IN VITRO EMBRYOGENESIS IN PLANTS

Current Plant Science and Biotechnology in Agriculture

VOLUME 20

Aims and Scope

The book series is intended for readers ranging from advanced students to senior research scientists and corporate directors interested in acquiring in-depth, state-of-the-art knowledge about research findings and techniques related to all aspects of agricultural biotechnology. Although the previous volumes in the series dealt with plant science and biotechnology, the aim is now to also include volumes dealing with animals science, food science and microbiology. While the subject matter will relate more particularly to agricultural applications, timely topics in basic science and biotechnology will also be explored. Some volumes will report progress in rapidly advancing disciplines through proceedings of symposia and workshops while others will detail fundamental information of an enduring nature that will be referenced repeatedly.

In Vitro Embryogenesis in Plants

Edited by

TREVOR A. THORPE

Plant Physiology Research Group, Department of Biological Sciences,
The University of Calgary, Calgary, Alberta, Canada

Kluwer Academic Publishers

DORDRECHT / BOSTON / LONDON

Library of Congress Cataloging-in-Publication Data

In vitro embryogenesis in plants / edited by Trevor A. Thorpe.
        p.   cm. -- (Current plant science and biotechnology in
    agriculture ; v. 20)
    Includes bibliographical references and index.
    ISBN 0-7923-3149-4 (HB : acid-free paper)
    1. Botany--Embryology. 2. Plant tissue culture. 3. Plant cell
culture. 4. Plant micropropagation.   I. Thorpe, Trevor A.
II. Series: Current plant science and biotechnology in agriculture ;
20.
QK665.I55   1995
581.3'3'0724--dc20   94-32558

ISBN 0-7923-3149-4

Published by Kluwer Academic Publishers,
P.O. Box 17, 3300 AA Dordrecht, The Netherlands.

Kluwer Academic Publishers incorporates
the publishing programmes of
D. Reidel, Martinus Nijhoff, Dr W. Junk and MTP Press.

Sold and distributed in the U.S.A. and Canada
by Kluwer Academic Publishers,
101 Philip Drive, Norwell MA 02061, U.S.A.

In all other countries, sold and distributed
by Kluwer Academic Publishers Group,
P.O. Box 322, 3300 AH Dordrecht, The Netherlands.

Table of Contents

Preface

The capacity of somatic plant cells in culture to form embryos by a process resembling zygotic embryogenesis is one of the most remarkable features of plants. This best exemplifies the concept of totipotency: that all normal living cells possess the potential to regenerate entire organisms. This concept, first proposed by the German plant physiologist Haberlandt in 1902, remained unproven until the 1950s when Reinert in Germany and Steward in the USA independently succeeded in regenerating carrot plantlets from cultured cells (see Chapter 1 for a historical perspective). Researchers using cell cultures have brought somatic embryogenesis to the forefront of plant tissue culture activity, in part because of the importance of micropropagation to agriculture, horticulture and forestry. Of the methods used for clonal propagation, somatic embryogenesis is potentially the most important, as it is capable of providing a larger number of plants in a shorter period of time than organogenic approaches. Although the phenomenon has been reported in numerous species (see Chapters 10-12), there is as yet no commercial production of plantlets by this method. Somatic embryogenesis is not restricted to in vitro conditions, as it is a phenomenon (apomixis) that has been known for a long time, and is widespread in nature. As a matter of fact, apomixis has been recorded in 94 families in the angiosperms. Therefore, a discussion on this topic has been included (Chapter 2).

Somatic embryogenesis is of fundamental biological interest and has often been compared to zygotic embryogeny, particularly with reference to 'normal' development. As well, the culture of excised mature zygotic embryos has been possible since the early 1900s and has led to the development of important in vitro techniques, in addition to application to problems in agriculture and horticulture since the 1930s. Thus chapters on zygotic embryogenesis in higher vascular plants (Chapter 3) and on the culture of zygotic embryos (Chapter 4) are included. The discussion of these subjects, as well as of apomixis (Chapter 2) provides valuable information on aspects germane to the examination of somatic embryogenesis. In addition, they can aid in appropriate in vitro experimentation.

Since the 1960s much information has been gathered about the requirements for the manipulation of somatic embryogenesis in vitro. This has included, inter alia, knowledge of empirical selections of the inoculum, the medium and the culture environment; so that now, well-established principles are in place (Chapter 5). Several reproducible morphogenic systems currently exist, of which carrot, for historical reasons, is the most important. These systems have allowed for detailed examinations of the structural and developmental patterns of somatic embryogenesis (Chapter 6), as well as for studies on the physiological and biochemical (Chapter 7), and the molecular biological (Chapter 8) bases of the process. The development of the synchronous single cells embryogenic system in carrot and the advance in molecular biological techniques are allowing solutions to formerly unanswerable questions. Although much remains to be learned and understood about the phenomenon, much progress has been made on these various aspects of in vitro somatic embryogenesis, particulary during the last decade.

The role of haploidy in plant breeding and in the study of genetics is well accepted. Successful androgenesis and gynogenesis (Chapter 9) are playing an important role in plant improvement. Finally, the last three chapters in the book attempt to summarize, mainly in tabular form, all the information available on somatic embryogenesis in herbaceous dicots (Chapter 10), herbaceous monocots (Chapter 11) and woody plants, both angiosperms and gymnosperms (Chapter 12). The magnitude of this task can be grasped from the number of references cited in these chapters: 491, 340 and 485, respectively. These tables show that species hither-to considered recalcitrant, such as cereals and grasses, legumes and conifers, have all responded favourably since the mid 1980s. These chapters are destined to serve as excellent reference sources for data up to the early 1990s, on somatic embryogenesis in all plants.

The various chapters of this book present together in one place and for the first time detailed information on all aspects of embryogenesis in plants, whether zygotic or asexual, in vivo or in vitro. The detailed reviews are directed to professionals and graduate students in all areas of the plant sciences who are interested in basic or applied aspects of plantlet formation and regeneration. The book will be useful, therefore, for cell biologists, physiologists, biochemists, molecular biologists, geneticists, and breeders working in agriculture, horticulture, or forestry, or in developmental botany, and morphogenesis. Plant tissue culture is playing an important part in plant biotechnology directly for mass clonal propagation, and indirectly as a central tool or as an adjunct to other methods, such as genetic engineering, in plant modification and improvement. In all of this work reliable methods of plantlet regeneration from selected or altered cells are required. In vitro somatic embryogenesis is destined for a pivotal role in this work, and this book contributes to the pursuit and achievement of the goals.

On a personal note, I am pleased to acknowledge the encouragement received from two of my former Postdoctoral Fellows (Indra Harry and Kiran Sharma) and one of my former graduate students (Richard Joy IV) to undertake this project, when I first began to consider it. They reinforced my view of the value of and the need for this book, a view also endorsed by my colleagues, David Reid and Edward Yeung. Drs. Harry, Sharma and Joy further critiqued my initial attempts at an outline for the book. While the final table of contents, authors, etc. are my own selections, I am grateful to all of the above for their intellectual input. I also wish to thank the present members of my laboratory for their help in indexing this book. Finally, I am indebted to the various authors who have contributed their time and expertise to the production of the in-depth, comprehensive and critical reviews which make up In Vitro Embryogenesis in Plants.

1. *In Vitro* Embryogenesis: Some Historical Issues and Unresolved Problems

WALTER HALPERIN

Contents

I. Introduction

Readers interested in a general discussion of the history of plant tissue culture cannot do better than to read the first two chapters of Philip White's book, *The Cultivation of Animal and Plant cells* [1]. White was a pioneer in the 1930s and 1940s, along with R.J. Gautheret and P. Nobecourt, in developing tissue culture techniques.

The following discussion will concentrate on carrot cultures since most of the early work was on *Daucus carota* and it is still the most studied system. Additional early work of significance which merits some discussion is the discovery that haploid embryos can be derived from cultured anthers.

II. Early reports of plantlet regeneration in carrot cultures

The story of embryogenesis *in vitro* begins with observations by early workers then plantlets were often regenerated in carrot callus cultures. The circumstances under which plantlets appeared in these early studies were diverse and offer only a few hints as to the controlling factors in organ regeneration. In

1

T.A. Thorpe (ed.), In Vitro Embryogenesis in Plants, pp. 1–16.
© 1995 *Kluwer Academic Publishers, Dordrecht. Printed in the Netherlands.*

1950, Levine [2] reported that carrot callus grown in a medium containing indoleacetic acid (IAA) would regenerate roots and shoots, in that order, when IAA was removed from the medium. Since no histological details were given, we do not know whether preexisting primordia were merely released from auxin inhibition by the removal of IAA or were initiated by the treatment. Levine reported that the shoots bore "primitive" leaves of a cotyledonous type and that these were followed by pinnatifid leaves resembling the mature leaves of carrot. He did not know how entire plantlets were ultimately produced from these isolated organ systems. In 1954, Wiggans [3] also observed carrot plantlet formation, but under different circumstances. His medium contained adenine sulfate. When the tissue was transferred to a medium lacking adenine, "buds" appeared and these gave rise to plantlets. Wiggans did not describe leaf morphology, but did state that tissue grown in the adenine-free medium "exhibited much less auxin activity than the original tissue did, thereby making bud formation possible". Thus, Levine concluded that the removal of auxin led to root formation, followed by whole plantlets; whereas Wiggans concluded that lowering of the auxin content of the tissue led to bud formation, followed by plantlets. Without exactly repeating their work, we cannot explain why in the one case roots appeared first and in the other case shoots appeared first, or how entire plantlets were ultimately regenerated from these isolated organ systems.

A more detailed report on the formation of carrot plantlets in tissue culture was published in two papers in 1958 by Steward and his coworkers [4, 5]. These papers are frequently cited, but seldom read, judging by the errors which continue to be propagated regarding their content. In the first paper [4], they called attention to the multiplicity of growth patterns exhibited by cultured cells and pointed out that certain patterns violated "classical concepts of cell behaviour". This is a highly original and stimulating paper. In the second paper [5], the authors discussed plantlet regeneration in their cultures and called attention, in particular, to proembryo-like early stages. Multicellular filaments, apparently derived from single cells, resembled the very early stage of proembryo development in seed embryos of *Daucus*. However, those who cite this paper overlook the fact that the authors report early cessation of embryo-like behaviour as the filaments grew into vascularized clumps. A cambial region appeared and roots were initiated, exactly in the manner described by earlier workers such as Sterling [6], who studied tobacco stem explants, and Gautheret [7], who made an exhaustively detailed study of the development and anatomy of such vascularized nodules. Steward and coworkers go on to state that when their rooted clumps, growing in suspension culture, were transferred to agar-solidified medium, buds formed at a point in the nodule opposite the root, thus forming a plantlet. That the shoot originated in a bud was apparently inferred, since the published data did not show histological sections establishing the developmental sequence involved in shoot formation. A similar pattern of plantlet formation, involving adventitious roots and presumptive buds sequentially formed in a vascularized nodule, has been reported several other times [8, 9].

Despite the fact that Steward and coworkers, in their often-cited paper of 1958 [5], did not observe somatic embryos, this paper is invariably cited as a study of *in vitro* embryogenesis. This is a consequence of the fact that, focusing on the early proembryo-like forms and the demonstrable totipotency of somatic carrot cells, they formulated a hypothesis which came to dominate the discussion of embryogenesis in cell cultures for a number of years. Its value lay then, as now, in providing a set of ideas which could be tested experimentally. Unfortunately, certain aspects of the hypothesis persisted and were incorporated into textbooks after there was compelling evidence that they were wrong [10].

III. Early reports of embryogenesis in carrot cell cultures

In 1959, Reinert [11] proposed that the plantlets appearing in his carrot cultures arose from bipolar embryos which were in turn derived from single cells. His conclusions were based on histological sections of callus. Although he did not publish photographs showing bipolar embryos, but only embryo fragments and cell clumps, this was the first hint in the literature that the early proembryo-like forms seen by Stewart and coworkers [5] might be capable of continuing the process of embryogenesis. In 1963, Kato and Takeuchi [9], Nakajima [12] and Wetherell and Halperin [13] independently published reports in which the photographic data unambiguously illustrated the production of bipolar embryos. In these three studies, the embryos formed in cultures derived from mature organs of the carrot plant. Also in 1963, Stewart et al. [14] reported on the production of embryos in cultures derived from carrot seed embryos. The production of adventive embryos derived from proliferating seed embryos in culture had earlier been reported in *Datura* [15], barley [16], *Cuscuta* [17], and *Dendrophthoe* [18], although these earlier reports were not well-documented.

Early studies [19–23] of the developmental sequence in carrot embryogenic cultures revealed the following pattern. Explants from virtually any organ or tissue of the carrot plant would proliferate on a medium containing only minerals, sucrose, thiamine and an auxin. Shortly after callus appeared, the proliferating tissue contained multicellular embryogenic clumps (Fig. 1E), easily recognizable by the characteristic appearance of the starch-filled cells (Fig. 1F). These clumps were called *proembryogenic masses* (PEMs) by Halperin [20] based on the prior use of a similar term by Jeffrey [24] ("embryogenic mass") and Guerin [25] ("corps embryogene") to describe polyembryony in several species of *Erythronium* in which the zygote grows into a cellular mass from which several embryos ultimately originate. In the continued presence of auxin, particularly a strong phenoxyacetic acid such as 2,4-D, organized development of PEMs was inhibited. The PEMs underwent cycles of growth and fragmentation (Fig. 1G) until such time as the auxin was removed (in a subculture) or the concentration fell (in an old culture) to the point where development was not inhibited. Differential development in the presence and

4

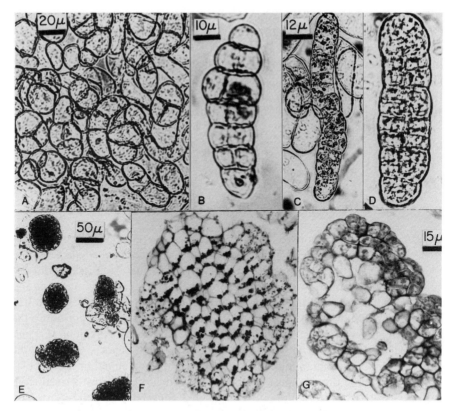

Fig. 1. (A-D) Development of highly-organized multicellular filaments from single cells in callus derived from primary explants. Cross walls tend to intersect mother cell walls at right angles. Filaments are packed with starch. (E-G) Proembryogenic clumps which probably result from continuation of the segmentation processes shown in A-D. (E) Whole mount of PEMs from suspension culture. (F) Periodic acid-Schiff stained section of PEM, revealing starch and distinction between inner and outer cells. (G) Hematoxylin-stained section of PEM showing clump disintegration as walls of innermost cells break down. (Reproduced from Halperin [22].)

absence of auxin has been illustrated diagrammatically by Wetherell [26] (Fig. 2). Embryo development proceeds rapidly on low or no auxin, with polarity of the shoot-root axis clearly established by the three-dimensionality of the proembryogenic clumps [21]. The inner cells were different cytologically and metabolically [21, 22] from the outermost cells which produced the embryos. The innermost cells did not utilize their starch (Fig. 3A) and did not synthesize DNA (Fig. 3D) as the outermost cells divided rapidly to produce globular embryos. The root pole always formed toward the inside of the clump and the shoot pole toward the outside. This was evident in large clumps where numerous embryos radiate outward, joined at the root end by nondividing cells which did not participate in embryo formation (Fig. 3B, C). The nondividing cells at the root pole were suspensor-like topographically, but little could be said

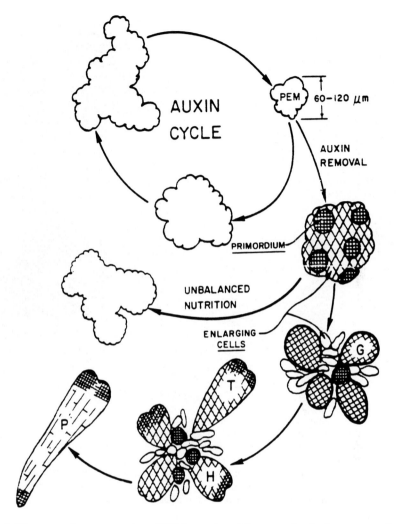

Fig. 2. Diagrammatic representation of PEM growth and fragmentation in the presence of auxin and embryo development in the absence of auxin. (Reproduced from Wetherell [26].)

about their function. They seem to be metabolically inert, judging by the fact that in addition to the absence of DNA synthesis, starch persists in such cells even after the attached embryos have reached the torpedo stage (Fig. 3B).

IV. Controversies and unresolved problems

In the following discussion, I will refer to the *initial* appearance of embryogenic cells as "induction" – for lack of a better word. This word has been used in a

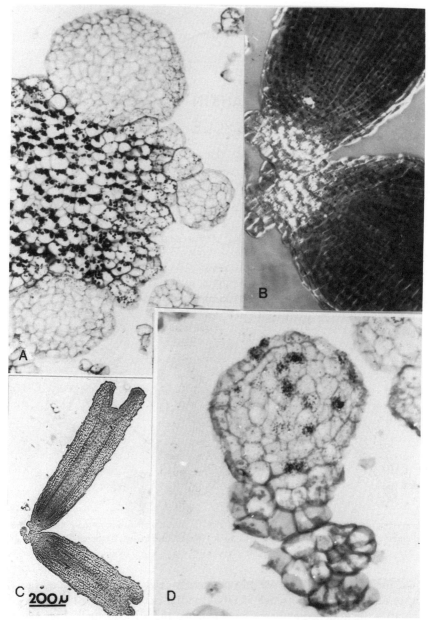

Fig. 3. (A) Globular stage embryos developing at the periphery of PEMs from cells which divide rapidly and utilize their starch, with internal suspensor-like cells retaining starch. (B) Enlargement of the root pole region of the embryos in Fig. 3C, viewed in polarized light to reveal starch retention in the suspensor-like cells. (C) Two torpedo-stage embryos joined at the root pole by suspensor-like cells derived from the original PEM. (D) Autoradiograph of embryo incubated in tritiated thymidine, showing the absence of DNA synthesis in the suspensor-like cells retained from the original PEM. ((C) reproduced from Halperin [20]; (D) reproduced from Halperin [22].)

variety of ways in the literature and is an unfortunate choice, since at present we do not know of unambiguous "inducing" factors. Embryogenesis is usually a spontaneous feature of those species in culture which are capable of forming somatic embryos and may not be "induced" by the experimental treatment as the term implies.

Embryogenic cells of carrot are usually recognizable and the time of their appearance can be reasonably well documented. In carrot cultures, they can usually be detected as multicellular filaments (Fig. 1A-D) or starch-filled clumps (Fig. 1E, F) in the callus which forms on primary explants during the first stage of culture. Transfer of such callus to low auxin or no auxin medium allows the second stage, development of organized embryos, to proceed. Frequent failure to recognize the existence of these two stages is a prime source of confusion, not just in the published literature, but in experimental design itself. Moreover, when the two stages *are* recognized, too great a significance may be attributed to them. In order to encourage the growth of explants, or maintain the growth of subcultures, the presence of an auxin is usually required in the medium. An unavoidable effect of the auxin, if the level is high enough, is to suppress the development of organized embryos from PEMs which may be present – but not detected. Self-perpetuating embryogenic cells can be carried along un-recognized through numerous subcultures. The ultimate development of organized embryos when the auxin level is lowered, either by design or by default (old cultures), does not, in my opinion, represent "induction" of embryogenesis, but release from auxin inhibition. While some workers in the field may dispute this view, it certainly remains true that cultures should be systematically examined for the presence of embryogenic cells *before* experimental treatments designed to "induce" them are applied. This is seldom done. In the case of carrot cultures, at least, testing is simple. One disperses the callus (or suspension) through a relatively coarse screen (for example, with 350 micron pores) and uses the effluent to start suspension cultures on embryo test media (basic medium without auxin). The purpose of screening is to provide reproducible aliquots for inoculating test cultures and to allow easy counting of any embryos which may develop. The inoculum as well as the material trapped on the screen should be examined cytologically. If the inoculum has a minimum of about 20,000 cells/cc, even tiny PEMs with fewer than 5 cells will develop into embryos with cotyledons within 10 days [27]. A schematic of this process is shown in Fig. 4.

A. Are there special chemical factors which promote embryogenesis?

Steward and coworkers claimed that the coconut milk used in their media was specifically responsible for embryogenesis *in vitro*. This view was codified by Bonner in the text, *Plant Biochemistry*, in 1965 [28], with the following statement ... "Two conditions must be satisfied for this to occur [embryogenesis]. The specialized cell must be separated from its neighbours, that is, it must be a single cell. In addition, the cell must be surrounded by medium which contains the

8

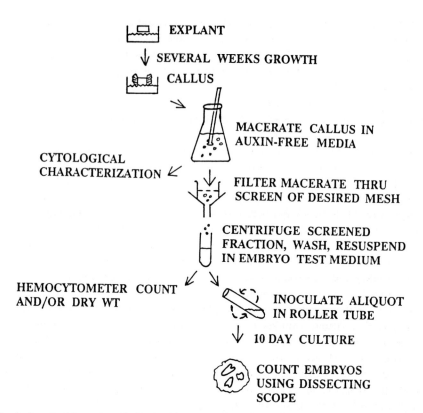

EXPLANT

↓ SEVERAL WEEKS GROWTH

CALLUS

MACERATE CALLUS IN
AUXIN-FREE MEDIA

CYTOLOGICAL
CHARACTERIZATION

FILTER MACERATE THRU
SCREEN OF DESIRED MESH

CENTRIFUGE SCREENED
FRACTION, WASH, RESUSPEND
IN EMBRYO TEST MEDIUM

HEMOCYTOMETER COUNT
AND/OR DRY WT

INOCULATE ALIQUOT
IN ROLLER TUBE

↓ 10 DAY CULTURE

COUNT EMBRYOS
USING DISSECTING
SCOPE

Fig. 4. A method for systematically checking suspension cultures for the presence of PEMs. See text for explanation.

nutrients needed for embryo growth. The liquid endosperm of coconut or horse chestnut contains the required substances. If either of the two conditions is not satisfied, embryos are not produced".

While the above hypothesis had an inherent logic, based on the fact that zygotes are more-or-less isolated cells which usually develop in the presence of endosperm, review of the relevant papers shows that experimental controls were lacking. Data from other labs made it clear that the hypothesis was incorrect with respect to a need for liquid endosperms. As far as I know, there has never been a controlled experiment demonstrating that coconut milk or other endosperms are required for the induction of embryogenic behaviour in cultured cells. A more complete discussion of this issue is available elsewhere [10].

Although embryogenesis in carrot cultures is initiated in simple, defined media containing only minerals, sucrose, thiamine and an auxin, one cannot conclude that other substances are not involved. Explants or subcultured tissues obviously contain a plethora of organic molecules, many of which are released into the medium. The possible role of such substances in triggering embryogenesis remains unknown. In 1922, Haberlandt [29] attempted to explain the induction of nucellar embryos by invoking the presence of "necrohormones" produced by dying cells in the vicinity of expanding embryo sacs, but others were unable to find evidence for this hypothesis. The production of somatic embryos in ovules, a potentially useful model for their production in cell cultures, remains unexplained, although Bonga [30] has made the interesting suggestion that it owes something to the juvenilizing influence of nearby meiotic tissue.

In 1965, Halperin and Wetherell [31] proposed that the ammonium ion was required for induction of carrot embryogenesis *in vitro*, but not for later stages of embryo development. It was later shown by Halperin that casein hydrolysate was nearly as effective as ammonium as a nitrogen source in promoting the appearance of embryogenic cells in explants [22]. Tazawa and Reinert [32] subsequently concluded that embryos would form in carrot cultures grown on nitrate only, but that a certain level of *intracellular* ammonium was essential. Their technique involved counting the percentage of callus cultures which formed embryos, a method which was not capable of detecting quantitative differences in the number of embryos per culture and did not avoid conditioning of media. In an investigation using suspension cultures in which the number of embryos per culture can be determined, and methods which attempted to minimize the effect of conditioned media, Wetherell and Dougall [33] came to the conclusion that some form of reduced nitrogen was essential in the medium. In additional work with the same experimental system, Brown, Wetherell and Dougall [34] concluded that potassium also had a specific developmental role to play in carrot embryogenesis *in vitro*.

A major problem encountered in reconciling the above sometimes contradictory results lies in the fact that different experimental systems were used and different stages of embryogenesis were studied. Halperin's work on the efficacy of various nitrogen sources involved petiole *primary explants* and the results referred to the initial appearance of embryogenic cells. The work of Wetherell et al. involved the role of ammonium and other factors in promoting embryogenesis when *proembryogenic masses* in cell suspension were released from auxin inhibition. The work of Reinert and coworkers involved *callus* cultures which were not characterized cytologically, so it is not clear whether they were dealing with mature cells undergoing the transition to the embryogenic condition or proembryogenic cells making the transition to the organized state.

It is clear that a great deal more work will have to be done to distinguish qualitative from quantitative effects of the various forms of nitrogen and other chemicals studied. It is crucial in future work that subcultured tissue be characterized cytologically and "behaviourally" (Fig. 4) before experimental treatments are begun.

Other authors, too numerous to cite, have stated that an auxin is essential to carrot somatic embryogenesis. Although the efficiency of the process seems to vary with the auxin source, 2,4-D clearly being the most effective, I do not know of experimental evidence for the view that an *exogenous* auxin plays some specific role in somatic embryo induction or development *in vitro*. Because one cannot obtain a culture from most explants, or maintain tissue in subculture, without the presence of auxin in the medium, it is difficult to separate the role of auxin in stimulating growth from a possible role in embryogenesis.

While the role of *exogenous* auxin in stimulating the initial appearance of embryogenic cells is uncertain, it does seem clear that embryogenic cells, once formed, are characterized by a substantial level of endogenous auxin synthesis. PEMs washed and placed in auxin-free media embark on a process of embryogenesis which is characterized by a cell division rate more rapid than that of controls on 2,4-D [22] a result which might be due to either 2,4-D inhibition of controls or endogenous growth factor synthesis in embryogenic cells. That the latter is probably the case was shown by Fujimura and Komamine [35] who found significant levels of endogenous auxin (IAA) at every stage of somatic embryogenesis in carrot suspension cultures. Smith and Krikorian [36] recently showed that under some circumstances embryogenic carrot cultures can be initiated and maintained in the absence of auxin, again pointing to an auxin biosynthetic capacity in such cells.

B. Is cell isolation required to induce embryogenesis?

An essential part of the original hypothesis framed by Steward and coworkers holds that conversion of a cultured cell to the embryogenic condition requires physical isolation, i.e. the single cell condition. The hypothesis was formulated at a time when the authors had seen only proembryo-like early stages (multicellular filaments), but had not yet witnessed the formation of somatic embryos. In subsequent studies in other labs, it was demonstrated that embryos are almost entirely derived from multicellular proembryogenic masses (PEMs) as described above. Thus the developmental history of PEMs becomes an important issue. Are they derived from single cells? Do they go through a single cell stage as they perpetuate themselves from subculture to subculture? The answer to the second question seems to be no. They reproduce by growth and fragmentation, although there is no question that single cells do appear in the cultures.

Callus and suspension cultures contain single cells of various sizes and developmental potentialities. The large, highly vacuolate cells which can be found in such cultures rarely divide, but in a few cases they can be induced to divide and produce clumps from which embryos may subsequently be derived [37, 38]. In suspensions containing PEMs, small, single cells about 12 μm in diameter are also numerous. Like the large, vacuolate cells, these small cells seldom divide. However, unlike the large cells, up to 90% of the small cells can be stimulated by special culture conditions to divide and give rise to embryos [39]. These small cells have been designated "type 1" cells by Nomura and Komamine, who point out that such cells appear to have "already been determined" as embryogenic cells in the PEMs from which they were derived.

Thus the question as to whether or not single, type 1 cells in proembryogenic cultures can produce embryos may be a trivial matter. The situation with respect to the large, highly vacuolate cells, present in a number of different kinds of cultures, is more problematic since the state of determination of such cells is unknown.

The question remains, what is the ultimate origin of PEMs? There is no direct evidence on this point, but strong circumstantial evidence indicates that at least some PEMs which appear in a proliferating explant are derived from the multicellular filaments which are common and which appear to be of single cell origin (Fig. 1A-D).

The ultimate origin of PEMs from single cells is thus consistent with the hypothesis of Steward and coworkers that a single cell stage is involved, but does not constitute evidence that such a stage is *required*. The hypothesis would be disproved by evidence indicating that two or more original cells of the explant can collaborate in the formation of PEMs. In fact this does occur [40], but possibly only in the special case known as "direct" embryogenesis – a situation which occurs when embryos or immature tissues are used as explants (see Section IV, C below). Proliferation of such primary explants can result in the outgrowth of somatic embryos which clearly originate in the coordinated activity of multiple cells of the explant. Such initiating cells do not seem to be isolated from neighbouring cells, as determined by the existence of plasmodesmata which connect them to neighbouring cells [41]. It has also been observed that the origin of embryos in PEMs, when the auxin content of the medium is reduced, is also associated with initiating cells which remain connected to other cells of the PEM by plasmodesmata [42]. Thus, cells which initiate the process of organized embryo development, whether in primary explants or in often subcultured proembryogenic clumps, may do so without being physically isolated from surrounding cells. Nothing can be said at present about possible physiological isolation.

C. Single vs. multiple cell origin of embryos and its bearing on the concept of isolation

An explanation for the circumstances controlling single vs multiple cell origin of embryos was proposed by Williams and Maheshwaran [43], building on the ideas of Sharp and coworkers [44]. In 1980, Sharp and coworkers attempted to distinguish between "direct" embryogenesis and "indirect" embryogenesis. The term "direct" was applied to explants which undergo a minimum of proliferation before forming somatic embryos. The term "indirect" was applied to explants which undergo an extensive period of disorganized proliferation before somatic embryos can develop. Sharp and coworkers suggested, in the paper cited above as well as in later papers, that direct embryogenesis is characteristic of explants in which all or some cells are "predetermined" as embryogenic cells. Predetermination was thought to be a consequence of having retained some of the properties of the parental meristematic cells from which such cells are derived. This would explain the tendency of direct embryogenesis to occur in explants consisting of young plants such as embryos or seedlings, or in tissues newly derived from meristems. Indirect embryogenesis was thought to be characteristic of mature organs in which cells must go through several cell cycles in order to achieve the embryogenic or "determined" condition.

Because we lack unambiguous biochemical or cytological markers defining embryogenic competence, there is circular reasoning involved in the concept of "predetermination". It is detected to begin with by the fact that direct embryogenesis has occurred. However, the terms "direct" and "indirect" are still useful in describing cases where either very little or a great deal of explant proliferation precedes embryogenesis, but do not necessarily indicate fundamental differences in the cells involved. The need for caution in applying these terms stems from a variety of data which seem inconsistent with the hypothesis. For example, Hanning and Conger [45] and Conger et al. [46] report that indirect embryogenesis is characteristic of explants from the *Dactylis* leaf base, whereas direct embryogenesis occurs in more distal regions of the leaf. This is not what one would expect from a grass leaf where the hypothetical predetermined state, allowing direct embryogenesis, should occur in the basal regions which are nearer to the intercalary meristem and thus younger. Another apparent contradiction is seen in the report by Dubois et al. that "mature cortical cells, with vacuoles and chloroplasts" were involved in direct embryogenesis from various organs of Cichorium [47]. In other cases, direct vs. indirect bud or embryo development clearly seem to be determined by correlative factors in tissues or the type of medium, rather than the cytological state of the cells at the time of excision. For example, in *Torenia* explants [48], there are two patterns of bud formation which can occur, direct or indirect, depending on the structure of the explant. But *the same cells serve as bud progenitors in both cases.* Clearly the surrounding tissue influences behavior of the bud-forming cells, not some intrinsic state of determination at the time of excision. If one considers the delayed embryogenesis which can occur when PEMs are involved, a phenomenon attributable to high auxin levels in the medium and not to delayed embryogenic determination, it is apparent that other cases of "indirect" embryogenesis may be similarly controlled.

Williams and Maheswaran [43] extended the ideas of Sharp and coworkers in a proposal which seems useful in explaining *multicellular vs. single* cell origin of somatic embryos when so called *direct* embryogenesis is involved. In the case of direct embryogenesis, they are of the opinion that the difference between multiple and single cell origin of embryos may be a trivial matter. A somatic embryo originating directly from multiple initials is assumed to be a consequence of the presence in the explant of several neighbouring cells in the same state of "predetermination" and thus capable of acting in coordinated fashion (as might be the case if the explant was an embryo itself). Single initials, according to Williams and Maheswaran, are characteristic of older plantlets or seedlings in which extensive cell maturation has occurred and only a few isolated cells are retained in the embryogenic or predetermined condition. If markers become available for the "predetermined" state, this proposal can be tested.

Thus, physical isolation, in the sense of single cells lacking contacts with other cells, is clearly not always characteristic of the initiating cells in somatic embryogenesis. Multiple cells may cooperate in forming an embryo. Single initiating cells may remain connected to other cells by plasmodesmata, although they later disappear. The early stages of direct embryogenesis are thus similar to axillary bud formation, in which a new branch forms from cells which remain fully integrated into the parental plant. But, unlike buds, the developing embryo eventually becomes isolated from the parental structure [41].

In the case of indirect embryogenesis, possibly characteristic of cells derived from mature tissue, there is little evidence for or against the concept that a required first step is physical, or at least physiological, isolation. It is logical to assume that the physical or physiological isolation which single cells or small groups of cells experience in a proliferating explant allows them to escape from correlative restraints, freeing them to embark on a new developmental program. In this sense, I believe that Steward and coworkers are correct in emphasizing the concept of isolation in such cases. However, the original hypothesis states that such isolation is a unique stimulus to embryogenic behaviour; the isolated cultured cell finding itself in the same situation as a zygote. But this cannot be the case, since few such isolated cells become embryogenic. Therefore, isolation of mature cells may be required before they can embark on an embryogenic program, but isolation *per se* is not a stimulus to embryogenic behaviour. It is essentially a permissive condition which can lead to a number of developmental outcomes. The fact that Umbellifers so readily produce embryogenic cells, whereas other taxa may resist doing so, seems to indicate that there is a genetic factor allowing some genotypes to take advantage of the *in vitro* environment.

D. Do buds form in carrot cultures?

In early papers, the formation of "buds" has often been noted in carrot cultures. In some cases, it has been assumed that adventitious root formation in vascularized callus was followed by later formation of a bud in the same callus – with eventual integration of the two systems to form a plantlet [5, 8, 9]. The formation of adventitious roots in carrot cultures is well-documented [7, 20], but as far as I know there are no histological sections illustrating the formation of buds in carrot callus. Reports that "buds" were present were simply based on observation of leafy shoots. The origin of the shoots was actually unknown.

There is an important difference between ontogeny of shoot apices in buds and in embryos. In bud development, a shoot apical meristem forms and is *followed by* the initiation of leaves. In an embryo, the first leaves (cotyledons) are initiated in the *absence of* a shoot apical meristem. In those early papers where "buds" were reported, it is noteworthy that the authors often described the first leaves as cotyledonary in form (or at least "entire", i.e. without the usual highly dissected form of carrot leaves). It seems possible, even likely, that the so called "buds" were actually shoots derived from embryos, however deformed or unrecognizable they may have been, and that carrots have a weak or even nonexistent capacity to form buds in callus. The reason may be that they do not normally form such buds in the intact plant. Carrots grow as rosettes and the axillary branches which grow out upon bolting may be florally determined from the outset. If carrots and other umbellifers have a weak genetic capacity to form vegetative buds, this may have something to do with the apparent absence of buds in cultures and the propensity to form embryos. This is, of course, wild speculation. We await better ideas.

V. Embryos from pollen grains

Guha and Maheshwari discovered in 1966 that the embryos they had earlier observed in cultured anthers of *Datura* were derived from pollen [49]. Their cytological studies showed "multicellular globular masses" within the exine. Chromosome counts of the embryos which ultimately formed from such masses revealed them to be haploid.

Subsequent studies by Nitsch and Nitsch with several species of *Nicotiana* established for the first time the conditions under which pollen-derived embryos could be obtained [50]. Anthers cultured at the tetrad stage were too young and anthers with mature pollen were too old. It appeared in this pioneering work that microspores were most sensitive to the embryo-inducing conditions of the culture tube when they were "fully individualized, uninucleate, and devoid of starch". A very simple medium was used, containing only minerals, vitamins, sucrose and indoleacetic acid. Later studies with a variety of species in many labs largely confirmed the conclusions reached by Nitsch and Nitsch, although very young pollen grains (i.e. binucleate) proved capable of producing embryos in certain species – largely from the vegetative cell only [51].

In retrospect, the fact that microspores or pollen grains are not rigidly limited in their development is not surprising. There were published reports a half-century ago that seed-derived plants occasionally showed strict paternal inheritance, indicating that pollen cells entering the embryo sac were stimulated to form embryos. More remarkably, it was also discovered at about the same time that microspores under certain circumstances will enlarge and undergo 3 successive mitoses to form what appears to be a perfectly normal 8 nucleate embryo sac – with a trio of nuclei at either end and 2 nuclei in the center [52]. Thus microspore nuclei, or one or more nuclei of the pollen tube, are not so rigidly programmed that they cannot occasionally express the developmental program expected of zygotes or even megaspore nuclei.

References

1. White, P.R., *The Cultivation of Animal and Plant Cells*, 2nd ed., Ronald Press, New York, 1963.
2. Levine, M., The growth of normal plant tissue *in vitro* as effected by chemical carcinogens and plant growth substances. I. The culture of the carrot tap root meristem, *Am. J. Bot.*, 37, 445, 1950.
3. Wiggans, S.C., Growth and organ formation in callus tissues derived from *Daucus carota*, *Am. J. Bot.*, 41, 321, 1954.
4. Steward, F.C., Mapes, M.O., and Smith, J., Growth and organized development of cultured cells. I. Growth and division of freely suspended cells, *Am. J. Bot.*, 45, 693, 1958.
5. Steward, F.C., Mapes, M.O., and Mears, K., Growth and organized development of cultured cells. II. Organization in cultures grown from freely suspended cells, *Am. J. Bot.*, 45, 705, 1958.
6. Sterling, C., Histogenesis in tobacco stem segments cultured *in vitro*, *Am. J. Bot.*, 27, 464, 1950.
7. Gautheret, R.J., *La Culture des Tissus Végétaux: Technique et Réalisations*, Masson et Cie., Paris, 1959.
8. Pilet, P.E., Culture *in vitro* de tissus de Carotte et organogenese, *Ber. Schweiz. Bot. Ges.*, 71, 189, 1961.
9. Kato, H., and Takeuchi, M., Morphogenesis *in vitro* starting from single cells of carrot root, *Plant Cell Physiol.*, 4, 243, 1963.
10. Halperin, W., Single cells, coconut milk, and embryogenesis *in vitro*, *Science* 9, 1287, 1966.
11. Reinert, J., Über die Kontrolle der Morphogenese und die Induktion von Adventivembryonen an Gewebekulturen aus Karotten, *Planta* 53, 318, 1959.
12. Nakajima, T., On the plant tissue culture with special reference to embryogenesis, in *Gamma Field Symposium no. 2, Cell Differentiation and Somatic Mutation*, 1963, 25.
13. Wetherell, D.F., and Halperin, W., Embryos derived from callus tissue cultures of wild carrot, *Nature*, 200, 1336, 1963.
14. Steward, F.C., Blakely, L.M, Kent, A.E., and Mapes, M.O., Growth and organization in free cell cultures, in *Brookhaven Symposium no. 16, Meristems and Differentiation*, 1963, 73.
15. Sanders, M., Development of self and hybrid Datura embryos in artificial culture, *Am. J. Bot.*, 37, 6, 1950.
16. Norstog, K., The growth and differentiation of cultured barley embryos, *Am. J. Bot.*, 48, 876, 1961.
17. Maheshwari, P., and Baldev, B., Artificial production of buds from the embryos of *Cuscuta reflexa*, *Nature* 191, 197, 1961.
18. Johri, B.M., and Bajaj, Y.P.S., Behaviour of mature embryo of *Dendrophthoe falcata* (L.f.) Ettings, *in vitro*, *Nature*, 193, 194, 1962.
19. Halperin, W., and Wetherell, D.F., Ontogeny of adventive embryos of wild carrot, *Science* 147, 756, 1965.
20. Halperin, W., Alternative morphogenetic events in cell suspensions, *Am. J. Bot.*, 53, 443, 1966.
21. Halperin, W., and Jensen, W.A., Ultrastructural changes during growth and embryogenesis in carrot cell cultures, *J. Ultrastruc. Res.*, 18, 428, 1967.
22. Halperin, W., Embryos from somatic plant cells, in *Control Mechanisms in the Expression of Cellular Phenotypes, Int'l Soc. for Cell Biology Symposium no. 9*, Padykula, H., Ed., Academic Press, New York, 1970, 169.
23. McWilliam, A.A., Smith, S.M., and Street, H.E., The origin and development of embryoids in suspension cultures of carrot (*Daucus carota*), *Ann. Bot.*, 38, 243, 1974.
24. Jeffrey, E.C., Polyembryony in *Erythronium americanum*, *Ann. Bot.*, 9, 537, 1895.
25. Guerin, P., L'ovule et la graine des *Erythronium* et des Calochortus, *Ann. Sci. Nat. Bot. Ser.* 10, Vol. 19, 225, 1937.
26. Wetherell, D.F., *In vitro* embryoid formation in cells derived from somatic plant tissues, in *Propagation of Higher Plants Through Tissue Culture*, Hughes, K.W., Henke, R., and Constantin, M., Eds., Conf. 780411, U.S. Tech. Inf. Serv., Springfield, VA, 1979, 102.

16

27. Halperin, W., Population density effects on embryogenesis in carrot cell cultures, *Exp. Cell Res.*, 48, 170, 1967.
28. Bonner, J., Development, in *Plant Biochemistry*, Bonner, J., and Varner, J.E., Eds., Academic Press, New York, 1965, 862.
29. Haberlandt, G., Über Zellteilungshormone und ihre Beziehungen zur Wundheilung, Befruchtung, Pathenogenesis und Adventivembryonie, *Biol. Zentbl.*, 42, 145, 1922.
30. Bonga, J.M., Vegetative propagation in relation to juvenility, maturity, and rejuvenation, in *Tissue Culture in Forestry*, Bonga, J.M., and Durzan, D.J., Eds., Martinus Nijhoff/Dr. W. Junk Publ., The Hague, 1982, 387.
31. Halperin, W., and Wetherell, D.F., Ammonium requirement for embryogenesis *in vitro*, *Nature* 205, 519, 1965.
32. Tazawa, M., and Reinert, J., Extracellular and intracellular chemical environments in relation to embryogenesis *in vitro*, *Protoplasma*, 68, 157, 1969.
33. Wetherell, D.F., and Dougall, D.K., Sources of nitrogen supporting growth and embryogenesis in cultured wild carrot tissue, *Physiol. Plant.*, 37, 97, 1976.
34. Brown, S., Wetherell, D.F., and Dougall, D.K., The potassium requirement for growth and embryogenesis in wild carrot suspension cultures, *Physiol. Plant.* 37, 73, 1976.
35. Fujimura, T., and Komamine, A., Involvement of endogenous auxin in somatic embryogenesis in a carrot cell suspension culture, *Z. Pflanzenphysiol.*, 95, 13, 1979.
36. Smith, D.L., and Krikorian, A.D., Somatic embryogenesis of carrot in hormone-free medium: external pH control of morphogenesis, *Am. J. Bot.*, 77, 1634, 1990.
37. Backs-Husemann, D., and Reinert, J., Embryobildung durch isolierte Einzelzellen aus Gewebekulturen von *Daucus carota*, *Protoplasma* 70, 49, 1970.
38. Jones, L.H., Factors influencing embryogenesis in carrot cultures (*Daucus carota* L.), *Ann. Bot.* 38, 1077, 1974.
39. Nomura, K., and Komamine, A., Identification and isolation of single cells that produce somatic embryos at a high frequency in a carrot suspension culture, *Plant Physiol.* 79, 988, 1985.
40. Dunstan, D.I., Short, K.C., and Thomas, E., The anatomy of secondary morphogenesis in cultured scutellum tissues of *Sorghum bicolor*, *Protoplasma* 97, 251, 1978.
41. Konar, R.N., Thomas E., and Street, H.E., Origin and structure of embryoids arising from epidermal cells of the stem of *Ranunculus sceleratus* L., *J. Cell Sci.* 11, 77, 1972.
42. Thomas, E., Konar, R.N., and Street, H.E., The fine structure of the embryogenic callus of *Ranunculus sceleratus* L., *J. Cell Sci.*, 11, 95, 1972.
43. Williams, E.G., and Maheswaran, G., Somatic embryogenesis: Factors influencing coordinated behaviour of cells as an embryogenic group, *Ann. Bot.*, 57, 443, 1986.
44. Sharp, W.R., Sendahl, M.R., Caldas, L.S., and Maraffa, S.B., The physiology of *in vitro* asexual embryogenesis, *Hort. Rev.* 2, 268, 1980.
45. Hanning, G.E., and Conger, B.V., Embryo and plantlet formation from leaf segments of *Dactylis glomerata* L., *Theor. Appl. Genet.* 63, 155, 1982.
46. Conger, B.V., Hanning, G.E., Gray, D.J., and McDaniel, J.K., Direct embryogenesis from mesophyll cells of orchardgrass, *Science* 221, 850, 1983.
47. Dubois, T., Guedira, M., Dubois, J., and Vasseur, J., Direct somatic embryogenesis in roots of *Cichorium*: Is callose an early marker? *Ann. Bot.* 65, 539, 1990.
48. Chlyah, H., Inter-tissue correlations in organ fragments: Organogenetic capacity of tissue excised from stem fragments of *Torenia fournieri* Lind. cultured separately *in vitro*, *Plant Physiol.* 54, 341, 1974.
49. Guha, S., and Maheshwari, S.C., Cell division and differentiation of embryos in the pollen grain of *Datura in vitro*, *Nature* 212, 97, 1966.
50. Nitsch, J.P., and Nitsch, C., Haploid plants from pollen grains, *Science* 163, 85, 1969.
51. Raghavan, V., *Experimental Embryogenesis in Vascular Plants*, Academic Press, New York, 1976.
52. Stow, I., Experimental studies on the formation of embryo sac-like giant pollen grains in the anther of *Hyacinthus orientalis*, *Cytologia* 1, 417, 1930.

2. Asexual Embryogenesis in Vascular Plants in Nature

KIRAN K. SHARMA and TREVOR A. THORPE

Contents

I. Introduction

In the flowering plants, sexual reproduction involves "double fertilization" where one sperm nucleus unites with the egg cell to form a zygote, which gives rise to the "embryo". Meiosis precedes the formation of gametes, and fertilization restores the somatic chromosome number. Another sperm nucleus unites with the two polar nuclei, which have usually fused to a central nucleus. The resulting "triploid" cell gives rise to the endosperm [1]. However, in many vascular plants, sexual reproduction is regularly combined with, or more or less replaced by, different kinds of asexual reproduction (Fig. 1). Asexual reproduction in seed plants (apomixis) can be divided into two main classes; vivipary or vegetative reproduction and agamospermy. Agamospermy involves the production of fertile seeds in the absence of sexual fusion between gametes or "seeds without sex", whereby, a full reductional division is usually absent, and thus chromosomes do not segregate. Agamospermy includes "adventitious embryony" and "gametophytic apomixis" and results in the formation of one or

17

T.A. Thorpe (ed.), In Vitro Embryogenesis in Plants, pp. 17–72.
© 1995 *Kluwer Academic Publishers, Dordrecht. Printed in the Netherlands.*

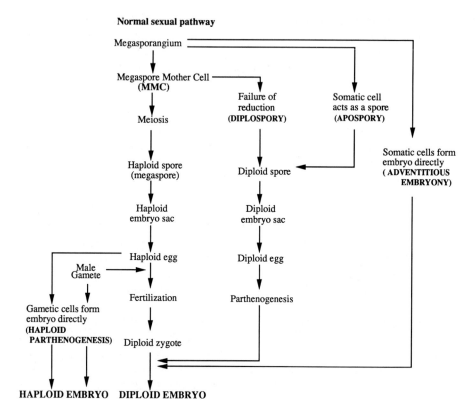

Fig. 1. Asexual embryogenic developmental pathways in vascular plants in nature as compared with the normal sexual cycle. (Modified from J. Heslop-Harrison, 1972 [2].)

more asexual embryos along with or without the sexually produced zygotic embryo leading to "polyembryony". Flowers designed for efficiency in the sexual process, underwent an evolutionary adjustment either by producing parthenogenetic seed (agamospermy) or by producing vegetative propagules instead, or in the addition of normal flowers (vivipary or vegetative reproduction). Since Leeuwenhoek in 1719 observed the presence of more than one embryo in the seeds of orange [1], and subsequent demonstration of the nature of polyembryony in *Citrus, Coelebogyne, Funkia* and *Nothoscordum*, many other plant species belonging to both Angiosperms and Gymnosperms have been reported to exhibit this phenomenon (see Table 1).

Adventitious embryony includes the formation of an embryo directly from a somatic cell in the ovule of the mother sporophyte as nucellar or integumental outgrowths without any intervening formation of embryo-sacs and egg cells. Thus in addition to the zygotic embryo, asexual non-zygotic embryo(s) may also develop in the same seed. On the other hand gametophytic apomixis involves the formation of the female gametophyte from an unreduced embryo-sac initial (megaspore mother-cell; MMC) or from somatic cells in the nucellus or chalaza.

This phenomenon (also known as apomeiosis) includes two variations. (1) diplospory, and (2) apospory. In diplospory, unreduced embryo-sacs are formed from a generative cell (female archesporial cell, megaspore mother cell), either directly by mitosis, or indirectly by modified meiosis, resulting in an unreduced restitution nucleus. In apospory, somatic cells of the ovule (usually nucellus) give rise to embryo sacs [298, 299]. The gametophyte has an unreduced gametophytic number in both cases, and an alteration of generations occurs, at least in the morphological sense [13]. The egg-cell may form an embryo without fertilization (parthenogenesis) or one of the other cells of the gametophyte may do so (apogamety). In a broad sense, asexual embryogeny may also include haploid parthenogenesis by the development of a haploid embryo from the egg (gynogenesis) or microspore (androgenesis) (see Chapter 10). This chapter is mainly confined to asexual embryogenesis in higher plants in nature whereby, apomictic embryos are produced in the seeds with or without the zygotic embryo. Table 1 includes a list of angiosperm and gymnosperm species where various forms of asexual embryogenesis have been reported so far. Various aspects of asexual embryogenesis in nature have been reviewed previously in both angiosperms [1, 10, 13, 20, 54, 299–306] and gymnosperms [13, 254, 290, 291, 307–309].

Although polyembryony may include (1) embryos produced by various types of apomixis, (2) embryos produced by cells other than the egg cell, and (3) plural embryos resulting from multiple embryo sacs in an ovule, it is customary to refer to a seed as polyembryonate if two or more embryos occur therein, regardless of their origin. According to Johansen [13], polyembryony should include only the accessory embryos arising from the zygotic proembryo. In accordance to this view, Maheshwari and Sachar [54] state that a truely polyembryonate seed is one in which the embryos are produced only from the zygotic proembryo, or from a mass of embryogenic cells. They also include embryos which develop from the synergids, antipodal cells, and sporophytic nucellar or integumentary cells. Yakovlev [310] prefers to include under polyembryony all the cases concerning the formation of adventive embryos which arise either from the sexual elements of the gametophyte or from the cells of the sporophyte. This latter view is maintained during the discussion in this chapter. Two groups of polyembryony can be ditinguished among higher plants according to their peculiarities in ontogeny: gametophytic and sporophytic. Both may be subdivided into types depending on their patterns and sources of the production of supernumerary embryos. Gametophytic polyembryony can arise at the expense of multiple archegonia, macrosporocytes and female gametes. Sporophytic polyembryony is a result of proliferation of the zygote, proembryo, endosperm and the initial cells of the sporophyte itself (nucellus, integument). Polyembryony that originates from the cells of the sporophyte is classified as embryoid polyembryony. Cleavage of the proembryo or the embryonal suspensor is a common cause of polyembryony in the gymnosperms. Other forms of asexual embryogenesis have generally not been observed in gymnosperms, and hence, asexual embryos are invariably present alongwith the zygotic embryos.

Table 1. Asexual embryogenesis in angiosperms and gymnosperms *in vivo*

Family/Genus/Species	Type[1]	Tissue/Organ[1]	Reference[2]
Angiosperms			
Acanthaceae			
Acanthus mollis L.	AE	–	3
Dipteracanthus patulus (Jacq.) Nees	CP	Su	4
Aceraceae			
Acer spicatum Lam.	–	–	5
Actinidiaceae			
Actinidia chinensis Planch.	CP	Su	6
polygama Franch. & Sav.	CP	Su	7
Alismaceae			
Limnocharis emarginata Humb. & Bonpl.	––	Su	8
Sagittaria graminea Mich.	AE	Sy	9
Amarantaceae			
Aerva tomentosa Forsk.	D + P	MMC	10
Amaryllidaceae			
Calostoma cunninghamii Ait.	(AE?)	–	10
purpureum R. Br.	(AE?)	Nu	10
Cooperia drummondii Herb.	D + P	Nu	11
pedunculata Herb.	D + P	Nu	11
Zephyranthes texana Herb.	D + P	MMC	10
Anacardiaceae			
Anacardium occidentale L.	AE	Nu	8
Lannea coromandelica (Houtt.) Merr.	Ha P	Sy	8
Mangifera indica L.	AE	Nu	12–19
odorata Griff.	AE	Nu	17
Apiaceae (see Umbelliferae)			
Apocynaceae			
Amsonia tabernaemontanae Walter	(AE?)	–	10
Rauwolfia tetraphylla L.	AE	Sy	8
Aracaceae			
Spathiphyllum patini (Hoog.) N.E. Br.	AE	Nu	8, 13
Aglaonema commutatum	A + P	––	10
psedobractaetum	A + P	––	10
roebelinii	A + P	––	10
Asclepiadaceae			
Vincetoxicum cretaceum Pobed.	–	–	20
medium Decne.	–	Sy	8
nigrum Moench	CP	Em	8
officinale Moench	AE	In	8
scandens Somm. et. Lev.	––	–	20
slepposum Pobed.	––	–	20
Aspleniaceae			
Onoclea sensibilis	PE	–	21
Asteraceae			
Antennaria media	PE	–	22
Balanophoraceae			
Balanophora elongata Blume	AE	En	23, 337
globosa	D + P	––	23
japonica	AE (D + P)	––	23

Table 1. Continued

Family/Genus/Species	Type[1]	Tissue/Organ[1]	Reference[2]
Betulaceae			
Alnus rugosa (Du Roi) Spreng.	AE (D + P)	En/Nu	8
Betula pubescens ssp. *tortuosa*	CP	Em	24
Bombacaceae			
Bombacopsis glabra (Pasq.) A. Robyns	AE	Nu	8
Pachira oleaginea Decne.	AE	Nu	25
Burmanniaceae			
Burmannia coelestis	D + P	--	10
Burseraceae			
Garuga pinnata Roxb.	AE	Nu	8
Buxaceae			
Sarcococca hookeriana Baill.	AE	Nu	10
humilis Stapf.	AE	Nu	10, 20, 347
pruniformis Lindl.	AE	Nu	8
ruscifolia Stapf.	AE	Nu	10, 13
zeylanica Baill.	AE	Nu	10
Cactaceae			
Opuntia aurantiaca Lindl.	AE	Nu	27
dillenii (Ker-Gawl.) Haw.	AE	Nu	28
elata Linke et Otto	AE	Nu	29
ficus-indica (L.) Mill.	AE	Nu	8
glaucophylla Wendl.	(AE?)	Nu	8
leucantha Link	AE	Nu	8
rafinesquii Engelm.	AE	Nu	8
tortispina Engelm.	(AE?)	Nu	8
vulgaris Mill	AE	Nu	8, 13
Mammillaria prolifera	AE	–	30
tenuis DC.	AE	Nu	8
zeilmanniana	AE	–	30
Caesalpiniaceae			
Bauhinia monandra	AE	Sy	31
Calycanthaceae			
Calycanthus fertilis L.	(D + P?)	--	10
floridus L.	(AE?, D + P?)	Nu	10, 13, 20
occidentalis Hook. and Arn.	(AE?, D + P?)	Nu	10, 13
Chimonanthus praecox Lindl.	--	--	13, 20
Campanulaceae			
Adenophora latifolia Fisch.	–	–	10
liliifolia	–	–	10
potaninii	–	–	10
utriculata	–	–	10
Isotoma longiflora Pres. Ind.	CP	Su	32
Lobelia syphilitica L.	CP	Su	33
Capparidaceae			
Capparis frondosa Jacq.	AE	Su, Nu	8
Isomeris arborea Nutt.	AE	En?	34
Caparifoliaceae			
Sambucus racemosa	A ± ?	–	35
Caricaceae			
Carica papaya × *C. cauliflora*	AE	Nu	36

Table 1. Continued

Family/Genus/Species	Type[1]	Tissue/Organ[1]	Reference[2]
Casuarinaceae			
Casuarina distyla Vent.	D + P	Sy	10
Celastraceae			
Celastrus scandens L.	AE	In	13
Euonymus alatus (Thumb.) Regel	AE	Nu, In, En	8
americanus L.	AE	Nu, In, En	8
dielsianus Loes.	AE	–	8
europeus L.	AE	In	8
japonica L.	AE	In	37
latifolius (L.) Mill.	AE	In, Nu, En	8
macroptera Rupr.	AE	Nu	8
maximowicziana Prokh.	AE	Nu	8
planipes (Koehne) Koehne	AE	Nu	8
oxyphyllus Miq.	AE	–	8
sachalinensis Mixim.	AE	–	8
vagans Wall. ex. Roxb.	AE	–	8
verrucosoides Loes.	AE	–	8
Chenopodiaceae			
Beta lomatogona	A + P	–	38
trigyna	D + P	–	38
vulgaris	A + P	–	39
Clusiaceae (Guttiferae)			
Clusia minor	AE	–	40
rosa Jacq.	AE	–	40
Garcinia cowa Roxb. (*G. Kydia* Roxb.)	––	––	41
livingstonii T. Anders	––	––	42
mangostana L.	––	Nu	8
parvifolia (Miq,) Miq.	D?	In	43
scortechinii (Miq.) Miq.	D?	In	43
treubii Pierre	––	––	44
Combretaceae			
Combretum paniculatum Vent.	––	––	20
pincianum Hook .	––	––	20
Poivrea coccinea DC.	––	Sy	45
Rudbeckia sullivantii Boynton et Beadle	AE	An	8
Cornaceae			
Garrya veatchii Kell.	––	Su	8
Compositae			
Ageratum conyzoides	A + P	–	46
Antennaria alpina	D + P	––	10
brainerdii	D + P	––	10
canadensis	D + P	––	10
carpathica R. Br. Reg.	A, D + P	––	10
fallax	D + P	––	10
glabrata	D (?) + P (?)	––	10
groenlandica	–	––	10
intermedia	D + P	––	10
megellanica	D + P	––	10
neodicica	D + P	––	10
occidentalis	D + P	––	10

Table 1. Continued

Family/Genus/Species	Type[1]	Tissue/Organ[1]	Reference[2]
parlinii	D + P	--	10
petaloidea	D + P	--	10
porsildii	D + P	--	10
Arnica alpina Rupp.	D + P	--	10
chamissonis Rupp.	D + P	--	10
diversifolia Rupp.	D + P	--	10
Artemisia nitida Bertol.	A, D (−P)	--	10
Carthamus tinctorius	AE	In	47
Centaurea cyanus L.	A (−P)	--	10
Chondrilla acantholepis	D + P	--	10
brevirostris	D + P	--	10
coronifera	(D + P)	--	10
graminea	D + P	--	10
juncea L.	D + P	--	10
kouzneezowii	(D + P)	--	10
lejosperma	(D + P)	--	10
ornata	(D + P)	--	10
pauciflora	D + P	--	10
Chichorium intybus L.	A + ?	--	10
Coreopsis bicolor L.	A (−P)	--	10
Crepis acuminata	A + P	--	10
atribarba	A + P	--	10
bakeri	A + P	--	10
barbigera	A + P	--	10
capillaris (L.) Wallr.	AE	Sy	48
intermedia	A + P	--	10
modocensis	A + P	--	10
monticola	A + P	--	10
occidentalis	A + P	--	10
pleurocarpa	A + P	--	10
Erigeron annuus Pers.	D + P	--	10
Karvinskianus var. *mucronatus*	D + P	--	10
ramosus	D + P	--	10
Eupatorium bupleurifolium	A + P	--	49
callilepsis	D + P	--	49
glandulosum	D + P	--	50
riparium Reg. Nota	D + P	--	51
tanacetifolia Gill. ex. H. et A (*Gyptis pinnatifida* Cass.)	D + P	Nu	52
Hieracium aurantiacum L.	AE	En	8
flagellare Reichb.	AE	An, En	53
ramosum Waldst. and Kit.	AE	Sy	54
vulgatum Fores.	AE	Sy	54
Hieracium Subg. Euhieracium (apomict)	D + P	--	10
pilosella (apomict)	A + P	--	10
Ixeris dentata	D + P	--	10
Leontodon hispidus L.	A (−P)	--	10
Leontopodium alpinus	D + P	--	10
Melampodium divaricatum DC	AE	In	55
Parthenium argentatum (apomict)	A, D + P	--	10

Table 1. Continued

Family/Genus/Species	Type[1]	Tissue/Organ[1]	Reference[2]
incanum	A, D + P	--	10
Picris hieracioides L.	A (− P)	--	10
Rudbeckia californica	D + P	--	10
deamii	D + P	--	10
lanciniata L.	D + P	--	10
speciosa	D + P	--	10
sullivantii	D + P	--	10
triloba L.	D + P	--	10
Taraxacum arctica	D + P	--	10
brachyglossum	D + P	–	56
ceratophora	D + P	--	10
crassiceps Hagl.	D + P	–	57
erythrosperma	D + P	--	10
hamatiforme Dahlst.	D + P	–	58
maculigera	D + P	--	10
obligua	D + P	--	10
officinale Wiggers	--	Sy	8
palustria	D + P	--	10
punctatum/austriacum	D + P	–	56
spectabilis	D + P	--	10
subcyanolepis	D + P	–	56
vulgaria	D + P	--	10
Tithonia rotundifolia	A + ?	Sy	59
Tridax procumbens L.	--	In/ES	60
Youngia Sect *crepidopsis*	–	--	10
Crassulaceae			
Sedum fabaria	AE	An	54
Cruciferae			
Arabis halleri L.	--	Sy	8
hirsuta Scop.	--	Sy	8
holboellii Hornem.	D + P	--	10
lyalli S. Wats.	Ha P	Sy	61
Erysimum inconspicuum	--	--	10
Raphanobrassica	A + P	Nu, In	62
Syrenia cana	PE	–	63
Cucurbitaceae			
Cucumis anguria	PE	–	64
ficifolius	PE	–	64
Melothria maderaspatana L.	AE	Nu	65
Momordica charantia L.	AE	Nu	8
Cyanastraceae			
Cyanella capensis L.	AE	Nu	66
Daphniphyllaceae			
Daphniphyllum himalayense Muell. Arg.	AE	Su	67
Dipterocarpaceae			
Hopea subalata Roxb.	AE	Nu	68
Shorea agami (Korth.) Bl.	AE	Nu?	68
ovalis (Korht.) Bl.	AE	Nu?	68
Euphorbiaceae			
Alchornea ilicifolia Muell.	AE	Nu	10

Table 1. Continued

Family/Genus/Species	Type[1]	Tissue/Organ[1]	Reference[2]
Euphorbia dulcis L.	AE	Nu	13, 69, 70
Fabaceae			
Coronilla emerus	PE	–	71
Fagaceae			
Quercus macrolepis Kotschy	AE	En	54
Garryaceae			
Garrya veatchii Dougl. ex Lindl.	CP	Su	72
Gentianaceae			
Cotylanthera tenuis	(D + P?)	--	10
Erythrae centaurium Pers.	Ha P	Sy	73
Gentiana carpatice Wettst.	--	Nu	74
livonica Fschh.	--	Nu	74
lutea L.	--	Nu	74
punctata L.	--	Nu	74
Swertia angustifolia	PE	–	75
Gramineae (Poaceae)			
Agropyron sp.	CP	Em	8
scaber Beauv.	A + P	--	76
Anthephora pubescens	A + P	--	10
Avena strigosa Vogler × *A. fatua* Viv.	--	Sy	8
Bothriochloa decipiens	A + P	--	10
ewartiana	A + P	--	10
intermedia	A + P	--	10
ischaemum	A + P	--	10
odorata	A + P	Nu	77
Bouteloua curtipendula	D + P	--	10
gracilis	--	--	10
hirsuta	--	--	10
Brachiaria brizantha	A + P	--	10
serrata	A + P	--	10
setigera	A + P, AE	Eg, En, Sy	78, 79
Buchloe dactyloides	(A + P)?	--	10
Calamagrostis canadensis	D + P	--	10
chalybaea	D + P	--	10
crassiglumis	D + P	--	10
inexpansa	D + P	--	10
langsdorffi	D + P	--	10
lapponica	D + P	--	10
purpurascens	D + P	--	10
purpurea	D + P	--	10
stricta subsp. *inexpansa*	D + P	In	80
Callipedium huegelii	A + P	Nu	77
parviflorum	A + P	Nu	10, 77
spicigerum	A + P	--	10
Chenchrus ciliaris	A + P	--	10
ciliaris × *Pennisetum americanum*	A + P	--	81
glaucus	A + P	–	82
pennisetiformis	A + P	–	83
setigerus	A + P	Nu	10, 83

Table 1. Continued

Family/Genus/Species	Type[1]	Tissue/Organ[1]	Reference[2]
Chloris andropogonoides	A + P	--	10
cucullata	A + P	--	10
gayana	A + P	--	10
verticilliata	A + P	--	10
Cortaderia jubata (Lem.) Stapf.	A (−P)	Nu	84, 85
rudiuscula	A (−P)	Nu	85
Dichanthium annulatum (Forssk.) Stapf.	A + P	Nu	10
aristatum (Poir.) Hub.	A + P	Nu	86
caricosum	A + P	--	10
nodosum	A + P	--	10
Echinochloa frumentacea occ.	A + P	--	10
stagnina (Retz.) P. Beav.	A + P	Nu	87
Eragrostis chloromelas	(D + P)?	--	10
curvula (Schrad.) Nees	D + P	An?	88, 89
Eremopogon foveolatus L. (= *Dichanthium foveolatum*)	A + P	Nu	77, 90
Eriochloa procera	AE	En	91
Heteropogon contortus	A + P	--	10
Hierochloe alpina (*Swartz*)	D + P	An	92
australis (*Schrader*)	D + P	An	93
monticola (*Bigelow*)	A + P, D + P	An	92, 94
odorata (L.) Beauv.	A + P	An	10
Hyparrhenia hirta	A + P	--	10
Nardus stricta	D + P	--	10
Panicum deustum	A + P	--	10
maximum Jacq.	A + P	--	10, 95
maximum hybrids	--	--	95
obtusum	A + P	--	10
virgatum	A + P	--	10
Paspalum chromorrhizon	A + P	-	96
dilatatum	A + P	Nu	10
distichum	A + P	Nu	77
guaraniticum	A (−P)	Nu	97
hartwegianum	A + P	--	10
macrostachyum (Brough) Trin.	A + P	-	98
macrourum Trinn.	A + P	-	98
malacophyllum	A + P	--	10
mandiocanum	A (−P)	Nu	99
notatum	A + P	--	10
paspaloides	A + P	Nu	100
polystachion (L.) Schult.	A + P	-	98
proliferum	A + ?	-	96
quadrifarium Lam.	A (−P)	Nu	97
scrobiculatum L.	A + P	Sy	10
secans	A + P	--	10
simplex	A + P	Nu	101
squamulatum Fresen.	A + P	-	98
Pennisetum ciliare	A + P	Nu	10
clandestinum	A + P	--	10
dubium	A + P	--	10

Table 1. Continued

Family/Genus/Species	Type[1]	Tissue/Organ[1]	Reference[2]
orientale	A + P	Nu	10, 77, 102
mezianum	AE	En, Nu	91
pendicellatum	A + P	Nu	103
polystachyon	A + P	--	104
purpureum	A + P	--	10
ramosum occ.	A + P	--	10
setaceum (Forssk.) Choiv.	A + P	--	10, 105
villosum	A + P	--	10
Poa alpina Guss.	D + P	Sy	10, 331
ampla	A + P	--	10
arctica subsp. *caespitans*	A + P	--	10
subsp. *depauperata*	A + P	--	10
subsp. *elongata*	A + P	--	10
subsp. *microglumis*	A + P	--	10
different forms	A + P	--	10
arida	A + P	--	10
compressa	A + P	--	10
glauca	D + P	--	10
granitica	A + P	--	10
herjedalica (= *alpina* × *pratensis*)	A + P	--	10
nemoralis	(D + P)	Sy	10, 106
nervosa	D + P	--	10
palustris	D + P	Sy	10, 106
pratensis L.	AE, D + P	Nu, Sy	13, 106
	AE, A + P	Nu, Sy	107
subsp. *alpigena*	A + P	--	10
subsp. *angustifolia*	A + P	--	10
subsp. *eupratensis*	A + P	--	10
subsp. *irrigata*	A + P	--	10
scabrella	A + P	--	10
secunda	A + P	Nu?	108
Schizachyrium sanguineum (Retz.) Alst.	A + P	--	109
Schmidtia bulbosa	--	--	10
kalahariensis	--	--	10
Setaria leucopila	A + P	--	10
villosissima	A + P	--	10
Sorghum bicolor (L.) Moench.	A + P	Nu	110–112
hybrids	A + P	Nu	112, 113
Themeda spp.	A + P	--	104
quadrivalvis	A + P	--	114
triandra	A + P	--	10, 114
Tricholaena monachne	A + P	--	114
Tripsacum dactyloides Willd.	D + P	Nu, Sy	10, 114, 115
Urochloa bulbodes	A + P	--	10, 114
mosambicensis	A + P	--	10, 114
pullulans	A + P	--	10, 114
trichopus	A + P	--	10, 114
Zea mays L.	CP	Em	8
	Ha P	Sy	116, 117

Table 1. Continued

Family/Genus/Species	Type[1]	Tissue/Organ[1]	Reference[2]
Guttiferae- (see Clusiaceae)			
Hamamelidaceae			
Hamamelis virginiana	PE	–	118
Hypericaceae			
Hypericaceae tetrapterum Fries	CP	Su	8
Hypericum perforatum	A + P	––	10
Iridaceae			
Belamcanda chinensis	Ae	In	119
Crocus speciosus Bieb	D + P?	–	120
Eleutherine plicata Herb.	AE	In	119
Iris sibirica L.	––	Sy	8
Juglandaceae			
Juglans sp.	––	––	10
regia	A + P	–	121
Pterocarya fraxinifolia Lam.	AE	Nu	122
Leguminosae			
Acacia farnesiana Willd.	PE	–	123
Albizzia lebbeck Benth.	PE	–	124, 125
Astragalus vaucasicus Pall.	PE	–	126
Cassia artemisioides Gaudich	––	––	20
australis Sims	––	––	20
desolata R. Muell.	––	––	20
eremophila A. Cunn.	––	––	20
Glycine max (L.) Merrill	Ha P	At?	127
Leucaena leucocephala Lam. De Vet	AE?	–	128
Medicago sativa L.	(AE?)	Nu/In	129
Melilotus alba × messanensis	CP	Em	130
messanensis × alba	Ha P	Sy	130
messanensis × (polonica × alba)	Ha P	Sy	130
(polonica × alba) × messanensis	Ha P	Sy	130
Mimosa denhartii Tenore.	Ha P	Sy	8
Phaseolus vulgaris	Ha P	Sy	54
Trifoloium ataphaxis	A + P	Nu	8
pratense L.	A + P	Nu	8
hybrids	A + P	–	131
Liliaceae			
Allium amplectens	(D + P)?	––	10
giganteum Regal.	AE	Sy	8
nutans L.	AE	Nu	10
odorum L.	AE	An, In, Nu, Sy	8, 13
roseum L.	AE	In	13
schoenoprasum L.	––	Sy	8
senescens L.	––	An	20
Asparagus officinalis L.	AE	–	132
Colchicum autumnale L.	––	Nu	13
speciosum Stev.	PE	Sy	133
Dianella tasmanica	––	––	10
coerulea	––	––	10
Erythronium americanum Ker-Gawl.	CP	Em	134
dens-cansis	CP	Em	135

Table 1. Continued

Family/Genus/Species	Type[1]	Tissue/Organ[1]	Reference[2]
Funkia (*Hosta*) *ovata* Spreng.	AE	Nu	18, 136
ovata Spreng.	AE	In, Nu	13
Hosta coerulea (Andrews) Tratt.	AE	Nu	8
Iphigenia indica Kunth.	--	Sy	137
Lilium elegans	AE	Sy	138
florum	AE	Sy	138
hansoni	AE	Sy	138
henryi	AE	Sy	138
longiflorum	AE	Sy	138
martagon	AE	Sy	138
regale	CP, AE	Em, Sy	8, 138
superbum	AE	Sy	138
usitatissum × *pallescens*	AE	Sy	8
Nothoscordum fragrans Kunth.	AE	Nu	8, 13
Smilacina racemosa (L.) Desf.	AE	Nu	8, 13
Trillium sp.	AE	En	139
undulatum Willd.	AE	Nu	54
Tulipa gesneriana (Ker-Gawl.)	AE	Nu, Su	8, 13
Linaceae			
Linum usitatissimum	Ha P	–	140
Loranthaceae			
Loranthus europaeus Jacq.	CP	Em	8
Malphigiaceae			
Hiptage madablota gaertn.	AE	Nu	10
Malvaceae			
Gossypium barbadense L.	Ha P	–	141
hirsutum L.	AE?	–	141
barbadense × *davidsonii*	Ha P	Sy	142
Melastomaceae			
Memecylon fasciculare	PE	–	143
Sonerila wallichii Benn.	CP	Em	144
Meliaceae			
Aphanamixis polystachya (Wall) Parker	AE	Nu	8
Azadirachta indica A. Juss	Ha P	Sy	145
Lansium domesticum	AE	Nu	146
Moraceae			
Morus indica	AE	Nu, In	147
laevigata	AE	Nu, In	147
Streblus taxoides (Heyne) Kurz.	--	Em	8
Myrsinaceae			
Ardisia crispa A.D.C.	AE	In	13
Myrtaceae			
Callistemon lanceolatus Sweet.	--	--	20
Eugenia sp.	AE	Nu	8
cumingii Hook.	(AE?)	–	20
grandis Duthia.	AE	Nu	8
hookeri Steud.	AE	Nu	8
jambolana Lam.	AE	Nu	8
jambos Blanco	AE	Nu, In	148, 149
malaccensis L.	AE	In	8

Table 1. Continued

Family/Genus/Species	Type[1]	Tissue/Organ[1]	Reference[2]
myrtifolia Sims.	--	--	13, 20
ugni Hook.	--	--	13, 20
Myricaria cauliflora	--	--	150
Syzygium cumini (L.) Skeels	AE	Nu	151
caryophyllifolium D.C.	AE	Nu	152
Najadaceae			
Najas major All.	AE	Sy	153
Nyctaginaceae			
Boerhaavia repanda Willd.	--	Sy	8
Nymphoeaceae			
Nymphaea advena Dryland	--	Em	8
Ochnaceae			
Ochna cerrulata Walp.	AE	Nu	20
Onagraceae			
Clarkia elegans Dougl.	AE	Nu	13
Epilobium angustifolium L. × *montanum* L.	AE	Nu	13
hirsutum L. × *dodanaei* Vill.	AE	Nu	13
biennis × *muricata*	AE	Nu	13
Oenothera lamarckiana Seringe.	AE	In, Nu	13, 154
Orchidaceae			
Bletilla striata var. *gebina*	Ha P	--	10
Bulbophyllum mysorense	CP	Em	155
Cephalantheria damasonium	Ha P	--	1
latifolia	Ha P	--	1
longifolia	Ha P	--	1
Coelogyne ilicifolia	--	Nu	18
Cymbidium bicolor	CP	Zy	156
ensifolium	CP	Zy	157
pendulum	CP	Zy	157
Dactylorchis sp.	--	--	20
Dulophia nuda	CP	Em	155
Epidendrum nocturnum	AE	--	158
Epipactis helleborine (L.) Crantz.	AE	Nu	159
veratrifolia Boiss & Hohn	AE	Nu	159
Eulophia epidendraea	CP	Zy	160
Geodorum densiflorum	CP	Zy	161
Granorum densiflorum	CP	Em	155
Gymnadenia conopsea (L.) R. Br.	AE	Nu	18
Habenaria platyphylla	CP	Zy	54
Leuxine sp.	AE, CP	In, Nu, Su	155
Listera ovata R. Br.	Ha P	Sy	54
Maxillaria cleistogama Brieg. et Illg	AE	In, Nu	162
Nigritella nigra (L.) Reichend.	AE	Nu	163, 164
widderi	AE	Nu	165, 166
Odontoglossum crispum Lindl.	AE	Sy	13
Orchis maculatus	Ha P	Sy	167
strictifolius	Ha P	Sy	10
Platanthera chlorantha	Ha P	Sy	10
Spathoglottis aurea	CP	Em	168
plicata	CP	Zy	161

Table 1. Continued

Family/Genus/Species	Type[1]	Tissue/Organ[1]	Reference[2]
Spiranthes australis Lindl.	AE	In	20
cernua (L.) Rich.	AE	In, Nu, Su	169, 170
Zeuxine longilabris	AE	Nu	171
sulcata Lindl.	AE	Nu	172
strateumatica (L.) Schltr.	AE	Nu	173
Zygopetalum mackayi Hook.	AE	Nu	54
mackaii Hook. ×			
Palmae			
Cocos nucifera L.	AE, CP	Em, Zy	174, 175
Rhapidophyllum hystrix	PE	–	176
Papaveraceae			
Argemone mexicana L.	AE	Sy	177
Pittosporaceae			
Pittosporum floribundum Wight et Arn.	AE	Su	178
heterophyllum Franch.	AE	Su	178
Plantaginaceae			
Plantago lanceolata L.	Ha P	Sy	54
Plumbaginaceae			
Statice olaeefolia var. *confusa*	D + P	--	10
Limonium binervosum	--	--	10
lychnidifolium	--	--	10
Poaceae (see Graminaea)			
Polygonaceae			
Acetosa arifolia	--	--	10
pratensis	--	--	10
thyrsiflora	--	--	10
Atraphaxis frutescens	A + P	Nu	179
Portulaceae			
Portulaca oleracea L.	AE	En?	54
Primulaceae			
Primula auricula L.	CP	Em	8
japonica × *chungensis*	--	--	10
japonica × *cockburnia*	--	--	10
japonica × *pulverulenta*	--	--	10
wilsoni × *cockburnia*	--	--	10
Ranunculaceae			
Eranthis hiemalis Salisb.	--	Em	8, 180
Nigella arvensis L.	AE	An	54
Ranunculus acris	--	--	10
aemulans Schwarz.	A + P	–	181
auricomus coll.	A + P	--	10
bulbosus	--	--	10
opimus Schwarz.	A + P	–	181
palmularis Schwarz.	A + P	–	181
pseudopimus Schwarz.	A + P	–	181
varicus Schwarz.	A + P	–	181
vertimualis Schwarz.	A + P	–	181
Rhamnaceae			
Zizyphus rotundifolia Lamk.	--	Wm, Sy	182

Table 1. Continued

Family/Genus/Species	Type[1]	Tissue/Organ[1]	Reference[2]
Rosaceae			
Alchemilla acutangula Buser	--	--	20
subg. *Aphanes arvensis*	A + P	--	10
subg. *Eualchemilla alpina* coll.	(A + P)	--	10
alpestris Schmidt.	A + P	–	183
alpina L.	--	--	20
acutiloba	--	--	10
glabra	--	--	10
glomerulans	--	--	10
micans	--	--	10
pastoralis Buser	AE	Nu	10, 183
subcreanata	--	--	10
vulgaris coll.	(A + P)	--	10
Amelanchier laevis Wiegand	A + P	Nu	184
Chaenomeles sp.	PE	Sy	185
Cotoneaster acutifolia var. *villosula*	A + P	--	10
bullata	A + P	--	10
obscura	A + P	--	10
var. *nova*	A + P	--	10
racemiflora var. *soongorica*	A + P	--	10
rosea	A + P	–	10
Crataegus (apomicts)	--	--	10
calpodendron (Ehrh.) Medic.	A + P	Nu	186
chrysocarpa Ashe	A + P	Nu	186
dissona Sarg.	D + P	Nu	187
dodgei Ashe.	A + P	Nu	186
flavida Sarg.	A + P	Nu	186
C. sp. aff. *Macrosperma* Ashe	A + P	Nu	186
lindmanii Hrabet.-Uhr.	A + P	–	188
macrocarpa Hagetschw.	A + P	–	188
palmstruchii Lindm.	A + P	–	188
pruinosa (Wend.) K.Koch	A + P	Nu	186
Schueltei Ashe var. *basilica*	A + P	Nu	186
var. *cuneata*	A + P	Nu	186
Fragaria grandiflora	A + P	Nu	54
vesca L.	AE	Sy	189
Geum rivale L.	--	Sy	8
Malus hupehensis	A + P	Nu	10
sargentii var. rosea	(A + P)?	--	10
sieboldii (Reg.) Rehd.	A + P	Nu	190, 191
sieboldii Hybrid	A + P	Nu	190
sikkimensis	(A + P)?	--	10
toringoides	(A + P)?	--	10
var. *macrocarpa*	A + P	--	10
zumi (Mats.) or [*sieboldii* × *baccata* (Max.)]	A + P	Nu	192
Potentilla arenaria	D + P, A + P	--	10
argentea	A + P	--	10
arguta	--	--	10
argyrophylla	--	--	10

Table 1. Continued

Family/Genus/Species	Type[1]	Tissue/Organ[1]	Reference[2]
aurea L.	--	--	10
breweri	--	--	10
canescens	P	--	10
chrysantha	--	--	10
collina	P	--	10
crantzii	(D + P)?	--	10
dealbata	--	--	10
degeni	--	--	10
diversifolia	--	--	10
dissecta	--	--	10
drummondii	--	--	10
erecta	PE	–	193
flabelliformis	--	--	10
gelida	--	--	10
geoides Beid	AE	Nu	13, 20
gracilis	--	--	10
hirta	--	--	10
intermedia	--	--	10
kurdica	--	--	10
levieri	--	--	10
multifida	--	--	10
nepalensis Hook. × *splendens* Ram.	AE	Nu	13
nepalensis Hook. × *argyrophylla* Wall.	AE	Nu	13
nevadensis	--	--	10
praecox	A + P	--	10
pulcherrima	--	--	10
recta	P	--	10
reptans L.	AE	In, Nu	194
verna (including) *P. tabernaemontani*	A + P, D + P	--	10
Rubus subg. *Eubatus* sect. *morifera*			
American apomicts	A + P	--	10
European apomicts	A + P	--	10
idaeeus	A + P	--	10
Sanguisorba minor complex	A + P	Nu	195
Sorbus aria coll.	--	--	10
(*obtusifolia*)	--	--	10
(*rupicola*)	--	--	10
chamaemespilus	A + P	--	10
eximia Kovanda (*aria* × *torminalis*)	A + P, D + P, AE	Nu	196
hybrida (= *fennica*)	A + P	--	10
intermedia	--	--	10
lancifolia	--	--	10
meinichii	--	--	10
minima	--	--	10
mougeoto	--	--	10
sudetica	AE, Ha P	--	10, 197
Rutaceae			
Aegle marmelos (L.) Corr.	AE	Nu	8, 198
Citrus aurantifolia (Christm.) Swing.	AE	Nu	199

34

Table 1. Continued

Family/Genus/Species	Type[1]	Tissue/Organ[1]	Reference[2]
aurantium L.	AE	Nu	8, 13, 18, 200, 201
bigaradia Risso & Poit.	AE	Nu	200
grandis (L.) Osb.	AE	Nu	8
jambhiri Lush.	AE	Nu	199
karna Raf.	AE	Nu	199
limon (L.) Burm.	AE	Nu	8, 199, 202
limomum Risso	AE	Nu	8
microcarpa	AE	Nu	203
nobilis Lour.	AE	Nu	8, 200, 204
paradisi Macfadyn	AE	Nu	8, 199, 201
pennivesiculata (Lush.) Tan.	AE	Nu	8
pseudoparadisi Hort. ex Y. Tan.	AE	Nu	202
reticulata Blanco	CP	Em, Su	8, 199
sinensis (L.) Osb.	AE	Nu	8, 199
sulcata Hort.	AE	Nu	205
tamurana Hort. ex Tan.	AE	Nu	8
trifoliata	AE	Nu	13
unshiu (Mak.) Marc.	AE	Nu	205, 206
Citrus species & Hybrids	AE	Nu	8
Citrus × *Poincirus* hybrids	PE	–	207
Eremocitrus glauca (Lindl.) Swing.	AE	Nu	208
Fortunella crassifolia Swing.	AE	Nu	208
Fortunella sp.	AE	Nu	8, 202
Murraya exotica Blanco	AE	Nu	8
koenigii (L.) Spreng.	AE	Nu	8
Poncirus trifoliata (L.) Raf.	AE	Nu	8
Ptelea trifoliata L.	AE	Nu	8
Skimmia japonica	A + P	--	10
lavellei	A + P	--	10
Toddalia asiatica Lamb.	AE	Nu	209
Triphasia aurantiola Lour.	AE	Nu	10
Xanthoxylum alata Wall.	AE	Nu	8
alutum Wall.	AE	Nu	10
americanum Mill.	AE	Nu	210
bungei Hance	AE	Nu	8, 210
planisoinum Seib.	AE	Nu	210
simulans Hance	AE	Nu	210
Salicaceae			
Populus deltoides L. hybrids	CP	Em	211
Santalaceae			
Exocarpus bidwillii Hook	CP	Su	212
sparea (= *sparteus*) R. Br. Prod.	CP	Su	213
Sapindaceae			
Allophylus alnifolius	A + P, AE	–	214
Soururaceae			
Houttuynia cordata	D + P	–	10
Saxifragaceae			
Bergenia delavayi	Ha P	Sy	215
Mitella ovalisfolia Greene	--	--	20

Table 1. Continued

Family/Genus/Species	Type[1]	Tissue/Organ[1]	Reference[2]
Tellima grandifolia R. Br.	AE	Sy	216
Solanaceae			
Capsicum annuum L.	AE?	–	217
frutescens L.	CP, Ha P	Em, Sy	218
Lycopersicon esculentum	CP	Em	219
Nicotiana glutinosa L. × *Tabacum* L.	Ha P	Sy	8
rustica L.	AE	Em, Nu	13, 54
rustica L. var. *brasilia* × *Petunia* sp.	CP	Em	8
Nicotiana tabacum	Ha P	Sy	54
violaceae	AE	Nu	13
Scopolila carniolica Jacq.	AE	In	13
Withania somnifera Don.	--	--	13, 20
Stachyuraceae			
Stachyurus chinensis Franch	AE, CP	Em, Sy	220
Symplocaceae			
Symplocos cochinchinensis	Ha P	Sy	221
foliosa	AE	Sy	221
paniculata	AE, Ha P	Sy	221
Symploeos klotzschii Bland.	--	--	20
Tamaricaceae			
Tamarix ericoides Rottl.	--	Sy	222, 223
Theaceae			
Thea sinensis L.	AE	Sy	13, 20, 224
Thymelaeaceae			
Wikstroemia indica C.A. Mey.	D + P	–	13, 20
viridifolia	D + P	--	225
Ulmaceae			
Ulmus americana L.	AE	An	8
glabra Huds.	AE	An	8
Umbelliferae (Apiaceae)			
Ammi majus Walt.	--	Nu	8
majus Walt.	AE	In	13
Coriandrum sativum	CP	Em	226
Foeniculum vulgare Mill.	CP	Em	226
Pimpinella candolleana W. & A.	AE, CP	Em, Sy	227
Urticaceae			
Elatostema acuminatum Brongn.	A + P	Nu	20, 228
eurhynchum Mig.	D + P	Nu	20, 228
eusinatum	AE	An	54
latifolium	A + P	--	10
machaerophyllum	D + P	--	10
peltifolium	D + P	--	10
penniverve	--	--	10
repens	--	--	10
rostratum	--	--	10
sinuatum	D + P	--	10
Procris frutescens	--	--	10
Violaceae			
Viola incognita	A + P	–	229
pallens	A + P	–	229

Table 1. Continued

Family/Genus/Species	Type[1]	Tissue/Organ[1]	Reference[2]
Miscellaneous Families			
Peltiphyllum peltatum	Ha P	Sy	230
Cardiocrinum giganteum	Ha P	Sy	54
Lumnitzera racemosa	PE	–	231
Gymnosperms			
Cephalotaxaceae			
Amentotaxus sp.	CP	PS	232
Cephalotaxus drupacea Sieb. et Zucc.	CP	Su	233
fortunei	CP	PS	234
Cupressaceae			
Biota orientalis Endl.	CP	Su	235
Callistris calcarata	CP	Su	236
drummondii	CP	Su	236
macleayana	CP	Su	236
muelleri	CP	Su	236
robusta	CP	Su	236
roei	CP	Su	236
spp.	CP	Su	237
Chamaecyparis obtusa	CP	PS	238
nootkatensis	CP	PS	239
Cupressus arizonica	CP	PS	240
funebris Endlicher	CP	PS	241
sempervirens L.	CP	Su	242
Juniperus communis	CP	Su	243
verginiana L.	CP	Su	244
Cycadaceae			
Encephalartos friderici guilielmi	AE	–	245
villosus	AE	–	245
Ephedraceae			
Ephedra distachya L.	CP	PS	246
Gnetaceae			
Gnetum funiculare	CP	PS	247
gnemon L.	CP	Su/PB	248
molulcense	CP	PS	247
Pinaceae			
Abies amabalis	CP	PB	249, 250
balsamea	CP	PB	249, 251
grandis	CP	PB	249
pindrow	CP	PB	252
Cedrus deodara (Roxb.) Loud.	CP	Su	253
libani	CP	Su	254
Keteleeria davidiana	CP	Su	255
Larix sp.	CP	Su	256
Pseudotsuga menziesii (Mirb.) Franco	AE?	–	257, 258
Pinus banksiana	CP	Su	254
cembroides Zucc. var. *monophylla* Voss	CP	–	259
contorta	CP	Su	260
coulteri Don.	CP	–	261
edule	CP	Su	254
gerardiana	CP	PB	262

Table 1. Continued

Family/Genus/Species	Type[1]	Tissue/Organ[1]	Reference[2]
jeffreyi (Murr.) Vasey	CP	–	263
lambertiana Dougl.	CP	–	263, 264
laricio	CP	Su	254
monophylla	CP	PB	259
monticola Don.	CP	PB	261, 265
muricata Don.	CP	–	263
nigra var. anstiaca	Ha P?	–	266
pinaster	Ha P?	–	267
pondorosa	CP	Su	254
resinosa	CP	Su	254
roxburghii Sarg.	CP	PB	268
sabiniana Dougl.	CP	–	259
strobus L.	CP	Su	254, 263
sylvestris	CP	Su	254
tabulaeformis	CP	Su	269
thunbergii Parl.	CP	–	270
torreyana Parry	CP	–	259
wallichiana Jack.	Ha P?	–	262
Pseudolarix sp.	CP	Su	271
Tsuga canadensis (L.) Carr.	CP	PS	254
mertensiana (Bong.) Carr.	CP	PS	272
Podocarpaceae			
Dacrydium bidwilii	CP	PS	273
colensoi	CP	PS	274
cupressinum	CP	PS	275
laxifolium	CP	PS	276
Pherosphaera hookeriana	CP	Su	277
Phyllocladus alpinus	CP	PS	278
Podocarpus acutifolius T. Kirk.	CP	PB	279
amarus	CP	PS	278
blumei	CP	PS	278
coriceus	CP	PS	278
dacrydioides	CP	PS	278
glomeratus	CP	PS	278
gracilior Pilger	CP	PS	278, 280
hallii T. Kirk.	CP	PB, PS	278, 279
imbricatus	CP	PS	278
macrophyllus maki (= *P. chinensis*)	CP	PS	278
matudai	CP	PS	278
nagi	CP	PS	278
nankoensis	CP	PS	278
purdeanus	CP	PS	278
nivalis Hook.	CP	PB	279
totarra D. Don.	CP	PB, PS	278
		PS	279
urbanii	CP	PS	278
usambarensis	CP	PS	279
Taxaceae			
Taxus cuspidata	CP	PS	281
Torreya nucifera	CP	Su	282

Table 1. Continued

Family/Genus/Species	Type[1]	Tissue/Organ[1]	Reference[2]
Taxodiaceae			
Cryptomeria japonica D. Don	CP	PS	283
		Su	284, 285
Cunninghamia lanceolata Hooker	CP	Su	285, 286, 287
konishii Hayata	CP	Su	288
Glyptostrobus sp.	CP	PB	289
Metasesequoia	CP	Su	290, 291
Sequoia gigantea	CP	PS	292
semipervirens	CP	PS	293
Sciadopitys sp.	CP	PB/PS?	294
Sequoiadendron sp.	CP	PB	293
Taxodium mucronatum Tenore	CP	Su	284, 295
Taiwania cryptomerioides Hayata	CP	PB	296
Welwitschiaceae			
Welwitschia mirabilis	CP	–	297

[1] Abbreviations: A = Apospory; AE = Adventive Embryony; An = Antipodals; At = Anthers; CP = Cleavage Polyembryony; D = Diplospory; Eg = Egg; Em = Zygotic embryo; ES = Embryo Sac; En = Endosperm; Ha = Haploid; In = Integument; Nu = Nucellus; OW = Ovary Wall; P = Parthenogenesis; PB = Proembryo; PE = Polyembryony (not certain about origin); Su = Suspensor; Sy = Synergid; Zy = Zygote; – indicates that the information has not been provided by the authors; –– indicates that the original references were not available for confirmation.
[2] This table is based in part on previously published reviews [8, 10, 20, 54].

II. Asexual embryogenesis in angiosperms *in vivo*

In nature, agamospermy includes the formation of seeds by apomictic means and primarily consists of three phenomena, viz., diplospory followed by parthenogenesis; apospory followed by parthenogenesis, and adventitious embryogenesis (see below). In the first two instances, gametophytes are formed, either from an MMC or its derivatives, or from a somatic cell in the ovule, and the new sporophyte arises from cells in the gametophyte. In the third case, the sporophytic generation (somatic cells in the ovule) give rise directly to a new sporophyte (an embryo). Unlike adventive embryony, all other forms of agamospermy involve an embryo-sac (female gametophyte). Apomictics may be obligate, in which case heterozygotic genotypes are preserved at the cost of evolutionary flexibility. More often, apomixis is facultative, in which case apomictic and sexual reproduction coexist. Pollination may or may not be required for the production of functional parthenogenetic seed.

A. Diplospory

In diplospory, a diploid embryo-sac is formed from a megaspore mother cell (MMC), without a regular meiotic division. The cytological events are often complex, and classification of various subtypes have been discussed elsewhere [300, 301]. Essentially the variation is in the way meiosis fails in the ovule. Diplospory may be divided into two groups depending on the type of division the MMC undergoes:

Type 1: MMC undergoes division begining like a meiotic prophase; some degree of pairing of the homologous chromosome may also take place. The dissociation of the chromosomes occurs with the formation of a restitution nucleus. The MMC with the restitution nucleus may directly develop into an embryo-sac, or it may divide into two cells by a mitotic division of which one degenerates and the other forms the embryo-sac. In triploid species like *Ixeris dentata* the MMC divides by a semi-heterotypic division where there is no pairing [311]. At anaphase the chromosomes are scattered on the spindle and, finally, a restitution nucleus is formed. The latter undergoes three mitotic divisions leading to the formation of an 8-nucleate embryo-sac. The diploid species of *Taraxacum* are sexually reproducing while the polyploid species are apomictic. In the latter, the MMC divides by a heterotypic division resulting in the formation of a restitution nucleus. The MMC, with the restitution nucleus, divides by a regular mitosis and forms a dyad. Usually the chalazal cell of the dyad gives rise to the embryo sac while the micropylar cell degenerates. This type of diplospory is typical to *Balanophora* [312], *Chondrilla* [313, 314], *Elastostoma* [228], *Enhieracium* species [315] and *Wikstroemia viridifolia* [225].

Type 2: In this case, the nucleus of the MMC directly undergoes the normal mitotic divisions to form an unreduced embryo-sac as in *Eupatorium glandulosum* [50] and *Antennaria* species [316, 317]. The resting stage of the nucleus in the MMC is very long and the division of MMC normally occurs when the meiosis in the pollen mother cells have reached the tetrad stage. This type is also referred to as the *Antennaria* type [10].

B. Apospory

The MMC undergoes the usual meiotic divisions and forms a tetrad. The embryo-sac may arise either from a cell of the archesporium (generative apospory) or from some other part of the nucellus (somatic apospory). There is no reduction in the number of chromosomes and hence, all the nuclei of the embryo-sac are diploid. The embryo finally may arise either from the egg (diploid parthenogenesis) or from some other cell of the gametophyte (diploid apogamy).

In the genus *Hieracium*, the megaspore mother cell goes through the usual meiotic divisions but at an early stage, a somatic cell situated in the chalazal region begins to enlarge and becomes vacuolated. This cell gradually increases in volume, encroaching upon the megaspore and finally crushing them. The aposporic embryo-sac arising from it has the unreduced chromosome number and is able to function without fertilization. In *H. aurantiacum*, the aposporic embryo-sac usually originates from a cell of the nucellar epidermis [1, 53]. While in the aposporic members of most angiosperm families only one nucellar cell acts as the mother cell and gives rise to a structurally normal, 8-nucleate embryo-sac, in the grasses, on the other hand, more than one aposporic embryo-sac may develop in the same nucleus, and the organization of the mature embryo-sac is 4-nucleate (3-celled egg apparatus and a single polar nucleus). In *Poa arctica* and *P. pratensis* a number of aposporic embryo-sacs may co-exist with the sexual embryo-sac formed by a haploid megaspore. The somatic cells do not start aposporous development until the true archespore has degenerated after completing meiosis. The cells destined for apospory show the typical cellular and nuclear growth accompanied by vacuolization of cytoplasm [301]. Just as the nucellar embryos of *Citrus* allow the sexual embryo-sac to undergo regular sexual fertilization and embryony, so do the nucellar embryo-sacs of apospory allow the archesporium to undergo a reductional meiosis which can result in nuclei of a sexual and agamospermous origin co-existing in the ovule. Thus the mature seed may have both the types (as is often the case in *Poa*), or only one (most frequently the agamospermous one) may survive. Hence, in aposporous mechanisms the archesporium, by definition, remains intact, and there is always the potential for sexuality. In such plants, agamospermy must always be considered at least potentially facultative rather than obligate. In the three aposporous groups most thoroughly investigated, the *Poa pratensis, P. alpina* complex, the *Potentilla verua* and *P. argentia* complexes, and the *Ranunculus auricomus* aggregate, agamospermy is very often facultative. Polyembryony is well-known in all three groups, and nucellar agamospermy co-exists with archesporial sexuality within the population, the individual, or even the ovule.

The aposporic embryo-sac is usually 8-nucleate whether formed by generative or by somatic apospory, although some variations have been observed; less than 8-nuclei in *Ochna serrulata* [318], and *Atraphaxis frutescens* [179] or more than eight as in *Elastema eurrhynchum* [228]. Very often, the various elements of the embryo-sac show disturbed polarity and lack of proper organization. Frequently, the egg and synergids are indistinguishable from one another, and occasionally there may be more than two polar nuclei, while the remaining elements of the apomictic embryo-sac remain undifferentiated [1]. In the *Atraphaxis* some embryo-sacs lacked an egg apparatus and some antipodal cells or both [179].

Theoretically the new sporophyte may arise from any cell or nucleus of the diploid embryo-sac, but usually only the egg is capable (diploid parthenogenesis). Occasionally one or both of the synergids, or antipodal cell may give rise to embryos (diploid apogamy). Most often, embryo development requires the stimulus of pollination or triple fusion (pseudogamy), while in some plants embryogenesis may occur without the stimulus of pollination (autonomous apomixis) [1].

C. Adventitious embryony

In adventitious embryony the embryos arise in the ovules but outside the embryo-sac. Hence the sporophyte generation gives rise directly to a new sporophytic generation. Adventitious embryony is often combined with true polyembryony. This phonemonan is found in a number of species mainly belonging to the families of Buxaceae, Cactaceae, Myrtaceae, Orchidaceae and Rutaceae (see also Table 1). Typically, pollination is followed by the double fertilization of a conventional reduced sexual embryo-sac and the embryo development results in the growth of further apomictic embryos in the nucellus or some other cell in the female gametophyte other than the egg cell (pollination by itself may also act as a stimulus).

Various fates of the sexual and the apomictic embryos are possible, but typically only one of the embryo and the apomictic embryos invades the embryo-sac, outcompeting the other apomictic and sexual embryos, and commandeers the sexually produced endosperm. Alternatively, the sexual embryo and one of the apomictic embryos co-exist, sharing the same endosperm. Thus the mature seed may have one sexual embryo, or two (or rarely, more) embryos, one of which is sexual and the other apomictic in origin. In common with aposporous agamosperms, generally there remains a dependence on the production of an endosperm by conventional sexual means (pseudogamy).

The species which undergo this process can be divided into two groups, the first of which need no fertilization for the formation of an endosperm, e.g. *Alchornea ilicifolia* [10], *Euphorbia dulcis* [69, 319], *Opuntia aurantia* [27]. Here the egg cell degenerates at an early stage, while adventitious embryos are formed from nucellus cells close to the embryo-sac. The endosperm develops autonomously just as in many diplosporous species. The second group contains species which require fertilization for the formation of an endosperm, e.g. the genus *Citrus*, where many species form adventitious embryos. Here the EMC forms an embryo-sac in a normal way, the egg cell of which is fertilized. In proximity to the egg apparatus adventitious initial cells arise which later grow out into embryos. The development of the normal embryo may be completed or it may degenerate.

D. Adventive embryogenesis in specific tissues

According to Johansen [13], adventive embryogeny includes all instances in which an embryo develops directly from a nucellar or integumental cell or from a group, isolated or not, of such cells. Hence, occasionally this phenomenon has also been referred to as nucellar embryony. With regards to the development of one or both synergids, antipodal cells and polar or primary endosperm nuclei into embryos, he designates under the term apogamety. In the strict sense, when a cell or nucleus develops into an embryo without actual fertilization it involves apogamety. However, for the sake of convenience, in this chapter, the term adventive embryogenesis has been used to all cases where the asexual embryos arise from cells outside the embryo sac (nucellus and integument) and from cells of the embryo sac other than the egg (synergids, antipodals and endosperm). Cleavage polyembryony from the zygote and suspensor has also been included under this section.

1. Nucellus

The inception of nucellar embryos generally takes place outside the embryo sac but they are gradually pushed into the embryo sac cavity where they divide and differentiate into mature embryos (see Fig. 2). Due to the lack of their synchronous development, a seed may show embryos at various developmental stages. In the Rutaceae, as well as in other families of angiosperms, it has not been unequivocally resolved whether or not pollination and/or fertilization are essential for adventitious embryogenesis *in vivo*. It has been suggested that pollination and fertilization are prerequisites for the initiation of adventive embryos [15, 17, 18, 203, 320, 321]. In some cases pollination but not subsequent fertilization has been required, e.g. *Citrus* sp. [204], *Coelobogyne ilicifolia* [18]

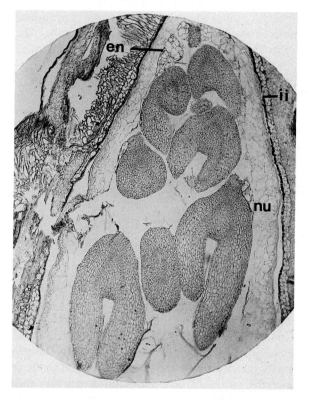

Fig. 2. Longisection of a part of the seed of *Citrus microcarpa* showing nucellar embryos in various stages of development; en = endosperm; ii = inner integument; nu = nucellus. (Courtesy: Professor N.S. Rangaswamy, Botany Department, University of Delhi, Delhi.)

Sarcococca pruniformis [322], and *Xanthoxylem* sp. [210], while even pollination is not necessary in *Nigritella nigra* and *Zygopetalum mackayi* [323], *Opuntia aurantiaca* [27], *Spiranthes cernua* [169], and *Citrus* [324]. However, Johri and Ahuja [198] reported that adventitious embryos develop in seeds of *Aegle marmelos* in which only triple fusion and subsequent endosperm development had occurred and 75% to 80% of the mature embryo sacs had degenerated completely. In the remaining embryo sacs the egg apparatus and the antipodals degenerate. From their interesting studies on the occurrence of adventive embryogenesis in the unfertilized seeds from control-pollinated fruits of apomictic *Citrus* cultivars, Wakana and Uemoto [324] made three important conclusions: (1) without pollination or fertilization, adventive embryos are autonomously initiated and develop to consume most of the nucellus; (2) adventive embryos inititiated in the chalazal end of the nucellus are larger than those in the micropylar end (see also Fig. 2); (3) adventive embryo development is slightly, but significantly, suppressed by pollination and fertilization of other seeds within the same fruit. The adventive embryos of *Citrus* may however initiate *in vivo* without fertilization, but are dependent on the endosperm for their continued development [325]. In *Eugenia jambos* the nucellar embryos are formed prior to pollination, but they attain full size only after pollination, fertilization, and endosperm formation [148, 149]. Similar observations have also been reported in *Syzygium cumini* [151] and *S. caryophyllifolium* [152] where nucellar embryos develop only after fertilization of the egg cell, development of endosperm and degeneration of the zygote.

The embryo initial cells divide to form clusters of meristematic cells, often called nucellar buds [12, 210, 326]. The nucellus degenerates eventually and serves as nourishment for the developing embryos. In *Opuntia aurantiaca* [27] and *O. dillenii* [28], nucellar embryogenesis occurs without pollination where the embryo sac degenerates and is displaced by the enlarging nucellus. The elongated cells of the nucellar-cap (the micropylar region cells), bordering the embryo sac and characterized by unevenly thickened walls have been identified as the initials. In the orchids *Nigritella nigra* and *Zygopetalum mackaii* asexual embryogenesis begins after the zygote proliferates into a four-celled embryo and the nucellar embryos originate in a single apical cell that contains a prominant nucleus [163, 164, 323]. In *Mangifera indica* the process of asexual embryogeny begins after the zygotic embryo has started to develop and the nucellus has started to degenerate [14, 15, 17, 19]. In *Cyanella capensis*, the accessary embryos are produced from the persistent nucellar cells, and the zygotic embryo is superseded by the faster growing nucellar embryos [66]. As many as 15 nucellar embryos may be formed in a single ovule of *Pachira oleaginea*, but only 5 or 6 attain maturity [25]. Almost 50 embryos in a seed were observed in this species [17]. In *Toddalia asiatiaca* the nucellar embryos develop in the middle region of the embryo sac when the zygotic embryo is 32- to 64-celled. The growth of the zygotic embryo is arrested at this stage, whereas the nucellar embryo enlarges considerably and shows well-differentiated procambial strands [209].

2. Integument

The inner epidermis of the inner integument of the bitegmic ovules and that of the single integument in the unitegmic ovules becomes differentiated as a specialized layer of the radially stretched cells with dense cytoplasm and prominent nuclei. This is also referred to as "endothelium" [327]. In general, only the inner integument and not the outer, has been shown to be involved in asexual embryogenesis [13, 37], and is mainly confined to the micropylar end [170]. The development of such embryos is not dependent on the fertilization of the egg cell, and there are no interpolation of the gametophytic phase. In *Spiranthes cernua* [170], the terminal cell and the lower adjoining cells enlarge and become richly cytoplasmic and divide to produce 2–6 proembryos while the sexual embryo sac degenerates. The formation of a single asexual embryo was reported in *Potentilla reptans* by Soueges [194]. In *Eugenia malaccensis* and *Scopolia carniolica* [13], a callus-like tissue serves as an intermediary while the embryos usually emerge directly from the integument cells. In *Flavaria repanda* failure of fertilization resulted in the development of globular structures, simulating embryos, from the endothelium [328]. In incompatible matings of *Solanum demissum* × *S. gibberulosum*, Walker [329] also found that the endothelial cells, following fertilization, proliferated into finger-like projections which extended into the embryo sac. Occasionally these cells divided and formed two-rowed structures which resembled the proembryos. Several developmental stages of the proembryo which resemble earlier stages of the zygotic embryo could be observed in *Melampodium divaricatum* [55]. In about 54% of the ovules the egg and secondary nuclei show signs of degeneration but the endothelial cells increase in size, become densely cytoplasmic and vacuolated. Some of these appear egg-like and divide by a transverse wall into two cells which subsequently undergo divisions to form structures resembling zygotic embryos. However, neither these nor the supernumerary embryos of *Carthamus tinctorius* developing from the endothelial cells during post-fertilization stages reached maturity [47]. Acording to Lakshmanan and Ambegaokar [303], so far, not a single instance of polyembryonic seed due to differentiated integumentary or endothelial embryos is known and they question the consideration of such multicelled and globular structures as proembryos.

3. Synergids

Synergids are a common source of asexual embryos which frequently become egg-like and are capable of developing into embryos with or without fertilization. Identical twins can be produced as a result of fertilization of the egg and one of the synergids by the two male gametes as in *Najas major* [153], *Tellima grandiflora* [216], *Peltiphylum peltatum* [230], and *Fragaria vesca* [189]. The twin embryos however abort if only two sperms enter an ovule resulting in no triple fusion and consequently no endosperm. In *Peltiphyllum peltatum*, Lebegue [230] claims that the twin embryos are able to develop even in the absence of endosperm. In some other cases, entry of an additional pollen tube, or the presence of additional sperm in the same pollen tube have resulted in the fertilization of the egg and one or both synergids and formation of endosperm. These include *Cuscuta reflexa* [330], *Atraphaxis frutescence* [179], *Crepis capillaris* [48], and *Sagittaria graminea* [9]. In *Tamarix ericoides* [222] and *Pennisetum squamulatum* [332] diploid synergid embryos developed along with the zygotic embryo. The suspensor of the zygotic embryo of the latter is typically multicellular as in Gramineae and that of the synergid is partly uniseriate and partly biseriate. Synergids can also produce haploid embryos without being fertilized. Several cases of haploid-diploid embryos are known in *Lilium* [138], *Arabis lyalli* [61], *Bergenia delavayi* [215], *Argemone maxicana* [177], and *Aristolochia bracteata* [333]. The suspensor of the zygote as well as of the synergid embryo of *Arabis lyalli* elongates into similar filamentous structures with a small embryonal mass at the apical end [61]. Both embryos of *Erythraea centaurium* have reduced suspensor and develop up to the heart-shaped stage [73]. This way, the zygotic and the synergid embryos recieve nourishment from the endosperm simultaneously. In one of the ovules of *Orchis* the egg as well as one of the synergids had commenced to divide even before the pollen tube had entered the embryo sac, thus indicating the possibility of production of twin haploid embryos [167]. While in *Azadirachata indica* [145] twin haploid embryos were formed, in *Carthamus tinctorius* [47] triplets were formed due to fertilization of the egg and both the synergids.

4. Antipodals

Antipodals which contain the haploid number of chromosomes are usually ephemeral and show signs of degeneration shortly before or after fertilization. However, sometimes the antipodals not only persist but also divide and produce embryos (or rather proembryo-like structures) as in *Paspalum scrobiculatum* [334], *Sedum fabaria* [54], *Hieracium flagellare* [53], and *Allium odorum* [335] for example. In most of these cases, since only young embryos were observed, it is not certain whether they reached maturity. According to Johanson [13], it is important to distinguish between mere proliferation of antipodal cells or nuclei and true antipodal embryos. In fact, Johri and Ambegaokar [336] consider the concept of antipodal embryos to be completely erroneous.

5. Endosperm

Some instances of asexual embryos arising from the cells or nuclei of the endosperm have been reported [34, 239, 154], but reinvestigations of most of these studies have yielded only negative results [54, 170, 338]. According to Maheshwari and Sachar [54], embryos may be wrongly considered to have originated from the endosperm, either because of their displaced position in the embryo sac, or during adventitious embryony because of their close association with the endosperm after the destruction of the nucellar cells from which they probably arose [13]. More recently, Muniyamma [79] claimed to have observed the formation of endosperm embryos in the grass *Bracharia setigera* which is also a facultative apomict, utilizing apospory where polyembryony is due to the formation of embryos from egg, synergids and endosperm [78]. However, Johri and Ambegaokar [336] consider these observations to be unjustified since in their view, except for the outgrowth of a cellular mass from the endosperm, Muniyamma did not observe any comparable development to that of a monocot embryo. Nevertheless, it is likely that on rare occassions, one of the synergids of an aposporous embryo sac may also be fertilized, and develop into an embryo which could account for a triploid seedling.

6. Zygote and suspensor

Like nucellar embryos, identical twins and super-twins result from a cleavage of the zygote. Such cleavage polyembryony is common in gymnosperms (see Section III) but is much less frequent in angiosperms. Supernumerary embryos develop from the cleavage of the apical cells of the embryonic mass of cells produced by the zygote as in *Erythronium americanum* [134], *E. dens-canis* [135], *Tulipa gesnariana* [339], *Cocos nucifera* [175], and *Primula auricula* [340]. The zygote may divide vertically or obliquely to form two indepenedent proembryos as in *Cymbidium bicolor* [156], or the zygote may produce an irregular mass of cells where some of the apical cells develop simultaneously into multiple proembryos as in *Eulophia epidendra* [160], *Habenaria platyphylla* [172], and *Geodorum densiflorum* [155]. These proembryos may also produce additional embryos by budding as in *Hamamelis virginiana* [118]. Due to incomplete cleavage of the zygotic proembryo in *Stachyurus chinensis* [220], two proembryos may develop side by side. In maize, of the 49,903 seeds examined, Sarkar and Coe [116] observed 32 pairs of twin embryos which were either due to cleavage of the proembryo or due to fertilization of two reduced female cells in the embryo sac by two sperms. Besides these diploid-diploid twins, 15 were haploid-haploid, and 2 haploid-diploid.

Suspensor polyembryony is of common occurrence in *Dipteracanthus patulus* [4] and *Exocarpus* species [212, 213]. In *Exocarpus* as many as five or six proembryos may develop simultaneously but, only one of them takes the lead and reaches maturity. The cells of the uniseriate suspensor develop buds in *Zygophyllum fabago* [341], *Actinidia chinensis* [33], *Bulbophyllum mysorense* and *Dulophia nuda* [155], and *Sonerila wallichii* [144]. In some plants, cleavage polyembryony occurs only as an abnormality. Additional embryos may arise from cells of the suspensor in *Actinidia chinensis* [6] and *Libelia syphilitica* [33]. Similar observations have been made in *Isotoma longifolia* [32], *Actinidia polygona* [7], and *Garrya veatchii* [72].

Thus a variety of cells and tissues both within and outside the embryo sac can give rise to asexual embryos in angiosperms in nature.

III. Asexual embryogenesis in gymnosperms *in vivo*

In gymnosperms, the mature seeds generally contain one embryo while only a very small percentage (1–3%) of the seeds show more than one embryo [342].

However, during late embryogenesis, the presence of more than one embryo in a young seed is common amongst most of the gymnosperms. This is mostly due to the fertilization of more than one egg and the development of multiple zygotes. This is known as "simple polyembryony". Frequently, a single zygote may also form multiple embryos by the cleavage or splitting of the cells of the embryonal tier into several embryonal units (see Fig. 3); thus resulting in "cleavage polyembryony" [254, 260, 309].

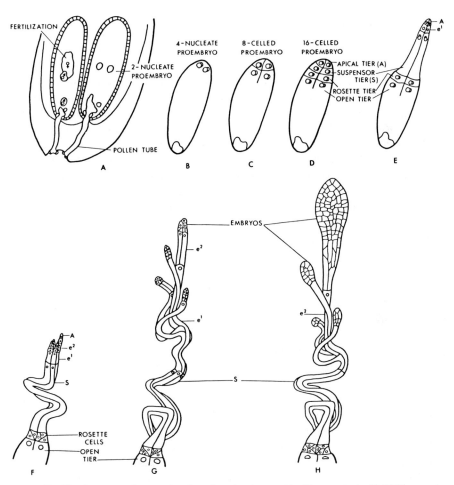

Fig. 3. Fertilization, proembryo and early embyo development in *Pinus contorta.* (A,B) The zygote nucleus divides forming two free nuclei which in turn divide to form free nuclei. These nuclei migrate to the distal end of the archegonium. (C,D) Division and cell wall formation occur forming the 16-celled proembryo. (E) The suspensor tier (S) elongates, forcing the apical tier into the female gametophyte. The apical tier divides to form apical cells (A) and embryonal tubes. (F) Cleavage polyembryony occurs when the apical cells and embryonal tubes (e^1, e^2) separate to form four files of cells. G,H. Apical cells divide forming multicellular embryos which are pushed into the female gametophyte by the elongating embryonal tubes and suspensors. (Courtesy: Professor J.N. Owens, Biology Department, University of Victoria, B.C., Canada.)

According to Buchholz [275], cleavage polyembryony is further divisible into two types, viz., "determinate cleavage polyembryony" and "indeterminate cleavage polyembryony". In the former, it becomes apparent at a very early stage as to which of the several embryos in a seed will be successfull. Usually it is the terminal embryo which is more favourably situated, as in *Dacrydium*. In the latter, there is no indication in the begining as to which embryo is going to survive during the struggle for supremacy. Any one of the many embryos derived from a zygote may survive, as in *Pinus* and *Cedrus*. Out of the 12 families belonging to Gymnospermae, at least 10 families show cleavage polyembryony. (see Table 1). In the Ginkgoaceae and Auraucariaceae there are no reports indicating any cleavage of the zygote. All nine genera belonging to the Pinaceae show simple polyembryony, but only four show cleavage polyembryony regularly (*Pinus*, *Cedrus*, *Tsuga*, and *Keteleeria*), while in *Abies* it is infrequent and the remaining four (*Pseudolarix*, *Picea*, *Larix* and *Pseudotsuga*) it is absent [308]. In the Taxaceae (*Taxus cuspidata*), although the cleavage of the embryo system is neither as regular nor as precise as that in the Pinaceae, it occurs in a significant percentage (30%) of the embryos [281]. In *Pseudotsuga menziesii*, Allen [257] and Orr-Ewing [258] have reported probable embryo formation in the unpollinated ovules. According to Orr-Ewing [258], this is due to agamospermy which is of rare occurrence, but when present, it is probably due to adventitious embryony or to the fusion of two large nuclei (possibly the egg and ventral canal nuclei observed in unpollinated ovules). Similar observations have also been reported in *Pinus pinaster* [267], *Pinus wallichiana* (10% of the cases) and in *P. nigra* var. *australis* (5% of the cases) where the egg nucleus divides parthenogenetically without the participation of the male nucleus [266]. According to Dogra [266] the possibility of survival of such embryos if formed is remote and based on the example of *Pinus pindrow*, he concluded that even if the ventral canal nucleus and egg nucleus came together and involve themselves in the act of near fertilization (parthenogenesis), it would take place in a non-recurrent fashion and would not be a fixed and regular feature of the species.

Independent behaviour of the embryonal units in terms of their segmentation and development of separate suspensor systems brings about cleavage polyembryony [309]. The separation of the polyembryos may occur at different stages of proembryo development. Cleavage may occur at wall formation and as the walls develop in the young proembryo, the different cells behave independently and each cell forms an embryonal mass and suspensor as is common in *Gnetum*, *Ephedra* and *Sequoia* [292]. Cleavage at the suspensor is uncommon, but has been reported in some genera like *Cryptomeria* [283, 284], *Taxodium* [295], *Glyptostrobus* [289], and *Taxus* [281]. The separation of suspensor from each other may occur either at the tip or farther down whereby each of the suspensor cells may carry one or more embryonal units [309]. The involvement of the embryonal suspensor in cleavage polyembryony has been reported in the Cupressaceae, Pinnaceae, and in *Sequoiadendron*, *Cunninghamia* and *Metasequoia* [290, 291]. However, cleavage between the embryonal units is most common amongst the conifers and occurs in the podocarps, taxads, taxodiads, cupressads and *Cedrus deodara*. Budding of the embryonal mass has been found in several gymnosperms, but it is not a regular phenomenon and it has been considered as a type of false cleavage [262].

In polyembryonic gymnospermous ovules the numerous embryos are gradually eliminated during development and only one survives in the seed. In *Sciadopitys* [294], the cleavage polyembryony is of an extreme type and the potential output per zygote is 12–28 embryos or more which is probably the highest number amongst all the conifers, although, eventually only one embryo matures. Mostly the physiologically superior embryo occupies the most suitable position, in regard to nourishment, in the corrosion region of the megagametophyte. In more than 98% of pine seeds, only one embryo matures [343], while the development of two pine seedlings from a single seed has been reported in over 20 species, but the occurrence is very infrequent [342]. The dominant growing embryos come in contact and stimulate the corrosion-region-tissue. The cells in contact with the embryo-tips are broken down and absorbed by the embryonal cells. This way, the supply of nutrition to the embryos shift from the egg cytoplasm (via the suspensor system) to direct absorption of broken down cells of the corrosion region of the gametophyte.

IV. Basis of *in vivo* asexual embryogenesis

Most of the studies dealing with the process of *in vivo* adventive embryogenesis has been carried out with *Citrus*, which undergoes nucellar embryony. The process of nucellar embryogenesis can be divided into three steps, viz., (1) formation of the initial cell, (2) division of the initial cells, and (3) development of nucellar embryos [344–346]. The process of pollination and fertilization do not seem to be essential for the formation of initial cells since these are already present in the nucellus at the time of pollination in the mature flower. Wilms et al. [346] have shown the presence of nucellar cells with a specific structure in the micropylar region of the embryo-sac of a polyembryonate cultivar of *Citrus sinensis*. The cytoplasm of these cells consists of homogenous matrix, a large spherical nucleus with globular nucleolus, some irregularly shaped proplastids with single thylakoids and some plastoglobuli, numerous oval shaped mitochondria with short cristae, some single strands of rough endoplasmic reticulum (RER) together with many free ribosomes and some polysomes. Walls of these cells show a specific character in the sense that in between the original primary wall and the plasma membrane, heterogenously dispersed, electron-dense material is present. Such type of cells were completely absent in the ovules of the monoembryonate cultivar of *Citrus limon*. The overall morphology of the adventive embryo initial cells is similar to that of an individual zygote [325]. Similar observations have also been noted in *Sarcococca humilis* where the micropylar part of the nucellus consists of the parietal tissue

[26]. Most of the parietal cells later become initial cells and develop to form embryos, while others degenerate. Such cells were characterized by a high metabolic activity before cell division and proembryo formation. The abundance of ribosomes, polysomes, and size, shape and position of mitochondria and plastids were specific characteristics of these cells. It seems that the closed plasmodesmata and degenerative processes in the tissue were a prerequisite for the differentiation of proembryos. Hence, these micropylar cells may perform the function of reproduction and produce an aposporic embryo or nucellar embryo [26].

In fact, it has been proposed that adventive embryogenesis is an inherent potential of all cells and that an embryogenic suppressant produced as early as after the first zygotic division is the key to the differences seen in the degree of polyembryony among species of the Rutaceae [208]. This suppressant may be responsible for the differences in degree of adventitious embryogenesis observed between the micropylar and chalazal regions of the nucellus in this family. More recently, it has been shown that in *Citrus* all AEICs (adventive embryo initial cells) are initiated prior to anthesis and whether or not the AEICs develop successfully into adventive embryos is dependent upon their position in the seed [348]. The farther the AEICs are located from the micropylar end, the more adventive embryogenesis is suppressed by the endosperm and the degree of adventive embryogenesis in the chalazal half is affected by time and extent of malfunction of the endosperm. Under natural conditions, these regulatory systems of adventive embryogenesis contribute to high production of zygotic seedlings in apomictic *Citrus* species and cultivars. According to another view, adventive embryos are the products of a "somatic fertilization" of nucellar cells. This might occur when two or more pollen tubes enter an ovule, the male gametes are discharged into the nucellar cells and also fertilize them. However, this view remains unconfirmed [203].

In fertilized seeds of *Citrus* in general, the development of adventive embryos is restricted to the micropylar half of the nucleus adjacent to the micropylar end and lateral flanks of the embryo-sac [18, 200, 201, 203, 206]. Wakana and Uemoto [348] have drawn two conclusions about nucellar embryogenesis: (1) appearance of adventive embryos may be at the flower bud stage [345, 346], and (2) the initials also exist at the chalazal end of the nucellus adjacent to the embryo-sac but rarely develop into preglobular embryos [208, 325]. The time of initiation and development of the adventive embryos in relation to the development of the zygote has been in question. Osawa [200] and Bacchi [201] reported that initiation *in vivo* coincided with the first division of the zygote. In contrast, Rangan et al. [349] reported that there was no evidence of initiation and adventive embryos in *Citrus aurantium* 100–120 days after pollination, when the zygotic embryo was at the globular or early heart stage. However, 28 out of 30 seeds of *Citrus reticulata* excised 47 days after pollination already had adventive embryos at the one-celled stage, thus suggesting that the initiation of the adventive embryos was apparently dependent on the presence of a zygote because one-celled adventive embryos were found in small ovules without embryo sacs or evidence of pollen tube entry [325].

In many cases, adventitious embryogenesis can also be triggered with certain chemical treatments, irradiation or even mechanical injury. While injecting the young *Citrus* fruits with maleic hydrazide repressed nucellar embryogenesis [202], auxin sprays on emasculated *Hosta coerulea* flowers enabled nucellar embryos to form even in the absence of pollination [350]. Administration of citric acid-phosphate buffer at pH 3.5 to 4.5 to *Eranthis hiemalis* seeds induced proliferation of the developing zygotic embryos into multiple embryos. Twinning of the zygotic embryo could be produced by treatment with auxins, while X-ray treatment of the ovary of *Eranthis* produced several meristematic centres and eventual development of multiple embryos from the zygotic embryos [180]. Fagerlind [136] reported that if a 1% lanolin paste of heteroauxin was applied to unpollinated pistils of *Hosta* they produced young adventitious embryos which however failed to develop further due to the lack of an endosperm. From the above, it is apparent that few recent studies have been carried out on this topic.

V. Causes of *in vivo* asexual embryogenesis

The causes of polyembryony in nature are largely unknown. Haberlandt's "wound hormone" and "necrohormone theory" [154, 351] is no longer tenable, and subsequent workers have failed to confirm his findings [54, 302]. The association of meiotic abnormalities, endoduplication of chromosomes in haploid synergid/egg, parthenogenetic development of unreduced/reduced synergid/egg, cleavage/budding of zygotic proembryo, budding of suspensor, occurrence of multiple embryos in hybrids, and nucellar and integumentary polyembryony have not provided any significant clues to the causes of this phenomenan [303]. Webber [306] considers polyembryony as (a) a gradual replacement of sexual reproduction, caused by the weakening of sexuality, (b) early stages of the development of something new, (c) a primitive feature, (d) a development by the plant to increase its capacity for dispersal, and (e) merely somatic buds which due to stimulation develop into embryos. Polyembryonate plants have been frequently associated with meiotic irregularities [352, 353] polyploidy [141, 354, 355], and hybridization [353, 356-358]. However, Pijl [148] referes to nucellar embryony of *Eugenia jambos* as a reduced form of normal reproduction, and to apogamy, apospory and other similar occurrences as intermediate forms. Kappert [140] suggests that although polyembryony is constitutional, environment determines the degree of expression. Polyembryony has also been suggested to be a recessive character conditioned by a series of multiple factors which are merely brought together in suitable recombinations following hybridization [140, 357, 358, 381]. Hybridization has been often thought to be a factor leading to the production of plural embryos [25, 70].

Several other factors like the climatic conditions and the physiological status of the plants have been shown to be indirectly associated with asexual embryogenesis in nature, but there seem to be differences between species. While very low temperatures resulted in polyembryony of *Genetiana* due to a failure of fertilization [74], excessively high temperatures caused nucellar embryony of *Opuntia aurantiaca* associated with a lack of fertilization [27]. Even within the same species of *Citrus*, Furusato et al. [205] observed that on the northern exposure, trees contained more embryos per seed than those on the southern side where temperature is often slightly warmer. However, *Citrus* polyembryony could not be enhanced by culturing the plants under warmer temperatures of 30 °C [202]. Nevertheless, Ueno and Nishiura [359] found the *Citrus* fruits that develop during the warmer months do manifest higher nucellar embryony than those from the cooler season. In *Mangifera indica*, the variation in rates of nucellar embryogeny is due probably to genetic differences [17], rather than climatic differences as suggested by Juliano [14, 15].

The balance between sexual and asexual reproductive modes in facultative apomicts has been shown to be influenced by the environment. The incidence of apomixis could be afffected by temperature and photoperiod. In the grass *Dicanthium aristatum* [86] and many other apomictic grasses of the tribe Andropogonea [114], the aposporous embryo sacs are 4-nucleate and compete in the ovule with sexually produced 8-nucleate sacs and the species flower only after short-day experience. Here, under continuous short days, 68–79% of the embryo sacs were aposporous, but with the minimal inductive period of 40 short days, only 27–46% were aposporous. A marked association between the length of the day during inflorescence development and degree of apomixis was observed in *Dicanthium aristatum* when exposed to natural conditions in the fields distributed over a span of 27° latitude in Australia [360]. In another grass species, *Themeda australis*, similar evidence of environmental control of apomixis was observed leading to the conclusion that short-day experience increased the degree of apospory, regardless of the day length requirement for flowering in the different genotypes [361]. However, due to the lack of detailed studies in other families, it is very difficult to draw general conclusions on the role of environmental factors on the apomictic behaviour.

Studies with *Citrus* have indicated that the nutritional status of trees do not affect the process of asexual embryogenesis [202, 359, 362]. But older *Citrus* trees have been observed to produce more nucellar embryos per seed than young trees [205]. Minessy [363] grafted a normally polyembryonate *Citrus* cultivar onto a monoembryonic rootstock and observed a reduced rate of nucellar embryony. During their *in vitro* studies on nucellar embryos of *Citrus* and *Mangifera*, Maheshwari and Rangaswamy [203] raised the following questions: (1) what factors determine the inception of nucellar embryos? (2) can the stimulus of fertilization be replaced, or overcome by any hormone, or suitable chemical treatment? (3) can the adventive embryos be induced in plants which are not naturally polyembryonate? However, these questions still remain to be answered suitably [54].

Genetic studies have revealed that one or more recessive genes are involved in nucellar embryogeny of *Linum* [140] and *Mangifera* [16, 19]. In the nucellar embryony of Rutaceae the involvement of a single dominant gene has been suggested [205, 364]. Genes controlling a high or obligate level of apomixis have been found in the wild relatives of some cultivated crops like *Elymus rectisetus, Triticum aestivum, Tripsacum dactyloides* [365]. However, these observations remain to be extended to most other plant species. With regards to gametophytic apomixis, according to Asker [298, 366], the genes that control successful agamospermy when they occur in combination, can be found singly as aberrations in many conventionally sexual populations, where they are likely to be unsuccessful. Hence, according to this view, all the functional features of various forms of agamospermy can be found exceptionally in sexual populations.

VI. Use of markers for isolation of apomictic individuals

One of the main obstacles in *Citrus* breeding arises from the fact that most varieties are apomictic and polyembryonic. This implies that in the nucellus of the mother plant asexual embryos develop together with, or instead of, zygotic embryos. This produces a large proportion of nucellar offsprings, genetically identical to their female progenitor. It is, therefore, of great interest to the breeder to be able to identify at an early stage of seedling development those zygotic plants that carry a combination of the desirable characters of both progenitors. Although asexual and zygotic embryos are sometimes distinguishable by their morphology and physiology, their distinction usually can be made only on their histogenic origin. Thakur and Bajwa [199] noted differences in shape of nucellar and hybrid *Citrus* seeds, while Furr and Reece [367] could separate them by the Almen reagent color reaction. Syakudo et al. [368] and Watanabe et al. [369] suggested the use of radioactively labelled pollen to enable separation between the two types of embryos of *Citrus*. However, it is often necessary to grow seedlings for five or more years to fruiting before the nucellar offsprings can be distinguished from the zygotic ones making it an expensive procedure [370]. Consequently, cultivars which are known not to produce nucellar embryos generally are used as the maternal parent, or in the case of root stock, the dominant trifoliate leaf gene of the closely related *Poncirus trifoliata* is used as a marker. These measures, while mostly effective, limit the possible crosses that can be attempted. A reliable set of markers available from vegetative tissue which would allow seedlings to be unequivocally scored as to parental origin would be more effective.

Although several attempts have been made by using biochemical markers to solve this problem [371–375], all have had their limitations, ascribable largely to a lack of understanding of the underlying genetic control of the compounds studied. As an alternative, genetic studies and genetic markers as a basis for taxonomic conclusions as well as for facilitating breeding programs might be more rewarding. According to Torres et al. [370] several conditions should be satisfied for devising genetic markers useful in breeding. The markers should be controlled by codominant alleles with each genotype producing a distinct phenotype, rather than dominants, so that genetic studies can be carried out with extant F_1s. The markers should be available from vegetative tissue, especially leaves, so that progeny can be scored while very young to aid in distinguishing zygotic from nucellar seedlings. The screening procedures should be simple and rapid to obviate the necessity for elaborate and laborious extraction techniques.

At present however, there are exceedingly few markers known for *Citrus* and relatives and some of these are not fully understood. The trifoliate leaf character of *Poncirus trifoliata* is apparently controlled by a single dominant gene [373], but some plants are heterozygous thus partly nullifying the usefulness of this marker. The gene for nucellar embryony is also apparently dominant in *Citrus* in which it occurs widely and contributes heavily to the problems of *Citrus* breeding [207]. Attempts have been made to distinguish nucellar from zygotic seedlings in progenies of *Citrus* hybrids through the infrared analysis of essential oils in leaves [376], or analysis of leaf morphology of the progeny plants [377]. Spot chromatography of 2,2-dimethoxy propane extracts of bark tissue from roots was used for rootstock identification [375]. Analysis of isozymes to differentiate nucellar from zygotic seedlings (20-day-old) in progenies of *Citrus* crosses has also been used [374]. In their studies, however, there were some doubtful situations that occurred when the peroxidase isozyme patterns present in seedlings were identical in the female and male progenitors. The authors concluded that the use of peroxidase isoenzymes as markers is limited by the low number of bands present in any given parental plant and by the narrow range of mobility that these bands have on the zymogram. Another serious limitation encountered in this study was that there were variations in the zymograms for peroxidase between plants of some varieties, and in some cases even between seedlings derived from one seed.

Since in polyembryonate seeds, the zygotic embryos develop earlier than nucellar embryos, Rangan et al. [349] were able to isolate and culture *in vitro* the zygotic embryos before they degenerated (120 days after pollination). However, Lima [378] found that under subtropical conditions, there is an acceleration of all the processes leading to the development of both the zygotic and nucellar seedlings and by the time the embryos reach the heart stage (60 days after pollination), in which they may be excised for successful culture, it is not possible to differentiate between them.

The presence of two distinct young shoot extract color phenotypes-browning and nonbrowning-in *Citrus* has been shown [379]. The browning taxa had one, possibly more, unknown phenolic substrates oxidized by polyphenol oxidase upon tissue homogenation, while the nonbrowning taxa contained neither the substrate nor any detectable enzyme activity. The authors further showed that browning was dominant to nonbrowning and was under single gene control. Their genetic studies based on a survey of 34 taxa, various F_1 hybrids and an F_2

population of 76 individuals indicated the possibility of distinguishing nucellar and zygotic progeny from non-browning X browning crosses at an early seedling stage which required only 0.2–0.5 g tissue [373, 379].

VII. Utilization of *in vivo* asexual embryogeny

In general, the asexual seed reproduction in *Citrus* by the formation of embryos from cells of the nucellus results in a marked and persistent rejuvenation, or vegetative invigoration also called as "neophyosis" (Greek " causing to grow new") and this declines with increasing age of the tree or clone from seed [380]. This invigoration is greatly enhanced in contrast to the less pronounced invigoration produced by ordinary nursery propagation and is definitely propagable and seems to persist for a more extended period [381, 382]. In their studies on the comparison of the Paper Rind orange variety before and after sexual seed production, the major differences noted were greater vigor, more upright habit of growth, thorniness and lower seed content in the young clone, and earlier and heavier bearing in the old clone.

Agamospermy assures reproduction even in the absence of pollination. This should favour propagation even in extreme climatic conditions. However, most species with apospory or adventitious embryony are pseudogamous requiring pollination for sexual seed development. Since meiosis and fertilization are not involved in development of the embryo, the progeny of apomictic plants are exact replicas of the female parent. Thus apomixis provides a method for cloning plants through seed and has important implications for use as a tool in plant breeding. Hence, apomictic seeds are of agricultural importance by (1) providing for uniformity in seed propagation of root stocks (e.g. in *Citrus*, where the sexual seedling is eliminated by weakness) [364, 383], (2) for true-breeding F_1 hybrids (e.g. in *Poa*, where up to 85% of the seeds are apomictics) [384], and (3) for removal of virus from clones of vegetatively propagated plants as in *Citrus* [364]. Stimulation of apomictic seed development occurs in *Citrus* spp., *Eugenia* spp. (adventitious embryony), and in *Parthenium argentatum* (guayule), *Rubus* spp. (raspberry), *Pyrus* spp. (apple), some *Poa* spp. (blue grass), as well as a result of interspecific pollination (e.g. in *Brassica* species crosses). The possibility of genetic manipulation of agamospermy as indicated in *Sorghum* and other Gramineae [110, 386], as well as in some flower crops of the Compositeae (e.g. *Centaurea cyanas*; cornflower), provides the way for stable maintenance and multiplication of superior hybrid cultivars of field, garden and forage crops. The development and release of apomictic cultivars of Kentucky bluegrass, *Poa pratensis* [387] and bufflegrass, *Cenchrus ciliaris* [388] has demonstrated the potential for use of apomixis in a breeding program.

Another significance of apomixis is that many virus diseases are not transmitted by seed. Consequently, growing apomictic seedlings provides a means of rejuvenating an old clone that has become infected with virus disease [385]. This procedure has been developed most extensively in improving citrus cultivars. These nucellar seedlings are in a juvenile state and are extremely vigorous and thorny; thus, such plants are not suitable horticulturally until they attain an adult state. Nucellar seedlings provide evidence that no permanant genetic change is involved since they are apomictic and, therefore, asexual, but such seedlings also go through the same juvenile-to-adult phase change [381].

Obligate apomixis offers an opportunity to clone plants through seed propagation and, when gene manipulation is possible, can be used effectively as a breeding tool [365]. Development of cross-compatible sexual or partially sexual (facultative apomictic) plants or strains allows for hybridization that can break the apomictic barriers of gene transfer. Obligately apomictic intra- and interspecific hybrids from such crosses breed true and are usually fertile. Their progenies represent seed propagated clones of the hybrid genotype [389]. Hybridization between sexual and asexual crops and their closest apomictic wild relatives have been performed with a view to obtaining homozygous lines, fixing heterosis and getting non-segregating populations from hybrids with a unique combination of beneficial characters from the parental forms [298]. One of the main advantages of apomixis is that it permits the development of hybrids or genotypes that breed true regardless of homozygosity, seed of any superior obligate apomictic could be increased through open-pollination for an unlimited number of generations without loss of vigor or change in genotype.

In spite of the many theoretical and practical advantages offered by apomictic reproduction, it has not really been possible to realize the complete potential of this natural phenomenon. Factors such as facultative behaviour, number of genes involved, modifiers, environmental factors, ploidy, and seed sterility have to be better understood and/or stabilized before apomixis can be usefully applied to produce new cultivars [365]. Discovery of sexual or partially sexual plants in apomictic species could provide an efficient mode for manipulation of plant breeding [386, 390]. However, the problem could be more complex in sexual species since, first apomixis has to be found in the species or in a cross-compatible wild relative. According to Hanna and Bashaw [365], transfer of apomixis from a wild relative would require establishment of phylogenetic relationships and to overcome differences in ploidy level, genome relationships, and gene pool. Moreover, high levels of apomictic reproduction have not been found generally in the cultivated species.

VIII. Conclusions

So far agamospermy has been recorded in at least 94 families of the angiosperms and no less than 60% of the agamospermous species belong to six families, viz., Compositeae, Gramineae, Liliaceae, Orchidaceae, Rosaceae and Rutaceae. From a total of 280 genera, more than 40% belong to these six families. Figure 4 shows 56 out of 94 families where at least two species of each family have been shown to exhibit some form of asexual embryogenesis. Apospory is a much commoner method of circumventing meiosis in the Gramineae and Rosaceae

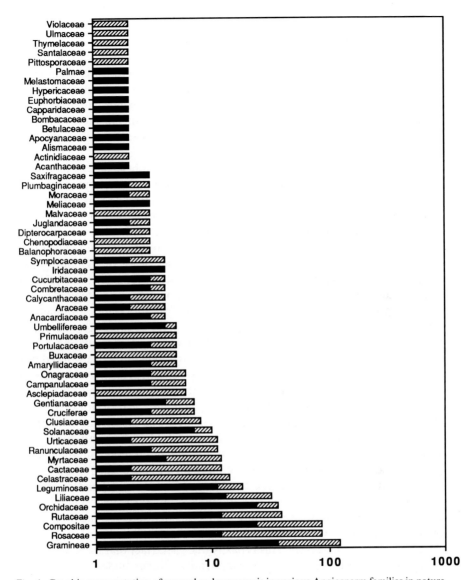

Fig. 4. Graphic representation of asexual embryogenesis in various Angiosperm families in nature. Only those families where at least two species have been reported to demonstrate some mode of asexual embryogenesis have been included. Against each family, the dark portion of the bar shows the number of genera while the total length of the bar (dark and/or the shaded portion) shows the total number of species.

than it is the Compositeae, in which diplospory prevails. The other three families of Liliaceae, Orchidaceae, and Rutaceae primarily exhibit adventive embryony. The pattern of distribution is certainly consistent with a hypothesis that suggests some "preadaptation" to a certain sort of agamospermous mechanism within a family. In gymnosperms, along with sexual reproduction, cleavage polyembryony due to cleavage of the proembryo or embryonal suspensor is the most common form of reproduction. In no case is asexual embryogenesis an obligate character. Apomictic embryo formation in the unpollinated ovules have been suspected to occur in a few taxa but never proven beyond doubt. Even if such a phenomenon is present, then it would be of rare occurrence and would take place in a non-recurrent fashion. Even the possibility of survival of such embryos, if formed, has been thought to be remote.

The advancement to our knowledge of asexual embryogenesis *in vivo* has been very slow and fragmentary, mainly due to the fact that except in relatively few genera, this phenomenon is very limited in occurrence and rare in most genera. Most of the detailed work on the basis of asexual embryogenesis has mainly concentrated on *Citrus* species. Artificial induction of adventive embryogenesis by certain chemical and mechanical treatments has not produced consistent results. Some correlations on the causes of asexual embryogenesis have provided some important insights into this phenomenon, including the genetic nature of apomixis. Meiotic abnormalities, polyploidy, and hybridization have been thought to cause polyembryony, although the environment has been shown to determine the degree of expression. Hormonal factors have been suspected to be strong candidates for regulating asexual embryogenesis, but their exact nature and control remains to be worked out.

The phenomenon of asexual embryogenesis seems to present itself with a vast number of potential applications, but so far not many have been fully realized. In *Citrus*, nucellar polyembryony has been used to initiate root stocks which are invariably uniform and vigorous in contrast to the occasional weaker and variable hybrid zygotic seedlings. In Kentucky bluegrass and bufflegrass, apomictic cultivars have demonstrated the potential uses of apomixis in breeding programs. Several apomixis mechanisms have been shown to be genetically controlled and therefore subject to genetic manipulation in plant breeding programs. Hence, obligate apomixis offers far reaching implications for genetic manipulation and cloning plants through seed propagation. So far the successes in introducing apomixis into sexual plants have been low inspite of the great efforts undertaken. This is partly due to the fact that there is as yet no other way than that of proceeding empirically. Proper understanding of the apomictic phenomenon has precluded its proper use. Absence of high levels of apomictic reproduction and inability to induce asexual embryogenesis in taxa where it is not present naturally are some of the bottlenecks which have been preventing the realization of the full potential of apomixis. Apart from this, identification of asexual from sexual seedlings is also difficult and needs further improvements in the selection strategies. The applications of modern techniques of physiology, biochemistry, genetics, microscopy, and plant tissue culture may provide useful information for proper understanding and utilization of natural asexual embryogenesis.

References

1. Maheshwari, P., *An Introduction to the Embryology of Angiosperms*, McGraw-Hill Book Co., New York, 1950.
2. Heslop-Harrison, J., Sexuality of angiosperms, in *Plant Physiology: A Treatise, vol. VI C, Physiology of Development: From Seeds to Sexuality*, Steward, F.C., Ed., Academic Press, New York, 1972, Chapter 9.
3. Gigante, R., Embriologia dell "*Acanthus mollis*" L., *Nuovo G. Bot. Ital. N.S.*, 36, 5, 1929.
4. Maheshwari, P., and Negi, V., The embryology of *Dipteracanthus patulus* (Jacq.) Nees., Phytomorphology, 5, 456, 1955.
5. Sullivan, J.R., Comparative reproductive biology of *Acer pensylvanicum* and *A. spicatum* (Aceraceae), *Amer. J. Bot.*, 70, 916, 1983.
6. Crété, P., Polyembryonie chez l'*Actinidia chinensis* Planch., *Bull. Soc. Bot. Fr.*, 91, 89, 1944.
7. Vijayaraghavan, M.R., Morphology and embryology of *Actinidia polygona* Franch. & Sav. and Systematic position of the family Actinidiaceae, *Phytomorphology*, 15, 224, 1965.
8. Tisserat, B., Esan, E.B., and Murashige, T., Somatic embryogenesis in angiosperms, *Hort. Rev.*, 1, 1, 1979.
9. Johri, B.M., Studies in the family Alismaceae – iV. *Alisma plantago* L., *A. plantago-aquatica* L., and *Sagittaria graminea* Mich., *Proc. Indian Acad. Sci. B.*, 4, 128, 1936.
10. Nygren, A., Apomixis in angiosperms, in, *Encyclopedia of Plant Physiology*, vol. XVIII, Ruhland, W., Ed., Springer-Verlag, Berlin, 1967, 551.
11. Coe, G.E., Cytology of reproduction in *Cooperia pedunculata*, *Amer. J. Bot.*, 40, 335, 1953.
12. Belling, J., Report of the Assistant in Horticulture: Mango, *Fla. Agr. Expt. Sta. Annu. Rpt.*, CX--CXXV, Plates 7–10, 1908.
13. Johansen, D.A., *Plant Embryology: Embryology of the Spermatophyta*, Chronica Botanica Co., Waltham, U.S.A., 1950.
14. Juliano, J.B., Origin of embryos in the Strawberry mango, *Philip. J. Sci.*, 54, 553, 1934.
15. Juliano, J.B., Embryos of Carabao mango, *Mangifera indica* L., *Philip. J. Agric.*, 25, 749, 1937.
16. Leroy, J.F., Sur un "Complexe agamique" des manguiers et sur l'origine et la physiologénie des variétiés cultivées. *Rev. Bot. Agr. Trip.*, 27, 304, 1947.
17. Sachar, R.C., and Chopra, R.N., A study of the endosperm and embryo in *Mangifera* L., *Indian J. Agric. Sci.*, 27, 219, 1957.
18. Strasburger, E., Über Polyembryonie, *Jena Z. Naturw.*, 12, 647, 1878.
19. Sturrock, T.T., Nucellar embryos of the Mango, *Proc. Fla. State Hort. Soc.*, 81, 350, 1968.
20. Naumova, T.N., and Yakovlev, M.S., Adventive embryony in Angiosperms, *Bot. Zh.*, 57, 1006, 1972.
21. Saus, G.L., and Lloyd, R.M., Experimental studies on mating systems and genetic load in *Onoclea sensibilis* (Aspleniaceae, Athyriodeae), *Bot. J. Linn. Soc.*, 72, 101, 1976.
22. Bayer, R.J., Ritland, K., and Purdy, B.G., Evidence of apomixis in *Antennaria media* (Asteraceae: Inuleae) detected by the segregation of genetic markers, *Amer. J. Bot.*, 77, 1078, 1990.
23. Fagerlind, F., Bildung und Entwicklung des Embryosacks bei sexuellen und agamospermischen *Belanophora*-Arten, *Svensk Bot. Tidskr.*, 39, 65, 1945.
24. Sulkinoja, M., and Valanne, T., Polyembryony and abnormal germination in *Betula pubescens* ssp. *tortuosa*, *Ann. Univ. Turk. Ser. A II Biol-Geogr. Geol.*, 63, 31, 1980.
25. Baker, H.G., Apomixis and polyploidy in *Pachira oleagina* (Bombacaceae), *Amer. J. Bot.*, 47, 296, 1960.
26. Naumova, T.N., and Willemse, M.T.M., Nucellar polyembryony in *Sarcococca humilis* – ultrastructural aspects, *Phytomorphology*, 32, 94, 1982.
27. Archibald, E.E.A., The development of the ovule and seed of jointed cactus (*Opuntia aurantiaca* Lindley), *S. African. J. Sci.*, 36, 195, 1939.
28. Maheshwari, P., and Chopra, R.N., The structure and development of the ovule and seed of *Opuntia dillenii* Haw., *Phytomorphology*, 5, 112, 1955.

60

29. Naumova, T.N., Developmental characteristics of nucellar tissue and nucellar polyembryony in *Opuntia elata*, Cactaceae, *Bot. Zh.*, 63, 344, 1978.
30. Ross, R., Chromosome counts cytology and reproduction in the Cactaceae, *Amer. J. Bot.*, 68, 463, 1981.
31. Birader, N.V., and Bhise, R.V., Studies on Caesalpiniaceae. 1. Embryology of *Bauhinia monandra*, *Bio-Vigyanam* (India), 8, 19, 1982.
32. Kaushik, S.B., and Subramanyam, K., A case of polyembryony in *Isotoma longifolia* Presl., *Curr. Sci.* (Bangalore), 15, 257, 1946.
33. Crété, P., La polyembryonic chez le *Lobelia syphilitica* L., *Bull. Soc. Bot. Fr.*, 85, 580, 1938.
34. Billings, F.H., Some new features in the reproductive cytology of angiosperms, illustrated by *Isomeris arborea*, *New Phytol.*, 36, 301, 1937.
35. Tokc, E., Embryology of *Sambucus racemosa*, *Acta Biol. Cracov. Ser. Bot.*, 22, 173, 1981.
36. Manshardt, R.M., and Wenslaff, T.F., Zygotic polyembryony in interspecific hybrids of *Carica papaya* and *Carica cauliflora*, *J. Am. Soc. Hort. Sci.*, 114, 684, 1989.
37. Copeland, H.F., Morphology and embryology of *Euonymus japonica*, *Phytomorphology*, 16, 326, 1966.
38. Jassem, B., Embryology and genetics of apomixis in the section *Corollinae* of the genus *Beta*, *Acta Biol. Cracov Ser. Bot.*, 19, 149, 1976.
39. Seilova, L.B., Abdurakhmanov, A.A., and Khailenko, N.A., Embryology of induced apomixis in sugarbeets, *Tsitol. Genet.*, 18, 90, 1984.
40. Maguire, B., Apomixis in the genus *Clusia* Clusiaceae – A preliminary report, *Taxon*, 25, 241, 1976.
41. Treub, M., Le sac embryonnaire et l'embryon dans les Angiospermes. Nouvelle serie des recherches, *Ann. du Jardin Bot. des Sci. Natur., Bot.*, 9, 1, 1911.
42. Puri, V., Studies in the order Parietales. I. A contribution to the morphology of *Tamarix chinensis* Lour., *Beih. Bot. Club.*, 59A, 355, 1939.
43. Ha, C.O., and Sands, V.E., Reproductive patterns of selected understory trees in the Malaysian rain forest: the apomictic species, *Bot. J. Linn. Soc.*, 97, 317, 1988.
44. Treub, M., Le sac embryonnaire et l'embryon dans les Angiosperms, *Ann. des Sci. Natur., Bot.*, 9, 1, 1911.
45. Venkateswarlu, J., Contributions to the embryology of Combretaceae: I – *Poivrea coccinea* DC., *Phytomorphology*, 2, 231, 1952.
46. Pullaiah, T., Studies in the embryology of Compositeae. 2. The Eupatorieae tribe, *Indian J. Bot.*, 5, 183, 1982.
47. Maheshwari Devi, H., and Pullaiah, T., Embryological abnormalities in *Carthamus tinctorius* Linn., *Acta Bot. Indica*, 5, 8, 1977.
48. Gerassimova, H., Fertilization in *Crepis capillaris*, *Cellule*, 42, 103, 1933.
49. Coleman, J.R., and Coleman, M.A., Apomixis in 2 triploid Brazilian species of *Eupatorium*, *Eupatorium bupleurifolium* and *Eupatorium callilepsis*, *Rev. Bras. Genet.*, 7, 549, 1984.
50. Holmgren, I., Zytologische Studien über die Fortpflanzung bei den Gattungen Erigeron und Eupatorium Kung., *Svensk Vet. Acad. Handl.*, 59, 1, 1919.
51. Sparvoli, E., Osservazioni cito-embriologiche in *Eupatorium riparium* Reg. Nota II. Megasporogenesi e sviluppo del gametofito femminile, *Ann. di Bot.* (Rome), 26, 481, 1961.
52. Rozenblum, E., Maldonado, S., and Waisman, C.E., Apomixis in *Eupatorium tanacetifolium* (Compositae), *Amer. J. Bot.*, 75, 311, 1988.
53. Rosenberg, O., Cytological studies on the apogamyin *Hieracium*, *Svensk. Bot. Tidskr.*, 28, 143, 1908.
54. Maheshwari, P., and Sachar, R.C., Polyembryony, in *Recent Advances in The Embryology of Angiosperms*, Maheshwari, P., Ed., Intl. Soc. Plant Morphologists, Univ. Delhi, Delhi, 1963, Chapter 9.
55. Maheshwari Devi, H., and Pullaiah, T., Embryological investigations in the Melampodinae. I. *Melampodium divaricatum*, *Phytomorphology*, 26, 77, 1976.
56. Richards, A.J., Eutetraploid facultative agamospermy in *Traxacum*, *New Phytol.*, 69, 761, 1970.

57. Malecka, J., Further embryological studies in the genus *Taraxacum*, *Acta Biol. Cracov Ser Bot.*, 24, 143, 1982.
58. Mogie, M., Latham, J.R., and Warman, E.A., Genotype-dependent aspects of seed ecology in *Taraxacum*, *Oikos*, 59, 175, 1990.
59. Pullaiah, T., Embryology of *Tithonia*, *Phytomorphology*, 28, 437, 1978.
60. Maheshwari, P., and Roy, S.K., The embryo sac and embryo of *Tridax procumbens* L., *Phytomorphology*, 2, 245, 1952.
61. Lebegue, A., Embryologie des Cruciferes: Polyembryonie chez l'*Arabis lyallii* S. Wats., *Bull Soc Bot., Fr.*, 95, 250, 1948.
62. Ellerstrom, S., and Zagorcheva, L., Sterility and apomictic embryo-sac formation in *Raphanobrassica*, *Hereditas*, 87, 107, 1977.
63. Belyayeva, L.Y., Chaika, K.A., and Fursa, M.S., Fertilization and development of seeds in *Diplotaxis tenuifolia*, *Ukr. Bot. Zh.*, 35, 484, 1978.
64. Zagorcheva, L., Autonomous apomictic propagation of *Cucumis ficifolius* and *Cucumis anguria*, *Genet. Sel.*, 18, 460, 1985.
65. Bhuskute, S.M., Generative apospory in *Momordica charantia* var. *muricata*, *Indian Bot. Rep.*, 6, 131, 1987.
66. Vos, M.P. de, Seed development in *Cyanella capensis* L.; a case of polyembryony, *S. African. J. Sci.*, 46, 220, 1950.
67. Bhatnagar, A.K., and Kapil, R.N., Seed development in *Daphniphyllum himalayense* with a discussion on taxonomic position of Daphniphyllaceae, *Phytomorphology*, 32, 66, 1982.
68. Kaur, A., Jong, K., Sands, V.E., and Seopadmo, E., Cytoembryology of some Malaysian dipterocarps, with some evidence of apomixis, *Bot. J. Linn. Soc.*, 92, 75, 1986.
69. Carano, E., Ulteriori osservazioni su *Euphorbia dulcis* L., in rapporto col suo compartamento apomittico, *Ann. Bot.*, 17, 50, 1926.
70. Cesca, G., Ricerche cariologiche ed embriologiche sulle Euphorbiaceae. I. Su alcuni biotipidi *Euphorbia dulcis* L. della toscana, *Caryologia*, 14, 79, 1961.
71. Krusheva, R.M., Comparative cytoembryological study of species of family Fabaceae. IV. Megasomatogenesis gametogenesis and embryogenesis in *Coronilla emerus* spp. *emeroides*, *Coronilla varia* and *Coronilla scorpioides*, *Kliment Okhridski Biol. Fak.*, 76, 3, 1985.
72. Kapil, R.N., and Mohana Rao, P.R., Studies of the Garryaceae. II. Embryology and systematic position of *Garrya* Douglas ex Lindley, *Phytomorphology*, 16, 564, 1966.
73. Crété, P., Un cas de polyembryonie chez une Gentianacee, l'*Erythrea centaurium* Pers., *Bull. Soc. Bot. Fr.*, 96, 113, 1949.
74. Rudenko, F.E., Apomixis in certain high mountain plants of Ukrainian Carpathians, *Ukr. Bot. Zh.*, 18, 24, 1961.
75. Maheshwari Devi, H., and Lakshminarayana, K., Embryological studies in Gentianaceae, *J. Indian Bot. Soc.*, 56, 182, 1977.
76. Hair, J.B., Subsexual reproduction in *Agropyron*, *Heredity*, 10, 129, 1956.
77. Bhanwra, R.K., Embryology in relation to systematics of Gramineae, *Ann. Bot.*, 62, 215, 1988.
78. Narayan, K.N., and Muniyamma, M., Endosperm embryos in *Brachiaria setigera* Retz., in *Proc. Intl. Bot. Congr.*, Washington, 1969.
79. Muniyamma, M., Triploid embryos from endosperm *in vivo*, *Ann. Bot.*, 41, 1077, 1977.
80. Greene, C.W., Sexual and apomictic reproduction in *Calamagrostis* (Gramineae) from eastern North America, *Amer. J. Bot.*, 71, 285, 1984.
81. Read, J.C., and Bashaw, E.C., Intergeneric hybrid between pearl millet and buffel grass, *Crop Sci.*, 14, 401, 1974.
82. Shanthamma, C., Apomixis in *Cenchrus glaucus*, *Proc. Indian Acad. Sci. Plant Sci.* (Plant Sci.), 91, 25, 1982.
83. Das, A., and Islam, A.S., Embryological studies of 3 species of *Cenchrus* and their relationship, *Bangladesh J. Bot.*, 6, 123, 1977.
84. Costas-Lippmann, M., Embryogeny of *Cortaderia selloana* and *Cortaderia jubata*, Graminaea, *Bot. Gaz.*, 140, 393, 1979.
85. Philipson, M.N., Apomixix in *Cortaderia jubata* (Gramineae), *N.Z. J. Bot.*, 16, 45, 1978.

86. Knox, R.B., and Heslop-Harisson, J., Experimental control of aposporous apomixis in a grass of the Andropogoneae, *Bot. Notiser*, 116, 127, 1963.

87. Muniyamma, M., Variations in microsporogenesis and the development of embryo sacs in *Echinochloa stagnina* (Gramineae), *Bot. Gaz.*, 139, 87, 1978.

88. Rabau, T., Longly, B., and Louant, B.P., Ontogeny of unreduced embryo sacs in *Eragrostis curvula*, *Can. J. Bot.*, 64, 1778, 1986.

89. Voigt, P.W., and Bashaw, E.C., Facultative apomixis in *Eragrostis curvula*, *Crop Sci.*, 16: 803, 1976.

90. Bhanwra, R.K., and Choda, S.P., Apomixis in *Eremopogon foveolatus* (Gramineae), *Nord. J. Bot.*, 1, 97, 1981.

91. Shanthamma, C., and Narayan, K.N., Studies in Poaceae Gramineae, *J. Mysore Univ. Sect. B.*, 27, 302, 1976.

92. Weimarck, G., Apomixis and sexuality in *Hierochloe alpina* (Gramineae) from Finland and Greenland and in *H. monticola* from Greenland, *Bot. Notiser*, 123, 495, 1970.

93. Weimerck, G., Apomixis in *Hierochloe monticola* (Gramineae), *Bot. Notiser*, 120, 448, 1967.

94. Weimerck, G., Apomixis and sexuality in *Hierochloe australis* and in Swedish *H. odorata* on different polyploid levels, *Bot. Notiser*, 120, 209, 1967.

95. Savidan, Y.H., Embryological analysis of facultative apomixis in *Panicum maximum* Jacq., *Crop Sci.*, 22, 467, 1982.

96. Quarin, C.L., Hanna, W.W., and Fernandez, A., Genetic studies in diploid and tetraploid *Paspalum* species, *J. Hered.*, 73, 254, 1982.

97. Burson, B.L., and Bennet, H.W., Cytology and reproduction of three *Paspalum* species, *J. Hered.*, 61, 129, 1970.

98. Dujardin, M., and Hanna, W., Microsporogenesis reproductive behavior and fertility in 5 *Pennisetum* species, *Theor. Appl. Genet.*, 67, 197, 1984.

99. Burson, B.L., and Bennet, H.W., Chromosome number, microsporogenesis, and mode of resproduction of seven *Paspalum* species, *Crop. Sci.*, 11, 292, 1971.

100. Srivastava, A.K., Apomixis in *Paspalum paspaloides*, *Acta Bot. Indica*, 10, 111, 1982.

101. Caponio, I., and Quarin, C.L., The genetic system of *Paspalum simplex* and of an interspecific hybrid with *Paspalum dilatatum*. *Kurtziana*, 19, 35, 1987.

102. Chatterji, A.K., and Timothy, D.H., Apomixis and tetraploidy in *Pennisetum orientale* Rich., *Crop Sci.*, 9, 796, 1969.

103. Kalyane, V.L., and Chatterji, A.K., Reproductive characteristics of *Pennisetum pedicellatum*, *Indian J. Genet. Plant Breed.*, 41, 384, 1981.

104. Birari, S.P., Mechanism of apomixis in *Pennisetum polystachyon*, *J. Maharashtra Agric. Univ.*, 6, 208, 1981.

105. Inamuddin, M., and Faruqi, S.A., Studies in Libyan grasses 8. Apomixis in *Pennisetum divisum* Sensu lato and *Pennisetum setaceum*, *Pak. J. Bot.*, 14, 69, 1982.

106. Khristov, M.A., and Terzijski, D.P., Origin of twin plants in some species of the genus Poa with apomictic reproduction, *Fitologiya*, 11, 3, 1979.

107. Batygina, T.B., and Freiberg, T.E., Polyembryony in *Poa pratensis* Poaceae, *Bot. Zh.*, 64, 793, 1979.

108. Kellogg, E.A., Apomixis in the *Poa secunda* complex, *Amer. J. Bot.*, 74, 1431, 1987.

109. Carman, J.G., and Hatch, S.L., Aposporus apomixis in *Schizachyrium* (Poaceae: Andropogoneae), *Crop Sci.*, 22, 1253, 1982.

110. Hanna, W.W., Schertz, K.F., and Bashaw, E.C., Apospory in *Sorghum bicolor* (L.) Moench., *Science*, 170, 338, 1970.

111. Rao, N.G.P., Narayana, L.L., and Reddy, R.N., Apomixis and its utilization in grain *Sorghum* – I. Embryology of two apomictic parents, *Caryologia*, 31, 427, 1978.

112. Tang, C.Y., Schertz, K.F., and Bashaw, E.C., Apomixis in *Sorghum bicolor* lines and their F_1 progenies. *Bot. Gaz.*, 141, 294, 1980.

113. Reddy, R.N., Narayana, L.L., and Rai, N.G.P., Apomixis and its utilization in Grain sorghum 2. Embryology of F_1 progeny of reciprocal crosses between R-473 and 302, *Proc. Indian Acad. Sci. Sect. B.*, 88, 455, 1979.

114. Brown, W.V., and Emry, W.H.P., Apomixix in the Gramineae: Panicoideae, *Amer. J. Bot.*, 45, 253, 1958.
115. Burson, B.L., Voigt, P.W., Sherman, R.A., and Dewald, C.L., Apomixis and sexuality in eastern Gamagrass, *Crop Sci.*, 30, 86, 1990.
116. Sarkar, K.R., and Coe, E.H. Jr., A genetic analysis of the origin of maternal haploids in maize, *Genetics*, 54, 453, 1966.
117. Zverzhanskaya, L.S., Grishina, E.V., and Komarova, P.I., Cytoembryological study of haploidy and embryo multiplicity in restitution maize lines, *Tsitol. Genet.*, 14, 3, 1980.
118. Mathew, C.J., Embryological studies in Hamamelidaceae, Development of female gametophyte and embryogeny in *Hamamelis virginiana*, *Phytomorphology*, 30, 172, 1980.
119. Venkateswarlu, J., Devi, P.S., and Nirmala, A., Embryological studies in *Eleutherine plicata* herb. and *Belamcanda chinensis* Lem., *Proc. Indian Acad. Sci.* (Plant Sci.), 89: 361, 1980.
120. Rudall, P.J., Owens, S.J., and Kenton, A.Y., Embryology and breeding systems in *Crocus* (Iridaceae) – A study in causes of chromosome variation, *Plant. Syst. Evol.*, 148, 119, 1984.
121. Terzijski, D., and Stefanova, A., On the nature of apomixis in some nut cultivars, *Rasteniev Dui Nauki*, 27, 73, 1990.
122. Bouman, F., and Boesewinkel, F.D., On a case of polyembryony in *Pterocarya fraxinifolia* (Juglandaceae) and on polyembryony in general, *Acta Bot. Neerl.*, 18, 50, 1969.
123. Kumar, P., Polyembryony in *Acacia farnesiana* Willd, *Indian For.*, 112, 742, 1986.
124. Kumar, A., and Gupta, B.B., Twin seedlings in *Albizzia lebbeck* L. Benth., *Indian J. For.*, 10: 313, 1987.
125. Shrivas, R.K., and Bajpai, S.P., A note on twin seedlings in *Albizzia lebbeck* Benth., *Indian For.*, 114, 292, 1988.
126. Akhalakatsi, M.S., Beridze, M.V., and Gvaladze, G.E., Embryologenesis and endospermogenesis in *Astragalus caucasicus* Pall., *Soobshch Akad. Nauk. Gruz. SSR*, 132, 601, 1988.
127. Beversdorf, W.D., and Bingham, E.T., Male-sterility as a source of haploids and polyploids of *Glycine max* , *Can. J. Genet. Cytol.*, 19, 283, 1977.
128. Dhawan, V., and Bhojwani, S.S., Polyembryony in *Leucaena leucocephala*, *Phytomorphology*, 35, 147, 1985.
129. Greenshields, J.E.R., Polyembryony in Alfalfa, *Sci. Agr.*, 31, 212, 1951.
130. Jaranowski, J., Haploid-diploid twin embryos in *Melilotus*, *Genet. Pol.*, 2, 129, 1961.
131. Kazimierska, E.M., Embryological studies of cross compatibility of species within the genus *Trifolium*. 3. Development of the embryo and endosperm in crossing *Trifolium repens* with *Trifolium hybridum* and *Trifolium fragiferum*, *Genet. Pol.*, 21, 37, 1980.
132. Randall, T.E., and Rick, C.M., A cytogenetic study of polyembryony in asparagus, *Amer. J. Bot.*, 32, 560, 1945.
133. Gvaladze, G.E., and Krialashvili, L.G., Polyembryony in *Colchicum speciosum* Stev., *Soobshch Akad. Nauk. Gruz. SSR*, 132, 117, 1988.
134. Jeffrey, E.C., Polyembryony in *Erythronium americanum*, *Ann. Bot.*, 9, 537, 1895.
135. Guerin, P., Le dévelopment de l'oeuf et la polyembryonie chez l'*Erythronium dens-canis* L. (Liliacée), *C.R. Acad. Sci., Paris*, 191, 1369, 1930.
136. Fagerlind, F., Hormonale Substanzen als Ursache der Frucht und Embryobildung bei pseudogamne *Hosta* Biotypen., *Sv. Bot. Tidskr.*, 40, 230, 1946.
137. Sulbha, The embryology of *Iphigenia indica* Kunth., *Phytomorphology*, 4, 180, 1954.
138. Cooper, D.C., Haploid-diploid twin embryos in *Lilium* and *Nicotiana*, *Amer. J. Bot.*, 30, 408, 1943.
139. Jeffrey, E.C., and Haertl, E.J., Apomixis in *Trillium*. *Cellule*, 48, 79, 1939.
140. Kappert, H., Erbliche Polyembryonie bei *Linum usitatissimum*, *Biol. Zentralbl.*, 53, 276, 1933.
141. Harland, S.C., Haploids in polyembryonic seeds of sea island cotton, *J. Hered.*, 27, 229, 1936.
142. Lee, J.A., On origin of haploid/diploid twinning in cotton, *Crop. Sci.*, 10, 453, 1970.
143. Jacqes-Felix, H., Seed and embryo in the African *Memecylon* (melastomataceae), *Adansonia*, 17, 193, 1977.

64

144. Subramanyam, K., A contribution to the life history of *Sonerila wallichii* Benn., *Proc. Indian Acad. Sci. B*, 19, 115, 1944.
145. Nair, N.C., and Kanta, K., Studies in Meliaceae 4. Floral morphology and embryology of *Azadirachta indica* A. Juss.: A reinvestigation, *J. Indian Bot. Soc.*, 40, 382, 1961.
146. Prakash, N., Lim, A.L., and Manurung, R., Embryology of Duku and Langsat varieties of *Lansium domesticum*, *Phytomorphology*, 27, 50, 1977.
147. Srivastava, B.L., and Tandon, S., Adventive embryony in *Morus* species (Mulberry), *Phytomorphology*, 35, 181, 1985.
148. Pijl, L.V., Über die Polyembryonie bei *Eugenia*, *Rec. Trav. Bot. Beerl.*, 31, 113, 1934.
149. Roy, S.K., Embryology of *Eugenia jambos* L., *Curr. Sci.* (Bangalore), 22, 249, 1953.
150. Traub, H.P., Polyembryony in *Myricaria cauliflora*, *Bot. Gaz.*, 101, 233, 1939.
151. Narayanaswami, S., and Roy, S.K., Embryo sac development and polyembryony in *Syzygium cumini* (Linn.) Skeels, *Bot. Not.*, 113, 273, 1960.
152. Roy, S.K., and Sahai, R., The embryo sac and embryo of *Syzygium caryophyllifolium* DC., *J. Indian Bot. Soc.*, 41, 45, 1962.
153. Guignard, L., La double fécondation dans le *Najas major*, *J. Bot.*, Paris, 15, 205, 1901.
154. Haberlandt, G., Über experimentelle Erzeugung von Adventiveembryonen bei *Oenothera lamarkiana*, *Sitz-ber. Acad., Berlin*, 40, 695, 1921.
155. Swamy, B.G.L., Embryological studies in the Orchidaceae. II. Embryogeny, *Amer. Midl. Nat.*, 41, 202, 1949.
156. Swamy, B.G.L., Female gametophyte and embryogeny in *Cymbidium bicolor* Lindli., *Proc. Indian Acad. Sci. B.*, 15, 194, 1942.
157. Singh, F., and Thimmappaiah, M., Polyembryony in orchid seeds, *Seed Sci. Technol.*, 10, 29, 1982.
158. Stort, M.N.S., and Pavanelli, E.A.D.S., Formation of multiple or adventive embryos in *Epidendrum nocturnum* Orchidaceae, *Ann. Bot.*, 55, 331, 1985.
159. Vij, S.P., and Sharma, M., Embryological studies in Orchidaceae V. *Epipactis* Adams, *Phytomorphology*, 37, 81, 1987.
160. Swamy, B.G.L., Gametogenesis and embryogeny in *Eulophia epidendra* Fischer., *Proc. Indian Acad. Sci. B*, 9, 59, 1943.
161. Ansari, R., Observation on the occurrence of polyembryony in two species of Orchids, *Curr. Sci.* (Bangalore), 46, 607, 1977.
162. Illg, R.D., Reproduction in *Maxillaria brasiliensis* and *Maxillaria cleistogama* (Orchidaceae), *Rev. Bras. Biol.*, 37, 267, 1977.
163. Afzelius, K., Die Embryibildung bei *Nigritella nigra*, *Sv. Bot. Tidskr.*, 22, 82, 1928.
164. Afzelius, K., Zur kenntnis der Fortpflanzungsverhaltnisse und Xhromosomenzahlen bei *Nogritella nigra*, *Sv. Bot. Tidskr.*, 26, 365, 1932.
165. Rossi, W., Capineri, R., Teppner, H., and Klein, E., *Nigritella widderi* Orchidaceae – Orchideae in the Apennines Italy, *Phyton*, 27, 129, 1987.
166. Teppner, H., and Klein, E., *Nigritella widderi* new species Orchidaceae orchideae. *Phyton*, 25, 317, 1985.
167. Hagerup, O., On fertilization, polyploidy and haploidy in *Orchis maculatus* L., *Dansk Bot. Ark.*, 11, 1, 1944.
168. Chua, L.G., and Rao, A.N., Polyembryony and suspensor characteristics in *Spathoglottis*, *Flora* (Jena), 167, 399, 1978.
169. Leavitt, R.G., Polyembryony in *Spiranthes cernua*, *Rhodora*, 2, 227, 1900.
170. Swamy, B.G.L., Agamospermy in *Spiranthes cernua*, *Lloydia*, 11, 149, 1948.
171. Karanth, K.A., Swamy, B.G.L., and Arekal, G.D., Embryogenesis in sexual and asexual species of *Zeuxine* Orchidaceae, *Proc. Indian Acad. Sci.* (Plant Sci.), 90, 1, 1981.
172. Swamy, B.G.L., The embryology of *Zeuxine sulcata* Lindl., *New Phytol.*, 45, 132, 1946.
173. Vij, S.P., Sharma, M., and Shekhar, N., Embryological studies in Orchidaceae. II: *Zeuxine strateumatica* complex, *Phytomorphology*, 32, 257, 1982.
174. Haccius, B., and Philip, V.J., Embryo development in *Cocos nucifera*: general understanding of palm embryogenesis, *Plant Syst. Evol.*, 132, 91, 1979.

175. Whitehead, R.A., and Chapman, G.P., Twinning and haploidy in *Cocos nucifera* Linn., *Nature*, 195, 1228, 1962.
176. Clancy, K.E., and Sullivan, M.J., Some observations on seed germination the seedling and polyembryony in the needle palm *Rhapidophyllum hystrix*, *Principes*, 32, 18, 1988.
177. Sachar, R.C., The embryology of *Argemone mexicana* L. – A reinvestigation, *Phytomorphology*, 5, 200, 1955.
178. Narayan, L.L., and Sundri, K.T., Embryology of Pittosporaceae 4, *Indian J. Bot.*, 4, 123, 1981.
179. Edman, G., Apomeiosis und Apomixis bei *Atraphaxis frutescens* C. Koch., *Acta Hort. Berg.*, 11: 13, 1931.
180. Haccius, B., and Reichert, R., Restitutionserscheninungen an pflanzlichen meristemen nach rontgenbestrahlung. II. Adventive-embryonie nach samenbestrahlung von *Eranthis hiemalis*, *Planta*, 62, 355, 1964.
181. Izmailow, R., Problem of apomixis in the *Ranunculus auricomus* group, *Acta Biol. Cracov. Ser. Bot.*, 19, 15, 1976.
182. Arora, N., The embryology of *Zizyphus rotundifolia* Lamk., *Phytomorphology*, 3, 88, 1953.
183. Mandryk, V.Y., Embryological studies of some species of the genus *Alchemilla*, *Ukr. Bot. Zh.*, 33, 476, 1976.
184. Campbell, C.S., Greene, C.W., Neubauer, B.F., and Higgins, J.M., Apomixis in *Amelanchier laevis*, shadbush (Rosaceae, Maloideae), *Amer. J. Bot.*, 72, 1397, 1985.
185. Samushia, M.D., and Mosashvili, V.A., Features of female gametophyte development and polyembryony in some *Quince* cultivars, *Soobshch Akad. Nauk Gruz. SSR*, 106, 385, 1982.
186. Muniyamma, M., and Phipps, J.B., Studies in *Crataegus*. XI. Further cytological evidence for the occurrence of apomixis in North American hawthorns, *Can. J. Bot.*, 62, 2316, 1984.
187. Muniyamma, M., and Phipps, J.B., Studies in Craaegus. X. A note on the occurrence of diplospory in *Crataegus dissona* Sarg. (Maloideae, Rosaceae), *Can. J. Genet. Cytol.*, 26, 249, 1984.
188. Ptak, K., Cyto-embryological investigations on the Polish representatives of the genus *Crataegus* L. II. Embryology of triploid species, *Acta Biol. Cracov. Ser. Bot.*, 31, 99, 1990.
189. Lebegue, A., Recherches sur le polyembryonie chez les Rosaceea (g. *Fragaria* et *Alchemilla*), *Bull. Soc. Bot. Fr.*, 99, 273, 1952.
190. Hjelmqvist, H., The apomictic development in *Malus sieboldii*, *Bot. Not.*, 110, 455, 1957.
191. Olden, E.J., Sexual and apomictic seed formation in *Malus sieboldi* Rehd., *Bot. Not.*, 106, 105, 1953.
192. Hjelmqvist, H., On the embryology of two *Malus* hybrids, *Bot. Not.*, 112, 453, 1959.
193. Mandryk, V.Y., and Mentkovskara, E.A., Cytoembryological study of some *Potentilla erecta* populations in the Ukrainian carpathian mountains in USSR: Fertilization and development of endosperm and embryo, *Bot. Zh.*, 63, 1326, 1978.
194. Souèges, R., Polyembryonie chez le *Potentilla reptans* L., *Bull. Soc. Bot. Fr.*, 82, 381, 1935.
195. Nordborg, G., Embryological studies in the Sanguisorba minor complex (Rosaceae), *Bot. Notiser*, 120, 109, 1967.
196. Jankun, A., and Kovanda, M., Apomixis at the diploid level in *Sorbus eximia*: Embryological studies in *Sorbus* 3, *Preslia* (Prague), 60, 193, 1988.
197. Jankun, A., and Kovanda, M., Apomixis in *Sorbus sudetica*. Embryological studies in *Sorbus* 1, *Preslia*, 58, 7, 1986.
198. Johri, B.M., and Ahuja, M.R., A contribution to the floral morphology and embryology of *Aegle marmelos* Correa, *Phytomorphology*, 7, 10, 1957.
199. Thakur, D.R., and Bajwa, B.S., Extent of polyembryony in some species and varieties of *Citrus*, *Indian J. Hort.*, 28, 25, 1971.
200. Osawa, J., Cytological and experimental studies in *Citrus*, *J. College Agric.*, Tokyo, 4, 83, 1912.
201. Bacchi, O., Cytological observations in *Citrus*. III. Megasporogenesis, fertilization, and polyembryony, *Bot. Gaz.*, 105, 221, 1943.
202. Furusato, K., and Ohta, Y., Induction and inhibition of nucellar embryo development in *Citrus*, *Seiken Ziho*, 21, 45, 1969.

66

203. Maheshwari, P., and Rangaswamy, N.S., Polyembryony and *in vitro* culture of embryos of *Citrus* and *Mangifera*, *Indian J. Hort.*, 15, 275, 1958.
204. Toxopeus, H.J., De polyembryonie van *Citrus* en haar beteenkensis voor de cultuur, *Vereen. Landbouw Neederl. Indie Landbouw Tijdksche.*, 6, 807, 1930.
205. Furusato, K.Y., Ohta, Y., and Ishibashi, K., Studies on polyembryony in *Citrus*, *Seiken Ziho*, 8, 40, 1957.
206. Yang, H.J., Fertilization and development of embryo on Satsuma orange (*Citrus unshiu* Marc.) and Natsudaidai (*C. natsudaidai* Hayata), *J. Japan. Soc. Hort. Sci.*, 37, 102, 1968.
207. Cameron, J.W., and Soost, R.K., Sexual and nucellar embryony in F_1 hybrids and advenaced crosses of *Citrus* with *poincirus* hybrid parent, *J. Amer. Soc. Hort. Sci.*, 104, 408, 1979.
208. Esan, E.B., A detailed study of adventive embryogenesis in the Rutaceae, Ph.D. Dissertation, Univ. California, Riverside, 1973.
209. Narmatha Bai, V., and Lakshmanan, K.K., Occurrence of polyembryony in *Toddalia asiatica* Lamb., *Curr. Sci.* (Bangalore), 51, 374, 1982.
210. Desai, S., Polyembryony in *Xanthoxylum* Mill., *Phytomorphology*, 12, 184, 1962.
211. Rajora, O.P., and Zsuffa, L., Atypical seedlings of *Populus* L.: Their genetic significance and value in breeding, *Silvae Genet.*, 35 122, 1986.
212. Bhatnagar, S.P., and Joshi, P.C., Morphological and embryological studies in the family Santalaceae. VII. *Exocarpus bidwillii* Hook. f., *Proc. Nat. Inst. Sci. India* (Part B), 31, 34, 1965.
213. Ram, M., Morphological and embryological studies in the family Santalaceae. II. *Exocarpus*, with a discussion on its systematic position, *Phytomorphology*, 9, 4, 1959.
214. Mathur, S., and Gulati, N., Embryology and taxonomy of *Allophylus alnifolius* Sapindaceae, *Indian J. Bot.*, 3, 103, 1980.
215. Lebègue, A., Embryologie des Saxifragacées: Polyembryonie chez le *Bergenia delavayi* Engl., *Bull. Soc. Bot., Fr.*, 96, 38, 1949.
216. Lebègue, A. Polyembryonie chez le *Tellima grandiflora* R. Br., *Bull Soc. Bot. Fr.*, 98, 114, 1951.
217. Christensen, H.M., and Bamford, R., Haploids in twin seedlings of pepper (*Capsicum annuum* L.), *J. Hered.*, 34, 98, 1943.
218. Morgan, D.T. Jr., and Rappleye, R.D., A cytogenetic study of the origin of multiple seedlings of *Capsicum frutescens*, *Amer. J. Bot.*, 41, 576, 1954.
219. Marshall, H.H., Polyembryony in tomatoes, *Can. J. Plant Sci.*, 38, 67, 1958.
220. Mathew, C.J., and Chapekar, M., Development of female gametophyte and embryogeny in *Stachyurus chinensis*, *Phytomorphology*, 27, 68, 1977.
221. Ravishankar, S., and Gulati, N., Polyembryony in Symplocaceae, *Indian J. Bot.*, 8, 13, 1985.
222. Johri, B.M., and Kak, D., The embryology of *Tamarix* Linn., *Phytomorphology*, 4, 230, 1954.
223. Sharma, B.D., Possible occurrence of polyembryony in Pentoxyleae, *Phytomorphology*, 39, 199, 1989.
224. Mikatadze-Pantsulaya, T.A., Polyembryony of some forms of *Thea sinensis* growing in the Georgian-SSR USSR, *Soobshch Akad. Nauk. Gruz. SSR*, 89, 173, 1978.
225. Fagerlind, F., Zytologie und Gametophyten bildung in der Gattung *Wikstroemia*, *Hereditas*, 26, 23, 1940.
226. Gupta, S.C., The embryology of *Coriandrum sativum* and *Foeniculum vulgare*, *Phytomorphology*, 14, 530, 1964.
227. Maheshwari Devi, H., and Santa Ram, A., Embryology of the Apiaceae – *Pimpinella*, *Phytomorphology*, 35, 183, 1985.
228. Fagerlind, F., Die Samenbildung und die Zytologie bei Agamospermischen und sexuellen Arten von *Elastostema* und einiger damit zusammenhangender Probleme, *Kl. Sv. Vet.-Akad. Handl. III*, 21, 1, 1944.
229. Kuta, E., Embryological observations on two Canadian species of the *Viola* L. genus section Plagiostigma Godr., *Acta Biol. Vracov Se. Bot.*, 30, 39, 1988.
230. Lebègue, A., Trois exemples de polyembryonie chez le *Peltiphyllim peltatum* Engl., *Bull. Soc. Bot. Fr.*, 99, 254, 1952.
231. Lakshmana, K.K., and Bai, V.N., A case of true polyembryony in *Lumnitzera racemosa*, *Curr. Sci.* (Bangalore), 55, 159, 1986.

232. Chen, Z.K., and Wang, F.X., Early embryogeny and its starch distribution in *Amentotaxus*, *Acta Bot. Sin.*, 26, 359, 1984.

233. Singh, H., The life history and sytematic position of *Cephalotaxus drupacea* Sieb. et Zucc., *Phytomorphology*, 11, 153, 1961.

234. Buchholz, J.T., The embryogeny of *Cephalotaxus fortunei*, *Bull. Torrey Bot. Club*, 52, 311, 1925.

235. Singh, H., and Oberoi, Y.P., A contribution to the life history of *Biota orientalis* Endl., *Phytomorphology*, 12, 373, 1962.

236. Baird, A.M., The life history of *Callitris*, *Phytomorphology*, 3, 258, 1953.

237. Looby, W.J., and Doyle, J., New observations on the life history of *Callitris*, *Sci. Proc. Roy. Dublin Soc.*, 22, 241, 1940.

238. Buchholz, J.T., The embryogeny of *Chamaecyparis obtusa*, *Amer. J. Bot.*, 19, 230, 1932.

239. Owens, J.N., and Molder, M., Pollination, female gametophyte; and embryo and seed development in yellow cedar (*Chamaecyparis nootkatensis*), *Can. J. Bot.*, 53, 186, 1975.

240. Doak, C.E., Morphology of *Cupressus arizonica*: gametophytes and embryogeny, *Bot. Gaz.*, 98, 808, 1937.

241. Sugihara, Y., The embryogeny of *Cupressus funebris* Endlicher., *Bot. Mag.* (Tokyo), 69, 820, 1956.

242. Suguhara, Y., The embryogeny of *Cupressus sempervirens* Linnaeus, *Sci. Rep. Tohoku Univ.*, 22 (Biol.), 1, 1956.

243. Cook, P.L., A new type of embryogeny in the conifers, *Amer. J. Bot.*, 26, 138, 1939.

244. Mathews, A.C., The morphological and cytological development of the sporophylls and seed of *Juniperus virginiana* L., *J. Elisha Mitchell Sci. Soc.*, 55, 7, 1939.

245. Sedgwick, P.J., Life history of *Encephalartos*, *Bot. Gaz.*, 77, 300, 1924.

246. Lehmann-Baerts, M., Étude sur les Gnétales. XII. Ovule, gamétophyte femelle et embryogenèse chez *Ephedra distachya* L., *Cellule*, 67, 53, 1967.

247. Haining, H.I., Development of embryo of *Gnetum*, *Bot. Gaz.*, 70, 436, 1920.

248. Sanwal, M., Morphology and embryology of *Gnetum gnemon* L., *Phytomorphology*, 12, 243, 1962.

249. Hutchinson, A.H., Embryogeny of *Abies*, *Bot. Gaz.*, 77, 280, 1924.

250. Owens, J.N., and Molder, M., Sexual reproduction of *Abies amabilis*, *Can. J. Bot.*, 55, 2653, 1977.

251. Buchholz, J.T., Polyembryony among Abietineae, *Bot. Gaz.*, 69, 153, 1920.

252. Dogra, P.D., Embryology of the Taxodiaceae, *Phytomorphology*, 16, 125, 1966.

253. Choudhury, C.R., The morphology and embryology of *Cedrus deodara* (Roxb.) Loud., *Phytomorphology*, 11, 283, 1961.

254. Buchholz, J.T., Embryo development and polyembryony in relation to the phylogeny of conifers, *Amer. J. Bot.*, 7: 125, 1920.

255. Sugihara, Y., Embryological observations on *Keteleeria davidiana* Biossner var. formosana Hayata, *Sci. Rep. Tohoku Univ.*, 17 (Biol.), 215, 1943.

256. Schopf, J.M., The embryology of *Larix*, *Illinois Biol. Monog.*, 19, 1, 1943.

257. Allen, G.S., Parthenocarpy, parthenogenesis, and self-sterility of Douglas-fir, *J. Forestery*, 40, 642, 1942.

258. Orr-Ewig, A.L., A cytological study of the effects of sef-pollination on *Pseudotsuga menziesii* (Mirb.) Franco, *Silvae Genet.*, 6, 179, 1957.

259. Clare, T.S., and Johnstone, G.R., Polyembryony and germination of polyembryonic coniferous seeds, *Amer. J. Bot.*, 18, 674, 1931.

260. Owens, J.N., Molder, M., and Simpson, S., Sexual reproduction in *Pinus contorta*. II. Postdormancy ovule, embryo and seed development, *Can. J. Bot.*, 60, 2071, 1982.

261. Johnstone, G.R., Further studies on polyembryony and germination of polyembryonic pine seeds, *Amer. J. Bot.*, 27, 808, 1940.

262. Dogra, P.D., Seed sterility and disturbances in embryogeny in conifers with particular reference to seed testing and tree breeding in Pinaceae, *Stud. Forest. Suecica*, 45, 1, 1967.

263. Johnstone, G.R., Multiple pine seedlings, *Science*, 84 (2180), 330, 1936.

68

264. Jacobs, A.W., Polyembryony in sugar pine, *J. Forestry*, 22, 573, 1924.
265. Owens, J.N., and Molder, M., Seed-cone differentiation and sexual reproduction in western white pine (*Pinus monticola*), *Can. J. Bot.*, 55, 2574, 1977.
266. Dogra, P.D., Observations on *Abies pindrow* with a discussion on the question of occurrence of apomixis in gymnosperms, *Silvae Genet.*, 15, 11, 1966.
267. Saxton, W.T., Preliminary account of the ovule, gametophytes, and embryo of *Widdringtonia cupressoides*, *Bot. Gaz.*, 48, 161, 1909.
268. Konar, R.N., The morphology and embryology of *Pinus roxburgii* Jack., *Phytomorphology*, 10, 305, 1960.
269. Mu, X.J., Chen, Z.K., and Wang, F.H., Variations in the starch of *Pinus tabulaeformis* during embryogenesis, *Acta Bot. Sin.*, 21, 117, 1979.
270. Toumey, J.W., Multiple pine embryos, *Bot. Gaz.*, 76, 426, 1923.
271. Buchholz, J.T., The pine embryo and the embryo of related genera, *Trans. Illinois State Acad. Sci.*, 23, 117, 1931.
272. Owens, J.N., and Molder, M., Sexual reproduction in mountain hemlock (*Tsuga mertensiana* (Bong.) Carr.), *Can. J. Bot.*, 53, 1811, 1975.
273. Quinn, C.J., Gametophyte development and embryogeny in the Podocarpaceae. II. *Dacrydium laxifolium* , *Phytomorphology*, 15, 37, 1965.
274. Quinn, C.J., Gametophyte development and embryogeny in the Podocarpaceae. III. *Dacrydium bidwillii* , *Phytomorphology*, 16, 81, 1966
275. Buchholz, J.T., Determinate cleavage polyembryony, with special reference to *Dacrydium*, *Bot. Gaz.*, 94, 579, 1933.
276. Quinn, C.J., Gametophyte development and embryogeny in the Podocarpaceae. VI. *Dacrydium colensoi* : General conclusions, *Phytomorphology*, 16, 199, 1966.
277. Elliot, C.G., A further contribution to the life history of *Pherosphaera*, *Proc. Linn. Soc. N.S.W.*, 75, 320, 1950.
278. Buchholz, J.T., Embryogenesis of the Podocarpaceae, *Bot. Gaz.*, 103, 1, 1941.
279. Brownlie, G., Embryology of the New Zealand species of the genus *Podocarpus*, section *Eupodocarpus*, *Phytomorphology*, 3, 295, 1953.
280. Konar, R.N., and Oberoi, Y.P., Studies on the morphology and embryology of *Podocarpus gracilior* Pilger., *Beitr. Biol. Pflanzen*, 45, 329, 1969.
281. Sterling, C., Proembryo and early embryogeny in *Taxus cuspidata*, *Bull Torrey Bot. Club.*, 75, 469, 1948.
282. Buchholz, J.T., The embryogeny of Torreya, with a note on *Austrotaxus*, *Bull. Torrey Bot. Club*, 67, 731, 1940.
283. Buchholz, J.T., The suspensor of *Cryptomeria japonica*, *Bot. Gaz.*, 93, 221, 1932.
284. Singh, H., and Chatterjee, J., A contribution to the life history of *Cryptomeria japonica* D. Don., *Phytomorphology*, 13, 428, 1963.
285. Dogra, P.D., Embryogeny of the Taxodiaceae, *Phytomorphology*, 16, 125, 1966.
286. Buchholz, J.T., The embryogeny of *Cunninghamia*, *Amer. J. Bot.*, 27, 877, 1940.
287. Sugihara, Y., The embryogeny of *Cunninghamia lanceolata* Hooker, *Sci. Rep. Tohoku Univ.*, 16 (Biol), 187, 1941.
288. Sugihara, Y., The embryogeny of *Cunninghamia konishii* Hayata, *Bot. Mag.* (Tokyo), 60, 53, 1947.
289. Wang, F.H., The early embryogeny of *Glyptostrobus*, *Bot. Bull. Acad. Sinica*, 2, 1, 1948.
290. Doyle, J., and Brennan, M., Cleavage polyembryony in conifers and taxads – a survey I. Podocarps, taxads and taxodiods, *Sci. Proc. Roy. Dublin Soc.*, 4A, 57, 1971.
291. Doyle, J., and Brennan, M., Cleavage polyembryony in conifers and taxads – a survey II. Cupressaceae, Pinnaceae and conclusions, *Sci. Proc. Roy. Dublin Soc.*, 4A, 137, 1972.
292. Buchholz, J.T., The morphology and embryogeny of *Sequoia gigantea*, *Amer. J. Bot.*, 26, 93, 1939.
293. Buchholz, J.T., The embryogeny of *Sequoia sempervirens* with a comparison of the sequoias, *Amer. J. Bot.*, 26, 248, 1939.
294. Buchholz, J.T., The suspensor of *Sciadopitys*, *Bot. Gaz.*, 92, 243, 1931.

295. Vasil, V., and Sahni, R.K., Morphology and embryology of *Taxidium mucronatum* Tenore, *Phytomorphology*, 14, 369, 1964.

296. Sugihara, Y., Embryological observations on *Taiwania cryptomerioides* Hayata, *Sci. Rep. Tohoku Iniv.*, 16 (Biol.), 291, 1941.

297. Martens, P., and Waterkeyn, L., Sur l'embryogenese de *Welwitschia mirabilis*, *Rev. Cytol. Biol. Veg.*, 32, 331, 1969.

298. Asker, S., Progress in apomixis research, *Hereditas*, 91, 231, 1979.

299. Nogler, G.A., Gametophytic apomixix, in *Embryology of Angiosperms*, Johri, B.M., Ed., Springer-Verlag, Berlin, 1984, Chapter 10.

300. Battaglia, E., Apomixis, in *Recent Advances in the Embryology of Angiosperms*, Maheshwari, P., Ed., Intl. Soc. Plant Mosphologists, Univ. Delhi, 1963, Chapter 8.

301. Gustafsson, H.V., Apomixis in higher plants – I. The mechanism of apomixis, *Lunds Univ. Arsskr, N.F. Adv. 2*, 42, 1, 1946.

302. Gustafsson, A., Apomixis in higher plants – II. The causal aspects of apomixis, *Lunds Univ. Arsskr. N.F. Adv. 2*, 43, 71, 1947.

303. Lakshmanan, K.K., and Ambegaokar, K.B., Polyembryony, in *Embryology of Angiosperms*, Johri, B.M., Ed., Springer-Verlag, Berlin, 1984, Chapter 9.

304. Richards, A.J., in *Plant Breeding Systems*, George Allen & Unwin, Boston, 1986, Chapter 11.

305. Stebbins, G.L., Apomixis in angiosperms, *Bot. Rev.*, 7, 507, 1941.

306. Webber, J.M., Polyembryony, *Bot. Rev.*, 6, 575, 1940.

307. Chamberlain, C.J., *Gymnosperms: Structure and Evolution*, Chicago Univ. Press, Chicago, 1935.

308. Choudhury, C.R., The embryogeny of conifers: A review, *Phytomorphology*, 12, 313, 1962.

309. Singh, H., in *Embryology of Gymnosperms*, Gebruder Borntraeger, Stuttgart, 1978, Chapter 11.

310. Yakovlev, M.S., Polyembryony in higher plants and principles of its classification, *Phytomorphology*, 17, 278, 1967.

311. Okabe, S., Parthenogenesis bei *Ixeris dentata*, *Bot. Mag.* (Tokyo), 46, 518, 1932.

312. Fagerlind, F., Bildung und Entwicklung des Embryosacks bei sexuellen und agamospermischen *Balanophora* Arten, *Svensk. Bot. Tidskr*, 39, 65, 1945.

313. Poddubnaja-Arnoldi, W.A., Geschlechtliche und ungeschlrchtliche Fortpflanzung bei einigen *Chondrilla* – Arten, *Planta*, 19, 46, 1933.

314. Bergman, B., Meiosis in two different clones of the apomictic *Chondrilla juncea*, *Hereditas*, 36, 297, 1950.

315. Bergman, B., Studies on the embryo sac mother cell and its development in *Hieracium*, subg. *Archieracium*, *Svensk. Bot. Tidskr.*, 35, 1, 1941.

316. Stebbins, G.L., Cytology of *Antennaria* II. Parthenogenetic species, *Bot. Gaz.*, 94, 322, 1932.

317. Bergman, B., On the formation of reduced and unreduced gametophytes in the females of *Antennaria carpatica*, *Hereditas*, 37, 501, 1951.

318. Chiarugi, A., and Francini, E., Apomissia in *Ochna serrulata* Walp., *Nuovo G. Bot. Ital.* (Nuova Ser.), 37, 1, 1930.

319. Carano, E., Sul particolare sviluppo del gametofito O di *Euphorbia dulcis* L., *R.C. Acad. Lincei, CL. Sci. Fis., Mat. e nat. Ser. VI a*, 1, 633, 1925.

320. Furusato, K., Studies on the polyembryony in *Citrus*, *Jap. J. Genet.*, 28, 165, 1953.

321. Frost, H.B., and Soost, R.K., Seed reproduction: Development of gametes and embryos, in *The Citrus Industry*, Reuther, W., Batchelor, L.D., and Webber, H.J., Eds., Univ. California, Riverside, 2, 1968, 290.

322. Wiger, J., Ein neuer Fall von autonomer Nuzellarembryonie, *Bot. Not.*, 83, 368, 1930.

323. Afzelius, K., Apomixis and polyembryony in *Zygopetalum mackayi* Hook., *Acta Hort. Bergiani.*, 19, 7, 1959.

324. Wakana, A., and Uemoto, S., Adventive embryogenesis in *Citrus*. I. The occurrence of adventitious embryos without pollination or fertilization, *Amer. J. Bot.*, 74, 517, 1987.

325. Esen, A., and Soost, R.K., Adventive embryogenesis in citrus and its relation to pollination and fertilization, *Amer. J. Bot.*, 64, 607, 1977.

70

326. Ueno, I., Iwamasa, M., and Nishiura, M., Embryo number of various varieties of *Citrus* and its relatives, *Bul. Hort. Res. Sta. Japan, Ser. B*, 7, 11, 1969.
327. Kapil, R.N., and Tiwari, S.C., The integumentary tapetum, *Bot. Rev.*, 44, 457, 1978.
328. Misra, S., Floral morphology of the family Compositae. II. Development of the seed and fruit in *Flavaria rependa, Bot. Mag.* (Tokyo), 77, 290, 1964.
329. Walker, R.I., Cytological and embryological studies in *Solanum*, section *Tuberarium, Bull. Torrey Bot. Club*, 82, 87, 1955.
330. Johri, B.M., and Tiagi, B., Floral morphology and seed formation in *Cuscuta reflexa* Roxb., *Phytomorphology*, 2, 162, 1952.
331. Hakkansson, A., Die Entwincklung des Embryosackes und die Befruchtung bei *Poa alpina, Hereditas*, 29, 25, 1943.
332. Shinde, A.N.R., Swamy, B.G.L., and Govindappa, D.A., Synergid embryo in *Pennisetum squamulatum, Curr. Sci.* (Bangalore), 49, 914, 1980.
333. Johri, B.M., and Bhatnagar, S.P., A contribution to the morphology and life history of *Aristolochia, Phytomorphology*, 5, 123, 1955.
334. Narayanaswami, S., The structure and development of the caryopsis in some Indian millets. II. *Paspalum scrobiculatum* L., *Bull. Torrey Bot. Club.*, 81, 288, 1954.
335. Modilewski, J., Die Embryobildung bei *Allium odorum* L., *Bull. Jard. Bot. Kieff*, 12/13, 27, 1931.
336. Johri, B.M., and Ambegaokar, K.B., Embryology: then and now, in *Embryology of Angiosperms*, Johri, B.M., Ed., Springer-Verlag, Berlin, 1984, Chapter 1.
337. Treub, M., L'organe femelle et l'apogamie du *Balanophora elongata, Ann. Jard. Bot.*, Buitenz., 15, 1, 1898.
338. Rosenberg, O., Apogamie und Parthenogenesis bei Pflanzen, in *Handbuch Vererbungswiss.*, Baur, E. and Hartman, M., Eds., Borntraeger, Berlin, 12, 1930, 1–66.
339. Ernst, A., Beitrage zur Kenntnis der Entwicklung des Embryosackes und des Embryos (Polyembryonie) von *Tulipa gesneriana* L., *Flora* (Jena), 88, 37, 1901.
340. Veillet-Bartoszewska, M., La polyembryonie chez le *Primula auricula* L., *Bull. Soc. Bot. Fr.*, 104, 473, 1957.
341. Masand, P., Embryology of *Zygophyllum fabago* Linn., *Phytomorphology*, 13, 293, 1963.
342. Berlyn, G.P., Developmental patterns in pine polyembryony, *Amer. J. Bot.*, 49, 327, 1972.
343. Buchholz, J.T., Volumetric studies of seeds, endosperms, and embryos in *Pinus pondorosa* during embryonic differentiation, *Bot. Gaz.*, 108, 232, 1946.
344. Kobayashi, S., Ikeda, I., and Nakatami, M., Studies on the nucellar embryogenesis in Citrus. I. Formation of nucellar embryo and development of ovule, *Bull. Fruit Tree Res. Sta.*, E2, 9, 1978.
345. Kobayashi, S., Ikeda, I., and Nakatani, M., Studies on nucellar embryogenesis in Citrus. II. Formation of the primordium cell of the nucellar embryo in the ovule of the flower bud, and its meristematic activity, *J. Jap. Soc. Hort. Sci.*, 48, 179, 1979.
346. Wilms, H.J., Van Went, J.L., Cresti, M., and Ciampolini, F., Adventive embryogenesis in *Citrus, Caryologia*, 36, 65, 1983.
347. Naumova, T.N., and Willemse, M.T.M., Ultrastructural aspects of nucellar polyembryony in *Sarcococca humilis* Buxaceae nucellus and initial cells of nucellar proembryos, *Bot. Zh.*, 68, 1044, 1983.
348. Wakana, A., and Uemoto, S., Adventitious embryogenesis in *Citrus* (Rutaceae) II. postfertilization development, *Amer. J. Bot.*, 75, 1033, 1988.
349. Rangan, T.S., Murashige, T., and Bitters, W.P., *In vitro* studies of zygotic and nucellar embryogenesis in *Citrus, Proc. First Intl. Citrus Symp.*, 1, 225, 1969.
350. Hu, S.Y., Studies in the polyembryony of *Hosta caerulea* Tratt. II. Observations on the development of adventive embryo under hormone-treatments, *Acta Bot. Sinica*, 11, 16, 1963.
351. Haberlandt, G., Über Zellteilungshormone und ihre Beziehungen zur Wundeheilung, befrüchtung, Parthenogenesis und Adventivembryonie, *Boil. Zentralbl.*, 42, 145, 1922.
352. Armstrong, J.M., A cytological study of the genus *Poa, Can. J. Res.*, 15, 281, 1937.
353. Lamm, R., A contribution to the embryology of the potato, *Svensk. Bot. Tidskr.*, 31, 217, 1937.

354. Bergstrom, I., Tetraploid apple seedlings obtained from the progeny of triploid varieties, *Hereditas*, 24, 210, 1938.

355. Cook, R.C., A haploid marglobe tomato. Practical application of a "short cut" for making pure line, *J. Hered.*, 27, 433, 1936.

356. Kasparyan, A.S., Haploids and haplo-diploids among hybrid twin seedlings of wheat, *Compt. Rend. Acad. Sci., (Doklady) U.S.S.R.*, 20, 53, 1938.

357. Ramaiah, K., Parthasarthi, N., and Ramanujam, S., Polyembryony in rice (*Oryza sativa*), *Indian J. Agr. Sci.*, 5, 119, 1935.

358. Yamamoto, Y., Twin and triple seeded plants and chromosome changes. *Kagaku (Sci.) Tokyo*, 7, 147, 1939 (Plant Breed. Abstr., 9, 288, 1939).

359. Ueno, I., and Nishimura, M., Studies on polyembryony in *Citrus* and its relatives, I. Influence of environmental condition on the number of embryos per seed, *Bull. Hort. Res. Sta. Japan, Ser. B*, 9, 11, 1969.

360. Knox, R.B., Apomixis: seasonal and population differences in a grass, *Science*, 157, 325, 1967.

361. Evans, L.T., and Knox, R.B., Environmental control of reproduction in *Themeda australis*, *Aust. J. Bot.*, 17, 375, 1969.

362. Minessy, F.A., and Higazy, M.K., Polyembryony in different *Citrus* varieties in Egypt, *Alexandria J. Agr. Res.*, 5, 89, 1957.

363. Minessy, F.A., Effect of rootstock on polyembryony in *Citrus*, *Alexandria J. Agr. Res.*, 1, 83, 1953.

364. Parlevliet, J.E., and Cameron, J.W., Evidence on the inheritance of nucellar embryony in *Citrus*, *Proc. Amer. Soc. Hort. Sci.*, 74, 252, 1959.

365. Hanna, W.W., and Bashaw, E.C., Apomixis: its identification and use in plant breeding, *Crop Sci.*, 27, 1136, 1987.

366. Asker, S., Gametophytic apomixis elements and genetic regulation, *Hereditas*, 93, 277, 1980.

367. Furr, J.R., and Reece, P.C., Identification of hybrid and nucellar seedlings by a modification of the rootstocks color test, *Proc. Amer. Soc. Hort. Sci.*, 46, 141, 1946.

368. Syakudo, K., Yamagat, H., and Watanabe, H., Studies on the *Citrus* generic polyembryony in relation to breeding. 1. Labelling of the nuclear DNA of pollen with ^{3}H and ^{14}C, *Japan. J. Breed.*, 19, 79, 1969.

369. Watanabe, H., Yamagat, H., and Syakudo, K., Studies on the *Citrus* generic breeding in relation to breeding. II. Discrimination of embryo fertilized by ^{3}H-labeled pollen grains, *Japan. H. Breed.*, 20, 141, 1970.

370. Torres, A.M., Soost, R.K., and Diedenhofen, U., Leaf isozymes as genetic markers in *Citrus*, *Amer. J. Bot.*, 65, 869, 1978.

371. Button, J., Vardi, A., and Spiegel-Roy, P., Root peroxidase isozymes as an aid in *Citrus* breeding and taxonomy, *Theor. App. Genet.*, 47, 119, 1976.

372. Esen, A., and Soost, R.K., Polyphenol oxidase catalyzed browning of young shoot extracts of *Citrus*, *J. Amer. Hort. Sci.*, 99, 484, 1974.

373. Esen, A., Scora, R.W., and Soost, R.K., A simple and rapid screening procedure for identification of zygotic *Citrus* seedlings among crosses of certain taxa, *J. Amer. Soc. Hort. Sci.*, 100, 558, 1975.

374. Iglesias, L., Lima, H., and Simon, J.P., Isoenzyme identification of zygotic and nucellar seedlings in *Citrus*, *J. Hered.*, 65, 81, 1974.

375. Selle, R.M., Spot chromatography identification of *Citrus* rootstocks, *Nature*, 181, 506, 1958.

376. Pieringer, A.P., and Edwards, G.J., Identification of nucellar and zygotic *Citrus* seedlings by infra-red spectroscopy, *Proc. Amer. Soc. Hort. Sci.*, 86, 226, 1967.

377. Teich, A.H., and Spiegel-Roy, P. Differentiation between nucellar and zygotic citrus seedlings by leaf shape, *Theor. Appl. Genet.*, 42, 314, 1972.

378. Lima, H., Embriogénesis cigotica nucelar en *Citrus*: Estudios *in vitro*, M.Sc. Thesis, Faculty of Science, University of Havana, 1973, 32.

379. Esen, A., and Soost, R.K., Inheritance of browning of young shoot extracts of *Citrus*. *J. Hered.*, 65, 97, 1974.

72

380. Swingle, W.T., Recapitulation of seedling characters by nucellar buds developing in the embryo sac of *Citrus*, *Proc. 6th Intl. Congr. Genet.*, 2, 196, 1932.
381. Frost, H.B., Nucellar embryony and juvenile characters in clonal varieties of citrus, *J. Hered.*, 29, 423, 1938.
382. Hodgson, R.W., and Cameron, S.H., Effects of reproduction by nucellar embryony on clonal characteristics in citrus, *J. Hered.*, 29, 417, 1938.
383. Cameron, J.W., and Soost, R.K., Size, yield, and fruit characters of orchard trees of citrus propagated from young nucellar-seedling lines and parental old lines, *Proc. Amer. Soc. Hort. Sci.*, 60, 255, 1952.
384. Brittingham, W.H., Type of seed formation as indicated by the nature and extent of variation in Kentucky bluegrass, and its practical implications, *J. Agric. Res.*, 67, 225, 1943.
385. Cameron, J.W., Soost, R.K., and Frost, H.B., The horticultural significance of nucellar embryogeny in citrus, in *Citrus* virus diseases, Wallace, J.M., Ed., Univ. Calif. Div. Agr. Sci., Berkeley, 1959, 191.
386. Taliaferro, C.M., and Bashaw, E.C., Inheritance and control of obligate apomixis in breeding buffelgrass, *Pennisetum ciliare*, *Crop Sci.*, 6, 473, 1966.
387. Pepin, G.W., and Funk, C.R., Interspecific hybridization as a method of breeding Kentucky bluegrass (*Poa pratensis* L.) for turf, *Crop Sci.*, 11, 445, 1971.
388. Bashaw, E.C., Registration of nueces and Liano buffelgrass, *Crop Sci.*, 20, 112, 1980.
389. Bashaw, E.C., and Hignight, K.W., Gene transfer in apomictic Buffelgrass through fertilization of an unreduced egg, *Crop Sci.*, 30, 571, 1990.
390. Hanna, W.W., Powell, J.B., Millot, J.C., and Burton, G.W., Cytology of obligate sexual plants in *Panicum maximum* Jacq. and their use in cotrolled breeding, *Crop Sci.*, 13, 695, 1973.

3. Zygotic Embryogenesis in Gymnosperms and Angiosperms

V. RAGHAVAN and KIRAN K. SHARMA

Contents

I. Introduction

Essentially all gymnosperms and angiosperms (phanerogams), like other groups of plants, begin their lives from a single cell, the fertilized egg or the zygote. What makes the zygote so unique a cell is that it is the product of sexual recombination between two gametes, the sperm contributing the paternal genome and the egg providing the maternal genome. The point in time when the egg and sperm fuse together to be woven into a new plant marks the ontogeny of the species. The phase of ontogeny concerned with the progressive division of the zygote to fabricate the embryo is known as embryogenesis. Triggered by fertilization, embryogenesis ranks as one of the most marvelous of all natural

73

T.A. Thorpe (ed.), In Vitro Embryogenesis in Plants, pp. 73–115.
© 1995 *Kluwer Academic Publishers, Dordrecht. Printed in the Netherlands.*

phenomena involving growth, differentiation, cell specialization and tissue formation. Simple though these processes may sound, embryogenesis is no doubt complex as the cells of the embryo go through the same biochemical and physiological transformations as other living cells. Our present ontogenetic concepts of embryogenesis have been given a new twist by the discovery that somatic cells of many phanerogams and pollen grains of certain angiosperms can display with a high degree of fidelity an embryogenic type of development leading to the formation of plants in full multicellularity, sexuality and structure. Embryogenic episodes from somatic cells and pollen grains are known as somatic embryogenesis and pollen embryogenesis, respectively. To avoid semantic confusion, embryogenesis from the zygote is nowadays referred to as zygotic embryogenesis; for the same reason, the term embryo is reserved to designate the product of fusion of the egg and sperm, while embryo-like structures generated from other cells of the plant, commonly referred to as embryoids previously, are now generally called somatic embryos.

It is well-known that successful completion of embryogenesis is the basis for the production of seeds and fruits in plants. Agricultural productivity is based largely on the control of the processes of pollination, fertilization, embryogenesis, seed formation and fruit development in our crop plants. Considering this, it is not surprising that various aspects of embryo development in angiosperms ranging from morphological to molecular continue to be investigated with enthusiasm and ingenuity. To a limited extent this is also true of some gymnosperms. The goal of this chapter is to outline the process of embryogenesis in these two groups of plants from the various angles that it has been studied in the hope of providing a sense of the past accomplishments and the outstanding future problems.

II. Patterns of embryo development

A. Historical background

The history of the study of embryo development in phanerogams may be separated into four periods, each of which is marked by certain fundamental contributions to the field. Early studies going back to the Greek philosophers up to about 1870 were chiefly concerned with the concept of sexuality in plants and were devoted to an understanding of the basic structural features of the male and female reproductive parts of the flower. The second epoch began with the publication of Hanstein's (1870) work on the development of embryos of certain dicotyledons and monocotyledons. With rare hindsight, Hanstein chose for his investigation *Capsella bursa-pastoris* which has since then received wide acceptance as an illustrative material to follow cleavage patterns during early embryogenesis in a dicot. In *C. bursa-pastoris*, Hanstein correctly identified the position of walls during successive segmentations of the zygote and introduced such terms as quadrants and octants to designate the four-celled and eight celled stages, respectively, of the globular embryo; this investigator also christened the cell at the lower end of the globular embryo as the hypophysis, a term which is used even to this day. This work as well as others that immediately followed received much impetus from the discovery of fertilization in plants by Strasburger in 1877 and the more unique phenomenon of double fertilization in angiosperms independently by Nawaschin and Guignard in 1899.

The study of embryo development in gymnosperms has been slow to begin with and is generally credited to the publication of Strasburger's books on the subject in 1872 ("Die Coniferen und die Gnetaceen") and 1879 ("Die Angiospermen und die Gymnospermen").

The third period concerned with the study of embryo development in phanerogams comprises about 40 years from 1910 to 1950 when the field of descriptive embryology encompassing studies of micro- and mega-sporogenesis and embryo and endosperm development was born and nourished and attained full bloom. A driving force behind much of the progress in this area was the introduction of microtomy, which allowed for the making of serial sections. It might be stated here parenthetically that the information gathered during this period provided material for the formulation of the laws of embryogenesis and the system of classification of embryo development types in angiosperms by the famous French embryologist, Souèges. Finally, the accumulation of information relating to the pattern of embryo development in a great number of gymnosperms and angiosperms paved the way for spectacular new developments in the fourth epoch, beginning about 1950 to the present time. These developments are aptly described under the term experimental embryogenesis, as they seek to determine the physical, physiological, biochemical and genetic factors that control embryogenic processes. Investigations during this period were greatly aided by the development of tissue culture techniques initially using angiosperms and later extending into gymnosperms, resulting in the ability to culture progressively small embryos in isolation in defined media, demonstration of the concept of cellular totipotency and the discovery of somatic embryogenesis and pollen embryogenesis. In recent years, while there has been a slackening of effort on the purely descriptive aspect of embryo development, molecular investigations are playing a major role in advancing our knowledge of the way embryo development in gymnosperms and angiosperms is accomplished. For references to the history of plant embryogenesis, see Johansen [1] and Maheshwari [2].

B. Classification of embryo types

Within the zygote lies the potential to form an entire plant by a series of genetically determined events; the zygote accomplishes this feat by extensive changes in form in defined and dramatic ways and by a progressive change from the undifferentiated to the differentiated state. Common to the changes that occur during progressive transformation of the zygote into an embryo are such processes as cell division, cell expansion, cell maturation, cell differentiation, tissue formation and organization of meristems. In a very young embryo, all cells are programmed to produce a new generation of daughter cells by simple mitotic divisions. As the embryo continues to grow, cell divisions are segregated to certain parts predictable by their position in the cell lineage to produce specialized cells, tissues and organs. Very little is currently known about the mechanisms that contribute to the restriction of functional activities in the developing embryo.

With a few exceptions, a major point of difference between embryogenesis of gymnosperms and angiosperms is that the former generally show a free-nuclear phase, while in the angiosperms, the first and subsequent divisions of the zygote are invariably followed by wall formation [1]. One of the salient features of the later phases of embryogenesis in many gymnosperms is the generation of multiple embryos, a phenomenon known as polyembryony. Common among cycads is embryogenic development of zygotes from several archegonia (simple polyembryony), while in conifers, secondary embryos are formed by cleavage of a single embryo (cleavage polyembryony). Multiple embryos however do not become part of the mature seed, as only one embryo survives the competition [3]. In both gymnosperms and angiosperms, the term proembryo is generally used to designate the early stages of development of the embryo up to the formation of the cotyledons.

1. Embryo development in gymnosperms

In the current state of our knowledge, it is convenient to consider embryogenesis in gymnosperms under three rubrics: (i) proembryogenesis, comprising the stages before suspensor elongation; (ii) early embryogenesis, including the stages after suspensor elongation and formation of the root generative meristem; and (iii) late embryogenesis, involving establishment of the polar meristems of the root and shoot and the attainment of form by the embryo. In the study of gymnosperm embryogenesis, Cycad and Ginkgo type, Conifer type, Ephedra and Sequoia type and Gnetum and Welwitschia type of embryos have been recognized [4]. The differences between these embryo types are best illustrated by brief accounts of their ontogeny.

a. Cycad and Ginkgo type

Early events of embryogenesis in cycads and *Gingko* are centered around a protracted free nuclear phase before wall formation sets in. The number of free nuclei formed vary from as low as 64 in *Bowenia* [3] to as high as 1024 in *Dioon edule* [5], and everything in between, as for example, 256 in *Gingko* [1] and 512 in *Cycas circinalis* [6]. During cellularization, wall formation is initiated at the lower end of the archegonium and progresses upward; although the newly formed cells fill the entire archegonium as in *Ginkgo* and *Macrozamia*, in most cycads, segmentation is confined to the lower portion of the archegonium. As to the fate of the remaining nuclei, they ultimately disintegrate without contributing to the mass of the embryo. Following wall formation, the cells at the lower end divide and function as the embryonal cells. The upper cells elongate and form a massive suspensor as in *Ginkgo* or differentiate into a row of suspensor cells, just above the embryonal cells and an upper row of buffer cells as in cycads [1, 4, 7]. In some cycads, the superficial embryonal cells elongate to form a short-lived layer of cells constituting a cap [8].

b. Conifer type

The significant feature of the Conifer type of embryogenesis is the brief period of free nuclear divisions of the proembryo and consequently, the limited number of free nuclei formed. The maximum number of free nuclei recorded is 64 in the proembryo of *Agathis australis* [9]. Wall formation is followed by the organization of a lower group of primary embryonal cells and an upper group of cells, the primary upper tier, arranged in a single layer [10]. The cells of the primary upper tier have no walls on the upper side and are therefore continuous with the general cytoplasm of the proembryo. The establishment of these two groups of cells is a signal for them to divide, resulting in a doubling of the number of cells. At this stage the proembryo consists of four tiers of cells. The lower two tiers comprise the embryonal tier; next is the suspensor tier and the uppermost is the upper tier of cells open at the top. With continued growth of the proembryo, the cells of the suspensor tier elongate and those of the embryonal tier divide to form a conspicuous mass of cells. The distal cells of this mass are transformed into secondary suspensors to which are later added more cells. Together with this, the whole embryo-suspensor complex becomes an intertwined mass of tubular and isodiametric cells.

Embryogenesis of Araucariaceae seems to differ from the basic Conifer type plan outlined above. Here, the free nuclei occupy the center of the archegonium where the proembryo takes shape. Wall formation results in a central group of cells surrounded by a jacket of peripheral cells. The proximal cells of the jacket develop into the cap cells and the distal cells into the suspensor. The central cells function as the embryonal cells [1].

c. Ephedra and Sequoia type

Although *Ephedra* and *Sequoia* are grouped together, there is much that is different in the embryogenic sequences of the two genera. In *Ephedra*, the limited number of free nuclei (generally eight) that are generated become proembryonic cells. These cells are transformed into independent embryos with their own embryonal tubes [11]. On the other hand, in *Sequoia*, the first and the subsequent division of the zygote are accompanied by wall formation which generates four independent embryos. Embryogenesis in *Sequoia* is also different from the rest of the conifers in the absence of tier formation as the proembryo develops the first walls [1]. Further development of the embryo in the two genera seems to follow more or less similar pathways.

d. Gnetum and Welwitschia type

An unusual feature of this type of embryogenesis is the absence of a free nuclear stage. There is also some variability in the behavior of the proembryos in the two genera. In *G. gnemon*, the zygote puts out branched tubes with a nucleus in each branch. The tubes, termed suspensor tubes, are very tenuous and grow in all directions in the female gametophyte. Finally, a terminal cell is cut off at the tip of the tube and by repeated divisions, it forms a globular mass of embryonal cells. As in certain other gymnosperm embryo types, the proximal cells of this mass elongate into embryonal tubes [12, 13]. In *Welwitschia*, the zygote divides to form a long suspensor cell and a small embryonal cell. The latter is segmented transversely to produce a succession of embryonal suspensors that grow towards the female gametophyte. During further growth, the terminal cell of the embryonal suspensor divides to form a group of embryonal cells. Later, some of these embryonal cells also produce embryonal tubes analogous to those found in other gymnosperms [14].

The formation of the mature embryo from the embryonal mass of cells follows a uniform pattern in the different embryo types described above. Generally, the cells in the proximal part of the embryonal mass elongate to form massive secondary suspensors, while the distal cells continue to add to the embryonal mass (Fig. 1A). The distal cells keep on adding to the length of the secondary suspensors so that the embryonal mass is pushed deep into the tissues of the female gametophyte. During the final stages of differentiation of the embryo, a variable number of cotyledons, ranging from two in cycads, *Ginkgo*, *Ephedra*, *Gnetum* and *Welwitschia* to many in conifers, arise laterally at the plumular pole. A root meristem is organized by the activity of a group of meristematic cells at the opposite end [1]. A distinctive feature of the mature embryos of *Gnetum* and *Welwitschia* is the presence of a lateral protuberance on the hypocotyl, known as the feeder [12, 14].

In summary, the most impressive embryogenic episodes in gymnosperms are the formation of free nuclei and the generation of multiple embryos from a single zygote. These are separated in time and space during embryogenesis. In consequence, proembryo formation must be interpreted and explained in terms of interactions of the free nuclei within a common cytoplasm, while the evolution of the final form of the embryo is probably due to intracellular events within the embryonal cells.

2. Embryo development in angiosperms

In angiosperms, the zygote is poised to divide within a few hours after fertilization and at this time it appears as a highly polarized cell. Polarity in the zygote is attributed to two factors. First, the orientation of this cell is polar in the sense that its one end is cemented to the embryo sac wall next to the micropyle (micropylar end), while the other end projects into the central cell facing the chalazal part of the embryo sac (chalazal end). Polarity is also bestowed on the zygote by its intrinsic structure, obvious in the light microscope and characterized by the dense concentration of cytoplasm at the chalazal end. The division of the zygote is the next milestone in the establishment of polarity. This division is almost invariably asymmetric and transverse, cutting off a large vacuolate basal cell towards the micropyle and a small, densely cytoplasmic

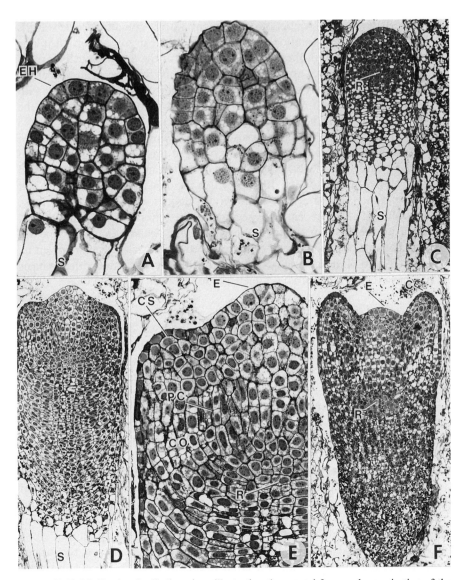

Fig. 1. (A-F) Median longitudinal sections illustrating the general form and organization of the zygotic embryo of *Pinus contorta* during later stages of embryogeny. The micropyle is towards the bottom of the page. [C = cotyledon; CO = cortex; CS = cotyledonary shoulder; E = epicotyl; EH = embryonal head; PC = procambium; R = root initials; S = suspensor]. (A) Early stage of embryo-suspensor complex. (B) Establishment of the proximal-distal axis of the embryo. (C) A cylindrical embryo. (D) Formation of the epicotyl and the origin of the pith in the hypocotyl-shoot axis. (E) A magnified view of the embryo shown in D. (F) A nearly mature embryo. The developing megagametophytes at different stages of growth were processed for glycol methacrylate sectioning and 3 μm thick sections were stained with Periodic acid Schiff's reagent (PAS) and counterstained with Amido black 10B. Magnifications: A: × 251; B: × 266; C: × 108; D: × 178; E: × 252; F: × 104. (Sharma and Thorpe [46].)

terminal cell. As we will see below, subsequent partitioning of these cells leads gradually towards the formation of a mature embryo. For this reason, the plane of division of one or both these cells and the fate of the daughter cells formed have an important bearing on the classification of embryo types in angiosperms.

The organogenetic part of the embryo is derived from the terminal cell with little or no contribution from the basal cell; on its way to form the embryo, the first division of the terminal cell may be transverse or longitudinal. These considerations have served as a basis for classifying the different types of embryos of angiosperms. A widely used system of classification has identified two major groups, one in which the division of the terminal cell of the two-celled embryo is longitudinal and the other in which the division wall is oriented in a transverse plane [1]. Rarely, the terminal cell divides obliquely. Within these major groups, different embryo segmentation types are recognized and identified by the name of the family in which many examples of the type are found. Accounts of the ontogeny of representatives of each group are given by Johansen [1] and Maheshwari [2], and so in the following account with emphasis on the basic division patterns, ontogenetic information is passed over.

a. Longitudinal division

Depending upon the extent of contribution of the basal cell of the two-celled proembryo to the formation of the organogenetic part of the mature embryo, Crucifer (or Onagrad) and Asterad types of embryo development have been recognized in this group. In the Crucifer type, exemplified by *Capsella bursa-pastoris* (Brassicaceae), the basal cell contributes very few derivatives to the formation of the organogenetic part of the embryo, which is thus formed almost entirely from the terminal cell. The cells derived from the basal cell form a filamentous suspensor. This type of embryogeny has also been described in plants belonging to Onagraceae, Bignoniaceae, Lamiaceae, Lythraceae, Papilionaceae, Ranunculaceae, Rutaceae and Scrophulariaceae among dicotyledons and in Juncaeae and Liliaceae among monocotyledons. In contrast, in the Asterad type the embryo is generated by the division of both terminal and basal cells of the two-celled proembryo [15]. Based on the description of embryogenesis in lettuce (*Lactuca sativa*) [16], there is a nice division of labor between the terminal and basal cells in the formation of the embryo, the former giving rise to the cotyledons and the latter to the root, root cap and hypocotyl; a small suspensor is also derived from the basal cell. Asterad type of embryo development or its variations have been described in members of Asteraceae, Geraniaceae, Lamiaceae, Oxalidaceae, Polygonaceae, Rosaceae, Urticaceae (dicots), Liliaceae and Poaceae (monocots) [15].

b. Transverse division

Here, three embryo development patterns (Solanad, Caryophyllad and Chenopodiad) have been recognized on the basis of the participation of the terminal and basal cells of the two-celled proembryo in the formation of the mature embryo. Solanad type found in several species of *Nicotiana* is characterized by the formation of the embryo from the terminal cell, while the basal cell divides to form a suspensor. Besides Solanaceae, members of Hydnoraceae, Linaceae, Papaveraceae and Rubiaceae display the Solanad type of embryo formation [15]. In the Caryophyllad type, the basal cell does not divide after it is cut off and the organogenetic part of the embryo is derived exclusively by the division of the terminal cell. These characteristics are clearly evident during early embryogenesis of *Sagina procumbens* described by Souèges [17]. A suspensor is not a regular feature of the Caryophyllad type of embryo and if one is present, it is also derived from the terminal cell with contributions from the basal cell as seen in *Saxifraga granulata* [18]. Caryophyllad type of embryo development has been described in plants included in Crassulaceae, Droseraceae, Fumariaceae, Holoragaceae, Portulacaceae, Pyrolaceae (dicots), Alismataceae, Araceae, Ruppiaceae and Zannuchelliaceae (monocots) [15]. A recent addition to the list is *Monotropa uniflora* (Monotropaceae) in which the basal cell collapses after it is cut off [19]. The Chenopodiad type has the distinction of having the division products of both the terminal and basal cells integrated into the organogenetic part of the embryo. Generally, the hypocotyl of the mature embryo has its origin in the basal cell, which also forms a short suspensor. Besides members of Chenopodiaceae, this type of embryo development is also found in plants included in Amaranthaceae and Polymoniaceae [15].

c. Oblique division of the zygote
This type of embryo development (Piperad type) was slow to be recognized and was also thought to be found only in a few families of the dicotyledons. Even to this day, the species assigned to this type are not many, although it has been identified in an occasional monocot such as the orchid *Pogonia* [20]. Besides Piperaceae, other families in which representatives of this type are found are Balanophoraceae, Dipsacaceae and Loranthaceae [15]. Embryo development types described above are summarized in Fig. 2.

It is clear from the above that the descendants of the basal cell of the two-celled proembryo defy a neat packaging because in some cases part or whole of this cell population becomes enmeshed in the organogenetic part of the embryo, the rest forming a suspensor while in others they wholly form the suspensor or simply wilt away. In other words, while the terminal cell is already determined as the progenitor of the embryo, the determinative state of the basal cell is at best described as being labile.

Mention should also be made here of the elaborate system of classification of embryo types developed by Souèges. The main features of this classification are summarized by Crété [21]; briefly, angiosperm embryos are considered to adhere to certain laws concerned with the origin of cells at a given stage, the number of cells produced in each generation, the role of cells formed and the determinative fate of the early formed cells. These laws no doubt form the basis for the orderly development of the diverse types of embryos, but when considered along with the descriptive formulae that go with the laws, the whole

82

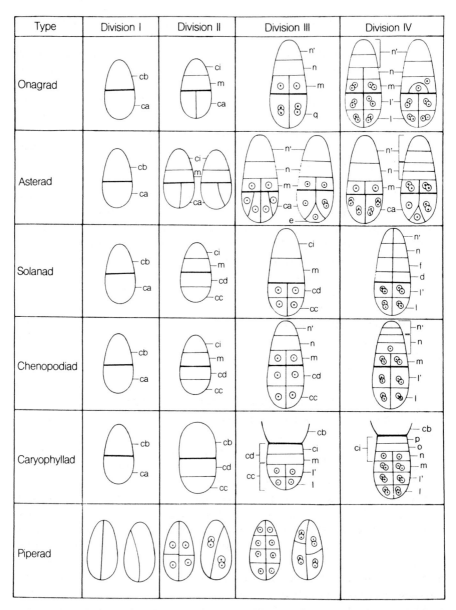

Fig. 2. A diagrammatic summary of the division patterns of embryo types in angiosperms. The micropyle is towards the top of the page. ca, cb = products of a transverse division of the zygote; cc, cd = products of division of ca; ci, m = products of division of cb; e = a wedge-shaped cell cut off at the embryonal apex known as epiphysis; l, l' = two tiers of cells at the octant stage; n, n' = products of division of ci; o, p = products of division of n'; q = quadrant (Natesh and Rau [15]).

system of classification has become very complex indeed and so is seldom used by embryologists.

Embryo development described in several species of *Paeonia* does not fall into the classification of embryo types described above. As was first shown by Yakovlev and Yoffe [22], this is due to the rampant failure of wall formation during the first few rounds of division of the zygote and so embryogenesis in this plant starts with a bunch of free nuclei suspended in a cellular bag. Later the nuclei migrate towards the periphery of the cell where wall formation is initiated to form a multicellular proembryo. Soon after, groups of marginal cells differentiate as embryo primordia. Although several primordia are born out of a proembryonal cluster, only one outlives the others and differentiates into a viable embryo. Independent investigations by other workers [23, 24] on different species of *Paeonia* have essentially confirmed the unusual features of embryogenesis in this genus. As noted earlier, a free nuclear phase is typical of embryogenesis in gymnosperms and its occurrence in *Paeonia* appears to be the angiosperm equivalent of a gymnosperm feature.

C. Comparative embryogenesis of dicots and monocots

The essence of embryogenesis is the establishment of tissue systems and their subsequent organization into recognizable primordial organs. Of necessity this requires an almost precise sequence of divisions of the zygote and early stage embryos as well as a high degree of inductive interactions between the growing parts of the embryo. The early division sequences of embryos of dicots and monocots are basically similar. It should not be surprising, therefore, that the modes of embryo development generally transcend the boundaries of dicots and monocots.

To understand how the zygote progresses from a single cell to a complex multicellular embryo in which fundamental tissues and organs are demarcated, we will consider embryogenesis of representatives from dicots and monocots. Any one of the several dicot and monocot embryos that have been studied over the years will illustrate the general principles of embryogenesis easily, but some plants have been favorites for this purpose. Embryogenesis in *Capsella bursa-pastoris* is almost a text-book example of the Crucifer type of embryogenesis, in the sense that the first division of the terminal cell of the two-celled proembryo is longitudinal. Each of the two cells thus formed again divides by longitudinal walls to yield a quadrant, followed by an octant stage by division of the four cells by transverse walls. Although the embryo is populated at this stage by a relatively small number of homogeneous cells, there are functional differences between the upper and lower tiers of cells. For example, the stem tip and cotyledons in the adult embryo are formed from the upper tier of cells and the hypocotyl is generated from the lower tier. The next round of division of cells of the octant embryo is periclinal and results in the formation of a globular embryo of 16 cells. This is phased out into the heart-shaped stage, due to a lateral expansion of the distal pole giving rise to a pair of cotyledons. The

84

elements of the future shoot apex are carved out by a few cells lying between the cotyledons and which remain essentially in a dormant state throughout the life of the embryo. As the cotyledons are initiated, division and differentiation of cells in the basal tier of the embryo give rise to the hypocotyl. There is considerable elongation of the cotyledons and hypocotyl at this stage, giving rise to a torpedo-shaped embryo. A root apex delimited in the torpedo-shaped embryo is the progenitor of the embryonic radicle. During progressive embryogenesis, continued growth of the embryonic organs, compounded by spatial restrictions within the ovule causes the embryo to curve at the tip to assume the walking-stick shaped stage and finally to take on the shape of a horseshoe (Fig. 3) [25, 26]. Although the mature embryonic organs differentiate as early as the heart-shaped or torpedo-shaped stage, subtle physiological and biochemical changes occur in the embryos and in the surrounding tissues before they become mature.

It will be recalled that a hallmark of the Crucifer type of embryo development is the ambiguity about the role of the descendants of the basal cell of the two-celled proembryo. That some of these cells are ultimately incorporated into the embryo of *C. bursa-pastoris* has been established by careful cell lineage studies [25, 26]. Initially, the basal cell of the two-celled proembryo is partitioned transversely. These cells give rise to the suspensor. Further divisions in the newly formed pair of cells are however restricted to the terminal cell which

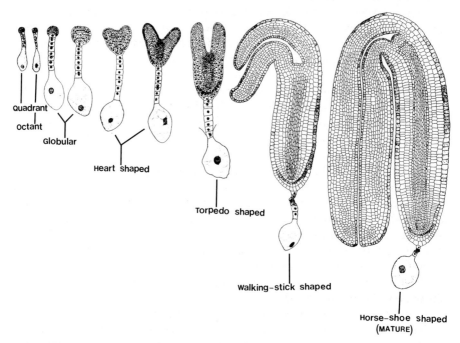

Fig. 3. Stages in the embryogenesis of *Capsella bursa-pastoris* in longitudinal sections. The lower end of the embryo is directed towards the micropyle (Schaffner [25]).

produces by transverse walls eight to ten cells held together as a filament. The suspensor is terminated at the chalazal end by the organogenetic part of the embryo and at the micropylar end by the basal cell which has now become disproportionately large. All the cells of the suspensor, except one, are eventually crushed and obliterated by the growing embryo; this fortunate cell known as the hypophysis is the one in immediate contact with the embryo and plays an important role in the formation of the root apex. This will be described later. It is as if a single cell from a group of about ten cells born out of simple mitotic divisions of a mother cell is funneled into a specialized channel while others perish along the way. Practically nothing is known about the mechanisms that bring about these changes in the developmental options of cells of the suspensor. Although the genetically permissible number of suspensor cells in the embryo of wild type *Arabidopsis thaliana* is six to eight, in recessive embryo lethal mutants, the suspensor grows to form a chain of two or more files of 15–150 cells [27]. This observation is suggestive of another type of relationship between the embryo and suspensor, namely, the presence of an actively growing embryo wields an inhibitory influence on the continued growth of the suspensor.

In their morphological appearance, mature embryos of monocots are strikingly different from those of dicots because of the presence of only one cotyledon. Studies on the comparative ontogeny of monocot and dicot embryos was bedeviled by controversy for some time, but there is now general agreement that the development of the embryo up to the octant stage is almost identical in monocots and dicots. The main point of contention is with regard to the origin of the shoot apex and the cotyledon in the monocots. For a period of time the dominant view was that in monocots the shoot apex is lateral and the cotyledon is terminal in origin. Extensive ontogenetic studies by Swamy and others [28–32] however seem to support the view that both the shoot apex and cotyledon share a common origin from the terminal cell of a three-celled proembryo. According to these investigators, what is perceived as a lateral shoot apex is a shoot apex displaced from its terminal position by the aggressive growth of the single cotyledon.

Sagittaria sagittaefolia exhibits a pattern of embryogenesis which is typical of many monocots. Here the differentiative potencies of the cells are established after the first transverse division of the zygote. The terminal cell formed from this division gives rise to the cotyledon and shoot apex, whereas the hypocotyl-root axis and the entire suspensor are derived from the basal cell. The cells destined to form the shoot apex are initially sluggish, so that the rapidly growing cotyledon dwarfs the shoot apex. As the hypocotyl is organized in the lower part of the embryo, there is a concomitant differentiation of the procambium and formation of root initials [33].

In their mature structure and ontogeny, embryos of Poaceae do not have much in common with other monocots. The main features of the grass embryo are the development of an absorptive organ known as the scutellum, presence of cap-like tissues covering the radicle (coleorhiza) and the plumule (coleoptile) and a flap-like outgrowth called the epiblast on one side of the coleorhiza.

We will consider embryo development in wheat (*Triticum aestivum*) as an example, based on the detailed work of Batygina [34]. Here, following an transverse-oblique division of the zygote, both cells divide by oblique walls to form a four-celled proembryo. Repeated divisions of these cells yield the mature embryo. Following a globular stage of 16 to 32 cells, the embryo becomes club-shaped. It is at this stage that the scutellum appears as a vaguely outlined elevation in the apical-lateral region of the embryo. At the same time, the opposite side of the embryo begins to enlarge to mark the differentiation of the shoot apex and leaf primordia. The final major event of embryogenesis is the differentiation of the radicle, which has an endogenous origin in the central zone of the embryo. In some other genera of Poaceae such as maize (*Zea mays*), the pattern of cell divisions in the proembryo becomes unpredictable even as early as the second round [35].

D. Reduced embryo types

It is clear from the above account that embryogenesis in angiosperms is a closely orchestrated process involving great many cells whose activities are interlaced. As a result, the mature embryo attains a basic organization consisting of a bipolar axis with one or two cotyledons attached at a node. The embryo is terminated at each pole by an apical meristem and the point of attachment of the cotyledons separates the embryonic axis into a hypocotyl-root region and an epicotyl-plumule region. The former has at its lower end the primordial root (radicle), while the primordial shoot (plumule) is attached to the stem part called the epicotyl. From the earliest investigations of embryogenesis, it has become apparent that embryo development does not proceed to completion in certain angiosperms. These plants thus shed seeds which harbor underdeveloped or reduced embryos. According to a list compiled by Natesh and Rau [15], about 15 families among dicots and 5 among monocots produce seeds which at the shedding stage contain embryos generally devoid of well-defined organs, although the level of organization varies in the embryos of different plants. Balanophoraceae, Ranunculaceae, Orobanchaceae and Pyrolaceae among dicots and Orchidaceae, Burmanniaceae and Pandanaceae among monocots are noteworthy in this respect. A paradigm of a seed with a reduced embryo is *Monotropa uniflora* (Fig. 4B). Here the embryo is generally embedded in the endosperm and consists of no more than two cells separated by a transverse wall [19, 36]. Equally striking for its reduced embryo is the seed of *Burmannia pusilla* in which embryogenesis terminates at the quadrant stage (Fig. 4A) [37]. A higher level of embryo differentiation is found in the seeds of various orchids [38] and of root parasites like *Orobanche aegyptiaca* and *Cistanche tubulosa* [39]; at the time of shedding, seeds harbor globular embryos which do not show any differentiation into radicle, hypocotyl, cotyledons, epicotyl and plumule. Mature seeds of *Alectra vogelii* and *Striga gesnerioides* contain a heart-shaped embryo with two rudimentary cotyledons and a radicular pole (Fig. 4C,D) [40]. Despite the lack of morphological differentiation, it has become apparent that

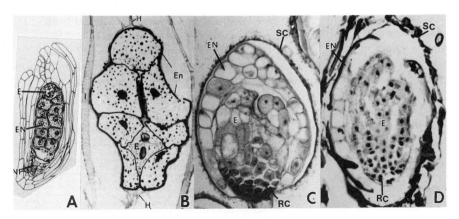

Fig. 4. (A-D) Seeds with reduced embryos. E = embryo; EN = endosperm; H = haustorium; N = nucellar pad; RC = root cap; SC = seed coat. (A) *Burmannia pusilla* (Arekal and Ramaswamy [37]). (B) *Monotropa uniflora* (Olson [36]). (C) *Alectra vogelii* (Okonkow and Raghavan [40]). (D) *Striga gesnerioides* (Okonkwo and Raghavan [40]).

undifferentiated embryos of certain plants display subtle differences between a shoot and a root pole. For example, in both *O. aegyptiaca* and *C. tubulosa*, cells at one end of the embryo identified as the radicular pole are small, while those at the opposite plumular pole are large and vacuolate [39]. Although embryos of *Eriocaulon robusto-brownianum* [41] and *E. xeranthemum* [42] are typically considered as reduced embryos, they show some differential cell division activity resulting in a relatively quiescent region and an actively dividing sector. The former represents the locus of differentiation of the epicotyl while the cotyledons arise from the latter sector. The embryo however does not differentiate a radicle or a hypocotyl. These examples serve to illustrate the occurrence of developmental blocks at programmed stages of embryogenesis. What causes the arrest of embryo development in these plants requires further investigation. By identifying where and when embryogenesis is blocked, it is possible to determine the causal factors involved in the process.

E. Nutrition of the embryo

We conclude this section with some comments about the nutrition of developing embryos. To a large extent, continued development of the embryo depends upon its interactions with the surrounding cells and with the cells of the ovule. Thus, in contrast to the static images conjured by diagrams of embryos in isolation, the latter maintain a channel of communication with their external milieu.

1. Gymnosperms

In gymnosperms, the female gametophyte serves the dual function of bearing gametes and nourishing the embryo. However, in its early stages of development, the embryo is nourished by the egg cytoplasm through the suspensor system and it is only later that it begins to draw upon the cells of the female gametophyte. The latter function is facilitated by the differentiation, in the gametophytic tissue below the archegonium, of a corrosion region consisting of densely packed linear cells. The embryo grows into this zone digesting, and apparently absorbing, the cellular metabolites. As the embryo begins to differentiate, the cells of the female gametophyte accumulate enormous reserves of starch, proteins and fats [43–46]. Admittedly, only a fraction of this is used by the developing embryo and the large part of the female gametophyte is held over in the seed for use of the germinating embryo.

2. Angiosperms

The function of nurturing the angiosperm embryo has traditionally been assigned to the endosperm, a tissue formed in the embryo sac following double fertilization. It is generally noted that the primary endosperm nucleus begins to divide ahead of the zygote, thus ensuring that a mass of cells charged with stored reserves is in place when the zygote begins to divide. In looking for structural adaptations in the early-division phase embryos that mediate in the transfer of nutrients from the endosperm, much attention has centered on the cells of the suspensor. Electron microscopic investigations have shown that the outer wall of the suspensor cells have an array of plasma membrane-lined invaginations that abut into the endosperm. In other cases, the surface area of the inner wall of the suspensor cells is considerably extended by the formation of rows of wall labyrinths. The suspensor cells thus display morphological features of transfer cells serving to facilitate the absorption of metabolites from the endosperm [47–56]. That the suspensor acts as a supplier of metabolites to the developing heart-shaped embryos has become evident from the work of Yeung [57]. This work demonstrated that if ^{14}C-sucrose is administered to the pods or isolated embryos of *Phaseolus coccineus* and French bean (*P. vulgaris*), much of the radioactivity appears in the suspensor and in the suspensor end of the embryo. When once the cotyledons have been mapped out, they function as the major absorptive organs of the embryo. Rapid uptake and translocation of labeled putrescine, a polyamine, by the suspensor cells of *P. coccineus* has also been demonstrated recently [58]. Since the amount of stored materials present in the newly formed endosperm cells is scanty, some questions have been raised as to whether this tissue can adequately nurture the embryo during its heterotrophic phase of growth [59]. A strong case can however be made for the possible utilization of endosperm reserves for later development of the embryo. It is well-known that during late embryogenesis, the cells of the endosperm begin to accumulate an acervate complex of reserve substances mainly composed of proteins, starch and lipids and that the embryo maintains an extended period of intimacy with this tissue. Continued growth of the embryo at this stage essentially involves the digestion of the endosperm cells as indicated by the appearance of a clear zone around the embryo [60] and by the presence of remnants of digested endosperm cells in this area [61]. According the Singh and

Mogensen [62], in *Quercus gambelii*, the degenerating endosperm cells probably transport metabolites from other parts of the ovule or from the vascular supply to the chalaza.

Morphological and ultrastructural observations of developing embryos of certain plants have provided additional evidence for the possible utilization of materials from parts of the ovule by the developing embryo. Thus, plants belonging to Rubiaceae, Orchidaceae and Trapaceae among others which do not have an endosperm, develop elaborate haustorial structures from the suspensor which come in contact with the cells of the integument, nucellus and placenta [63]. At the ultrastructural level, invaginations of the inner wall of the embryo sac interfacing with the endosperm and other sites of nutrient flux have been recorded in pea (*Pisum sativum*) [49], *Stellaria media* [50], *Vigna sinensis* [53], *Medicago sativa* [54], sunflower (*Helianthus annuus*) [64], cotton (*Gossypium hirsutum*) [65], *Haemanthus katherinae* [66], soybean (*Glycine max*) [67–69] and rice (*Oryza sativa*) [70], while in barley (*Hordeum vulgare*) [71], invaginations have been seen on the outer wall of the endosperm cells. Accentuating the changing nutritional needs of the proembryo, in soybean, soon after fertilization wall invaginations appear at additional sites in the embryo sac wall [68, 69]. The wall labyrinths are thought to facilitate the passage of cytoplasmic nutrients from the cells of the ovule into the embryo sac.

In summary, there is considerable indirect evidence in support of the view that embryos of phanerogams come equipped with adequate resources for their continued growth as part of the developing seed. These include provision of cells containing storage reserves in the immediate vicinity of the growing embryos and structural adaptations for drawing upon the metabolites of the surrounding cells.

III. Tissue and meristem differentiation during embryogenesis

Progressive embryogenesis is manifest in the changing shape of the embryo. To give legitimacy to the embryo as the future sporophyte, these changes are accompanied by the mapping out of primary meristems that evolve into the protoderm, ground tissue and procambium as well as of the shoot and root apices. An overview of the anatomical changes associated with meristematic organization in gymnosperm and angiosperm embryos is presented in this section.

A. Gymnosperm embryogenesis

A survey of the literature shows that some differences exist between the different embryo types in regard to the plan of meristem organization, although it is remarkably uniform in the Conifer type of embryos [1]. In most gymnosperms, a proximal region (close to the micropyle) and a distal region (close to the base of the ovule) become established in the young embryonal mass (Fig. 1B). During late embryogenesis, coincident with the embryo becoming massive, growth activity is centered around a small group of initial cells at the surface of the free (distal) apex of the embryo. From these initials, all other cells of the embryo appear to diverge in periclinal rows. The anticlinal walls are envisaged as concentric arcs radiating from the initials, indicating successive growth increments at the apex. Thus, the free apex of the embryo is organized and functions like a shoot apical meristem. There is a clear demarcation between the cells of the proximal apex and those of the suspensor. The apical region consists of small cells with densely staining cytoplasm and large nuclei (Fig. 1C). In contrast, cells of the suspensor are large and vacuolate with lightly staining cytoplasm [72, 73].

1. Organization of primary meristems

Preparatory to the establishment of the primary meristems, a new center of cell division is formed a short distance below the summit of the apex. At this stage the embryo appears as an elongate cylindrical body with a hemispherical apex at the distal end and with a proximal part continuous with the suspensor (Fig. 1C). The configuration of cells at the distal region of the embryo is quite different from that of the cells in the proximal region. In the latter, cell divisions occur predominantly in a transverse plane (anticlinal to the long axis of the embryo), whereas in the distal region, divisions seem less restricted to a particular plane; yet a sharp anatomical delimitation does not occur between these regions. Cells of the proximal region which are not added to the secondary suspensor are the forerunners of the root cap even though the root initials are not established at this stage. The distal portion becomes the hypocotyl, shoot apex and cotyledons.

a. Origin of the pith

The first indication of anatomical organization in the embryo is seen when the cells in the central region of the hypocotyl-shoot axis tend to divide rapidly in the transverse plane. This represents an early expression of a developing pith; it is evident in embryos in which the shoot apex has attained a low rounded form and is well-established in embryos in which an epicotyl is present. Thus, as shown in embryos of *Larix* [44] and *Pinus strobus* [72] the pith is the first tissue of the hypocotyl-shoot axis to become clearly defined (Fig. 1D).

b. Origin of the procambium

The pith is laterally surrounded by a region of densely staining cells which divide both transversely and longitudinally. These cells are the progenitors of the procambium and appear in embryos before cotyledonary primordia are formed. As is well-known, the procambium represents the vascular meristem from which the primary xylem and phloem, and when present, the vascular cambium, are derived. In the embryo of *P. strobus*, the procambium is not only more densely staining than the cells of the pith and cortex, but its outer limits can be closely approximated as young secretory elements can be easily recognized [72].

c. Origin of the epidermis

Epidermis is the outer layer of the embryo, whether actively dividing or not, except in the regions of localized meristems such as the stem apex and the apical and marginal meristems of the leaf. In *P. strobus*, the cells of the outer layer of the distal portion of the early stage embryos are restricted to divisions which are either anticlinal or periclinal. In embryos of advanced stages, the cells of this outer limiting layer divide for the most part by anticlinal divisions which are transverse to the long axis of the embryo. Occasional periclinal divisions occur along the flanks of the hypocotyl and cotyledons to form a continuous layer of the epidermis [72].

2. Organization of shoot and root apices

The organization of shoot and root apical meristems in the embryo involves the formation of new growth centers in a homogeneous mass of cells. This primary event is followed by the elongation of the embryo axis to segregate the growth centers to opposite ends.

a. Origin of the shoot apex

For purposes of this discussion, in the embryo of gymnosperms the apical meristem of the shoot is considered to give rise to the cotyledons and to the epicotyl. During late stages of embryogenesis, the distal portion of the embryo increases considerably in diameter at the same time as the apex becomes rounded with rather well-defined shoulders, representing the beginning of the epicotyl (Fig. 1D,E). With further growth, the characteristic conical epicotyl is formed (Fig. 1F). At this stage the cotyledons appear as slight prominences on the upper flanks or shoulders of the embryo. The region between the bases of the cotyledons represents a portion of the original shoot apex of the embryo which eventually attains a conical form [46, 72].

In the mature embryo, the nuclei in the cells of the outer portion of the epicotyl are notably lightly staining and relatively large; in contrast, cells of the inner region are rather deeply staining with somewhat smaller nuclei (Fig. 5A,B). These latter cells thus correspond closely in organization and staining qualities to those of the pith. Cell divisions in the developing epicotyl are rather infrequent as its growth is characterized by marked increases in the size of its cells [46, 72, 73].

The embryonic region concerned with cotyledonary development is characterized by small, densely staining cells and is established before distinct primordia arise. In embryos in which the epicotyl is just beginning to differentiate, the cotyledons develop below the incipient epicotyl. The appearance of discrete cotyledonary primordia is preceded by a period in which the hypocotyl-shoot axis shows a greater increase in diameter than in length. During this period, the embryo apex becomes more rounded as the shoulders are established and is closely followed by the development of the epicotyl. With minor variations, this ontogenetic pattern has been described in embryos of *Larix* [44], *Pinus strobus* [72], *Taxus cuspidata* [73] and *Pseudotsuga* [74].

92

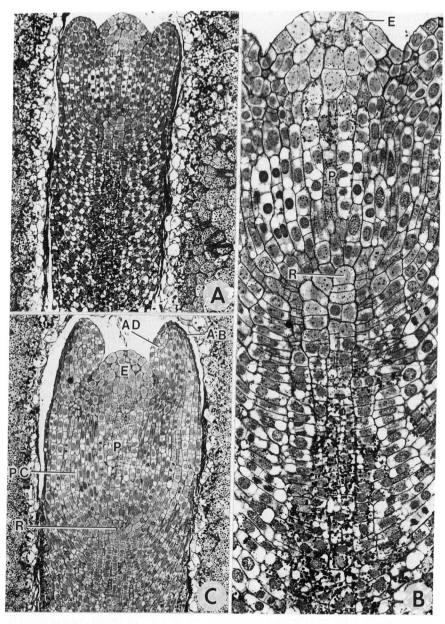

Fig. 5. (A-C) Median longitudinal sections of the zygotic embryos of *Pinus contorta* during post-cotyledonary stages of development [AB = abaxial side; AD = adaxial side; E = epicotyl; P = pith; PC = procambium; R = root initials]. (A) A fully mature embryo (× 102). (B) Magnified view of the embryo in A showing the formation of a distinct root initial and pith. Note that the cells of the outer portion of the epicotyl are lightly staining and relatively large (× 245). (C) The embryo at this stage has a fully differentiated tissue systems including the extension of procambial strands into the cotyledons. Further maturation of the embryo mainly involves the elongation of the hypo-cotyledonary region (× 104) (Sharma and Thorpe [46]).

Once the various meristems have been demarcated and the cotyledonary primordia appeared, further development in the embryo involves an increase in size and tissue differentiation. Although the cotyledons elongate somewhat, their growth in length does not match the growth in length of the hypocotyl. The embryo apex which is a remnant of the broad apex, is now definitely the shoot apex of the embryo (Fig. 5C). Very little mitotic activity is displayed in the shoot apex of the mature embryo. However, as the cotyledons elongate, the shoot apex slowly enlarges to form a conical mound of tissue between them.

Development of the hypocotyl involves a great deal of cell division and cell elongation to produce the rather long body of the embryo. The procambial tissue in the center is composed of rows of narrow, elongate cells which are traceable down to the root generative meristem. Following the pattern observed before cotyledon initiation, some transverse and many longitudinal divisions occur in the procambial core. Thus, a cell which is cut off by the meristem towards the procambial area by a transverse wall elongates along the long axis of the embryo and undergoes longitudinal divisions.

b. Origin of the root apex

With further development of the embryo, cell divisions in the potential root cap region continue largely in a transverse plane. The cells of the proximal region are arranged for the most part in vertical rows. The upper limits of this organization become well-defined since immediately above this limit cell divisions are in vertical or steeply oblique planes. This discontinuity is the first anatomical indication of the establishment of root initials and the root apex. The region below the level of the root initials constitutes the root cap and the suspensor system (Fig. 1D) [72].

As the root cap develops, the cells in its outer region assume an oblique orientation. This condition persists until in advanced stage embryos the long axis of the more peripheral cells are steeply inclined or almost vertical. The oblique orientation of the cells gradually decreases as the center of the root cap is approached. Those cells in the central region form a column, a characteristic feature of gymnosperm roots. The portion of the root cap surrounding the column is referred to as the peripheral region [74].

The cells produced by the peripheral initials of the root meristem divide tangentially; this results in an increase in the number of cells at each successive level above the meristem. These rows of cells are transitional between the procambium and cortex in their histology and mode of origin. The root generative meristem furnishes cells not only for the body of the root, but eventually also for most of the root cap-suspensor tissue. After the initiation of the cotyledon primordia, there is a noticeable increase in the activity of the root apical initials in the direction of the suspensor (Fig. 5A, B). This results in the production of longitudinal rows of cells directly below the meristem constituting the root cap [73].

B. Angiosperm embryogenesis

As we saw earlier, in angiosperms, the divisions of the zygote up to the stage of the globular embryo are in predictable planes. This makes it possible to trace the primary meristems to their progenitor cells to a more precise level than in gymnosperms. In *Capsella bursa-pastoris*, the first meristem to differentiate is the protoderm, which is formed by a periclinal division of the cells of the octant embryo. Because of the nature of this division, the packet of cells of the globular embryo can be neatly separated into eight external cells of the protoderm, sheltering an internal group of eight cells which later form the ground tissue and procambium [75]. The cells of the dermal layer thus blocked out acquire the characteristics of the epidermis. According to Bruck and Walker [76], in *Citrus jambhiri*, the commitment to differentiate an epidermis is made by the zygote itself by the development of a cuticle around its outer surface. The epidermal trait is thus perpetuated in the external layer of cells as far back in embryogenesis as the zygote. Failure of epidermal reformation in embryos whose epidermis was surgically removed has further shown that both epidermal and subepidermal cell layers are committed along divergent pathways early during embryogenesis [77]. Incompetence of two embryos to form a graft between surface cells has also been cited as evidence of epidermal determination [78].

The differentiation of the inner cells of the 16-celled embryo of *C. bursa-pastoris* into the progenitor cells of the ground tissue is signaled by their decreased stainability and increased vacuolation [75]. The cells of the ground meristem of the embryo may differentiate exclusively into the cortex as in *Phlox drummondii* [79], *Dianthus chinensis* [80], *Sphenoclea zeylanica* [81], *Downingia bacigalupii, D. pulchella* [82], and *Stellaria media* [83]; in some cases the rigid control on cell differentiation may be relaxed as the ground meristem differentiates into both the cortex and pith as in *Juglans regia* [84], *Pisum sativum* [85], *Nerium oleander* [86] and *Sesamum indicum* [87].

Procambial cells are cut out of the cortex or pith of globular embryos before the primordia of cotyledons are formed. This is clearly described in the embryos of *P. drummondii* [79], *Micropiper excelsum* [88], *N. oleander* [86], *S. indicum* [87], and *Downingia* [82], while in the embryos of *Cassipourea elliptica* [89] and *Bruguiera exaristata* [90], the initiation of procambial cells may be postponed until the cotyledons appear. As shown in *N. oleander*, a procambium is first seen in the innermost part of the embryonic cortex of the early heart-shaped embryo at the level of the presumptive cotyledons. More initials are laid down on either side of the original file leading to the formation of a complete ring of procambial cells. After the procambium is established in the hypocotyl axis of the developing embryo it extends acropetally into the cotyledons and shoot apex and basipetally into the root apex. Thus, by the time the embryo is phased into the torpedo-shaped stage, it has a solid file of procambium extending through its entire length and into the cotyledons [86]. In *Rhizophora mangle*, the procambium files linking the shoot and root apices are apparently each initiated independently as they appear discontinuous for a brief period until they are connected to the hypocotyl procambium by cotyledonary strands [91]. With regard to the maturation of the first vascular elements, sieve tubes are the first to differentiate in mature embryos of *J. regia* [84], *Lupinus albus*, *Helianthus annuus*, *Ricinus communis* and *Mirabilis jalapa* [92], while xylem elements appear ahead of the phloem in embryos of *P. drummondii* [79], *C. elliptica* [89] and *B. exaristata* [90]. The vascular elements may first appear in the hypocotyl-root axis [92] or at the base of the cotyledonary node [84] or in the cotyledons [79, 89, 90]. On the whole, the pattern of tissue differentiation is a crucial process of embryogenesis and one of enduring histogenetic interest, but one about whose mechanism we know very little. Based on the beautiful work done on the control of vascular tissue differentiation in other systems [93], one way of getting information on this question would be to study how hormonal gradients in the embryo modify the process.

1. Origin of the apical meristems

Careful anatomical studies on the embryos of *Nerium oleander* [86] and *Downingia bacigalupii* and *D. pulchella* [82] have provided some basic information on the origin of the shoot and root apical meristems. In the former, zones of partially differentiated meristematic cells at either ends of the globular embryo foreshadow the shoot and root apices. These zones are interpreted as direct descendants of the octant stage embryo rather than new cells generated as meristems. In all three species, differentiation of the shoot apex is also augured by a change from the random distribution of cell divisions to their concentration in the apical half and by cytohistological differentiation into peripheral and central zones. It is to be noted that these changes occur before the appearance of the cotyledons, suggesting that the latter are the first formative organs produced by the shoot apex. The attainment of a dome-like shape by the embryonic shoot apex is coincident with the change in the shape of the embryo from globular to heart-shaped and has been attributed to the continued meristematic activity in the protoderm and in the underlying layer of cells in the apical part of the embryo. The formation of the central zone of the embryonic shoot apex is foreshadowed by the cytological differentiation of three to four layers of cells below the protoderm. Differentiation of the shoot apex is complete when it assumes the shape of a rounded dome and lapses into a state of quiescence or dormancy in the seed.

The architecture of the embryonic root apex, which consists of the root meristem and root cap is generally established in the globular stage of the embryo. Depending upon the type of embryo ontogenesis, the root meristem and the root cap have their origin in the terminal or the basal cell of the two-celled proembryo. *D. pulchella* will serve as our illustrative example in which the embryonic root apex is derived from the terminal cell of a two-celled proembryo. Limited transverse and longitudinal divisions in this cell lead to the formation of an embryo consisting of four superimposed tiers of one to four cells. The first sign of initiation of the root apex is an oblique division in the two cells of the most basal tier which give rise to the root cap. Protodermal cells of the next tier towards the apical part of the embryo also undergo periclinal divisions to contribute to the root cap. This latter tier functions as the apical initials that generate the rest of the root apex [82]. In *Capsella bursa-pastoris*, the embryonic root is derived from the hypophysis. This cell divides transversely to form two cells. By further transverse and longitudinal divisions, the daughter cell close to the suspensor gives rise to the root cap and the root epidermis and the one close to the organogenetic part of the embryo generates the rest of the root apex including the root meristem [25, 26].

2. Origin of the quiescent center

The quiescent center is a packet of cells of the angiosperm root apical meristem which divide rather infrequently or not at all. These cells are also characterized by low DNA, RNA and protein synthetic activities leading to the suggestion that the quiescent center originates as a result of the decreasing metabolic activity of certain cells of the embryonic root meristem. A variety of methods have been used to monitor the formation of this tissue in developing embryos. Based on the tempo of mitotic activity, Sterling [94] showed that the radicle meristem of lima bean (*Phaseolus lunatus*) embryo has a zone of mitotically inactive cells, analogous to a quiescent center, surrounded by actively dividing cells. In the embryo of *Alyssum maritimum* [95], a group of cells that stains lightly with methyl-green pyronin is considered to represent the earliest indication of a quiescent center, while in *Petunia hybrida* [96], cell lineage studies have shown that cells comparable to the quiescent center are present throughout embryogenesis beginning with the globular stage. Using ^{3}H-thymidine as a marker to monitor autoradiographically detectable DNA synthesis, a quiescent center has been identified in the embryos of maize [97] and pea [98] with well-developed root cap initials or a primary root cap. As a model for the earliest seen quiescent center, in the embryo of pea, it is represented by a band of no more than four to seven cells. As the embryo matures, the quiescent center disappears at the same time as meristematic activity in the root apex is curtailed [98]. Based *on situ* hybridization with ^{3}H-polyuridylic acid, a recent study has presented evidence for the single-celled origin of the quiescent center in the embryo of *Capsella bursa-pastoris* and a role for the hypophysis in it (Figs. 6 and 7) [99]. Overall, these results indicate that the quiescent center is a product of embryogenic development and originates as soon as the root meristem is delimited in the globular embryo.

IV. Cellular and biochemical embryogenesis

Embryogenesis involves not only the division of the zygote according to a predetermined plan, but also correlated ultrastructural and biochemical changes in the cells leading to the accumulation of storage reserves. Although ultrastructural and biochemical studies have been slow to get off the ground,

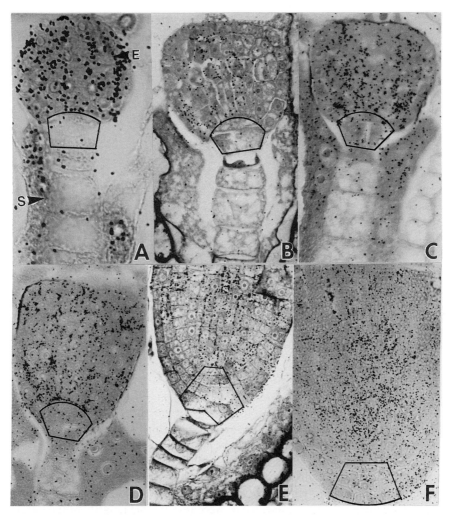

Fig. 6. (A-F) Autoradiographs of sections of embryos of different ages of *Capsella bursa-pastori* hybridized *in situ* with ³H-polyuridylic acid. Lines are drawn on the prints to indicate the boundaries of the hypophysis and cells cut off by the hypophysis. E = embryo; S = suspensor. (A) An early globular embryo. (B) A late globular embryo. (C) Another late globular embryo. (D) An early heart-shaped embryo. (E) A torpedo-shaped embryo. (F) Radicle end of a mature seed embryo (Raghavan [99]).

some of the results generated have succeeded in identifying the changes associated with functional differentiation in the embryo. As will be seen below, most of the work concerns angiosperms. Embryos of gymnosperms have hardly been examined at the ultrastructural or biochemical levels, but the challenge is great.

98

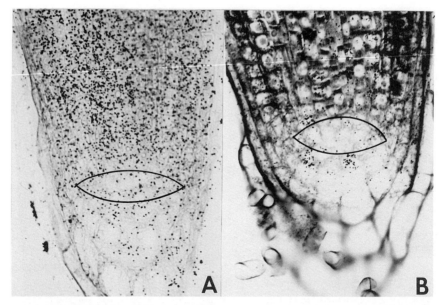

Fig. 7. (A,B) Localization of the quiescent center (bounded by lines drawn on prints) in the seedling root of *Capsella bursa-pastoris*. (A) Localization by *in situ* hybridization with ^{3}H-polyuridylic acid. (B) Localization by ^{3}H-thymidine labeling (Raghavan [99]).

A. Ultrastructural changes

The simplicity of the structure of the zygote of various angiosperms as seen in the electron microscope almost betrays its potential as the progenitor of a whole plant. Present indications are that compared to the unfertilized egg, the zygote displays a polarized distribution of organelles, most strikingly of the endoplasmic reticulum, ribosomes, plastids, and mitochondria which cluster around the nucleus at the chalazal end. Accentuating the polarity of the zygote is a gradient of decreasing density of plastids and mitochondria towards the micropylar end. Conversely, both the size and number of vacuoles increase markedly towards the micropylar end of the zygote [62, 75, 100–103]. Rarely, as in *Papaver nudicaule* which has a micropylar nucleus and a large chalazal vacuole in the unfertilized egg, the nucleus and the vacuole are found in opposite orientations in the zygote [104].

Other ultrastructural observations indicate that the zygote is a metabolically active cell. While in many plants the chalazal part of the unfertilized egg lacks a cell wall, laying down and consolidation of a new wall is an important step in the structural evolution of the zygote; increased dictyosome activity coincident with cell wall regeneration seems to concur with this observation [105]. Increase in the number of plastids and in the degree of their internal differentiation and in the ribosome configuration of the zygote could also have a role in its localized metabolic functions. Usually changes in the ribosome profile in the fertilized egg are rather dramatic and in the different species investigated they include formation of new population of ribosomes, their aggregation into polysomes and lengthening of polysomes already present [75, 106–109]. Besides an increase in stainable RNA in the zygote of *Capsella bursa-pastoris* [75], *in situ* hybridization with ^{3}H-polyuridylic acid has shown a gradient in the accumulation of mRNA in this cell, with a high concentration in the chalazal end [110]. In general, these results suggest that fertilization signals the production of specific mRNAs and their engagement by ribosomes to form polysomes.

The polar distribution of organelles and metabolites in the zygote naturally affects the ultrastructure of the daughter cells formed from its first division because the partition wall lies close to the chalazal end. As a result, the terminal cell of the two-celled embryo has a dense cytoplasm enriched in organelles, whereas the basal cell is relatively improvished of them. Among the organelles affected in the two-celled embryo of cotton are plastids and mitochondria which are numerous in the terminal cell and vacuoles of various sizes and shapes in the basal cell [100]. If the localized distribution of metabolites are of developmental significance in the two-celled embryo, it is likely that they will be nucleic acids and proteins. As shown in cotton [100] and *Vanda* [111], a functional differentiation between the terminal and basal cells is signaled by a higher concentration of RNA in the former, while in *C. bursa-pastoris* [75], and *Stellaria media* [112], RNA concentration is higher in the basal cell than in the terminal cell.

Several important ultrastructural and histochemical differences separate the two-celled proembryo from its subsequent division stages. As shown in *C. bursa-pastoris*, the events of this period are an increased concentration of ribosomes in the terminal cell of the three-celled embryo and an abundance of ER and dictyosomes in the basal and middle cells. The ribosome density remains unchanged as the terminal cell forms the globular embryo. The earliest stage of embryogenesis at which differences between the cells are noted preparatory to tissue and organ differentiation is the heart-shaped embryo; here the cells of the procambium and ground meristem appear more highly vacuolate than those of the protoderm [75]. From the heart-shaped to the mature stages of the embryo, there are no characteristics ultrastructural changes associated with further differentiation of cells and tissues and formation of organs. Although there are differences in detail, ultrastructural changes auguring cell specialization during embryogenesis in barley [113], sunflower [61] and *Quercus gambelii* [62] are somewhat similar to those described in *C. bursa-pastoris*.

B. Biochemical changes

We possess only a limited understanding of the biochemical trends that prevail during embryogenesis and in no case has there been a protracted study of one or two species to provide secure generalizations applicable widely. A number of investigators [114–117] have shown that in embryos of various legumes and in particular in the cotyledons which initially grow by cell division, the nuclei continue to duplicate DNA by endoduplication after the cell number has stabilized. The result is that cells of full-grown cotyledons attain DNA values as high as 64C in *Pisum arvense* [114]. With regard to changes in RNA content, a rising level of RNA occurs during most periods of growth of the embryos of *Vicia faba* [118] and other plants [119]. When the organogenetic part of the embryo and suspensor of *Phaseolus coccineus* were analyzed separately, RNA and protein contents and the rates of synthesis of these macromolecules per cell are found to be high in the early heart-shaped embryos and to decline thereafter, but on a per cell basis, at comparable stages of development, the biosynthetic capacity of the cells of the embryo is much less than that of the nondividing cells of the suspensor [120, 121]. The relationship between RNA and protein metabolism during embryogenesis is not clearly elucidated. However, in embryos of *P. coccineus* it has been shown that proteins continue to accumulate even after cessation of RNA synthesis. This observation supports the notion that these proteins are coded on mRNA synthesized at an earlier period in the ontogeny of the embryo and held over for later use [122]. A series of observations on RNA metabolism of cotyledons of developing embryos have generally supported the results from whole embryos, although some variations in the timing of peak accumulation of RNA have been noted in the cotyledons of various species [119]. Present evidence also shows that protein synthetic activity declines precipitously in the cotyledons of mature and dehydrated embryos. Since the rate of *in vitro* protein synthesis using mRNA extracted from cotyledons of mature embryos is much lower than that primed by mRNA from cotyledons at the stage of rapid growth, it was naturally thought that mature cotyledons lack translatable mRNA. This inference has been fortified by the demonstration that addition of an artificial messenger completely restores the protein synthetic activity of mRNA of mature embryos [123].

1. The suspensor

A number of observations have been made on the nuclear cytology and biochemical properties of the suspensor cells of various angiosperms. A recurrent theme from these investigations indicative of a transcriptionally active nature of the suspensor cells is endoduplication of their nuclei coupled with the presence of polytene chromosomes in them [119]. On the basis of nuclear volume or quantitative Feulgen microspectrophotometry, values for nuclear DNA content so far reported range from 16C for *Sophora flavescens* [124] to 8192C for *Phaseolus coccineus* [125]. In *P. coccineus* [124, 125] and *Eruca sativa* [126], there is a gradient of DNA values beginning with low values in the suspensor cells close to the embryo, intermediate values in the neck region of the suspensor and high values in the cells in the basal region. Although the chromosomal constitution of endoduplicated suspensor cells of several plants has revealed polyteny, the phenomenon has most extensively been studied in *P. coccineus* [127] and *P. vulgaris* [128]. Like polytene chromosomes in the salivary glands of dipteran larvae, those of *P. vulgaris* [129] and *P. coccineus* [130] also display puffing. Since puffs represent sites at which DNA synthesis and transcription occur, it is clear that suspensor cells are transcriptionally very active. This has been verified in experiments in which the changes in RNA and protein metabolism in the suspensor of *P. coccineus* surgically removed from the embryo were monitored [120, 121]. These studies showed that on a per cell basis, RNA content and the rate of RNA synthesis are low during the early development of the suspensor, increase to a maximum at the late heart-shaped or early cotyledon stage of the embryo and then decline. Protein content and the rate of protein synthesis are also low in the suspensor cells of early stage embryos, but increase substantially in the cells of late stage embryos. Compared to the cells of the embryo, cells of the suspensor contain more RNA and proteins and synthesize them more efficiently than embryo cells of the same age. However, further work is necessary to gain insight into the mechanism by which gene action is regulated in the suspensor and the ways by which the products of gene action contribute to suspensor function.

Another indication of the biochemical activity of the suspensor is the production of growth hormones such as auxins [131, 132], gibberellins [131, 133], cytokinins [134] and abscisic acid (ABA) [135]. In the early stage embryos of *P. coccineus*, both gibberellins and cytokinins are present in higher concentrations in the suspensor cells than in the cells of the organogenetic part of the embryo; a complete reversal of the status of active gibberellins and cytokinins occurs in the embryos at cotyledon stage. Changes in the auxin content of embryos have drawn little attention, although it appears that the suspensor cells contain more auxins than the cells of the embryo. Very little ABA was present in the suspensor of *P. coccineus* during the initial stages of its development, although some increase with two peaks was noted during later stages of embryogenesis. An idea for a role for hormones in the growth of the suspensor has come from experiments in which embryos of *P. coccineus* deprived of the suspensor were found to achieve very little growth unless the medium was supplemented with gibberellins or cytokinins [136–138]. This suggests that hormones necessary for the early growth of the embryo are made by the suspensor.

C. Storage protein synthesis

In recent years, the study of storage protein accumulation and synthesis has opened up the question of protein metabolism in developing embryos in an impressive way. The termination of the early proliferative phase of growth of embryos marks the beginning of the synthesis and accumulation of storage proteins, leading to increases in fresh and dry weight of embryos. Although virtually all cells of the embryo accumulate varying amounts of storage protein reserves, the latter appear in enormous quantities in the cotyledons.

Many aspects of the metabolism of storage proteins during embryogenesis have been investigated in embryos of agronomically important plants such as peas, beans, cotton and rape (*Brassica napus*). Among legumes, one of the most thoroughly studied examples is soybean (*Glycine max*) whose major storage proteins are a 7S glycoprotein, conglycilin and an 11S nonglycosylated protein, glycinin. Both conglycinin and glycinin have several subunits which appear at different times during the ontogeny of the cotyledons [139, 140]. Legumin and vicilin are two high molecular weight proteins found in the cotyledons of *Vicia faba* and pea. In the cotyledons of *V. faba*, although the synthesis of vicilin generally precedes that of legumin, legumin production overtakes that of vicilin, accounting for the ability of the cotyledons to store nearly four times more legumin than vicilin [141, 142]. In pea, vicilin subunits are present at earlier stages of embryogenesis than legumin, whereas the latter and convicilin (a minor storage protein fraction) are prominent at later stages [143]. In *Phaseolus vulgaris*, a globulin G1 (also known as phaseolin), along with a smaller amount of another globulin G2, form the predominant components of the storage proteins of cotyledons [144]. Unlike in the legumes, in cotton, major increases in storage protein content occur during the cell division phase of growth of cotyledons [145]. Storage proteins of rape are constituted of a legumin-like 12S neutral globulin designated as cruciferin and a bunch of basic proteins designated as napin, both of which are distributed in the embryo axis as well as in the cotyledons. Although both proteins begin to accumulate at about the same rate in the embryo during its cell expansion phase, there is however a higher ratio of cruciferin to napin in the mature embryo [146]. The major storage protein in the embryos of sunflower is a globulin, helianthinin. The synthesis of this protein is developmentally regulated, maximum synthesis occurring between 14 to 20 days after flowering [147]. Embryo proteins have also been characterized in cereals like rice [148], and palms like *Phoenix dactylifera* [149] and *Washingtonia filifera* [150] but their developmental regulation during embryogenesis awaits investigation.

Much less is known about the storage proteins of gymnosperms. The major storage proteins of several species of *Pinus* [151] and *Picea* [152] are identified as crystalloids with solubility, structural and size characteristics similar to crystalloids found in angiosperms. When comparisons were made between early stage and maturation phase embryos, storage proteins were found to accumulate in abundance in the latter [153, 154]. According to Dodd et al. [155], embryos of *Podocarpus henkelii* continue to accumulate storage proteins and metabolites even after the seeds attain maturity. This has been interpreted as a developmental strategy to account for uninterrupted embryo growth without dormancy. At this stage, these studies are of interest as a first step to more extensive studies on developmental gene expression during embryogenesis in gymnosperms.

V. Gene activity during embryogenesis

It is clear by now that dramatic morphological, physiological and biochemical changes occur as the zygote is transformed into an embryo within the confines of the ovule. These involve repeated mitoses, establishment of polarity and the morphological pattern of the adult plant, histodifferentiation and accumulation of complex metabolites and storage reserves. Since these changes are generally attributed to alterations in gene expression patterns, a brief account of the mechanism of regulation of genes specifically expressed during embryogenesis is germane to the theme of this chapter. Fascinating though are the changes in gene expression patterns as an explanation of progressive form change, its significance in plant embryogenesis was not appreciated until recently. Several recent reviews [156–159] published on this subject are intended to show that gene activation during plant embryogenesis is not simply an inference enshrined in a dogma, but is a demonstrable phenomenon.

A. Fertilization and early phase of embryogenesis

Progressive division of the zygote into an embryo results from the temporal expression of genes originally present in the egg and further endowed at the time of fertilization. There has been very little investigation on the molecular biology of fertilization and the mechanism of gene activation during the early division phase of the zygote. A major impediment to these studies is that events of fertilization and embryogenesis occur in the relatively inaccessible environments of the archegonium in gymnosperms and the embryo sac in angiosperms. In one study, based on *in vitro* translation of poly (A) + RNA isolated from ovules of *Pinus strobus* before and after fertilization, evidence was obtained for the transcription of a prominent mRNA at the time of fertilization and the beginning of embryogenesis. However, no attempt was made to separate the archegonia from the rest of the gametophytic and sporophytic tissues of the ovule. Moreover, the transient mRNA was found in the integument and nucellar tissue, but not in the female gametophyte [160]. This casts considerable doubt on the significance of the new mRNA in fertilization and embryogenesis.

It is not established whether mature eggs of gymnosperms and angiosperms contain stored mRNAs which become available after fertilization to code for the first proteins of the dividing zygote. However, there are many claims for an independent role for the maternal genome in directing early embryo development based largely on the end result – activation of the egg in the absence of the sperm (parthenogenesis). These include the division of the egg in the absence of a stimulus from the male gametophyte [161], physical and chemical agents which prevent the growth and penetration of the pollen tube [162], haploids that arise by chromosome elimination in certain interspecific crosses [163] and the purported origin of embryos from the egg of cultured unfertilized ovules or ovaries containing unfertilized ovules [164, 165]. The point to be emphasized is that partially or fully developed embryos arise in the absence of fertilization because of the ability of the maternal genome to support continued division of the egg. Although this is not a rigorous demonstration of the presence of stored mRNA, there is probably some truth in its validity.

Another line of evidence has come from an autoradiographic investigation of RNA and protein synthesis in early stage embryos of rice and barley. When florets of these plants were cultured in a medium containing ^{3}H-uridine as a precursor of RNA synthesis, incorporation of the label was not detected in embryos of less than 100 cells. At the same time, ^{3}H-leucine, a precursor of protein synthesis, was readily incorporated into embryos of all ages. Since the eggs are fertilized after the florets are placed in culture, it has been reasoned that protein synthesis in the absence of concurrent RNA synthesis probably involves stored maternal mRNA [166].

B. Structural and storage protein gene expression

Induction of mutations at discrete stages of embryo development has provided an elegantly simple means of establishing the necessity of certain genes for completion of embryogenesis. By disturbing embryogenesis in *Arabidopsis thaliana* [167] and maize [168] by mutagenesis, embryo-lethal mutants, that is, embryos arrested in developmental or nutritional functions have been obtained. Arrest of development at a particular stage by a mutation implies that the embryo is unable to complete a metabolic reaction or synthesize a specific nutrient substance required for its developmental program due to a defective gene. More to the point is the fact that in some cases the embryos can be rescued by culturing them in a nutrient medium that supplies the missing substance. A striking example of a nutritional mutant is a proline-requiring mutant of maize in which the defect has been traced to a genetic block in the pathway of proline biosynthesis [169, 170]. In a similar way, in a viviparous mutant of *A. thaliana* characterized by a low level of ABA, the embryo was found to germinate in the seed (vivipary), skipping the normal developmental arrest. This is what one would expect of embryos in which the onset of natural dormancy is correlated with a high ABA content [171]. Another embryo-lethal mutant of *A. thaliana* was recently shown to be a biotin auxotroph, since it was rescued by supplementation of the culture medium by biotin or by watering the mutant plants with biotin [172].

In recent years a substantial body of research has provided molecular evidence supporting an underlying program of sequential gene activation during embryogenesis. In one of the first investigations of its kind, Goldberg et al. [173] constructed a cDNA library of mRNA sequences for certain subunits of the soybean storage proteins glycin and conglycinin and showed that the cloned transcripts are either absent or present in very low levels at early stages of embryogenesis but accumulate dramatically at the mid-maturation phase when protein synthesis is rapid. This is followed by a decay prior to desiccation of the seed and coincident with the arrest of protein accumulation. Later studies showed that accumulation and decay of lectin and trypsin inhibitor mRNAs during embryogenesis also follow a course analogous to that of genes for glycinin and conglycinin [174, 175]. In addition, both trypsin inhibitor mRNA and seed protein messages are found to be more prevalent in the embryonic cotyledons than in the embryo axis, indicating the existence of mechanisms for both temporal and spatial regulation of seed protein gene expression. Since the storage proteins are the most visible and abundant reserves of dicotyledonous embryos, they serve as a paradigm of a highly regulated embryo-specific gene set.

Equally impressive molecular evidence for the existence of gene regulation during embryogenesis has come from the work of Galau and Dure on cotton [176–179]. These investigators have identified several major mRNA species which are more abundant in mature embryos than in young embryos or seedlings of cotton. The concentrations of some of these mRNA sequences designated as late embryogenesis abundant (*lea*) [180] are found to increase dramatically during late embryogenesis and decline precipitously during the early phase of germination [181]. Storage protein mRNA sequences and mRNAs analogous to *lea* have also been identified in cDNA libraries of embryos of other plants such as *A. thaliana* [182], radish (*Raphanus sativus*) [183] and *Vigna unguiculata* [184]. We know very little about the functions of proteins encoded by the various embryo-specific mRNAs, although there is some evidence that *lea* mRNAs code for proteins involved in seed desiccation.

The function of proteins encoded by chloroplast genes is, of course, well-known. In the embryos of cotton there is a temporal pattern of accumulation of mRNA sequences for chlorophyll a/b binding protein (*cab*) and for the large subunit of ribulose bisphosphate carboxylase (RuBisCo) [185]. A temporal pattern was also characteristic of the accumulation of RuBisCo in the developing embryos of *Phaseolus vulgaris* [186]. Since these embryos are not exposed to light during their development, expression of the genes and their proteins is associated with developmental signals present in the embryos.

A general feature underlying the expression of embryo-specific mRNA sequences is that they appear and decay at various times during embryogenesis. This indicates that the expression of each set of genes is controlled by specific regulatory signals. What causes the activation of specific genes at specific stages of embryogenesis is not well understood, but recent studies have assigned a role for ABA in this process. For example, in cotton, all *lea* mRNAs are induced by culturing immature embryos in a medium containing ABA [177, 180]. Regulation of ABA-induced mRNAs and proteins has also been analyzed in embryos of other plants such as rape [187], wheat [188], soybean [189, 190], barley [191, 192], corn [193], sunflower [194] and pea [186]. ABA also controls the synthesis and localization of lectin wheat germ agglutinin [195, 196] and the accumulation of lectin in rice embryos [197]. From this it appears that under natural conditions, continuation of the normal program of embryogenesis including the accumulation of specific proteins is geared to the changes in the concentration of ABA in the embryo and in the surrounding tissues of the ovule.

All of the studies on gene expression during embryogenesis described above were undertaken with midmaturation or mature stage embryos. Thus, there is a serious lack of molecular information concerning gene expression during early stages of embryogenesis when the polarity of the embryo is established and when the major events of cell specialization occur. In an effort to fill this void, attempts have been made to follow the gene expression pattern during early embryogenesis by *in situ* hybridization localization of mRNAs. With regard to the expression of a rice histone gene [198], the gene for glutelin [199], the major storage protein of rice, and genes for soybean seed storage protein [200] and storage proteins of *Brassica napus* [201] these studies have shown that transcriptional changes are initiated almost concomitant with organ initiation in the embryo and not at earlier stages (Fig. 8A,B). A case for transcription of mRNAs of these genes on the sporophytic (post-fertilization) rather than on the gametophytic (pre-fertilization) genome is implied in this observation.

VI. Concluding comments

The major achievements in the study of embryology of phanerogams fall into two general areas. Firstly, the detailed examination of the pattern of embryo development in scores of gymnosperms and angiosperms laid the foundation for future experimental studies on this system; secondly, the spectacular new developments in molecular biology and genetic engineering have made it possible to examine the regulatory factors that control the temporal, spatial and cell-specific gene expression patterns in developing embryos of normal and transformed plants. Sandwiched between these areas is the experimental study of embryos, designed to determine the physiological cues that control their orderly growth in space and time.

Yet much remains to be learned regarding the interactions of various factors in embryo growth and there is a need for a fresh approach to study the interrelationships between the embryo and its source of nutrition. It is easy to consider the early division phase of the embryo as one when it increases its endowment of cells drawing upon the food materials of the female gametophyte as in gymnosperms or of the endosperm in angiosperms. It is thus heterotrophic in nature during the early period, but becomes progressively autotrophic. Increase in cell number goes hand in hand with an increase in the amount of DNA the embryo has at its disposal. This is followed by the synthesis of unique RNA and proteins which account for the specificity in cytodifferentiation and tissue and organ formation. How the upsurge of cellular and biochemical activities contributes to the characteristic division patterns displayed by the zygote remains elusive at this time. Basic to our understanding of the role of these divisions in the generation of form in the embryo are the factors which determine the plane of divisions. When all these are taken into account, it seems certain that cellular, molecular and biochemical techniques will assume major importance in gaining deeper insight into embryogenesis in phanerogams.

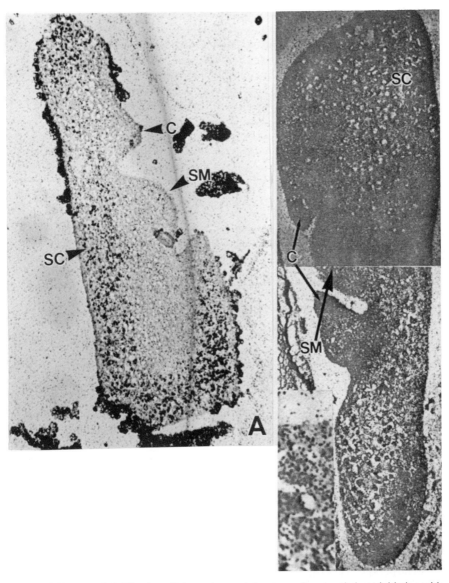

Fig. 8. (A,B) *In situ* hybridization of rice embryos at the stage of root and shoot initiation with cloned genes. C = coleoptile; SC = scutellum; SM = shoot meristem. (A) Hybridization with ^{3}H-labeled rice histone H$_3$ gene. Label is found mostly in the scutellum (Raghavan and Olmedilla [198]). (B) Hybridization with ^{35}S-labeled rice glutelin gene. Label is found in all parts of the embryo (Ramachandran and Raghavan [199]).

References

1. Johansen, D.A., *Plant Embryology. Embryogeny of the Spermatophyta*, Chronica Botanica Co., Waltham, 1950.
2. Maheshwari, P., *An Introduction to the Embryology of Angiosperms*, McGraw-Hill Book Co., New York, 1950.
3. Gifford, E.M., and Foster, A.S., *Morphology and Evolution of Vascular Plants*, third ed., W.H. Freeman & Co., New York, 1988.
4. Singh, H., *Embryology of Gymnosperms (Handbuch der Pflanzenanatomie, Band 10, Teil 2)*, Gebrüder Borntraeger, Berlin, 1978.
5. Chamberlain, C.J., *Gymnosperms: Structure and Evolution*, Chicago University Press, Chicago, 1935.
6. Rao, L.N., Life history of *Cycas circinalis* L. II. Fertilization, embryogeny and germination of the seed, *J. Indian Bot. Soc.*, 42, 319, 1963.
7. Brough, P., and Taylor, M.H., An investigation of the life cycle of *Macrozamia spiralis* Miq., *Proc. Linn. Soc. NSW*, 65, 494, 1940.
8. Bryan, G.S., The cellular proembryo of *Zamia* and its cap cells. *Amer. J. Bot.*, 39, 433, 1952.
9. Roy Chowdhury, C., The embryogeny of conifers: A review, *Phytomorphology*, 12, 313, 1962.
10. Dogra, P.D., Morphology, development and nomenclature of conifer embryo, *Phytomorphology*, 28, 307, 1978.
11. Lehmann-Baerts, M., Études sur les Gnetales. XII. Ovule, gametophyte femelle et embryogenèse chez *Ephedra distachya* L., *Cellule*, 67, 51, 1967.
12. Maheshwari, P., and Vasil, V., *Gnetum*, Council of Scientific & Industrial Research, New Delhi, 1961.
13. Martens, P., *Les Gnétophytes (Handbuch der Pflanzenanatomie, Band 12, Teil 2)*, Gebrüder Borntraeger, Berlin, 1971.
14. Martens, P., and Waterkeyn, L., Études sur les Gnétales. XIII. Recherches sur *Welwitschia mirabilis*. V. Évolution ovulaire et embryogenèse, *Cellule*, 70, 163, 1974.
15. Natesh, S., and Rau, M.A., The embryo, in *Embryology of Angiosperms*, Johri, B.M., Ed., Springer-Verlag, Berlin, 1984, Chapter 8.
16. Jones, H.A., Pollination and life history studies of lettuce (*Lactuca sativa*), *Hilgardia*, 2, 425, 1927.
17. Souèges, R., Développement de l'embryon chez le *Sagina procumbens* L., *Bull. Soc. Bot. Fr.*, 71, 590, 1924.
18. Souèges, R., Embryogénie des Saxifragacées. Développement de l'embryon chez le *Saxifraga granulata* L., *C.R. Acad. Sci. Paris*, 202, 240, 1936.
19. Olson, A.R., Postfertilization changes in ovules of *Monotropa uniflora* L. (Monotropaceae), *Amer. J. Bot.*, 78, 99, 1991.
20. Tohda, H., Development of the embryo of *Pogonia* (Orchidaceae), *Sci. Rep. Tohoku Univ. Fourth Ser. (Biol.)*, 37, 89, 1974.
21. Crété, P., Embryo, in *Recent Advances in the Embryology of Angiosperms*, Maheshwari, P., Ed., International Society of Plant Morphologists, Delhi, 1963, Chapter 7.
22. Yakovlev, M.S., and Yoffe, M.D., On some peculiar features in the embryogeny of *Paeonia* L., *Phytomorphology*, 7, 74, 1957.
23. Cave, M.S., Arnott, H.J., and Cook, S.A., Embryogeny in the California peonies with reference to their taxonomic position. *Amer. J. Bot.*, 48, 397, 1961.
24. Carniel, K., Über die Embryobildung in der Gattung *Paeonia*, *Österr. Bot. Z.*, 114, 4, 1967.
25. Schaffner, M., The embryology of the Shepherd's Purse, *Ohio Naturl.*, 7, 1, 1906.
26. Souèges, R., Les premières divisions de l'oeuf et les différenciation du suspenseur chez le *Capsella bursa-pastoris* Moench. (1), *Ann. Sci. Naturl. X Bot.*, 1, 1, 1919.
27. Marsden, M.P.F., and Meinke, D.W., Abnormal development of the suspensor in an embryo-lethal mutant of *Arabidopsis thaliana*, *Amer. J. Bot.*, 72, 1801, 1985.
28. Swamy, B.G.L., Embryogenesis in *Cheirostylis flabellata*, *Phytomorphology*, 29, 199, 1979.

29. Swamy, B.G.L., and Lakshmanan, K.K., The origin of the epicotylary meristem and cotyledon in *Halophila ovata* Gaudich, *Ann. Bot.*, 26, 243, 1962.

30. Swamy, B.G.L., and Lakshmanan, K.K., Contributions to the embryology of the Najadaceae, *J. Indian Bot. Soc.*, 41, 422, 1962.

31. Guignard, J.-L., The development of cotyledon and shoot apex in monocotyledons, *Can. J. Bot.*, 62, 1316, 1984.

32. Ba, L.T., Cavé, G., Henry, M., and Guignard, J., Embryogénie des Potomogetonacées. Étude en microscopie électronique à balayage de l'origine du cotylédon chez *Potomogeton lucens* L., *C.R. Acad. Sci. Paris*, 286D, 1351, 1978.

33. Swamy, B.G.L., Embryogenesis in *Sagittaria sagittaefolia*, *Phytomorphology*, 30, 204, 1980.

34. Batygina, T.B., On the possibility of separation of a new type of embryogenesis in angiospermae, *Rev. Cytol. Biol. Vég.*, 32, 335, 1969.

35. Randolph, L.F., Developmental morphology of the caryopsis in maize, *J. Agric. Res.*, 53, 881, 1936.

36. Olson, A.R., Seed morphology of *Monotropa uniflora* L. (Ericaceae), *Amer. J. Bot.*, 67, 968, 1980.

37. Arekal, G.D., and Ramaswamy, S.N., Embryology of *Burmannia pusilla* (Wall. ex Miers) Thw. and its taxonomic status, *Beitr. Biol. Pflanzen*, 49, 35, 1973.

38. Arditti, J., Aspects of the physiology of orchids, *Adv. Bot. Res.*, 7, 421, 1979.

39. Rangaswamy, N.S., Morphogenesis of seed germination in angiosperms, *Phytomorphology*, P. Maheshwari Mem. Vol., 17, 477, 1967.

40. Okonkwo, S.N.C., and Raghavan, V., Studies on the germination of seeds of the root parasites, *Alectra vogelii* and *Striga gesnerioides*. I. Anatomical changes in the embryos, *Amer. J. Bot.*, 69, 1636, 1982.

41. Ramaswamy, S.N., Swamy, B.G.L., and Govindappa, D.A., From zygote to seedling in *Eriocaulon robusto-brownianum* Ruhl. (Eriocaulaceae), *Beitr. Biol. Pflanzen*, 55, 179, 1981.

42. Ramaswamy, S.N., and Arekal, G.D., Embryology of *Eriocaulon xeranthemum* Mart. (Eriocaulaceae), *Acta Bot. Neerl.*, 31, 41, 1982.

43. Doyle, J., and Looby, W.J., Embryogeny in *Saxegothaea* and its relation to other podocarps, *Sci. Proc. Roy. Dublin Soc.*, 22NS, 127, 1939.

44. Schopf, J.M., The embryology of *Larix*, *Illinois Biol. Monogr.*, 19, 1, 1943.

45. Hakansson, A., Seed development in *Picea abies* and *Pinus sylvestris*, *Medd. Stat. Skogsforskinst.*, 46, 1, 1956.

46. Sharma, K.K., and Thorpe, T.A., Unpublished observations.

47. Schulz, P., and Jensen, W.A., *Capsella* embryogenesis: The suspensor and the basal cell, *Protoplasma*, 67, 139, 1969.

48. Schnepf, E., and Nagl, W., Über einige Strukturbesonderheiten der Suspensorzellen von *Phaseolus vulgaris*, *Protoplasma*, 69, 133, 1970.

49. Marinos, N.G., Embryogenesis of the pea (*Pisum sativum*). I. The cytological environment of the developing embryo, *Protoplasma*, 70, 261, 1970.

50. Newcomb, W., and Fowke, L.C., *Stellaria media* embryogenesis: The development and ultrastructure of the suspensor, *Can. J. Bot.*, 52, 607, 1974.

51. Simoncioli, C., Ultrastructural characteristics of "*Diplotaxis erucoides* (L.) DC" suspensor, *Gior. Bot. Ital.*, 108, 175, 1974.

52. Yeung, E.C., and Clutter, M.E., Embryogeny of *Phaseolus coccineus*: The ultrastructure and development of the suspensor, *Can. J. Bot.*, 57, 120, 1979.

53. Hu, S., Zhu, C., and Zee, S.Y., Transfer cells in suspensor and endosperm during early embryogeny of *Vigna sinensis*, *Acta Bot. Sinica*, 25, 1, 1983.

54. Sangduen, N., Kreitner, G.L., and Sorensen, E.L., Light and electron microscopy of embryo development in perennial and annual *Medicago* species, *Can. J. Bot.*, 61, 837, 1983.

55. Prabhakar, K., and Vijayaraghavan, M.R., Histochemistry and ultrastructure of suspensor cells in *Alyssum maritimum*, *Cytologia*, 48, 389, 1983.

56. Bohdanowicz, J., *Alisma* embryogenesis: The development and ultrastructure of the suspensor, *Protoplasma*, 137, 71, 1987.

110

57. Yeung, E.C., Embryogeny of *Phaseolus*: The role of the suspensor, *Z. Pflanzenphysiol.*, 96, 17, 1980.
58. Nagl, W., Translocation of putrescine in the ovule, suspensor and embryo of *Phaseolus coccineus*, *J. Plant Physiol.*, 136, 587, 1990.
59. Vijayaraghavan, M.M., and Prabhakar, K., The endosperm, in *Embryology of Angiosperms*, Johri, B.M., Ed., Springer-Verlag, Berlin, 1984, Chapter 7.
60. List, A. Jr., and Steward, F.C., The nucellus, embryo sac, endosperm, and embryo of *Aesculus* and their interdependence during growth, *Ann. Bot.*, 29, 1, 1965.
61. Newcomb, W., The development of the embryo sac of sunflower *Helianthus annuus* after fertilization, *Can. J. Bot.*, 51, 879, 1973.
62. Singh, A.P., and Mogensen, H.L., Fine structure of the zygote and early embryo in *Quercus gambelii*, *Amer. J. Bot.*, 62, 105, 1975.
63. Raghavan, V., *Experimental Embryogenesis in Vascular Plants*, Academic Press, London, 1976.
64. Newcomb, W., and Steeves, T.A., *Helianthus annuus* embryogenesis: Embryo sac wall projections before and after fertilization, *Bot. Gaz.*, 132, 367, 1971.
65. Schulz, P., and Jensen, W.A., Cotton embryogenesis: The early development of the free nuclear endosperm, *Amer. J. Bot.*, 64, 384, 1977.
66. Newcomb, W., The development of cells in the coenocytic endosperm of the African blood lily *Haemanthus katherinae*, *Can. J. Bot.*, 56, 483, 1978.
67. Tilton, V.R., Wilcox, L.W., and Palmer, R.G., Postfertilization wandlabrinthe formation and function in the central cell of soybean, *Glycine max* (L.) Merr. (Leguminosae), *Bot. Gaz.*, 145, 334, 1984.
68. Folsom, M.W., and Cass, D.D., Changes in transfer cell distribution in the ovule of soybean after fertilization. *Can. J. Bot.*, 64, 965, 1986.
69. Dute, R.R., Peterson, C.M., and Rushing, A.E., Ultrastructural changes of the egg apparatus associated with fertilization and proembryo development of soybean, *Glycine max*, *Ann. Bot.*, 64, 123, 1989.
70. Jones, T.J., and Rost, T.L., Histochemistry and ultrastructure of rice (*Oryza sativa*) zygotic embryogenesis, *Amer. J. Bot.*, 76, 504, 1989.
71. Norstog, K., Nucellus during early embryogeny in barley: Fine structure, *Bot. Gaz.*, 135, 97, 1974.
72. Spurr, A.R., Histogenesis and organization of the embryo in *Pinus strobus* L., *Amer. J. Bot.*, 36, 629, 1949.
73. Sterling, C., Proembryo and early embryogeny in *Taxus cuspidata*, *Bull. Torrey Bot. Cl.*, 76, 116, 1948.
74. Allen, G.S., Embryogeny and the development of the apical meristems of *Pseudotsuga*. III. Development of the apical meristems, *Amer. J. Bot.*, 34, 204, 1947.
75. Schulz, R., and Jensen, W.A., *Capsella* embryogenesis: The early embryo, *J. Ultrastr. Res.*, 22, 376, 1968.
76. Bruck, D.K., and Walker, D.B., Cell determination during embryogenesis in *Citrus jambhiri*. I. Ontogeny of the epidermis, *Bot. Gaz.*, 146, 188, 1985.
77. Bruck, D.K., and Walker, D.B., Cell determination during embryogenesis in *Citrus jambhiri*. II. Epidermal differentiation as one-time event, *Amer. J. Bot.*, 72, 1602, 1985.
78. Bruck, D.K., and Walker, D.B., Cell determination during embryogenesis in *Citrus jambhiri*. III. Graft formation and nonformation in embryonic tissues, *Can. J. Bot.*, 64, 2057, 1986.
79. Miller, H.A., and Wetmore, R.H., Studies in the developmental anatomy of *Phlox drummondii* Hook. I. The embryo, *Amer. J. Bot.*, 32, 588, 1945.
80. Buell, K.M. Developmental morphology in *Dianthus*. I. Structure of the pistil and seed development, *Amer. J. Bot.*, 39, 194, 1952.
81. Swamy, B.G.L., and Padmanabhan, D., Embryogenesis in *Sphenoclea zeylanica*, *Proc. Indian Acad. Sci.*, 54B, 169, 1961.
82. Kaplan, D.R., Seed development in *Downingia*, *Phytomorphology*, 19, 253, 1969.
83. Ramji, M.V., Histology of growth with regard to embryos and apical meristems in some angiosperms. I. Embryogeny of *Stellaria media*, *Phytomorphology*, 25, 131, 1975.

84. Nast, C., The embryogeny and seedling morphology of "*Juglans regia*" L., *Lilloa*, 6, 163, 1941.
85. Reeve, R.M., Late embryogeny and histogenesis in *Pisum*, *Amer. J. Bot.*, 35, 591, 1948.
86. Mahlberg, P.G., Embryogenesis and histogenesis in *Nerium oleander* L. I. Organization of primary meristematic tissues, *Phytomorphology*, 10, 118, 1960.
87. Hanawa, J., Late embryogeny and histogenesis in *Sesamum indicum* L., *Bot. Mag. Tokyo*, 73, 369, 1960.
88. Balfour, E., The development of the vascular system in *Macropiper excelsum* Forst. I. The embryo and the seedling, *Phytomorphology*, 7, 354, 1957.
89. Juncosa, A.M., Embryogenesis and seedling development in *Cassipourea elliptica* (Sw.) Poir. (Rhizophoraceae), *Amer. J. Bot.*, 71, 170, 1984.
90. Juncosa, A.M., Embryogenesis and developmental morphology of the seedling in *Bruguiera exaristata* Ding Hou (Rhizophoraceae), *Amer. J. Bot.*, 71, 180, 1984.
91. Juncosa, A.M., Developmental morphology of the embryo and seedling of *Rhizophora mangle* L. (Rhizophoraceae), *Amer. J. Bot.*, 69, 1599, 1982.
92. Dauphiné, A., and Rivière, S., Sur la présence de tubes criblés dans des embryons de graines non germées, *C.R. Acad. Sci. Paris*, 211, 359, 1940.
93. Sachs, T., The control of the patterned differentiation of vascular tissues, *Adv. Bot. Res.*, 9, 151, 1981.
94. Sterling, C., Embryogeny in the lima bean, *Bull. Torrey Bot. Cl.*, 82, 325, 1955.
95. Rondet, P., L'organogènse au cors l'embryogenèse chez *Alyssum maritimum* Lamk., *C.R. Acad. Sci. Paris*, 255, 2278, 1962.
96. Vallade, J., Structure et fonctionnement du meristémè lors de la formation de la jeune racine primaire chez un *Petunia hybrida* hort., *C.R. Acad. Sci. Paris*, 274D, 1027, 1972.
97. Clowes, F.A.L., Origin of the quiescent centre in *Zea mays*, *New Phytol.*, 80, 409, 1978.
98. Clowes, F.A.L., Origin of quiescence at the root pole of pea embryos, *Ann. Bot.*, 42, 1237, 1978.
99. Raghavan, V., Origin of the quiescent center in the root of *Capsella bursa-pastoris* (L.) Medik., *Planta*, 181, 62, 1990.
100. Jensen, W.A., Cell development during plant embryogenesis, *Brookhaven Symp. Biol.*, 16, 179, 1964.
101. D'Alascio-Deschamps, R., Le sac embryonnaire du lin après la fécondation, *Botaniste*, 50, 273, 1972.
102. Vallade, J., Données cytologiques sur la proembryogenèse du *Petunia*; intérêt pour une interprétation morphogénétique du développement embryonnaire, *Bull. Soc. Bot. Fr. Actual. Bot.*, 127, 19, 1980.
103. van Lammeren, A.A.M., Early events during embryogenesis in *Zea mays* L., *Acta Soc. Bot. Pol.*, 50, 289, 1981.
104. Olson, A.R., and Cass, D.D., Changes in megagametophyte structure in *Papaver nudicaule* L. (Papaveraceae) following *in vitro* placental pollination, *Amer. J. Bot.*, 68, 1333, 1981.
105. Jensen, W.A., Cotton embryogenesis: The zygote, *Planta*, 79, 346, 1968.
106. Diboll, A.G., Fine structural development of the megagametophyte of *Zea mays* following fertilization, *Amer. J. Bot.*, 55, 797, 1968.
107. Cocucci, A., and Jensen, W.A., Orchid embryology: Megagametophyte of *Epidendrum scutella* following fertilization, *Amer. J. Bot.*, 56, 629, 1969.
108. Deschamps, R., Premiers stades du développement de l'embryon et de l'albumen du lin: Étude au microscope électronique, *Rev. Cytol. Biol. Vég.*, 32, 379, 169.
109. D'Alascio-Deschamps, R., Embryologie du *Linum catharticum* L. Le zygote: étude ultrastructurale, *Bull. Soc. Bot. Fr. Lett. Bot.*, 128, 269, 1981.
110. Raghavan, V., Spatial distribution of mRNA during pre-fertilization ovule development in *Capsella bursa-pastoris*, *Sex. Plant Reprod.*, 3, 170, 1990.
111. Alvarez, M.R., and Sagawa, Y., A histochemical study of embryo development in *Vanda* (Orchidaceae), *Caryologia*, 18, 251, 1965.
112. Pritchard, H.N., A cytochemical study of embryo development in *Stellaria media*, *Amer. J. Bot.*, 51, 472, 1964.

113. Norstog, K., Early development of the barley embryo: Fine structure, *Amer. J. Bot.*, 59, 123, 1972.

114. Scharpé, A., and van Parijs, R., The formation of polyploid cells in ripening cotyledons of *Pisum sativum* L. in relation to ribosome and protein synthesis, *J. Exp. Bot.*, 24, 216, 1973.

115. Smith, D.L., Nucleic acid, protein, and starch synthesis in developing cotyledons of *Pisum arvense* L., *Ann. Bot.*, 37, 795, 1973.

116. Millerd, A., and Whitfeld, P.R., Deoxyribonucleic acid and ribonucleic acid synthesis during the cell expansion phase of cotyledon development in *Vicia faba* L., *Plant Physiol.*, 51, 1005, 1973.

117. Walbot, V., and Dure, L.S., III. Developmental biochemistry of cotton seed embryogenesis and germination. VII. Characterization of the cotton genome, *J. Mol. Biol.*, 101, 503, 1976.

118. Wheeler, C.T., and Boulter, D., Nucleic acids of developing seeds of *Vicia faba* L., *J. Exp. Bot.*, 18, 229, 1967.

119. Raghavan, V., *Embryogenesis in Angiosperms, A Developmental and Experimental Study*, Cambridge University Press, New York, 1986.

120. Walbot, V., Brady, T., Clutter, M., and Sussex, I., Macromolecular synthesis during plant embryogeny: Rates of RNA synthesis in *Phaseolus coccineus* embryos and suspensors, *Develop. Biol.*, 29, 104, 1972.

121. Sussex, I., Clutter, M., Walbot, V., and Brady, T., Biosynthetic activity of the suspensor of *Phaseolus coccineus*, *Caryologia Suppl.*, 25, 261, 1973.

122. Walbot, V., RNA metabolism during embryo development and germination of *Phaseolus vulgaris*, *Develop. Biol.*, 26, 369, 1971.

123. Beevers, L., and Poulson, R., Protein synthesis in cotyledons of *Pisum sativum* L. I. Changes in cell-free amino acid incorporation capacity during seed development and maturation, *Plant Physiol.*, 49, 476, 1972.

124. Nagl, W., *Endopolyploidy and Polyteny in Differentiation and Evolution*, North-Holland Publishing Co., Amsterdam, 1978.

125. Brady, T., Feulgen cytophotometric determination of the DNA content of the embryo proper and suspensor cells of *Phaseolus coccineus*, *Cell Diffn.*, 2, 65, 1973.

126. Corsi, G., Renzoni, G.C., and Viegi, L., A DNA cytophotometric investigation on the suspensor of *Eruca sativa* Miller, *Caryologia*, 26, 531, 1973.

127. Nagl, W., Die Riesenchromosomen von *Phaseolus coccineus* L.: Baueigentümlichkeiten, Strukturmodifikationen, zusätzliche Nukleolen und Vergleich mit den mitotischen Chromosomen, *Österr. Bot. Z.*, 114, 171, 1967.

128. Nagl, W., Banded polytene chromosomes in the legume *Phaseolus vulgaris*, *Nature*, 221, 70, 1969.

129. Nagl, W., Puffing of polytene chromosomes in a plant (*Phaseolus vulgaris*), *Naturwissenschaften*, 56, 221, 1969.

130. Tagliasacchi, A.M., Forino, L.M.C., Cionini, P.G., Cavallini, A., Durante, M., Cremonini, R., and Avanzi, S., Different structure of polytene chromosomes of *Phaseolus coccineus* suspensors during early embryogenesis. 3. Chromosome pair VI, *Protoplasma*, 122, 98, 1984.

131. Alpi, A., Tognoni, F., and D'Amato, F., Growth regulator levels in embryo and suspensor of *Phaseolus coccineus* at two stages of development, *Planta*, 127, 153, 1975.

132. Przybyllok, T., and Nagl, W., Auxin concentration in the embryo and suspensors of *Tropaeolum majus*, as determined by mass fragmentation (single ion detection), *Z. Pflanzenphysiol.*, 84, 463, 1977.

133. Picciarelli, P., Alpi, A., Pistelli, L., and Scalet, M., Gibberellin-like activity in suspensors of *Tropaeolum majus* L. and *Cytisus laburnum* L., *Planta*, 162, 566, 1984.

134. Lorenzi, R., Bennici, A., Cionini, P.G., Alpi, A., and D'Amato, F., Embryo-suspensor relations in *Phaseolus coccineus*: Cytokinins during seed development, *Planta*, 143, 59, 1978.

135. Perata, P., Picciarelli, P., and Alpi, A., Pattern of variations in abscisic acid content in suspensors, embryos, and integuments of developing *Phaseolus coccineus* seeds, *Plant Physiol.*, 94, 1776, 1990.

136. Cionini, P.G., Bennici, A., Alpi, A. and D'Amato, F., Suspensor, gibberellin and *in vitro* development of *Phaseolus coccineus* embryos, *Planta*, 131, 115, 1976.
137. Bennici, A., Cionini, P.G., and D'Amato, F., Callus formation from the suspensor of *Phaseolus coccineus* in hormone-free medium: A cytological and DNA cytophotometric study, *Protoplasma*, 89, 251, 1976.
138. Yeung, E.C., and Sussex, I.M., Embryogeny of *Phaseolus coccineus*: The suspensor and the growth of the embryo-proper *in vitro*, *Z. Pflanzenphysiol.*, 91, 423, 1979.
139. Gayler, K.R., and Sykes, G.E., β-conglycinins in developing soybean seeds, *Plant Physiol.*, 67, 958, 1981.
140. Meinke, D.W., Chen, J., and Beachy, R.N., Expression of storage-protein genes during soybean seed development, *Planta*, 153, 130, 1981.
141. Millerd, A., and Spencer, D., Changes in RNA-synthesizing activity and template activity in nuclei from cotyledons of developing pea seeds, *Austr. J. Plant Physiol.*, 1, 331, 1974.
142. Millerd, A., Simon, M., and Stern, H., Legumin synthesis in developing cotyledons of *Vicia faba* L., *Plant Physiol.*, 48, 419, 1971.
143. Gatehouse, J.A., Evans, I.M., Bown, D., Croy, R.R.D., and Boulter, D., Control of storage-protein synthesis during seed development in pea (*Pisum sativum* L.), *Biochem. J.*, 208, 119, 1982.
144. Sun, S.M., Mutschler, M.A., Bliss, F.A., and Hall, T.C., Protein synthesis and accumulation in bean cotyledons during growth, *Plant Physiol.*, 61, 918, 1978.
145. Dure, L., III, and Chlan, C, Developmental biochemistry of cotton seed embryogenesis and germination. XII. Purification and properties of principal storage proteins, *Plant Physiol.*, 68, 180, 1981.
146. Crouch, M.L., and Sussex, I.M., Development and storage-protein synthesis in *Brassica napus* L. embryos *in vivo* and *in vitro*, *Planta*, 153, 64, 1981.
147. This, P., Goffner, P.D., Raynal, M., Chartier, M., and Delseny, M., Characterization of major storage proteins of sunflower and their accumulation, *Plant Physiol. Biochem.*, 26, 125, 1988.
148. Chen, L.-J., and Luthe, D.S., Analysis of proteins from embryogenic and non-embryogenic rice (*Oryza sativa* L.) calli. *Plant Sci.*, 48, 181, 1987.
149. Chandra Sekhar, K.N., and DeMason, D.A., Quantitative ultrastructure and protein composition of date palm (*Phoenix dactylifera*) seeds: A comparative study of endosperm vs. embryo, *Amer. J. Bot.*, 75, 323, 1988.
150. Chandra Sekhar, K.N., and DeMason, D.A., A comparison of endosperm and embryo proteins of the palm *Washingtonia filifera*, *Amer. J. Bot.*, 75, 338, 1988.
151. Gifford, D.J., An electrophoretic analysis of the seed proteins from *Pinus monticola* and eight other species of pine, *Can. J. Bot.*, 66, 1808, 1988.
152. Misra, S., and Green, M.J., Developmental gene expression in conifer embryogenesis and germination. I. Seed proteins and protein body composition of mature embryo and the megagametophyte of white spruce (*Picea glauca* [Moench] Voss.), *Plant Sci.*, 68, 163, 1990.
153. Hakman, I., Stabel, P., Engström, P., and Eriksson, T., Storage protein accumulation during zygotic and somatic embryo development in *Picea abies* (Norway spruce), *Physiol. Plantarum*, 80, 441, 1990.
154. Roberts, D.R., Flinn, B.S., Webb, D.T., Webster, F.B., and Sutton, B.C.S., Characterization of immature embryos of interior spruce by SDS-PAGE and microscopy in relation to their competence for somatic embryogenesis, *Plant Cell Rep.*, 8, 285, 1989.
155. Dodd, M.C., van Staden, J., and Smith, M.T., Seed development in *Podocarpus henkelii*: An ultrastructural and biochemical study, *Ann. Bot.*, 64, 297, 1989.
156. Higgins, T.J.V., Synthesis and regulation of major proteins in seeds. *Ann. Rev. Plant Physiol.*, 35, 191, 1984.
157. Dure, L., III, Embryogenesis and gene expression during seed formation, *Oxford Surv. Plant Mol. Cell Biol.*, 2, 179, 1985.
158. Casey, R., Domoney, C., and Ellis, N., Legume storage proteins and their genes, *Oxford Surv. Plant Mol. Cell Biol.*, 3, 1, 1986.

114

159. Goldberg, R.B., Barker, S.J., and Perez-Grau, L., Regulation of gene expression during plant embryogenesis, *Cell*, 56, 149, 1989.

160. Whitmore, F.W., and Kriebel, H.B., Expression of a gene in *Pinus strobus* ovules associated with fertilization and early embryo development, *Can. J. For. Res.*, 17, 408, 1987.

161. Battaglia, E., Apomixis, in *Recent Advances in the Embryology of Angiosperms*, Maheshwari, P., Ed., International Society of Plant Morphologists, Delhi, 1963, Chapter 8.

162. Lacadena, J., Spontaneous and induced parthenogenesis and androgenesis, in *Haploids in Higher Plants. Advances and Potential*, Kasha, K.J., Ed., University of Guelph, Guelph, 1974, pp. 13–32.

163. Kasha, K.J., and Kao, K.N., High frequency haploid production in barley (*Hordeum vulgare* L.), *Nature*, 225, 874, 1970.

164. Puri, P., *In vitro* culture of floral organs of an apomict, *Aerva javanica* (Brum. F.) Spreng, *Phytomorphology*, 14, 564, 1964.

165. Yang, H.Y., and Zhou, C., *In vitro* induction of haploid plants from unpollinated ovaries and ovules. *Theor. Appl. Genet.* 63, 97, 1982.

166. Nagato, Y., Incorporation of ^{3}H-uridine and ^{3}H-leucine during early embryogenesis of rice and barley in caryopsis culture. *Plant Cell Physiol.*, 20, 765, 1979.

167. Meinke, D.W., Embryo-lethal mutants and the study of plant embryo development, *Oxford Surv. Plant Mol. Cell Biol.*, 3, 122, 1986.

168. Sheridan, W.F., and Clark, J.K., Maize embryogeny: A promising experimental system, *Trends Genet.*, 3, 3, 1987.

169. Racchi, M.L., Gavazzi, G., Monti, D., and Manitto, P., An analysis of the nutritional requirements of the *pro* mutant in *Zea mays*, *Plant Sci. Lett.*, 13, 357, 1978.

170. Sheridan, W.F., and Neuffer, M.G., Defective kernel mutants of maize. II. Morphological and embryo culture studies, *Genetics*, 95, 945, 1980.

171. Karssen, C.M., Brinkhorst-van der Swan, D.L.C., Breekland, A.E., and Koornneef, M., Induction of dormancy during seed development by endogenous abscisic acid: Studies on abscisic acid deficient genotypes of *Arabidopsis thaliana* (L.) Heynh., *Planta*, 157, 158, 1983.

172. Schneider, T., Dinkins, R., Robinson, K., Shellhammer, J., and Meinke, D.W., An embryo-lethal mutant of *Arabidopsis thaliana* is a biotin auxotroph, *Develop. Biol.*, 131, 161, 1989.

173. Goldberg, R.B., Hoschek, G., Ditta, G.S., and Breidenbach, R.W., Developmental regulation of cloned superabundant embryo mRNAs in soybean, *Develop. Biol.*, 83, 218, 1981.

174. Goldberg, R.B., Hoschek, G., and Vodkin, L.O., An insertion sequence blocks the expression of a soybean lectin gene, *Cell* 33, 465, 1983.

175. Walling, L., Drews, G.N., and Goldberg, R.B., Transcriptional and post-transcriptional regulation of soybean seed protein mRNA levels, *Proc. Natl. Acad. Sci. USA*, 83, 2123, 1986.

176. Galau, G.A., and Dure, L., III, Developmental biochemistry of cotton seed embryogenesis and germination: Changing messenger ribonucleic acid populations as shown by reciprocal heterologous complementary deoxyribonucleic acid-messenger ribonucleic acid hybridization, *Biochemistry*, 20, 4169, 1981.

177. Dure, L., III, Greenway, S.C., and Galau, G.A., Developmental biochemistry of cottonseed embryogenesis and germination: Changing messenger ribonucleic acid populations as shown by *in vitro* and *in vivo* protein synthesis, *Biochemistry*, 20, 4162, 1981.

178. Galau, G.A., and Hughes, D.W., Coordinate accumulation of homeologous transcripts of seven cotton *Lea* gene families during embryogenesis and germination, *Develop. Biol.*, 123, 213, 1987.

179. Hughes, D.W., and Galau, G.A., Temporally modular gene expression during cotyledon development, *Genes Develop.*, 3, 358, 1989.

180. Galau, G.A., Hughes, D.W., and Dure, L., III, Abscisic acid induction of cloned late embryogenesis-abundant (*Lea*) mRNAs, *Plant Mol. Biol.*, 7, 155, 1986.

181. Galau, G.A., Bijaisoradat, N., and Hughes, D.W., Accumulation kinetics of cotton late embryogenesis-abundant mRNAs and storage protein mRNAs: Coordinate regulation during embryogenesis and the role of abscisic acid, *Develop. Biol.*, 123, 198, 1987.

182. Pang, P.P., Pruitt, R.E., and Meyerowitz, E.M., Molecular cloning, genomic organization, expression and evolution of 12S seed storage protein genes of *Arabidopsis thaliana*, *Plant Mol. Biol.*, 11, 805, 1988.

183. Raynal, M., Depigny, D., Cooke, R., and Delseny, M., Characterization of a radish nuclear gene expressed during late seed maturation, *Plant Physiol.*, 91, 829, 1989.

184. Ishibashi, N., and Minamikawa, T., Molecular cloning and characterization of stored mRNA in cotyledons of *Vigna unguiculata*, *Plant Cell Physiol.*, 31, 39, 1990.

185. Borroto, K.E., and Dure, L., III, The expression of chloroplast genes during cotton embryogenesis, *Plant Mol. Biol.*, 7, 105, 1986.

186. Medford, J.I., and Sussex, I.M., Regulation of chlorophyll and Rubisco levels in embryonic cotyledons of *Phaseolus vulgaris*, *Planta*, 179, 309, 1989.

187. Finkelstein, R.R., Tenbarge, K.M., Shumway, J.E., and Crouch, M.L., Role of ABA in maturation of rapeseed embryos, *Plant Physiol.*, 78, 630, 1985.

188. Williamson, J.D., and Quatrano, R.S., ABA-regulation of two classes of embryo-specific sequences in mature wheat embryos, *Plant Physiol.*, 86, 208, 1988.

189. Bray, E.A., and Beachy, R.N., Regulation by ABA of β-conglycinin expression in cultured developing soybean cotyledons, *Plant Physiol.*, 79, 746, 1985.

190. Eisenberg, A.J., and Mascarenhas, J.P., Abscisic acid and the regulation of synthesis of specific seed proteins and their messenger RNAs during culture of soybean embryos, *Planta*, 166, 505, 1985.

191. Mundy, J., Hejgaard, J., Hansen, A., Hallgren, L., Jorgensen, K.G., and Munck, L., Differential synthesis *in vitro* of barley aleurone and starchy endosperm proteins, *Plant Physiol.*, 81, 630, 1986.

192. Olsen, O.-A., Jakobsen, K.S., and Schmelzer, E., Development of barley aleurone cells: Temporal and spatial patterns of accumulation of cell-specific mRNAs, *Planta*, 181, 462, 1990.

193. Goday, A., Sánchez-Martinez, D., Gómez, J., Puigdomènech, P., and Pagès, M., Gene expression in developing *Zea mays* embryos: Regulation by abscisic acid of a highly phosphorylated 23- to 25-kD group of proteins, *Plant Physiol.*, 88, 564, 1988.

194. Goffner, D., This, P., and Delseny, M., Effects of abscisic acid and osmotica on helianthinin gene expression in sunflower cotyledons *in vitro*, *Plant Sci.*, 66, 211, 1990.

195. Triplett, B.A., and Quatrano, R.S., Timing, localization, and control of wheat germ agglutinin synthesis in developing wheat embryos, *Develop. Biol.*, 91, 491, 1982.

196. Mansfield, M.A., and Raikhel, N.V., Abscisic acid enhances the transcription of wheat-germ agglutinin without altering its tissue-specific expression, *Planta*, 180, 548, 1990.

197. Stinissen, H.M., Peumans, W.J., and De Langhe, E., Abscisic acid promotes lectin biosynthesis in developing and germinating rice embryos, *Plant Cell Rep.*, 3, 55, 1984.

198. Raghavan, V., and Olmedilla, A., Spatial patterns of histone mRNA expression during grain development in rice, *Cell Diffn. Develop.*, 27, 183, 1989.

199. Ramachandran, C., and Raghavan, V., Intracellular localization of glutelin mRNA during grain development in rice, *J. Exp. Bot.*, 41, 393, 1990.

200. Perez-Grau, L., and Goldberg, R.B., Soybean seed protein genes are regulated spatially during embryogenesis, *Plant Cell*, 1, 1095, 1989.

201. Fernandez, D.E., Turner, F.R., and Crouch, M.L., *In situ* localization of storage protein mRNAs in developing meristems of *Brassica napus* embryos, *Development*, 111, 299, 1991.

4. Culture of Zygotic Embryos

MICHEL MONNIER

Contents

T.A. Thorpe (ed.), In Vitro Embryogenesis in Plants, pp. 117–153.
© 1995 *Kluwer Academic Publishers, Dordrecht. Printed in the Netherlands.*

I. Introduction

A. History of embryo culture

There is a profound difference in the historical development of plant tissue culture and that of embryo culture. Cultures of embryos preceded a long time before plant tissue culture. Tissue culture was definitely settled when two researchers Nobécourt and Gautheret, had the idea to supplement the medium with a hormone in 1937 [1, 2], but this addition had been delayed until Kögl et al. in 1934 discovered this hormone: indole-3-acetic acid [3]. The early beginning of immature embryo culture (1904) can be easily explained since the culture of embryos was achieved with a hormone-free medium containing only nutrients [4]. Comparing the two research developments, we must emphasize that embryo culture has been known for nearly a century and has slowly progressed with time while tissue culture, after a long period of limited progress, rapidly developed after 1937. From this date an explosion of discoveries has carried the technique all over the world, whereas embryo culture has kept the same steady pace in its advancement.

Plant embryology was formerly a descriptive science. The first investigators examined the sequence of segmentation within the embryo by means of histological sections. But rapidly numerous researchers wished to know more about the live embryo. They wanted to examine the step by step development of the embryo. Since the embryo is enclosed in parental tissues during its entire embryological life, and thus hidden from view, it became clear that the only way to obtain further information was to excise and grow it on an artificial medium.

Plant physiologists wondered if it was possible to grow immature embryos, on an entirely artificial medium, without any link with the mother plant. The earliest experimental embryologist was probably a Frenchman Charles Bonnet who excised and cultured decotylated mature embryos of beans in 1754 [5]. The first attempts with immature embryos were made by Hannig in 1904 who successfully grew cruciferous embryos on a simple medium and some of his results are still valid today [4]. He emphasized the importance of a high concentration of sugar to prevent the embryo from germinating before maturity. He described for the first time the precocious germination which leads to the formation of a weak plantlet which retains embryological characteristics.

In 1906 Brown studied the effect of various organic nitrogenous compounds and found that glutamine was particularly effective in promoting the shoot growth of barley embryos [6]. An application of embryo culture was found for the first time by Laibach who demonstrated the possibility of using this technique in genetic studies [7]. In certain interspecific crosses the fruit was found shrunken and the seeds non-viable, but when the embryos of such seeds were excised and culture *in vitro* they were able to give plants. LaRue, who attempted to grow embryos of different plants, encountered precocious germination in numerous species, especially in monocotyledonous embryos

[8, 9]. A significant application of embryo culture is the breakage of seed dormancy in order to shorten breeding cycles. Dormancy is circumvented by embryo excision and culture. In 1945 Randolph was able to shorten the *Iris* breeding cycle from years to months [10].

In 1934, after observing the development of the *Ginkgo* embryo in the seed at the endosperm's expense, Li added an extract of the reserves to the medium in order to stimulate the growth of embryos [11]. It became obvious that young embryos could benefit from plant extract supplements. On the basis of such considerations, Van Overbeek et al. added coconut milk (more correctly coconut water), which itself is a liquid endosperm, to the culture medium of *Datura* embryos [12]. They showed that it was necessary to sterilize this liquid by filtration to prevent heat-labile substances from being damaged [12]. Mention must be made of Tukey (1934), who found an original application for embryo culture. In early-ripening species of fruit trees and particularly some varieties of stone fruits, it seems that embryo growth cannot keep pace with the fast growth of the fruit. The embryos are not able to complete their development, the nucellus and integuments collapse, seeds appear shrivelled and are not viable. If the embryos are inoculated in a medium they can complete their embryogenesis *in vitro* and give plants [13]. Rijven in his cultures of immature *Capsella* embryos showed the importance of osmotic pressure *in vitro*, which must be similar to that *in situ* for proper development [14]. After much experimentation with amino acids, he concluded that glutamine was the best. For a comprehensive review of the effects of nitrogen compounds, reference can be made to Raghavan [15].

Norstog and Smith, in analysing the endosperm of bean ovules, discovered the presence of a high quantity of malic acid. By adding this organic acid to the medium they stimulated the growth of embryos of barley [16]. Monnier examined the mineral requirements of embryos considering at the same time their effects on growth and survival. He evolved a new well-balanced solution which had less toxicity and promoted growth of embryos [17]. The composition of the medium can actually modify the pattern of embryo development. When the mineral solution meets the requirements of the embryo, its differentiation is closer to the embryo developed *in situ* [18]. Monnier observed that during culture, there is not only a transformation in the morphology of the embryo but it is also accompanied by drastic physiological changes. Indeed the requirements of embryos vary with time. Thus, the culture of embryos is complex, and it differs profoundly from tissue or cell culture, since there is an important change of nutritional requirements from early to mature embryos [18, 19].

Monnier and Norstog also reported that the differentiation of embryos varies with the period of time spent in the ovule [20]. In addition, cultured embryos can undergo surgical operations similar to those performed with great success on animal embryos. Thus the destiny of embryological territories can be explored.

B. Advancement in the technique

The former research workers after successful culture of large and almost mature embryos, expected to obtain the development of smaller and younger embryos. They attempted to elucidate such problems as nutrition in order to obtain by careful adjustment of the medium fast growth and normal embryogenesis. Considerable progress has been made from that early start. On the whole, if we examine the history of the evolution of research, we notice that techniques and composition of media have been constantly improved, allowing investigators to grow smaller and thus younger embryos. One criterion which allows us to appreciate this advancement is to consider the minimum size of the embryo that it is possible to cultivate.

The improvement of the conditions of culture allowed LaRue in 1936 to successfully cultivate 500 μm long embryos [8]. In 1942, Van Overbeek et al. supplemented the medium with coconut milk and obtained plantlets from 150 μm embryos [21]. In 1963 Raghavan and Torrey [22] and in 1965 Norstog [23] reported the beginning of development of 60 μm embryos. In 1970, Monnier observed that the toxicity of the medium was caused by too high a concentration of some mineral salts [24]. After alterations of the concentration of some elements, he succeeded to develop a new medium in which the concentration of each element was carefully adjusted in order to increase survival of early embryos. In that way development of 50 μm embryos into plants was obtained [18–25].

C. Interest

By trying to reproduce *in vitro* the normal development of embryos, embryo culture appears as a tool to elucidate some fundamental *in situ* problems such as nutrition of the embryo in the ovule, control of embryogenesis, utility of the suspensor and change from developing to resting embryo in the seed. The interest is also based on the multiple applications of embryo culture in the improvement of plants.

1. Nutrition and embryogenesis

If the regular improvement of the composition of the culture medium in course of history is examined, it is observed that the number of components is increasing so that the medium meets better and better the requirements of the embryo. During the constant advancement in the composition of the medium, it has become progressively clear that, if a modification of the experimental conditions of embryo culture, provokes a more rapid growth, a higher survival and an improved embryogenesis, it can be assumed that the new conditions are closer to those that exist in the embryo sac. The culture of zygotic embryos, therefore, permits us to determine the conditions in which the embryo grows in the ovary. In fact these experimental results are confirmed by the data provided by the analysis of the composition of the ovular liquid surrounding the embryo [26].

2. Utility of the suspensor

The difficulty of excising small embryos and the lack of suitable culture media for them have delayed the successful culture of early-staged embryos. One of the major problems that have puzzled research workers for a long time is why an embryo of a few cells cannot survive in culture whereas a cell of the same size, isolated from a differentiated tissue, is easily cultured. This question has not yet been completely solved. However it is evident that the suspensor of the *in situ* embryo has not only a mechanical part to play but also a nutritive role which is essential in the early life of the embryo [27]. The functioning of the suspensor of the embryo in culture is probably arrested shortly after the inoculation.

3. Resting period

Recently Kononwicz and Janick tried to compare the development *in vitro* to the *in situ* growth where the embryo, during formation of seed, becomes dehydrated and enters dormancy. They observed that in a certain medium they can reproduce the accumulation of reserves as in a seed [28]. Finkelstein and Crouch studied the possibility to monitor the embryogeny and germination processes by modifying the concentration of sucrose in the medium. The transition is appreciated by looking at the morphology and also analysing cotyledonary reserves [29].

4. Applications

The main application of embryo culture is certainly the rescue of hybrid embryos which abort on the mother plants after particular crosses. The method introduced by Laibach in 1925 has remained the most powerful process to obtain new hybrid plants [8].

Embryo culture has several recent new applications. For instance the production of haploids. Kasha and Kao have developed a technique to produce barley haploids. In the embryo resulting from interspecific crosses there is an elimination of one of the genomes, and a haploid is obtained [30].

Culture of immature embryos in a medium supplemented with hormones at a high concentration disturbs organogenesis and rejuvenates the cells which acquire the potentialities of the zygote. After transfer of the tissue onto a hormone-free medium, a multitude of embryos appear. The immature embryo becomes a good source of somatic embryos [31, 32].

Hybridization is a method which unfortunately has drawbacks. It can bring in the new plant numerous characters but a number of them can be uninteresting or disadvantageous. Modification of the genome by means of transfer of genes seems a better alternative [33].

II. Culture technique

A. Basic method

1. Plant material

a. Choice of plant

The most interesting plant for making fundamental research is certainly *Capsella bursa-pastoris*. It is a classical material since the embryo has been well studied by Hanstein [34], Schaffner [35], Souèges [36], and more recently the ultrastructural aspect was reported by Pollock and Jensen [37].

Qualities of this plant are multiple. Initially, when Hanstein wanted to study aspects of the embryo *in toto*, he chose *Capsella* because embryos are easy to extract since the endosperm stays liquid for a long time [34]. Taking this important work as a basis, Souèges accurately described embryogenesis of the *Capsella* embryo by means of histological sectioning and put forward the remarkable homogeneity of development from one plant to another [36–38]. Later when the development of tissue culture stimulated embryo culture studies,

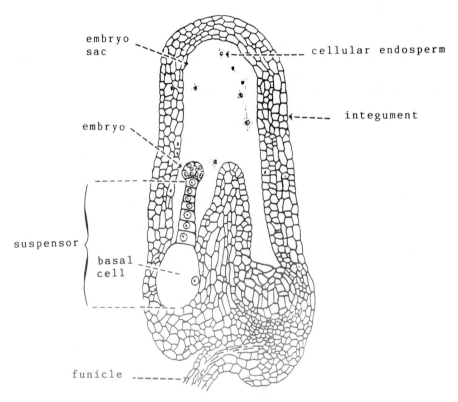

Fig. 1. Median section of a *Capsella bursa-pastoris* ovule.

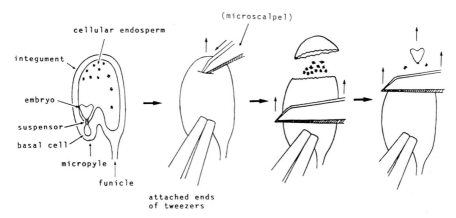

Fig. 2. Dissection of an ovule of *Capsella* to remove a heart-shaped embryo prior to culture. The dissection is made in a drop of sucrose solution.

numerous research workers selected this plant because of its well described embryogenesis and the long lasting nature of the liquid endosperm of its ovules (Fig. 1).

b. Inoculation

The advantage of *Capsella* is that the flowers are arranged along a long terminal inflorescence and that, from the top to the bottom, the fruits offer different stages of maturity. Therefore it is possible to choose the right stage of embryo by considering the size of the fruit. The fruit is then collected, sterilized with a calcium hypochlorite solution and washed in sterile water. Further operations are carried out under a dissecting microscope, on a slide, in a few drops of a sucrose solution to prevent desiccation. The fruit is opened with forceps and then ovules can be removed. The tissue within the ovule, in which the embryo is embedded, is already sterile. The ovule is isolated in a drop of sucrose solution and with the help of a sharp mounted blade (microscalpel), the ovule is cut at the top (Fig. 2). A slight pressure with the attached ends of tweezers and the blade breaks the suspensor, separates the embryo from the surrounding ovular tissues and liberates it in the sucrose solution. The excised embryo is transferred by a micropipette to a Petri dish containing agar medium (see Monnier) [18]. This Petri dish is engraved to spot the embryos.

2. Nutrient medium

a. Mineral solution

Growth and survival of embryos are considerably affected by the composition of the medium.

– Mineral solution: All plant growth media cannot be utilized for culture of embryos. Monnier has pointed out that embryos are quite sensitive to the mineral solution used (Fig. 3) [24, 25]. One of the first used mineral solutions

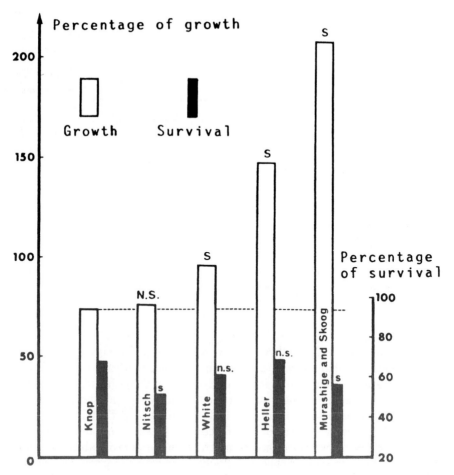

Fig. 3. Effect of different mineral solutions on the growth and survival of immature *Capsella* embryos. The Murashige and Skoog solution which enhances considerably growth has a certain toxicity for smaller embryos. S means significantly different from the Knop solution (risk 5%) and n.s. non-significantly different.

was Knop's solution by Dieterich [39]. The mineral solution of Knop [40] was generally diluted by half for plant tissue culture use, and at this concentration it is not perfectly suitable to cultivate small embryos (Fig. 4). It induces generally only limited growth which is slow in comparison to the growth *in situ* [41]. Besides, this growth is aberrant because the differentiation of the embryo is delayed when compared to an *in situ* embryo of the same size, and in addition, the smallest embryos generally die before they reach maturity. This slowness in their differentiation becomes particularly evident when the ratio length/width of the embryo and the histological development is considered [18].

The study of mineral nutrition shows that growth and survival do not exactly tally (Fig. 3) [18–24]. The mineral solutions which promote growth are toxic,

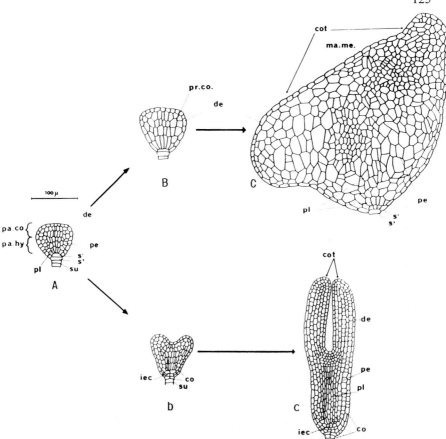

Fig. 4. Comparison of the development of a late globular-staged embryo of *Capsella in vitro* and *in situ*. (A,B,C) Embryo cultivated *in vitro* on a medium containing Knop mineral solution. (A) Embryo excised at the time of inoculation. (B) Enlargement of the cells. (C) Multiplication of the cells and formation of two unequal cotyledons. It can be noticed the massive and minimum differentiation of the embryo at the end of the culture. The structure is different from the regular, symmetrical morphology of the embryo developed *in situ*, which has the same size but is thinner and possesses two well developed cotyledons (c). The differentiation of the C embryo can be compared to the b one. (A,b,c) Evolution of an embryo *in situ*. (A) Late globular embryo. (b) Early heart-shaped embryo. (C) Claw-shaped embryo; pa. co. = cotyledonary region; pa. hy. = hypocotyledonary region; de. = dermatogen; pe. = periblem; pl. = plerome; s₁, s₂ = hypophysis; su. = suspensor; pr. co. = cotyledonary primordium; cot. = cotyledons; ma. me. = meristem; iec = position of initials of root cortex; co. = cap.

and on the contrary non-toxic solutions, such as Knop's solution, are able to induce normal development (Fig. 4). To obtain a new solution which promoted the best growth with minimum toxicity the following method was employed. After the choice of a particularly growth-promoting solution the concentration of every salt or element was altered to examine the effect of such a variation. In this way a new mineral solution was developed, whose action upon growth was

126

Table 1. Composition of the new solution (concentration in mg/l)

KNO_3	=	1990	H_3BO_3	=	12.4
$CaCl_2 \cdot 2H_2O$	=	880	$MnSO_4 \cdot H_2O$	=	33.6
NH_4NO_3	=	825	$ZnSO_4 \cdot 7H_2O$	=	21.0
$MgSO_4 \cdot 7H_2O$	=	370	KI	=	1.66
KCl	=	350	$Na_2MoO_4 \cdot 2H_2O$	=	0.5
KH_2PO_4	=	170	$CuSO_4 \cdot 5H_2O$	=	0.05
Na_2 EDTA	=	14.9*	$CoCl_2 \cdot 6H_2O$	=	0.05
$FeSO_4 \cdot 7H_2O$	=	11.1*			

* 2 ml of a stock solution containing 5.57 g $FeSO_4 \cdot 7H_2O$ and 7.45 g of Na_2EDTA/l. Generally stock solutions are prepared in advance at $100\times$ concentration.

identical to the best of the original solutions, but which happened to be less toxic than the other solutions commonly used for tissue culture [18].

Murashige and Skoog's salt formulation [42] was chosen for study as it provided the best growth for culture of embryos, but it appeared to possess a certain toxicity for the smallest or youngest embryos (Fig. 3). Two elements particularly influence the rate of embryo survival: iron and calcium. Iron is

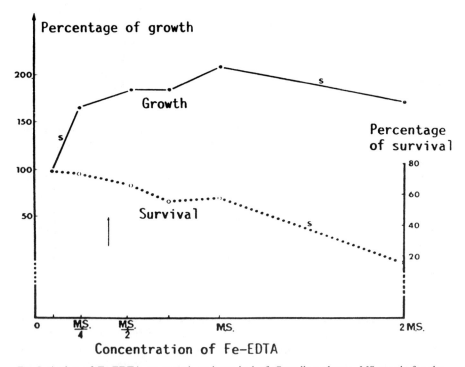

Fig. 5. Action of Fe-EDTA on growth and survival of *Capsella* embryos. MS stands for the concentration of the substance in the Murashige and Skoog solution. The toxicity of iron is manifest.

always added as Fe-EDTA which is more easily absorbed by plants (Fig. 5). The concentration marked MS on the abscissa is the one used in the solution of Murashige and Skoog. The doubling of this concentration (2 MS) provoked a significant decrease of survival (S). When the quantity of Fe-EDTA was lowered, it can be observed that survival increased regularly. Iron cannot be completely removed from the solution because this element is essential for the growth of embryos. In the new solution which was evolved, the Fe-EDTA concentration amounts to 2/3 of Murashige and Skoog's. Such a diminution of the concentration caused a reduction of toxicity without an excessive decrease of growth. The other element calcium exerts a positive action on survival (Fig. 6). The effect is observed by using $CaCl_2$, but since it is generally considered that Cl^- has a negligible effect on plant growth, it is probable that the action of $CaCl_2$ on survival is mainly due to the ion Ca^{++}. Figure 6 shows that survival was greater with increasing Ca^{++} concentrations. This increase of survival is significant (S). The ion calcium is generally considered to have antitoxic properties. It has a protective action on embryos and it decreases cellular permeability. The concentration chosen to develop the new solution is twice that contained in the Murashige and Skoog solution. The study of the effect of

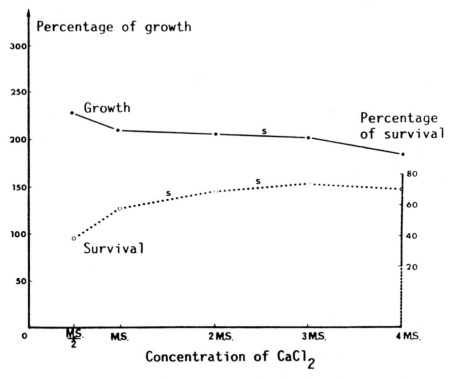

Fig. 6. Action of $CaCl_2$ on growth and survival of *Capsella* embryos. Ca^{++} has a significant (= S) beneficial effect on survival.

the total quantity of microelements showed that this solution can be easily employed at a double concentration without any toxic effect.

The examination of the other elements or salts of the Murashige and Skoog solution provided useful information on growth and survival. Different observations were made:
– The increase of the concentration of potassium enhances the growth to a large extent without effect on survival.
– Nitrate appeared to be toxic and its concentration was reduced.

The final composition of this new solution is shown in Table 1.

b. Other elements

This solution is supplemented with sucrose (120 g/l). This high concentration was recommended by Rijven who demonstrated that it equilibrated the osmotic pressures of embryo cells [14]. Examination of embryo sucrose requirement showed that this concentration was perfectly suitable [19]. Glutamine was added at the concentration of 400 mg/l. This amino acid was found very effective by Rijven [14]. Two vitamins B_1 and B_6, both at the concentration of 1 mg per liter, are put in the medium. These two vitamins were used by Raghavan and Torrey [22]. Their concentrations have been multiplied by 10 because it was noticed by Linsmaier and Skoog that after sterilisation by autoclaving, they were partly degraded [43]. Difco Bacto-Agar at the dose of 7 g/l is used for two reasons. First, the agar is transparent enough to see the embryos through the bottom of the Petri dish when the lid is dimmed with vapour. Secondly the survival is generally better with that partially purified agar. This new solution happened to promote the growth of the embryos as well as Murashige and Skoog's, but its toxicity was considerably reduced (Fig. 7, improved medium) [18].

The method of sterilization of the medium is of importance. It has been observed that the temperature of sterilization has an effect on the development of embryos. A low temperature does not alter the elements and ensures better growth and survival of embryos [44]. Sterilization of the medium is carried out by heating twice at 100 °C (tyndallisation), with an interval of 24 h separating these two heatings. When the dissection is made in soft agar (1.75 g/l) another improvement of growth and survival is obtained (Fig. 7, improved culture) [18].

3. Measurements

The action of this improved mineral solution was tested on growth and survival of immature *Capsella* embryos (Fig. 7, improved medium) [18]. The length of the embryos was measured after 6 days with a micrometer gauge. The embryos were cultured in the dark at 25 °C. Generally, when the effect of a factor on the development of embryos is examined, it is necessary to inoculate two types of embryos: large embryos whose growth follows statistical laws and small embryos that are very fragile and whose survival varies according to the toxicity of the medium [24]. For example, in studies on *Capsella* embryos 160 μm long embryos were inoculated to examine growth and 65 μm long embryos to have

PERCENTAGE OF GROWTH

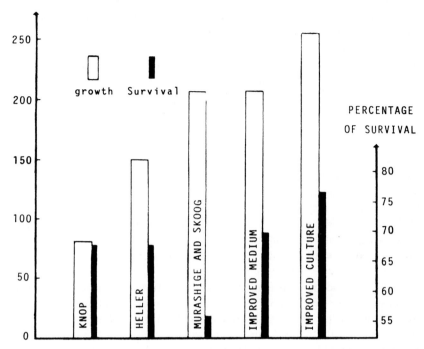

Fig. 7. Effect of various mineral solutions on growth and survival of immature embryos of *Capsella*. The term "improved medium" corresponds to the new mineral solution. The expression "improved culture" includes the improved medium and a change in the culture technique (agar coating of embryos and tyndallisation of the medium).

information about survival. Growth is measured by calculating the increase of size according to the initial length, that is to say the ratio $L_6 - L_0/L_0$, where L_6 = length after 6 days, L_0 = length at the inoculation time. This ratio is multiplied by 100 to convert to a percentage of growth. The percentage of survival is obtained by the ratio: number of embryos grown/number of transplanted embryos × 100.

B. Variation with the basic method

1. Inoculation of different explants

a. Immature hybrid embryos

When culture of hybrid embryos is attempted, emasculation and bagging are necessary prior to maturation of the flowers. Pollination is made by hand, and excision of the embryos is performed after a certain time when the embryo has reached a sufficient size. This period of necessary growth of the embryo inside the fruit varies considerably with the plant and the development must be followed by a series of successive dissections. Embryos are found to be more ontogenetically advanced as time progresses, and simultaneously, prospective survival in culture *in vitro* increases with their size. But another factor must be considered, the number of aborted embryos inside the fruit increases over time. Generally, the best compromise is when the embryo reaches the heart-shaped stage. If abortion occurs too early when the embryos are still globular, at a stage impossible to rescue, a preliminary culture of the entire ovary can sustain development of the embryos up to the heart stage.

b. Seed embryos

The excision of embryos from seeds offers two advantages.

First, as the behaviour of embryos cultured *in vitro* depends essentially on the size of embryos when inoculated, their stage must be carefully chosen. One of the methods allowing for control of the right stage of embryos is to work with species whose embryos are continually growing in the seed, e.g. *Gingko biloba* (Gymnosperm) [45]. Angiosperm embryos like *Eranthis hiemalis* also exhibit the same feature which permitted Haccius to examine the effect of several substances on undifferentiated embryos without culturing them on a medium [46].

Secondly, when it is difficult to obtain immature embryos on a regular basis because they are supplied only seasonally by flowering plants, seeds constitute a good material providing embryos all the year round. However these embryos are mature, and it is necessary to excise from the seed, the embryo without cotyledons. The decotylated embryos which are also called embryonic axes are really dependent on the nutrient elements and can testify to the qualities of the medium [62].

2. Alternative media

a. Support

Rijven cultured immature embryos in a drop of liquid medium placed between two slides [14], but agar-solidified medium gives better results. It is preferable to use the Difco Bacto-Agar which stimulates the growth of the embryos and is transparent enough to see them through the medium. Other types of support were tested. Polyacrylamide gel, used generally for electrophoresis, can be employed after having been thoroughly washed [47]. This support is particularly transparent.

b. Carbohydrate source

The best energy source for embryo culture is certainly sucrose. Embryos are considered to need a concentration of sugar much higher than what was generally used in the media for plant tissue culture. High concentrations of sucrose (8–12%) are used which approximate the high osmotic potential of the intracellular environment of the young embryo sac. Rijven demonstrated that this high osmolarity prevents precocious germination, which is characterized by a too early development of the apical meristem and formation of rootlets [14]. If the concentration of sucrose is too low, normal embryogenesis is deeply disturbed. The use of mannitol has been proposed as an osmoticum [14]. The optimal concentration of sucrose for the fastest growth of embryos varies according to the length of the embryo when it is excised and inoculated [19]. The concentration suitable for heart-shaped embryos is about 120 g/l.

c. Mineral solution

Using the above information many useful studies examining various morphogenetic, physiological and biochemical aspects have been carried out with embryos from a variety of species (Table 2). These studies have shown that certain basic requirements are indispensable. Although, the effects of these various media are certainly different on growth, survival and differentiation of embryos, until now it has not been proved that the requirements of embryos differ profoundly from one species to another.

Table 2. Compilation of different media developed for various species.

Plant Species	Purpose of Study	Medium Used	References
Brassica napus	Storage-protein synthesis	Monnier's medium, sucrose 12%	Crouch and Sussex [48]
	Precocious germination	Monnier's medium, sucrose 1% and 12% + sorbitol	Finkelstein and Crouch [29]
Capsella bursa-pastoris	Improvement of growth	Olsen's minerals, sucrose 12%	Rijven [14]
	Survival of globular embryos	Modified Robbins and Schmidt's medium, sucrose 2%	Raghavan and Torrey [22]
	Comparison *in vitro* and *in situ*	Knop's medium, sucrose 12%	Monnier [41]
	Growth and survival	Monnier's medium, sucrose 12%	Monnier [17]
	Action of oxygen	Monnier's medium, sucrose 12%	Monnier [49]
	Effect of various factors	Monnier's medium, sucrose 12%	Monnier [18]
	Variation of requirements	Monnier's medium, sucrose 12%	Monnier [19]
	Frost-resistance	Monnier's medium, sucrose 30%	Monnier and Leddet [50]
	Culture of ovules	Monnier's medium, sucrose 2%	Lagriffol and Monnier [51]
Citrus microcarpa	Growth of proembryos	Modified White's medium, sucrose 10%	Rangaswamy [52]
Corchorus olitorius	Action of IAA and GA	White's medium and yeast extract, sucrose 5%	Iyer et al. [53]
Cucumis sativus	Effect of endosperm substances	White's medium, sucrose 5%	Nakajima [54]

Table 2. Continued

Plant Species	Purpose of Study	Medium Used	References
Datura stramonium	Effect of coconut milk	Tukey's medium, dextrose 1%	Van Overbeek et al. [21]
	Effect of quantity of sucrose	Modified Randolph and Cox's medium, sucrose 8–12%	Rietsema et al. [55]
Datura tatula	Effect of diffusates from seeds	Modified White's medium, sucrose 8%	Matsubara [56]
	Effect of different factors	Modified White's medium, sucrose 8%	Matsubara and Nakajira [57]
Gossypium hirsutum	Effects of malic acid	Modified White's medium, sucrose 2%	Mauney et al. [58]
Hordeum vulgare	Effect of endosperm	Randolph and Cox's medium, sucrose 12%	Ziebur and Brink [59]
	Growth and differentiation	White's medium, sucrose 12%	Norstog [60]
	Effect of coconut milk	White's medium with phosphate, sucrose 9%	Norstog and Smith [16]
	Composition of a new medium	Norstog's medium, sucrose 3%	Norstog [61]
Phaseolus vulgaris	Effect of extract of cotyledons	Monnier's medium, sucrose 3%	Monnier [62]
Pinus mugo	Effect of conditioning	Halperin's medium, sucrose 2%	Thomas [63]
Prunus avium	Rescue of embryos of fruit-tree	Tukey's medium, glucose 0.5%	Tukey [64]
Trifolium repens	Effect of *in vitro* pollinisation	Gamborg's B5 medium, sucrose 3%	Leduc et al. [65]
Triticum × *Aegilops*	Overcoming inviability	Monnier's medium, sucrose 12%	Chueca et al. [66]
Zamia pumila	Action of surgical operations	Norstog's medium, sucrose 12%	Monnier and Norstog [67]
Zea mays	Effect of coconut milk and amino acids	Tukey's medium, 5% sucrose	Haagen-Smit [68]

d. Supplementary elements

Among various explants cultivated *in vitro*, the embryo is probably the most sensitive to different substances added to the medium. Possibly its capacity to grow very quickly makes it a perfect testimony of the nutritive quality of the medium.

– Nitrogen compounds. It seems that very young immature embryos do not have the active enzyme system to reduce nitrate to nitrite and then to ammonium nitrogen. This explains the prominent part played by reduced nitrogen in embryo nutrition. Glutamine and asparagine have proved to be very efficient sources of nitrogen [69]. Glutamine was reported being the best by many

authors. Recently other amino acids, such as proline and serine were tested and it was shown that they stimulated somatic embryo growth [70]. Sometimes, inorganic nitrogen such as ammonium sulfate can produce growth increments of the embryo of the same magnitude as those observed on embryos growing in a mixture of amino acids [71]. When the embryos grow older they reach a metabolic autonomy by acquiring nitrate reductase. In that case potassium nitrate can be the sole nitrogen source [72]. Nevertheless, since the embryo develops in culture and that its mechanism of nitrogen utilization can be changing, it is preferable to supplement the medium with different sources of nitrogen. It is also possible to meet the nitrogen requirement of the embryo by the amino acid complex of casein hydrolysate [18].

– *Hormones.* In many cases, exogenously supplied hormones are not required for embryo culture. Embryos can be considered as a plant with its own endogenous hormones so it is thought that it is not necessary to supplement the medium with hormonal substances [18]. When growth substances are added to the medium at usual concentrations, they induce modification in the ontogenic pattern of embryos; including suppression of root growth and precocious leaf expansion with kinetin, and longer and thinner embryos with gibberellins [73, 74]. Nevertheless, there are some cases in which low concentrations of hormones in the medium have facilitated embryo culture. It seems that for certain species, natural endosperms do contain hormones and so the culture medium, to better reproduce the conditions of the *in ovulo* environment, may be supplemented with hormones at a very low level.

– *Vitamins.* Commonly added vitamins are thiamine, myo-Inositol, ascorbic acid, nicotinic acid, pyridoxine and pantothenic acid. The first ones have been proved to be the most essential [43].

– *Organic acids.* Organic acids, such as malic, pyruvic and citric acids, were found to be effective when used as supplements together in the medium, especially when decotylated mature embryos of beans are cultivated (Monnier, unpublished).

– *Plant extracts.* When embryo development, in the cultivated ovule, was compared with or without the surrounding ovary tissues still attached, it was observed that with the tissues, growth and survival was increased [51]. This fact is probably due to the presence of elements in the tissue which are not in the medium. These substances cannot be easily extracted because they may be labile and destroyed during the extraction. In addition, inhibitors are likely to appear by oxidation of certain compounds in contact with the oxygen of the air and prevent the extract from stimulating embryo development. Nevertheless, some extracts promote growth of embryos. Coconut milk and its action on the embryo of *Datura* [12] and barley are well-known [75]. Matsubara tried alcohol diffusates of young seeds [56]. He found that the activity in extracts of *Lupinus* seeds was twice as effective as coconut milk. Also cotyledons of beans appeared to be a good source of growth-promoting substances, if they were previously cultivated on a medium composed of only agar and water. After a 12-day culture period, their extract considerably enhanced the growth of either mature ecotyledonous bean embryos [62] or immature *Capsella* embryos [76]. It seems that embryo-growth factors accumulate in the isolated cotyledon since their transport is precluded as the embryonic axis had been previously removed [62]. After 45 days of culture on a medium supplemented with an extract of cotyledons, decotylated embryos of beans weighed 6 times more than the embryos on Murashige and Skoog medium [62].

All these results suggest three conclusions: first of all, embryos are very sensitive to the action of plant extracts, and they can be used as a tool to determine the effective substances in a plant. Secondly, the enhancement of growth obtained with plant extracts proves that completely synthetic media lack certain substances [77]. Thirdly, it can be assumed that the better development of embryos obtained with a medium supplemented with plant extracts is probably due to the fact that, by this addition, the composition of the medium is closer to that of the ovular sap.

In that way, embryo culture allows us to acquire a good knowledge of the conditions of the natural environment which surrounds the embryo. For example, embryos greatly differ from tissues in culture which require only a low level of sugars. They need to be cultivated in a medium which contains a large concentration of sucrose. Indeed, the analysis of the composition of the liquid in the embryo sac revealed that the amount of sugars was very high and thus provided a high osmotic pressure [26].

3. Variable medium

a. Change of requirements during culture

If a mature embryo in a seed is completely self-sufficient, it can be surmised that previously, when it starts developing in the ovule, it is largely food-dependent on maternal tissue, and that it becomes progressively autonomous during seed formation. This explains the difficulty in finding a suitable culture medium for very early embryos and shows that they are probably supplied *in situ* with special substances, which are still unknown presently. In Figure 8 the optimal concentration of sucrose for each type of embryo inoculated is plotted. As it can be seen, the longer the inoculated embryos are, the lower the concentration of sucrose they require. It can be thought that, *in vitro*, in the same manner, the increasing length is also accompanied by physiological change. Indeed as embryos grow in the medium, there is a continuous change in their requirements, so it is useful to adjust the concentration of elements over the course of time. They need a constant modification in the concentration of mineral salts, sucrose and amino acids [19].

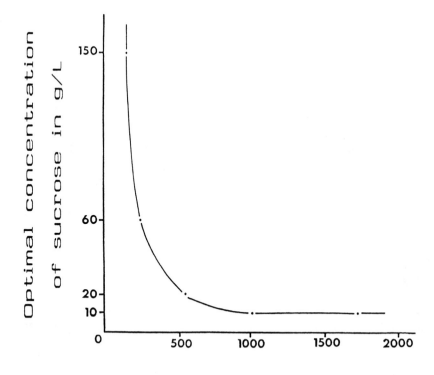

Fig. 8. Variation of the optimal concentration of sucrose according to the initial length of the immature embryo of *Capsella* at the time of inoculation.

b. Development of a culture device

Monnier engineered a device which eliminates the need for a sequential transfer from one medium to another [18]. This device allows the juxtaposition of two media of distinct composition (Fig. 9). The first agar medium is liquefied by

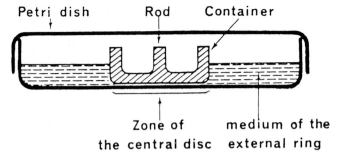

Fig. 9. Device allowing the juxtaposition of two media with different composition. The first medium is poured around the glass container when it is hot (external ring). After solidification, the central container is removed and a second medium of a different composition is poured in the hole. Embryos are cultivated on the area of the central disk and grow on a variable medium.

heating and then poured around a small central glass container in the center of a Petri dish. This medium constitutes the external ring. Upon solidification of the first medium, the central container is removed and a second medium of a different composition is poured into the hole. The central medium contains a large amount of sucrose (180 g/l), of calcium (protection of the embryo), a low concentration of mineral salts, no iron but amino acids in large quantity. The external ring contains mineral salts at a high concentration, a large amount of iron but no sucrose and no amino acids. Embryos are cultured on the medium in the center of the Petri dish, and as a result of diffusion, they are gradually subjected to the action of a variable medium (Fig. 10) [18]. This device makes the transition from heterotrophy to autotrophy easier for the embryo, while biochemical mechanisms are progressively being organized to sustain physiological activities. Thanks to this new method of culture, for the first time, the complete development of very young embryos (from about 50 μm long) up to germination was obtained. At the beginning of the culture, the embryo is globular and the appearance of the cotyledons is an event of great interest.

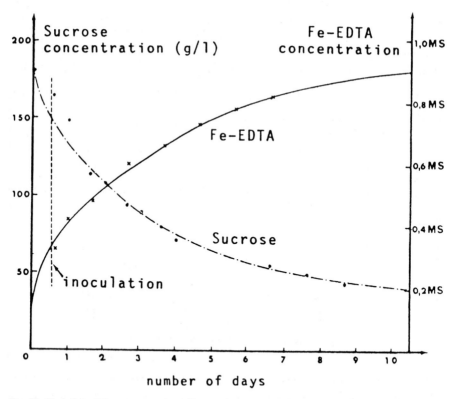

Fig. 10. Variation of the concentration of iron and sucrose in the central disc when two media have been juxtaposed. The sucrose in the disc diffuses towards the external ring and on the contrary iron goes into the central area.

4. Mimicking the normal development

a. Precocious germination

When the embryo in culture reaches the size of a mature embryo, it must be inoculated onto a medium containing a low concentration of sucrose. If this transfer is made before the state of maturity the embryo germinates, although it has not completed its embryogenesis. This particular germination is called precocious germination and is characterized by a premature elongation of hypocotyl cells and development of the shoot apex, while the radicle elongates. But as was reported in *Brassica*, the plantlet retains embryonic characteristics. Secondary cotyledons are produced and, in them, the accumulation of a specific storage protein of *Brassica* (cruciferin) can be detected [29].

b. Rest period

The use of the device (Fig. 9) ensures the complete development of the embryo up to a plant, without any period of rest. As there is no appearance of dormancy, germination immediately follows the embryo development. This property is used to shorten the breeding cycle especially in research laboratories looking to obtain the rapid growth of a plant, in order to verify quickly the effect of some experiments. However this development without a rest period is abnormal. Generally, *in situ*, embryogenesis is stopped and dehydration occurs (Fig. 11). During this period of maturation, polymeric reserves accumulate, the

Stages	Embryogenesis	Rest period	Germination
Nature of embryo	immature	mature	seedling
Embryo change	multiplication of small cells accumulation of reserves	presence of: dehydrated cells, reserves, dormancy.	hydration, enlargement and multiplication of cells. mobilisation of reserves, acquisition photo and geo-tropism.

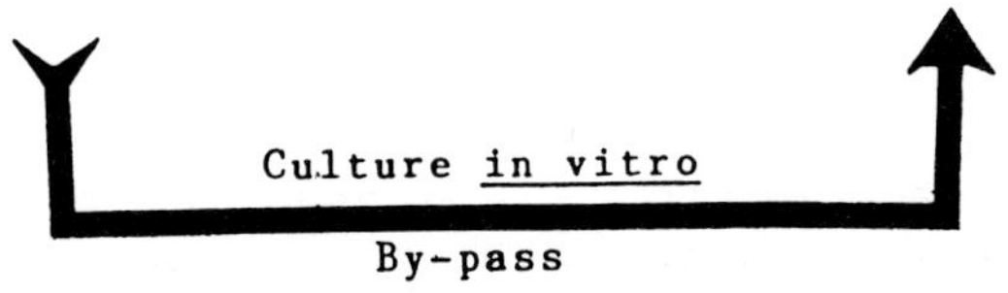

Culture in vitro

By-pass

Fig. 11. Embryo passes through different stages *in situ* which are not completely reproduced in culture *in vitro*. The rest period of the embryo in the seed is suppressed. The *in vitro* culture of the embryo in a high osmotic pressure can induce the formation of reserves and the acquisition of frost resistance characteristic of the embryo in the seed.

138

integuments of the ovule harden, and the seed is constituted. At the same time dormancy may occur. Embryo culture avoids this stage. The state of the embryo in the seed can be reproduced in embryo culture by cultivating the embryo in a high concentration of sucrose (300 g/l) when it becomes mature [78]. This treatment induces transformation in the embryo that mimics the developmental pattern characteristic of *in situ* embryo during formation of the seed. There is a decrease of growth [78], an accumulation of reserves for instance anthocyanins, alkaloids and lipids in cacao embryos [28] and an acquisition of frost resistance nearly equal to that of the seed (Fig. 12) [50]. In the same way *Brassica* embryos cultivated on a medium with high osmoticum (sorbitol) mimic the normal development, i.e. there is suppression of germination, continuation of synthesis of a storage cotyledonary protein called cruciferin and maintenance of cruciferin mRNA at its normal level [79].

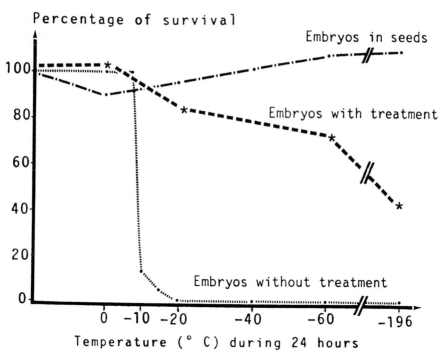

Fig. 12. Acquisition of frost resistance of embryos of *Capsella*. To measure it, the embryos are submitted to a low temperature during a 24-h period. After action of a high concentration of sucrose (embryos with treatment) their resistance is superior to the embryos in the ovule (embryos without treatment) and mimics that of embryos in seeds.

III. Survival of embryos

At the present state of knowledge, it is not possible to culture small isolated embryos composed of only a few cells. They do not survive in culture and generally die as soon as they are inoculated. The problem of their culture has not yet been solved, and study of their survival shows that it varies with different conditions of culture.

A. Effect of the size

The culture of early embryos like globular embryos is much more difficult than older stages (heart and torpedo stage). This problem was studied on immature *Capsella* embryos [80]. The action of the new mineral solution was tested on the survival of embryos of various sizes. It was observed that the survival of embryos increases with their size (Fig. 13, isolated embryo). When the embryo is longer than 200 μm, it is not influenced any more by the culture conditions. The new solution supplies the best survival for the embryos (Fig. 7) but, in spite of the improvement in the composition of the nutritive solution, globular embryos of 25–30 μm do not develop (Fig. 13, isolated embryo). In order to

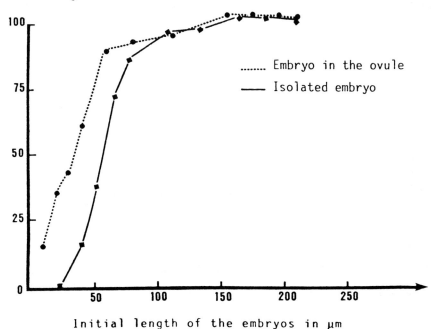

Fig. 13. Survival rate of immature embryos of *Capsella* cultured either isolated (embryo culture) or growing in ovule cultured *in vitro* (ovule culture). The better survival of the embryo in the ovule may be due to the integrity of the suspensor.

explain this fact, we tried to compare the survival of embryos cultivated isolated on the medium and embryos inside ovules cultured *in vitro*.

B. Action of ovular environment

Ovules of different sizes were inoculated on the medium supplemented with 80 g/l of sucrose. This concentration was found optimal for the culture of ovules [81]. It is observed (Fig. 13) that the survival rate of embryos growing inside the ovules is always higher than that of isolated embryos. Although it is not possible to cultivate isolated embryos of 25 μm, these embryos survive in ovules cultured *in vitro*. In this case the survival rate reaches 40% [80]. This technique is particularly useful when hybrid embryos are to be cultivated. Quite often abortion occurs very early, when the embryo is too small to be cultivated directly on the medium with any chance of success. One way to rescue the embryo is to culture the whole ovule. Along the same line, it was reported that a preliminary culture of the ovary can delay the abortion of hybrid embryos, and the embryo culture can be attempted later with larger embryos whose survival is greater [82].

C. Conditioning dependence

We examined the effect of a previous conditioning of the medium with four nurse-cultures (Fig. 14) on the survival of 30 μm *Capsella* embryos. In Table 3 it can be seen that in a medium, conditioned by tobacco tissue, the embryos have a significant higher survival (Monnier, unpublished). This observation had already been reported by Thomas working on pine embryos [63]. He found that very young undifferentiated pine embryos could only survive in a medium in which pine tissues were previously cultivated [63]. These nurse-cultures have conditioned the medium to support the growth of the delicate embryos. In this case (pine embryos) a considerable lengthening of the suspensor was observed, thus mimicking the development *in situ*.

Table 3. Effect of the presence of nurse cultures to condition the medium. Three media have been tested. The reduced deviation is calculated between conditioned and non-conditioned medium (risk 5%). When it is superior to 1.96 the difference is significant.

Concentration of Sucrose	Without Conditioning	Reduced Deviation	With Conditioning
30 g/l	$\frac{0}{47} = 0\%$	5.40	$\frac{26}{56} = 46\%$
120 g/l	$\frac{0}{55} = 0\%$	2.42	$\frac{6}{59} = 10\%$
30 g/l without Fe	$\frac{0}{50} = 0\%$	3.02	$\frac{10}{60} = 16\%$

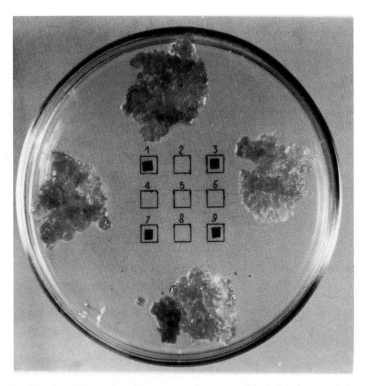

Fig. 14. Conditioning of the medium. Four nurse cultures are cultivated during a week to condition the medium. The location of the inoculated embryos is indicated by black squares. The bottom of the Petri dish is engraved to spot the embryos.

D. Causes of mortality

As was reported, embryos of 25 μm or less cannot be cultivated, as they die within the few hours following their inoculation [80].

Three hypotheses can be formulated to explain this observation. At first the endosperm supplies the embryo with special substances which are lacking in the medium. In this case, rescuing very small embryos requires media refinements that must still be found. The second explanation is based on the protective effect of the mass of tissue that represents the ovule. As a matter of fact, it has been said that a previous culture of tissue prepares the medium to culture very fragile embryos. In the same manner, in ovule culture, there is probably an adjustment of the medium to the needs of the embryo by the endosperm, at least at the beginning of the culture, because it degenerates in course of time. The third hypothesis is that the mortality is due to the rupture of the suspensor. It seems that the suspensor must be intact for survival of early-staged embryos. If the suspensor is broken when the embryo is very small, the wound results in embryo death. It is possible, that through the rupture of the suspensor, cellular substances indispensable for embryonic development leak into the medium, or

embryo

suspensor

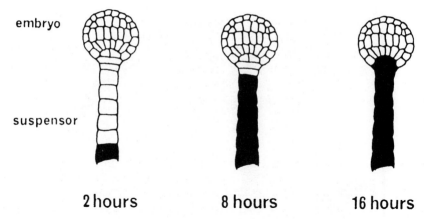

2 hours   **8 hours**   **16 hours**

Fig. 15. Globular embryos of *Capsella* after 2, 8 and 16 h of culture. The advancement of necrosis is detected by Evan's blue which stains the dead cells.

that the sudden penetration of mineral salts into the suspensor induces the necrosis of a great number of embryonic cells. A certain number of observations favour that explanation [83]. After culture, the base of the embryo where the suspensor was inserted is brown. This brown spot is composed of a group of cells which have kept the same embryological size and, as generally the cells increase in size at the beginning of the culture, it shows that these cells have died very soon after inoculation. The ultrastructure of these cells confirms this observation [83]. The contracted protoplasm of the hypophysis cells reveals that they are dead and that some of them have increased considerably the thickness of their cell walls [83]. The suspensor and the base of the embryo constitute the most fragile parts. By using a stain which colours the proteins of dead cells, the necrosis of the suspensor can be traced from its basal cell to the hypophyseal cell of the embryo base in a few hours (Fig. 15) [80].

IV. Applications

Embryo culture methods promise to be useful in studying some very fundamental problems of embryo and ovule growth. They provide an excellent opportunity for studies in basic embryogenesis including the possibility to change the development of the embryo and its metabolism by modification of nutrition, environmental conditions and phytohormones. The isolation of the whole embryo or of parts of it and the study of development in culture provide useful information on the induction received by the cells during their *in ovulo* stay.

A. Study of zygotic embryogenesis

Embryo development in culture *in vitro* changes according to three factors: the size of the embryo when it is inoculated, the nature of the nutrient solution, and the more or less prolonged stay in the ovule before inoculation. However, the complex interaction of these and other factors is unclear and requires additional study.

1. Effect of size

When globular embryos are inoculated numerous abnormal aspects are observed after culture. A small percentage of these early embryos develop in a normal way and can survive whereas the others stop growing after a certain period. These small embryos are very sensitive to the composition of the mineral solution of the medium. Embryogenesis of a more differentiated embryo in culture, like a heart-shaped or torpedo-shaped embryo, is very similar to the growth pattern of an embryo *in situ* and it is little affected by the nature of the medium.

2. Effect of the nature of medium

Generally a globular embryo exhibits an original and variable development, especially in the way the cotyledons appear. This depends on the kind of mineral medium used. The study of this critical moment may be very helpful to acquire more knowledge about differentiation. The emergence of cotyledons is more or less complete (Fig. 16) depending on the composition of the mineral solution employed [18]. Knop's solution with a low level of mineral salts does not induce normal cotyledon formation. The embryo stays massive and poorly differentiated. In the new solution (Table 1), better balanced and adapted to the

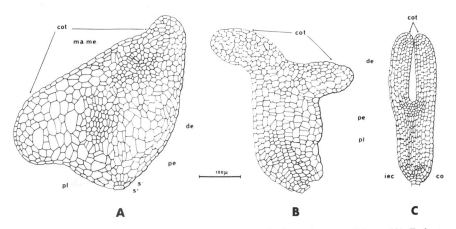

Fig. 16. Development of a globular embryo of *Capsella* in various conditions. (A) Embryo cultivated *in vitro* on a medium with Knop solution. (B) Embryo cultivated *in vitro* on the new solution (Table 1). (C) Embryo developed *in situ*.

144

requirements of the embryos, the embryo is made up of smaller cells, the cotyledons are more protruding and histological differentiation appears in the axis of the embryo. Nevertheless, *in situ* embryos exhibit longer cotyledons and thinner hypocotyls (Fig. 16).

3. Action of the in ovulo period

Embryo development in culture varies according to the time spent by the embryo in the ovule. The duration of the *in ovulo* period has an effect on the differentiation and regulation capacity of the embryos in culture even if they are in the same undifferentiated state when they are inoculated. Depending upon excision chronology, the embryos exhibit two types of behaviour when cultured *in vitro* (Fig. 17). Early embryos, excised shortly after fertilization, grow isodiametrically without organogenesis. Late embryos, although also undifferentiated at the time of explantation, directly enter a period of organogenesis with rapid initiation of two cotyledons. These embryos, therefore, seem to have received an inductive stimulus during the more prolonged stay in the ovule which permits subsequent differentiation. This organogenetic impulse is also expressed even when the embryo is longitudinally bisected. Each half-embryo regenerates a complete embryo having two cotyledons [20]. In this instance, the independent development of each embryo

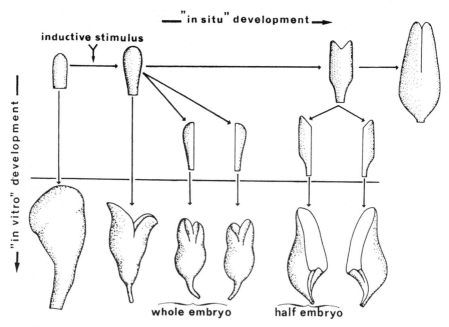

Fig. 17. Development of embryos and parts of embryos of *Zamia*. The development *in situ* is represented horizontally. Embryos, after a more or less prolonged stay in the ovule, are inoculated either complete or in parts. Growth *in vitro* is depicted vertically.

was a phenomenon of regulation. However, when the longitudinal bisection is done on an old and well-differentiated embryo, which already possesses two cotyledons, each half-embryo continues to develop as if it remained attached to the missing half. Thus, well-advanced embryos do not exhibit regulation to the same degree as do younger embryos, and bisection of the embryos results in the formation of two-half embryos. These experiments were carried out on embryos of the cycad *Zamia* which were large enough to stand dissection [20].

The nature of the stimulus induced by the ovule is not very well-known. As indicated earlier, the embryogenesis of an early embryo in culture differs very much from that of an *in situ* embryo. But the morphology of the embryo growing in an ovule cultured *in vitro* is closer to that of an *in situ* embryo [84]. Wardlaw suggested that in the ovule some bilateral gradients could determine the formation and localization of cotyledons on the surface of the embryo [85]. Growth of any particular part of the embryo is probably controlled by intercellular and extracellular reaction to a particular factor, and embryo culture is a means to appraise the contribution of each individual factor.

B. Shortening of the breeding cycle

For example, *Iris* seeds require a long period of dormancy which can reach several years. The use of embryo culture shortens the breeding cycle by overcoming the seed dormancy [10]. In a general way, dormancy can be caused by numerous factors including dry storage requirements, low temperatures, specific light requirements and even endogenous inhibitors. All these factors can be circumvented by embryo excision and culture.

C. Overcoming incompatibility

1. Rescue of growth-arrested embryos
Two cases are interesting.

a. Underdeveloped embryos
The first case concerns early-ripening trees whose fruit grow ripe so quickly that embryos fail to develop completely and are thus nonviable. Embryos from seeds of such fruits germinate when cultured *in vitro* and produce normal plants provided embryos are not too small [13].

b. Hybrid embryos
In horticultural and breeding practices, when plants are distantly related, crosses are often unsuccessful because, although fertilization occurs, the hybrid embryo aborts on the mother plant. The embryo appears to possess the potential for initiating development but is somehow prevented from reaching the adult size with normal differentiation. The cause of this abortion lies in the fact that, in a normal ovule, endosperm formation precedes and supports embryo development nutritionally. The endosperm does not develop or

degenerates, probably on account of the failure of the second fertilization, which does not occur because of a genomic incompatibility between the two parents. This absence or degeneration of endosperm generally results in starvation of the embryo and provokes its abnormal development. There are other causes which can provoke embryo abortion. In *Datura* crosses the embryo sac is invaded by a proliferating tissue which leads to a final digestion of the endosperm and abortion of the undernourished embryo [86]. In any event, isolation and culture of the hybrid embryo prior to starvation on a synthetic medium can offer to the embryo a substitute for the missing endosperm. The first successful interspecific hybrid rescued through embryo culture was with *Linum* by Laibach in 1925 [7]. These embryos were unable to be reared to maturity by conventional methods. By using embryo culture Laibach laid the foundations of a method to surmount barriers to crossability in plants. Since that time, techniques and knowledge of media requirement have been improved making it possible to cross two more distantly related plants. These crossings can provide new genotypes by gathering different and interesting characteristics into a unique plant. For instance, hybrids between *Triticum* and *Aegilops* were attempted in order to introduce the resistance to leaf rust in wheat [66]. However, pollination of wheat with *Aegilops* pollen yielded only very few mostly abortive seeds. A high percentage of hybrid plants could nevertheless be obtained by means of culture of the embryo on Monnier's medium [66]. Recently utilizing the same medium a cross between *Phaseolus coccineus* and *Phaseolus acutifolius* was achieved and provided useful drought resistant hybrids [87].

The method of embryo culture appears sometimes of limited value since the hybrid plants are sterile or with very low fertility. Indeed chromosomes are often too different for pairing during meiosis. Doubling the chromosome number by the use of colchicine provides tetraploid plants with normal gametogenesis and restores fertility in many cases.

But embryo culture cannot solve all problems of crossability, there are definite limitations to the method. The impossibility can occur earlier, e.g. the fertilization does not happen because the pollen cannot reach the female gamete. There is a pollen-stigma incompatibility.

2. Pollen-stigma incompatibility

It happens that the barrier to crossability is placed well before fertilization. In that case sterility between two plants originates from an incompatibility between the pollen of one plant and the stigma of the other. *In vitro* pollination and fertilization are potential approaches to overcome this incompatibility.

– *Stigmatic pollination in vitro*. In *Trifolium repens* a gametophytic self-incompatibility prevents inbreeding. Inbred lines are necessary for the development of hybrid varieties. Production of homozygotes by the diploidization of plants derived from haploid cells has not yet succeeded in the Leguminosae. Culture of flowers *in vitro*, pollination *in vitro* and the culture of the resulting embryos can increase the number of self-pollinated plants by more than ten fold [65].

– *Non-stigmatic pollination in vitro*. By culturing flowers *in vitro* and depositing pollen grain directly on the ovule, incompatibility can be by-passed. Although the percentage of fertilization decreases dramatically in comparison with the fertilization *in situ*, hybrids can be obtained that have never been observed *in situ* [88].

D. Production of haploids

In some plants even if the embryo develops, via embryo culture, one of the genomes can be eliminated, in the course of the successive cell divisions of the embryo, and thus the hybrid plant is not obtained. The most remarkable case is the cross between species of barley (*H. vulgare* × *H. bulbosum*) where there is a gradual elimination of the chromosomes of *H. bulbosum*. This property was utilized in the production of haploids [30], and is now used routinely in barley breeding.

E. Production of somatic embryos

Somatic embryos are generally obtained from cells of differentiated tissue which are first submitted to the action of an auxin and secondly transferred to a new medium without auxin. It seems that the influence of a hormone, rejuvenates the cell so much that it reacquires the capacity of a zygote to produce an embryo. It is understandable that the younger the original somatic tissue is, the more juvenile the result is. When culture is started from an immature embryo, somatic embryos are generally more easily obtained [31]. In that way efficient regeneration can be obtained in "recalcitrant" families like the Leguminosae [32].

V. Future trends of research

As has been shown, multiple difficulties occur at different levels of a crossing: incompatibility between pollen and stigma, plant sterility, abortion of the hybrid embryo and finally if the embryo survives rejection of one of the chromosome sets. All these difficulties come from the fact that one is attempting to merge two half genomes whose characteristics are too different. This experimental procedure is certainly ill-chosen especially because generally the transfer of only one character is needed as, for example, a disease resistance. So it appears that it would be certainly preferable to transfer only one gene at a time. Recently numerous experiments demonstrated that it was possible to transfer genes at various stages in the course of fertilization [89] and embryo development [33].

The improvement of the qualities of plants of agricultural importance needs the direct transfer of some interesting genes from one plant to another. Two problems must be solved, namely the isolation or construction of the necessary genes and the penetration of genes into the plant. The first question has already been explored, whereas the second is on its way towards solution.

A. Construction of genes

In order to examine the penetration of genes inside plants the best manner is probably to use a solution containing DNA from a plasmid. On the plasmid several genes have been inserted whose expression in the plant can be easily detected. In the plasmid different parts of the genome of a plant can be inserted, generally the replicative regions to increase the integration of the foreign gene and also the replication of the plasmid [89].

DNA directly extracted from plants can also be employed. In maize, DNA was extracted from plant leaves crushed in liquid nitrogen and after a special preparation added to a sucrose solution [90].

B. Transfer of genes

The introduction of genes can be done at three main stages in the reproduction of the plant.

1. In pollen grains

Shortly after pollination the solution containing the plasmid, carrying an antibiotic resistant gene, is added on the stigma of each flower. The solution contains also agents to increase permeability of the membrane. The pollen grains germinate on the stigma and the pollen tubes enter the style. At the same time the plasmid penetrates into the pollen tube and, using it as a vector, slides down to the ovule and to the female gamete. Germination of the seeds on a solution containing an antibiotic permits the screening for the transformed plants. By using a radioactive probe, corresponding to the plasmid, it has been proved that the plants were entirely transgenic [89]. The same approach has been used with DNA from plants [90].

2. In early embryos

When the embryo is very young the suspensor is rather large compared to the size of the embryo. Probably this is the way in which the nutritive substances are conveyed from the maternal tissue to the embryo. Schulz and Jensen [91] using an electron microscope showed that the suspensor cells largely communicate between themselves with numerous plasmodesmata which considerably facilitate the passage of dissolved substances. Probably these communications between cells would permit DNA to enter the embryo and to transform it, but until now, to our knowledge, no such experiment has been performed on that subject.

3. In mature embryos

Dry seeds are opened and the embryonic axis (decotylated embryo) is plunged into the solution containing the plasmid and agents increasing permeability. Embryos have to be in a desiccated condition since DNA uptake happens during the initial period of imbibition of the embryos. Indeed membranes in dry embryos have unique properties, they are discontinuous and disorganized, large pores are opened and molecules of a certain size can pass through. At the same time, in the opposite direction, some substances can leak out to the exterior. But this period is very short during imbibition, since the membranes quickly become intact and uniform [33]. The feasibility of inserting DNA into dry embryo cells during imbibition was recently confirmed using somatic embryos [92].

Probably, in the future, these methods will offer the possibility to transfer useful foreign genes to plants and, in that way, improve their agricultural qualities.

VI. Conclusion

Gradually a better understanding of the requirements and of the special physiology of embryos has dawned. The principal application of embryo culture, that is to say the improvement of plants by crossing them and production of hybrids, was known rather early, and has not changed in the course of time. But research workers are more and more interested in cultivating smaller embryos and scrutinizing embryological cells. Study of embryo differentiation by means of molecular biology and transfer of genes are probably the next steps. In one way or another, embryo culture will certainly remain a promising method for developing new plants tailored to withstand unfavourable environmental conditions, to give higher yields and to offer resistance to diseases and pests.

References

1. Nobécourt, P., Cultures en série de tissus végétaux sur milieu artificiel, C.R. Acad. Sc. 205, 521, 1937.
2. Gautheret, R.J., Nouvelles recherches sur la culture de tissu cambial, C.R. Acad. Sc. 20, 5, 572, 1937.
3. Kögl, F., Haagen-Smit, A.J., and Erxleben, H., Über den Einfluss der Auxine auf das Wurzelwachstum und über die chemische Natur des Auxins der Graskoleoptilen, Z. Physiol. Chem. 228, 104, 1934.
4. Hannig, E., Zur Physiologie pflanzlicher Embryonen. Ueber die Kultur von Cruciferen-Embryonen ausserhalb des Embryosacks, Bot. Zeit., 62, 45, 1904.
5. Bonnet, C., Recherches sur l'Usage des Feuilles dans les Plantes et sur Quelques Autres Sujets Relatifs à l'Histoire de la Végétation, Elie Luzac, Gottingue and Leide, 1754.
6. Brown, H.T., On the culture of excised embryos of barley on nutrient solutions containing nitrogen in different forms, Trans. Guinness Res. Lab. 1, 288, 1906.

7. Laibach, F., Das Taubwerden von Bastardsamen und die künstliche Aufzucht früh absterbender Bastard-Embryonen, *Zeist. f. Bot.*, 17, 417, 1925.
8. LaRue, C.D., The growth of plant embryos in culture, *Bull. Torrey Bot. Club*, 63, 365, 1936.
9. LaRue, C.D., and Avery, G.S. Jr., The development of the embryo of *Zizania aquatica* in the seed and in artificial culture, *Bull. Torrey Bot. Club*, 65, 11, 1938.
10. Randolph, L.F., Embryo culture of *Iris* seed. *Bull. Am. Iris Soc.*, 97, 33, 1945.
11. Li, T.T., The development of the embryo of *Ginkgo biloba in vitro*, *Sci. Rept. Nat. Tsing. Hua Univ.*, Ser. B2, 29, 1934.
12. Van Overbeek, J., Conklin, M.E., and Blakeslee, A.F., Factors in coconut milk essential for growth and development of very young *Datura* embryos, *Science*, 94, 350, 1941.
13. Tukey, H.B., Growth of the peach embryo in relation to growth of fruit and season of ripening, *Proc. Amer. Soc. Hort. Sci.*, 30, 209, 1933.
14. Rijven, A.H.G.C., *In vitro* studies on the embryo of *Capsella bursa-pastoris*, *Acta. bot. Neerl.*, 1, 157, 1952.
15. Raghavan, V., *Experimental Embryogenesis in Vascular Plants*, Academic Press, London/New York/San Francisco, 1976.
16. Norstog, K., and Smith, J., Culture of small barley embryos on defined media, *Science*, 142, 1655, 1963.
17. Monnier, M., Croissance et développement des embryons globulaires de *Capsella bursa-pastoris* cultivés *in vitro* dans un milieu à base d'une nouvelle solution minérale, Colloque de Morphologie végétale, *Bull. Soc. Bot. Fr., Mém.*, 179, 1973.
18. Monnier, M., Culture *in vitro* de l'embryon immature de *Capsella bursa-pastoris* Moench., *Rev. Cytol. Biol. Vég.*, 39, 1, Thesis, 1976.
19. Monnier, M., Variations des besoins nutritifs des embryons immatures de *Capsella bursa-pastoris* au cours de leur culture *in vitro*, *Congrès Nat. Soc. Sav. Lille Sciences*, 1, 595, 1976.
20. Monnier, M., and Norstog, K., Effect of *in ovulo* period on the differentiation and regulation of immature embryos of *Zamia* cultured *in vitro*, *J. Exp. Bot.*, 37, 1633, 1986.
21. Van Overbeek, J., Conklin, M.E., and Blakeslee, A.F., Cultivation *in vitro* of small *Datura* embryos, *Amer. J. Bot.*, 29, 472, 1942.
22. Raghavan, V., and Torrey, J.G., Growth and morphogenesis of globular and older embryos of *Capsella* in culture, *Amer. J. Bot.*, 50, 540, 1963.
23. Norstog, K., Development of cultured barley embryos. 1. Growth of 0.1–0.4 mm embryos, *Amer. J. Bot.*, 52, 538, 1965.
24. Monnier, M., Croissance et survie des embryons de *Capsella bursa-pastoris* cultivés *in vitro* dans diverses solutions minérales, *Rev. Gén. Bot.*, 77, 73, 1970.
25. Monnier, M., Action de divers sels minéraux sur le développement *in vitro* des embryons immatures de *Capsella bursa-pastoris*, 99ème Congrès Nat. Soc. Sav., Toulouse, Sciences, 4, 257, 1971.
26. Ryczkowski, M., Sugars and osmotic value of the sap surrounding the embryo in developing ovules (dicotyledonous perennial plants), *Acta Soc. Bot. Polon.*, 33, 397, 1964.
27. Yeung, E.C., and Sussex, I.M., Embryogeny of *Phaseolus coccineus*: the suspensor and the growth of the embryo-proper *in vitro*, *Z. Pflanzenphysiol.* 91, 423, 1979.
28. Kononowicz, A.K., and Janick, J., *In vitro* development of zygotic embryos of *Theobroma cacaco*, *J. Amer. Soc. Hort. Sci.*, 109, 266, 1984.
29. Finkelstein, R.R., and Crouch, M.L., Precociously germinating rapeseed embryos retain characteristics of embryogeny, *Planta*, 162, 125, 1984.
30. Kasha, K.J., and Kao, K.N., High frequency haploid production in barley (*Hordeum vulgare* L.), *Nature*, 225, 874, 1970.
31. Vasil, I.K., Plant regeneration and genetic variability, in *Cell Culture and Somatic Cell Genetics of Plants*, Vol. 3, Vasil, I.K., Ed., Academic Press, Orlando, FL, 1986, 757.
32. Tétu, T., Sangwan, R.S., and Sangwan-Noreel, B.S., Direct somatic embryogenesis and organogenesis in cultured immature zygotic embryos of *Pisum sativum* L., *J. Plant Physiol.*, 137, 102, 1990.

33. Töpfer, R., Gronenbom, B., Schell, J., and Steinbiss, H.-H., Uptake and transient expression of chimeric genes in seed-derived embryos, *Plant Cell*, 1, 133, 1989.
34. Hanstein, J., Die Entwicklung des Keimes der Monokotylen und Dikotylen. *Bot. Abh.*, 1, Hanstein, 1, 1870.
35. Schaffner, M., The embryology of the shepherd's purse, *Ohio Naturalist*, 7, 1, 1906.
36. Souèges, R., Nouvelles recherches sur le développement de l'embryon chez les Crucifères. *Ann. Sci. nat., Bot.*, 9 Sér., 19, 311, 1914.
37. Pollock, E.G., and Jensen, W.A., Cell development during early embryogenesis in *Capsella* and *Gossypium*, *Amer. J. Bot.*, 51, 915, 1964.
38. Souèges, R., Les premières divisions de l'oeuf et les différenciations du suspenseur chez le *Capsella bursa-pastoris* Moench., *Ann. Sci. Nat., Bot.*, 10 Sér., 1, 1, 1919.
39. Dieterich, K., Über die Kultur von Embryonen ausserhalb des Samens, *Flora*, 117, 379, 1924.
40. Knop, W., Bereitung einer concentrirten Nährstofflösung für Pflanzen. *Landw. versuchs-Stat.*, 30, 292, 1884.
41. Monnier, M., Comparaison du développement des embryons immatures de *Capsella bursa-pastoris in vitro* et *in situ*, *Bull. Soc. bot. Fr.*, 115, 15, 1968.
42. Murashige, T., and Skoog, F., A revised medium for rapid growth and bioassays with tobacco tissue cultures, *Physiol. Plant*, 15, 473, 1962.
43. Linsmaier, E.M., and Skoog, F., Organic growth factor requirements of tobacco tissue cultures, *Physiol. Plant*, 18, 100, 1965.
44. Monnier, M., Action des conditions de stérilisation sur la valeur nutritive des milieux utilisés pour la culture des embryons isolés de *Capsella bursa-pastoris*, *Rev. Gén., Bot.*, 78, 57, 1971.
45. Camefort, H., and Boué, H., *Reproduction et Biologie des Végétaux Supérieurs*. Doin, Paris, 1979.
46. Haccius, B., Über die Beeinflussung der Morphogenese pflanzlicher Embryonen durch Lithium-Ionen, *Ber. Dtsch. Bot. Ges.* 69, 87, 1956.
47. Monnier, M., Action d'un gel de polyacrylamide employé comme support pour la culture de l'embryon immature de *Capsella bursa-pastoris*, *C.R. Acad. Sc. Paris*, 280, 705, 1975.
48. Crouch, M.L., and Sussex, I.M., Development and storage-protein synthesis in *Brassica napus* L. embryos *in vivo* and *in vitro*, *Planta*, 153, 64, 1981.
49. Monnier, M., Action de la pression partielle d'oxygène sur le développement de l'embryon de *Capsella bursa-pastoris* cultivé *in vitro*, *C.R. Acad. Sc. Paris*, 282, 1009, 1976.
50. Monnier, M., and Leddet, C., Sur l'acquisition de la résistance au froid des embryons immatures de *Capsella bursa-pastoris*, *C.R. Acad. Sc. Paris*, 287, 615, 1978.
51. Lagriffol, J., and Monnier, M., Effects of endosperm and placenta on development of *Capsella* embryos in ovules cultivated *in vitro*. *J. Plant Physiol.* 118, 127, 1985.
52. Rangaswamy, N.S., Experimental studies on female reproductive structures of *Citrus microcarpa* Bunge., *Phytomorphology*, 11, 109, 1961.
53. Iyer, R.D., Sulbha, K., and Ramanujam, S., Embryo culture studies in jute and tomato, *Memoirs, Indian Bot. Soc.* 2, 30, 1959.
54. Nakajima, T., Physiological studies of seed development, especially embryonic growth and endosperm development, *Bull. Univ. Osaka Pref. Ser.* B 13, 13, 1962.
55. Rietsema, J., Satina, S., and Blakeslee, A.F., The effect of sucrose on the growth of *Datura stramonium* embryos *in vitro*, *Amer. J. Bot.* 40, 538, 1953.
56. Matsubara, S., Studies on the growth-promoting substance "embryo factor" necessary for the culture of young embryos of *Datura tatula in vitro*, *Bot. Mag. Tokyo*, 75, 10, 1962.
57. Matsubara, S., and Nakahira, R., Some factors affecting the growth of young embryo *in vitro*. *Sci. Rep. Kyoto Pref. Univ.* (Nat. Sci., Liv. Sci. & Welf. Sci.) Ser. A 16, 1, 1965.
58. Mauney, J.R., Chappell, J., and Ward, B.J., Effects of malic acid salts on growth of young cotton embryos *in vitro*, *Bot. Gaz.* 128, 198, 1967.
59. Ziebur, N.K., and Brink, R.A., The stimulative effect of Hordeum endosperms on the growth of immature plant embryos *in vitro*, *Amer. J. Bot.* 38, 253, 1951.
60. Norstog, K., The growth and differentiation of cultured barley embryos, *Amer. J. Bot.*, 48, 876, 1961.

152

61. Norstog, K., New synthetic medium for the culture of premature barley embryos, *In Vitro*, 8, 307, 1973.
62. Monnier, M., Culture of mature ecotyledonous embryos of *Phaseolus vulgaris* and the nutritional role of cotyledons, *Amer. J. Bot.*, 69, 896, 1982.
63. Thomas, M.J., Comportement des embryons de trois espèces de Pins (*Pinus mugo* Turra, *Pinus silvestris* L. et *Pinus nigra* Arn.), isolés au moment de leur clivage et cultivés *in vitro* en présence de cultures-nourrices, *C.R. Acad. Sc.*, Paris, 274, 2655, 1972.
64. Tukey, H.B., Growth patterns of plants developed from immature embryos in artificial culture, *Bot. Gaz.* 99, 630, 1938.
65. Leduc, N., Douglas, G.C., Monnier, M., and Connolly, V., Pollination *in vitro*: effects on the growth of pollen tubes, seed set and gametophytic self-incompatibility in *Trifolium pratense* L. and *T. repens* L., *Theor. Appl. Genet.*, 80, 657, 1990.
66. Chueca, M.C., Cauderon, Y., and Tempé, J., Technique d'obtention d'hybrides Blé tendre × *Aegilops* par culture *in vitro* d'embryons immatures, *Ann. Amélior. Plantes*, 27, 539, 1977.
67. Monnier, M., and Norstog, K., Developmental aspects of immature *Zamia* embryos cultured *in vitro*, *Z. Pflanzenphysiol.* 113, 105, 1984.
68. Haagen-Smit, A.J., Siu, R., and Wilson, G., A method for the culturing of excised immature corn embryos *in vitro*, *Science* 101, 234, 1945.
69. Rijven, A.H.G.C., Effects of glutamine, asparagine and other related compounds on the *in vitro* growth of embryos of *Capsella bursa-pastoris*, *Koninkl. Nederl. Akad. Wetensch. Proc.* C 58, 368, 1955.
70. Nuti-Ronchi, V., Caligo, M.A., Nozzolini, M., and Luccarini, G., Stimulation of carrot embryogenesis by proline and serine, *Plant Cell Rep.*, 3, 210, 1984.
71. Harris, G.P., Amino acids and the growth of isolated oat embryos. *Nature*, 172, 1003, 1953.
72. Mitra, J., and Datta, C., Nutritional malformations in excised embryos of jute, *Sci. and Cult.*, 16, 531, 1951.
73. Raghavan, V., Interaction of growth substances in growth and organ initiation in the embryos of *Capsella*, *Plant Physiol.*, 39, 816, 1964.
74. Raghavan, V., and Torrey, J.G., Effects of certain growth substances on the growth and morphogenesis of immature embryos of *Capsella* in culture, *Plant Physiol.*, 39, 691, 1964.
75. Norstog, K.J., The growth of barley embryos on coconut milk media, *Bull. Torrey Bot. Club*, 83, 27, 1956.
76. Monnier, M., and Clippe, A., Effect of plant extracts on development of *Capsella* embryos in ovules cultured *in vitro*, *Biol. Plant* 34, 31, 1992.
77. Monnier, M., Culture des embryons adultes de *Phaseolus vulgaris*: rôle des réserves. *C.R. Acad. Sci. Paris*, 286, 461, 1978.
78. Monnier, M., and Leddet, C., Action du saccharose sur la résistance au gel des embryons immatures de Capselle, *Bull. Soc. bot. Fr.*, *Actual. Bot.*, 127, 71, 1980.
79. Finkelstein, R.R., and Crouch, M.L., Rapeseed embryo development in culture on high osmoticum is similar to that in seeds, *Plant Physiol.*, 81, 907, 1986.
80. Monnier, M., Survival of young immature *Capsella* embryos cultured *in vitro*, *J. Plant Physiol.*, 115, 105, 1984.
81. Lagriffol, J., and Monnier, M., Étude de divers paramètres en vue de la culture *in vitro* des ovules de *Capsella bursa-pastoris*, *Can. J. Bot.* 61, 3471, 1983.
82. Scemama, C., and Raquin, C., An improved method for rescuing zygotic embryos of *Pelargonium* × *hortorum* Bailey, *J. Plant Physiol.* 135, 763, 1990.
83. Monnier, M., Culture of zygotic embryos, in *Frontiers of Plant Tissue Culture*, Thorpe, T.A., Ed., IAPTC, Calgary, Canada, 1978, 277.
84. Monnier, M., and Lagriffol, J., Effect of ovular tissue on the development of *Capsella* embryos cultivated *in vitro*, *J. Plant Physiol.*, 122, 17, 1986.
85. Wardlaw, C.W., General physiological problems of embryogenesis in plants, in *"Encyclopedia of Plant Physiology"*, Ruhland, W., Ed., Springer-Verlag, Berlin, 1965, 15, 424.
86. Sansome, E.R., Satina, S., and Blakeslee, A.F., Disintegration of ovules in tetraploid-diploid and in incompatible species crosses in *Datura*, *Bull. Torrey Bot. Club*, 69, 405, 1942.

87. Ben Rejeb, R., and Benbadis, A., Fertile allotetraploid from the cross between *Phaseolus coccineus* L. and *Phaseolus acutifolius* A. Gray, *Plant Cell Rep*, 8, 178, 1989.
88. Zenkteler, M., Test-tube fertilization of ovules in *Melandrium album* Mill. with pollen grains of *Datura stramonium* L., *Experientia*, 26, 661, 1971.
89. Piccard, E., Jacquemin, J.M., Granier, F., Bobin, M., and Forgeois, P., Genetic transformation of wheat (*Triticum aestivum*) by plasmid DNA uptake during pollen tube germination, in *7th Intl. Wheat Genet. Symposium of Cambridge*, Miller, T.M., and Kobner, R.M.D., Eds., 1988, 1, 779.
90. Ohta, Y., High-efficiency genetic transformation of maize by a mixture of pollen and exogenous DNA, *Proc. Natl. Acad. Sci. USA*, 83, 715, 1986.
91. Schulz, P., and Jensen, W.A., *Capsella* embryogenesis: the suspensor and the basal cell, *Protoplasma*, 67, 139, 1969.
92. Senaratna, T., McKersie, B.D., Kasha, K.J., and Procunier, J.D., Direct DNA uptake during the imbibition of dry cells, *Plant Sci.* 79, 223, 1991.

5. Morphogenic Aspects of Somatic Embryogenesis

S.A. MERKLE, W.A. PARROTT and B.S. FLINN

Contents

T.A. Thorpe (ed.), In Vitro Embryogenesis in Plants, pp. 155–203.
© 1995 *Kluwer Academic Publishers, Dordrecht. Printed in the Netherlands.*

I. Introduction

Somatic embryogenesis is the formation of an embryo from a cell other than a gamete or the product of gametic fusion. This is not an artificial phenomenon, and is known in nature as a form of apomixis called adventitious embryony, first described by Strasburger in 1878 [1] (see Sharma and Thorpe, this volume). Although Steward [2] and Reinert [3] are generally given credit for the first descriptions of somatic embryogenesis in 1958, such credit might more properly belong to Levine [4], who in 1947 reported the recovery of carrot "seedlings" from tissues exposed to low levels of α-naphthaleneacetic acid, via a process whose description sounds very much like somatic embryogenesis.

Seed-bearing plants represent an ancient lineage, having evolved during the Paleozoic era from pterophytes or "progymnosperms". The origin of naked seeds dates at least to the Devonian period, and gymnosperms, in the form of the seed ferns were well established by the Mississippian. By the mid-Cretaceous, the angiosperms had become widespread [5], although they did not become the dominant vegetation until much later [6]. Today, the major features of embryogenesis remain conserved among the seed-bearing plants, with the most obvious modification being a progressive reduction in cotyledon number. Sufficient developmental and biochemical similarities remain between embryogenesis in the seed-bearing plants that information obtained from one group can be successfully applied to another. Comparisons between groups of plants reveal fundamental processes associated with embryogenesis, and one goal of this chapter is to demonstrate how knowledge of these processes may be exploited to develop or optimize *in vitro* protocols for somatic embryo production, maturation, and conversion to plantlets.

A substantial amount of information has become available in the 35 years since somatic embryogenesis was first recognized as such in the laboratory. Nevertheless, the successful induction of somatic embryos and subsequent recovery of viable plants is not routine or efficient for the majority of species. Currently the trend is towards the recognition that embryo development is comprised of a variety of different stages involving specific patterns of gene expression [7–9]. The standard use of simple two-step media sequences to promote the induction and development stages of embryogenesis, originally identified by Kohlenbach [10], is proving inadequate to accommodate the multiple and distinct phases that somatic embryos undergo in the course of their ontogeny and subsequent development. Therefore, the more closely the pattern of somatic embryo gene expression matches that of zygotic embryos, the greater the chance of obtaining highly efficient regeneration systems. Such normalization of gene expression patterns will be achieved through the optimization of media and culture protocols for each individual stage of embryo development. It is the overall goal of this chapter to review the current concepts of embryogenesis as a morphogenic phenomenon, and describe culture protocols or manipulations that permit the normal development of a somatic embryo.

tomato species [58]. Soybean zygotic embryos reach a developmental point, associated with the end of growth through cell division, beyond which somatic embyrogenesis may no longer be induced [59–61]. Likewise zygotic embryos of interior spruce lose their embryogenic capacity when the accumulation of storage proteins begins [62], and the loss of embryogenic capacity in yellow-poplar correlates with the appearance of a 55 kDa polypeptide, which is a pututative storage product [63].

Similar observations on age and developmental stage have been made on coconut leaves [64]. In grasses, embryogenic capacity appears to be limited to embryos, and inflorescences and leaves that are still meristematic, as long as these have been physically separated from mature, differentiated tissues [37]. In clovers, seedling tissue or tissues associated with reproduction are more readily induced to become embryogenic than mature tissues [65]. In many tissues and genotypes, the non-EC status is irreversible using currently available techniques for embryo induction. One possible explanation is that this phenomenon might be associated with the absence of soluble auxin receptors in nonresponsive explant tissues [66], while another is that the proper signal molecules have not yet been identified.

The health and vigor of the explant tissue also affects embryogenesis. In white spruce, embryogenic capacity is inversely correlated with the age and germinability of seeds used as explant tissue [67]. When zygotic embryos are used as explants, the entire embryo may not be embryogenic, as is the case in pea, in which only the axis, and not the cotyledons, is responsive [51]. In soybean, the orientation of the explant is important, with the highest number of somatic embryos forming when excised cotyledons are placed with their abaxial sides on the medium [38].

b. Medium components

A minimum concentration of Ca^{2+} was necessary to induce somatic embryos from pollen of horsenettle, and the use of an ionophore to increase the intracellular free calcium ions doubled the number of pollen grains undergoing embryogenesis, perhaps by mediating the necessary signal transductions [68]. Calcium also increased the frequency of carrot somatic embryogenesis [69]. Sucrose and auxin concentrations have been found to interact with each other [70], with the optimal concentration of one being dependent on the concentration of the other. Glucose was reported to be superior to sucrose for the induction of somatic embryos of the scarlet runner bean [71]. Galactose and lactose were reported to be superior to sucrose for the induction of somatic embryos from nucellar callus of citrus, while sucrose was more effective than glucose or fructose [72]. Maltose, in the presence of NH_4^+, improved the number of primary alfalfa embryos recovered [73].

The importance of ammonium and nitrate, and reduced nitrogen in general, has also been reported in other somatic embryogenic systems, such as carrot [74] and spruce [75]. Amino acid supplements have also been used as nitrogen sources. Proline and serine interactions were noted in orchardgrass, and

12.5 mM concentrations of the two were optimal for increasing embryo number. While some amino acids are neutral or even inhibitory, the addition of 50–300 mM proline or proline analogues is also effective in increasing embryo number in alfalfa [76], as long as NH_4^+ is present in the medium [77]. In general, the induction of somatic embryogenesis usually requires at least 5–12.5 mM of ammonium in the medium [78–80], but very high levels of ammonium may inhibit embryogenesis [81]. The importance of nitrogen supply during embryogenesis may be a reflection of the nitrogen requirement for continued protein, nucleic acid and reserve substance synthesis. However, some of the effects may be associated with pH. Proline was suggested to buffer the growth medium within the appropriate range and stimulate embryogenesis in orchardgrass [82], and nitrate helped to counter medium acidification due to ammonium ions during carrot embryo development [74].

c. Culture environment

Very little information is available on the effect of culture environment on embryogenesis. Somatic embryo induction in eggplant is reported to have an absolute requirement for light [83]. In contrast, somatic embryo induction of poplar [84] and *Podophyllum* [85] have an absolute requirement for darkness. In Norway spruce, embryogenesis was possible only in complete darkness if ammonium is present in the medium. The removal of ammonium was necessary before embryogenesis is possible in the light [86]. Red light enhanced while blue light decreased the induction of date palm [87] somatic embryos. Green light, red light, or darkness were equally effective, while blue light or white light at higher intensities were inhibitory for induction of carrot somatic embryos [88]. Darkness, or white light from fluorescent tubes (75 μE m^{-2} s^{-1}), tended to lower somatic embryogenesis of soybean, while lower light intensities (10 μE m^{-2} s^{-1}), or light provided by Grolux® fluorescent tubes, which provide more light in the red spectrum, resulted in higher frequencies of embryogenesis [89].

In terms of the gaseous environment, the addition of ethylene inhibitors, such as cobalt, nickel, and salicylic acid, increased embryogenesis in carrot [90, 91] and rubber [92]. Inhibition of ethylene action with silver nitrate or the use of culture vessels that permit free gaseous exchange was also effective [92]. Simulation of the *in ovulo* environment, especially providing a reduced oxygen tension, was effective for improving embryogenesis in wheat [93, 94].

d. Bacterial compounds

Tantalizing hints that additional signal molecules remain to discovered come from the use of compounds of bacterial origin. Cocultivation of soybean callus with *Pseudomonas maltophilia* can induce the formation of somatic embryos [95]. Prior to this report, soybean somatic embryos had only been recovered directly from immature cotyledons. Extracts from the marine cyanobacteria *Nostoc*, *Anabaena*, *Synechococus*, and *Xenococcus* increased embryo formation following exposure to 2,4-D in carrot cell cultures [96].

III. Proliferation and mass propagation

A. Factors influencing proliferation of embryogenic tissues

Regardless of whether ECs originate as an IEDC or a PEDC, once they have been produced, the potential fates of these cells are functionally identical. Cells from both sources are able to either develop into somatic embryos or enter into another cycle of new EC production [97]. One of the most powerful aspects of somatic embryogenesis for such applications as mass propagation and gene transfer is the ability of embryogenic cultures of many species to proliferate indefinitely. This proliferative process has been variously termed secondary, recurrent or repetitive embryogenesis. Proliferation of embryogenic cells takes a number of forms and is apparently influenced by a variety of factors, some of which can be controlled during the culture process, and some of which are yet undefined. Factors which have been investigated include many of the same factors which have been associated with induction of embryogenesis, such as the effects of plant growth regulators, reduced nitrogen, plant species, and genotype of the cultured material. In the case of proliferation, however, these factors not only determine whether or not an embryogenic culture will continue to perpetuate the embryogenic state, but also influence the character of the repetitive cycle. Thus these factors appear to be responsible for controlling variation in the stage of somatic embryo development at which proliferation of new embryogenic material is triggered or the area of the primary embryo from which repetitive embryos are produced.

1. Plant growth regulators and reduced nitrogen

The most broadly documented factor associated with continuous proliferation of embryogenic cells is auxin (Fig. 2). However, it has become apparent that the effects of auxin cannot be considered independently from those of reduced nitrogen, since there is strong evidence of interaction between the two [98]. Once a group of ECs has been obtained, not only does the continued presence of auxin inhibit normal development and maturation of somatic embryos, but, if the auxin level is high enough, a new round of somatic embryo production is initiated. The exact role of auxin in triggering proliferation as opposed to development is unknown. Furthermore, the level of auxin necessary to maintain repetitive embryogenesis varies among species. Carrot embryogenic cultures, for example, can be maintained in a state of repetitive embryogenesis in the form of cell clumps known as proembryogenic masses (PEMs) [99] or more accurately, embryogenic clusters [100], in medium supplemented with 2,4-D concentrations ranging from $5 \times 10^{-5} \mu M$ to $2 \times 10^{-4} \mu M$, depending on the concentration of reduced nitrogen in the medium [98]. Once 2,4-D concentrations are reduced below a certain threshold, the repetitive cycles of embryogenesis cease, and development is permitted, although an even lower threshold must be crossed for the embryos to develop normally [101].

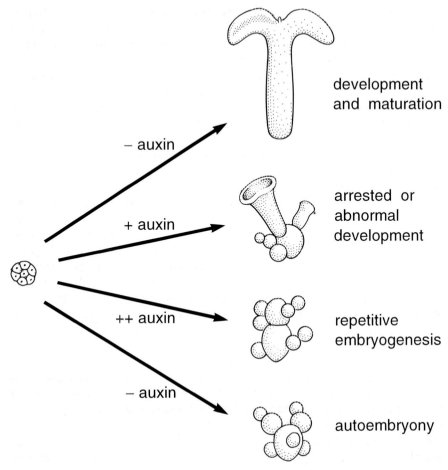

− auxin   development and maturation

+ auxin   arrested or abnormal development

++ auxin   repetitive embryogenesis

− auxin   autoembryony

Fig. 2. The role of exogenous auxin during histodifferentiation. The most normal histo-differentiation is usually achieved through the complete removal of exogenous auxins from the medium (top). Auxins in the medium may lead to abnormal histodifferentiation, or if the levels are high enough, arrest the histodifferentiation process at a globular or heart stage (2nd from top). Once the level of auxin surpasses a threshold, repetitive embryogenesis may result (3rd from top). Finally, repetitive embryogenesis may occur even in the absence of exogenous auxins (bottom), in a process termed "autoembryony".

    Another group of species, including a number of legumes such as alfalfa [102] and black locust [103], and some woody species such as walnut [104] and pecan [105], require only a pulse of auxin during a critical period, which can be as short as a few days, for cycles of repetitive embryogenesis to be initiated and maintained on growth-regulator-free medium indefinitely. Day lily cultures required 12 weeks of exposure to 2,4-D before they could be maintained without exogenous growth regulators by using a medium with a low pH [106]. Such systems might be thought of as maintaining a state of "autoembryony", since no exogenous growth regulators are required for maintenance and proliferation of the embryogenic state. In order for a somatic embryo produced in such a system to be released from this repetitive cycle and allowed to develop and mature, it must usually be removed from the repetitive culture environment and be "reprogrammed" by a stimulus such as individual transfer to soil [102]. In some cases, no effective treatment has been found to completely break the repetitive cycle. Although the underlying mechanism for autoembryony has not been determined, in the case of oilseed rape, repetitive embryogenesis in microspore-derived cultures was attributed to high endogenous levels of cytokinin, since repetitive embryogenesis was halted by application of exogenous cytokinin, which putatively restored internal control of cytokinin production through a feed-back mechanism [107].

At the far end of the spectrum from auxin-maintained repetitive embryogenesis are those reports in which no auxin-mediated induction was required for repetitive cycles of embryogenesis to be initiated. Such auxin-less induction has been reported for carrot. In this case, once a proembryonic culture was established by breaking or wounding zygotic embryos, a medium containing 1–5 mM NH_4^+ as the sole nitrogen source allowed continuous formation and multiplication of proembryos [26]. Later, the role of low pH in maintenance of continuous somatic proembryo proliferation in carrot was demonstrated [108].

2. Plant genotype

Given the genotypic effects on induction of the embryogenic state discussed earlier, genotype would also be expected to have impact on culture proliferation rates, although few reports have actually analyzed variation among genotypes for this trait. For example, significant variation in proliferation of somatic embryos was detected among 15 embryogenic Norway spruce lines, but it was not determined what proportion of this variation was due to genetic differences among the lines [109]. Two alfalfa genotypes monitored continuously for repetitive somatic embryo production over almost 2 years showed not only that embryo production remained stable for each of the two lines, but that one line consistently produced significantly more somatic embryos than the other over this period [110]. Similarly, the frequency of explants displaying proliferative embryogenic tissue was significantly different among 8 soybean genotypes, as was embryo yield over cycles of proliferation [111].

3. Species

In addition to the interactive effects between auxin level and species noted above, variation among species in the nature of repetitive embryogenesis can also be found in the developmental stage at which entry into repetitive cycles of embryo production is triggered. In carrot and yellow-poplar, for example, the repetitive cycle is continually expressed as proliferation of PEMs [16, 99], with no further development until auxin levels are lowered below a critical level. The ultrastructure of carrot PEMS in suspension cultures is composed of an inner region of highly vacuolate, nondividing cells and a peripheral region of small meristematic cells [112]. When continuously exposed to 2,4-D, PEMs undergo repeated cycles of fragmentation and growth. It is only when the 2,4-D level is reduced below 0.01 mg/l that globular embryos form [112]. In *Citrus* and soybean, somatic embryos advance to the globular stage before initiating a new round of repetitive embryogenesis [113, 114], while in alfalfa and black locust, proliferation continues from cotyledon-stage or even germinated embryos [102, 103, 110, 115]. Black locust embryogenic cultures are apparently unique in that although they will continue to proliferate indefinitely as later-stage embryos on auxin-free medium following a brief pulse of 2,4-D, they can also be maintained as PEMs by continuous culture in the presence of 2,4-D [116]. Variation among species has also been noted in the portion of the primary embryo from which repetitive embryos are derived. In pecan and black locust, radicles were the site of repetitive embryo production, while in walnut, repetitive embryos formed on cotyledons, hypocotyls, and roots [103–105].

Proliferation of embryogenic cultures of coniferous species is of particular interest because the mechanism of proliferation is very distinctive from the analogous process in angiosperms. Embryogenic conifer cultures appear most frequently to proliferate as embryonal masses with their associated suspensors, which have been termed embryonal suspensor masses (ESMs) [117–119], but have also been described as mixtures of unaggregated suspensor-like cells and small dense clusters of embryogenic cells [120]. While it may be tempting to describe ESMs as homologous to the PEMs which characterize proliferating embryogenic cultures of some angiosperms, the polyembryonic nature of proliferating conifer cultures [121] does not at all correspond, for example, to the ultrastructural descriptions of embryogenic carrot cultures [112] (see above). Even though the proliferation processes are divergent, however, the primary factors influencing this proliferation appear to be remarkably similar. For example, with most conifers, the 2,4-D and BA levels used to maintain embryogenesis (usually about 10 μM and 5 μM, respectively [122, 123]) are similar to the levels effective with angiosperms. In the case of silver fir however, embryogenic cultures were both initiated and maintained using cytokinin as the sole growth regulator [124].

4. Culture environment

Increasing attention has been given to the effects of the culture environment with regard to its effects on embryogenic culture proliferation. One aspect of the culture environment which is of particular importance to proliferation in bioreactor cultures (see below) is the gaseous environment. Carrot cell proliferation and embryo differentiation have different optimum levels of dissolved oxygen, with high levels promoting proliferation, while lower levels promote embryo differentiation [125]. Similarly, increasing oxygen concentration to 60% decreased the cell doubling time of poinsettia embryogenic cultures to 40 h, compared to 57 h at 10% [126]. Proliferation of embryogenic Norway spruce cultures was stimulated by increasing partial pressures of both oxygen and carbon dioxide [127], while growth of embryogenic white spruce cultures was apparently inhibited not by lack of oxygen, but by the accumulation of high levels of ethylene and carbon dioxide [128].

B. Embryo cloning

According to Vasil [129], current micropropagation methods are inadequate for the scale of plant production required to satisfy the large volume needs of the agriculture and forest industries, and reduction of costs per propagule cannot be achieved without total automation of the most labor intensive (and hence most expensive) components of the system. The fact that embryogenic cultures of many economically important species have the potential for continuous proliferation and virtually unlimited production of propagules has stimulated an interest in adapting these cultures for mass propagation via embryo cloning in bioreactors.

1. Bioreactors

While bioreactors are just one aspect of the scale-up and automation strategies under development for tissue-culture-based propagation, they are an aspect that many researchers believe to be particularly well-suited for application to proliferative embryogenic cultures. However, while much has been written about the potential of bioreactors for large-scale culture of somatic embryos (e.g. the recent review by Denchev et al. [130]), as yet there has been only a limited number of tests with model systems.

The first reported large-scale culture of embryogenic cells employed 20-L carboys to grow carrot cells, and only resulted in the formation of a few somatic embryos [131]. Since this early effort, a number of bioreactor configurations have been tested for somatic embryo production. The design which has been most frequently tested for plant cell culture is the stirred-tank bioreactor, which was originally developed for microbial fermentation. Although most of the research conducted with this design has been directed at large-scale culture of plant cells for production of secondary metabolites (e.g. Ten Hoopen et al. [132]), they have also been adapted for production of propagules such as somatic embryos. One of the earliest applications of this design employed a 5-L fermenter to determine the effects of dissolved oxygen on embryogenic carrot cultures (see above section on gaseous culture environment) [125].

Because stirred tank bioreactors have the disadvantage of generating high shear forces which are damaging to plant cells [133], researchers have investigated alternative bioreactor designs. One such alternative is the air-lift bioreactor, which has been demonstrated to support the growth of several plant cell types [134]. When an air-lift design was compared to a propeller-stirred design and to shaken flasks for production of alfalfa somatic embryos, slightly higher yields of embryos were obtained using the air-lift design, but conversion rates of these embryos were low [135]. More recent results from experiments established to compare different bioreactor designs for production of convertible embryos have indicated that some mechanically agitated designs may actually be more suitable for somatic embryo production than air-lift designs. An air-lift design was superior to stirred tank and vibro-mixer designs for growing alfalfa cells, but was not suited for producing alfalfa somatic embryos, possibly due to high dissolved oxygen levels [136]. A hanging stir bar fermentor produced embryos similar in quality to those produced in shaken flasks. Similarly, when air-lift, bladed turbine, and magnetic stir bar bioreactor designs were compared, the highest yields of normal carrot somatic embryos were obtained using the magnetic stir bar design at a speed of 50–200 rpm [137]. A vibro-mixer design equipped with silicone tubes for bubble-free aeration was capable of producing 100,000 poinsettia somatic embryos per liter of suspension, if the suspension was plated on filter paper and allowed to develop for four weeks [138].

Most bioreactor systems reported to date are batch culture designs which are aimed primarily at producing populations of mature somatic embryos, and as such, fail to take advantage of the proliferative nature of repetitive embryogenic cultures. There seems to be little recognition in most bioreactor studies of the fact that the functions of proliferation, development and maturation are clearly separate steps that are physiologically incompatible. If bioreactor technology is to meet its potential for efficient somatic embryo production, systems should be developed in which each step is separated from the other steps so that each can be monitored and controlled independently. Bapat et al. [139] separated these functions to some extent by generating populations of embryogenic sandalwood cells in a 7-L bioreactor while promoting embryo maturation in a separate 1-L bioreactor. The bioreactor configurations with most promise for taking advantage of the potential of proliferative embryogenic cultures, however, are those designed for continuous culture [140]. In a continuous culture bioreactor, the tank is initially filled and inoculated, as with a batch type bioreactor, but when cell proliferation reaches the log phase, fresh medium is introduced at a low rate while the same volume of spent medium and cells is removed. Maintenance of cells at the desired density can be accomplished by employing a spin filter, such that the bioreactor operates as a perfusion device, allowing removal of spent medium from the bioreactor without cell loss [140, 141]. In one test, when cell proliferation medium was replaced with embryo differentiation medium in the spin filter bioreactor, a constant number of PEMs was produced, each of which increased in cell number [141]. However, even this relatively advanced design does not truly provide continuous culture, since at some point, the bioreactor must be stopped, the mature somatic embryos removed, and a new culture cycle initiated.

The approach to truly continuous culture designs involves the establishment of multi-stage systems in which bioreactor units, each with its own function (i.e. cell proliferation, embryo development, embryo maturation, quiescence), are linked together in series [140]. Operational deployment of such a system will probably depend on the development of a complex automated design in which each step can be monitored. An automated commercial micropropagation system capable of handling somatic embryos has been described in which a bioprocessor controlled separation, sizing, and distribution of propagules into a culture vessel, but few details were provided [142, 143]. Progress in building the technology necessary for a truly continuous culture system has been made in the form of a kinetic model of carrot somatic embryo development in suspension culture, in which substrate utilization, culture growth, and embryo development are monitored over the time course of an embryogenic culture [144]. The integration of such a model with image analysis hardware and software for monitoring embryo concentration and development will be a key to establishing an automated system for continuous somatic embryo production in bioreactors [145].

2. Alternative scale-up strategies

Some embryogenic cultures, even if they can be maintained in liquid medium, simply cannot be grown as true suspension cultures. For example, although soybean somatic embryos can be multiplied in liquid culture, they proliferate not as cell suspensions or PEMs, but as clumps of globular embryos [114], limiting the potential for their production using a completely automated bioreactor system. Other embryogenic systems are not suited to bioreactor culture because high-quality embryos cannot be produced in liquid medium. For example, while embryogenic alfalfa cultures grow prolifically as suspensions, somatic embryos grown on semisolid medium were found to convert at a frequency 30 times higher than those grown in an air-lift bioreactor [135]. A protocol whereby alfalfa embryogenic suspensions are fractionated on nylon mesh followed by plating of the desired fraction on semisolid medium was therefore developed to produce large numbers of roughly synchronous alfalfa somatic embryos for subsequent maturation and desiccation treatments [146–148]. A similar system of fractionation and plating has been applied to promote high frequency production of yellow-poplar somatic embryos [149].

Conifer embryogenic cultures present a special problem with regard to potential bioreactor production, since they often proliferate as ESMs [119]. Although some testing of bioreactors for coniferous species has apparently been conducted [123, 150], results have not been published to date. Since cotyledonary-stage conifer somatic embryos have yet to be generated in submerged liquid culture, proembryos formed in suspension are usually plated on solid medium or an artificial support system to allow development and maturation [123]. One innovative approach to scaled-up production of conifer somatic embryos allows spruce somatic embryos to develop and mature on a flat absorbent bed above the surface of liquid medium which is continuously pumped into one end of a chamber, while spent medium flows from the opposite end [151]. Although this configuration does not fit the general description of a bioreactor, such alternative scale-up strategies will be of great value for high-frequency embryo production from culture systems which behave differently from carrot.

IV. Histodifferentiation

As a somatic embryo continues to form following induction, it proceeds through ontogenetic stages similar to those of their zygotic counterparts. Following the proembryonal stage, coniferous and dicotyledonous embryos (Fig. 3) go through globular, heart, torpedo, and cotyledonary stages. Monocotyledonous embryos proceed through globular, coleoptilar, and scutellar stages. The process of organ formation through which a globular-stage embryo develops into a cotyledon-stage embryo has been termed histodifferentiation [15]. Media manipulations that have a detrimental effect during this stage ultimately have a detrimental effect on plant recovery.

Of the histodifferentiation process, the most information is available with regard to the transition between the globular and the heart stages, largely because of the availability of temperature-sensitive mutants that cannot proceed beyond the globular-stage of development at non-permissive temperatures [152]. Such mutants can be rescued by a glycoprotein with endochitinase activity excreted into the culture medium by non-mutant cell lines [153], or by the addition of rhizobial *nod* factors (N-acetylglucosamine-containing lipooligo-saccharides). These latter could be an analogue of the plant compound actually produced through the action of the endochitinase [154].

While the addition of amino acids can be beneficial during the induction or maturation stages, proline and serine can act as inhibitors if added at the histodifferentiation stage [155]. Serine blocks cell division beyond a certain point, permanently arresting the development of globular-stage embryos. The

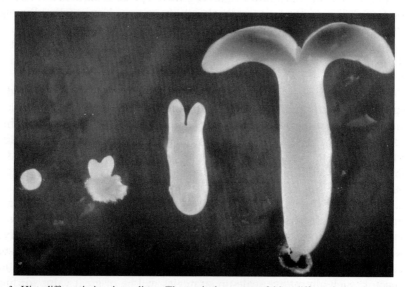

Fig. 3. Histodifferentiation in a dicot. The typical process of histodifferentiation in a dicot is illustrated by these soybean somatic embryos. Globular-stage embryos (left) proceed through heart, torpedo, and cotyledon stages. Photo courtesy of M.A. Bailey.

effects of proline are more varied, and the use of proline to arrest embryo development, along with esterase staining to detect the formation of stele tissue, has indicated two distinct phases in globular-stage embryos. The first manifests no evidence of polarity, while the latter exhibits esterase staining at the future shoot pole. Likewise, the heart stage also shows two distinct phases, with esterase revealing the beginning of vascular tissue formation in the cotyledonary area in the early phase, and in the hypocotyl during the later phase. Addition of proline at a late globular or early heart stages, followed by its removal from the medium, leads to the recovery of embryos which can germinate, but lack root meristems, highlighting the sensitivity of this developmental stage to disruptions [155].

In general, continued embryo histodifferentiation beyond the globular stage and subsequent maturation requires the removal of growth regulators from the medium, or at the least, a reduction of the growth regulator concentrations associated with induction and proliferation, to a level that will permit proper embryo development. While examples of continued embryo development under the same conditions used for induction have been reported [156, 157], such treatments usually result in somatic embryos of poor quality. Other factors can affect histodifferentiation as well. For example, a high osmolarity stress to globular-stage somatic embryos stimulated further embryo development [158]. The inclusion of abscisic acid or its analogues has been shown to promote the morphological normalcy of embryos [159, 160].

To the extent it has been traditional for embryogenic protocols to use a single medium for histodifferentiation and the subsequent maturation phases, it is difficult to review the literature and determine if a given treatment affected the histodifferentiation stage or the maturation stages. Nevertheless, enough information is available to make the following observations.

A. Role of auxins

Innumerable published protocols routinely expose somatic embryos undergoing histodifferentiation to auxins. Nevertheless, while exposure to auxin is essential for induction of the embryogenic state, continued exposure to auxin is detrimental to histodifferentiation (Fig. 2). Halperin and Wetherell [101] first noticed that the maintenance of globular-stage carrot somatic embryos on 2,4-D inhibited their histodifferentiation or led to abnormal development of the apical meristem. Similar observations have been made for crownvetch [161], and the development of soybean somatic embryos was inversely proportional to length of auxin exposure [162]. Up to 50% of carrot somatic embryos induced with cadmium were reported to germinate, while only 15% of somatic embryos induced with 2,4-D were able to do so [19], illustrating the point that exposure to auxins can have a deleterious effect that eventually affects germination capacity. The negative effect of auxins on meristem development is probably not surprising given the normal role that auxins play in the suppression of lateral meristems *in planta*. Even in grasses, in which it is thought that low levels of exogenous auxin are required to achieve continued cell division during histodifferentiation [163], enough protocols omitting the use of auxins during this phase have been published (e.g. [164, 165]) to suggest that exogenous auxins are not a universal requirement for proper somatic embryo development in this group of plants. The inclusion of indole-3-butyric acid (0.1–10 μM) in combination with ABA enhanced cotyledon development and morphology of interior spruce somatic embryos [166], suggesting species exist which may not have sufficient endogenous auxin for normal development. If this is the case, application of low levels of exogenous auxin is necessary for proper development.

1. Polarity

Low levels of endogenous auxin are required to allow the establishment of polarity and allow bipolar growth [167]. Electrical polarity preceding differentiation has been measured around cellular aggregates of carrot, with ionic currents flowing in and out of the future shoot and root poles, respectively [168]. Somatic embryos undergoing histodifferentiation continue to exhibit electrical polarity, with ionic currents detected around developing embryos from tobacco microspores [169], and a K^+ influx at the shoot apex and an H^+ efflux at the root apex has been detected in heart and torpedo-stage somatic embryos of carrot [170]. The addition of auxin transport inhibitors to somatic embryos of carrot has been associated with the inability of the embryos to make the transition between a globular and a heart stage, or from a heart stage to a torpedo stage [171].

Other indications of polarity are an asymmetry in DNA [172] and RNA [173] synthesis, and an uneven distribution of activated calmodulin, with the greatest concentrations occurring at the site of the future root pole [174]. Imposing polarity on cultured cells may also assist in the induction of embryogenesis. Anchoring cellular masses of sweet potato in alginate beads was reported to help establish a physiological polarity prior to the formation of somatic embryos [175]. Imposition of polarity through the application of a low voltage field to alfalfa protoplast-derived callus stimulated the formation of somatic embryos [176].

Likewise, polar auxin transport is necessary to establish the bilateral symmetry associated with dicotyledonous embryos. Zygotic embryos of *pin1–1* mutants of *Arabidopsis*, which have reduced auxin transport, tend to develop a continuous cone of cotyledon tissue instead of two distinct cotyledons [177]. Zygotic embryos of Indian mustard cultured *in vitro* fail to form two separate cotyledons when exposed to auxin transport inhibitors, and instead exhibit this phenotype [178]. Such fused cotyledons are characteristic of the horn-shaped somatic embryos commonly recovered from cell culture, and it may be that the exogenous auxin present in the medium may sufficiently perturb the normal auxin transport within a globular-stage embryo that normal cotyledonary development is prevented.

Simply placing somatic embryos on medium devoid of exogenous auxins for the histodifferentiation stage may not guarantee an auxin-free environment for proper histodifferentiation. Somatic embryos have the ability to conjugate auxins, which are subsequently excreted into the medium during histodifferentiation [167]. Hence, treatments such as inclusion of activated charcoal in the medium have led to the recovery of somatic embryos with a more normal morphology and increased germination ability [179–181], presumably because the charcoal adsorbs auxins released from developing tissues [182].

B. Role of cytokinins

The role of cytokinins during the histodifferentiation phase is more difficult to assess. Endogenous cytokinin levels have been observed to decrease as development occurs in both zygotic embryos of soybean [183] and somatic embryos of celery [184]. Furthermore, exogenous cytokinins can lead to abnormal development or suppression of the main embryo axis [43]. Alternatively, the inclusion of a cytokinin during the histodifferentiation phase can compensate for auxin-induced detrimental effects on meristem development. Globular-stage soybean somatic embryos, when exposed to a cytokinin, forgo further development while the apical meristem elongates to form multiple shoots [185]. Cytokinins in the medium may multiply the apices of somatic embryos, resulting in structures that are difficult to characterize as embryogenic or organogenic based on appearance alone.

V. Maturation

A. Factors affecting maturation

Following cell division and histodifferentiation, the period of embryo development in which cell expansion and reserve deposition occurs is considered the maturation phase [186]. Analyses of reserve accumulations in developing somatic embryos have revealed striking similarities or striking differences in comparison to zygotic embryos. These differences may primarily be attributed to the *in vitro* maturation conditions used. Many reports on somatic embryo reserve accumulations have not been carried out with optimal maturation protocols. Hence, manipulation of cultural conditions to prolong and improve embryo maturation, and to prevent precocious germination, will probably increase the similarity observed between zygotic and somatic embryos. Furthermore, the monitoring of markers such as storage protein transcript levels or the storage compounds themselves may provide the means to gauge the degree of maturation or onset of precocious germination. This would allow the determination of the appropriate time for somatic embryo removal from maturation to pre-germination treatments, prior to the onset of precocious germination.

1. Nitrogen

A variety of different medium formulations have been used for embryo maturation, involving various salt formulations and strengths, often varying in their nitrogen content. The presence of nitrogen, supplied as nitrate, ammonium, individual amino acids and/or casein hydrolysate is a requirement for continued embryo histodifferentiation and maturation. Walker and Sato [78] reported that no embryo development occurred in alfalfa in the absence of ammonium or nitrate. Furthermore, ammonium was required continuously during the first 6 days of embryo regeneration for embryo development. Certain amino acids were neutral or inhibitory for alfalfa embryo development, while others were stimulatory, increasing embryo size and/or number, and their eventual conversion [76]. In relation to this, the application of glutamine during alfalfa embryo maturation significantly enhanced protein synthesis and dry weight accumulation [187]. Proline and serine interactions were noted in orchardgrass, and 12.5 mM concentrations of the two were optimal for reduced precocious germination [82].

2. Carbon

Carbohydrate supply during embryo maturation also appears to be important for both embryo quality and number. Sucrose is the most commonly used carbohydrate, usually supplied at levels between 3–6%. A variety of other sugars support carrot embryo development, but certain sugars such as galactose, raffinose and stachyose result in a marked delay in the rate of embryo growth and production [188]. Different stages of embryo maturation were promoted by different carbohydrates during European silver fir embryo development [189]. Early maturation was promoted by a variety of carbohydrates, but lactose was superior to the others for further development and led to more normally structured embryos. With Nordmans fir, maltose was better for embryo maturation than sucrose at equimolar concentrations [41]. Similarly, maltose rather than sucrose enhanced alfalfa somatic embryo production and conversion [73], and the most recent maturation protocols for soybean [190] and alfalfa [191] somatic embryos use maltose. Furthermore, maturation of white clover on 6% maltose significantly enhanced conversion over maturation on 3% sucrose [192]. Different sugars were found to affect the total number of embryos that developed during spruce embryo maturation [193]. In addition to effects on somatic embryo number and development, Anandarajah and McKersie [194] reported that increased sucrose levels during alfalfa embryo development and maturation doubled their subsequent vigor.

The primary role attributed to the supplied carbohydrate is that of a carbon source. However, part of the carbohydrate effect on development and maturation is osmotic, as a portion of the supplied carbohydrate can be replaced by osmoticum [193, 195]. Further exploration of the role of osmoticum on somatic embryo development was provided by Roberts [196]. Low levels of mannitol (2–6%) promoted globular embryo formation, and osmoticum in combination with ABA doubled the production of mature embryos. High mannitol levels prevented precocious germination and prolonged the maturation phase. The requirement for high osmolarity may mimic changes in osmolarity that may occur in the environment surrounding the seed embryo.

3. Other factors

In addition to the different compounds mentioned previously, factors that affect the redox status of somatic embryos can affect their maturation. Addition of buthionine sulfoximine, an inhibitor of reducing agents, to Norway spruce cultures enhanced the number of maturing embryos [197] and high antioxidant levels were shown to be detrimental to the completion of carrot somatic embryogenesis [198].

The culture environment during maturation is also critical. Studies of the gaseous environment revealed that low oxygen levels significantly increased the number of developing wheat somatic embryos, and inhibited their precocious germination [94]. In Norway spruce, maturation was affected by the oxygen and carbon dioxide partial pressures, although gas phase effects varied between different cell lines [127]. In addition to oxygen and carbon dioxide, ethylene affected embryo maturation. Ethylene is usually inhibitory to development, and the use of ethylene inhibitors stimulates embryo development [91, 92]. However, ethylene was found to be essential for alfalfa embryo maturation [199]. It has been suggested that altering the gaseous environment in culture may help to simulate the *in ovulo* environment normally encountered during embryo development [94].

One final environmental factor that is not normally explored during embryo maturation is the light environment. Although somatic embryo maturation has been reported in both the dark and the light, the maturation of spruce embryos in the dark reduced the total number of mature embryos, and the number that eventually germinated, when compared to maturation using a 16-h photoperiod [75]. Furthermore, during carrot somatic embryo development, light quality and quantity affected both the total number of developing embryos, as well as their endogenous ABA content [88]. Hence, alterations in light spectra may be used to optimize maturation protocols.

B. Reserve substances as indicators of maturation

The accumulation of reserve substances represents a key stage of zygotic embryogenesis [186], providing compounds that are utilized by the germinating embryo until the development of autotrophy. Since the accumulation of these substances is a normal stage of embryo development, these provide excellent markers with which to compare the quality and fidelity of somatic embryogenesis. The correct accumulation of reserves and corresponding increases in dry weight in somatic embryos may indicate a high degree of vigor for subsequent germination [200]. In addition, differences in patterns of reserve substance accumulation in somatic embryos may lead to modifications of differentiation protocols that could enhance development and subsequent plant yield. Several studies have used reserve substances as markers to gauge the quality of somatic embryo development, with most concentrating on storage protein and lipid accumulation.

1. Storage proteins

Somatic embryos accumulate storage proteins that exhibit the same size and the same biochemical (solubility, isoelectric variants and disulfide linkage) characteristics as those of their zygotic counterparts [201–205]. In addition, storage proteins are targeted to their correct subcellular location, the storage protein body, in somatic embryo systems [206, 207]. These results suggest that the tissue culture process does not alter the regions of storage protein genes encoding structural and targeting information, allowing the expression of the same storage proteins as found in zygotic embryos.

Studies have reported that storage protein levels in somatic embryos of rapeseed [201, 208], cotton [204], alfalfa [203, 205], soybean [209], and white spruce [210] were lower than those of their zygotic counterparts, approximately 10% of the zygotic levels [201, 205]. In addition, the developmental timing of storage protein accumulation has been reported to be abnormal [201, 203, 204], with storage proteins accumulating at earlier stages of development than in zygotic embryos. These differences have been attributed, in part, to reduced storage protein transcript levels [203, 211], translational repression of storage protein transcripts during specific developmental stages [212], and post-translational regulation [212] in the somatic embryos.

The abnormal cases of storage protein expression in somatic embryos may be attributable to the type of maturation protocols utilized during somatic embryo development, and not an inherent problem associated with somatic embryo-genesis. The above mentioned studies used little or no ABA during somatic embryo maturation. Others, such as the abnormal reserve accumulation documented in soybean somatic embryos by Dahmer et al. [209], may be the result of embryo development in the presence of relatively high auxin levels. Due to the previously described beneficial effects of ABA on somatic embryo development, and known effects on zygotic embryo storage protein expression [213, 214], the lack of sufficient ABA may account for the above results. In support of this, the application of ABA to microspore-derived rapeseed embryos stimulated storage protein transcript accumulation [208] and promoted the correct developmental induction of storage protein transcripts to levels comparable to equivalent stage zygotic embryos [211]. Developing Norway and interior spruce somatic embryos matured on ABA displayed storage protein levels similar to those of zygotic embryos [202, 215, 216], with similar developmental patterns of storage protein transcript appearance [217] and protein accumulation [202, 217], as developing zygotic embryos. Furthermore, reduced levels of ABA resulted in lower levels of storage protein accumulation [166].

While most somatic embryo systems have reported stimulation of storage protein expression when ABA has been utilized, Lai et al. [187] reported that ABA imposed quiescence on developing alfalfa somatic embryos, and inhibited reserve substance deposition. However, the application of 50 mM glutamine with or without low levels of ABA, during the first half of their somatic embryo maturation phase enhanced dry weight accumulation and storage protein synthesis. Application of ABA during the second half of the maturation period then prepared the embryos for desiccation. Hence, the timing of ABA application may determine if ABA has a beneficial or detrimental effect on embryo maturation.

In addition to ABA, osmoticum plays a role in obtaining correct storage protein expression in developing somatic embryos. The application of 13–20% mannitol alone could replace 40 μM ABA during the later stages of interior spruce embryo maturation, preventing precocious germination and allowing storage protein accumulation [196, 217]. A similar stimulation of storage protein transcript accumulation by sorbitol was observed in rapeseed microspore-derived embryos [211]. While mRNA profiles could be normalized in somatic embryos of white spruce through the addition of ABA, proper translation of the mRNAs into storage proteins did not occur unless an osmoticum, in the form of 7.5% polyethylene glycol, was used in combination with 16 μM ABA [206].

The above reports have dealt with media additions to enhance storage protein accumulations. However, Stuart et al. [205] noted the importance of removing the induction growth regulators for storage protein expression. In this study, alfalfa somatic embryos induced on 50 μM 2,4-D exhibited little storage protein accumulation, while those induced on 10 μM 2,4-D exhibited higher storage protein levels, suggesting that the high auxin levels during induction led to carryover that suppressed storage protein gene expression and deposition.

The reduced storage protein levels that have been observed in somatic embryos may also be a reflection of culture protocols in which the embryos were allowed to precociously germinate before sufficient maturation occurred, as is often described. Interior spruce embryos matured on 10 μM ABA displayed initial accumulations of storage protein transcripts and proteins, but these declined rapidly during the early cotyledonary stages as precocious germination commenced [217]. In contrast, embryos matured on 40 μM ABA developed well into the cotyledonary stage without precocious germination, and exhibited a prolonged period of storage protein transcript abundance and major accumulations of storage proteins [202, 217]. Therefore, one of the roles played by ABA and osmoticum may be the prevention of precocious germination, allowing the continued expression of the embryo maturation program and storage protein accumulation.

2. Lipids

In addition to storage proteins, storage lipids (triglycerides) and their constituent fatty acids have been analyzed in developing somatic embryos.

Similar to the reports for storage proteins, storage lipids accumulate in their correct subcellular organelle, the lipid body, in somatic embryos [21, 210, 218]. Storage lipid accumulation occurs primarily during the maturation phase of embryo development and is often coincident with storage protein deposition [186].

Triglyceride contents have been shown to increase during the appropriate stages of somatic embryo development [210, 219, 220], as does the incorporation of ^{14}C-acetate into the lipid fraction [219, 221, 222], similar to observed patterns in developing zygotic embryos [186]. In addition, qualitative and quantitative changes in lipid fatty acid composition have been noted during somatic embryo development [208, 220, 221]. Pomeroy et al. [220] reported that the pattern of triglyceride fatty acid composition changed during rapeseed microspore-derived embryo development in a similar manner to that observed during zygotic embryo development, and that by the late cotyledonary stage, was similar to that of the dry seed. Taylor et al. [208] also reported a similar fatty acid composition of rapeseed microspore embryos to that of the dry seed, although differences in the levels of 22:1 and 18:1 fatty acids were evident between the two embryo types. White spruce somatic embryos also contained a triglyceride fatty acid composition similar to that of zygotic embryos [223]. In addition, the R_1 progeny of somatic embryo soybean regenerants displayed little variation in fatty acid composition, and were somewhat similar to non-regenerated, seed-derived material [224]. In contrast, Shoemaker and Hammond [225] reported that although the most prominent and least prominent fatty acids were the same in soybean zygotic and somatic embryos, the other fatty acids did not show the same trend. Furthermore, celery somatic embryos contained lower fatty acid levels and substantially lower proportions of 18:1 to 18:2 fatty acids compared to mature zygotic embryos [226].

In addition to fatty acid composition studies, total triglyceride levels have been measured. Rapeseed microspore embryos contained triglyceride levels that were similar to those of the dry seed, based on dry weight [208] and percentage of total lipid [220]. Interior spruce somatic embryos contained similar triglyceride levels to zygotic embryos on a per embryo basis, but genotype differences were noted between the somatic embryos [215]. White spruce embryos contained five times the triglyceride levels on a per embryo basis as compared to zygotic embryos [227]. In contrast to these reports, somatic embryos of Norway spruce [228] and white spruce [210] contained significantly less triglycerides than zygotic embryos, while in soybean, total neutral lipid levels (containing triglycerides) were significantly lower than mature zygotic embryos [229].

Again, differences in lipid and fatty acid compositions between zygotic and somatic embryos probably reflect the type of maturation protocols used. Dahmer et al. [229] reported that the type and level of auxin and/or cytokinin in the maturation medium greatly affected the total lipid level and fatty acid composition in soybean somatic embryos. Both ABA and osmoticum promoted fatty acid accumulation in cultured zygotic embryos [230], while ABA in combination with proline stimulated fatty acid accumulation in celery somatic embryos [226]. Application of ABA promoted triglyceride accumulation in spruce somatic embryos [228] and increasing sucrose levels promoted triglyceride accumulation in *Brassica* somatic embryos [218], with somatic embryos cultured on 20% sucrose accumulating triglyceride levels higher than those found in mature seed [218]. The ability to obtain somatic embryos with similar lipid and fatty acid compositions to those of zygotic embryos probably requires culture modifications that help to prolong the embryo maturation phase, and prevent precocious germination, a process which leads to the induction of enzymes associated with triglyceride degradation [219]. The large levels of storage lipids reported in spruce somatic embryos [223] resulted from the prolonged culture period (8 weeks) in the presence of 16 μM ABA and 7.5% polyethylene glycol, which prevented precocious germination. The large discrepancies in spruce storage lipid levels reported [210, 223] probably reflect the low ABA level and the short maturation period used, as well as the apparent precocious germination of the somatic embryos in the study by Joy IV et al. [210]. Apart from the influence of the maturation regime, Taylor et al. [208] suggested that spontaneous diploidization of microspore-derived rapeseed embryos may lead to alterations in gene expression and storage product deposition due to differences in gene dosage.

3. Carbohydrates (starch)

Very few studies have used starch accumulation to gauge somatic embryo development. However, those carried out have reported increases in starch content during somatic embryo maturation [200, 210, 219] and in all cases, the somatic embryos contained significantly higher starch levels than their zygotic counterparts. These results suggested that the somatic embryos were metabolically different from zygotic embryos and unable to efficiently convert carbohydrates into lipid and protein reserves. However, the type of maturation conditions, including such factors as sucrose concentration, exogenous growth regulators, osmoticum and others may play a role. In support of this, spruce somatic embryos matured on low ABA levels [206, 210] show large levels of starch compared to other reserves. In contrast, when the embryos are matured on high levels of ABA or ABA plus osmoticum, they accumulate substantial levels of protein [202, 206] and lipids [206], relative to starch, which is similar to the pattern observed in zygotic embryos.

VI. Post-maturation/pre-germination

Embryos removed from maturation conditions for further development often display poor or aberrant subsequent germination, growth and vigor [148, 162, 231]. These results suggest that further post-maturation treatments are required. Certain aspects of the culture environment, such as residual ABA from the maturation treatment, may be inhibiting further development. Attree et al. [232]

utilized a growth regulator-free, post-maturation treatment in the dark with spruce somatic embryos. During the treatment, continued cotyledon enlargement, but no precocious germination, occurred. This treatment stimulated subsequent embryo germination and development, and the enhancement was attributed to the loss of residual ABA from the maturation treatment, before exposure of the embryos to true germination conditions. The culture regime during maturation may also artificially induce dormancy [233]. The use of dormancy-breaking treatments such as cold treatments or gibberellic acid application has been found to enhance somatic embryo germination and growth following maturation in grape [233], rapeseed [234], and barley [235]. Other growth regulator treatments, such as cytokinins, may enhance development from the embryos when transferred to subsequent germination conditions [179, 235].

One of the fundamental aspects of zygotic embryo development not normally encountered during somatic embryo development is desiccation, which leads to embryo quiescence. It has been proposed that desiccation is required for the correct transition from an embryo maturation program to a germination program [9]. Hence, some of the problems associated with continued somatic embryo development following maturation may be due to the absence of a programmed drying regime. In support of this, germination of zygotic embryos prior to their desiccation is considered to be precocious and is characterized by aberrant development [236]. The imposition of a drying or partial drying treatment on somatic embryos enhances their subsequent germination and growth [147, 231, 237–239] (Fig. 4), and gives a better synchronization of root and shoot growth [231, 240]. Desiccation tolerance in somatic embryos can be induced by application of ABA, heat stress or proline [147, 226, 241, 242]. The stage of embryo development at which ABA is applied for the induction of desiccation tolerance is critical, and in alfalfa, the treatment of torpedo to cotyledonary stage embryos with ABA was optimal [148]. In addition, the rate of embryo drying is important for their subsequent survival on imbibition, with slow, gradual drying regimes better than more rapid drying [122, 147, 234]. In spruce, the duration of the partial drying treatment required for optimal subsequent development was related to the ABA levels used during maturation [240]. Maturation on high ABA levels required longer periods of partial drying to acquire a high frequency and quality of subsequent development. This suggests that desiccation following maturation may enhance germination by allowing a reduction in endogenous ABA levels, decreasing embryo sensitivity to ABA, or both, without the occurrence of germination during the process. Therefore, maturation of embryos on high ABA levels to promote reserve substance accumulation, followed by an appropriate duration of drying may be required to obtain good subsequent growth and development. While the above studies show a role for some type of desiccation treatment, desiccation may not be required for all species. In support of this, somatic embryos of birdsfoot trefoil, crown vetch [50] and white clover [243] germinated well in the absence of a desiccation treatment.

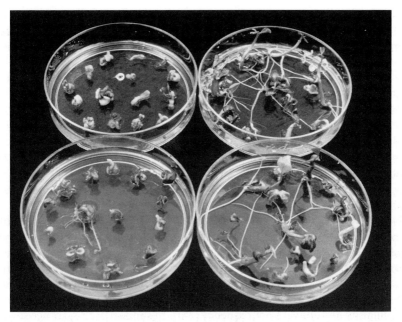

Fig. 4. Effect of desiccation on germination of somatic embryos. Desiccation following maturation helps normalize and synchronize germination, as illustrated by these somatic embryos of soybean. All embryos are of the same genotype and age, and received the same maturation treatment. Those on the right were desiccated for one week, while those on the left were not.

VII. Germination and conversion

While most studies report the development of roots and the germination of their somatic embryos, little distinction is made between germination and conversion. Germination refers to root and/or shoot development, while conversion, as defined by Stuart and Strickland [76], refers to the survival and development of these propagules in an *ex vitro* soil environment. While the ability to obtain somatic embryo germinants is often reported, little mention is made of their convertability. The ability to obtain rooted plants *in vitro* is not necessarily an indicator of continued growth and vigor [148, 194, 200, 241]. In alfalfa, somatic-embryo-derived plants show lower vigor than seed-derived embryos [148]. This may be related to the lack of abundant nutritional reserves in the embryo.

A. Storage reserves and embryo quality

Little biochemical or physiological analysis has been done to compare somatic and zygotic embryo germination patterns. Storage protein and lipid degradation occurred in dried spruce somatic embryos following imbibition

[206, 215, 223]. Protein and lipid reserves declined substantially within one day following imbibition, well before radicle elongation [215]. This was more rapid than in seed embryos, and suggested that the kinetics of reserve degradation differed between zygotic and somatic embryos, probably as a result of the lack of surrounding nutritive seed tissues in the somatic embryo. Subsequent root and hypocotyl lengths were somewhat lower in the somatic-embryo-derived plants than seedlings, which may have been due to the rapid reserve depletion in the somatic embryos and the absence of megagametophyte tissue. These results point out the need for the germination medium to substitute for the surrounding nutritive seed tissues, and may indicate that in addition to stimulating correct reserve accumulations in the developing somatic embryos, it may be necessary to try and mimic the normal kinetics of reserve utilization during germination to achieve normal post-germinative growth.

Large dry weight and reserve accumulations may be necessary for high conversion and vigor in the soil environment. Fujii et al. [200] reported that a 25 day ABA treatment during maturation, which stimulated dry weight and starch accumulation, increased conversion rates from 5% to 80%. Similarly, a high sucrose concentration (9%) during the first half of maturation, followed by a lower sucrose concentration (6%) during the second half increased dry weight accumulation and conversion frequency [241]. In support of this, spruce somatic embryos matured under conditions that promote abundant reserve accumulations [223, 244] exhibit good conversion rates and vigor.

B. Germination media and environment

Germination usually occurs *in vitro* on agar-solidified media or on liquid-saturated supports such as filter paper, Sorbarods® or Kimpaks®. Various medium formulations are used, often at levels less than full-strength. The use of lower salt concentrations during germination suggests that different stages of embryogenesis may require different types or concentrations of basal salts. For example, while white clover somatic embryogenesis was best induced and maintained on EC6 medium, Murashige and Skoog salts were better for the conversion of embryos into plants [245]. Germination media also require sucrose, or some other carbon source. Spruce somatic embryos were not capable of autotrophic growth immediately, and required 2% sucrose for root development [246].

Other compounds may also aid germination. The addition of 100–400 mg/l of an extract from a *Synechococcus* strain greatly increased the germination of carrot somatic embryos [247], again suggesting that additional molecules with growth regulator activity may yet be discovered in plants. In this case, the active component of the extract was removable by dialysis. As mentioned previously, the use of cytokinins during germination can help counteract deleterious effects produced by auxin exposure during the histodifferentiation stage. Cytokinins may also be effective when maturation treatments were ineffective, or omitted altogether from the regeneration protocol.

In embryos that have been desiccated, the rate of imbibition during the initial stages of germination may be important, since rapid imbibition may lead to abnormal germination and vitrification. The overly rapid rate of imbibition may be due to the absence of surrounding seed coat and other tissues, which serve to control the rate of imbibition in zygotic embryos [186]. The imbibition of dried interior spruce embryos on agar-solidified medium, rather than exposure to free liquid, gave somatic embryos whose radicles appeared similar to those of non-dried embryos, suggesting a lack of damage [246].

Embryo orientation during germination has also been shown to be important. Spruce embryos on agar slants [180, 244], or with their cotyledons immersed into the medium and the tubes inverted, had better germination than when the radicles were embedded in the medium [180]. This may be attributed to the normal absorptive function of the cotyledons and the requirement for good root aeration and reduced resistance for root growth.

Somatic embryo germination is commonly carried out in the light, and germination of interior, black and white spruce was promoted under low light intensities [232, 244]. However, von Arnold and Hakman [248] reported that germination in the dark was slightly better than in the light with Norway spruce embryos. Photoperiod may also be a critical factor for day length-sensitive plants. The use of a 23-h photoperiod in soybean was necessary to prevent the premature induction of flowering [162], a process which results in the cessation of all vegetative growth, regardless of the size of the seedling.

Finally plantlets grown in the water-saturated *in vitro* environment exhibit reduced cuticular wax development and aberrant stomatal function [249]. Therefore, removal of plants from the *in vitro* to the *ex vitro* environment may lead to reduced conversion rates if not protected from transpirational water loss and stress during the acclimatization phase. The transfer of spruce somatic embryos from culture to soil:peat or vermiculite and growth under high humidity and low light intensity allowed good conversion [232, 244]. However, some terminal bud set and dormancy was observed and was attributed to water stress during the acclimatization process [232]. Somatic embryos that had been dried prior to germination did not exhibit any terminal bud set, abnormal swelling or vitrification [246], and 80% of individuals converted for most interior spruce genotypes tested [244].

While the above medium addenda or environmental manipulations can affect germination capacity, one aspect that cannot be manipulated as readily is the genetic background of the embryo. Studies with plants such as soybean [111, 250] and interior spruce [244] have indicated that germination capacity is greatly influenced by genotype.

VIII. Factors affecting success of delivery systems

The preceding sections have illustrated how advances in understanding of the factors controlling induction and proliferation of embryogenic cultures, and histodifferentiation and maturation of somatic embryos, have led to enhanced somatic embryo production for many agronomic, horticultural and forestry species. As a result of this progress, there has been a corresponding rise in the level of interest in developing efficient methods for delivering the resulting propagules to the greenhouse, nursery and field. The current explosion of published research in this area is far too extensive to be adequately reviewed here. Indeed, an extensive review [233] and an entire book [251] devoted to the subject of synthetic seeds have been published. Here, we select published examples of this technology that illustrate how increased knowledge of critical factors controlling somatic embryo development, maturation, dormancy, quiescence, and conversion has been applied to the development of somatic embryo delivery systems.

A. Artificial seeds

As noted earlier, if tissue culture-derived plants are to compete with seedlings, per unit labor costs must be reduced, and one of the more labor-intensive steps is the transfer of plantlets to *ex vitro* conditions. Fortunately, somatic embryos have a distinct advantage over other *in vitro*-derived propagules in that they correspond in a number of aspects to nature's own propagule-delivery system, the seed. Like seed embryos, somatic embryos are equipped with preformed root and shoot tissues, as well as the "program" to develop into a complete plant. Also, somatic embryos, unlike axillary shoots or adventitious buds, by definition have no vascular connection to source tissue. Thus somatic embryos, unlike other vegetative propagation products, have no requirement for shoot elongation, excision, or rooting steps that must be carried out before the propagule can survive transfer to soil. However, somatic embryos do differ from their zygotic counterparts in that: (1) they do not develop inside a protective seed coat, (2) they do not undergo a controlled desiccation mediated by the maternal tissues, and (3) they are not integrated into a system of storage reserves (endosperm, megagametophyte or storage cotyledons) to provide the necessary energy and other factors needed by the embryo until it begins to photosynthesize. Thus, research directed at designing artificial seeds based on somatic embryos has focused on such areas as development of encapsulation methods, testing of desiccation treatments and improvement of embryo "quality" [252].

In fact, a recurring theme in research with synthetic seeds is that the success of a delivery system depends less on such factors as the engineering of the capsules than on the quality of the embryos themselves. Progress in the area of improving somatic embryo quality, such as maturation treatments, enhancing accumulation of storage products and desiccation treatments, has been summarized above, but since somatic embryo quality (in comparison to that of zygotic embryos) is still the primary barrier to the operational use of somatic embryos for artificial seeds for most species [251], these manipulations bear further discussion in this context.

Although the concept of generating synthetic seeds by encapsulating somatic embryos was first introduced by Murashige [253], the first laboratory testing of this concept was not carried out until years later [254, 255], and the required properties for an effective synthetic seed coat were first summarized by Redenbaugh et al. [256, 257]. Some of the requirements for the encapsulation material noted by these authors were that it be durable enough to protect the embryo during transport and planting, yet still allow conversion. It would also be required to hold and deliver nutrients, carbohydrates and other chemical factors to promote conversion. Redenbaugh et al. [252] divided synthetic seeds into 4 categories: (1) uncoated, desiccated synthetic seeds, (2) coated, desiccated synthetic seeds, (3) coated, hydrated synthetic seeds, and (4) uncoated, hydrated synthetic seeds.

With recent advances in the control and understanding of desiccation of somatic embryos as an effective pretreatment for germination, efforts to develop synthetic seed technology have concentrated on the use of desiccated somatic embryos [252]. As discussed earlier, following pioneering research by Gray [237, 238] with grape and orchardgrass somatic embryos, desiccation has been used to enhance conversion for a number of species. In addition to its effects on conversion, dehydration can be applied to induce somatic embryo quiescence for storage and handling [238]. Early studies reported that viability of desiccated somatic embryos dropped quickly. Orchardgrass somatic embryos, for example, converted at a frequency of 32% after being dehydrated for 24 h to a water content of 13%, but conversion decreased to 4% after 21 days of storage [237]. More recent studies, however, have demonstrated the ability of desiccated somatic embryos to survive long-term storage. Quiescent grape somatic embryos converted at a rate of 34% following 21 days of storage at 70% relative humidity [258]. Alfalfa somatic embryos treated with ABA and air dried to moisture contents of 15% could be stored at room temperature and humidity for 8 months without loss of viability [146], and after a year of storage, 60% of alfalfa somatic embryos given this treatment converted to plantlets [148].

The majority of coated, desiccated synthetic seed research has employed the water-soluble resin polyethylene oxide as the coating agent. Clumps of carrot somatic embryos were encapsulated with polyethylene oxide which was subsequently dried to form wafers [259, 260]. Embryo "hardening" treatments, which included culture with 12% sucrose or 10 μM ABA, increased survival of encapsulated and desiccated somatic embryos up to 58%, but conversion to plantlets was not reported. Individual celery somatic embryos coated with the same resin survived for up to 9 days of desiccation, while uncoated embryos failed to survive beyond 4 days [261]. Some coated celery embryos converted to plantlets [262].

Experiments to test hydrated coatings were originally undertaken because of the low tolerance of somatic embryos for desiccation [252]. Although recent advances in controlling desiccation have shifted the emphasis to desiccated synthetic seeds, many studies testing coated, hydrated synthetic seeds have been published, and the only published field tests of synthetic seeds to date have employed this type of coating [263]. Redenbaugh et al. [255–257] pioneered the use of hydrogels for encapsulating embryos of alfalfa, celery and cauliflower by mixing them with sodium alginate and dropping them into a solution of calcium chloride to form calcium alginate capsules around each embryo (Fig. 5). *In vitro* conversion frequencies of encapsulated embryos ranged as high as 29% and 55% for randomly picked alfalfa somatic embryos and visually selected celery somatic embryos, respectively, while conversion on sand or peat plugs was considerably lower [256]. Thirty-four percent of encapsulated alfalfa somatic embryos matured with 1–5 μM ABA converted following sowing in trays of potting soil in the greenhouse [264] and 23% converted following field planting under styrofoam cups [263]. Somatic embryos of a number of other species have been encapsulated using alginate capsules, including barley [265], carrot [266], sandalwood [267–269], loblolly pine [117], and interior and black spruces [270]. One common observation with regard to hydrogel-encapsulated somatic embryos is that conversion rates for encapsulated embryos are often lower than for nonencapsulated embryos, when conversion is attempted *in vitro* [252, 271]. Encapsulated alfalfa somatic embryos planted in the field or in potting mix in the growth chamber also converted at lower frequencies than naked embryos [263]. Redenbaugh et al. [251] attributed the lowered conversion frequencies of encapsulated embryos to poor embryo respiration due to restricted gas flow through the alginate gel. When stored for 3 months at 4 °C, however, more than 95% of encapsulated carrot somatic embryos germinated while less than 5% of nonencapsulated embryos germinated [272].

Fig. 5. Synthetic seeds of black locust. These were obtained by encapsulating hydrated somatic embryos in calcium alginate gel. Photo courtesy of I. Arrillaga.

The procedure for employing uncoated, hydrated synthetic seeds in a viscous carrier gel, known as fluid drilling, was originally developed as a means of sowing pregerminated seeds to improve seedling emergence [273]. Carrot somatic embryos pretreated with ABA and sucrose, and sowed in a gel supplemented with growth regulators, sucrose, and nutrients onto soil mix in the greenhouse, produced no plantlets [274]. Sweet potato somatic embryos sown in Natrosol® 250 HHR supplemented with Murashige and Skoog medium with sucrose, maltose or glucose converted at frequencies of 20–25% [275]. An interesting variation on the use of uncoated, hydrated synthetic seeds circumvents the use of a gel carrier by pipetting alfalfa somatic embryos directly from suspension culture onto an agar nutrient cap covering a plug of potting medium [276]. When the embryos germinate, their roots grow through the agar cap and establish in the potting medium, and once the plantlets are established they can be removed from aseptic conditions and placed in a mist bed. The authors claim a plantlet recovery rate of up to 97% using this technique.

Clearly, advances in manipulating somatic embryo development, maturation and quiescence have made great advances possible in the area of synthetic seed technology over the past ten years, at least for some important species. In fact, Redenbaugh [251] believes that although embryo quality is still a significant issue, the production of robust and vigorous somatic embryos in such species as alfalfa has become sufficiently routine that new synthetic seed research with these species should focus primarily on development of improved synthetic seed coats.

B. Factors affecting performance of somatic embryo-derived plants

Given that embryogenic systems are capable of producing substantial numbers of propagules for a number of commercially important species, it might be expected that field testing of somatic embryo-derived plants would be well underway. However, problems with somatic embryo quality have limited the number of published reports of ex vitro performance of somatic embryo-derived plants. Until somatic embryos can be consistently fortified to the point where they resemble zygotic embryos with regard to robustness and vigor, there is little point in deploying them in the field. It is probably safe to say that many of the same factors that determine the conversion and early growth of somatic embryos used as artificial seeds will influence their behavior once they are deployed. However, the ability of somatic embryo-derived plantlets to survive and grow in the field is not the only aspect of their performance which is of consequence, as there are likely to be some behaviors associated with their origin that are not apparent until after field establishment. Of these behaviors, one which is of the most concern is the appearance of unexpected variation among regenerants, which may either be epigenetic or heritable (somaclonal). Although it is generally believed that plants regenerated via somatic embryogenesis would be less likely to display within-clone variation than plants regenerated via other in vitro routes [37], this has not been supported by data.

No significant differences were found in frequencies of possible mutations for plantlets derived from embryogenic versus organogenic cultures [277], and embryogenic callus of maize was actually shown to produce a higher frequency of phenotypic variants than organogenic callus [278].

The presence of 2,4-D for somatic embryo induction and proliferation may affect the genetic stability of the embryogenic cells, since this growth regulator is believed responsible for some of the somaclonal variation observed in regenerated plants [279]. In support of this, studies with carrot [280] and yellow-poplar [281] have shown a reduction in ploidy in embryogenic cultures. Cytological analysis of carrot embryogenic cells revealed that a prophase chromosome reduction mechanism was operating, leading to haploid regenerated embryos [280]. Altered ploidy levels are also associated with loss of embryogenicity. Non-embryogenic carrot lines exhibited ploidy reduction via somatic meiosis [23] and reduced embryogenicity in yellow-poplar suspension cultures was associated with a high frequency of haploid cells [281]. Apart from ploidy reduction, Kott et al. [282] noticed an increase in ploidy levels in embryogenic callus that originated from haploid barley embryos. Each embryogenic line displayed its own unique polyploidization rate, with the accumulation of diploid cells a common characteristic.

An extensive review of variability in somatic embryo-derived plants is beyond the scope of this chapter, and we would refer the interested reader to a recent review of the subject by Caligari and Shohet [279]. We would, however, like to call attention to the idea that the same conditions applied for the induction and maintenance of somatic embryogenesis, especially the presence of 2,4-D, are also thought to affect the incidence of both somaclonal and epigenetic variation in plants regenerated from embryogenic cultures [279]. Given the evidence that 2,4-D can induce mutation and other genetic changes in plants (reviewed in Shoemaker et al. [283]), it would be expected that the incidence of somaclonal variation would increase with increasing concentration of 2,4-D employed in the medium and duration of exposure to 2,4-D. However, as was the case with the supposed genetic stability of somatic embryo-derived plants, studies supporting this concept are lacking. In fact, heritable variation in soybean actually decreased with increasing 2,4-D concentration [283]. One group of species where one would expect the long-term effects of the somatic embryo origin of propagules to manifest themselves is woody perennials. To date, however, there are few field plantings of somatic embryo-derived trees and even these have only been established for less than 4 years. Over 1200 interior spruce regenerants tested for nursery performance displayed growth rates, shoot and root morphology, and frost hardiness similar to those of seedlings following the first growing season [244]. The incidence of phenotypic variants was not reported. The best example of long term results for field performance of somatic embryo-derived plants is that of oil palm, since somatic embryogenesis has been used for operational production of oil palm clones from selected ortets for over 10 years [284, 285]. Data collected from clone comparison trials established in three countries showed that somatic embryo-derived material from selected ortets displayed oil productivity which averaged 11% higher than that from seed-derived material. Although one report described flowering abnormalities on somatic embryo-derived oil palms [286], Durand-Gasselin [284] described the phenomenon as rare and characterized by reversion, even on seriously abnormal trees. With regard to the idea that duration in culture affects the incidence of variation, these investigators found no relationship between time spent *in vitro* and abnormality rate.

IX. Conclusions

In the more than three decades since the discovery that somatic embryogenesis could be induced *in vitro*, protocols have evolved to encompass a great diversity of species. Sufficient information on the induction of an embryogenic state, along with the subsequent histodifferentiation, maturation, and germination stages, is available to provide a reasonable starting point when developing or optimizing a regeneration system based on somatic embryogenesis. At this point, additional gains in efficiency – whether in a laboratory petri dish or in a commercial bioreactor – will be made through the adoption of multi-step regeneration systems. In these, each step will entail protocols specifically designed to permit optimal proliferation of embryogenic cells or growth of somatic embryos at each stage of their development. Such a multi-step regeneration system for alfalfa has been presented by McKersie and Bowley [287], which recognizes seven stages in the embryogenic process and nine medium or culture manipulations.

Table 1 presents a convenient way to separate somatic embryogenesis into stages and summarizes the factors controlling events at each stage. However, to make maximum use of the available knowledge on each of these processes for such goals as mass propagation, not only must researchers be prepared to consider all of the factors that influence each stage, they must also understand how the factors active in one stage will influence behavior during subsequent stages. That is to say that although this chapter has artificially divided the morphogenic events of somatic embryogenesis into neat stages, the changes occurring in culture are not always so cleanly segregated. By recognizing the critical factors involved at each stage, and those that exert their influence throughout the process, the protocols at each stage can be tailored to more closely simulate the conditions *in planta*. With informed manipulation of these factors, not only will proliferative embryogenic cultures realize their potential for virtually unlimited propagule production, but the somatic embryos produced will come to have the vigor and germination associated with their zygotic counterparts.

Table 1. Summary of the biochemical, cytological, histological, and morphological events observed during the various stages of somatic embryogenesis, and the manipulations or factors known to influence each stage.

	Induction	Proliferation	Histodifferentiation	Early Maturation	Late Maturation	Desiccation	Germination	Conversion
Biochemical Events:	• mRNA • DNA methylation	• embryogenic tissue	• polarity: CA++ gradients, calmodulin, auxin • extracelluar proteins		• accumulation of: starch, lipids, protein	• quiescence dormancy[3] • germination-specific proteins	• reserve breakdown • endogenous ABA↓	
Cytology, Histology, and Morphology:	• small isodiametric cells • richly cytoplasmic • high starch content callose deposition plasmodesmata severed	• ploidy cycling • same as induction	• cell division • organ differentiation	• cell expansion	• chlorophyll loss[2]		• greening • cell division • cell expansion	• concomitant shoot and root development • autotrophy
Culture Manipulations and Controlling Factors:	• auxins • cytokinins • heavy metals • Cl⁻ ions • osmotic stress • bacterial extracts • wounding/disruption • donor plant conditions • explant status • genotype • gasses • carbohydrate type • nitrogen source • light quality	• auxins • gasses • bioreactors • fractionation • synchronization • genotype	• removal of auxins • removal of cytokinins • activated charcoal • osmoticum • ABA[1]	• ABA[1] • osmoticum	• ABA • osmoticum	• desiccation • thermal treatments • encapsulation	• eliminate osmoticum • salt formulation • cyanobacterial extracts • GA$_3$ • cytokinins[4] • genotype • ethylene inhibitors	• humidity control • light quality/intensity • photoperiod[5] • genotype

[1] Conifers.
[2] In species with chlorophyll in young embryos.
[3] Probable in some species.
[4] Effective in cases where maturation and/or desiccation treatments not applied or ineffective.
[5] Photoperiod-sensitive species.

Acknowledgements

Figures 1 and 2 were drawn by C.G. Hahn. In addition, the authors wish to express their gratitude to Drs. M.A. Bailey, H.Y. Wetzstein, and H.D. Wilde for their critical review of the manuscript.

References

1. Strasburger, E., Über Polyembryonie, *Jenaische Z. Naturwiss.*, 12, 647, 1878.
2. Steward, F.C., Growth and development of cultivated cells. III. Interpretations of the growth from free cell to carrot plant, *Am. J. Bot.*, 45, 709, 1958.
3. Reinert, J., Morphogenese und ihre Kotrolle an Gewebekulturen aus Carotten, *Naturwiss.*, 45, 344, 1958.
4. Levine, M., Differentiation of carrot root tissue grown in culture, *Bull. Torrey Bot. Club*, 74, 321, 1947.
5. Burns, G.W., *The Plant Kingdom*, Macmillan Publishing Co., Inc., New York, 1974.
6. Wing, S.L., Hickey, L.J., and Swisher, C.C., Implications of an exceptional fossil flora for Late Cretaceous vegetation, *Nature*, 363, 342, 1993.
7. Goldberg, R.B., Barker, S.J., and Perez-Grau, L., Regulation of gene expression during plant embryogenesis, *Cell*, 56, 149, 1989.
8. Hughes, D.W., and Galau, G.A., Temporally modular gene expression during cotyledon development, *Genes Dev.*, 3, 358, 1989.
9. Kermode, A.R., Regulatory mechanisms involved in the transition from seed development to maturation, *Crit. Rev. Pl. Sci.*, 9, 155, 1990.
10. Kohlenbach, H., Comparative somatic embryogenesis, in *Frontiers of Plant Tissue Culture*, Thorpe, T.A., Ed., University of Calgary, Calgary, 1978, 59.
11. Goldberg, R.B., Regulation of plant gene expression, *Phil. Trans. R. Soc. Lond. B*, 314, 343, 1986.
12. Sharp, W.R., Evans, D.A., and Sondahl, M.R., Application of somatic embryogenesis to crop improvement, in *Plant Tissue Culture 1983. Proceedings of the Fifth International Congress of Plant Tissue Culture*, Fujiwara, A., Ed., Japanese Association for Plant Tissue Culture, 1982, 759.
13. Evans, D.A., Sharp, W.R., and Flick, C.E., Growth and behaviour of cell cultures. Embryogenesis and organogenesis, in *Plant Tissue Culture: Methods and Applications in Agriculture*, Thorpe, T.A., Ed., Academic Press, New York, 1981, 45.
14. Williams, E.G., and Maheswaran, G., Somatic embryogenesis: factors influencing coordinated behaviour of cells as an embryogenic group, *Ann. Bot.*, 57, 443, 1986.
15. Carman, J.G., Embryogenic cells in plant tissue cultures: occurrence and behavior, *In Vitro Cell. Dev. Biol.*, 26, 746, 1990.
16. Merkle, S.A., Parrott, W.A., and Williams, E.G., Applications of somatic embryogenesis and embryo cloning, in *Plant tissue culture: Applications and limitations*, Bhojwani, S.S., Ed., Elsevier Science Publishers, Amsterdam, 1990, 67.
17. LoSchiavo, F., Pitto, L., Giuliano, G., Torti, G., Nuti-Ronchi, V., Marazziti, D., Vergara, R., Orselli, S., and Terzi, M., DNA methylation of embryogenic carrot cell cultures and its variations as caused by mutation, differentiation, hormones and hypomethylating drugs, *Theor. Appl. Genet.*, 77, 325, 1989.
18. Pechan, P.M., Bartels, D., Brown, D.C.W., and Schell, J., Messenger-RNA and protein changes associated with induction of *Brassica* microspore embryogenesis, *Planta*, 184, 161, 1991.
19. Kamada, H., Kobayashi, K., Kiyosue, T., and Harada, H., Stress induced somatic embryogenesis in carrot and its application to synthetic seed production, *In Vitro Cell. Dev. Biol.*, 25, 1163, 1989.

192

20. Imamura, J., and Harada, H., Stimulation of tobacco pollen embryogenesis by anaerobic treatments, *J. Plant Physiol.*, 103, 259, 1981.
21. Hakman, I., and Von Arnold, S., Plantlet regeneration through somatic embryogenesis in *Picea abies* (Norway spruce), *J. Plant Physiol.*, 121, 149, 1985.
22. Herberle-Bors, E., *In vitro* haploid formation from pollen: a critical review, *Theor. Appl. Genet.*, 71, 361, 1985.
23. Nuti Ronchi, V., Giorgetti, L., Tonelli, M., and Martini, G., Ploidy reduction and genome segregation in cultured carrot cell lines. II. Somatic meiosis, *Plant Cell Tissue Organ Cult.*, 30, 115, 1992.
24. Bonnelle, C., Lejeune, F., Fournier, D., and Tourte, Y., Remaniements infrastructuraux des cellules cotylédonaires en culture chez deux légumineuses; relations avec l'acquisition de potentialités embryogènes, *C.R. Acad. Sci. Paris*, 310, 657, 1990.
25. Vergara, R., Verde, F., Pitto, L., LoSchiavo, F., and Terzi, M., Reversible variations in the methylation pattern of carrot DNA during somatic embryogenesis, *Plant Cell Rep.*, 8, 697, 1990.
26. Smith, D.L., and Krikorian, A.D., Release of somatic embryogenic potential from excised zygotic embryos of carrot and maintenance of proembryonic cultures in hormone-free medium, *Am. J. Bot.*, 76, 1832, 1989.
27. Trigiano, R.N., Gray, D.J., Conger, B.V., and McDaniel, J.K., Origin of direct somatic embryos from cultured leaf segments of *Dactylis glomerata*, *Bot. Gaz.*, 150, 72, 1989.
28. Dubois, T., Guedira, M., Dubois, J., and Vasseur, J., Direct somatic embryogenesis in roots of *Cichorium*: Is callose an early marker? *Ann. Bot.*, 65, 539, 1990.
29. Dubois, T., Guedira, M., and Vasseur, J., Direct somatic embryogenesis in leaves of *Cichorium*, *Protoplasma*, 162, 120, 1991.
30. Wetherell, D.F., Enhanced adventive embryogenesis resulting from plasmolysis of cultured wild carrot cells, *Plant Cell Tiss. Org. Cult.*, 3, 221, 1984.
31. Tsukahara, M., and Hirosawa, T., Simple dehydration treatment promotes plantlet regeneration of rice (*Oryza sativa* L.) callus, *Plant Cell Rep.*, 11, 550, 1992.
32. Chuang, M.-J., and Chang, W.-C., Somatic embryogenesis and plant regeneration in callus culture derived from immature seeds and mature zygotic embryos of *Dysosma pleintha* (Hance) Woodson, *Plant Cell Rep.*, 6, 484, 1987.
33. Gharyal, P.K., and Maheshwari, S.C., *In vitro* differentiation of somatic embryoids in a leguminous tree – *Albizzia lebbeck* L, *Naturwiss.*, 68, 379, 1983.
34. Dhanalakshmi, S., and Lakshmanan, K.K., *In vitro* somatic embryogenesis and plant regeneration in *Clitoria ternatea*, *J. Exp. Bot.*, 43, 213, 1992.
35. Street, H.E., and Withers, L.A., The anatomy of embryogenesis in culture, in *Tissue Culture and Plant Science*, Street, H.E., Ed., Academic Press, London/New York, 1974, 71.
36. Haccius, B., Question of unicellular origin of non-zygotic embryos in callus cultures, *Phytomorphology*, 28, 74, 1977.
37. Vasil, I.K., Developing cell and tissue culture systems for the improvement of cereal and grass crops, *J. Plant Physiol.*, 128, 193, 1987.
38. Hartweck, L.M., Lazzeri, P.A., Cui, D., Collins, G.B., and Williams, E.G., Auxin-orientation effects on somatic embryogenesis from immature soybean cotyledons, *In Vitro Cell. Dev. Biol.*, 24, 821, 1988.
39. Sato, S., Newell, C., Kolacz, K., Tredo, L., Finer, J., and Hinchee, M., Stable transformation via particle bombardment in two different soybean regeneration systems, *Plant Cell Rep.*, 12, 408, 1993.
40. Polito, V.S., McGranahan, G., Pinney, K., and Leslie, C., Origin of somatic embryos from repetitively embryogenic cultures of walnut (*Junglans regia* L.): implications for *Agrobacterium*-mediated transformation, *Plant Cell Rep.*, 8, 219, 1989.
41. Norgaard, J.V., and Krogstrup, P., Cytokinin induced somatic embryogenesis from immature embryos of *Abies nordmanniana* Lk, *Plant Cell Rep.*, 9, 509, 1991.
42. Maheswaran, G., and Williams, E.G., Direct somatic embryoid formation on immature embryos of *Trifolium repens*, *T. pratense* and *Medicago sativa*, and rapid clonal propagation of *T. repens*, *Ann. Bot.*, 54, 201, 1984.

43. Maheswaran, G., and Williams, E.G., Origin and development of somatic embryoids formed directly on immature embryos of *Trifolium repens in vitro*, *Ann. Bot.*, 56, 619, 1985.

44. Mo, L.H., and Von Arnold, S., Origin and development of embryogenic cultures from seedlings of Norway spruce, *J. Plant Physiol.*, 138, 223, 1991.

45. Saunders, J.W., and Bingham, E.T., Growth regulator effects on bud initiation in callus cultures of *Medicago sativa*, *Am. J. Bot.*, 62, 850, 1975.

46. Walker, K.A., Yu, P.C., Sato, S.J., and Jaworski, E.G., The hormonal control of organ formation in callus of *Medicago sativa* L. cultured *in vitro*, *Am. J. Bot.*, 65, 654, 1982.

47. Saxena, P.K., and King, J., Morphogenesis in lentil: plant regeneration from callus cultures of *Lens culinaris* Medik. via somatic embryogenesis, *Plant Sci.*, 52, 223, 1987.

48. Rajasekaran, K., Hein, M.B., Davis, G.C., Carnes, M.G., and Vasil, I.K., Endogenous growth regulators in leaves and tissue cultures of *Pennisetum purpureum* Schum, *J. Plant Physiol.*, 130, 13, 1987.

49. Wenck, A.R., Conger, B.V., Trigiano, R.N., and Sams, C.E., Inhibition of somatic embryogenesis in orchardgrass by endogenous cytokinins, *Plant Sci.*, 88, 990, 1988.

50. Arcioni, S., and Mariotti, D., Tissue culture and plant regeneration in the forage legumes *Onobrychis viciaefolia* Scop., *Coronilla varia* and *Lotus corniculatus* L, in *Plant Tissue Culture*, Fujiwara, A., Ed., Japanese Association for Plant Tissue Culture, Tokyo, 1982, 707.

51. Kysely, W., and Jacobsen, H.-J., Somatic embryogenesis from pea embryos and shoot apices, *Plant Cell Tiss. Org. Cult.*, 20, 7, 1990.

52. Lippmann, B., and Lippmann, G., Induction of somatic embryos in cotyledonary tissue of soybean, *Glycine max* L. Merr., *Plant Cell Rep.*, 3, 215, 1984.

53. Lazzeri, P.A., Hildebrand, D.F., and Collins, G.B., Soybean somatic embryogenesis: effects of hormones and culture manipulations, *Plant Mol. Biol. Rep.*, 10, 197, 1987.

54. Nadolska-Orczyk, A., Somatic embryogenesis of agriculturally important lupin species (*Lupinus angustifolius*, *L. albus*, *L. mutabilis*), *Plant Cell Tiss. Org. Cult.*, 28, 19, 1992.

55. Parrott, W.A., Merkle, S.A., and Williams, E.G., Somatic embryogenesis: potential for use in propagation and gene transfer systems, in *Advanced Methods in Plant Breeding and Biotechnology*, Murray, D.R., Ed., CAB International, Wallingford, 1991, 158.

56. Komatsuda, T., Kaneko, K., and Oka, S., Genotype × sucrose interactions for somatic embryogenesis in soybean, *Crop Sci.*, 31, 333, 1991.

57. Pretova, A., and Williams, E.G., Direct somatic embryogenesis from immature embryos of flax (*Linum usitatissimum* L.), *J. Plant Physiol.*, 126, 155, 1986.

58. Young, R., Kaul, V., and Williams, E.G., Clonal propagation *in vitro* from immature emryos and flower buds of *Lycopersicon peruvianum* and *L. esculentum*, *Plant Sci.*, 52, 237, 1987.

59. Ghazi, T.D., Cheema, H.V., and Nabors, M.W., Somatic embryogenesis and plant regeneration from embryogenic callus of soybean, *Glycine max* L., *Plant Cell Rep.*, 5, 452, 1986.

60. Ranch, J.P., Oglesby, L., and Zielinski, A.C., Plant regeneration from embryo-derived tissue cultures of soybean by somatic embryogenesis, *In Vitro Cell. Dev. Biol.*, 21, 653, 1985.

61. Barwale, U.B., Kerns, H.R., and Widholm, J.M., Plant regeneration from callus cultures of several soybean genotypes via embryogenesis and organogenesis, *Planta*, 167, 473, 1986.

62. Roberts, D.R., Flinn, B.S., Webb, D.T., Webster, F.B., and Sutton, B.C.S., Characterization of immature embryos of interior spruce by SDS-PAGE and microscopy in relation to their competence for somatic embryogenesis, *Plant Cell Rep.*, 8, 285, 1989.

63. Sotak, R.J., Sommer, H.E., and Merkle, S.A., Relation of the developmental stage of zygotic embryos of yellow-poplar to their somatic embryogenic potential, *Plant Cell Rep.*, 10, 175, 1991.

64. Karunaratne, S., Gamage, C., and Kovoor, A., Leaf maturity, a critical factor in somatic embryogenesis, *J. Plant Physiol.*, 139, 27, 1991.

65. Williams, E.G., Collins, G.B., and Myers, J.R., 2. Clovers (*Trifolium* spp.), in *Biotechnology in Agriculture and Forestry, Vol. 10, Legumes and Oilseed Crops, I.*, Bajaj, Y.P.S., Ed., Springer-Verlag, Berlin/Heidelberg, 1990, 242.

66. Jacobsen, H.-J., Somatic embryogenesis in seed legumes: The possible role of soluble auxin receptors, *Israel J. Bot.*, 40, 139, 1991.

194

67. Tremblay, F.M., Somatic embryogenesis and plantlet regeneration from embryos isolated from stored seeds of *Picea glauca*, *Can. J. Bot.*, 68, 236, 1990.

68. Reynolds, T.L., Interactions between calcium and auxin during pollen androgenesis in anther cultures of *Solanum carolinense* L, *Plant Sci.*, 72, 109, 1990.

69. Jansen, M.A.K., Booij, H., Schel, J.H.N., and De Vries, S.C., Calcium increases the yield of somatic embryos in carrot embryogenic suspension cultures, *Plant Cell Rep.*, 9, 221, 1990.

70. Lazzeri, P.A., Hildebrand, D.F., Sunega, J., Williams, E.G., and Collins, G.B., Soybean somatic embryogenesis: interactions between sucrose and auxin, *Plant Cell Rep.*, 7, 517, 1988.

71. Genga, A., and Allavena, A., Factors affecting morphogenesis from immature cotyledons of *Phaseolus coccineus* L, *Plant Cell Tiss. Org. Cult.*, 27, 189, 1991.

72. Kochba, J., Spiegel-Roy, P., Neumann, H., and Saad, S., Effect of carbohydrates on somatic embryogenesis in subcultured nucellar callus of *Citrus* cultivars, *Z. Pflanzenphysiol.*, 105, 359, 1982.

73. Strickland, S.G., Nichol, J.W., McCall, C.M., and Stuart, D.A., Effect of carbohydrate source on alfalfa somatic embryogenesis, *Plant Sci.*, 48, 113, 1987.

74. Wetherell, D.F., and Dougall, D.K., Sources of nitrogen supporting growth and embryogenesis in cultured wild carrot tissue, *Physiol. Plant.*, 37, 97, 1976.

75. Tremblay, L., and Tremblay, F.M., Effects of gelling agents, ammonium nitrate, and light on the development of *Picea mariana* (Mill) B.S.P. (black spruce) and *Picea rubens* Sarg. (red spruce) somatic embryos, *Plant Sci.*, 77, 233, 1991.

76. Stuart, D.A., and Strickland, S.G., Somatic embryogenesis from cell cultures of *Medicago sativa* 1. the role of amino acid additions to the regeneration medium, *Plant Sci. Lett.*, 34, 165, 1984.

77. Stuart, D.A., and Strickland, S.G., Somatic embryogenesis from cell cultures of *Medicago sativa* 2. The interaction of amino acids with ammonium, *Plant Sci. Lett.*, 34, 175, 1984.

78. Walker, K.A., and Sato, S.J., Morphogenesis in callus tissue of *Medicago sativa*: the role of ammonium ion in somatic embryogenesis, *Plant Cell Tiss. Org. Cult.*, 1, 109, 1981.

79. Meijer, E.G.M., and Brown, D.C.W., Role of exogenous reduced nitrogen and sucrose in rapid high frequency somatic embryogenesis in *Medicago sativa*, *Plant Cell Tissue Organ Cult.*, 10, 11, 1987.

80. Greinwald, R., and Czygan, F.-C., Regeneration of plantlets from callus cultures of *Chamaecytisus purpureus* and *Chamaecytisus austriacus* (Leguminosae), *Bot. Acta*, 104, 64, 1991.

81. Trigiano, R.N., Beaty, R.M., and Graham, E.T., Somatic embryogenesis from immature embryos of redbud (*Cercis canadensis*), *Plant Cell Rep.*, 7, 148, 1988.

82. Trigiano, R.N., and Conger, B.V., Regulation of growth and somatic embryogenesis by proline and serine in suspension cultures of *Dactylis glomerata*, *J. Plant Physiol.*, 130, 49, 1987.

83. Gleddie, S., Keller, W., and Setterfield, G., Somatic embryogenesis and plant regeneration from leaf explants and cell suspensions of *Solanum melongena* (eggplant), *Can. J. Bot.*, 61, 656, 1983.

84. Michler, C.H., and Bauer, E.O., High frequency somatic embryogenesis from leaf tissue of *Populus* spp, *Plant Sci.*, 77, 111, 1991.

85. Arumugan, N., and Bhojwani, S.S., Somatic embryogenesis in tissue cultures of *Podophyllum hexandrum*, *Can. J. Bot.*, 68, 487, 1990.

86. Verhagen, S.A., and Wann, S.R., Norway spruce somatic embryogenesis: high-frequency initiation from light-cultured mature embryos, *Plant Cell Tiss. Org. Cult.*, 16, 103, 1989.

87. Calero, N., Actions de radiations rouges et bleues sur l'embryogenèse somatique du Palmier dattier (*Phoenix dactylifera* L.) en culture *in vitro* et sur sa teneur en leucoanthocyanes, *C.R. Soc. Biol.*, 183, 307, 1989.

88. Michler, C.H., and Lineberger, R.D., Effects of light on somatic embryo development and abscisic levels in carrot suspension cultures, *Plant Cell Tiss. Org. Cult.*, 11, 189, 1987.

89. Lazzeri, P.A., Hildebrand, D.F., and Collins, G.B., Soybean somatic embryogenesis: effects of nutritional, physical and chemical factors, *Plant Mol. Biol. Rep.*, 10, 209, 1987.

90. Roustan, J.P., Latche, A., and Fallot, J., Stimulation of *Daucus carota* somatic embryogenesis by inhibitors of ethylene synthesis: cobalt and nickel, *Plant Cell Rep.*, 8, 182, 1989.

91. Roustan, J.-P., Latche, A., and Fallot, J., Inhibition of ethylene production and stimulation of carrot somatic embryogenesis by salicylic acid, *Biol. Plant.*, 32, 273, 1990.

92. Auboiron, E., Carron, M.-P., and Michaux-Ferrière, M., Influence of atmospheric gases, particularly ethylene, on somatic embryogenesis of *Hevea brasiliensis*, *Plant Cell Tiss. Org. Cult.*, 21, 31, 1990.

93. Carman, J.G., The *in ovulo* environment and its relevance to cloning wheat via somatic embryogenesis, *In Vitro Cell. Dev. Biol.*, 25, 1155, 1989.

94. Carman, J.G., Improved somatic embryogenesis in wheat by partial simulation of the in-ovulo oxygen, growth-regulator and desiccation environments, *Planta*, 175, 417, 1988.

95. Yang, Y.-S., Wada, K., Goto, M., and Futsuhara, Y., *In vitro* formation of nodular calli in soybean (*Glycine max* L.) induced by cocultivated *Pseudomonas maltophilia*, *Japan J. Breed.*, 41, 595, 1991.

96. Wake, H., Umetsu, H., Ozeki, Y., Shimomura, K., and Matsunaga, T., Extracts of marine cyanobacteria stimulated somatic embryogenesis of *Daucus carota* L, *Plant Cell Rep.*, 9, 655, 1991.

97. McCoy, T.J., Tissue culture evaluation of sodium chloride tolerance in *Medicago* sp cellular versus whole plant response, *Plant Cell Rep.*, 6, 31, 1987.

98. Mohan Ram, H.Y., Mehta, U., Ramanuja Rao, I.V., and Narasimham, M., Haploid induction in legumes, in *Plant Tissue Culture*, Fujiwara A., Ed., Japanese Association for Plant Tissue Culture, Tokyo, 1982, 541.

99. Halperin, W., Alternative morphogenetic events in cell suspensions, *Am. J. Bot.*, 53, 443, 1966.

100. McWilliam, A.A., Smith, S.H., and Street, H.E., The origin and development of embryoids in suspension culutres of carrot (*Daucus carota*), *Ann. Bot.*, 38, 243, 1974.

101. Halperin, W., and Wetherell, D.F., Adventive embryony in tissue cultures of the wild carrot, *Daucus carota*, *Am. J. Bot.*, 51, 274, 1964.

102. Lupotto, E., Propagation of an embryogenic culture of *Medicago sativa* L., *Z. Pflanzenphysiol.*, 111, 95, 1983.

103. Merkle, S.A. and Wiecko, A.T., Regeneration of *Robinia psuedoacacia* via somatic embryogenesis, *Can. J. For. Res.*, 19, 285, 1989.

104. Tulecke, W., and McGranahan, G., Somatic embryogenesis and plant regeneration from cotyledons of *Juglans regia* L, *Plant Sci. Lett.*, 1985.

105. Wetzstein, H.Y., Ault, J.R., and Merkle, S.A., Further characterization of somatic embryogenesis and plantlet regeneration in pecan (*Carya illinoensis*), *Plant Sci.*, 64, 193, 1989.

106. Smith, D.L., and Krikorian, A.D., Growth and maintenance of an embryogenic cell culture of daylily (*Hemerocallis*) on hormone-free medium, *Ann. Bot.*, 67, 443, 1991.

107. Loh, C.-S., Ingram, D.S., and Hanke, D.E., Cytokinins and the regeneration of plantlets from secondary embryoids of winter oilseed rape, *Brassica napus* ssp. *oleifera*, *New Phytol.*, 95, 349, 1983.

108. Smith, D.L., and Krikorian, A.D., Somatic proembryo production from excised, wounded zygotic carrot embryos on hormone-free medium: evaluation of the effects of pH, ethylene and activated charcoal, *Plant Cell Rep.*, 9, 34, 1990.

109. Becwar, M.R., Noland, T.L., and Wann, S.R., A method for quanitification of the level of somatic embryogenesis among Norway spruce callus lines, *Plant Cell Rep.*, 6, 35, 1987.

110. Parrott, W.A., and Bailey, M.A., Characterization of recurrent somatic embryogenesis of alfalfa on auxin-free medium and its control by sugars and nicotinic acid, *Plant Cell Tiss. Org. Cult.*, 32, 69, 1993.

111. Bailey, M.A., Boerma, H.R., and Parrott, W.A., Genotype effects on repetitive embryogenesis and plant regeneration of soybean, *In Vitro Cell. Dev. Biol.*, 29P, 102, 1993.

112. Halperin, W., and Jensen, W.A., Ultrastructural changes during growth and embryogenesis in carrot cell cultures, *J. Ultrastructure Res.*, 18, 428, 1967.

113. Button, J., Kochba, J., and Bornman, C.H., Fine structure of and embryoid development from embryogenic ovular callus of "Shamouti" orange (*Citrus sinensis* Osb.), *J. Exp. Bot.*, 25, 446, 1974.

114. Finer, J.J., and Nagasawa, A., Development of an embryogenic suspension culture of soybean (*Glycine max* Merrill.), *Plant Cell Tiss. Org. Cult.*, 15, 125, 1988.

115. Lupotto, E., The use of single somatic embryo culture in propagating and regenerating lucerne (*Medicago sativa* L.), *Ann. Bot.*, 57, 19, 1986.

116. Merkle, S.A., Somatic embryogenesis in black locust, in *Black Locust: Biology, Culture and Utilization*, Hanover, J.W., Miller, K., and Plesko, S., Eds., Michigan State University, East Lansing, 1992, 136.

117. Gupta, P.K., and Durzan, D.J., Biotechnology of somatic polyembryogenesis and plantlet regeneration on loblolly pine, *Bio/Technol.*, 5, 147, 1987.

118. Durzan, D.J., and Gupta, P.K., Somatic embryogenesis and polyembryogenesis in Douglas-fir suspension cultures, *Plant Sci.*, 52, 229, 1987.

119. Gupta, P.K., and Durzan, D.J., Somatic polyembryogenesis from callus of mature sugar pine embryos, *Bio/Technol.*, 4, 643, 1986.

120. Becwar, M.R., Nagmani, R., and Wann, S.R., Initiation of embryogenic cultures and somatic embryo development in loblolly pine (*Pinus taeda*), *Can. J. For. Res.*, 20, 810, 1990.

121. Durzan, D.J., and Gupta, P.K., Somatic embryogenesis and polyembryogenesis in conifers, in *Biotechnology in Agriculture*, Anonymous, Ed., Alan R. Liss, Inc, 1988, 53.

122. Attree, S.M., Moore, D., Sawhney, V.K., and Fowke, L.C., Enhanced maturation and desiccation tolerance of white spruce [*Picea glauca* (Moench) Voss] somatic embryos: Effects of a non-plasmolysing water stress and abscisic acid, *Ann. Bot.*, 68, 519, 1991.

123. Gupta, P.K., Timmis, R., Pullman, G., Yancey, M., Kreitinger, M., Carlson, W., and Carpenter, C., Development of an embryogenic system for automated propagation of forest trees, in *Scale-Up and Automation in Plant Propagation. Cell Culture and Somatic Cell Genetics, Vol. 8*, Vasil, I.K., Ed., Academic Press, Inc., San Diego, 1991, 75.

124. Schuller, A., Reuther, G., and Geier, T., Somatic embryogenesis from seed explants of *Abies alba*, *Plant Cell Tiss. Org. Cult.*, 17, 53, 1989.

125. Kessell, R.H.J., and Carr, A.H., The effect of dissolved oxygen concentration on growth and differentiation of carrot (*Daucus carota*) tissue, *J. Exp. Bot.*, 23, 996, 1972.

126. Priel, W., Florek, P., Wix, U., and Beck, A., Towards mass propagation by use of bioreactors, *Acta Hort.*, 226, 99, 1988.

127. Kvaalen, H., and Von Arnold, S., Effects of various partial pressures of oxygen and carbon dioxide on different stages of somatic embryogenesis in *Picea abies*, *Plant Cell Tiss. Org. Cult.*, 27, 49, 1991.

128. Kumar, P.P., Joy, R.W. IV, and Thorpe, T.A., Ethylene and carbon dioxide accumulation, and growth of cell suspension cultures of *Picea glauca* (white spruce), *J. Plant Physiol.*, 135, 592, 1989.

129. Vasil, I.K., Rationale for the scale-up and automation of plant propagation, in *Scale-Up and Automation in Plant Propagation. Cell Culture and Somatic Cell Genetics of Plants, Volume 8*, Vasil, I.K., Ed., Academic Press, Inc., San Diego, 1991, 1.

130. Denchev, P.D., Kuklin, A.E., and Scragg, A.H., Somatic embryo production in bioreactors, *J. Biotech.*, 26, 99, 1992.

131. Backs-Hüsemann, D., and Reinert, J., Embryobildung durch isolierte Einzelzellen aus Gewebekulturen von *Daucus carota*, *Protoplasma*, 70, 49, 1970.

132. Ten Hoopen, H.J.G., Van Gulik, W.M., and Meijer, J.J., Possibilities, problems, and pitfalls of large-scale plant cell cultures, in *Progress in Plant Cellular and Milecular Biology*, Nijkamp, H.J.J., Van der Plas, L.W.H., and Van Aatrijk, J., Eds., Kluwer Academic Publishers, Dordrecht, 1990, 673.

133. Fowler, M.W., Process systems and approaches for large scale plant cell culture, in *Plant Tissue and Cell Culture*, Green, C.E., Somers, D.A., Hackett, W.P., and Biesboer, D.D., Eds., Alan R. Liss, Inc., New York, 1987, 459.

134. Fowler, M.W., Large-scale cultures of cells in suspension, in *Cell Culture and Somatic Cell Genetics of Plants, Vol. 1*, Vasil, I.K., Ed., Academic Press, New York, 1984, 167.

135. Stuart, D.A., Strickland, S.G., and Walker, K.A., Bioreactor production of alfalfa somatic embryos, *HortSci.*, 22, 800, 1987.

136. Chen, T.H.H., Thompson, B.G., and Gerson, D.F., *In vitro* production of alfalfa somatic embryos in fermentation systems, *J. Ferment. Technol.*, 65, 353, 1987.

137. Terashima, T., and Nishimura, S., Mass propagation of somatic embryos in carrot (*Daucus carota*) using bioreactor, *Japan J. Breed.*, 41, 234, 1991.

138. Preil, W., Florek, P., and Wix, U., Towards mass propagation by use of bioreactors, *Acta Hort.*, 226, 99, 1988.

139. Bapat, V.A., Fulzele, D.P., Heble, M.R., and Rao, P.S., Production of sandalwood somatic embryos in bioreactors, *Curr. Sci.*, 59, 746, 1990.

140. Ammirato, P.V., and Styer, D.J., Strategies for large-scale manipulation of somatic embryos in suspension culture, in *Biotechnology in Plant Science*, Zaitlin, M., Day, P., and Hollaender, A., Eds., Academic Press, Inc., Orlando, 1985, 161.

141. Styer, D.J., Bioreactor technology for plant propagation, in *Tissue Culture in Forestry and Agriculture*, Henke, R.R., Hughes, K.W., Constantin, M.J., and Hollaender, A., Eds., Plenum Press, New York, 1985, 117.

142. Levin, R., Gaba, V., Tal., Hirsch, S., DeNola, D., and Vasil, I.K., Automated plant tissue culture for mass propagation, *Bio/Technol.*, 6, 1035, 1988.

143. Levin, R., and Vasil, I.K., An integrated and automated tissue culture system for mass propagation of plants, *In Vitro Cell. Dev. Biol.*, 25, 21, 1989.

144. Cazzulino, D.L., Pedersen, H., Chin, C.K., and Styer, D., Kinetics of carrot somatic embryo development in suspension culture, *Biotech. Bioeng.*, 35, 781, 1990.

145. Cazzulino, D., Pedersen, H., and Chin, C.-K., Bioreactors and image analysis for scale-up and plant propagation, in *Scale-Up and Automation in Plant Propagation. Cell Culture and Somatic Cell Genetics of Plants, Vol. 8*, Vasil, I.K., Ed., Academic Press, Inc., San Diego, 1991, 147.

146. McKersie, B.D., Senaratna, T., Bowley, S.R., Brown, D.C.W., Krochko, J.E., and Bewley, J.D., Application of artificial seed technology in the production of hybrid alfalfa (*Medicago sativa* L.), *In Vitro Cell. Dev. Biol.*, 25, 1183, 1989.

147. Senaratna, T., McKersie, B.D., and Bowley, S.R., Desiccation tolerance of alfalfa (*Medicago sativa* L.) somatic embryos. Influence of abscisic acid, stress pretreatments and drying rates, *Plant Sci.*, 65, 253, 1989.

148. Senaratna, T., McKersie, B.D., and Bowley, S.R., Artificial seeds of alfalfa (*Medicago sativa* L.). Induction of desiccation tolerance in somatic embryos, *In Vitro Cell. Dev. Biol.*, 26, 85, 1990.

149. Merkle, S.A., Wiecko, A.T., Sotak, R.J., and Sommer, H.E., Maturation and conversion of *Liriodendron tulipifera* somatic embryos, *In Vitro Cell. Dev. Biol.*, 26, 1086, 1990.

150. Walker, C.C., *Growth of embryogenic Norway spruce cultures in a bioreactor. A - 190 Independent Study. The Institute of Paper Science and Technology, Atlanta, GA*, 1989.

151. Attree, S.M., Pomeroy, M.K., and Fowke, L.C., Production of vigorous, desiccation tolerant white spruce [*Picea glauca* (Moench) Voss.] synthetic seeds in a bioreactor, *Plant Cell Rep.*, 13, 601, 1994.

152. LoSchiavo, F., Giuliano, G., De Vries, S.C., Genga, A., Bollini, R., Pitto, L., Cozzani, F., Nuti-Ronchi, V., and Terzi, M., A carrot cell variant temperature sensitive for somatic embryogenesis reveals a defect in the glycosylation of extracellular proteins, *Mol. Gen. Genet.*, 223, 385, 1990.

153. De Jong, A.J., Cordewener, J., Lo Schiavo, F., Terzi, M., Vandekerckhove, J., Van Kammen, A., and De Vries, S.C., A carrot somatic embryo mutant is rescued by chitinase, *Plant Cell*, 4, 425, 1992.

154. De Jong, A.J., Heidstra, R., Spaink, H.P., Hartog, M.V., Meijer, E.A., Hendriks, T., Lo Schiavo, F., Terzi, M., Bisseling, T., Van Kammen, A., and De Vries, S.C., *Rhizobium* lipooligosaccharides rescue a carrot somatic embryo mutant, *Plant Cell*, 5, 615, 1993.

155. Caligo, M.A., Nuti Ronchi, V., and Nozzolini, M., Proline and serine affect polarity and development of carrot somatic embryos, *Cell Differentiation*, 17, 193, 1985.

156. Fobert, P.R., and Webb, D.T., Effect of polyamines, polyamine precursors, and polyamine biosynthetic inhibitors on somatic embryogenesis from eggplant (*Solanum melongena*) cotyledons, *Can. J. Bot.*, 66, 1734, 1988.

198

157. Conger, B.V., Hovanesian, S.C., Trigiano, R.N., and Gray, D.J., Somatic embryo ontogeny in suspension cultures of orchardgrass, *Crop Sci.*, 29, 448, 1989.

158. Litz, R.E., Effect of osmotic stress on somatic embryogenesis in *Carica* suspension cultures, *J. Amer. Soc. Hort. Sci.*, 111, 969, 1986.

159. Ammirato, P.V., The effects of abscisic acid on the development of somatic embryos from cells of caraway (*Carum carvi* L.), *Bot. Gaz.*, 135, 328, 1974.

160. Ammirato, P.V., Hormonal control of somatic embryo development from cultured cells of caraway. Interactions of abscisic acid, zeatin, and gibberellic acid, *Plant Physiol.*, 59, 579, 1977.

161. Duskova, J., Opantrny, Z., Sovova, M., and Dusek, J., Somatic embryogenesis and plant regeneration in *Coronilla varia* L. (crownvetch) long-term tissue cultures, *Biol. Plant*, 32, 8, 1990.

162. Parrott, W.A., Dryden, G., Vogt, S., Hildebrand, D.F., Collins, G.B., and Williams, E.G., Optimization of somatic embryogenesis and embryo germination in soybean, *In Vitro Cell. Dev. Biol.*, 24, 817, 1988.

163. Vasil, I.K., Somatic embryogenesis and its consequences in the gramineae, in *Tissue Culture in Agriculture and Forestry, Vol. 31*, Henke, R.R., Hughes, K.W., Constantin, M.J., and Hollaender, A., Eds., Plenum Press, New York, 1985, 31.

164. Kearney, J.F., Parrott, W.A., and Hill, N.S., Infection of somatic embryos of tall fescue with *Acremonium coenophialum*, *Crop Sci.*, 31, 979, 1991.

165. Zaghmout, O.M.F., and Torello, W.A., Enhanced regeneration from long-term callus cultures of red fescue by pretreatment with activated charcoal, *HortSci.*, 23, 615, 1988.

166. Roberts, D.R., Flinn, B.S., Webb, D.T., Webster, F.B., and Sutton, B.C.S., Abscisic acid and indole-3-butyric acid regulation of maturation and accumulation of storage proteins in somatic embryos of interior spruce, *Physiol. Plant.*, 78, 355, 1990.

167. Michalczuk, L., Cooke, T.J., and Cohen, J.D., Auxin levels at different stages of carrot somatic embryogenesis, *Phytochem.*, 31, 1097, 1992.

168. Gorst, J., Overall, R.L., and Wernicke, W., Ionic currents traversing cell clusters from carrot suspension cultures reveal perpetuation of morphogenetic potential as distinct from induction of embryogenesis, *Cell Differentiation*, 21, 101, 1987.

169. Overall, R.L., and Wernicke, W., Steady ionic currents around haploid embryos formed from tobacco pollen in culture, in *Ionic Currents in Development*, Nuccitelli, R., Ed., Alan R. Liss, Inc., New York, 1986, 139.

170. Rathore, K.S., Hodges, T.K., and Robinson, K.R., Ionic basis of currents in somatic embryos of *Daucus carota*, *Planta*, 175, 280, 1988.

171. Schiavone, F.M., and Cooke, T.J., Unusual patterns of somatic embryogenesis in the domesticated carrot: developmental effects of exogenous auxins and auxin transport inhibitors, *Cell Differentiation*, 21, 53, 1987.

172. Nomura, K., and Komamine, A., Molecular mechanisms of somatic embryogenesis, *Oxford Surv. Plant Molec. Cell Biol.*, 3, 456, 1986.

173. Nomura, K., and Komamine, A., Physiological and biochemical aspects of somatic embryogenesis from single cells, in *Somatic Embryogenesis*, Terzi, M., Pitto, L., and Sung, Z.R., Eds., Consiglio Nationale delle Recherche, 1985, 1.

174. Timmers, A.C.J., De Vries, S.C., and Schel, J.H.N., Distribution of membrane-bound calcium and activated calmodulin during somatic embryogenesis of carrot (*Daucus carota* L.), *Protoplasma*, 153, 24, 1989.

175. Chée, R.P., and Cantliffe, D.J., Embryo development from discrete cell aggregates in *Ipomoea batatas* (L.) Lam. in response to structural polarity, *In Vitro Cell. Dev. Biol.*, 25, 757, 1989.

176. Dijak, M., Smith, D.L., Wilson, T.J., and Brown, D.C.W., Stimulation of direct embryogenesis from mesophyll protoplasts of *Medicago sativa*, *Plant Cell Rep.*, 5, 468, 1986.

177. Okada, K., Ueda, J., Komaki, M.K., Bell, C.J., and Shimura, Y., Requirement of the auxin polar transport system in early stages of *Arabidopsis* floral bud formation, *Plant Cell*, 3, 677, 1992.

178. Liu, C.-m., Xu, Z.-H., and Chua, N.-H., Auxin polar transport is essential for the establishment of bilateral symmetry during early plant embryogenesis, *Plant Cell*, 5, 621, 1993.

179. Ho, W.-J., and Vasil, I.K., Somatic embryogenesis in sugarcane (*Saccharum officinarum* L.) I. The morphology and physiology of callus formation and the ontogeny of somatic embryos, *Protoplasma*, 18, 169, 1983.

180. Becwar, M.R., Noland, T.L., and Wyckoff, J.L., Maturation, germination and conversion of Norway spruce (*Picea abies* L.) somatic embryos to plants, *In Vitro Cell. Dev. Biol.*, 25, 575, 1989.

181. Buchheim, J.A., Colburn, S.M., and Ranch, J.P., Maturation of soybean somatic embryos and the transition to plantlet growth, *Plant Physiol.*, 89, 768, 1989.

182. Ebert, A., and Taylor, H.F., Assessment of the changes of 2,4-dichlorophenoxyacetic acid concentrations in plant tissue culture media in the presence of activated charcoal, *Plant Cell Tiss. Org. Cult.*, 20, 165, 1990.

183. Aung, L.H., Buss, G.R., Crosby, K.E., and Brown, S.S., Changes in the hormonal levels of soybean fruit during ontogeny, *Øyton*, 42, 151, 1982.

184. Danin, M., Upfold, S.J., Levin, N., Nadel, B.L., and Van Standen, J., Polyamines and cytokinins in celery embryogenic cell cultures, *Plant Growth Regulation*, 12, 245, 1993.

185. Wright, M.S., Launis, K.L., Novitzky, R., Duesing, J.H., and Harms, C.T., A simple method for the recovery of multiple fertile plants from individual somatic embryos of soybean [*Glycine max* (L.) Merrill], *In Vitro Cell. Dev. Biol.*, 27P, 153, 1991.

186. Bewley, J.D., and Black, M., Seeds: germination, structure and composition, in *Physiology of Development and Germination*, Bewley, J.D., and Black, M., Eds., Plenum Press, New York, 1985, 1.

187. Lai, F., Senaratna, T., and McKersie, B.D., Glutamine enhances storage protein synthesis in *Medicago sativa* L. somatic embryos, *Plant Sci.*, 87, 69, 1992.

188. Verma, D.C., and Dougall, D.K., Influence of carbohydrates on quantitative aspects of growth and embryo formation in wild carrot suspension cultures, *Plant Physiol.*, 59, 81, 1977.

189. Schuller, A., and Reuther, G., Response of *Abies alba* embryonal-suspensor mass to various carbohydrate treatments, *Plant Cell Rep.*, 12, 199, 1993.

190. Finer, J.J., and McMullen, M.D., Transformation of soybean via particle bombardment of embryogenic suspension culture tissue, *In Vitro Cell. Dev. Biol.*, 27P, 175, 1991.

191. Denchov, P., Velcheva, M., and Atanassov, A., A new approach to direct somatic embryogenesis in *Medicago*, *Plant Cell Rep.*, 10, 338, 1991.

192. Kielly, G.A., and Bowley, S.R., Genetic control of somatic embryogenesis in alfalfa, *Genome*, 35, 474, 1992.

193. Tremblay, L., and Tremblay, F.M., Carbohydrate requirements for the development of black spruce (*Picea mariana* (Mill.) B.S.P.) and red spruce (*P. rubens* Sarg.) somatic embryos, *Plant Cell Tiss. Org. Cult.*, 27, 95, 1991.

194. Anandarajah, K., and McKersie, B.D., Enhanced vigor of dry somatic embryos of *Medicago sativa* L. with increased sucrose, *Plant Sci.*, 71, 261, 1990.

195. Lu, C.-Y., and Thorpe, T.A., Somatic embryogenesis and plantlet regeneration in cultured immature embryos of *Picea glauca*, *J. Plant Physiol.*, 128, 297, 1987.

196. Roberts, D.R., Abscisic acid and mannitol promote early development, maturation and storage protein accumulation in somatic embryos of interior spruce, *Physiol. Plant.*, 83, 247, 1991.

197. Jain, S.M., Newton, R.J., and Soltes, E.J., Enhancement of somatic embryogenesis in Norway spruce (*Picea abies* L.), *Theor. Appl. Genet.*, 76, 501, 1988.

198. Earnshaw, B.A., and Johnson, M.A., Control of wild carrot somatic embryo development by antioxidants, *Plant Physiol.*, 85, 273, 1987.

199. Kepczynski, J., McKersie, B.D., and Brown, D.C.W., Requirement of ethylene for growth of callus and somatic embryogenesis in *Medicago sativa* L, *J. Exp. Bot.*, 43, 1199, 1992.

200. Fujii, J.A.A., Slade, D., Olsen, R., Ruzin, S.E., and Redenbaugh, K., Alfalfa somatic embryo maturation and conversion to plants, *Plant Sci.*, 72, 93, 1990.

201. Crouch, M.L., Non-zygotic embryos of *Brassica napus* L. contain embryo-specific storage proteins, *Planta*, 156, 520, 1982.

202. Flinn, B.S., Roberts, D.R., and Taylor, I.E.P., Evaluation of somatic embryos of interior spruce. Characterization and developmental regulation of storage proteins, *Physiol. Plant.*, 82, 624, 1991.

203. Krochko, J.E., Pramanik, S.K., and Bewley, J.D., Contrasting storage protein synthesis and messenger RNA accumulation during development of zygotic and somatic embryos of alfalfa (*Medicago sativa* L.), *Plant Physiol.*, 99, 46, 1992.

204. Shoemaker, R.C., Christofferson, S.E., and Galbraith, D.W., Storage protein accumulation patterns in somatic embryos of cotton (*Gossypium hirsutum* L.), *Plant Cell Rep.*, 6, 12, 1987.

205. Stuart, D.A., Nelsen, J., and Nichol, J.W., Expression of 7S and 11S alfalfa seed storage proteins in somatic embryos, *J. Plant Physiol.*, 132, 134, 1988.

206. Misra, S., Attree, S.M., Leal, I., and Fowke, L.C., Effect of abscisic acid, osmoticum, and desiccation on synthesis of storage proteins during the development of white spruce somatic embryos, *Ann. Bot.*, 71, 11, 1993.

207. Tewes, A., Manteuffel, R., Adler, K., Weber, E., and Wobus, U., Long-term cultures of barley synthesize and correctly deposit seed storage proteins, *Plant Cell Rep.*, 10, 467, 1991.

208. Taylor, D.C., Weber, N., Underhill, E.W., Pomeroy, M.K., Keller, W.A., Scowcroft, W.R., Wilen, R.W., Moloney, M.M., and Holbrook, L.A., Storage protein regulation and lipid accumulation in microspore embryos of *Brassica napus* L, *Planta*, 181, 18, 1990.

209. Dahmer, M.L., Hildebrand, D.F., and Collins, G.B., Comparative protein accumulation patterns in soybean somatic and zygotic embryos, *In Vitro Cell. Dev. Biol.*, 28P, 106, 1992.

210. Joy, R.W. IV, Yeung, E.C., Kong, L., and Thorpe, T.A., Development of white spruce somatic embryos. I. Storage product deposition, *In Vitro Cell. Dev. Biol.*, 27, 32, 1991.

211. Wilen, R.W., Mandel, R.M., Pharis, R.P., Holbrook, L.A., and Moloney, M.M., Effects of abscisic acid and high osmoticum on storage protein gene expression in microspore embryos of *Brassica napus*, *Plant Physiol.*, 94, 875, 1990.

212. Pramanik, S.K., Krochko, J.E., and Bewley, J.D., Distribution of cytosolic mRNAs between polysomal and ribonucleoprotein complex fractions in alfalfa embryos, *Plant Physiol.*, 99, 1590, 1992.

213. Bray, E., and Beachy, R.N., Regulation by ABA of β-conglycinin expression in cultured developing soybean cotyledons, *Plant Physiol.*, 79, 746, 1985.

214. Finkelstein, R.R., Tenbarge, K.M., Shumway, J.E., and Crouch, M.L., Role of ABA in maturation of rapeseed embryos, *Plant Physiol.*, 78, 630, 1985.

215. Cyr, D.R., Webster, F.B., and Roberts, D.R., Biochemical events during germination and early growth of somatic embryos and seed of interior spruce (*Picea glauca engelmanii* complex), *Seed Sci. Res.*, 1, 91, 1991.

216. Hakman, I., Stabel, P., Engström, P., and Eriksson, T., Storage protein accumulation during zygotic and somatic embryo development in *Picea abies* (Norway spruce), *Physiol. Plant.*, 80, 441, 1990.

217. Flinn, B.S., Roberts, D.R., Newton, C.H., Cyr, D.R., Webster, F.B., and Taylor, I.E.P., Storage protein gene expression in zygotic and somatic embryos of interior spruce, *Physiol. Plant.*, 1993 (in press).

218. Avjioglu, A., and Knox, R.B., Storage lipid accumulation by zygotic and somatic embryos in culture, *Ann. Bot.*, 63, 409, 1989.

219. Hara, S., Falk, H., and Kleinig, H., Starch and triacylglycerol metabolism related to somatic embryogenesis in *Papaver orientale* tissue cultures, *Planta*, 164, 303, 1985.

220. Pomeroy, M.K., Kramer, J.K.G., Hunt, D.J., and Keller, W.A., Fatty acid changes during development of zygotic and microspore-derived embryos of *Brassica napus*, *Physiol. Plant.*, 81, 447, 1991.

221. Warren, G.S., and Fowler, M.W., Changing fatty acid composition during somatic embryogenesis in cultures of *Daucus carota*, *Planta*, 144, 451, 1979.

222. Turnham, E., and Northcote, D.H., The incorporation of $[1\text{-}^{14}C]$ acetate into lipids during embryogenesis in oil palm tissue cultures, *Phytochem.*, 23, 35, 1984.

223. Attree, S.M., Pomeroy, M.K., and Fowke, L.C., Manipulation of conditions for the culture of somatic embryos of white spruce for improved triacylglycerol biosynthesis and desiccation tolerance, *Planta*, 187, 395, 1992.

224. Hildebrand, D.F., Adams, T.R., Dahmer, M.L., Williams, E.G., and Collins, G.B., Analysis of lipid composition and morphological characteristics in soybean regenerants, *Plant Cell Rep.*, 7, 701, 1989.

225. Shoemaker, R.C., and Hammond, E.G., Fatty acid composition of soybean (*Glycine max* (L.) Merr.) somatic embryos, *In Vitro Cell. Dev. Biol.*, 24, 829, 1988.

226. Kim, Y.-H., and Janick, J., Abscisic acid and proline improved desiccation tolerance and increase fatty acid content of celery somatic embryos, *Plant Cell Tiss. Org. Cult.*, 24, 83, 1991.

227. Stuart, D.A., and McCall, C.M., Induction of somatic embryogenesis using side chain and ring modified forms of phenoxy acid growth regulators, *Plant Physiol.*, 99, 111, 1992.

228. Feirer, R.P., Conkey, J.H., and Verhagen, S.A., Triglycerides in embryogenic conifer calli: a comparison with zygotic embryos, *Plant Cell Rep.*, 8, 207, 1989.

229. Dahmer, M.L., Collins, G.B., and Hildebrand, D.F., Lipid content and composition of soybean somatic embryos, *Crop Sci.*, 31, 741, 1991.

230. Finkelstein, R., and Somerville, C., Abscisic acid or high osmoticum promote accumulation of long-chain fatty acids in developing embryos of *Brassica napus*, *Plant Sci.*, 61, 213, 1989.

231. Roberts, D.R., Sutton, B.C.S., and Flinn, B.S., Synchronous and high frequency germination of interior spruce somatic embryos following partial drying at high relative humidity, *Can. J. Bot.*, 68, 1086, 1990.

232. Attree, S.M., Tautorus, T.E., Dunstan, D.I., and Fowke, L.C., Somatic embryo maturation, germination and soil establishment of plants of black and white spruce (*Picea mariana* and *Picea glauca*), *Can. J. Bot.*, 68, 2583, 1990.

233. Gray, D.J., and Purohit, A., Somatic embryogenesis and development of synthetic seed technology, *Crit. Rev. Pl. Sci.*, 10, 33, 1991.

234. Kott, L.S., and Beversdorf, W.D., Enhanced plant regeneration from microspore-derived embryos of *Brassica napus* by chilling, partial desiccation and age selection, *Plant Cell Tiss. Org. Cult.*, 23, 187, 1990.

235. Kott, L.S., and Kasha, K.J., Initiation and morphological development of somatic embryoids from barley cell cultures, *Can. J. Bot.*, 62, 1245, 1984.

236. Finkelstein, R.R., and Crouch, M.L., Precociously germinating rapeseed embryos retain characteristics of embryogeny, *Planta*, 162, 125, 1984.

237. Gray, D.J., Conger, B.V., and Songstad, D.D., Desiccated quiescent somatic embryos of orchardgrass for use as synthetic seeds, *In Vitro Cell. Dev. Biol.*, 23, 29, 1987.

238. Gray, D.J., Quiescence in monocotyledonous and dicotyledonous somatic embryos induced by dehydration, *HortSci.*, 22, 810, 1987.

239. Hammatt, N., and Davey, M.R., Somatic embryogenesis and plant regeneration from cultured zygotic embryos of soybean (*Glycine max* L. Merr.), *J. Plant Physiol.*, 128, 219, 1987.

240. Roberts, D.R., Lazaroff, W.R., and Webster, F.B., Interaction between maturation and high relative humidity treatments and their effects on germination of sitka spruce somatic embryos, *J. Plant Physiol.*, 138, 1, 1991.

241. Anandarajah, K., and McKersie, B.D., Manipulating the desiccation tolerance and vigor of dry somatic embryos of *Medicago sativa* L. with sucrose, heat shock and abscisic acid, *Plant Cell Rep.*, 9, 451, 1990.

242. Anandarajah, K., Kott, L., Beversdorf, W.D., and McKersie, B.D., Induction of desiccation tolerance in microspore-derived embryos of *Brassica napus* L. by thermal stress, *Plant Sci.*, 77, 119, 1991.

243. Parrott, W.A., Auxin-stimulated somatic embryogenesis from immature cotyledons of white clover, *Plant Cell Rep.*, 10, 17, 1991.

244. Webster, F.B., Roberts, D.R., McInnis, S.M., and Sutton, B.C.S., Propagation of interior spruce by somatic embryogenesis, *Can. J. For. Res.*, 20, 1759, 1990.

245. Weissinger, A.K. II, and Parrott, W.A., Repetitive somatic embryogenesis and plant recovery in white clover, *Plant Cell Rep.*, 12, 125, 1993.

246. Roberts, D.R., Webster, F.B., Flinn, B.S., Lazaroff, W.R., and Cyr, D.R., Somatic embryogenesis of spruce, in *Synseeds. Applications of Synthetic Seeds to Crop Improvement*, Redenbaugh, K., Ed., CRC Press, Inc., Boca Raton, FL, 1993, 427.

247. Wake, H., Akasaka, A., Umetsu, H., Ozeki, Y., Shimomura, K., and Matsunaga, T., Enhanced germination of artificial seeds by marine cyanobacterial extract, *Appl. Microbiol. Biotechnol.*, 36, 684, 1992.

248. Von Arnold, S., and Hakman, I., Regulation of somatic embryo development in *Picea abies* by abscisic acid (ABA), *J. Plant Physiol.*, 132, 164, 1988.

249. Wetzstein, H.Y., and Sommer, H.E., Scanning electron microscopy of *in vitro* cultured *Liquidambar styraciflua* plantlets during acclimatization, *J. Amer. Soc. Hort. Sci.*, 108, 475, 1983.

250. Komatsuda, T., and Ohyama, K., Genotypes of high competence for somatic embryogenesis and plant regeneration in soybean *Glycine max, Theor. Appl. Genet.*, 75, 695, 1988.

251. Redenbaugh, K., *SynSeeds: Applications of Synthetic Seeds to Crop Improvement*, CRC Press, Boca Raton, FL, 1993.

252. Redenbaugh, K., Fujii, J.A., and Slade, D., Synthetic seed technology, in *Scale-Up and Automation in Plant Propagation. Cell Culture and Somatic Cell Genetics of Plants, Vol. 8*, Vasil, I.K., Ed., Academic Press, Inc., San Diego, 1991, 35.

253. Murashige, T., Plant cell and organ cultures as horticultural practices, *Acta Hort.*, 78, 17, 1977.

254. Kitto, S., and Janick, J., Polyox as an artificial seed coat for asexual embryos, *HortSci.*, 17, 448, 1982.

255. Redenbaugh, K., Nichol, J., Kossler, M.E., and Paasch, B., Encapsulation of somatic embryos for artificial seed production, *In Vitro Cell. Dev. Biol.*, 20, 256, 1984.

256. Redenbaugh, K., Paasch, B.D., Nichol, J.W., Kossler, M.E., Viss, P.R., and Walker, K.A., Somatic seeds: Encapsulation of asexual plant embryos, *Bio/Tech.*, 4, 797, 1986.

257. Redenbaugh, K., Slade, D., Viss, P., and Fujii, J.A., Encapsulation of somatic embryos in synthetic seed coats, *HortSci.*, 22, 803, 1987.

258. Gray, D.J., Effects of dehydration and exogenous growth regulators on dormancy, quiescence and germination of grape somatic embryos, *In Vitro Cell. Dev. Biol.*, 25, 1173, 1989.

259. Kitto, S.L., and Janick, J., Hardening treatments increase survival of synthetically-coated asexual embryos of carrot, *J. Amer. Soc. Hort. Sci.*, 110, 283, 1985.

260. Kitto, S.L., and Janick, J., Production of synthetic seeds by encapsulating asexual embryos of carrot, *J. Amer. Soc. Hort. Sci.*, 110, 277, 1985.

261. Kim, Y.-H., and Janick, J., ABA and Polyox-encapsulation or high humidity increases survival of desiccated somatic embryos of celery, *HortSci.*, 24, 674, 1989.

262. Janick, J., Kitto, S.L., and Kim, Y.-H., Production of synthetic seed by desiccation and encapsulation, *In Vitro Cell. Dev. Biol.*, 25, 1167, 1989.

263. Fujii, J.A., Slade, D., Aguirre-Rascon, J., and Redenbaugh, K., Field planting of alfalfa artificial seeds, *In Vitro Cell. Dev. Biol.*, 28P, 73, 1992.

264. Fujii, J.A.A., Slade, D., and Redenbaugh, K., Maturation and greenhouse planting of alfalfa artificial seeds, *In Vitro Cell. Dev. Biol.*, 25, 1179, 1989.

265. Datta, S., and Potrykus, I., Artificial seeds of barley: encapsulation of microspore-derived embryos, *Theor. Appl. Genet.*, 77, 820, 1989.

266. Shigeta, J., Mori, T., Toda, K., and Ohtake, H., Effects of capsule hardness on germination frequency of encapsulated somatic embryos of carrot, *Biotechnol. Tech.*, 4, 21, 1990.

267. Bapat, V.A., and Rao, P.S., Sandalwood plantlets from "synthetic seeds", *Plant Cell Rep.*, 7, 434, 1988.

268. Bapat, V.A., and Rao, P.S., Somatic seeds of sandalwood (*Santalum album* L.) and mulberry (*Morus indica* L.), in *Proceedings of the National Seminar on Advances in Seed Science and Technology*, Shetty, H.S., and Prakash, H.S., Eds., Univ. of Mysore, Mysore, 1989, 372.

269. Bapat, V.A., and Rao, P.S., Plantlet regeneration from encapsulated and nonencapsulated desiccated somatic embryos of a forest tree: sandalwood (*Santalum album* L.), *J. Plant Biochem. Biotechnol.*, 1, 109, 1992.

270. Lulsdorf, M.M., Tautorus, T.E., Kikcio, S.I., Bethune, T.D., and Dunstan, D.I., Germination of encapsulated embryos of interior spruce (*Picea glauca engelmannii* complex) and black spruce (*Picea mariana* Mill.), *Plant Cell Rep.*, 12, 385, 1993.

271. Rao, P.V.L., and Singh, B., Plantlet regeneration from encapsulated somatic embryos of hybrid *Solanum melongena* L, *Plant Cell Rep.*, 10, 7, 1991.

272. Shigeta, J., Mori, T., and Sato, K., Storage of encapsulated somatic embryos of carrot, *Biotechnol. Tech.*, 7, 165, 1993.

273. Currah, I.E., Gray, D., and Thomas, T.H., The sowing of germinating vegetable seeds using a fluid drill, *Ann. Appl. Biol.*, 76, 311, 1974.

274. Baker, C.M., Synchronization and fluid sowing of carrot, *Daucus carota*, somatic embryos. M.Sc. Thesis, University of Florida, Gainesville, 1985.

275. Schultheis, J.R., and Cantliffe, D.J., Plant formation of *Ipomea batatas* Poir. from somatic embryos in gel carriers with additives, *HortSci.*, 23, 812, 1988.

276. McElroy, A.R., and Brown, D.C.W., A transplant plug technique for production of alfalfa (*Medicago sativa* L.) plants from somatic embryos, *Can. J. Plant Sci.*, 72, 483, 1992.

277. Barwale, U.B., and Widholm, J.M., Somaclonal variation in plants regenerated from cultures of soybean, *Plant Cell Rep.*, 6, 365, 1987.

278. Armstrong, C.L., and Phillips, R.L., Genetic and cytogenetic variation in plants regenerated from organogenic and friable, embryogenic tissue cultures of maize, *Crop Sci.*, 28, 363, 1988.

279. Caligari, P.D.S., and Shohet, S., Variability in somatic embryos, in *Synseeds: Applications of Synthetic Seeds to Crop Improvement*, Redenbaugh, K., Ed., CRC Press, Boca Raton, FL, 1993, 163.

280. Nuti Ronchi, V., Giorgetti, L., Tonelli, M., and Martini, G., Ploidy reduction and genome segregation in cultured carrot cell lines. I. Prophase chromosome reduction, *Plant Cell Tiss. Org. Cult.*, 30, 107, 1992.

281. Rugh, C.L., Parrott, W.A., and Merkle, S.A., Ploidy variation in embryogenic yellow-poplar, in *Proc. 22nd Southern Forest Tree Improvement Conference*, Atlanta, GA, June 14–17, 1993, 493.

282. Kott, L.S., Flack, S., and Kasha, K.J., A comparative study of initiation and development of embryogenic callus from haploid embryos of several barley cultivars. II. Cytophotometry of embryos and callus, *Can. J. Bot.*, 64, 2107, 1986.

283. Shoemaker, R.C., Amberger, L.A., Palmer, R.G., Oglesby, L., and Ranch, J.P., Effect of 2,4-dichlorophenoxyacetic acid concentration on somatic embryogenesis and heritable variation in soybean [*Glycine max* (L) Merr.], *In Vitro Cell. Dev. Biol.*, 27P, 84, 1991.

284. Durand-Gasselin, T., Le Guen, V., Konan, K., and Duval, Y., Plantations en Côte-d'Ivoire de palmiers à huile (*Elaeis guineensis* Jacq.), obtenues par culture *in vitro*. Premiers résultats, *Oléagineux*, 45, 1, 1990.

285. Le Guen, V., Samaritaan, G., Othman, A.Z., Chin, C.W., Konan, K., and Durand-Gasselin, T., Oil production in young oil palm clones, *Oléagineux*, 46, 347, 1991.

286. Corley, R.H.V., Lee, C.H., Law, L.H., and Wong, C.Y., Abnormal flower development in oil palm clones, *Planter*, 62, 233, 1986.

287. McKersie, B.D., and Bowley, S.R., Synthetic seeds of alfalfa, in *Synseeds. Applications of Synthetic Seeds to Crop Improvement*, Redenbaugh, K., Ed., CRC Press, Boca Raton, FL, 1993, 231.

6. Structural and Developmental Patterns in Somatic Embryogenesis

EDWARD C. YEUNG

Contents

I. Introduction

Somatic embryogenesis is a process by which somatic cells undergo a developmental sequence similar to that seen in zygotic embryos. This process is an important plant propagation technique [1] and provides an essential tool for basic research into plant embryo development [2, 3] and other aspects of plant physiology [4]. Living organisms are known for the high degree of order in their constituent parts, so it is not surprising to find that embryos undergo characteristic morphological and anatomical changes during their development. Therefore, a study of the pattern of development will provide a

205

T.A. Thorpe (ed.), In Vitro Embryogenesis in Plants, pp. 205–247.
© 1995 *Kluwer Academic Publishers, Dordrecht. Printed in the Netherlands.*

better understanding of the order-generating processes. In the last decade, a number of reviews have dealt with different aspects of somatic embryogenesis [5–14]. However, a comprehensive review specifically on the structural aspects of somatic embryogenesis is not available. This chapter provides a synthesis of the information available concerning the structural and developmental patterns observed during somatic embryogenesis. Only selected, well-characterized systems will be discussed. Some speculations, hypotheses, and questions are provided to stimulate further experimentation in this area of research. Additionally, comparisons will be made with zygotic embryo development.

II. The early events leading to embryogenic cell formation

A. The initiation of somatic embryogenesis

The pattern of embryogenic cell formation varies according to the plant species studied. Single cells such as pollen and epidermal cells of a variety of explants have been shown to give rise directly to somatic embryos; while in other systems, especially those involving a liquid phase such as the carrot somatic embryo system, embryogenic cells are formed after a callus phase.

In order to gain a better understanding of the early events leading to somatic embryo formation, the concepts of competence, induction, and determination used by animal biologists are now being used by botanists to interpret their observations [5, 6, 15–19].

Competence of cells/tissues/organs is essentially cell reactivity. Namely, it is the capacity to respond to specific signals, e.g. environmental, chemical, or other manipulated treatments, and to do so in a consistent manner [17–19]. Induction occurs when a signal produces a unique developmental response from competent tissue [18]. Determination is a process by which the developmental fate of a cell or group of cells becomes fixed and is limited to a particular developmental pathway [5]. The developmental potential of cells/tissues/organs becomes more restricted as determination proceeds, and involves stable changes in the phenotype [17]. Although these events are usually tightly coupled and may be impossible to separate, the usefulness of these terms is that they allow for a systematic way of thinking about early events in somatic embryogenesis. Hopefully this will lead to a better understanding of the underlying cellular and molecular mechanisms. For a more in depth discussion on the concepts of competence, induction, and determination, readers are referred to the work of Meins [17], McDaniel [18], Lyndon [19], and Sachs [20]. It is important to point out that when compared with animal development, plant development is more plastic, and thus it is not surprising to see that once the system is perturbed, the stability of various "states" of development can be altered. When these concepts are applied, the experimental conditions should be clearly defined.

In the following discussion, somatic embryo formation is perceived as somatic cells having first to become competent and then undergo the process of induction. At the end of the induction process, some somatic cells become determined as embryogenic cells. Although this may be a simplistic way of looking at somatic embryo formation, it allows for comparisons to be made between different systems.

Is the embryogenic cell "determined"? Due to the developmental plasticity of embrogenic cells, Carmen [8] suggested that embryogenic cells are not determined. However, a "developmentally plastic" embryogenic cell could be the characteristic of a "determined" embryogenic cell. The fact that polarized development occurs early in somatic embryogenesis [21, 22] suggests pattern formation has begun and that process is essential to subsequent histo-differentiation within the embryo proper. The formation of a recognizable embryonic structure, albeit labile in its pattern of development, can be regarded as the expression step [23] of the determined embryogenic cell.

In the plant body, all cells have specific functions to play. From the study of messenger-RNA and protein changes associated with induction of canola [24] and tobacco [184] microspore embryogenesis, it appears that in order to change the developmental pathway of a cell, the existing developmental information must be stopped or altered. Thus cells have to change or dedifferentiate prior to becoming competent to respond to the new signal. The redirection of the developmental program and the acquisition of competence in the embryogenic process can be initiated by pretreatment procedures prior to excision, or by placing explants in a new environment subsequent to the excision procedure. Many of the procedures such as excision, low or high temperature treatments, gamma radiation, and pretreatment with mannitol and ethanol application [25–28] drastically alters the environment of the explants. These treatments can be regarded as forms of stress that make cells more responsive to new stimuli. In the zygotic embryo, the maturation drying process of the seed has been shown to shut down the embryo developmental program and initiate the germination program [29, 30]. A shift in the temperature of canola microspore culture could be associated with the interruption of the pollen developmental program [31] which is essential for the "dedifferentiation" process. The successful use of cadmium and other heavy metal ions to induce somatic embryo formation on the surface of carrot seedlings without visible callus formation is also regarded as a stress treatment [32, 33]. These examples clearly indicate that altering the growth conditions and subjecting tissues and/or organs to unusual conditions can be the trigger that enables cells to undergo changes in the developmental process and cause them to become competent to the inductive signals for somatic embryogenesis. At present, the tobacco and canola microspore system offers promise in the study of the physiological, biochemical and molecular events during the acquisition of competence proir to the onset of somatic embryogenesis. A number of mRNAs and proteins have been detected in the canola and tobacco microspores during the commitment phase of the inductive process [24, 184].

Once primed by altering growing conditions prior to excision, etc., the explant becomes more responsive to the inductive signal. The most important inductive signals in somatic embryogenesis are the plant growth regulators, especially auxins. Other agents such as heavy metal ions [32, 33] and changes in the pH of the explant environment [37] can also serve as inductive agents. It is unfortunate that these chemical agents can have multiple effects on somatic embryogenesis; effects which are often difficult to separate. The induction process leads to the formation of an embryogenic cell which becomes determined to embark on the embryogenic pathway. Little is known about the mechanisms and the biochemical and molecular changes associated with the inductive process. Using a careful cell separation procedure, Nomura and Komamine [38] demonstrated that different states of cells exist in carrot suspension cultures. Auxin is important in channelling cells into their final determined state with the capacity to form the somatic embryo. The work of Nomura and Komamine [38] clearly shows that auxin plays an important role in the inductive process. The presence of auxin is also essential for the formation of heat labile factors from an established embryogenic culture that can accelerate the acquisition of embryogenic potential [39, 40]. In a recent report, 2,4-D has been shown to influence IAA metabolism in carrot cells [41]. The switch in "IAA metabolism may affect the developmental transition from callus to embryo" [41]. There is no doubt that changes in auxin metabolism is essential to the process of carrot somatic embryogenesis. Additional auxin-regulated processes may be important in embryogenic cell induction, such as the formation of labile factors reported by de Vries [39, 40]. If detailed structural and biochemical comparisons can be made using the carefully sieved cells of carrot or other embryogenic systems, valuable information on the inductive processes leading to the formation of "determined" embryogenic cells may be gained. In carrot, further comparative studies using different agents in somatic embryo induction may enable us to determine whether a common mechanism exists for the induction of somatic embryos.

The final step in the study of the early events in somatic embryogenesis is the determination process. The product of cell determination can be judged indirectly through the formation of the characteristic embryogenic cells (see next section). Little is known about the physiological and biochemical properties of determined embryogenic cells. By comparing the molecular processes of embryogenesis, using the proembryogenic mass in the presence of auxin, it was found that the program was already established [39, 40, 42]. These studies indicate that a "determined" embryogenic mass has acquired properties of developing embryos. Maybe one can use this feature as one of the criteria to judge whether the embryogenic cell is "determined" or not. The ability to maintain a large number of preglobular stage proembryos [37, 43, 44] by simply altering the pH of the culture medium may provide a useful source of material to study the properties of embryogenically determined cells. A protein ECP31, which has characteristics resembling those of a late embryogenic abundant protein has been localized in stress-induced peripheral cells of embryo-bearing segments and it gradually disappears as the cells develop into somatic embryos

[34–36]. This protein can be a useful marker for the study of biochemical changes during early stages of embryogenesis in carrot cells. Recently, a novel carbohydrate epitope, JIM8 has been found to associate with carrot embryogenic cells in suspension cultures [185]. This cell wall epitope can be used as an additional marker for an early determinative event in carrot somatic embryogenesis [185, 186].

In addition, more work is needed to study how the "determined" state is maintained. In Gramineae species, a high concentration of auxin is required in order to maintain the embryogenic nature of the cultures [11]. These cultures may provide useful systems to study the effect of auxin on the maintenance of the "determined" state.

In the majority of systems studied, events relative to the acquisition of competence, induction, and determination are usually tightly coupled and thus difficult to separate for experimental studies. The period over which the formation of embryogenic cells takes place in some systems, such as *Brassica* microspores, is very short, e.g. only eight hours are needed, [24] making it difficult to separate the above events. Furthermore, one has to be careful in interpreting data obtained from embryogenic cultures generated from callus or liquid suspension systems. In these embryogenic cultures, cells with different developmental "states" are present. Therefore, it is difficult to study the early events related to somatic embryogenesis. If one can develop protocols to arrest specific "states" of early embryogenesis, an effective means of studying early events in somatic embryogenesis will become available. With improved sensitivities towards detection of various macromolecules and growth substances and manipulation of growing conditions prior to excision, it is hoped that significant progress can be made in the near future.

A detailed description of anatomical changes related to the inductive process that leads to the formation of embryogenic cells is not readily available in the literature. The initiation sequence of the "primary" embryogenic callus is usually ignored. Most published results deal with the "secondary" embryogenic callus, i.e. the maintenance and propagation of the original embryogenic callus. This type of study provides little uesful information concerning embryogenic cell initiation. The following is an account of several selected examples in the generation of the "determined" embryogenic cell. The purpose is to discover whether similar events occur during the early stages of somatic embryo formation.

1. Embryogenic cell formation from callus

a. The carrot suspension culture system
The best documented system concerning somatic embryogenesis using a liquid suspension is carrot. One of the advantages of the liquid suspension system is that a large volume of cells and embryos can be obtained for various uses. The general structural pattern of development is summarized in Plates 1 and 2 (Figs. 1–7).

In carrot suspension cultures, a number of ultrastructural studies have attempted to compare the fine structure of quiescent and growing carrot cells

210

Plates 1 and 2 illustrate the major stages of carrot somatic embryo development. The structural pattern of development will vary according to the cell line used and the method of culture. The following micrographs simply serve as a guide. Additional details of carrot somatic embryo development can be found in the paper by Schiavone and Cooke [140].

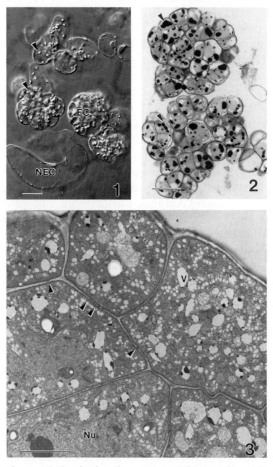

Plate 1. (1) A light micrograph showing interference contrast images of proembryogenic cell masses (EC) and a non-embryogenic cell (NEC). The proembryogenic cell mass consists of small, dense cytoplasmic cells with starch grains (arrowhead) while the non-embryogenic cell is large and highly vacuolated in appearance. Scale bar = 40 μm. (2) A light micrograph showing a section through proembryogenic cell masses. The cells have a dense cytoplasm and a prominent nucleus. Starch grains (arrowhead) are abundant within the cytoplasm. More vacuolated cells (*) are located in the centre of the proembryogenic cell mass. Scale bar = 25 μm. (3) An electron micrograph showing a portion of the cells in the peripheral region of the proembryogenic cell mass. The ultrastructural features concur with the light microscope observation in that the cells have numerous organelles, a prominent nucleus (Nu), vacuoles (V) are small in size and few in number. Plasmodesmata (arrowhead) are present in the walls between the surface cells and those beneath them. It appears that more plasmodesmata are present in the periclinal walls than the anticlinal walls. Scale bar = 5 μm.

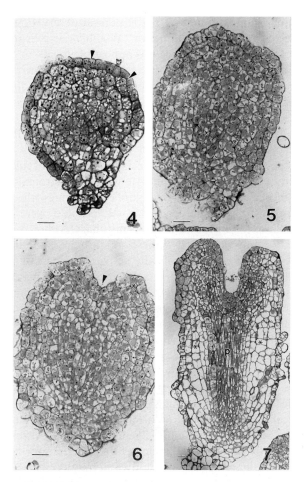

Plate 2. (4) Polarity is gradually established within the proembryogenic cell mass. Mitotic activities result in the formation of a globular shaped embryo. Different cell lines and culture conditions will influence the formation and structural development of the globular embryo from the proembryogenic cell mass. A globular stage embryo is illustrated in Fig. 4. One of the most important structural features at this stage of development is the differentiation of the protoderm (arrowhead). The protoderm reacts more strongly with the protein stain, amido black 10B. Scale bar = 25 μm. (5) The globular embryo begins to elongate into an oblong shaped embryo. This change in shape coincides with the differentiation of the procambial-like cells (*) in the central core towards the root pole. Scale bar = 25 μm. (6) The initiation of the cotyledon primordia denotes the beginning of the heart stage of development. Major events in histodifferentiation begin at this stage with the differentiation of the cotyledon primordia (*) and the apical meristems (arrowhead). From the structural standpoint, the differentiation of these respective structures is judged from their position, cell profiles and cytoplasmic density. Cotyledon initiation is readily detected by localized outgrowth of the primordia. For the apical meristems, although their locations are predictable, the exact timing of apical meristem differentiation is often more difficult to judge. At present, it is difficult to determine when the cells become the apical meristem cells. Procambium differentiation continues. Scale bar = 25 μm. (7) A light micrograph showing a well developed heart stage embryo. The cotyledons are well development. However, these cotyledons are relatively small when compared

[45, 46]. It is important to note that the structural observations reported are changes to selected cells in the population. Whether the cells in question will ultimately function as embryogenic cells is impossible to tell. In dormant roots, the carrot cells are highly vacuolated with only a thin layer of cytoplasm near the wall. Mitochondria are small and spherical in shape and few in number. Similarly, plastids exhibit little internal differentiation and are also few in number. Endoplasmic reticulum cisternae are sparse and appear to be discontinuous. In the presence of 2,4-D, cells grow, divide rapidly, and actively increase their cytoplasmic content and organelles. Plastids have a better developed internal membrane system after six to seven days of culture and starch grains of various sizes begin to accumulate within the stroma of the plastids. Similarly, Wilson et al. [47] observed a range of cytoplasmic inclusions during the induction of growth in quiescent cells, but microtubules were not prominent within the cytoplasm during the initial callusing phase. Although these cells may not produce the embryo directly, these studies indicate that during de-differentiation cells become metabolically active. The increase in cytoplasmic components and cellular activity suggests that these changes may be essential for the subsequent induction of competent cells to become determined.

With the establishment of a suspension culture, direct observation indicates that somatic embryos can form from isolated parenchymatous single cells of carrot [38, 48, 187]. Through careful isolation, small single cells have been isolated which have the ability to become fully determined [38]. Vacuolated single cells can be treated with a low concentration of auxin and cytokinin to bring about the final step in the determination process. Furthermore, Nomura and Komamine [22] have shown that the majority of cells undergo unequal cell division and that a polarized macromolecular synthesis pattern is established very early in the formation of the embryogenic clump [49]. A recent study using video cell tracking by Toonen et al. indicates that different forms of single cells can give rise to somatic embryos albeit at different frequencies [187]. It is possible to predict the developmental pathway leading to a somatic embryo by the initial morphology of the cell [187]. Through careful isolation and observation of cells, the work of Nomura and Komamine and Toonen et al. clearly indicate that it is possible to gain important information concerning the early events in somatic embryo formation. Detailed comparisons between these cells will provide new insights into the determination event. Further ultrastructural characterization may provide a structural basis for determination in carrot cells.

---

to the zygotic embryo. The protoderm cells are tightly packed and remain cytoplasmic. The cells of the ground tissue are highly vacuolated (*). The center core of the somatic embryo is occupied by the procambium (P) which bifurcates into the cotyledons at the shoot pole. The shoot apical meristem consists of a small group of cytoplasmic cells occupying the apical notch between the cotyledons. The root apical meristem occupies the region between the cap and the procambium pole. Structural features alone do not enable one to determine the precise location of the root apical meristem. Future studies using physiological, biochemical, and molecular markers will enable us to localize the structure and formation of the apical meristem. Scale bar = 25 μm.

b. Embryogenic cell formation in the Gramineae
The process of somatic embryo formation was reviewed recently by Vasil [11]. In many Gramineae species such as *Zea mays*, although somatic embryos can be formed directly from explants without going through a callus phase, the experimental conditions as such are often optimized for a continuous production of embryogenic callus. Embryogenic calluses have been derived from immature embryos, developing inflorescences, and the basal meristem of leaves. The induction of embryogenic calluses appears to be more stringent when compared to carrot, as only specific tissues within the original explant are capable of forming embryogenic callus. Different levels of endogenous growth substances have been found within different regions of the leaves of *Pennisetum purpureum* which correlate with their regenerative potential upon culture [50]. Thus, the differentiated state of the explant is important in acquiring competence to respond to the inductive signal.

The most common explants used for the production of embryogenic calluses are developing embryos. For example, in corn, the epidermis and the subepidermal cells of the scutellum tissue are primarily responsible for the formation of embryogenic callus. Within 2–3 days of culture, the internal scutellum cells enlarge considerably and are separated from one another by large intercellular air spaces. This enlargement results in the separation of the epidermis and the subepidermal layer from the rest of the embryonic axis. Some cells of the epidermis and subepidermal layers enlarge, become more cytoplasmic and undergo rapid division [51–53]. After 10 days in culture, many small cytoplasm-rich cells can be found which presumably are the embryogenic cells of the friable callus [53]. Similar observations have been reported in *Panicum maximum* [54]. Although general information is available concerning the generation of the embryogenic callus, detailed structural changes associated with the early inductive process has not been documented. Is the separation of the epidermal and the subepidermal layer from the remaining scutellum tissue essential to the formation of the embryogenically determined cells? Further detailed structural and developmental analysis is needed to characterize the early events related to somatic embryogensis in this important group of plants.

2. Embryogenic cell formation in explants without a callusing phase
One advantage associated with the use of solid medium for the study of somatic embryogenesis is the immobilization of explants. Through careful anatomical investigations, the ontogenetic sequence can be studied. Detailed structural information related to the process of somatic embryo formation can be obtained.

In a recent study, Jones and Rost [55] detailed the development and ultrastructure of somatic embryos from rice scutellum epithelial cells. At the time of culture, rice scutellum epithelial cells contain a centrally located nucleus, numerous lipid bodies, and small protein bodies in the cytoplasm. Plastids and mitochondria are present but are not well developed; endoplasmic reticulum cisternae and dictyosomes appear to be absent. Upon excision and placement on culture medium with and without 2,4-D, the epithelial cells begin to elongate and develop small profiles of rough endoplasmic reticulum cisternae. Mitochondria and plastids have a better developed internal membrane system. Under germination conditions, starch appears while lipid bodies break down. In contrast, cells undergoing the process of somatic embryogenesis retain numerous lipid bodies. Plastids do not accumulate starch grains and possess only a simple thylakoid system. A typical complement of organelles is present, and ribosome density increases. The work of Jones and Rost [55] clearly demonstrates that cytoplasmic reorganization is essential for the redirection of the developmental program of the cells, i.e. from an epithelial cell to a determined embryogenic cell. Alteration of metabolic activity between cells with and without 2,4-D is clearly demonstrated by the differences in the starch accumulation pattern and lipid metabolism.

In rice, some epithelial cells have been induced to divide by 24 h [55]. The first cell division signals the beginning of somatic embryo formation, i.e. the induction of an epithelial cell to become a determined embryogenic cell occurs within 24 h. A shift in the direction of cell division can be an important early indicator for somatic embryogenesis. In *Trifolium repens*, the first sign of somatic embryo induction is a shift from anticlinal to irregular, periclinal and oblique quantal divisions [56]. It is often noted that in shoot histogenesis from the cotyledon explants such as the radiata pine, the first positive signal for shoot induction is also a change in the direction of cell division [57]. Epidermal cells, once formed, rarely divide in the periclinal direction throughout development [58] and these cells are considered to be more highly determined than underlying parenchyma cells [59]. Thus, the observed change in the direction of cell division associated with somatic embryogenesis and shoot histogenesis is a clear signal that there has been a change in the developmental state of the cells. Additional experimental and ultrastructural studies on the cell division process, especially on the regulation of new wall formation, can provide important information concerning the early events of somatic embryogenesis.

3. Embryogenic cell formation from developing pollen grains

The use of pollen embryos is becoming an important tool in the study of the early events of somatic embryo formation. Recent reviews by Raghavan [9, 60] document the important cytological and biochemical events related to the early induction process. Our earlier information concerning structural changes associated with pollen embryo development is limited mainly to tobacco and *Datura* [61]. In these studies, changes in the organization of the cytoplasm in pollen prior to the commencement of embryogenesis have been detailed in tobacco during the initial stages of embryo formation [62–64]. Dunwell and Sunderland [63] reported that "ribosomes appeared to be virtually eliminated from the cell and many other organelles degraded; the few that remained, mainly plastids, were clustered around and in close contact with the vegetative nucleus". This change is considered to be rather drastic. Since these changes occur prior to the commencement of divisions in the vegetative cell, it is concluded that the gametophytic cytoplasm is destroyed prior to the initiation of a new developmental program [63]. However, in a similar study [65], such an observation could not be demonstrated. This discrepancy might be due to low temperature pretreatment of tobacco anthers in the latter study and different varieties used in the respective experiments.

One of the potential difficulties in this area of research is that only a small percentage of pollen grains will differentiate into embryos. One cannot be certain that the structural changes observed in a pollen grain is related to the differentiation process. Thus it is no surprise that controversies exist in this area of research regarding the early stages of pollen embryo formation. With the optimization of culture conditions, additional structural, biochemical, and molecular information is becoming available.

Recently, improvements have been made to the tobacco pollen embryo system and render it an attractive system to study the early inductive changes during pollen embryogenesis [66, 184]. A starvation treatment of isolated, immature pollen grains induces the formation of embryogenic cells that develop into embryos upon transfer to a sugar-containing medium. It has been shown that during the induction process of tobacco pollen embryogenesis, one of the most important events is the de-repression of the G1 arrest in the cell cycle of the vegetative cell [66]. Starvation treatment causes DNA replication in the vegetative cell of the developing pollen grain. The addition of hydroxyurea, a DNA replication inhibitor, to the "starvation" medium only postponed S phase entry, and does not affect subsequent embryo formation when pollen are transferred to a "rich" medium. Thus, cell cycle control is a key to the understanding of the inductive process during pollen embryogenesis [66]. Furthermore, specific mRNAs are synthesized during the starvation treatment and changes in protein kinase activity are also detected [184]. It is concluded that "the induction of tobacco pollen embryogenesis is the consequence of at least three processes triggered by starvation: degradation of the cytoplasm of the immature pollen grains isolated at a specific developmental stage (probably necessary to inhibit and eliminate the products of gametophytic gene expression), activation of the expression of specific genes required in the very early stages of embryogenesis and the activation/inactivation of specific protein kinases through a yet unknown signal transduction pathway" [184]. Further characterization of this system will yield important insights into the early inductive process of pollen embryogenesis.

Protocols have been developed to generate a high frequency of embryogenic microspores from canola [67, 67a]. After subjecting the microspores to high temperature (32.5 °C for 24 h) an embryogenic frequency of 21% was obtained [67]. This high percentage of embryogenic microspore induction provides a better confidence limit in interpreting the results regarding the process of pollen embryogenesis. This temperature treatment provides the signal to alter the developmental pathway. Several changes have been observed in association with the high temperature treatment. Microspores begin to divide in a symmetrical manner. A loss of cellular polarity in terms of organelle distribution and an increased division symmetry occurs after 24 h of treatment. It is important to note that the symmetrical division at high temperature is preceded by the formation of a preprophase band [69]. New patterns of microtubule arrangement have been observed and likely played an important role in the newly induced symmetrical division [69a]. Thus, similar to that reported by Zarsky et al. [66], during the high temperature treatment changes to cell cycle regulatory processes must have taken place in canola pollen grains, as the vegetative cell will divide soon after. Additional structural changes can be observed in association with the high temperature treatment, e.g. appearance of cytoplasmic granules [68, 70]. These granules appear to be the heat shock proteins. It is possible that heat shock proteins are involved in the dedifferentiation process. The preferential synsthesis of heat shock proteins may prevent the synthesis of regulatory proteins necessary for pollen development [68]. And it has been proposed that heat shock proteins could play a helper role in heat-induced proliferation of cells [68a]. However, these features are not absolute markers of embryogenic development, as some pollen fail to develop into somatic embryos. These structural changes may reflect changes during the inductive period.

One of the most significant observations reported by Telmer et al. [70] were differences in cell wall structure. Electron dense deposits and vesicle-like structures were present in the walls of potentially embryogenic cultured microspores. Although the significance of this newly synthesized wall is not known, this feature may represent one of the structural markers for the "fixation" of the embryogenic pathway. Different cell wall components may play an important role in pollen embryogenesis [70]. Recent studies in carrot indicate that extracellular glycoproteins can rescue arrested somatic embryos and one of these glycoproteins may regulate the cell expansion process [71, 72]. In *Citrus*, glycoproteins released from proembryogenic masses also play an important role in regulating the progression of embryo development [73]. All these studies clearly indicate the importance of cell wall components during embryogenesis. Although there are still many unanswered questions, the canola pollen embryo system has the potential to provide us with more definitive answers concerning the early events of somatic embryogenesis. Some structural changes associated with the early inductive phase in canola pollen embryo development are summarized in Plate 3 (Figs. 8–11).

In other similar studies, Zaki and Dickinson [74, 75] reported major changes in the cytoplasmic organization during canola pollen embryo formation. Within 24 h of culture in the embryo induction medium, the central vacuole becomes fragmented, allowing the nucleus to assume a central position within the cell. Starch synthesis commences in the plastids and the plastids begin to aggregate near the nucleus. This is followed by the synthesis of a new cell wall. It is also reported that experimental disruption of the microtubular cytoskeleton with colchicine promotes somatic embryo formation [75]. However, as pointed out earlier in this discussion, the low embryogenic frequency (1–2%), the significance of these findings awaits future confirmation.

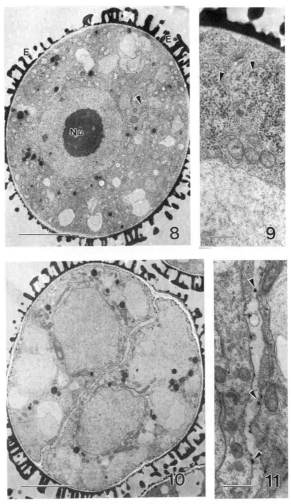

Plate 3 represents some of the events related to canola pollen embryo development during the first 24 h. For details of the structural changes and discussion, see Telmer et al. [68, 70]. (8) An electron micrograph showing a *Brassica napus* cv. Topas microspore after culture at 32.5 °C for 6 h. The microspore has not yet divided at this time. The cell consists of a prominent nucleus (Nu) and a cluster of cytoplasmic granules (arrowheads). Small vacuoles, proplastids, mitochondria, and lipid globules are present in the cytoplasm. The exine (E) is still present at this time. Scale bar = 10 μm. (9) One of the outstanding features at this time is the formation of the cytoplasmic granules (arrowhead). These granules are speculated to be the heat shock proteins. Although these granules are not exclusive to embryogenic microspores, the formation of these granules appear to show reaction to the stress treatment. Since stress treatments may serve as signals for the early inductive events, the formation of these granules can be a useful structural marker for pollen embryogenesis. Scale bar = 1 μm. (10) An electron micrograph showing a two celled structure after 24 h of culture. The cell division is symmetrical, resulting in two cells of similar size. Scale bar = 10 μm. (11) One of the most notable features at this time is the presence of electron-dense deposits (arrowhead) along the walls of the cells. These deposits are found in potentially embryogenic microspores. These could be used as a structural marker for the successful early inductive process. Scale bar = 1 μm. Micrograph courtesy of Dr. C. Telmer.

From the few case histories available, it appears that the state of differentiation of the pollen at the time of culture will determine the degree of cytoplasmic reorganization observed. A more dramatic change in the cytoplasm is likely to happen if the developing pollen has proceeded more towards the gametophytic pathway. Different methods of anther and pollen pre-treatments may also result in differences observed in the structural pattern of development. Correlative biochemical and molecular studies similar to that reported by Pechan et al. [24] will provide better insights into the structural changes observed. Furthermore, comparative studies using the tobacco and canola systems may enable us to determine whether common molecular events exist during the early inductive process of pollen embryogenesis.

4. Embryogenic cultures from conifer species

Because of the economic importance of Conifer species, tremendous progress has been made in the development of systems for in vitro propagation. Many excellent reviews are available in the literature dealing with different aspects of somatic embryogenesis in conifers [76, 77]. Therefore, this information will not be repeated.

In order to produce embryogenic cultures, explant selection is critical to the successful induction of somatic embryogenesis [76]. In a majority of reports, it is clear that embryogenic cultures from conifer species are derived from immature and mature embryos. This suggests that the early events leading to embryogenic culture formation are stringent in tissues other than developing or mature embryos. The successful utilization of embryos as explants suggests that information required for embryogenic culture formation is already present in the initial explants. Drastic physiological and biochemical changes may not be required during the initial stages of somatic embryo induction. The key to a successful manipulation is to induce the explants to divide while maintaining embryogenic properties.

Different patterns of somatic embryo initiation have been suggested and documented [76, 77]. Little information is available concerning the structural, physiological and biochemical changes during the initiation of embryogenic culture from zygotic embryo explants. This information will provide a better insight into the early events of embryogenic culture initiation.

5. Conclusions

Changes in metabolic activity seems to be a common process associated with embryogenic cell formation, especially for those explants with vacuolated cells such as carrot. The premitotic increase in cellular activity is reflected by an increase in the number of organelles and changes in nuclear morphology and vacuolar structure. These changes reflect premitotic enhancement of respiration, protein- and carbohydrate-synthesis, and a turnover of membranes similar to the cytoplasmic reactivation process in the root cortical cells of pea during vascular differentiation [78]. For those somatic embryos derived from more cytoplasmic cells, ultrastructural studies [55, 68, 70, 74, 75] clearly indicate changes in macromolecular synthesis such as changes in the pattern of storage product catabolism and synthesis. This observation suggests that alteration of the metabolic pathway is essential to the process of somatic embryogenesis. Changing metabolism has been suggested to play an important role in morphogenesis [79].

A structural manifestation of early changes in somatic embryogenesis is the alteration in cell division pattern as denoted by the newly formed cell wall. A change in the symmetry of cell division and/or reorientation of the cell wall appears to be important in the inductive/determinative process. In the study of pattern formation in *Arabidopsis*, mutations that result in the absence of defined tissue patterns can be traced to the first division within the zygote [79a]. Instead of dividing asymmetrically, the zygote divides equally into two cells. The absence of meristem formation in *Arabidopsis* can also be traced to the alteration or absence of a defined division plane [79b]. These studies clearly indicate the importance of the regulation of the mitotic process in cell differentiation.

The key structural components related to the mitotic process are the cytoskeletal elements. The information at hand clearly indicates the importance of these elements in the early inductive/determinative events in somatic embryogenesis. The formation of the preprophase band prior to cell division in canola pollen during the high temperature treatment is similar to that reported in the zygotic embryo of *Arabidopsis thaliana* [80]. A broad putative preprophase band develops in the elongated zygote prior to its first mitosis [80]. This may be a structural sign that signals the beginning of a new development pathway. In mesophyll protoplasts of *Medicago sativa* undergoing embryonic induction, an altered microtubule morphology is observed; this change is followed by an asymmetric first cell division and rapid cell division coupled with limited cell expansion [81]. In the vascular differentiation of root cortical cells of peas during the "reprogramming" process prior to cell divisions, microtubules organize into parallel arrays [78]. In the study of the wound reaction in pea root, Hush and Overall [82] also concluded that "microtubules play a critical role in establishing and maintaining cell polarity". Schulz [78] suggested that the microtubles may be responsible for the repositioning of the nucleus and that they might participate in the orientation of the plane of future division. Irrespective of the differences observed, the preceding discussion clearly gives evidence of the importance of cytoskeletal elements in relation to cell differentiation, and that they can serve as structural markers for the study of the early events in somatic embryogenesis. It is also important to note that the cytoskeletal elements can have distinct roles to play during the commitment/ induction/ determination/ and polarity establishment phases. For example, in the *Fucus* zygote, cytoskeletal elements and the extracellular matrix may be essential for polar axis fixation [83]. A careful selection and characterization of the system is essential in determining the role of the cytoskeletal elements and how they are regulated in the formation of embryogenic cells.

The processes of induction and determination are completed within a comparatively short period of time in the absence of a callus phase. It is interesting that cells of these systems have a relatively dense cytoplasm at the time of experimental manipulation. The initial high activity of the cells may have shortened the early events as the basic biosynthetic apparatus, such as ribosomes and endoplasmic reticulum, is intact.

B. Polarity establishment

Information concerning polarity establishment is essential to our understanding of somatic embryogenesis. Physical, electrical, ionic, and hormonal signals have been implicated as an important influence in polarity establishment and plant morphogenesis [84–86]. Little is known about the initial events that establish the polarity of cells destined to become somatic embryos. In the investigation of polarity establishment in algae, it is well-known that changes in ion fluxes, especially calcium ions, the generation of ion current, intracellular redistribution of organelles and macromolecules are important to the creation and the fixation of the polar axis of the cell [19]. Cytoskeletal elements and the extracellular matrix may also be essential for polar axis fixation in *Fucus* [83]. Whether any of these processes occur during the formation of embryogenic cells and the early stages of somatic embryo formation is not clear at present.

Some of the physical factors associated with plant growth are the generation of electric and ion currents by plant cells. Electric current has been detected in cultured plant cells [87–89]. Electrical polarity and ion currents have been observed in different stages of carrot somatic embryos [90–92] and in tobacco microspore embryos [93]. These observations clearly indicate that electrical polarity and ion currents are integral parts of the developmental process. Changes in electrical and ion currents may lead to alteration in the pattern of development. As an example, Dijak et al. [94] observed that the application of a low voltage electrical field greatly enhances the formation of somatic embryos from alfalfa. This mild electrical treatment may have perturbed the organization of the cell leading to the establishment of polarity and organized development.

It is well-known that calcium ions can mediate a large number of developmental processes. An increase in calcium ion concentration stimulates pollen embryo production in *Solanum carolinense* [95] and carrot [96]. A recent study indicates that a calcium ion flux precedes organogenesis in *Graptopetalum* [97], again suggesting that calcium ions can play a role in the establishment of polarity in higher plant tissues. In the case of somatic embryo induction, it is not clear whether calcium ions have any effect during the inductive phase. A more precise experimental design will provide greater insight into the role of calcium during somatic embryogenesis.

The formation of electric current in plant cells may be essential to the establishment of polar auxin transport or the "fixation" of a polar auxin movement pathway. The polarity generated by the observed electric current may prevent auxin diffusion from occuring within the developing embryo. Subsequently, these factors may reinforce one another as suggested by Sachs

[98], leading to histodifferentiation. This argument is in part supported by the work of Goldsworthy and Rathore [87]. The application of auxin inhibitors abolishes the electrically induced growth of plant cultures, even in the presence of auxin. In the carrot somatic embryo, a disruption of auxin flow at the globular stage results in the enlargement of the globular mass without further differentiation [99]. Rathore et al. [92] suggested that "an electrical field set up across the embryo might result in polar transport of indole-3-acetic acid [100] which in turn might control embryonic development" [101]. Although the importance of electric current in relation to plant growth and development has clearly been established, it is important to point out that there is no information concerning the initiation of electrical polarity establishment in plant cells, and how this polarity affects subsequent somatic embryo development.

From the structural stand point, some of the early detectable signs of polarity establishment are uneven distribution of cytoplasmic components and asymmetric cell divisions. Asymmetric cell division seems to be necessary for subsequent differentiation. However, as Lyndon [19] pointed out, if polarization of cytoplasm exists within the cell, a symmetrical cell division still results in daughter cells having different cytoplasmic constitutions. Unequal cell division is observed in the carrot suspension system prior to the formation of the embryogenic masses [22, 48]. In the formation of pollen somatic embryos, both equal and unequal cell division is observed. A careful evaluation is needed to determine whether cytoplasmic differentiation, which is presumably the first sign of cytoplasmic polarity, occurs prior to the first cell division. Furthermore, can the change in the direction of cell division such as the epithelium of rice scutellum be used as the indicator of polarity redirection in cells?

Embryogenic cells are developmentally plastic [8], and this may be due to a weak polar axis or the inability to stablize the polar axis once it is established within cells. In his study of vascular element differentiation, Sachs [102] suggested that "polarity is stable whenever a flux of inductive signals follows this polarity". Upon the transfer of the pro-embryogenic cell mass to the medium without auxin, somatic embryo development proceeds. During the early stages of development, the endogenous inductive signal within the proembryo is still developing and is most likely to be weak. According to Sachs' hypothesis [20, 102], a weak signal could de-stabilize the polarity and make the young embryos more susceptable to external influences. As embryo development progresses with the formation of the cotyledons, a strong polarity exists within the embryo. Surgical experiments performed by Schiavone and Racusen [103] clearly indicate that a strong regenerative power of the carrot somatic embryo and the principle polar signal is due to the polar movement of auxin. The observed electrical polarity found in the more mature embryos may be an important factor in the co-ordination and maintenance of polarity within the entire embryo. Further quantitative analysis of endogenous growth substances during somatic embryo development will provide a better insight into the nature of the inductive signals essential to the maintenance of polarity within the developing embryo.

C. Single vs multicellular origin of the somatic embryo

Systems such as the pollen embryos, and those which develop without a callus phase such as the rice epithelial cell system, show beyond a doubt that somatic embryogenesis has a unicellular origin. Other systems, especially those requiring a callus phase in liquid suspension are not as definitive and remain controversial (see Halperin, this volume). Halperin [104], McWilliams et al. [105], and Backs-Huseman and Reinert [48] have shown that carrot somatic embryos are derived from cellular aggregates and not directly from single, free floating cells [105]. However, we do not know whether all cells within the embryogenic mass are "embryogenically determined". Furthermore, in a liquid suspension system, the physical environment may have altered the physical configuration of the somatic embryo preventing the formation of a recognizable "compact-looking" proembryo from a single cell.

Once the cells are determined, can all embryogenically determined cells give rise to somatic embryos? Are all the cells in the peripheral region of an embryogenic clump capable of developing into somatic embryos? In the rice epithelium system, judging from similarities in their structure, many cells appear able to give rise to somatic embryos. However, only a few somatic embryos are present. Similarly, in carrot, only a single or a few somatic embryos are produced from a single proembryogenic mass. Jones and Rost [55] suggested that competition may have prevented further development of some embryogenic cells in favor of others. As embryogenic cells are tightly aggregated and have similar metabolic processes, metabolic co-operation between cells is a more efficient means to carry out a single task. Metabolic co-operation can also lead to fewer somatic embryos being formed. It is important to point out that the strategy involved in the regulation of continual embryo development may be quite different from that involved in induction and determination events. In different culture systems, it is likely that a single cell is capable of giving rise to a somatic embryo either directly or somewhat indirectly through a few division cycles, but culture conditions and neigbouring effects would certainly influence their subsequent development. In support of the above idea, plasmolysis of cell cultures, such as that performed by Wetherell [106], leads to a higher number of somatic embryos being formed. The plasmolysis experiment suggests that a single embryogenic cell has the ability to form a somatic embryo. Within an aggregate, symplastic communication, metabolic co-operation, and/or competition can take place, resulting in fewer somatic embryos. Careful structural and biochemical studies on the formation of somatic embryos from embryogenic masses will allow us to determine the dynamics within individual embryogenic masses.

D. Is isolation a necessary prerequisite for embryogenic cell induction?

During the formation of reproductive structures, such as the embryo sac and the zygote, plasmodesmata are usually absent and a direct symplastic connection is not present between the zygote and its maternal tissues. This and other observations lead to the suggestion that physiological isolation – the absence of plasmodesmata and hence symplastic transport – plays an important role in somatic embryo formation.

In the embryogenic ovular callus of "Shamouti" orange, somatic embryos are formed from single cells having a thick cell wall which lacks plasmodesmata [107]. Although this observation suggests that physiological isolation may be important in somatic embryo formation, it is important to note that these cells are derived from existing embryogenic callus. Plasmodesmata may be present during the induction phase of somatic embryogenesis. Recently, it has been reported that callose completely surrounds developing somatic embryos in roots and leaves of Cichorium [108, 109]. In flowering plants, the formation of callose during pollen development is associated with the severing of existing cytoplasmic connections between cells. This suggests that the embryogenic cells of Cichorium are isolated from their neighbours. However, in the absence of detailed ultrastructural information of the entire process, it is premature to conclude that "isolation" is essential to the induction process.

In Ranuculus, it is clear that the initial epidermal embryogenic cells are connected with their neighbours via plasmodesmata prior to further embryo development, indicating that these cells are not "isolated" during the inductive events [110]. However, plasmodesmatal connections appear to have been severed during subsequent somatic embryo development. This observation suggests that physiological isolation may not be a necessary prerequisite for the initial inductive events in somatic embryogenesis. Subsequent changes in the distribution and number of plasmodesmata may be essential to allow further development of the somatic embryo.

Can plasmodesmata be used as the marker for physiological isolation? In order to induce somatic embryogenesis, tissues and/or organs have to be excised and placed in a new environment. Although morphologically recognizable plasmodesmata are present, will the excision process and other pretreatment protocols affect normal plasmodesmatal function? Plasmodesmata are dynamic entities [111]; new ones are formed while others are rendered nonfunctional. An alteration in the rate of intercommunication between cells may have a profound influence on cell development. At present, information is not available on whether existing plasmodesmata remain functional during the early inductive events. Until further physiological data are available, structural data alone cannot provide an unequivocal answer as to whether physiological isolation is a prerequisite for somatic embryogenesis. Detailed ultrastructual studies, e.g. serial sections and physiological investigation, such as microinjection of dyes [112, 113], will allow us to evaluate this problem fully. Furthermore, do we need to have complete isolation (absence of plasmodesmata) or simply an alteration in plasmodesmatal function and/or distribution in order that cells can undergo the inductive process?

E. Direct and indirect embryogenesis

Two basic patterns of induction are recognized; direct and indirect embryogenesis [12]. In direct embryogenesis, it is suggested that pre-embryonic determined cells are present and require favorable inductive conditions to initiate embryo development. Indirect embryogenesis requires redetermination of differentiated cells and the acquisition of the embryogenic state prior to the initiation of embryo development [6, 13]. This pattern of formation is still open to interpretation [114]. In a recent article, Merkle et al. [1] provided an excellent discussion of this subject.

> Somatic embryogenesis may be "direct", with embryonic cells developing directly from explant cells, or it may be "indirect", with a number of unorganized, non-embryonic structures. The linguistic distinction can be misleading. In practice these terms define opposite ends of a continuum whose intermediate regions may be difficult to quantify. To distinguish patterns of embryogenesis as direct or indirect based simply on intercalation of mitotic cycles between explant and embryo organization is, in physiological terms, an oversimplification. The most meaningful way to define "direct" and "indirect" appears to be with reference to the epigenetic state of explant cells. Thus, somatic cells which are themselves embryonic, or not far removed from embryonic, are generally more easily induced to undergo somatic embryogenesis than differentiated vegetative cells. Highly differentiated cells appear to require major epigenetic changes, making the initiation of embryogenesis less direct. In these terms, the directness of embryogenesis is measured as epigenetic "distance" of explant cells from the embryonic state [1].

With this in mind, one of the important distinctions between direct and indirect embryogenesis depends on the timing of dedifferentiation and acquisition of competence. In the former case, the cells respond to experimental treatments and become determined in a short time without prior proliferation of cells. In the latter case, a longer time is needed to acquire the embryogenic competent state. This state is often preceded by cell proliferation.

III. Structural characteristics of embryogenic cells

The problems associated with the recognition and characterization of embryogenic cells of somatic cultures were detailed by Street and Withers [115]. In carrot, although peripheral cells of the proembryogenic mass have similar cytological features, it is not certain as to whether all these cells can continue to develop into somatic embryos. It is also difficult to observe significant structural changes occurring during the initial divisions, such as those similar to the first division of the zygote, to ascertain the identity of a embryogenic cell. In canola pollen embryos, the "determination" stage appears to occur only after periclinal divisions begin within the cell mass [70]. Little significant structural changes can be noted within the cells prior to the initiation of periclinal divisions. Because of the uncertainties in the developmental potential of the cells, it is difficult to interpret the published information.

In a number of systems studied todate, embryogenic cells share similar structural features. They are small, highly cytoplasmic, and often have an accumulation of starch within the plastids. The study by Halperin and Jensen [116] documents the ultrastructure of embryogenic carrot cells. In their culture, embryos are derived from cell clumps which perpetuate themselves by growth and fragmentation [116]. The peripheral cells are smaller with thin walls, and contain numerous small starch grains. These cells are highly cytoplasmic with numerous organelles and are capable of repeated cell divisions. It is interesting to note that a high density of free ribosomes are present in the cytoplasm of carrot embryogenic cells. The cells possess a large, diffuse nucleus and stain intensely with protein and RNA stains [105]. Plasmodesmata are present between embryogenic cells as well as between embryogenic and the underlying vacuolated cells, though they are reduced in number in this location [115]. Similar structural observations can be found in other species such as corn [53], *Ranunculus* [117] and other Gramineae species [118].

In the rice epithelial cells [55], which are presumed to be embryogenically determined cells, the cytoplasm is dense and numerous lipid bodies are present. Plastids do not accumulate starch grains and have a simple thylakoid system. Rough endoplasmic reticulum cisternae profiles are not as prominent as in the control treatment and the nucleus has distinct nucleoli.

In canola microspores, the cells within the cluster remain highly cytoplasmic and the cell walls remain thin. Small starch grains are present within the plastids. In contrast to the reports of Zaki and Dickinson [74], Telmer et al. [68] reported that starch and/or lipid accumulation only accompany non-embryogenic development.

The embryogenic cells in the embryonal region of a conifer species, *Picea glauca* consist of small, densely cytoplasmic cells. Numerous organelles are present within the cytoplasm. The embryonal region is subtended by a suspensor composed of long, highly vacuolated cells. Preprophase bands of microtubules can be detected in the densely cytoplasmic embryonal cells [76].

Judging from the ultrastructural studies, embryogenic cells are metabolically active. The abundance of organelles is an indicator of rapid cell growth [76]. Thus, mitogenic activity is an essential property of embryogenic cells.

Is the embryogenic cell similar to that of the zygote in flowering plants? Prior to fertilization, the egg cell of flowering plants is usually quite large and the nucleus is located near the micropylar end, while a large vacuole is located near the chalazal end of the cell. After fertilization, there is usually a reduction in the size of the zygote. The nucleus migrates to the chalazal end of the cell. Organelles begin to congregate around the nucleus. In most species, ribosomes aggregate to form polysomes, and endoplasmic reticulum cisternae become more abundant. Histochemical staining suggests that mRNA synthesis increases, as does metabolic activity within the zygote. After fertilization, a new cell wall is formed over the cell boundary. Other cytological changes include the accumulation of storage products, especially in the form of starch [119]. Thus, the embryogenic cells determined to give rise to somatic embryos are similar. The active appearance of the cytoplasm and the accumulation of storage products is a structural feature indicative of embryogenesis.

The most striking difference between the zygote of flowering plants and embryogenic cells destined to become somatic embryos is that the former have a polarized appearance: the vacuole is near the micropylar end while the chalazal end of the cell is highly cytoplasmic. Marked polarization is absent from the somatic embryogenic cell. This may account for variations in subsequent development of the somatic embryo, especially as somatic embryos lack a functional suspensor system during the early stages of development. This observation suggests that a stronger polarity may be present within the zygote.

IV. Somatic embryo development

A. Proembryo development

Upon the transfer of determined embryogenic carrot cells onto medium without 2,4-D, changes begin to occur. The originally smooth embryogenic cell clumps take on a more undulated appearance as superficial embryogenic cells begin to divide and give rise to somatic embryos. Ultrastructural observations indicate that there is an increase in the amount of endoplasmic reticulum cisternae and ribosomes. Organelles become even more abundant as compared to the embryogenic cells described in the previous section [115].

Based on the segmentation pattern, the carrot proembryo appears to have its origin in a single embryogenic cell within the cell mass. A periclinal division marks the beginning of somatic embryogenesis in carrot; the terminal cell gives rise to the embryo proper and the basal cell forms the suspensor [105]. Each group appears to give rise to an individual somatic embryo. However, the population dynamics of the surface cells and their ultimate fate is not clear. Can all surface cells give rise to somatic embryos, or do some of them have alternate developmental pathways? In carrot, different rates of cell division are observed within the cell mass, leading to the formation of the globular embryo. During the first three days after transfer onto auxin-free medium, the rate of cell division within the embryogenic cluster is slow. From day 3 to day 4, rapid cell division occurs in the determined locus of the cell cluster, leading to the formation of globular embryos [21]. Polarized DNA synthesis is observed during the initial stage of embryo development. A polarized distribution of calcium ions was noted by Nomura and Komamine [111] using fura-2 AM as the fluorescent calcium ion indicator. When chlorotetracycline is used as the calcium stain, marked polar distribution of calcium is not observed in the proembryogenic mass [120]. However, the background fluorescence could have obstructed the real distribution of calcium within the cell. The use of the confocal laser scanning microscope will allow a much clearer determination of calcium distribution. The calcium binding protein, calmodulin, has been found to be distributed unevenly within the proembryogenic mass [120].

When comparing the early stages of somatic and zygotic embryo formation, similar patterns of development have been observed in rice [121], *Vitis* [121a] and *Ranunculus* [110]. In the case of carrot, the early cell division pattern differs from that of the zygotic embryo. McWilliam et al. [105] clearly demonstrated that the carrot somatic embryo follows the Crucifer type and not the Solanad type of division found in the zygotic embryo. In the Crucifer (Onagrad) type, the terminal cell of a two celled embryo divides anticlinally while in the Solanad type, the terminal cell divides periclinally [119]. Similarly, in alfalfa, different patterns of somatic proembryo formation can be observed depending on the region of the explant in which the embryo is initiated [122]. However, this variation in the cell division pattern appears not to have a profound influence during subsequent development. Davidson [123] suggested that "developing embryos can integrate variations in cell size and orientation of division without undergoing developmental perturbations". The differences in planes of cell division may be due to physical constraints imposed on the embryonic cells. It is known that physical environment can alter the direction of cell plate formation [124]. In the light of the recent report by Mayer et al. [79a], it would be of interest to reinvestigate the cell division pattern and determine whether differing cell division patterns affect the subsequent histodifferentiation process.

In corn, a transitional structure analagous to the proembryo stage is observed [125]. This structure is more organized than the embryogenic units and consists of cells with different degrees of vacuolation. Intercellular spaces are present. Franz and Schel [125] suggest that the transition unit marks the change from unorganized to organized development of the somatic embryo. Further studies are needed to determine how the globular embryo develops from such a structure.

B. The suspensor

During zygotic embryo development, the suspensor exhibits structural specialization in that it has a transfer cell morphology, large amounts of endoplasmic reticulum cisternae, and numerous polysomes, mitochondria, and plastids [126, 127]. This structural specialization clearly indicates that a division of labour exists within the zygotic embryo. During zygotic embryo development, the suspensor is believed to play an active role [128, 129]. In the study of suspensor function of *Phaseolus*, the suspensor has been shown to be the active uptake site for the young developing embryo. The uptake process by the suspensor is an active process as it can be inhibited by metabolic inhibitors, while the embryo proper of the heart stage embryo is not sensitive to metabolic inhibitors [128]. This uptake pattern indicates that the embryo proper is heterotrophic in the early stages of development. When one examines the ultrastructure of the embryo proper of the zygotic embryo, the cells are packed with ribosomes; however, few other organelles are present within the cytoplasm. Ultrastructural work suggests that cells of the embryo proper are not as metabolically specialized as are the suspensor cells. The embryo proper cells are primarily geared to mitotic activity during early stages of development. For a more critical discussion on suspensors of zygotic embryos, see Yeung and Meinke [129a].

In species such as maize in which a structurally active suspensor is not present, the surrounding tissues, i.e. the nucellar epidermis and the endosperm [130, 131] may have functions similar to that of the *Phaseolus* suspensor. The organization of the embryonic environment and the structure of the developing zygotic embryo indicates that a division of labour exists within the seed. In the case of somatic embryogenesis of flowering plants, a morphologically distinct suspensor may or may not be present. Even when it is present, structural specialization has not been found. This pattern of development suggests that the suspensor in somatic embryos is not required or essential to its development. For those embryos formed on solid medium, the suspensor most likely functions primarily as a conduit for nutrients and may not have specialized synthetic functions such as those found in bean. If one compares the ultrastructure of the embryo proper, cells in the somatic embryo usually have a full complement of organelles, and the organelles are more abundant than cells in the embryo proper of the zygotic embryo. This ultrastructural pattern indicates that the somatic embryo proper is more metabolically active than its zygotic counterpart. The cytoplasm of the zygote is highly polarized prior to its first cell division, whereas a similar process in the somatic embryo cell has not been reported. This suggests that the organogenetic portion of the somatic embryo is autotrophic and does not rely on the supply of metabolites from the suspensor for its development. Further comparisons of metabolic activity between the embryo proper of zygotic and somatic embryos will allow for direct testing of this hypothesis.

Differences in suspensor morphology may also be due to the timing of polarity establishment [132]. The polarity of the zygote is well marked, which leads to the formation of a structurally defined suspensor. On the other hand, a strong polarity may not be present during the early induction and formation of the somatic proembryo, thus preventing the formation and development of a structurally defined suspensor. For those somatic embryos generated from the proembryogenic mass such as carrot, the remaining cells within the proembryo-genic mass may in fact function as a rudimentary suspensor [56]. If metabolic co-operation indeed exists, the cells forming the rudimentary suspensor may be important in the development of the somatic embryo. Histochemical studies may provide additional information about this suspensor system.

In conifers, many reports indicate that the suspensor is present together with the embryonal cell mass during somatic embryo formation [76, 77, 133]. Once formed, the somatic embryonal cell mass in *Larix* produces both suspensor and embryo. Thus, the suspensor is an integral part of conifer somatic embryos.

Structural specialization similar to that of *Phaseolus* has not been reported in the suspensor of conifer species. Suspensor cells are usually highly vacuolated. In contrast to cells in the cytoplasmic cells of the embryonal mass, plastids of the suspensor remain undifferentiated [76]. In white spruce, those suspensor cells that show active cytoplasmic streaming have cable-like actin filaments that can be detected to run parallel to the long axis of the cells [76, 134].

Although physiological and experimental data are not available concerning the potential function of the suspensor in conifer species, the fact that the suspensor is an integral part of the embryonal mass suggests that it may be essential to embryo development. Furthermore, in *Larix* [77], suspensor cells have been shown to give rise to organized meristematic centres, again suggesting a close interrelationship between the suspensor cells and the embryonal mass. This evidence indicates that a division of labor may have existed within the embryonal mass, and that a functional suspensor is essential to its development.

C. The globular embryo

Since the general morphology of somatic embryos is similar to that of their zygotic counterpart, little attention has been paid to the study of somatic embryo development beyond the proembryo stage. This may be due to the fact that somatic embryos are plastic in their development, as a range of embryonic forms can be found within a single culture [14], and morphological abnormalities that develop during somatic embryogenesis may not prevent subsequent development [135]. Thus this aspect of somatic embryogenesis appears not to be important.

One of the most fascinating aspects of somatic embryogenesis is the generation of forms. There is little physical constraint surrounding the somatic embryo and yet the progression of forms, i.e. globular, heart, etc., are remarkably similar to those of the zygotic embryo. The answer to this observation may lie in our understanding of protoderm differentiation and the regulation of the cell division pattern.

The globular stage of somatic embryogenesis clearly marks the beginning of structural differentiation. Histogenesis begins with the formation of a protoderm covering the globular embryo. Abnormal protoderm formation can lead to the arrest of the carrot somatic embryo. Several glycoproteins have been identified [71, 72, 136–138] and some of these proteins can rescue arrested somatic embryos. These observations clearly indicate the importance of the protoderm in the further development of the carrot somatic embryo. Changes in cell wall components can result in changes in the physical characteristics of the cells such as the regulation of cell expansion [71, 186]. This could lead to the generation of forces that are important to the morphogenesis of the embryo. Although little is known about the specific interactions at the surface of plant cells [139], the cell surface components are the most likely candidates for the generation of various shapes of the somatic embryo.

In carrot, the protoderm cells at the globular stage are not as tightly packed as those of the zygotic embryo and the protoderm itself has a tendency to form secondary embryos in some cases [105, 122]. The phenomenon of secondary embryogenesis may be a result of an absence of a fully differentiated protoderm.

In the carrot globular stage embryo, the outer cell layer tends to divide in an anticlinal direction, while the inner layers preferentially divide in a periclinal direction. Using geometric analysis on the growth pattern of the carrot somatic embryo, Schiavone and Cooke [140] observed that prior to the formation of the heart shaped embryo, axial elongation of the inner isodiametric cells of the globular embryo leads to the formation of a longitudinal extension near the lower end of the embryo. They termed this an oblong embryo. Schiavone and Cooke [140] further suggested that the formation of the oblong-shaped embryo is the first sign of incipient procambium formation. The change in the growth axis of the procambium is an important step in morphogenesis as it involves changes in cell size and shape and these changes most likely involves changes in the cytoskeletal pattern of cells. Wilson et al. [47] reported that microtubules are not conspicuous until the globular stage of somatic embryo development. Furthermore, the observed changes in microtubule pattern may be an important marker for the establishment of auxin polarity within the embryo. Schulz [78] suggested that there may be a correlation between the occurrence of stranded forms of microtubules as reported by Goosen-de Roo et al. [141] and the onset of channelled auxin transport. Thus, the formation of the oblong-shaped somatic embryo of carrot may provide a useful system to study such a postulate.

Differentiation of organelles also begins at the globular stage. Plastids begin to turn green in the terminal portion of the developing somatic embryo [115]. The ground meristem of the oblong embryo begins to vacuolate, a process which continues throughout embryo maturation [140]. When compared to a similar stage of zygotic embryo development, vacuolation occurs at an earlier stage in the somatic embryo [104].

Much more information is needed at the globular stage of development. In zygotic embryogenesis, histodifferentiation begins at the globular stage [142]. Through careful studies of selected experimental systems, additional information concerning the structural, biochemical, and molecular changes in the embryo can be obtained. For example, in a recent report, somatic embryos of *Ipomoea batatas* can be arrested at the globular stage of development using a liquid suspension system [143]. Upon transferring the embryos onto solid medium, structural polarity is observed with the formation of torpedo-shaped embryos. Through the utilization of temperature-sensitive carrot variants, stage specific polypeptides have been identified [144]. These and other systems may be ideal to study changes at the globular stage of development.

D. Cotyledons of the somatic embryo

The next important morphogenetic event in embryogenesis is the formation of the cotyledons and a clearly defined embryonic axis. This is usually referred as the heart stage of embryo development. In dicots and in conifer species, cotyledons arise as small protrusions from the peripheral region of the terminal end of the somatic embryo. Local activity results in the formation of small cotyledon primordia. In carrot somatic embryos, cotyledon primordia are composed of dense isodiametric cells. As development progresses, cotyledons normally grow in size, but seldom reach the same size as in the corresponding zygotic embryo. In zygotic embryo development, the pattern of cotyledon formation is a predictable event. In somatic embryogenesis, one of the most often reported abnormalities involves the cotyledon, i.e. a variable number and a variety of cotyledonary forms can be found in the same culture [145–147]. In soybean somatic embryos, a range of cotyledon morphologies have been observed, i.e. mono-, di-, poly, and fused forms [135].

Do cotyledons have a morphogenetic effect on somatic embryo development? The work of Schiavone and Racusen [2, 103, 148] suggests that the cotyledons are important in the regeneration of the somatic embryo. Using microsurgical procedures, different regions of carrot somatic embryos show strong regenerative power [103, 148] and spatially-specific proteins can be detected in the upper and lower half of the somatic embryo [149]. The upper half of the somatic embryo readily regenerates the root pole while the bottom half behaves as if it had been released from growth inhibition by forming root hairs and a root cap. This observation suggests a physiological gradient is established at the heart stage of embryo development. Auxin may play an important role in this process [99, 148]. The formation of the oblong-shaped embryo may well be the first sign of the establishment of a strong physiological gradient. Surgical studies indicate pattern restoration in tissue explants that do not possess well-defined meristems [103]. The physiological activity of the apical meristem is not known at present. Its role in the maintenance of polarity and regeneration requires further experimental study. Since the cotyledons constitute the major organ at this stage, they are the most likely candidates in the generation and/or the maintenance of the observed polarity in carrot somatic embryos.

The cotyledons may also influence the germination process. Buchheim et al. [135] demonstrated that there is a direct relationship between the amount of cotyledonary tissue and conversion days, i.e. the time required for recovery of plantlets or plants from somatic embryos. Although the mechanism of action is not clear at present, these observations clearly suggest that the cotyledonary tissue in the somatic embryo may actively affect embryo development, as well as subsequent germination events.

E. Apical meristems of somatic embryos

The formation of the apical meristems is one of the most important events in embryogenesis. The quality of the somatic embryo and their germinability depends on the proper development of a normal, functional meristem.

At or near the end of the globular stage, the zygotic embryo expands laterally, eventually giving rise to the shoot apical meristem and the cotyledon primordia. At the same time, the root meristem begins to be delimited. Upon germination, subsequent plant development depends on the activities of these apical meristems. In the case of somatic embryos, the formation of the apical meristems is generally ignored. Little detailed structural information is available on meristem formation in somatic embryos. One study of soybean somatic embryogenesis claims that apical meristems resemble zygotic embryos [150]. However, a detailed evaluation cannot be made from the published micrographs. In carrot somatic embryos, Street and Withers [115] showed the presence of well defined shoot apical meristems in their preparation. Upon close examination, a range of shoot apex organization can be found in carrot [151]. For embryos that do not convert, cells in the apical notch showed pronounced vacuolation. These embryos failed to form a meristem, and the entire shoot pole became necrotic, though healthy roots persisted. Embryos that successfully converted maintained a nucleus of densely cytoplasmic cells in the apical notch [151]. It has been shown recently that abscisic acid plays a role in shoot meristem development [188]. Shoot apical meristem organization in carrot is illustrated in Plate 2 (Figs. 4–7). Similarly, in white spruce, cells of the shoot apical meristem have cytological features that differ from those of zygotic embryos. In mature somatic embryos, the functional part of the shoot apical meristem is bi-layered. Large air spaces are present between meristem cells, while cells in the zygotic embryo are tightly packed. After partial drying and upon germination, cells of the epidermis and the subepidermal layers of the somatic embryo begin to divide and develop into a new shoot apical meristem which is comparable to its zygotic counterpart. The structural changes in the apical meristem of white spruce are summarized in Plates 4 and 5 (Figs. 12–19). These observations indicate that the development of the meristem in somatic embryos can be different from those of zygotic derived embryos. Without a careful structural study, one cannot assume that the structure is normal. Experimental manipulations such as the addition of abscisic acid, silver nitrate, removal of ethylene, osmotic stress, and a partial drying treatment may greatly improve on the organization of the apical meristem and its subsequent conversion rate [151, 152, 189, 190]. It is unfortunate that so little attention has been paid to the formation of the meristem in somatic embryos. In a discussion of determination, Wareing [16] indicated that the meristems are "determined" structures. Special conditions may be required for their formation. The large number of abnormalities observed in differing culture systems, and the low germination and conversion percentage indicates that the culture conditions may not be optimal for meristem differentiation. The low percentage of germination and conversion in species such as grape [153] may be due to the failure of shoot meristem formation.

Little is known concerning the organization and activity of the root apical meristem. In the zygotic embryo, the root meristem begins to differentiate at the late globular to heart stage of embryo development [141]. In *Arabidopsis*, a defined pattern is observed that results in root meristem initiation [79b]. A quiescent center begins to develop in the future root pole of the *Capsella* embryo as early as the globular stage [154]. Whether a quiescent center exists in the

Plates 4 and 5 illustrate structural changes in the shoot apical meristem of the white spruce somatic embryo. The purpose is to indicate that apical meristem development in the somatic embryo can be quite different from that of the zygotic embryo. For a detailed description, see Kong and Yeung [152].

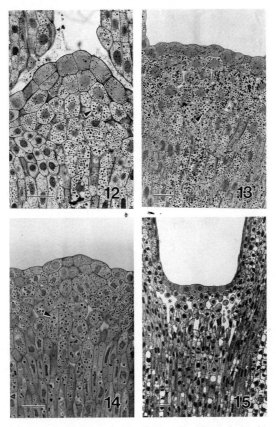

Plate 4. (12) Light micrograph showing a longitudinal section through the shoot apex of a mature zygotic embryo. Shoot apex is conical. The top layer of cells has more densely stained cytoplasm and prominent nuclei. Cells within the shoot meristem are tightly packed and filled with abundant protein bodies (arrowhead). Scale bar = 40 μm. (13) Light micrograph showing the shoot apex of a mature somatic embryo before partial drying. Cells within the first layer of the meristem are similar to the zygotic embryo in having dense cytoplasm with prominent nuclei. Relatively little starch accumulates in this layer. In contrast to the zygotic embryo, the cells of the shoot apex are not tightly packed; intercellular spaces (*) exist. Except for the first layer, other cells are vacuolated and contain abundant starch granules (arrowhead) within their cytoplasm. Scale bar = 40 μm. (14) After partial drying, prominent protein bodies are present (arrowhead) within the cells, whereas storage starch deposits have disappeared from the cytoplasm. Scale bar = 40 μm. (15) This figure illustrates the fact that shoot apical meristem development differs between cell lines. The shoot apex of the IMF-5 line has marked intercellular spaces (*). Subapical cells are loosely attached to the apical layer. Even after partial drying treatment, their germination percentage is extremely low. Scale bar = 40 μm.

234

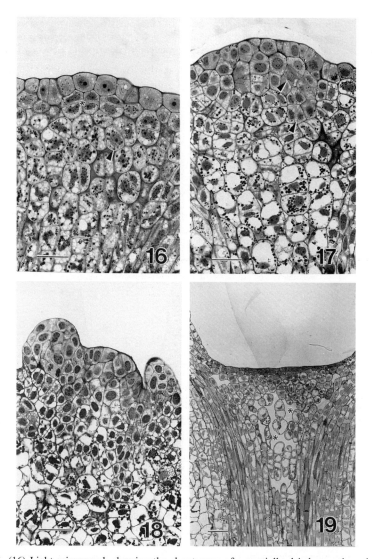

Plate 5. (16) Light micrograph showing the shoot apex of a partially dried somatic embryo on germination medium at Day 2. The subepidermal cells begin to divide (*). These initial cell divisions are mainly in the periclinal direction. Protein bodies disappear soon after the placement of embryos onto the gemination medium. Starch granules (arrowhead) are again becoming abundant in the cells of the subapical region. Scale bar = 40 μm. (17) At Day 3, apical layer cells and the cells newly derived from the subapical layer are mitotically active (arrowhead). This mitotic activity results in the formation of a small meristematic dome. The cells within the meristematic dome are tightly packed with dense cytoplasm, a prominent nucleus, and small vacuoles. Scale bar = 40 μm. (18) By Day 7, a shoot primordium begins to appear. Leaf primordia have been established. Scale bar = 40 μm. (19) In the absence of the partial drying treatment, large intercellular spaces (*) develop in the apical region of the somatic embryo. Although mitotic activities are present in the apical layer, organized development leading to the formation of a shoot primordium is absent. Scale bar = 40 μm.

somatic embryo is not known as present. In the study of regeneration patterns of the carrot somatic embryo after microsurgery, it was shown that the shoot pole can cause a regeneration of the root pole [103]. Furthermore, the shoot pole can regulate the growth of the root pole; once the root pole is separated from the shoot pole, rapid expansion begins. This suggests that the development of the root meristem may be tightly coupled to the differentiation of the shoot apical meristem. Much more work is needed related to meristem differentiation in somatic embryogenesis.

F. Embryo maturation

In zygotic embryos, the maturation stage is characterized by storage product accumulation and the acquisition of tolerance to desiccation. The desiccation process enhances subsequent germination [29, 155, 156].

Storage product deposition has received less attention in the study of somatic embryogenesis. It appears that all somatic embryos have the ability to synthesize storage products just like their zygotic counterparts [133, 157–160]. However, the pattern of storage product deposition, the amount deposited, and the chemical composition may not be the same. In white spruce, polysaccharides accumulate first, followed by lipid and lastly protein [133]. Quantitatively, cotyledonary stage somatic embryos have less lipid and protein and more starch when compared to zygotic embryos at the same developmental stage. SDS gels also reveal differences in the protein profiles, and qualitative differences are also observed. In mature somatic embryos, the cotyledons of the same individual are extremely rich in lipids and proteins while other plant parts have little. In mature zygotic embryos, storage lipid and protein bodies are densely packed within the cells and both are uniformly distributed throughout the entire embryo [133]. The variability observed in most systems is likely due to differences in the embryonic environment. The zygotic embryo develops within an enclosed environment. The physical and chemical environment of the endosperm, seed coat, and maternal tissues exert a combined influence on embryo development. In the somatic environment, the signals for storage product synthesis may not be frequent or significant enough to stimulate massive storage synthesis, or the composition of the culture medium may not be conducive or adequate to such massive synthesis.

Somatic embryos in culture are able to germinate precociously. Precocious germination often leads to abnormal seedling formation or the cessation of growth after cotyledon elongation. In zygotic embryos, physical enclosure by the seed coat, a gradual lowering of water content, and qualitative and quantitative changes in endogenous growth substances leads to developmental arrest [9, 156, 161] In the case of somatic embryos, such programmed and co-ordinated changes do not exist, as the embryonic environment is different from that found in zygotic embryos. Thus, it is not surprising to find that the somatic embryos germinate precociously. Their inability to prepare properly for the subsequent germination events may predispose them towards the production of abnormalities.

If the embryo is allowed to "mature" in culture prior to germination, a higher percentage of germination can be achieved. The somatic embryos of soybeans require a period of maturation before they acquire the capacity to convert to a normal developmental pattern [134]. Recent protocols developed by Slawinska and Obendorf [162] are successful in bringing the mature soybean somatic embryos to a quiescent state without precocious germination. Thus, if the process of precocious germination can be inhibited and the maturation process prolonged, more "normal" plantlets may be obtained.

A number of studies have documented the usefulness of abscisic acid, osmoticum, and/or dessication of the somatic embryo in the prevention of precocious germination and the production of a more "normal" looking somatic embryo. The importance of abscisic acid in zygotic embryo development has long been recognized [161, 163]. Abscisic acid can stimulate growth and protein accumulation of zygotic embryos in culture and appears to maintain embryos in a developmental mode [29, 155]. However, other factors also appear to be important. Recently, Xu et al. [164] indicated that osmotic stress can also prevent the germination of developing alfalfa zygotic embryos. While both abscisic acid and osmotic stress can prevent germination of the developing alfalfa embryo, only the osmoticum can maintain the synthesis of developmental proteins [164]. Although the mode of action of abscisic acid and osmotic stress are not clear at present, treatment of somatic embryos with either or both of these agents is beneficial to their subsequent maturation. Abscisic acid is required for improved somatic embryo development in a number of conifer species [165, 166, 190]. It has been shown to promote storage product accumulation in alfalfa somatic embryos, leading to improvements in the conversion frequency [167]. Osmoticum, such as a high sucrose content, is required for canola pollen embryoid development [168] and the maturation of soybean somatic embryos [135, 162]. Depending on the system, other agents such as indolebutyric acid in Norway spruce [165] can also play a role in the maturation process of somatic embryos. Although the mode of action of these chemical and physical agents is not clear at present, somatic embryos provide an excellent system for such investigations.

The final phase in seed development is desiccation. One of the possible roles of desiccation is that this process switches the pattern of gene expression from a maturation to a germination program [29, 30, 155]. Drying may affect the hormonal balance and decrease the sensitivity of the embryo to abscisic acid, releasing the embryo from the constraints of development and allowing germination to proceed [30]. The low percentage of germination observed in a number of somatic embryo systems may be due to the failure of the embryo to switch to a germination program even when structurally "normal" shoot and root apices are present [169]. A number of somatic embryo systems, including grape and orchardgrass, have successfully survived the desiccation treatment [170]. This treatment has been shown to greatly enhance the germination of interior spruce [171], white spruce [172], and wheat somatic embryos [173, 174]. There is no doubt that some form of dessication treatment plays an important role in the germinative development of some somatic embryos.

At present, more information is needed in relation to developmental, maturation, and germination processes of the somatic embryo. Information concerning the formation and activity of the apical meristems is essential to our understanding of the desiccation and germinative events. The "quality" of the somatic embryo can be greatly enhanced if we have a proper understanding of these processes. This information is also essential for the development of artificial seed technology.

V. Are somatic embryos similar to their zygotic counterparts?

Considering the differences in both the physical and chemical environment of the somatic and zygotic embryos, one of the most fascinating aspects of somatic embryogenesis is the similarity between somatic and zygotic developmental patterns. The question to be asked is just how similar are these two processes?

In the preceeding discussion, a number of differences have been pointed out to indicate that differences do exist between the two systems. The most important difference is the inductive process that leads to the formation of a determined embryogenic cell. The strategy utilized by somatic cells is certainly different from the controlled fusion of gametes that results in the formation of the zygote. Gene activity studies by Wilde et al. [42] and morphological comparisons clearly indicate that different developmental strategies are employed. During somatic embryogenesis, less division of labour exists within the embryogenic mass, and the embryogenic program is already present in "determined" embryogenic cells. On the other hand, during zygotic embryo-gensis, embryo development appears to be a highly regulated process; this is exemplified by the formation of the structually specialized suspensor and surrounding maternal tissues. From a theoretical standpoint, if the preceeding speculation is correct, information obtained from the study of the early events on somatic embryogenesis may not be applicable to that of zygotic embryo-genesis.

From the globular stage onwards, although differences still exist between the two systems, the basic developmental strategies are likely similar. The differences observed, such as cotyledon morphology and storage product deposition, are most likely due to differences in the embryonic environment. In the zygotic embryo of beans, the surrounding seed coat has a number of morphological specializations [175, 176]. Physiological studies indicate that the seed coat has an important nutritive function [177]. The zygotic embryo is enclosed within the seed coat and is in turn covered by the fruit wall. This arrangement indicates that the immediate embryonic environment is hypoxic. The high level of alcohol and lactate dehydrogenases [178, 179] observed within the seed indicate that the embryonic environment is indeed unique. With a better understanding of the embryonic environment and developmental sequence of the zygotic embryo, culture conditions can be improved to simulate the natural environment, and a more "normal" pattern of somatic embryo development can be achieved. Using this principle, Carmen [173, 174] has enhanced somatic embryo production in wheat by changing the oxygen concentration so that it resembled that of the ovules. Continual improvements in culture conditions may allow for the production of embryos that more and more resemble their zygotic counterparts.

VI. Future perspectives

The ultimate control in the observed developmental patterns rest in the genes. It is clear from recent genetic analyses of *Arabidopsis* zygotic embryo mutants that genes are responsible for tissue patterning and meristem formation [79a,b]. Although this review deals with somatic embryogenesis, it is important to note that study of zygotic embryogenesis will continue to aid in our understanding of how somatic embryos develop. The utilization of embryo mutants of *Arabidopsis* will provide additional information for all phases of zygotic embryo development [180, 181]. Temperature sensitive mutants that affects carrot somatic embryo development have been reported [182]. Thus, embryo mutants will provide important insight into the mechanism of gene action.

In order to understand gene action, detailed structural investigations are essential. The recent studies by Mayer et al. [79a] and Burton and Poethig [79b] clearly demonstrate the importance and the reward of detailed structural studies. The structural method is a powerful approach in the study of embryogenesis. Although a number of structural studies are available, more work is still needed, especially during early stages of embryogenic callus induction. In order that structural information be useful, it is essential that detailed ontogenetic investigations be carried out. At present, only fragmentary information is available for various systems. Most structural information that has been published provides only a superficial description of the systems. Little attention is paid to the establishment of the "primary" embryogenic callus. Without this information, it is impossible to gain a better understanding of the early inductive processes that lead to the formation of determined embryogenic cells. High resolution light microscopy using thin plastic sections [183] and detailed ultrastructural studies are required. These methods, when coupled with other biochemical and cellular methods such as autoradiography, immuno-cytochemistry, hybridization of defined nucleotide sequences, and additional experimental approaches, will provide valuable insight into this process. It is hoped that this review will both stimulate and facilitate further exploitation of this remarkable process, which is of both theoretical and practical interest.

Acknowledgements

I thank my colleagues who generously supplied reprints and preprints used in preparing this review, especially Drs. C. Telmer, P. Pechan, and S. Poethig. I also thank Mr. Todd Nickle for the critical reading of the manuscript. Research in the author's laboratory has been supported by the Natural Sciences and Engineering Research Council of Canada.

References

1. Merkle, S.A., Parrott, W.A., Williams, E.G., Applications of somatic embryogenesis and embryo cloning, in *Plant Tissue Culture: Applications and Limitations*, Bhojwani, S.S., Ed., Elsevier, Amsterdam, 1990, 67.
2. Racusen, R.H., and Schiavone, F.M., Positional cues and differential gene expression in somatic embryos of higher plants, *Cell Diff. Develop.*, 30, 159, 1990.
3. Dudits, D., Bogre, L., and Gyorgyey, J., Molecular and cellular approaches to the analysis of plant embryo development from somatic cells *in vitro*, *J. Cell Sci.*, 99, 475, 1991.
4. Jacobsen, H.-J., Somatic embryogenesis in seed legumes: the possible role of soluble auxin receptors, *Isreal J. Bot.*, 40, 139, 1991.
5. Ammirato, P.V., Patterns of development in culture, in *Tissue Culture in Forestry and Agriculture*, Henke, R.R., Hughes, K.W., and Constantin, M.J., Eds., Plenum Press, New York, 1985, 9.
6. Ammirato, P.V., Organization events during somatic embryogenesis, in *Plant Tissue and Cell Culture*, Green, C.E., Somers, D.A., Hackett, W.P., and Biesboer, D.D., Eds., A.R. Liss, Inc., New York, 1987, 57.
7. Ammirato, P.V., Recent progress in somatic embryogenesis, *Int. Assoc. Plant Tiss. Cult. Newslett.*, no. 57, 2, 1989.
8. Carman, J.G., Embryogenic cells in plant tissue cultures: occurrence and behavior, *In Vitro Cell. Dev. Biol.*, 26, 746, 1990.
9. Raghavan, V., *Embryogenesis in Angiosperm*, Cambridge University Press, Cambridge, 1986.
10. Tisserat, B., Esan, E.B., and Murashige, T., Somatic embryogenesis in angiosperms, *Hort. Rev.*, 1, 1, 1979.
11. Vasil, I.K., Somatic embryogenesis and its consequences in the Gramineae, in *Tissue Culture in Forestry and Agriculture*, Henke, R.R., Hughes, K.W., and Constantin, M.J., Eds., Plenum Press, New York, 1985, 31.
12. Sharp, W.R., Sondahl, M.R., Caldas, L.S., and Maraffa, S.B., The physiology of *in vitro* asexual embryogenesis, *Hort. Rev.*, 2, 268, 1980.
13. Williams, E.G., and Maheswaran, G., Somatic embryogenesis: factors influencing coordinated behavior of cells as an embryogenic group, *Ann. Bot.*, 57, 443, 1986.
14. Thorpe, T.A., *In vitro* somatic embryogenesis, *ISI Atlas Sci.: Animal Plant Sci.*, 1, 81, 1988.
15. Christianson, M.L., Causal events in morphogenesis, in *Plant Tissue and Cell Culture*, Green, C.E., Somers, D.A., Hackett, W.P., and Biesboer, D.D., Eds., A.R. Liss, Inc., New York, 1987, 45.
16. Wareing, P.F., Determination in plant development, *Bot. Mag. Tokyo*, Special Issue, 1, 3, 1978.
17. Meins, F. Jr., Determination and morphogenetic competence in plant tissue culture, in *Plant Cell Culture Technology*, Yeoman, M.M., Ed., Blackwell Scientific Publications, Oxford, 1986, 7.
18. McDaniel, C.N., Competence, determination, and induction in plant development, in *Pattern Formation: a Primer in Developmental Biology*, Malacinski, G.M., Ed., Macmillan Pub. Co., New York, 1984, 393.
19. Lyndon, R.F., *Plant Development: The Cellular Basis*, Unwin Hyman, London, 1990.

240

20. Sachs, T., *Pattern Formation in Plant Tissues*, Cambridge University Press, Cambridge, 1991.
21. Fujimura, T., and Komamine, A., The serial observation of embryogenesis in a carrot cell suspension culture, *New Phytol.*, 86, 213, 1980.
22. Nomura, K., and Komamine, A., Polarized DNA synthesis and cell division in cell clusters during somatic embryogenesis from single carrot cells, *New Phytol.*, 104, 25, 1986.
23. Sung, Z.R., Development states of embryogenic culture, in *Somatic Embryogenesis of Carrot*, Terzi, M., Pitto, L., and Sung, Z.R., Eds., Incremento Produttinita Risorse Agricole, Rome, 1985, 117.
24. Pechan, P.M., Bartels, D., Brown, D.C.W., and Schell, J., Messenger-RNA and protein changes associated with induction of *Brassica* microspore embryogenesis, *Planta*, 184, 161, 1991.
25. Pechan, P.M., and Keller, W.A., Induction of microspore embryogenesis in *Brassica napus* L. by gamma irradiation and ethanol stress, *In Vitro Cell. Develop. Biol.*, 11, 1073, 1989.
26. Gland, A., Lichter, R., and Schweiger, H.-G., Genetic and exogenous factors affecting embryogenesis in isolated microspore cultures of *Brassica napus* L., *J. Plant Physiol.*, 132, 613, 1988.
27. Keller, W.A., Arnison, P.G., and Cardy, B.J., Haploids from gametophytic cells – recent developments and future prospects, in *Plant Tissue and Cell Culture*, Green, C.E., Somers, D.A., Hackett, W.P., and Biesboere, D.D., Eds., Alan Liss, Inc., New York, 1987, 223.
28. Roberts-Oehlschlager, S.L., and Dunwell, J.M., Barley anther culture: pretreatment on mannitol stimulates production of microspore-derived embryos, *Plant Cell Tiss. Org. Cult.*, 20, 235, 1990.
29. Kermode, A.R., Regulatory mechanisms involved in the transition from seed development to germination, *Crit. Rev. Plant Sci.*, 9, 155, 1990.
30. Kermode, A.R., and Bewley, J.D., Developing seed of *Ricinus communis* L., when detached and maintained in an atmosphere of high relative humidity, switch to germinative mode without the requirement of complete desiccation, *Plant Physiol.*, 90, 702, 1989.
31. Pechan, P.M., and Schell, J., Molecular changes associated with the commitment phase of microspore embryogenesis, in *Progress in Plant Cellular and Molecular Biology*, Nijkamp, H.J.J., Van der Plas, L.H.W., and van Aartrijk, J., Eds., Kluwer Academic Publishers, Dordrecht, 1990, 407.
32. Kamada, H., Kobayashi, K., Kiyosue, T., and Harada, H., Stress induced somatic embryogenesis in carrot and its application to synthetic seed production, *In Vitro Cell. Dev. Biol.*, 25, 1163, 1989.
33. Kiyosue, T., Takano, K., Kamada, H., and Harada, H., Induction of somatic embryogenesis in carrot by heavy metal ions, *Can. J. Bot.*, 68, 2301, 1990.
34. Kiyosue, T., Satoh, S., Kamada, H., and Harada, H., Purification and immunohistochemical detection of an embryogenic cell protein in carrot, *Plant Physiol.*, 95, 1077, 1991.
35. Kiyosue, T., Nakayama, J., Satoh, S., Isogai, A., Suzuki, A., Kamada, H., and Harada, H., Partial amino-acid sequence of ECP31, a carrot embryogenic-cell protein, and enhancement of its accumulation by abscisic acid in somatic embryos, *Planta*, 186, 337, 1992.
36. Kiyosue, T., Satoh, S., Kamada, H., and Harada, H., Immunological detection of an embryogenic-cell protein (ECP31) during stress-induced embryogenesis in carrot, *Can. J. Bot.*, 70, 651, 1992.
37. Smith, D.L., and Krikorian, A.D., Release of somatic embryogenic potential from excised zygotic embryos of carrot and maintenance of proembryonic cultures in hormone-free medium, *Am. J. Bot.*, 76, 1832, 1989.
38. Nomura, K., and Komamine, A., Identification and isolation of single cells that produce somatic embryos at a high frequency in a carrot suspension culture, *Plant Physiol.*, 79, 988, 1985.
39. de Vries, S.C., Booij, H., Janssens, R., Vogels, R., Saris, L., LoSchiavo, F., Terzi, M., and Van Kammen, A., Carrot somatic embryogenesis depends on the phytohormone-controlled presence of correctly glycosylated extracellular proteins, *Genes Develop.*, 2, 462, 1988.

40. de Vries, S.C., Booij, H., Meyerink, P., Huisman, G., Wilde, H.D., Thomas, T.L., and van Kammen, Ab., Acquisition of embryogenic potential in carrot cell-suspension cultures, *Planta*, 176, 196, 1988.
41. Michalczuk, L., Ribnicky, D.M., Cooke, T.J., and Cohen, J.D., Regulation of indole-3-acetic acid biosynthetic pathways in carrot cell cultures, *Plant Physiol.*, 100, 1346, 1992.
42. Wilde, H.D., Nelson, W.S., Booij, H., de Vries, S.C., and Thomas, T.L., Gene-expression programs in embryogenic and non-embryogenic carrot cultures, *Planta*, 176, 205, 1988.
43. Smith, D.L., and Krikorian, A.D., Production of somatic embryos from carrot tissues in hormone-free medium, *Plant Sci.*, 58, 103, 1988.
44. Smith, D.L., and Krikorian, A.D., Growth and maintenance of an embryogenic cell culture of Daylily (*Hemerocallis*) on hormone-free medium, *Ann. Bot.*, 67, 443, 1991.
45. Israel, H.W., and Steward, F.C., The fine structure of quiescent and growing carrot cells: its relation to growth induction, *Ann. Bot.*, 30, 63, 1966.
46. Israel, H.W., and Steward, F.C., The fine structure and development of plastids in cultured cells of *Daucus carota*, *Ann. Bot.*, 31, 1, 1967.
47. Wilson, H.J., Israel, H.W., and Steward, F.C., Morphogenesis and the fine structure of cultured carrot cells, *J. Cell Sci.*, 15, 57, 1974.
48. Backs-Husemann, D., and Reinert, J., Embryobildung durch isolierte Einzelzellen aus Gewebekulturen von *Daucus carota*, *Protoplasma*, 70, 49, 1970.
49. Komamine, A., Kawahara, R., Matsumoto, M., Sunabori, S., Toya, T., Fujiwara, A., Tsukahara, M., Smith, J., Ito, M., Fukuda, H., Nomura, K., and Fujimura, T., Mechanisms of somatic embryogenesis in cell cultures: physiology, biochemistry, and molecular biology, *In Vitro Cell. Dev. Biol.*, 28P, 11, 1992.
50. Rakasekaran, K., Hein, M.B., and Vasil, I.K., Endogenous growth regulators in leaves and tissue cultures of *Pennisetum purpureum* Schum., *J. Plant Physiol.*, 130, 13, 1987.
51. Vasil, V., Lu, C., and Vasil, I.K., Histology of somatic embryogenesis in culture immature embryos of maize (*Zea mays* L.), *Protoplasma*, 127, 1, 1985.
52. Von Lammeren, A.A.M., Observations on the structural development of immature maize embryos (*Zea mays* L.) during *in vitro* culture in the presence or absence of 2,4-D, *Acta Bot. Neerl.*, 37, 49, 1988.
53. Fransz, P.F., and Schel, J.H.N., Cytodifferentiation during the development of friable embryogenic callus of maize, *Can. J. Bot.*, 69, 26, 1991.
54. Lu, C., and Vasil, I.K., Histology of somatic embryogenesis in *Panicum maximum* (Guinea grass), *Amer. J. Bot.*, 72, 1908, 1985.
55. Jones, T.J., and Rost, T.L., The developmental anatomy and ultrastructure of somatic embryos from rice (*Oryza sativa* L.) scutellum epithelial cells, *Bot. Gaz.*, 150, 41, 1989.
56. Maheswaran, G., and Williams, E.G., Origin and development of somatic embryoids formed directly on immature embryos of *Trifolium repens in vitro*, *Ann. Bot.*, 56, 619, 1985.
57. Yeung E.C., Aitken, J., Biondi, S., and Thorpe, T.A., Shoot histogenesis in cotyledon explants of radiata pine, *Bot. Gaz.*, 142, 494, 1981.
58. Stewart, R.N., and Dermen, H., Flexibility in ontogeny as shown by the contribution of the shoot apical layers to leaves of periclinal chimeras, *Amer. J. Bot.*, 62, 935, 1975.
59. Bruck, D.K., and Walker, D.B., Cell determination during embryogenesis in *Citrus jambhiri*. I. Ontogeny of the epidermis, *Bot. Gaz.*, 146, 188, 1985.
60. Raghavan, V., From microspore to embryoid: faces of the angiosperm pollen grain, in *Progress in Plant Cellular and Molecular Biology*, Nijkamp, H.J.J., Van der Plas, L.H.W., and Van Aartrijk, J., Eds., Kluwer Academic Publishers, Dordrecht, 1990, 213.
61. Sunderland, N., and Dunwell, J.M., Anther and pollen culture, in *Plant Tissue and Cell Culture, 2nd ed.*, Street, H.E., Ed., Blackwell Scientific Pub., Oxford, 1977, 223.
62. Dunwell, J.M., and Sunderland, N., Pollen ultrastructure in anther culture of *Nicotiana tabacum*, I. Early stages of culture, *J. Exp. Bot.*, 25, 352, 1974.
63. Dunwell, J.M., and Sunderland, N., Pollen ultrastructure in anther cultures of *Nicotiana tabacum*. II. Changes associated with embryogenesis, *J. Exp. Bot.*, 25, 363, 1974.

64. Dunwell, J.M., and Sunderland, N., Pollen ultrastructure in anther culture of *Nicotiana tabacum*. III. The first sporophytic division, *J. Exp. Bot.*, 26, 240, 1975.

65. Rashid, A., Siddiqui, A.W., and Reinert, J., Subcellular aspects of origin and structure of pollen embryos of *Nicotiana*, *Protoplasma*, 113, 202, 1982.

66. Zarsky, V., Garrido, D., Rihova, L., Tupy, J., Vicente, O., and Heberle-Bors, E., Derepression of the cell cycle by starvation is involved in the induction of tobacco pollen embryogenesis, *Sex. Plant Reprod.*, 5, 189, 1992.

67. Telmer, C.A., Simmonds, D.H., and Newcomb, W., Determination of developmental stages to obtain high frequencies of embryogenic microspores in *Brassica napus*, *Physiol. Plant.*, 84, 417, 1992.

67a. Pechan, P.M., and Keller, W.A., Identification of potentially embryogenic microspores in *Brassica napus*, *Physiol. Plant.*, 74, 377, 1988.

68. Telmer, C.A., Newcomb, W., and Simmonds, D.H., Microspore development in *Brassica napus* and the effect of high temperature on division *in vivo* and *in vitro*, *Protoplasma*, 172, 154, 1993.

68a. Pechan, P.M., Heat shock proteins and cell proliferation, *Fed. Eur. Biochem. Soc.*, 280, 1, 1991.

69. Simmonds, D.H., Gervais, C., and Keller, W.A., Embryogenesis from microspores of embryogenic and non-embryogenic lines of *Brassica napus*, in *Rapeseed in a Changing World, Proceedings of the Rapeseed Congress*, Saskatoon, Saskatchewan, Canada, 1991, 306.

69a. Hause, B., Hause, G., Pechan, P., and Van Lammeren, A.A.M., Cytoskeletal changes and induction of embryogenesis in microspore and pollen cultures of *Brassica napus* L., *Cell Biol. Intern. Rep.*, 17, 153, 1993.

70. Telmer, C.A., Newcomb, W., and Simmonds, D.H., Embryo induction and development from cultured microspores of *Brassica napus* cv. Topas, manuscript in preparation, personal communication.

71. Von Engelen, F.A., and De Vries, S.C., Extracellular proteins in plant embryogenesis, *Trends in Genet.*, 8, 66, 1992.

72. de Jong, A.J., Cordewener, J., Schiavo. F.L., Terzi, M., Vandekerckhove, J., van Kammen, A., and de Vries, S.C., A carrot somatic embryo mutant is rescued by chitinase, *Plant Cell*, 4, 425, 1992.

73. Gavish, H., Vardi, A., and Fluhr, R., Suppression of somatic embryogenesis in *Citrus* cell cultures by extracellular proteins, *Planta*, 186, 511, 1992.

74. Zaki, M.A.M., and Dickinson, H.G., Structural changes during the first divisions of embryos resulting from anther and free microspore culture in *Brassica napus*, *Protoplasma*, 156, 149, 1990.

75. Zaki, M.A.M., and Dickinson, H.G., Microspore-derived embryos in *Brassica*: the significance of division symmetry in pollen mitosis I to embryogenic development, *Sex. Plant Reprod.*, 4, 48, 1991.

76. Tautorus, T.E., Fowke, L.C., and Dunstan, D.I., Somatic embryogenesis in conifers, *Can. J. Bot.*, 69, 1873, 1991.

77. Von Aderkas, P., Bonga, J, and Owens, J., Comparison of Larch embryogeny *in vivo* and *in vitro*, in *Woody Plant Biotechnology*, Ahuja, M.R., Ed., Plenum Press, New York, 1991, 139.

78. Schulz, A., Vascular differentiation in the root cortex of peas: premitotic stages of cytoplasmic reactivation, *Protoplasma*, 143, 176, 1988.

79. Allsopp, A., The metabolic status and morphogenesis, *Phytomorphology*, 14, 1, 1964.

79a. Mayer, U., Buttner, G., and Jurgens, G., Apical-basal pattern formation in *Arabidopsis* embryo: studies on the role of *gnom* gene, *Development*, 117, 149, 1993.

79b. Barton, M.K., and Poethig, R.S., Formation of the shoot apical meristem in *Arabidopsis thaliana*: an analysis of development in the wild type and in the Shoot Meristemless mutant, *Development*, 119, 823, 1993.

80. Webb, M.C., and Gunning B.E.S., The microtubular cytoskeleton during development of the zygote, proembryo and free-nuclear endosperm in *Arabidopsis thaliana* (L.) Heynh, *Planta*, 184, 187, 1991.

81. Dijak, M., and Simmonds, D.H., Microtubule organization during early direct embryogenesis from mesophyll protoplasts of *Medicago sativa* L., *Plant Sci.*, 58, 183, 1988.
82. Hush, J.M., and Overall R.L., Re-orientation of cortical F-actin is not necessary for wound-induced microtubule re-orientation and cell polarity establishment, *Protoplasma*, 169, 97, 1992.
83. Quatrano, R.S., Brain, L., Aldridge, J., and Schultz, T., Polar axis fixation in *Fucus* zygotes: components of the cytoskeleton and extracellular matrix, *Development*, Suppl., 1, 11, 1991.
84. Wardlaw, C.W., *Morphogenesis in Plants*, Methuen, London, 1968.
85. Schnep, E., Cellular polarity, *Annu. Rev. Plant Physiol.*, 37, 23, 1986.
86. Osborne, D.J., Chemical signals in plant morphogenesis, in *Signals in Plant Development*, Krekule, J., and Seidlova, F., Eds., APB Academic Publishing, Hague, 1989, 1.
87. Goldsworthy, A., and Rathore, K.S., The electrical control of growth in plant tissue cultures: the polar transport of auxin. *J. Exp. Bot.*, 36, 1134, 1985.
88. Goldsworthy, A., and Mina, M.G., Electrical patterns of tobacco cells in media containing indole-3-acetic acid or 2,4-dichlorophenoxyacetic acid. Their relation to organogenesis and herbicide action, *Planta*, 183, 368, 1991.
89. Mina, M.G., and Goldsworthy, A., Changes in the electrical polarity of tobacco cells following the application of weak external currents, *Planta*, 186, 104, 1991.
90. Brawley, S.H., Wetherell, D.F., and Robinson, K.R., Electrical polarity in embryos of wild carrot precedes cotyledon differentiation, *Proc. Nat. Acad. Sci.*, 81, 6064, 1984.
91. Gorst, J., Overall, R.L., and Wernicke, W., Ionic currents traversing cell clusters from carrot suspension cultures reveal perpetuation of morphogenetic potential as distinct from induction of embryogenesis. *Cell Diff.*, 21, 101, 1987.
92. Rathore, K.S., Hodges, T.K., and Robinson, K.R., Ionic basis of currents in somatic embryos of *Daucus carota*, *Planta*, 175, 280, 1988.
93. Overton, R.L., and Wernicke, W., Steady ionic currents around haploid embryos formed from tobacco pollen in culture, in *Ionic Currents in Development*, Nuccitelli, R., Ed., Alan R. Liss, New York, 1986, 139.
94. Dijak, M., Smith, D.L., Wilson, T.J., and Brown, D.C.W., Stimulation of direct embryogenesis from mesophyll protoplasts of *Medicago sativa*, *Plant Cell Rep.*, 5, 468, 1986.
95. Reynolds, T.L., Interactions between calcium and auxin during pollen androgenesis in anther cultures of *Solanum carolinense* L., *Plant Sci.*, 72, 109, 1990.
96. Jansen, M.A.K., Booij, H., Schel, J.H.N., and de Vries, S.C., Calcium increases the yield of somatic embryos in carrot embryogenic suspension cultures, *Plant Cell Rep.*, 9, 221, 1990.
97. Hush, J.M., Overall, R.L., and Newman, I.A., A calcium influx precedes organogenesis in *Graptopetalum*, *Plant Cell Environment*, 14, 657, 1991.
98. Sachs, T., Cell polarity and tissue patterning in plants, *Development*, Suppl., 1, 83, 1991.
99. Schiavone, F.M., and Cooke, T.J., Unusual patterns of somatic embryogenesis in the domesticated carrot: developmental effects of exogenous auxins and auxin transport inhibitors, *Cell Diff.*, 21, 53, 1987.
100. Raven, J.A., The possible role of membrane electrophoresis in the polar transport of IAA and other solutes in plant tissues, *New Phytol.*, 82, 285, 1979.
101. Fry, S.C., and Wangermann, E., Polar transport of auxin through embryos, *New Phytol.*, 77, 313, 1976.
102. Sachs, T., Polarity changes and tissue organization in plants, in *International Cell Biology 1980-1981*, Schweiger, H.G., Ed., Springer-Verlag, Berlin, 1981, 489.
103. Schiavone, F.M., and Racusen, R.H., Microsurgery reveals regional capabilities for pattern reestablishment in somatic carrot embryos, *Dev. Biol.*, 141, 211, 1990.
104. Halperin, W., Embryos from somatic plant cells, in *Control Mechanisms in the Expression of Cellular Phenotypes*, Padykula, H.A., Ed., Symp. Int. Soc. Cell Biol. Vol. 9, Academic Press, New York, 1970, 169.
105. McWilliam, A.A., Smith, S.M., and Street, H.E., The origin and development of embryoids in suspension cultures of carrot (*Daucus carota*), *Ann. Bot.*, 38, 243, 1974.

106. Wetherell, D.F., Enhanced adventive embryogenesis resulting from plasmolysis of cultured wild carrot cells, *Plant Cell Tiss. Org. Cult.*, 3, 221, 1984.

107. Button, J., Kochba, J., and Bornman, C.H., Fine structure of and embryoid development from embryogenic ovular callus of "Shamouti" orange (*Citrus sinensis* Osb.), *J. Exp. Bot.*, 25, 446, 1974.

108. Dubois, T., Guedira, M., Dubois, J., and Vasseur, J., Direct somatic embryogenesis in roots of *Cichorium*: Is callose an early marker? *Ann. Bot.*, 65, 539, 1990.

109. Dubois, T., Guedira, M., Dubois, J., and Vasseur, J., Direct somatic embryogenesis in leaves of *Cichorium*, a histological and SEM study of early stages, *Protoplasma*, 162, 120, 1991.

110. Konar, R.N., Thomas, E., and Street, H.E., Origin and structure of embryoids arising from epidermal cells of the stem of *Ranunculus sceleratus* L., *J. Cell Sci.*, 11, 77, 1972.

111. Robards, A.W., and Lucas, W.J., Plasmodesmata, *Ann. Rev. Plant Physiol. Plant Mol. Biol.*, 41, 369, 1990.

112. Nomura, K., and Komamine, A., Embryogenesis from microinjected single cells in a carrot cell suspension culture, *Plant Sci.*, 44, 53, 1986.

113. Tucker, E.B., Translocation in the staminal hairs of *Setcreasea purpurea*. I. Study of cell ultrastructure and cell-to-cell passage of molecular probes, *Protoplasma*, 113, 193, 1982.

114. Halperin, W., Attainment and retention of morphogenetic capacity *in vitro*, in *Cell Culture and Somatic Cell Genetics of Plants, Vol. 3*, I.K. Vasil, Ed., Academic Press, Orlando, 1986, 3.

115. Street, H.E., and Withers, L.A., The anatomy of embryogenesis in culture, in *Tissue Culture and Plant Science 1974*, Street, H.E., Ed., Academic Press, London, 1974, 71.

116. Halperin, W., and Jensen, W.A., Ultrastructural changes during growth and embryogenesis in carrot cell cultures, *J. Ultrast. Res.*, 18, 428, 1967.

117. Thomas, E., Konar, R.N., and Street, H.E., The fine structure of the embryogenic callus of *Ranunculus sceleratus* L., *J. Cell Sci.*, 11, 95, 1972.

118. Karlsson, S.B., and Vasil, I.K., Morphology and ultrastructure of embryogenic cell suspension cultures of *Panicum maximum* (Guinea grass) and *Pennisetum purpureum* (Napier grass), *Amer. J. Bot.*, 73, 894, 1986.

119. Natesh, S., and Rau, M.A., The embryo, in *Embryology of Angiosperms*, Johri, B.M., Ed., Springer-Verlag, Berlin, 1984, 377.

120. Timmers, A.C.J., de Vries, S.C., and Schel, J.H.N., Distribution of membrane-bound calcium and activated calmodulin during somatic embryogenesis of carrot (*Daucus carota* L.), *Protoplasma*, 153, 24, 1989.

121. Jones, T.J., and Rost, T.L., Histochemisry and ultrastructure of rice (*Oryza sativa* L.) zygotic embryogenesis, *Amer. J. Bot.*, 76, 504, 1989.

121a. Altamura, M.M., Cersosimo, A., Majoli, C., and Crespan, M., Histological study of embryogenesis and organogenesis from anthers of *Vitis rupestris* du Lot cultured *in vitro*, *Protoplasma*, 171, 134, 1992.

122. Santos, A.V. dos, Cutter, E.G., and Davey, M.R., Origin and development of somatic embryos in *Medicago sativa* L. (Alfalfa), *Protoplasma*, 117, 107, 1983.

123. Davidson, D., Cell division, in *Plant physiology, a Treatise, Vol. 10: Growth and Development*, Bidwell, R.G.S., Ed., Academic Press, San Diego, 1991, 341.

124. Lintilhac, P.M., and Vesecky, T.B., Stress-induced alighment of division plane in plant tissues grown *in vitro*, *Nature*, 307, 363, 1984.

125. Fransz, P.F., and Schel, J.H.N., An ultrastructural study on the early development of *Zea mays* somatic embryos, *Can. J. Bot.*, 69, 858, 1991.

126. Yeung, E.C., and Clutter, M.E., Embryogeny of *Phaseolus coccineus*: growth and microanatomy, *Protoplasma*, 94, 19, 1978.

127. Yeung, E.C., and Clutter, M.E., Embryogeny of *Phaseolus coccineus*: the ultrastructure and development of the suspensor, *Can. J. Bot.*, 57, 120, 1979.

128. Yeung, E.C., Embryogeny of *Phaseolus*: the role of the suspensor, *Z. Pflanzenphysiol.*, 96, 17, 1980.

129. Yeung, E.C., and Sussex, I.M., Embryogeny of *Phaseolus coccineus*: the suspensor and the growth of the embryo-proper *in vitro*, *Z. Pflanzenphysiol.*, 91, 423, 1979.

129a. Yeung, E.C., and Meinke, D.W., Embryogenesis in angiosperms: development of the suspensor, *Plant Cell*, 1993 (in press).

130. Smart, M.G., and O'Brien, T.P., The development of the wheat embryo in relation to the neighbouring tissues, *Protoplasma*, 114, 1, 1983.

131. Schel, J.H.N., Kieft, H., and Van Lammeren, A.A.M., Interactions between embryo and endosperm during early developmental stages of maize caryopses (*Zea mays*), *Can. J. Bot.*, 62, 2842, 1984.

132. Haccius, B., and Bhandari, N.N., Delayed histogen differentiation as a common primitive character in all types of non-zygotic embryo, *Phytomorphology*, 25, 91, 1975.

133. Joy, IV, R.W., Yeung, E.C., Kong, L., and Thorpe, T.A., Development of white spruce somatic embryos: I. storage product deposition, *In Vitro Cell. Dev. Biol.*, 27P, 32, 1991.

134. Hakman, I., Rennie, P., and Fowke, L., A light and electron microscope study of *Picea glauca* (white spruce) somatic embryos, *Protoplasma*, 140, 100, 1987.

135. Buchheim, J.A., Colburn, S.M., and Ranch, J.P., Maturation of soybean somatic embryos and the transition to plantlet growth, *Plant Physiol.*, 89, 768, 1989.

136. Sterk, O., Booij, H., Schellekens, G.A., van Kammen, A., and de Vries, S.C., Cell-specific expression of the carrot EP2 lipid transfer protein gene, *Plant Cell*, 3, 907, 1991.

137. Von Engelen, F.A., Sterk, P., Booij, H., Cordewener, J.H.G., Rook, W., van Kammen, Ab., and de Vries, S.C., Heterogeneity and cell type-specific localization of a cell wall glycoprotein from carrot suspension cells, *Plant Physiol.*, 96, 705, 1991.

138. Satoh, S., Sturm, A., Fujii, T., and Chrispeels, M.J., cDNA cloning of an extracellular dermal glycoprotein of carrot and its expression in response to wounding, *Planta*, 188, 432, 1992.

139. Knox, P., Emerging patterns of organization at the plant cell surface, *J. Cell Sci.*, 96, 557, 1990.

140. Schiavone, F.M., and Cooke, T.J., A geometric analysis of somatic embryo formation in carrot cell cultures, *Can. J. Bot.*, 63, 1573, 1985.

141. Goosen-de Roo, L., Bakhuizen, R., Van Spronsen P.V., and Libbenga, K.R., The presence of extended phragmosomes containing cytoskeletal elements in fusiform cambial cells of *Fraxinus excelsior* L., *Protoplasma*, 122, 145, 1984.

142. Krishnamurthy, K.V., The Angiosperm embryo: correlative controls in development, differentiation, and maturation, in *Growth Patterns in Vascular Plants*, Iqbal, M., Ed., Dioscorides Press, Portland, Oregon, 1994, 372.

143. Chee, R.P., and Cantliffe, D.J., Embryo development from discrete cell aggregates in *Ipomoea batatas* (L.) Lam. in response to structural polarity, *In Vitro Cell. Dev. Biol.*, 25, 757, 1989.

144. Schnall, J., Hwang, C.H., Cooke, T.J., and Zimmerman, J.L., An evaluation of gene expression during somatic embryogenesis of two temperature-sensitive carrot variants unable to complete embryo development, *Physiol. Plant.*, 82, 498, 1991.

145. Santos, A.V.P. dos, and Machado, R.D., A scanning electron microscope study of *Theobroma cacao* somatic embryogenesis, *Ann. Bot.*, 64, 293, 1989.

146. Preece, J.E., Zhao, J., and Kung, F.H., Callus production and somatic embryogenesis from white ash, *HortSci.*, 24, 377, 1989.

147. Xu, N., and Bewley, J.D., Contrasting pattern of somatic and zygotic embryo development in alfalfa (*Medicago sativa* L.) as revealed by scanning electron microscopy. *Plant Cell Rep.*, 11, 279, 1992.

148. Schiavone, F.M., Microamputation of somatic embryos of the domestic carrot reveals apical control of axis elongation and root regeneration, *Development*, 103, 657, 1988.

149. Racusen, R.H., and Schiavone, F.M., Detection of spatially- and stage- specific proteins in extracts from single embryos of the domesticated carrot, *Development*, 103, 665, 1988.

150. Christou, P., and Yang, N., Developmental aspects of soybean (*Glycine max*) somatic embryogenesis, *Ann. Bot.*, 64, 225, 1989.

151. Nickle, T.C., and Yeung, E.C., Failure to establish a functional shoot apical meristem may be a cause of conversion failure in somatic embryo of *Daucus carota*, *Amer. J. Bot.*, 80, 1284, 1993.

152. Kong, L., and Yeung, E.C., Development of white spruce somatic embryos: II. Continual shoot meristem development during germination, *In Vitro Cell Dev. Biol.*, 28P, 125, 1992.

153. Gray, D.J., Somatic embryogenesis and plant regeneration from immature zygotic embryos of muscadine grape (*Vitis rotundifolia*) cultivars, *Amer. J. Bot.*, 79, 542, 1992.

154. Raghavan, V., Origin of the quiescent centre in the root of *Capsella bursa-pastoris* (L.) Medik, *Planta*, 181, 62, 1990.

155. Kermode, A.R., and Bewley, J.D., The role of maturation drying in the transition from seed development to germination. I. Acquisition of desiccation-tolerance and germinability during development of *Ricinus communis* L., *J. Exp. Bot.*, 36, 1906, 1985.

156. Kermode, A.R., Dumbroff, E.B., and Bewley, J.D., The role of maturation drying in the transition from seed development to germination. VII. Effects of partial and complete desiccation on abscisic acid levels and sensitivity in *Ricinus communis* L. seeds, *J. Exp. Bot.*, 40, 303, 1989.

157. Hakman, I., Stabel, P., Engstrom, P., and Eriksson, T., Storage protein accumulation during zygotic and somatic embryo development in *Picea abies* (Norway spruce), *Physiol. Plant.*, 80, 441, 1990.

158. Roberts, D.R., Flinn, B.S., Webb, D.T., Webster, F.B., and Sutton, B.C.S., Abscisic acid and indole-3-butyric acid regulation of maturation and accumulation of storage proteins in somatic embryos of interior spruce, *Physiol. Plant.*, 78, 355, 1990.

159. Dahmer, M.L., Hilderbrand, D.F., and Collins, G.B., Comparative protein accumulation patterns in soybean somatic and zygotic embryos, *In Vitro Cell. Dev. Biol.*, 28P, 106, 1992.

160. Wilen, R.W., Mandel, R.M., Pharis, R.P., Holbrook, L.A., and Moloney, M.M., Effects of abscisic acid and high osmoticum on storage protein gene expression in microspore embryos of *Brassica napus*, *Plant Physiol.*, 94, 875, 1990.

161. Quatrano, R.S., The role of hormones during seed development, in *Plant Hormones and their Role in Plant Growth and Development*, Davies, P.J., Ed., Martinus Nijhoff Publishers, Dordrecht, 1987, 494.

162. Slawinska, J., and Obendorf, R.L., Soybean somatic embryo maturation: composition, respiration and water relations, *Seed Sci. Res.*, 1, 251, 1991.

163. Hetherington, A.M., and Quatrano, R.S., Mechanisms of action of abscisic acid at the cellular level, *New Phytol.*, 119, 9, 1991.

164. Xu, N., Coulter, K.M., and Bewley, J.D., Abscisic acid and osmoticum prevent germination of developing alfalfa embryos, but only osmoticum maintains the synthesis of developmental proteins, *Planta*, 182, 382, 1990.

165. Becwar, M.R., Noland, T.L., and Wyckoff, J.L., Maturation, germination, and conversion of Norway spruce (*Picea Abies*, L.) somatic embryos to plants, *In Vitro Cell. Dev. Biol.*, 25, 575, 1989.

166. Roberts, D.R., Abscisic acid and mannitol promote early development, maturation and storage protein accumulation in somatic embryos of interior spruce, *Physiol. Plant.*, 83, 247, 1991.

167. Fujii, J.A., Slade, D., and Redenbaugh, K., Maturation and greenhouse planting of alfalfa artificial seeds, *In Vitro Cell. Dev. Biol.*, 25, 1179, 1989.

168. Dunwell, J.M., and Thurling, N., Role of sucrose in microspore embryo production in *Brassica napus* ssp. *oleifera*, *J. Exp. Bot.*, 36, 1478, 1985.

169. Faure, O., Embryons somatiques de *Vitis rupestris* et embryons zygotiques de Vitis sp.: morphologie, histologie, histochimie et developpement, *Can. J. Bot.*, 68, 2305, 1990.

170. Gray, D.J., Effects of dehydration and exogenous growth regulators on dormancy, quiescence and germination of grape somatic embryos, *In Vitro Cell. Dev. Biol.*, 25, 1173, 1989.

171. Roberts, D.R., Sutton, B.C.S., and Flinn, B.S., Synchronous and high frequency germination of interior spruce somatic embryos following partial drying at high relative humidity, *Can. J. Bot.*, 68, 1086, 1990.

172. Attree, S.M., Moore, D., Sawhney, V.K., and Fowke, L.C., Enhanced maturation and desiccation tolerance of white spruce [*Picea glauca* (Moench) Voss] somatic embryos: Effects of a non-plasmolysing water stress and abscisic acid, *Ann. Bot.*, 68, 519, 1991.

173. Carman, J.G., Improved somatic embryogenesis in wheat by partial simulation of in-ovulo oxygen, growth-regulator and desiccation environments, *Planta*, 175, 417, 1988.

174. Carman, J.G., The in ovulo environment and its relevance to cloning wheat via somatic embryogenesis, *In Vitro Cell. Dev. Biol.*, 25, 1155, 1989.

175. Yeung, E.C., Developmental changes in the branched parenchyma cells of bean seed coat, *Protoplasma*, 118, 225, 1983.

176. Yeung, E.C., and Cavey, M.J., Developmental changes in the inner epidermis of the bean seed coat, *Protoplasma*, 154, 45, 1990.

177. Murray, D.R., *Nutrition of the Angiosperm Embryo*. Research Studies Press, Somerset, England, 1988.

178. Boyle, S.A., and Yeung, E.C., Embryogeny of *Phaseolus*: developmental pattern of lactate and alcohol dehydrogenases, *Phytochemistry*, 22, 2413, 1983.

179. Yeung, E.C., and Blackman, S.J., Histochemical localization of alcohol dehydrogenase in developing bean seeds, *Amer. J. Bot.*, 74, 1461, 1987.

180. Meinke, D.W., Perspectives on genetic analysis of plant embryogenesis, *Plant Cell*, 3, 857, 1991.

181. Castle, L.A., and Meinke, D.W., Embryo-defective mutants as tools to study essential functions and regulatory processes in plant embryo development. *Sem. Develop. Biol.*, 4, 31, 1993.

182. Schnall, J.A., Hwang, C.H., Cooke, T.J., and Zimmermann, J.L., An evaluation of gene expression during somatic embryogenesis of two temperature-sensitive carrot variants unable to complete embryo development, *Physiol. Plant.*, 82, 498, 1991.

183. Yeung, E.C., and Law, S.K., Serial sectioning techniques for a modified LKB Historesin, *Stain Technol.*, 62, 147, 1987.

184. Garrido, D., Eller, N., Heberle-Bors, E., and Vicente, O., De novo transcription of specific mRNAs during the induction of tobacco pollen embryogenesis, *Sex. Plant Reprod.*, 6, 40, 1993.

185. Pennell, R.I., Janniche, L., Scofield, G.N., Booij, H., de Vries, S.C., and Roberts, K., Identification of a transitional cell state in the developmental pathway to carrot somatic embryogenesis, *J. Cell Biol.*, 119, 1371, 1992.

186. de Jong, A.J., Schmidt E.D.L., de Vries, S.C., Early events in higher-plant embryogenesis, *Plant Mol. Biol.*, 22, 367, 1993.

187. Toonen, M.A.J., Hendriks, T., Schmidt, E.D.L., Verhoeven, H.A., van Kammen, A., de Vries, S.C., Description of somatic-embryo-forming single cells in carrot suspension cultures employing video cell tracking, *Planta*, 194, 565, 1994.

188. Nickle, T.C., Yeung, E.C., Further evidence of a role for abscisic acid in conversion of somatic embryos of *Daucus carota*, *In Vitro Cell Dev. Biol.*, 30P, 96, 1994.

189. Kong, L., and Yeung, E.C., Effects of ethylene and ethylene inhibitors on white spruce somatic embryo maturation, *Plant Science*, 1994 (in press).

190. Kong, L., and Yeung, E.C., Effects of silver nitrate and polyethylene glycol on white spruce (*Picea glauca*) somatic embryo development: enhancing cotyledonary embryo formation and endogenous ABA content, *Physiol. Plant.*, 1994 (in press).

7. Physiological and Biochemical Aspects of Somatic Embryogenesis

KOJI NOMURA and ATSUSHI KOMAMINE

Contents

I. Introduction

Recent studies on the mechanisms of somatic embryogenesis are focused mainly on molecular approaches. However, physiological and biochemical aspects of somatic embryogenesis are also important through the use of high frequency and synchronous embryogenic systems, which are also necessary for investigation at the molecular level. Furthermore, understanding the physiological and biochemical basis of somatic embryogenesis is essential for the elucidation of its causal mechanisms.

In this chapter, the action of plant growth regulators, effects of nutrients, cell density and other factors on somatic embryogenesis, mainly in carrot, are described. In addition, changes in metabolism and chromatin during somatic embryogenesis, molecular markers for embryogenesis and other molecular approaches are also mentioned. Finally, an established high frequency and synchronous embryogenesic system is introduced.

T.A. Thorpe (ed.), In Vitro Embryogenesis in Plants, pp. 249–265.

II. Action of plant growth regulators

Somatic embryogenesis in carrot is readily achieved by transferring cultured cells or cell clusters, which have been induced and subcultured in a medium containing auxin, to an auxin-free medium [1-3]. Addition of auxin, 2,4-D or IAA, inhibited somatic embryogenesis from embryogenic cell clusters at concentrations higher than 10^{-7} or 10^{-8}M, respectively, in carrot suspension cultures [4].

The antiauxins, 2,4,6-trichlorophenoxyacetic acid and p-chlorophenoxy-isobutyric acid added to the medium also inhibited embryogenesis. Embryogenic cell clusters were most sensitive to both auxin and antiauxin, but auxin (2, 4-D) was less inhibitory when it was added after the globular stage of embryogenesis. Endogenous auxin, IAA, could be detected in embryos, however, the level did not change significantly during embryogenesis. From these results, a hypothesis can be proposed that endogenous auxin polarity in embryogenic cell clusters is essential for induction of embryogenesis; exogenously supplied auxin could cancel the polarity by diffusion into cell clusters, resulting in inhibition of embryogenic development. Antiauxin should also inhibit the action of endogenous auxin, resulting in the inhibition of embryogenic induction by auxin.

As regards auxin levels during somatic embryogenesis, Michalczuk et al. [5] reported that a carrot embryogenic cell line exhibited a rapid decline in both free and conjugated 2,4-D metabolites within seven days, while IAA levels expressed per fresh weight remained relatively steady for seven days in the preglobular stage after which the levels declined steadily in all subsequent stages of embryo development. But their findings are not contradictory to the hypothesis mentioned above.

Further investigations demonstrated that single cells or cell clusters smaller than 31μm would not differentiate to embryos by direct transfer into an auxin-free medium. Auxin pretreatment prior to transferring to an auxin-free medium was required for induction of embryogenesis. Thus, the role of auxin in embryogenesis is twofold. During the first phase, auxin is required for the transition from single cells to embryogenic cell clusters. During the second phase, auxin is inhibitory for the development of embryogenic cell clusters to embryos [6].

Indoleacetic acid, B-naphthoxyacetic acid, indolebutyric acid and NAA have been reported to induce roots at lower concentrations and somatic embryos at higher concentrations in hypocotyl segments of carrot. Short term application of 2,4-D stimulated the formation of somatic embryos and continuous application induced embryogenic calluses. Phenoxyacetic acid did not stimulate embryo formation, whereas o-chlorophenoxyacetic acid slightly stimulated root formation and 2,4,5-trichlorophenoxyacetic acid, 2-methyl-4-chloro-phenoxyacetic acid and p-chlorophenoxyacetic acid stimulated somatic embryo formation [7, 8]. Effects of aryloxyalkanecarboxylic acids have been examined by Chandra et al. [9]. When carrot cells cultured in the presence of 2,4-D were transferred to media containing α-(2-chlorophenoxy)isobutyric acid or 3,5-dichlorophenoxyacetic acid, the yield of embryos was enhanced, however 2-chlorophenoxyacetic acid suppressed embryo numbers when compared to cultures in auxin-free medium. Among the aryloxyalkanecarboxylic acids examined, 2,4-D was the only one not to cause a decline of embryo formation in carrot cultures.

The cytokinin, zeatin, was reported by Fujimura and Komamine [10] to promote embryogenesis from embryogenic cell clusters in a narrow concentration range around 10^{-7}M, but other synthetic cytokinins such as 6-benzyladenine (BA) and kinetin did not show promotive effects. Zeatin has also been found to promote embryogenesis from single cells to embryogenic cell clusters in the presence of auxin [6]. Cytokinins such as BA, zeatin, and N-phenyl-N'-(4-pyridyl)urea (4-PU) did not induce organogenesis. Application of BA plus 2,4-D induced soft friable calluses, suppressing somatic embryo formation in carrot [7].

Embryogenesis has been shown to be suppressed by ethephon and ethylene. Depression of embryogenesis by 2,4-D was unrelated to ethylene evolution [11].

Giberellic acid (GA_3) also inhibited somatic embryo formation [3, 7]. Embryos in a medium without 2,4-D contained low levels (0.2–0.3 $\mu g \bullet g^{-1}$ dry weight) of polar GA (GA_1 like), but 13–22 times higher levels of less polar GA ($GA_{4/7}$ like). Non-embryogenic cells and embryogenic cells in a medium containing 2,4-D showed high levels of polar GA (2.9–4.4 $\mu g \bullet g^{-1}$), and reduced levels of less polar GA [12]. Endogenous giberellin-like substances were also examined in suspension cultures of a hybrid grape (*Vitis vinifera* × *V. rupestris*) during somatic embryogenesis. Free and highly H_2O-soluble GA-like substances, expressed on a dry weight basis, decreased (however, they increased on an embryo basis) during development of embryos [13].

Abscisic acid (ABA) concentration increased during seed development and promoted the onset and maintenance of dormancy. Addition of ABA during embryo growth and development in caraway (*Carum carvi* L.) cells, suppressed abnormal embryo morphology. Growth in a medium with 10^{-7} M ABA, in the dark, produced normal embryos which closely resembled their zygotic embryo counterparts [14]. Ammirato [15] also observed the interaction of ABA, zeatin and giberellic acid on the development of somatic embryos from cultured cells of caraway and demonstrated that the balance between abscisic acid on the one hand and zeatin and giberellic acid on the other could effectively control somatic embryo development and either disrupt or ensure normal maturation.

Exogenously supplied ABA was found to be inhibitory to development of embryos from embryogenic cell clusters in carrot [3]. The amount of endogenous ABA in cultured cells and globular stage somatic embryos of carrot remained low, but concentrations increased during further development. However, the content of ABA increased continuously from the 7th until the 13th day and then decreased in a culture containing 2,4-D [16]. The ABA content in developing embryos reached a maximum on the 10th day and then decreased in a culture lacking 2,4-D. Addition of ABA to the medium suppressed the formation of abnormal embryos but also decreased the total number of somatic embryos. The effect of ABA was especially marked when it was applied during embryo development [16].

Abscisic acid levels in developing somatic embryos of grape (*Vitis vinifera* × *V. rupestris*) decreased from the globular stage to the mature stage. Chilling at 4°C induced normal germination of seeds and mature somatic embryos and precocious germination of immature somatic embryos. Chilling led to a marked reduction in endogenous ABA and exogenous ABA inhibited the germination of chilled somatic embryos [17]. The capacity for somatic embryogenesis was lost by inhibition of ABA synthesis with fluridone (1-methyl-3-phenyl-a-(3-[trifluor-omethyl] phenyl)-4-(1H)pyridinone) in *Pennisetum purpureum*. Inhibition of somatic embryogenesis by fluridone could be partially overcome by the addition of (±)-ABA to the culture medium which enhanced somatic embryogenesis and reduced the formation of non-embryogenic callus [17]. Endogenous ABA levels measured in chilled somatic embryos of carrot by immunoassay were found to be similar to non-chilled embryos [18]. Thus, the effects of ABA on somatic embryogenesis is still controversial and not conclusive.

Recently, Wake et al. [19] reported that hot water extracts from marine cyanobacteria, Nostoc, Anabaena, *Synochnococcus* and *Xenococcus* promoted carrot embryogenesis and increased plantlet numbers to more than 3.7 times of controls. Both dialysates and nondialysates of each of these extracts increased plantlet formation. The extracts also promoted embryo-like structure formation from one year old, carrot cell cultures which were unable to produce plantlets using conventional methods.

III. Effects of nutrients from carbon and nitrogen sources

Callus derived from carrot petioles grown in media containing only nitrate failed to produce embryos in either of two liquid media (nitrate of ammonium plus nitrate-containing). In the case of the callus cell fraction grown on media containing ammonium, many embryos were observable in cultures within two weeks of inoculation. While ammonium was required for initiation of embryogenesis in cultured cells, nitrate alone may be sufficient for the further development of pro-embryos (embryogenic cell clusters) in the undifferentiated state. It may be possible that young pro-embryos require reduced nitrogen but obtain it in these cultures by absorbing the nitrogen released in various molecular forms from the suspension by living and dead cells [20].

It has been shown that nitrogen in the medium, both qualitatively and quantitatively, was important in the induction of embryogenesis. No embryos were formed with low levels of nitrogen, but embryogenesis occurred when nitrogen compounds were added to the medium [21]. Among inorganic and organic nitrogenous compounds, ammonium nitrate and glutamine were especially efficient [22]. The importance of reduced nitrogen was clearly demonstrated by Wetherell and Dougall [23]. Nitrate at concentrations ranging from 5 to 95 mM KNO$_3$ supported only weak growth and very low embryogenesis. Ammonium chloride (0.1 mM) added to a nitrate medium allowed some embryogenesis and 10 mM NH$_4$CI was near optimal when KNO$_3$ was in the range of 12 to 40 mM.

Each of glutamine, glutamic acid, urea and alanine could partially replace NH$_4$CI in the medium. Glutamine, alanine and possibly glutamic acid could serve as sole sources of nitrogen supporting both good growth and embryogenesis. It was also shown that the pH of the culture medium was strongly influenced by the ratio of NH$_4$CI to KNO$_3$ supplied [23]. Dougall and Verma [24] also reported that carrot suspension cultures could grow and produce somatic embryos in the presence of NH$_4^+$ as a sole nitrogen source when the pH of the medium was controlled by continuous titration.

According to Kamada and Harada [25], the process of somatic embryogenesis in carrot seems to consist of two phases: the induction phase and the development phase. The former requires 2,4-D, but not any nitrogenous compounds, while the latter requires reduced nitrogen. Reduced nitrogenous compounds like ammonium nitrate and ammonium chloride with added nitrate such as potassium nitrate were very effective in inducing somatic embryogenesis, but ammonium sulfate was not. A high level of potassium nitrate in a medium without reduced nitrogen slightly stimulated somatic embryogenesis.

When various amino acids were added to a medium containing 20 mM potassium, nitrate, α-alanine was the most effective and glutamine, asparagine, aspartic acid, glutamic acid, arginine and proline were also found to be stimulatory, but lysine, valine, histidine, leucine and methionine were not effective. The addition of α-alanine accelerated cell division during the earlier stages of embryogenesis. Without any nitrogenous compounds, glutamine stimulated somatic embryogenesis the most, and with α-alanine and glutamic acid also being effective [8].

After uptake into cells, α-alanine was found to be quickly transformed to glutamic acid by alanine aminotransferase and utilized as a nitrogen source. Alanine aminotransferase activity was observed in cells and embryos, although the activity decreased during the culture period [25]. Serine and proline, which were added to the medium in the presence of 2,4-D, stimulated embryogenic development when the hormone was subsequently omitted [26]. Proline and serine were also found to affect polarity of somatic embryos in carrot [27].

In alfalfa, the production of somatic embryos was also increased by alanine and proline addition. However, utilization of nitrogen for synthesis of protein from alanine, proline and other amino acids was not qualitatively different. Overall metabolism of the nitrogen from proline, alanine, glutamate and glycine was similar in the regenerating and non-regenerating cell lines, except that the levels of free amino acids were consistently higher in the non-regenerating line. When regeneration was suppressed in regenerating lines, the level of intracellular free amino acids increased. This increased level of metabolites was the only direct evidence of differences between the regenerating and non-regenerating states provided by analysis of nitrogen metabolism [28].

IV. Population density and conditioned medium

It has been reported that somatic embryogenesis in wild carrot cells was influenced by the population density of the cells in culture. Cells cultured at low cell densities produced fewer embryos than the expected number as compared to cells at higher densities [29]. The effect of population density and of conditioned medium on embryogenesis indicated that one particular cell line of carrot exhibited a linear relationship between population density and total number of embryos produced at all densities examined. However, another cell line produced lower numbers of embryos than expected at low cell densities. At all densities the presence of conditioned medium increased the total number of embryos produced and the rate of development was also accelerated [30]. It was also reported in anise (*Pimpinella aniseum* L.) that the efficiency of somatic embryogenesis was affected by population density, and a close relationship existed between the rate of cell division in cell aggregates and the efficiency of embryogenesis [31].

The results mentioned above were mostly from experiments on somatic embryogenesis from cell clusters. Nomura and Komamine [6] reported the importance of cell density in somatic embryogenesis from single cells in carrot suspension cultures. The progression of the process from single cells to embryogenic cell clusters required auxin and rather high cell densities (10^6 cells/ml was the optimal cell density).

There are many reports illustrating the promotive effects on plant cell culture and somatic embryogenesis of conditioned medium, however, the substance or substances involved has remained obscure. De Vries et al. [32] reported that a small number of proteins was secreted into the medium within 2 days after embryo initiation. These proteins were either absent or present in reduced concentrations in cultures which were unable to produce somatic embryos. This suggested that these proteins were correlated with an early, preglobular stage of somatic embryo development. A cDNA corresponding to one of these proteins, a 10-KD protein, designated extracellular protein 2 (EP2) was obtained by screening expression libraries. The EP2 protein was identified as a lipid transfer protein and found to be present in cell walls and conditioned medium of cell cultures. The EP2 gene is expressed in embryogenic cell cultures, the shoot apices of seedlings, developing flowers, and maturing seeds. Based on the extracellular location of the EP2 protein and the expression pattern of the encoding gene, it was proposed that EP2 protein, a plant lipid transfer protein, plays a role in the transport of cutin monomers through the extracellular matrix to sites of cutin synthesis. It appears likely that early plant embryos are indeed covered with a cuticular layer as soon as the protoderm is formed [33].

V. Gravity

Experiments using carrot cell cultures were performed in space with Vostok biosatellite Kosmos 782 to test whether or not cells could develop to embryos normally under conditions of microgravity. Weightlessness seemed to have no adverse effects on the induction of somatic embryos or on the development of their organs [34].

VI. Dissolved oxygen

Dissolved oxygen has been shown to play an important role in carrot suspension culture growth and differentiation. When dissolved oxygen was below critical levels, dry weight increased linearly with time while cell numbers increased exponentially. This condition favored differentiation into plantlets via somatic embryogenesis, most likely from cell clusters. Above critical levels, both cell dry weight and cell numbers increased exponentially which favored plantlet differentiation by rhizogenesis [35]. Below the critical level, cyanide sensitivity of respiration and cellular concentrations of adenosine-5'-triphosphate (ATP) increased. The level of ATP in the cells could be increased by adding adenosine to the medium. Both dissolved oxygen and adenosine treatments produced more embryos than control cultures [36].

In addition, it has been reported that high levels of oxygen favored embryogenesis from single cells in carrot suspension cultures [6]. Competent single cells were cultured in the presence of auxin in flasks filled with 0, 20, 40 and 100% oxygen for 7 days and then transferred to a medium lacking auxin which was contained in flasks filled with air (21% oxygen). Cultures in 40% oxygen in the presence of auxin produced the greatest number of embryos.

VII. Polyamine metabolism

Polyamine metabolism during embryogenesis has been widely investigated. Since polyamines are thought to be involved in the regulation of transcription and translation in animal cells. It is important to investigate polyamine metabolism during embryogenesis where transcription and translation are very active.

Putrescine levels were found to be elevated two fold over control values after transfer of cells to auxin free medium. Spermidine levels were also elevated but spermine levels appeared to be lower [37]. The activity of arginine decarboxylase, which is important in putrescine synthesis in plants, was elevated two fold in cells during embryogenesis [37].

Ornithine carbamoyltransferase was analyzed and partially purified from carrot cultures of low embryogenic potential. The properties of ornithine carbamoyltransferase from cells possessing embryogenic ability were compared with those cells lacking embryogenic ability [38].

The regulation of polyamine metabolism during embryo development was compared with that of proliferative growth (control culture). Polyamine levels in developing embryo cultures were several times higher than in control cultures. The activities of arginine decarboxylase and S-adenosylmethionine decarboxylase were suppressed by auxin, but increased in cultures of developing embryos. These results indicate that increased polyamine levels are required for somatic embryogenesis in carrot cell cultures [39].

The importance of polyamine metabolism in embryo development also has been shown through the use of enzyme inhibitors such as α-difluoromethyl-arginine, an inhibitor of arginine decarboxylase. It has been suggested that polyamines are synthesized mainly via arginine decarboxylase and that they play an important role in embryogenesis [40]. Minocha et al. [41] reported that methylglyoxal bis (guanylhydrazone) (MGBG), an inhibitor of S-adenosyl-methionine (SAM) decarboxylase completely inhibited somatic embryogenesis in carrot. MGBG causes a significant reduction in the cellular levels of spermidine and spermine, while allowing increased accumulation of putrescine. Cellular levels of 1-aminocyclopropane-1-carboxylic acid, a precursor of ethylene, were considerably higher in the presence of MGBG. These results suggest that an interaction between ethylene and polyamine biosynthetic pathways through competition for the common precursor, SAM, plays an important role in the development of somatic embryos in carrot cell cultures.

VIII. Changes in DNA, RNA and protein metabolism during embryogenesis

DNA and RNA metabolism and protein synthesis have been investigated during development of somatic embryos. Synthesis of macromolecules has been investigated during embryogenesis and active DNA synthesis was observed during the formation of globular embryos. Prior to active DNA synthesis, the turnover rate of RNA and protein increased substantially [42]. It was also confirmed that the activity of de *novo* and *salvage* pathways for pyrimidine nucleotide biosynthesis increased during embryogenesis accompanying active RNA synthesis [43]. RNA synthesis first increased following transfer and laterdecreased [44]. The rate of rRNA synthesis was lower for embryogenesis than for undifferentiated growth in media containing auxin. On the other hand, the rate of poly(A)$^+$RNA synthesis was higher for embryogenesis than for undifferentiated growth [45]. Rates of protein synthesis also increased after transfer to auxin-free media.

Changes in endogenous amino acids also were investigated by Kamada and Harada [25], however, no significant differences were found. Free amino acid levels were found to be higher in non-embryogenic than in embryogenic cell lines of alfalfa. When embryogenesis was suppressed, the levels of intracellular free amino acids increased [28].

These observations indicate that both active cell division and protein synthesis occurred during embryogenesis. However, the identity and function of actively synthesized proteins were undetermined in most studies. Recently, functionally specific proteins have been investigated during embryogenesis; namely, protein glycosylation activities [46], protease activities [47], and the expression of tubulin genes [48].

More recently, a gene which was expressed preferentially in the globular stage of embryogenesis was isolated by differential screening and was found to encode the elongation factor 1a(EF1a) which is an essential protein for elongation of polypeptides in ribosomes of eukaryotic cells [49]. These results indicate that active protein synthesis occurs in the globular stage, associated with active cell division.

IX. Molecular markers for somatic embryos

Much effort has gone into trying to find specific molecular markers for somatic embryo genesis. Changes in the electrophoretic patterns of soluble proteins were analyzed during the formation of 'pro-embryoid' (clusters of densely cytoplasmic cells) in petiole explants of Chinese celery [50]. However, the 'pro-embryoid' was not precisely defined, and suspension cultures are considered to be more suitable for biochemical investigations than materials such as explants.

The first report on the translational profiles of somatic embryos and cells showing undifferentiated growth was by Sung and Okimoto [51]. Two-dimensional gel electrophoresis revealed two 'embryonic' proteins which appeared in somatic embryos but were undetectable in undifferentiated cells.

Fujimura and Komamine [52] also compared the two dimensional patterns of proteins synthesized using a wheat germ *in vitro* translation system in developing embryos and in control cultures. A high frequency embryogenesis system was used and selected embryogenic cell clusters were transferred to auxin-free medium to continue development. In control cultures, cell clusters were cultured in medium containing auxin and grew without differentiation. About 400 translated proteins could be detected but the patterns were very similar for both developing embryos and control cultures. Two proteins appeared and two disappeared in developing embryos prior to the globular stage (4 days after transfer to auxin-free medium) suggesting that these proteins may play important roles in the early stage of embryogenesis and could be regulated at the transcriptional level.

The above studies indicated the possibility of using specific proteins as markers for the early stages of embryogenesis. These 'embryonic' proteins could also be used to investigate gene expression during embryogenesis and undifferentiated growth. Sung and Okimoto [53] showed that undifferentiated cells produce 'callus-specific' proteins, and that these 'callus-specific' and 'embryonic" proteins are coordinately regulated [53, 54].

Using gel electrophoresis, another approach was used to find specific marker proteins in embryogenic cell lines of maize. Everett et al. [55] reported that the gel electrophoresis patterns of isozymes of esterase and glutamate dehydrogenase could be used to distinguish embryogenic cell lines from non-embryogenic cell lines.

The expression of certain proteins with known properties has also been investigated during embryogenesis. Heat shock proteins have been studied in several laboratories and one was reported to be a stage specific marker of somatic embryogenesis [56].

The goal of such investigations is to elucidate gene regulation during somatic embryogenesis, and to demonstrate the expression of totipotency in biochemical terms. The molecular aspects of somatic embryogenesis are discussed in another chapter (see Dudits et al., this volume).

X. Changes in chromatin and chromosomal proteins

Chromosomal histone and non-histone proteins were analyzed using the carrot system and a group of non-histone proteins with low molecular weight were found to disappear after the continuation of embryo development. In contrast, histone components were the same in embryos and unorganized cells [57]. The template activity of chromatin in embryos was higher than in cells during unorganized growth [57, 58]. Using a system of synchronized embryogenesis, a lower ratio of histone H1 to total histone, and a higher ratio of H2 + H3 to H4 were observed in developing embryo cultures [58].

XI. Polarity in embryogenesis

It is well-known that an endogenous electric current is associated with the initiation of polarization in embryos of fucoid algae. Similar endogenous currents traversing somatic embryos of carrot have also been shown. The current enters the apical pole and leaves the region near the presumptive radicle in globular embryos. This electrical polarity preceded differentiation of vascular tissue and cotyledon development. Inward currents were found at the cotyledonary end and outward currents at the radicle end; with exogenous IAA reversibly inhibiting these currents. The ionic gradients generated by these currents may be important in the accumulation of metabolites and in other developmental processes within the embryo [59].

Electrical polarity was found to be established in clusters undergoing apparently disorganized proliferation in the presence of 2,4-D. The electrical polarity was similar to that found in organized somatic embryos. This suggested that the potential to undergo embryogenesis was present even before 2,4-D was removed [60].

Removal of K^+ from the medium reversibly reduced the currents to about 25% of their original value at both the cotyledon and radicle. Deletion of Cl^- decreased the currents slightly and removal of Ca^{2+} resulted in a rapid doubling of currents. Addition of either N,N'-dicyclohexylcarbodiimide or tetraethyl ammonium chloride substantially reduced overall currents, and their removal resulted in partial recovery of the currents. These results suggested that the inward current at the cotyledon was due largely to a K^+ influx and the outward current at the radicle was mainly the result of active H^+ efflux [61].

Cell technology can also be applied to cells during embryogenesis. Microinjection into single cells and embryogenic cell clusters has been attempted. Single cells injected with fluorescent dye could differentiate to form embryos at high frequency [62]. This system might be used for analyzing gene expression during embryogenesis. If microinjection is applied to cell clusters, individual cells can be marked, and pattern formation and interactions between cells can be followed (Nomura and Komamine, unpublished). Other methods such as protoplast fusion [63] and efficient transformation of carrot protoplasts [64] may be helpful for introducing foreign genes and studying their expression during embryogenesis.

XII. High frequency and synchronous embryogenic system

As mentioned above, many contradictory findings have been obtained on the physiological and biochemical aspects of somatic embryogenesis. This is because investigators have used systems in which high frequency and synchronous somatic embryogenesis does not always occur. Thus, a model system, in which high frequency and synchronous embryogenesis occurred, if possible from single cells, should be a requirement for studies on the physiological, biochemical and molecular aspects of somatic embryogenesis. Fujimura and Komamine [65] and Nomura and Komamine [6] succeeded in establishing such a system in carrot suspension cultures and on which physiological findings have been based. Heterogeneous cell populations from carrot suspensions were fractioned by sieving through nylon screens and subsequent density gradient centrifugation (DGC) in Ficoll. From DGC, small, round and cytoplasm-rich competent single cells (State 0), or embryogenic cell clusters (State 1) which consisted of less than 10 cytoplasm-rich cells were isolated. State 0 single cells could divide to form State 1 embryogenic cell clusters in the presence of auxin (2,4-D at 5×10^{-8} M was optimal). State 1 embryogenic cell clusters could differentiate to globular, heart and torpedo shaped embryos synchronously and at high frequency when they were transferred to auxin-free medium containing zeatin at 10^{-7}M. Embryogenic frequency from single cells in this system was found to be between 80–90%. When State 0 single cells were cultured in the absence of auxin, they did not divide, but elongated. Elongated cells could not divide nor differentiate even if they were cultured in the presence of auxin and, therefore, appear to have lost their totipotent potential. Thus, the process from State 0 cells tot State 1 cells and embryos could be regarded as the process where totipotency was expressed, while the process from State 0 cells to elongated cells in the absence of auxin was that in which totipotency was lost.

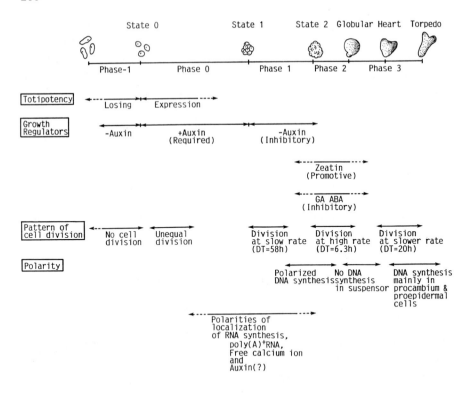

Fig. 1. A summary of the morphological and physiological aspects of somatic embryogenesis, arising from studies carried out with the high frequency and synchronous embryogenic single cell system of carrot.

Polarity of DNA and RNA synthesis, mRNA, free Ca^{2+} ion and probably auxin appeared in the late phase of State 0 cells to State 1 cell clusters. Polarity might play an important role during development of embryos from embryogenic cell clusters. Therefore, it could be considered that auxin was required to induce embryogenesis from competent single cells, while auxin was inhibitory during development of embryos from embryogenic cell clusters. Thus, it could be assumed that determination of embryogenesis had already occurred in State 1 cell clusters, because the fate of each cell was already determined. This was confirmed by following the distribution of labelled cells with fluororesent dye introduced by microinjection (Nomura and Komamine, unpublished data).

In the early stage of the process, from State 1 cell clusters to embryos, three phases could be recognised. In Phase 1, during the first three days after transfer to auxin-free medium, undifferentiated cell division occurred slowly. In Phase 2, between 3 and 4 days, very rapid cell division took place, giving rise to globular stage embryos which subsequently developed into heart and torpedo shaped embryos. Active turnover of RNA and protein occurred during Phase 1 and

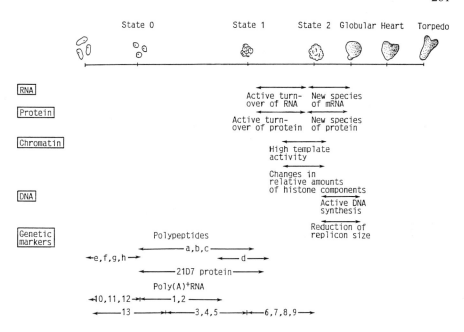

Fig. 2. A summary of the biochemical aspects of somatic embryogenesis, arising from studies carried out with the high frequency and synchronous embryogenic single cell system of carrot.

synthesis of new species of mRNA and protein appeared in Phases 1–2. In Phase 2, active DNA synthesis occurred which was at least partially due to reduction of replicon size. High template activity and relative changes in components of histone were also observed in Phases 1–2. The cell doubling times in Phase 2 were found to be only 6.3 hours.

Several molecular markers associated with the totipotent nature of cells have been isolated. Four polypeptides a, b, c and d, a protein (21D7) and two mRNAs 1 and 2 have been detected during the induction and development phases of somatic embryogenesis from single cells. If, however, single cells were cultured in the absence of auxin, producing elongated cells with concomitant loss of totipotency, these markers were no longer detectable. These results are summarized in Figs. 1 and 2.

XIII. Concluding Remarks

In this chapter, physiological and biochemical aspects on somatic embryogenesis are described mainly in carrot. Most investigations on mechanisms of somatic embryogenesis have recently focused on molecular aspects of it, which are described in another chapter (see Dudits et al.). However, different physiological situations of somatic embryogenesis have been well-known in other species; for instance, a different response to auxin from that in carrot was reported in somatic embryogenesis of rice suspension cultures [66]. Though mechanisms of somatic embryogenesis in plants may be the same in every species, auxin is inhibitory on somatic embryogenesis in some species, while it is required to induce embryogenesis in another species. It is still unclear whether such differences in auxin response is due to different endogenous auxin level in different species or tissues, or is due to different physiological situations in each species. Thus, the final goal of elucidation of physiological aspects of somatic embryogenesis, i.e. to know a common principle, remains a long way off. As a tool, it is important to establish high frequency and synchronous somatic embryogenesis systems in other species than carrot. When this is achieved, it will be possible to elucidate the common physiological, biochemical and molecular aspects of somatic aspects in higher plants.

References

1. Halperin, W., and Wetherell, D.F., Adventive embryony in tissue cultures of the wild carrot, Daucus carota., *Amer. J. Bot.*, 51, 274, 1964.
2. Halperin, W., Alternative morphogenetic events in cell suspensions, *Amer. J. Bot.*, 53, 443, 1966.
3. Fujimura, T., and Komamine, A., Effects of growth regulators on embryogenesis in a carrot suspension culture., *Plant Sci. Lett.*, 5, 359, 1975.
4. Fujimura, T., and Komamine, A., Involvement of endogenous auxin in somatic embryogenesis in a carrot cell suspension culture., *Z. Pflanzenphysiol.*, 95, 13, 1979.
5. Michalczur, L., Cooke, T.J., and Cohen, J. Auxin, levels at different atages of carrot somatic embryogenesis, *Phytochem.* 31, 1097, 1992.
6. Nomura, K., and Komamine, A., Identification and isolation of single cells that produce somatic embryos at a high frequency in a carrot suspension culture, *Plant Physiol.,* 79, 988, 1985.
7. Kamada, H., and Harada, H., Studies on the organogenesis in carrot tissue cultures. I. Effects of growth regulators on somatic embryogenesis and root formation, *Z. Pflanzenphysiol.*, 91, 255, 1979.
8. Kamada, H., and Harada, H., Studies on the organogenesis in carrot tissue cultures. II. Effects of amino acids and inoganic nitrogenous compounds and somatic embryogenesis, *Z. Pflanzenphysiol*, 91, 453, 1979.
9. Chandra, N., Lam, T.L., and Street, H.E., The effects of selected aryloxyalkanecarboxylic acids on the growth and embryogenesis of a suspension cultures of carrot (*Daucus carota* L.), *Z. Pflanzenphysiol.*, 86, 55, 1978.
10. Fujimura, T., and Komamine, A., Mode of action of 2,4-D and zeatin on somatic embryogenesis in a carrot cell suspension culture., *Z. Pflanzenphysiol.*, 99, 1, 1980.
11. Tisserat, B., and Murashige, T., Effects of ethephon, ethylene, and 2,4-dichlorophenoxy-acetic acid on asexual embryogenesis *in vitro, Plant Physiol.,* 60, 437, 1977.
12. Noma, M., Huber, J., Ernst, D., and Pharis, R.P., Quantitation of gibeerellins and the metabolism of [^{3}H]gibberellin A_1 during somatic embryogenesis in carrot and anise cell cultures, *Planta*, 155, 369, 1982.
13. Takeno, K., Koshioka, M., Pharis, R.P., Rajasekaran, K., and Mullins, M.G., Endogenous gibberellin-like substances in somatic embryos of grape (*Vitis vinifera* × *Vitis rupestris*) in relation to embryogenesis and the chilling requirement for subsequent development of mature embryos, *Plant Physiol.*, 73, 803, 1983.

14. Ammirato, P.V., The effects of abscisic acid on the development of somatic embryos from cells of caraway (Carum carvil, L), *Bot. Gaz.*, 135, 328, 1974.

15. Ammirato, P.V., Hormonal control of somatic embryo development from cultured cells of caraway, *Plan t Physiol.*, 59, 579, 1977.

16. Kamada, H., and Harada, H., Changes in the endogenous level and effects of abscisicacid during somatic embryogenesis of *Daucus carota* L., *Plant Cell Physiol*, 22, 1423, 1981.

17. Rajasekaran, K., Vine, J., and Mullins, M.G., Dormancy in somatic embryos and seed of *Vitis:* changes in endogenous abscisic acid during embryogeny and germination, *Planta*, 154, 139, 1982.

18. Spencer, T.M., and Kitto, S.L., Measurement of endogenous ABA levels in chilled somatic embryos of carrot by immunoassay, *Plant Cell Rep.*, 7, 352, 1988.

19. Wake, H., Umetsu, H., Ozeki, Y., Shinomura, K., and Matsunaga, T., Extracts of marine cyanobacteria stimulated somatic embryogenesis of *Daucus carota* L., *Plant Cell Rep.*, 9, 655, 1991.

20. Halperin, W., and Wetherell, D.F., Ammonium requirement for embryogenesis in vitro., *Nature*, 205, 519, 1965.

21. Tazawa, M., and Reinert, J., Extracellular and intracellular chemical environments in relation to embryogenesis in vitro, *Pnotoplasma*, 68, 157, 1969.

22. Reinert, J., and Tazawa, M., Wirkung von Strickstoffverbindungen und von Auxin auf die Embryogenese in Gewebekulturen, *Planta*, 87, 239, 1969.

23. Wetherell, D.F., and Dougall, D.K., Sources of nitrogen supporting growth and embryogenesis in cultured wild carrot tissue, *Physiol. Plant.*, 37, 97, 1976.

24. Dougall, D.K., and Verma, D., Growth and embryoformation in wild-carrot suspension cultures with ammonium ion as a sole nitrogen source, *In Vitro*, 14, 180, 1978.

25. Kamada, H., and Harada, H., Studies on nitrogen metabolism during somatic embryogenesis in carrot. I. utilization of a-alanine as a nitrogen source, *Plant Sci. Lett.*, 33, 7, 1984.

26. Nuti Ronchi, V., Caligo, M.A., Nozzolini, M., and Luccarini, G., Stimulation of carrot somatic embryogenesis by proline and serine, *Plant Cell Rep.*, 3, 210, 1984.

27. Caligo, M.A., Nuti Ronchi, V., and Nozzolini, M., Proline and serine affect polarity and development of carrot somatic embryos., *Cell Differ.*, 17, 193, 1985.

28. Skokut, T.A., Manchester, J., and Schaefer, J., Regeneration in alfalfa tissue culture. Stimulation of somatic embryoproduction by amino acids and N-15 NMR determination of nitrogen utilization, *Plant Physiol.*, 19, 579, 1985.

29. Halperin, W., Population density effects on embryogenesis in carrot-cell cultures, *Exp. Cell Res.*, 48, 170, 1967.

30. Hari, V., Effect of cell density changes and conditioned media on carrot cell embryogenesis, *Z. Pflanzenphysiol.*, 96, 227, 1980.

31. Huber, J., Constabel, F., and Gamborg, O.L., A cell counting procedure applied to embryogenesis in cell suspension cultures of anis (*Pimpinella anisum* L.), *Plant Sci. Lett.*, 12, 209, 1978.

32. De Vries, S.C., Booij, H., Janssen, R., Vogels, R., Saris, L., Lo Shiavo, F., Terzi, M., and Van Kammen, A., Carrot somatic embryogenesis depends on the phytohormone-controlled expression of correctly glycosylated extracellular proteins, *Gene Dev.*, 2, 462, 1988.

33. Steak, P., Booij, H., Schellenkens, G.A., van Kammen, A., and de Vries, S.C., Cell-specific expression of the carrot EP2 lipid transfer protein gene, *Plant Cell*, 3, 907, 1991.

34. Krikorian, A.D., and Steward, F.C., Morphogenetic response of cultured totipotent cells of carrot (*Daucus carota* var *carota*) at zero gravity, *Science*, 200, 67, 1978.

35. Kessel, R.H.J., and Carr, A.H., The effect of dissolved oxygen concentration on growth and differentiation of carrot (*Daucus carota*) tissue, *J. Exp. Bot.*, 23, 996, 1972.

36. Kessel, R.H.J., Goodwin, J.C., Philip, J., and Fowler, M.W., The relationship between dissolved oxygen concentration and embryogenesis in carrot (*Daucus carota*) tissue cultures, *Plant Sci. Lett.*, 10, 265, 1977.

37. Montague, M.J., Koppenbrink, J.W., and Jaworski, E.G., Polyamine metabolism in embryogenic cells of *Daucus carota*. I. Changes in intracellular content and rates os synthesis, *Plant Physiol.*, 62, 430, 1978.

38. Baker, S.R., and Yon R.J., Characterization of ornithine carvamoyltransferase from cultured carrot cells of low embryogenic potential, *Phytochemistry*, 22, 2171, 1983.

39. Fienberg, A.A., Choi, J.H., Lubich, W.P., and Sung, Z.R., Developmental regulation of polyamine metabolism in growth and differentiation of carrot culture, *Planta*, 162, 532, 1984.

40. Feirer, R.P., Mignon, G., and Litvay, J.D., Arginine decarboxylase and polyamines requirred for embryogenesis in the wild carrot, *Science*, 223, 1433, 1984.

41. Minocha, S.C., Papa N.S., Kahn, A.J., and Somuelson, A.E., 'Polyamines and somatic embryogenesis in carrot. III. Effects of methylglyoxal bis (guanylhydrazone)', *Plant Cell Physiol.*, 32, 395, 1991.

42. Fujimura, T., Komamine, A., and Matsumoto, H., Aspects of DNA, RNA and protein synthesis during somatic embryogenesis in a carrot cell suspension culture, *Physiol. Plant.*, 49, 255, 1980.

43. Ashihara, H., Fujimura, T., and Komamine, A., Pyrimidine nucleotide biosynthesis during somatic embryogenesis in a carrot cell suspension culture, *Z. Pflanzenphysiol.*, 104, 129, 1981.

44. Sengupta, C., and Raghavan, V., Somatic embryogenesis in carrot cell suspension. I. Pattern of protein and nucleic acid synthesis, *J. Exp. Bot.*, 31, 247, 1980.

45. Sengupta, C., and Raghavan, V., Somatic embryogenesis in carrot cell suspension. II. Synthesis of ribosomal and poly(A) +RNA, *J. Exp. Bot.*, 31, 259, 1980.

46. Lo Schiavo, F., Quesada-Allue, L., and Sung, Z.R., Synthesis of mannosylated dolichyl derivates in proliferating cells, in *Somatic Embryogenesis*, Terzi, M., Pitto, L., and Sung, Z.R., Eds., IPRA, Rome, 1985, 64.

47. Carlberg, I., Sonderhall, K., Glimelius, K., and Erikisson, T., Protease activities in non-embryogenic and embryogenic carrot cell strains during callus growth and embryo formation, *Physiol. Plant*, 62, 458, 1984.

48. Borkird, C., and Sung, Z.R., Expression of tublin genes during somatic embryogenesis, in *Somatic Embryogenesis,* Terzi, M., Pitto, L., and Sung, Z.R., Eds., IPRA, Rome, 1985, 14.

49. Komamine, A., Matsumoto, M., Tsukahara, M., Fujiwara, A., Kawahara, R., Ito, M., Smith, J., Nomura, K., and Fujimura, T., Mechanisms of somatic embryogenesis in cell cultures: physiology, Biochemistry, and Molecular Biology', *In Vitro Cell. Dev. Biol.,* 28 p, 179, 1992.

50. Zee, S.-Y., Wu, S.C., and Yue, B., Morphological and SDS-polyacrylamide gel electrophoretiic studies of pro-embryoid formation in the periole explants of chinese celery, *Z. Pflanzenphysiol.*, 95, 397, 1979.

51. Sung, Z.R., and Okimoto, R., Embryogenic proteins in somatic embryos of carrot, *Proc. Natl. Acad. Sci. USA*, 78, 3683, 1981.

52. Fujimura, T., and Komamine, A., Molecular aspects of somatic embryogenesis in a synchronous system, in *Plant Tissue Culture 1982*, Fujiwara, A., Ed., Maruzen, Tokyo, 1982, 105.

53. Sung, Z.R., and Okimoto, R., Coordinate gene expression during somatic embryogenesis in carrots, *Proc. Natl. Acad. Sci. USA*, 80, 2661, 1983.

54. Choi, J.H., and Sung, Z.R., Two dimensional gel analysis of carrot somatic embryogenic proteins, *Plant Mol. Biol. Rep.*, 2, 19, 1984.

55. Everett, N.P., Wach, M.J., and Ashworth, D.J., Biochemical markers of embryogenesis in tissue cultures of the maize inbred B73, *Plant Sci.*, 41, 133, 1985.

56. Pitto, L., Lo Schiavo, F., Giuliano, G., and Terzi, M., Analysis of the heat-shock protein pattern during somatic embryogenesis of carrot', *Plant Mol. Biol.*, 2, 231, 1983.

57. Matsumoto, H., Gregor, D., and Reinert, J., Changes in chromatin of *Daucus carota cells* during embryogenesis, *Phytochem.*, 14, 41, 1975.

58. Fujimura, T., Komamine, A., and Matsumoto, H., Changes in chromosomal proteins during early stages of synchronized embryogenesis in a carrot cell suspension culture, *Z. Pflanzenphysiol.*, 102, 293, 1981.

59. Brawley, S.H., Wetherell, D.F., and Robinson, K.R., Electrical polarity in embryos of wild carrot precedes cotyledon differentiation, *Proc. Natl. Acad. Sci. USA*, 81, 6064, 1984.

60. Gorst, J., Overall, R.L., and Wernicke, W., Ionic currents traversing cell clusters form carrot suspension cultures reveal perpetuation of morphogenetic potential as distinct from induction of embryogenesis, *Cell Differ.*, 21, 101, 1987.

61. Rathore, K.S., Hodges, T.K., and Robinson, K.R., Ionic basis of currents in somatic embryos of *Daucus carota.*, *Planta*, 175, 280, 1988.
62. Nomura, K., and Komamine, A., Embryogenesis from microinjected single cells in a carrot cell suspension culture, *Plant Sci.*, 44, 53, 1986.
63. Widholm, J.M., Hauptmann, R.M., and Dudits, D., Carrot protoplast fusion, *Plant Mol. Biol. Rep.*, 2, 26, 1984.
64. Langridge, W.H.R., Li, B.J., and Szalay, A.A., Electric field mediated stable transformation of carrot protoplasts with naked DNA, *Plant Cell Rep.*, 4, 355, 1985.
65. Fujimura, T., and Komamine, A., Synchronization of somatic embryogenesis in a carrot cell suspension culture, *Plant Physiol.*, 64, 162, 1979.
66. Ozawa, K., and Komamine, A., Establishment of a system of high-frequency embryogenesis from long-term cell suspension culture of rice *(Oryza sativa L.)*, *Theor. Appl. Genet.*, 77, 205, 1989.

8. Molecular Biology of Somatic Embryogenesis

DÉNES DUDITS, JÁNOS GYÖRGYEY, LÁSZLÓ BÖGRE and
LÁSZLÓ BAKÓ

Contents

I. Key molecular events during induction of embryo development from somatic plant cells: A general concept

Similar to other higher eukaryotic organisms, in flowering plants, embryo development is the consequence of fertilization events. Union of gametes as the male sperm nucleus and the female egg results in the zygote which later develops into an embryo within the ovule. During the sexual reproductive cycle, the egg cell is prepared for initiation of the embryogenic development that is triggered by signals after sperm-egg contact. *In vivo* the gametophytic and sporophytic cell differentiation is separated and the haploid gametes are specialized for sexual fusion and the fertilized egg has the potential to develop into a new

267

T.A. Thorpe (ed.), In Vitro Embryogenesis in Plants, pp. 267–308.

268

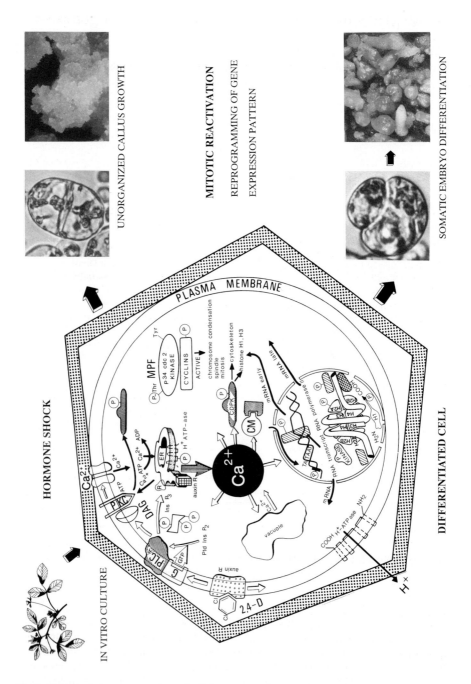

Fig. 1. Presumptive components in auxin (2,4-D)-activated signal transduction cascade involved in the reprogramming of the gene expression and induction of cell division leading to either callus growth or somatic embryo development. Callus tissue is derived from a cell dividing equally in the

organism. In most higher eukaryotes, the differentiation of totipotent embryogenic cells is controlled by a pre-set developmental program and terminally differentiated cells are formed. In early embryos, the cells have rapid division cycles and the chromatin becomes transcriptionally active after variable number of division cycles during embryo development.

Unlike other organisms, in plants, the cellular totipotency extends not only to the cells in zygotic embryos but also to the somatic cells. Flexibility of the differentiation program makes it possible to generate the embryogenic cell stage in a fully differentiated cell under defined conditions. The reset of the whole ontogenic program by initiation of somatic embryogenesis requires an essential reprogramming of the gene expression pattern. After abolishment of the previous differentiated cell functions, totipotent somatic cells with metabolic states similar to that of the fertilized egg cells are formed during the hormone-induced cell divisions. To understand the molecular and cellular basis of this dramatic developmental switch we might analyze possible analogies between activated egg cells and hormone-treated somatic cells.

A large number of empirical observations on various embryogenic tissue culture systems, and the results of biochemical, physical and structural studies support a general concept, which emphasizes the central role of hormone- or stress-induced activation of signal transduction systems. Consequently the internally transmitted signals trigger substantial changes in chromatin structure, alteration of transcription and induction of a series of cell divisions that leads to the formation of either dedifferentiated callus tissues or somatic embryos. Some potential components of this complex biochemical chain are summarized by Fig. 1. The reprogrammed gene expression is reflected by synthesis of new mRNA molecules during cell division after exposing the somatic cells or tissues to hormone treatment under *in vitro* conditions. Figure 1 also indicates that the initiation of the embryogenic pathway is restricted only to certain responsive cells with defined structure and characteristic patterns of cell division. Depending on the complex interaction between cells and external stimuli, e.g. hormone treatment, the subsequent cell divisions can result in either unorganized callus growth or polarized growth with well-coordinated pattern formation resulting in embryo development. Experimentally, these two developmental pathways should be separated. Figure 1 shows an example for the control of this developmental decision by the concentration of a synthetic

---

←

presence of a low concentration (0.1 mg/l) of 2,4-D. The dense cytoplasmic cell with unequal division is cultured in the presence of higher concentration (1 mg/l) of 2,4-D, and it has the potential to develop into a somatic embryo. R: receptor; G: GTP-binding protein; PLC: phospholipase C (phosphoinositidase); Ptd $InsP_2$: phosphatidylinositol 4,5-bisphosphate; $InsP_3$: inositol 1,4,5-trisphosphate; ER: endoplasmic reticulum; DAG: 1,2-diacylglycerol; PKC: protein kinase C; P-phosphoprotein; CDPK: Ca^{2+}-dependent protein kinase; CM: calmodulin; MPF: maturation promoting factor; histones: H1; H2A; H2B; H3; H4. References are cited in the text. Some components of this presentation are from Sussmann and Harper [229], Morse et al. [87], Csordás [230].

auxin, 2,4-D in the culture medium of alfalfa mesophyll protoplasts. Embryo development from protoplast-derived cells requires high (> 1 mg/l) concentration of 2,4-D in the case of A2 of *Medicago varia* L. genotype. As far as the molecular events are concerned, the process of somatic embryogenesis can be divided into induction and expression phases. Commitment of competent somatic cells towards embryogenic development is followed by the progression of embryogenesis with formation of globular, heart and torpedo shaped embryos.

Somatic embryogenesis as an unusual start of ontogenesis provides an exciting, special experimental system for developmental molecular biology. The first results in identification of embryo-specific proteins in cultured carrot cells and in cloning of the corresponding genes have been published in the early eighties [1–3].

Carrot suspension cultures are the most frequently used material in molecular studies focused on somatic embryogenesis [4–6]. Alternatively, callus or protoplast cultures from other species, e.g. from various *Medicago* genotypes offer experimental material for analysis of molecular changes especially during the early inductive phase (see the review by Dudits et al. [7]). The recent progress in establishment of embryogenic protoplast culture systems in monocot cereals such as rice [8] wheat [9] and maize [10] will further extend the range of available embryogenic cultures for molecular studies.

In this article, we describe a general concept about the central role of hormone-induced cell divisions in resetting the developmental program. The transition from somatic to embryogenic cell type will be analyzed as a hormonal or stress response that is activated via similar components of signal transduction as in the case of the egg cells after sperm-egg binding. As it is proposed by Fig. 1, we will attempt to link the hormone-induced molecular changes and synthesis of the specific mRNAs with the reactivation of the cell cycle and initiation of the embryogenic processes. Furthermore, characteristics of the early and late marker genes will be described. In addition to the transcriptional reprogramming, it is important to emphasize the significance of ultrastructural changes in cytoskeleton architecture during induction of somatic embryogenesis.

II. Initiation of the embryogenic pathway in cultured plant cells as a stress response

In a wider sense, somatic embryogenesis can be considered to be an extreme case of adaptation that is based on the phenotypic plasticity of individual somatic cells. Phenotypic plasticity as a characteristic of plants allows individuals to adapt or acclimate to a wide range of environments [11]. These can include instability in morphogenic or developmental programs and responses to damaging external conditions in order to insure survival and reproduction. Smith [11] summarizes three principal features of phenotypic plasticity, namely:

a. plasticity is manifested as the flexible expression of a constant genotype,
b. plasticity is intimately connected with environmental variations, and
c. plasticity confers adaptive value in respect to the capacity of the individual organism to adapt or acclimate to environmental conditions.

The embryogenic response in cultured plant cells can be correlated to the potential of single cells for developmental plasticity. A particular feature of the switch from somatic to an embryogenic cell type as an adaptation process is that it occurs *in vitro* under artificial growth conditions. Furthermore, the capability for embryogenic response as "survival strategy" depends on a specific metabolic state of the cells with defined genetic potential to activate the proper set of genes involved in generation of embryogenic cells during the early phase of ontogenesis.

Establishment of embryogenic tissue cultures includes several manipulation steps which severely injure and stress the inoculated tissues. During initiation of the cultures, the cells or tissues are separated from their original milieu. The wounded plant material is placed into *in vitro* growth conditions with unbalanced nutrition and hormone supply. Furthermore, the tissues can be exposed to various degrees of desiccation during culture on agar surfaces or anaerobic stress in liquid culture.

As will be discussed later, among the external factors the exogenously applied hormones, mainly auxins such as 2,4-dichlorophenoxy acetic acid (2,4-D), play a critical role in the reactivation of the cell cycle and the initiation of the embryo formation. Application of high concentrations of 2,4-D in the culture medium itself is a stress signal, since embryogenic induction requires the use of a physiological auxin concentrations that inhibit the callus growth (see also Fig. 4).

The inductive effect of a short auxin shock can be clearly demonstrated with the help of microcallus suspensions (MCS) from alfalfa (*Medicago sativa*) RA$_3$ genotype. As shown by Fig. 2, treatment of dedifferentiated cells grown in the presence of weak auxin: naphthalene acetic acid (NAA) with 100 μM 2,4-D for a few minutes up to a few hours is sufficient to induce embryo formation. This system also provides evidence for the key role of 2,4-D in the formation of embryogenic somatic cells. In addition, the use of these cultures allows the exact timing of the inductive phase. In contrast, the proembryogenic nature of carrot suspension cultures makes it difficult to determine the time of commitment of somatic cells towards embryogenesis. Differences between carrot and alfalfa embryogenic culture systems are summarized by Dudits et al. [7].

There are several tissue culture observations which convincingly indicate the significant influence of wounding on embryogenic response. On primary explants, the somatic embryos emerge first from cells around the cut surface. The number of somatic embryos formed can be increased by the use of tissue inocula chopped into smaller pieces. The wounding effect is the most pronounced if protoplasts are isolated and used for initiation of embryogenic cultures. Removal of the cell wall by enzyme treatment wounds each cell. The

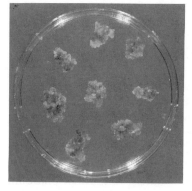

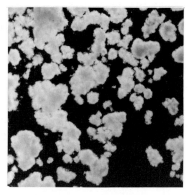

INITIATION OF CALLUS TISSUE   MICROCALLUS SUSPENSION

  2,4-D shock

treatment with 100 μM for 1 hour

EMBRYO DEVELOPMENT

under hormone free conditions

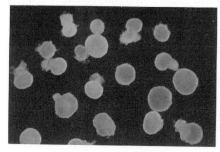

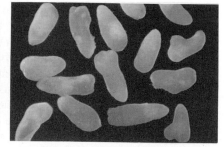

GLOBULAR EMBRYOS   TORPEDO-SHAPED EMBRYOS

Fig. 2. Induction of somatic embryos in alfalfa (*Medicago sativa cv.* RA3) microcallus suspension (MCS) by short auxin shock. The primary callus tissues are produced from leaf petiols in the presence of 15 μM naphthalene acetic acid, and 10 μM kinetin. The dedifferentiated callus cells grown as microcolonies are cultured in liquid suspension culture with the same hormones. After auxin shock, the tissues are transferred into hormone-free medium, where globular and torpedo-shaped somatic embryos can develop from totipotent cells (see also Dudits et al. [7]).

rapid cellular responses to auxin treatment and wounding share common features (see review by Brummel and Hall [12]; Davies [13]). Both of these external factors can play a role in the initiation of direct embryo formation in alfalfa protoplast-derived cells. In this case, embryos are formed from mesophyll cells without callus formation [14, 15].

The number of embryos developed from protoplast cultures can be increased by treatment of alfalfa cells with a low voltage electrical field [16].

In addition to the detection of the systemic signals released by wounding [17], a set of wound-inducible genes have been identified and cloned [18–23]. Wounding and 2,4-D treatment exhibit synergic effects in the stimulation of ethylene production [24]. Ethylene inhibitors such as $AgNo_3$, $CoCl_2$ and $NiCl_2$ were shown to enhance the number of somatic embryos in carrot suspension or in wheat and tobacco callus cultures [25–27]. High salt concentration in the culture medium or application of osmoticum as another stress stimulus can also increase the number of embryos in wheat culture [28] or in carrot suspension culture [29]. NaCl can prolong the embryogenic potential [30].

The consideration of somatic embryogenesis as a specific form of stress response related to an adaptation process is supported by experimental findings that show the involvement of the heat shock systems in this developmental reprogramming. First, Pitto et al. [31] observed characteristic differences in pattern of heat shock proteins at different developmental stages of carrot embryogenesis. Also in carrot cultures, the globular stage somatic embryos accumulated considerably less heat shock mRNA in comparison to embryos at later stages or to cultured callus cells after heat shock [32]. Analysis of the level of HS 17.5 mRNA indicated the lack of transcriptional induction of this heat shock gene in globular carrot embryos.

A possible link between embryogenic response and heat shock proteins is also proposed by the finding that one of the temperature sensitive non-embryogenic carrot mutant (Ts59) turned out to be defective in the phosphorylation of a heat shock protein (see review by Terzi and Lo Schiano [33]). Common elements in auxin and heat shock response can be predicted from molecular studies on different experimental systems. Czarnecka et al. [34] found the activation of heat shock genes by 2,4-D treatment of soybean hypocotyl tissues.

More interestingly, the 3'-intergenic element of an auxin regulated gene cluster (SAUR genes) in soybean shows high homology to the sequence motif located 150 bp downstream of the stop codon in the soybean heat shock gene 6834 [35]. The functional significance of this homology is not known.

In addition to the stress-activated defense function of HSPs, these proteins can play a role during cell proliferation, differentiation and embryogenesis (reviewed by Bond and Schlesinger [36]). In a variety of eukaryotic organisms, the heat shock genes are transiently expressed during early embryogenesis [37, 38]. In agreement with the previous studies on animal embryos, a small heat shock gene (Mshsp18) was also found to be expressed in early, globular and heart stage alfalfa embryos developed from somatic cells under normal culture conditions [39]. The authors proposed that this HSP may serve as a molecular chaperone (reviewed by Ellis [40]) with an assembly function during the developmental switch for initiation of the embryogenic cell stage. This hypothesis is supported by the fact that the GVLTV sequence motif is present in the amino acid sequence of this protein. Ingolia and Craig [41] suggest that this structural element can be responsible for the aggregation capability of HS proteins. The currently available data about the relationship between the start of embryogenic program in somatic cells and activation of heat shock genes open new directions in further studies on the role of stress in plant development.

The discussion of the stress-related molecular changes would be incomplete without the analysis of the structural reorganization of the genome or alteration of DNA methylation in cultured plant cells or tissues. Both of these mechanisms can contribute to the change of chromatin structure as a consequence of stresses caused by the *in vitro* culture conditions, especially by the hormonal effects. During the process of cell dedifferentiation – callus formation – and differentiation, the nuclear DNA may undergo quantitative variations generated by amplification, underreplication or elimination of specific sequences.

The early cytological studies on explants at callus initiation phase have revealed DNA extrusion into the cytoplasm, differential increase of nuclear DNA content, or nuclear fragmentation [42]. The existence of differential replication processes during callus induction is evident from various experiments [43–45]. The amplification processes are not restricted only to G+C-rich satellite DNA [46], but they can influence the copy number of defined genes [47].

Based on cytological and cytophotometric data, Nuti-Ronchi et al. [48] have suggested a particularly interesting hypothesis about the involvement of somatic meiosis in the process of somatic embryogenesis. According to this theory, the DNA diminution and formation of haploid nuclei may be an essential step in the restoration of totipotency. This model requires further experimental support with an explanation for the development of diploid embryos from the somatic cells with haploid chromosome sets. It should be mentioned that the flexibility of the plant genome is reflected by changes of DNA content in differentiating root cells [49, 50]. Both cytophotometric and biochemical results show changes in the amount and organization of the nuclear DNA at various developmental stages. The loss of sequences with ageing or extra DNA synthesis during initiation of a new differentiation step can be a component in the realization of the developmental program [51]. Therefore, the so-called phenotypic plasticity might also be based on the manifestation of genomic fluidity. Both of these can contribute to the potential of somatic cells to become embryogenic.

In addition to the quantitative changes in the nuclear DNA content in cell and tissue cultures, qualitative alterations, e.g. in the degree of DNA methylation can also influence the differentiation processes in plants [52, 53]. The relation of DNA methylation to gene activity has been proposed by experiments with a variety of eukaryotic organisms, including plants [54, 55]. Studies on the content of 5-methylcytosine in various plant organs and cultured cells or tissues have not yet provided a coherent view about the role of DNA methylation in developmental changes.

Phytohormones such as 2,4-D have inhibited the replicative methylation in tissues of wheat seedlings [56, 57]. Morrish and Vasil [58] also described a lower level of methylation in Napier grass callus than in leaves. The 2,4-D controlled changes in DNA methylation in embryogenic callus showed a complex pattern [59]. LoSchiavo et al. [59] found a significant hypermethylation after the application of the auxin. The removal of 2,4-D from the medium caused a rapid demethylation, and the somatic embryos exhibited 14% of methylcytosine at the early stage. At present, we still lack substantial evidence which supports the idea that the methylation status of plant DNA directly influences the differentiation processes in tissue cultures. Progress in this field will require the use of specific genes with known function in somatic embryogenesis or the application of other methodology e.g. genomic sequencing [60].

III. Similarities in signal transduction and molecular responses between fertilized egg cells and 2,4-D-treated somatic plant cells

It is a striking fact that auxin (2,4-D) treatment can reprogram differentiated somatic cells to become totipotent and to reach developmental potential similar to that of the egg cells after fertilization. In both cases the embryogenic differentiation relies on cell division-induced by the activation of the signaling system as a consequence of hormone or sperm binding. The major signal transduction events involved in the triggering of the development by fertilization in sea urchin embryos have been listed by Epel [61]. According to this overview (see Fig. 3), the primary signals in egg activation are the breakdown of phosphoinositol with a resultant increase in Ca_i and pH_i. The reconstructing of egg plasma membrane is followed by the formation of an actin-rich cortical cytoskeleton, and later new proteins are synthesized simultaneously with the accumulation of cyclin.

The activation of the cell cycle is shown by the initiation of DNA synthesis and subsequent cell division. Surprisingly, several of the above mentioned molecular changes do occur in plant cells treated with a synthetic auxin, such as 2,4-D. The considerable similarity becomes apparent from experimental evidence that can help to outline a general molecular concept of embryogenic induction in cultured plant cells grown in the presence of 2,4-D. Some essential components of this signaling are also indicated in Fig. 1.

A. Binding of auxins to receptors; auxin sensitivity as a limiting factor

As shown by several lines of evidence, the primary mechanism of auxin action is related to the binding of the hormone to proteins. Recent results have been reviewed by Palme et al. [62]. Membrane-associated auxin-binding sites were located on the endoplasmic reticulum (ER), the tonoplast and the plasmalemma [63]. Subsequently, auxin-binding proteins (ABP) have been purified from maize shoots [64–68]. As recently reviewed by Klämbt [69] progress has been

FERTILIZATION (sea urchin egg)	AUXIN TREATMENT (somatic cells)
SPERM/EGG BINDING	AUXIN/RECEPTOR BINDING
Membrane depolarization	+
PIP_2 hydrolysis	+
Ca^{2+} release	+
- cortical granule exocytosis	
- NAD kinase	
Na^+ - H^+ exchange	+
- pH_i increase	decrease
Oxygen consumption increase	+
-H_2O_2 production	
Actin polymerization	?
K^+ conductance	+
Transport changes	+
Endocytosis	?
Protein synthesis	+
Pronuclear movements	?
Cyclin accumulation	?
DNA synthesis	+
MPF activation	?
Nuclear envelope breakdown	+
Cell division	+

Fig. 3. Analogy in signal transduction and cellular responses between fertilized egg cells and auxin-treated somatic plant cells. The sequence of changes initiated by fertilization in the sea urchin embryo is cited from the review by Epel [61]. The references for data about auxin effects are listed in the text. "+" indicates similarity between the two cell systems. "?" represents response that has not yet been studied in plant cells.

made in elucidating the biological function of auxin-binding proteins with different cellular localizations by using specific antibodies [64, 66–70]. The auxin-induced modification of protoplast transmembrane potential difference was inhibited by antibodies directed against plasma membrane antigen and plasmalemma ATPase [71]. These experiments have provided evidence for the presence of a membrane receptor for auxin at the plasmalemma, and the binding of the hormone to this receptor likely can activate the proton-pumping ATPase. Consistent with the proposed ER location of auxin-binding proteins the sequence analysis of cDNAs encoding for ABPs has shown a deduced C-terminal tetrapeptide sequence Lys-Asp-Glu-Leu(-KDEL) that is characteristic of other proteins in endoplasmic reticulum [67, 72]. Expression of axr [1] gene was found to be organ-specific with high amounts of mRNA in maize ears and styles [67].

Despite the accumulating data about auxin-binding proteins, the direct or indirect involvement of these molecules in the mediation of auxin action has not yet been determined. Several models have been proposed that emphasize the functional significance of the auxin-protein complex in the plasmalemma or in the cytoplasm alternatively in the nucleus [12, 69, 73]. In the plasmamembrane, H^+-ATPase can serve as an effector-protein after auxin-binding. Modulation of the ATPase activity can induce H^+ extrusion, change of cytoplasmic pH and membrane alterations [74]. The first response to auxin (IAA) was the membrane depolarization which lasted for 5–12 min and after repolarization a hyperpolarized membrane was formed [75, 76].

Similarly to other auxin responses, the induction of embryogenesis is a concentration-dependent process which shows considerable variation among species and genotypes. Furthermore, the structural studies have revealed that only a few cells in the inoculated primary explant appeared to be competent for embryogenic induction. All these observed differences can originate from various degree of sensitivity of cells or tissues towards auxins. The possible relation between auxin sensitivity and embryogenic potential was indicated by comparative studies on embryogenic (A2) and non-embryogenic (R15) alfalfa genotypes [77]. As shown in Fig. 4A, in the responsive genotype (A2) somatic embryos were formed in the presence of 2,4-D at a concentration which already inhibited the growth of callus tissues. The root explants of R15 alfalfa plants produced large amounts of dedifferentiated callus tissues without embryogenesis. The observed differences between the two genotypes in embryogenic potential could be correlated with 2,4-D sensitivity of the plants (Fig. 4B,C). If shoot explants were rooted in culture medium with various concentration of 2,4-D the root formation of A2 plants was already severely inhibited by 0.01 mg/l 2,4-D. In the case of R15 plants a similar degree of inhibition was only observed at 0.1 mg/l 2,4-D.

The proposed significance of auxin sensitivity in somatic embryogenesis might help to explain differences between plant species, genotypes or cells in the same explants or in explants with different origin in their capability to become embryogenic. The role of auxin sensitivity might be studied by production of transgenic plants [78].

278

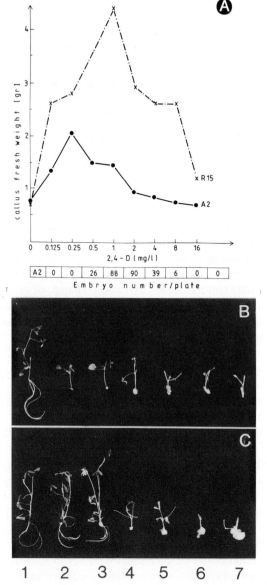

| A2 | 0 | 0 | 26 | 88 | 90 | 39 | 6 | 0 | 0 |

Embryo number/plate

Fig. 4. Differences in auxin sensitivity between embryogenic (A2) and non-embryogenic (R15) alfalfa clones from cultivar: Rambler (*Medicago varia*). (A) 2,4-D concentration-dependent *in vitro* responses in root explants. In addition to the callus induction A2 explants produce somatic embryos at auxin concentrations that inhibit the callus growth. The non-embryogenic R15 tissues respond with intensive callus growth also at a high 2,4-D concentration range. (B,C) Characterization of 2,4-D sensitivity of the clones on the basis of inhibition of root formation on shoot explants and callus initiation. (B) Explants from A2 embryogenic clone. (C) Explants from R15 non-embryogenic clone. 2,4-D concentration (mg/l): 1:0; 2:0,01; 3:0,05; 4:0,1; 5:0,5; 6:1; 7:2.

B. Phosphatidylinositol cycle, calcium messenger system

The bifurcating signal pathway based on inositol trisphosphate (InsP$_3$) and diacylglycerol (DAG) (see the review by Berridge [79]) can be involved in the initiation of a whole range of cellular processes including those events that take place at fertilization. The increase in polyphosphoinositide content is one of the earliest biochemical events following fertilization of the sea urchin egg [80]. Ciapa and Whitaker [81] have found a biphasic increase in the production of InsP$_3$ and DAG at the fertilization of the sea urchin eggs. Microinjection of InsP$_3$ activates sea urchin eggs by releasing calcium from the egg's intercellular stores and calcium stimulates InsP$_3$ production [82]. This wave of calcium release arises from the mutually reinforcing release of calcium and InsP$_3$. The cleavage of the membrane lipid phosphatidylinositol 4,5-bisphosphate (Ptd Ins(4,5)P$_2$) by inositol-specific phospholipase C (PLC) into InsP$_3$ and DAG is under the influence of guanyl nucleotide-binding proteins (G$_p$ proteins). According to the microinjection experiments with guanosine-5-0-(3-thio-triposphate (GTP-γ-S) the active form of the protein (with bound GTP) stimulates the production of InsP$_3$ in sea urchin eggs [83].

Miyazaki [84] has proposed that the sperm-egg interaction activates a GTP-binding protein that stimulates the production of InsP$_3$, causing the first one or two Ca^{2+} releases and also activates the pathway for the elevation of Ca^{2+} permeability. Diacylglycerol can be responsible for the activation of protein kinase C (PKC) resulting in a cascade of phosphorylation events (reviewed by Jaken [85]).

Experimental evidence for the role of the phosphatidylinositol (PtdIns) cycle in the activation of eggs at fertilization in animals encourages us to ask questions about the possible involvement of this signal pathway in mediation of a variety of hormone responses including induction of cell division and somatic embryogenesis in plants. As reviewed by Poovaiah et al. [86], Morse et al. [87], Guern et al. [88], and Palme [89] several studies have provided proofs for the occurrence in plant cells of various components of this cycle. More importantly, the PtdIns turnover is influenced by auxins. Indoleacetic acid (IAA) or 2,4-D can stimulate the fast breakdown of polyphosphoinositides, and the application of these auxins triggers a rapid, transient rise in InsP$_3$ and InsP$_2$ [90–92]. Furthermore, analogous to that in animal systems, InsP$_3$ releases Ca^{2+} from both vacuolar membranes and intact vacuoles [93–95].

The hormone-induced changes in cytosolic calcium via InsP$_3$-mediated calcium release can contribute to the activation of the second messenger system based on calcium (reviewed by Poovaiah et al. [86]). Generation of cytoplasmic Ca^{2+} signals through the auxin-mediated PtdIns response might depend on the phospholipases and GTP-binding proteins. However the various components with defined functions of this pathway have not been clearly identified. There are experiments that support the theory of the existence of this signaling in plants (see review by Guern et al. [88], Palme [89]). High-affinity GTP-binding proteins were detected in plant extracts [96, 97]. Further evidence for the presence of G-proteins in plants was provided by immunological detection experiments with antisera raised against a synthetic oligopeptide representing a highly conserved region of mammalian α-subunit of G-proteins [98].

Several plant genes encoding GTP-binding proteins have been identified so far [99–102]. The function of various phospholipases (A2; C; D) in the phosphatidylinositol cycle has been also discussed by several reviewers [87, 89, 103].

The involvement of auxin receptor function in auxin (e.g. 2,4-D)-induced activation of phospholipase A2, and the requirement for the presence of GTP were shown by Scherer et al. [104] and André and Scherer [105].

Experimental data support a link between the increase of cytosolic inositol trisphosphate and the mobilization of intra cellular calcium [93–95].

In addition to the release of Ca^{2+} from internal stores due to auxin promoted PtdIns turnover, the change in cytosolic Ca^{2+} level can result from opening the Ca^{2+} channels in the plasma membrane. (see the model by Elliott [106]). The cell wall and extracellular wall space can also be considered as a source of Ca^{2+} [107].

As a common feature of various models for the mode of action of auxins (see the review by Brummell and Hall [12]) and for the wound response [13], the elevated concentration of cytoplasmic Ca^{2+} is shown to have a major, yet complex role in the modulation of a set of cellular changes. Some of the responses can occur transiently, as e.g. the cessation of cytoplasmic streaming [108]. The long-lived responses can be mediated by Ca^{2+}-calmodulin complex [109] and by regulation of activity of various protein kinases [110].

Significant increase in Ca^{2+} level at fertilization has been shown in eggs of many species (reviewed by Jaffe [111]). Periodic oscillation of Ca^{2+} concentration during the initial few hours after fertilization was observed in ascidian and mammal eggs [112, 113]. During carrot somatic embryogenesis the elevation of $CaCl_2$ concentration in the growth medium caused an increase in number of embryos [114]. Nomura [115] described a polarized distribution of free Ca^{2+} in proembryogenic masses of carrot. However, the membrane-bound calcium showed diffuse and uniform distribution during carrot somatic embryogenesis according to fluorescence studies with chlorotetracycline [116].

C. K^+-h^+ exchange and pH changes

Studies on fertilization in animal eggs have indicated the critical role of pH change – increase of cytoplasmic pH – in activation of later fertilization events such as movement of pronucleus, initiation of protein and DNA synthesis (reviewed by Epel [117]). There is a minimum period of 10–20 min after fertilization during the pH level must be increased to trigger the cell division cycle [118]. In egg activation events two regulatory systems can be experimentally identified such as the earliest ionic changes regulated by Ca^{2+} and the later pH-related events.

As far as the auxin-stimulated changes are concerned, the analytical studies were primarily restricted to monitoring the prompt responses in shifting pH or altering Ca^{2+} level in tissues from both dicotyledonous and monocotyledonous plants. There are no experimental data available about cytosolic pH or Ca^{2+} concentration in auxin-treated cells prior to initiation of cell division. Therefore only limited comparison can be performed between the zygotic and somatic systems from this aspect. It is generally accepted that one of the most characteristic primary cellular responses to auxin is the decrease of cytosolic pH (reviewed by Brummell and Hall [12]). As a consequence of a pH drop, the ATPase H^+ pump and H^+ extrusion is stimulated [119]. The auxin-induced H^+ extrusion was coupled with K^+ influx after a lag period of 30 min [120]. The recent studies by Felle [121, 122] have provided a convincing evidence for auxin-induced oscillations of Ca^{2+} and pH. Measurements by using double-barrelled ion-sensitive microelectrodes have made it possible to suggest a sequential order in which the pH change is the primary response that can modulate cytosolic free calcium.

The influence of intercellular and external pH on somatic embryogenesis has been studied both in alfalfa and carrot tissue cultures. Based on phosphorus-31 NMR Schaefer [123] has shown that cells of non-embryogenic alfalfa callus tissues have a lower average intracellular pH, while the embryogenic cultures contain cells with high pH value.

Interestingly, there is an interaction between amino acid components of the culture medium and the actual pH, favoring, the embryogenic response. In the presence of proline, embryogenesis was enhanced with an increase of average pH with time in culture. In contrast, glutamate caused lower average intercellular pH with simultaneous inhibition of embryogenesis. A relation between the pH value of the culture medium and change in developmental state of the cells was suggested by experiments based on wounded carrot zygotic embryos [124, 125]. Continuous culture of preglobular stage proembryos could be established in hormone-free medium supplemented with 1 mM NH_4Cl. Under these conditions the pH of the medium dropped from pH 5.7 to $\sim$ pH 4. The low pH was essential for maintaining the proembryogenic stage, however embryo development proceeded when the pH was increased above 5.7.

D. Protein and DNA synthesis, induction of cell division

The early embryogenic cell cycle of higher eukaryotic organisms is accompanied by synthesis of several new proteins. The cyclins begin to accumulate shortly after fertilization and are destroyed during mitosis. In Xenopus oocytes fertilization triggers a rise in the intracellular Ca^{2+}, the inactivation of a cytostatic factor (CSF) and the reduction in activity of a maturation promoting factor (MPF). The transient elevation in level of active MPF leads to the induction of mitosis, while the fall of MPF activity allows the cells to enter in the next interphase. Murray and Kirschner [126] showed that the synthesis of sea urchin cyclin is sufficient to induce both the morphological and biochemical changes characteristic of mitosis. Epel [61] suggests two separate pathways initiated at fertilization – one leading to cyclin synthesis and the other one leading to DNA synthesis.

As reviewed by several authors [12, 73, 127, 128] the auxin treatment can significantly alter the synthesis of particular proteins via the stimulation or modification of gene expression.

Comparison of various *in vitro* translation products produced by mRNA from auxin-treated and untreated tissues revealed "early" as well as "late" auxin-inducible proteins. In higher plants, cross-reaction between anticyclin B(cdc13) antibodies and a 62KD protein has been just recently detected in alfalfa cell extract [129]. The role of this protein during zygotic or somatic embryogenesis in not known at this point.

According to experiments carried out by Evans (cited by Yeoman [130]) there is a considerable increase in the rate of oxygen uptake after excision of artichoke explant and before the start of DNA replication. Out of the multiple actions of auxins, initiation of DNA synthesis is one of the key functions during both callus formation and induction of totipotent embryogenic cells. The ability of auxins such as 2,4-D to activate DNA synthesis in non-dividing cells was already shown by the early studies (see review by Key [127]; Yeoman [130]). In Jerusalem artichoke tuber tissues, a frequently used experimental material, DNA synthesis was detected at a specific concentration of 2,4-D. This inductive effect requires protein and RNA synthesis [131]. In experiments with tissue explants, the wound reaction can interact with the hormones in the activation processes [132]. In potato tuber discs the cutting itself can initiate DNA synthesis without subsequent cell division [133]. Mitotic activation and initiation of cell division under the influence of auxins appear to be a general phenomenon in both tissue explants and single cells. Characteristics of division phase during early development in callus cultures are summarized by Yeoman [130]. Parameters of the first cell cycle in protoplast-derived tobacco cells indicated 60 % of colonies with two daughter cells 50 h after plating [134].

As seen, the described comparison has revealed a considerable similarity in a variety of molecular and cellular changes during fertilization that activates the development of eggs into zygotes and auxin initiated events that can trigger embryo development in somatic plant cells.

A hypothesis proposing the similarity is also strongly supported by *in vivo* experiments that showed development of embryos from unpollinated ovules after auxin treatment of influorescences of several gramineous species [135]. First Britten [136] reported induction of grain formation by applying auxin to maize. Marshall et al. [137] detected similar effects after 2,4-D treatment of wheat. It remains to study the molecular bases of auxin action in both systems that results in initiation of embryogenic program. The primary aim of this analysis was to provide theoretical basis for an overall strategy to design experimental approaches for better understanding the problem of somatic embryogenesis. Further progress in analysis of the molecular basis of transition from somatic to embryogenic cell type would be considerably stimulated with information about the early events of fertilization and zygotic development in higher plants (see Raghavan [138]). Because of the limited experimentation with plant egg cells and zygotes we were forced to rely on results from studies of animal systems. At the same time it is hoped that knowledge about somatic embryogenesis will contribute to the future development of experimental embryology of plants as well.

IV. Hormone- and stress-induced reactivation of cell cycle: Prerequisite for trigger of embryogenic program in somatic plant cells

One of the basic features in initiation of somatic embryogenesis is the reactivation of the cell cycle in differentiated plant cells under the influence of external stimuli. The artificially induced series of cell divisions open the way to switch from somatic to embryogenic cell type that requires coordinated expression of a set of genes and the post-translational modifications of the regulatory proteins involved also in the cell cycle control. Under defined circumstances, the reprogrammed transcription will generate an unique cellular state with specific functions that are needed to restart ontogenesis.

Based on these considerations molecular studies on somatic embryogenesis may be focused on the regulatory mechanisms in cell cycle with special emphasis on the mode of signal transduction. As documented by a large number of tissue culturists the hormone- or wound-induced cell proliferation does not necessarily lead to the formation of embryogenic cells, as frequently dedifferentiated callus tissue is formed. Unknown mechanisms in selected cells initiate specific additional molecular and cellular processes that generate the embryogenic state. Therefore the primary aim of molecular studies on various embryogenic culture systems is to identify these characteristic regulatory components.

At present, only limited information is available about the cell cycle parameters in embryogenic cells. Warren and Fowler [139] have shown that one of the consequences of the start of somatic embryogenesis is the decrease of cell doubling time in carrot cultures. A relatively short doubling time (6.3 h) was detected in early, globular stage carrot somatic embryos [140]. In addition, polarity and spatial distribution of cells at S phase were characteristic of carrot somatic embryos [5]. In general, only a defined fraction of the cells in suspension takes part in the division cycle that is maintained by the 2,4-D. The responses of embryogenic and non-embryogenic cells are different to the removal of the auxin from the medium. Cells in proembryogenic clusters continue the divisions with characteristic distribution of the daughter cells during embryo formation. In contrast, the non-embryogenic cells are as a rested at the G_1 phase by auxin starvation [141].

The use of primary explants allows the monitoring of the reactivation processes in mature tissues consisting of both proliferating and non-proliferating cells. As an example, the analysis of the auxin effect on various sectors of young wheat leaves has clearly indicated a gradient in cell cycle response as a function of the position of the cells and the concentration of applied auxin. The basal, meristematic explants required lower concentration of 2,4-D for G_1/S transition than the more mature tissue sectors [142].

Multifactorial control of the cell cycle in tissues cultured *in vitro* was suggested by the results of cell cycle studies that showed nuclear DNA replication in highly differentiated cells in both the absence and presence of exogenous auxin, but these cells could not complete the mitotic cycle. The differences between tissue sectors in cell cycle stage and in their potential to respond to stimuli can be correlated with the observations about the plant regeneration experiment in similar systems. These studies showed that the morphogenic or embryogenic potential is characteristic of tissue inocula from basal regions of cereal leaves [143–145].

The relation between the cell cycle reactivation and the embryogenic response can also be studied in the relatively uniform, synchronously responding cell populations obtained from leaf mesophyll protoplasts [146]. The majority of freshly isolated mesophyll protoplasts from tobacco was found to be in G_1 phase [134]. After resynthesis of the cell wall, the auxin-induced transition can be monitored by labeling experiments with ^{3}H-thymidine or flow cytometry. Bögre et al. [77] have compared the time course of ^{3}H-thymidine incorporation into mesophyll protoplasts-derived cells from leaves of embryogenic (A2) and non-embryogenic (R15) genotypes of alfalfa. In the embryogenic cultures, the maximum in the first incorporation peak was detected 2 days earlier than in cultures of the non-embryogenic genotype. Considering the available limited number of experimental results a link may be proposed between accelerated activation of cell cycle and embryogenic processes in somatic cells. Furthermore, there is a need for studies on cell cycle control in relation to auxin-induced division and embryogenic response.

The elucidation of the regulation of the cell division cycle in eukaryotic cells has evolved rapidly during the recent years. Several models summarize current understanding and interpretations of the general features of the key proteins implicated in cell cycle regulation [147–152]. Studies on yeast, Xenopus, starfish or sea urchin oocytes, fertilized eggs, chicken or mammalian cells have revealed common regulatory proteins including p34[cdc2] protein kinase as the product of cdc2 gene or its homologs and various cyclins.

Despite the vast evolutionary distance, in several eukaryotic cells the complex between p34[cdc2] and cyclins, as components of the maturation promoting factor (MPF), has regulatory function in the cell cycle.

This is achieved by alteration of the active or inactive form through phosphorylation/dephosphorylation of various constituents of the complex (see the recent model by Solomon et al. [150]; Fig. 1) The kinase activity of p34[cdc2] is regulated by its phosphorylation state on threonine and tyrosine residues and its association with other proteins e.g. cyclins. Broek et al. [153] suggest two states for the protein kinase p34[cdc2]: the active S form leads to entry into S phase and later p34[cdc2] is converted to its M form.

The p34[cdc2] kinase phosphorylates a variety of different substrates, as reviewed by Moreno and Nurse [154]. In addition to the modification of various nuclear proteins (histone H1, nucleolin, lamins) cdc2 kinase can be responsible for rearrangement of cellular structures [155] or changes in microtubule dynamics [156]. The possible role of cdc2 protein kinase in transcriptional regulation was proposed on the basis of phosphorylation of RNA polymerase II by CTD kinase that contains the cdc2 protein as a component [157]. From a series of studies, it is evident that the cyclins are directly involved in cell cycle regulation [149, 158, 159]. According to the proposed model of Solomon et al. [150], cyclin accumulation induces the activation of p34[cdc2]. The complex formation between various types of cyclins (A or B) and cdc2 protein has a differential role at each cell cycle phase (see Draetta [149]). Murray and Kirschner [126] have concluded that in early Xenopus embryos the concentration of cyclin is the only control in triggering of mitosis.

The overwhelming increase in the knowledge about regulation of cell cycle in variety of eukaryotic organisms has also stimulated research with higher plants and considerable progress has been achieved in the recent years(see reviews by Jacobs [152], and Dudits et al. [160]). The first, experimental data with plants are in agreement with an unified view of eukaryotic cell cycle control.

It has been demonstrated that several of the above mentioned conserved components of these regulatory pathways exist in plant cells and they play a role in the control of cell division. Evidence for the presence of a plant homolog of the cell cycle control protein p34[cdc2] has been provided by different experimental approaches.

A sequence comparison of p34[cdc2] from different species showed a conserved region of amino acids, the so-called PSTAIR region [161]. The use of antibodies raised against this internal peptide or Mab-J4 has made it possible to recognize a 34kDa protein in plant cell extracts [162–165].

The co-purification of the p34 with histone H1 kinase activity was found during chromatography of pea and alfalfa cell extracts [129, 163]. In experiments with alfalfa, anti-cyclin B (cdc13) antibodies recognized a 62kDa protein after immunoblotting. The first indication for the existence of p34[cdc2]-cyclin complex in higher plants has been provided by the use of p13[suc1] affinity binding and of antibodies against PSTAIR peptide and human cyclin A [166]. The suc1 gene encodes a 13kDa protein, p13[suc1] that binds *in vitro* to p34[cdc2] or its complexes with cyclins from various sources [167, 168]. In addition to the biological significance of this binding, the specific interaction between p13 and p34 can be used for p13-Sepharose affinity binding in studies on p34[cdc2]-cyclin complexes. The involvement of the p34[cdc2] kinase at various phases of the cell cycle can be concluded from the changes in H1 kinase activity that was associated with the complex. As an example, Fig. 5 presents the result of *in vitro* phosphorylation studies with cell extracts from partially synchronized alfalfa cells. In this experiment the suspension culture was exposed to aphidicolin treatment. A significant peak in incorporation of ^{3}H-thymidine can be observed after release of the block. Based on auto-radiography after histone H1 phosphorylation *in vitro* we can recognize

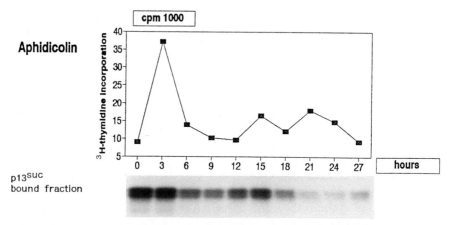

Fig. 5. Changes in p34^{cdc2}-related kinase activity in cell extracts from partially synchronized alfalfa cell suspension culture. The cultured alfalfa cells were synchronized by amphidicolin treatment (24 h; 20 µg/ml). Cells were collected at various time points after release of amphidicolin block, and labelled with ^{3}H-thymidine for 3 h. At 3 h intervals cell extracts were prepared, and loaded onto p13^{sucl}-Sepharose. The *in vitro* phosphorylation was carried out in the presence of histone H1 in the following reaction mixture: 25 mM Tris-MES pH 7,4; 15 mM MgCl$_2$; 5 mM EGTA; 1 mM DTT; 7,5 µCi/ml [γ^{32}-P] ATP; 1 mg/ml histone H1 (Sigma). The histone H1 phosphorylation by p13^{sucl}-Sepharose bound protein fractions was monitored by autoradiography after separation of proteins on SDS-PAGE.

changes in p34^{cdc2}-related kinase activity during the studied period of the cycle.

Elevated H1 kinase activity in protein fractions bound to p13^{sucl}-Sepharose was found in the first two samples, one with increased level of ^{3}H-thymidine incorporation and in samples collected 12–15 h after washing out the aphidicolin. In agreement with similar studies on cell extracts from leaf protoplast cultures (see Dudits et al. [7]), and cultured cells synchronized by hydroxyurea [166], this experiment also suggests p34^{cdc2} kinase activity associated with the G$_1$-S or G$_2$/M phases in plants. These results may suggest a role for the S and M form of p34^{cdc2} or different p34^{cdc2}-related kinases in cultured plant cells similar to other eukaryotes [153, 169]. The recent experiments by Magyar et al. [166] provide further information about complex formation between p34^{cdc2} kinase and cyclin like proteins. Immuno-precipitation with human anti-cyclin A antibodies allowed to detect histone kinase activity in cells at G$_1$/S phase. The present data suggest phase-specific p34^{cdc2}-cyclin complexes during cell cycle in higher plants.

As a complementary approach, isolation of cdc2/CDC28 homolog genes from plant cDNA libraries provides additional proof for the existence of p34^{cdc2} in plants. cDNAs related cdc2 were identified from pea [163], rice [170], alfalfa [164], *Arabidopsis* [171, 172], and maize [165]. Since the alfalfa cDNA (cdc2 Ms) encodes for a protein with all the characteristic amino acid sequence elements required for p34^{cdc2} kinase function, the expression of this protein in yeast could restore the temperature-sensitive cdc2 mutation [164]. The availability of the alfalfa cdc2 cDNA clone allowed more comprehensive studies on expression of this gene or genes in dividing cells and alfalfa somatic embryos. Northern analysis has revealed that the 2,4-D treatment significantly induced the expression of this gene. It is important to emphasize that the hybridization has indicated at least two transcripts with different patterns of expression [164]. This observation is in agreement with the results published by Hata [170], who also found two hybridizing bands in Northern analysis of RNA from rice cells.

Further Northern blot analysis of alfalfa cdc2 genes revealed considerably different expression patterns in cells of primary root explant, or from leaf mesophyll protoplasts exposed to hormone treatment, and of rapidly cycling suspension culture [166]. Transcription of al teast one of the cdc2 genes was activated by hormones (auxins, cytokinins) in root and mesophyll cells of alfalfa. In suspension culture of alfalfa cells, cdc2 transcripts were found to be at constitutively high level, independently from the actual cell cycle phase of cells. This might indicate the loss of normal cell cycle control mechanisms under the artificial conditions of *in vitro* cultures.

The p34[cdc2] is one of the representatives from the large number of protein kinases that are essential components of signal transduction during several cellular functions. For the illustration of the complexity of phosphoprotein patterns during the culture of alfalfa mesophyll protoplasts we show the results of *in vitro* phosphorylation in cell extracts from various time points (Fig. 6). By comparing the phosphorylation patterns from cells of embryogenic (A2) and non-embryogenic (R15) genotypes significant differences between the two cell systems were observed. In the presence of the same concentration of 2,4-D (1 mg/l), the embryogenic cell population exhibited a marked increase in phosphorylation of defined proteins at the 3rd and 5th day of culture, while in extracts from the R15 cells the differences in phosphorylation were much more reduced. In order to be able to understand the basis of these differences we have to identify the various protein kinases, their substrate-specificity and the influence of various external and internal factors on the status of the phosphorylated form of these proteins.

It is well established that protein phosphorylation represents a major mechanism for the stimulus-response coupling mediated by second messengers (see the review by Ranjewa and Boudet [173]). Ca^{2+} or calmodulin-dependent protein kinases have been described in a variety of plants. These activities were associated with plasma membranes [174], chloroplasts [175], soluble fractions [176, 177], and chromatin [178].

The pioneer work of Murray and Key [179] has shown the predominant effect of 2,4-D in the increase of the phosphorylation of nuclear proteins from soybean. There was a striking increase in the phosphorylation of a 48 kD protein after 24 h treatment. Characteristic changes in the amount of a major phosphoproteins (CDPK 52–54 kDa) were detected in alfalfa protoplasts grown in the presence of 2,4-D [160]. As the number of the dividing cells increased in culture, the Ca^{2+}-dependent phosphorylation of the 52–54 kDa protein doublet

288

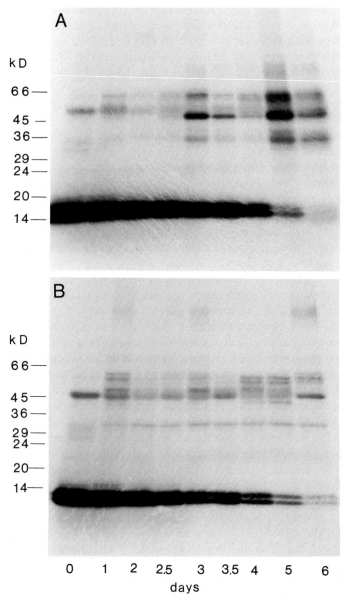

Fig. 6. Differences in *in vitro* phosphorylation of proteins from cultured alfalfa leaf mesophyll protoplasts and the derived cells of embryogenic and non-embryogenic genotypes. (A) Cell extracts from cultures of A2 embryogenic genotype. (B) Cell extracts from cultures of R15 non-embryogenic genotype. The extraction buffer for alfalfa cells composed of the followings: 25 mM Tris-HCl pH 7.4; 15 mM MgCl$_2$; 15 mM EGTA; 100 mM NaCl; 0.1% TWEEN20; 1 mM DTT; 0.5 mM PMSF; 20 μg/ml leupeptin; 20 μg/ml aprotinin; 1 mM Na-pyrophosphate; 1 mM Na-vanadate; 2 mM NaF and 10 mM α-naphthyl phosphate. After kinase reaction in the absence of histone H1, the proteins were separated on SDS-PAGE and the phosphoproteins were detected by autoradiography.

also showed elevated signals. After partial purification of these proteins it turned out that the 52–54 kDa protein was a Ca^{2+}-dependent protein kinase (CDPK) with autophosphorylation capability [177]. This kinase shares several common characteristics with the CDPK from soybean [180]. This alfalfa kinase can phosphorylate a synthetic nonapeptide (AAASFKAKK) that is a specific substrate of animal protein kinase C [181, 182]. Furthermore this CDPK from alfalfa can preferentially phosphorylate histone H3 in comparison to other histone substrates. Here it should be mentioned that the soybean CDPK isolated from suspension culture can phosphorylate gizzard myosin light chain and a synthetic myosin peptide suggesting a role for CDPK in cytoplasmic streaming [183].

Phosphoproteins can be detected among embryogenic proteins coded by abscisic acid (ABA)-inducible genes [184, 185]. The RAB-17 proteins were highly phosphorylated in mature maize embryos, while these proteins accumulated in low amounts in ABA-treated callus without considerable phosphorylation [185]. Involvement of the modification of the regulatory proteins through phosphorylation by various protein kinases in the alteration of the transcription can be concluded from several recent publications. Guilfoyle [186] has described a protein kinase from wheat germ with a native molecular weight of about 200 kD that phosphorylates the large subunit of RNA polymerase II. Such a change might modify the interaction of the RNA polymerase II with transcriptional factors.

The DNA-binding ability of the AT-1 nuclear protein to specific AT-rich elements within the promoter of photosynthetic genes is strongly affected by phosphorylation [187]. In view of the described experimental data about the protein phosphorylation-related molecular mechanisms in the mediation of various external stimuli or in control of cell division, the auxin-activated embryogenic program in somatic cells can be approached experimentally from new aspects and in a more defined manner.

Accepting the fact that the initiation of somatic embryogenesis is closely linked to the hormone-induced cell divisions, the molecular characterization can also be based on genes with cell cycle-dependent expression. Among others, histone genes can serve as molecular markers during both the inductive phase and the realization of the embryogenic program. Analysis of the expression of histone genes in relation to cell division and embryogenic differentiation may provide new insights into the transcriptional changes during a developmental switch that depends also on altered chromatin structure. Structural and functional characterization of histone genes has already revealed a multi-component regulatory mechanism that is responsible for replication-dependent, constitutive or tissue-specific expression of various histone genes [188, 189]. Although a considerable number of histone genes have been cloned and sequenced from higher plants (reviewed by Gigot [190]), our knowledge about the regulation of histone gene expression during plant cell division and differentiation is very limited. In germinating *Vicia faba* the histone synthesis was dependent on the transcription of new mRNAs [191]. Accumulation of H3 mRNA was detected by *in situ* hybridization in non-dividing cells during the development of rice embryos [192].

Figure 1 already indicated that considerable nucleosomal changes with possible functional consequences can occur at the end point of a signal cascade. This can originate from the post-translational modification of the protein constituents of the chromatin or alterations in the nucleosome structures mediated by histone variants.

As reviewed by Maxson et al. [193] histone genes are differentially expressed during sea urchin development that includes the synthesis of histone variants during both the rapid embryogenic cell division and tissue morphogenesis. Existence of histone variants was proven on the basis of electrophoretic characteristics of plant histone H3 proteins [194], as well as of nucleotide sequence of various histone H3 cDNAs (pH3c − 1; pH3c − 11) by Wu et al. [195].

In partially synchronized alfalfa cells or 2,4-D-treated protoplast cultures the amount of H3c − 1 transcripts was correlated with the maximum ^{3}H-thymidine incorporation. In contrast to the S-phase-dependent expression of histone H3c − 1 gene, the mRNA of H3c − 11 gene is detectable in all cell types [196, 197].

It is interesting to mention that the expression of the H3c − 11 variant gene is inducible by hormone or other stress treatments. Both genes provide abundant mRNA in somatic embryos, but the H3c − 11 shows differential expression. The amount of its transcript is higher in early stage embryos in comparison to those in the later developmental stages. The functional role of this H3c − 11 variant during embryogenesis is not known. Waterborg et al. [198] have detected high steady state level of acetylation of this alfalfa histone variant. This finding can be important in analyzing the role of this histone H3 variant in formation of the active chromatin required at early embryogenic stages.

V. Differential gene expression underlying embryogenic induction and embryo development

The previous discussion has emphasized that initiation of somatic embryo-genesis is a consequence of external stimuli, and depends on the induction of cell division with defined characteristics. Furthermore, a complex signaling system is responsible to bridge the effect of inducers and the coordinated alterations in the expression of a set of genes. The developmental switch in somatic cells at the time of embryogenic induction involves essential changes in the physiological state of the cells that require an overall reprogramming of gene expression. The success in the identification of key regulatory factors at different chech points during somatic embryogenesis highly depends on the tissue culture systems used. The controlling elements are expected to be considerably different during the processes of embryogenic commitment in primary explants or fully differentiated somatic cells from those that act in suspension cultures with proembryogenic structures as, e.g. in the case of carrot. The inductive phase and progression of embryo development rely on various molecular events in decision-making during development. Therefore isolation and characterization of so-called "early" and "late" marker genes are equally needed for comprehensive molecular analysis of this developmental pathway.

The early inductive molecular events can be monitored by detection of characteristic mRNA or protein molecules in tissues or protoplast-derived single cells that are synthesized under the influences of external stimuli that trigger the embryogenic state.

It is also recognized that at this stage the cellular responses are also influenced by "housekeeping" genes that are components of normal metabolic functions such as cell division, and stress response. It is also possible that activation of these genes may follow an unusual order or variation in the level of expression.

According to the general concept outlined before, the crucial step is the initiation of the same cascade of gene activation chain that is activated at sexual fertilization. Since in a majority of tissue culture systems, the auxins, typically 2,4-D treatment and wounding are the key inducers of embryogenic cell type, the auxin- and wound-responsive genes are the primary candidates for detecting specific changes at the level of gene expression. Preferentially, the molecular studies on auxin-regulated gene expression with the help of cloned cDNAs have focused on rapid responses related to e.g. cell elongation (see review by Theologis [73]; Brummel and Hall [12]). Significantly, a fewer number of auxin-responsive genes has been cloned that are linked to later cellular changes such as activation of cell division [199]. As an example, Bögre et al. [77] used auxin-responsive genes as molecular markers for detection of differences between embryogenic and non-embryogenic genotypes of *Medicago sativa* cv. Rambler. In these experiments Northern hybridization was carried out using RNA samples from alfalfa tissues treated with low or high concentrations of 2,4-D. Two soybean cDNA clones were used as heterologous hybridization probes. The pJCW1 cDNA represents an auxin-inducable gene called Aux28 [200], while the pTUO4 clone is a soybean gene with a developmentally controlled pattern of expression [201]. In auxin-treated tissues from plants with embryogenic potential, 2,4-D considerably induced the expression of the gene, homolog of Aux28, and reduced the amount of transcripts from the gene corresponding to pTUO4. Similar changes in the expression of the above genes were not detectable in tissues from the non-embryogenic genotype.

A variety of auxin-responsive genes has been identified with the help of specific tissue culture systems based on microcallus suspensions of alfalfa.

As shown by Fig. 2 a short pulse treatment with high concentration of 2,4-D is sufficient to trigger embryo differentiation in this culture. Differential screening of a cDNA library constructed from poly(A$^+$) RNA of 2,4-D-shocked tissues has revealed a set of genes with characteristic expression pattern during the transition to embryo differentiation and subsequently at different stages of embryo development. As an example, Fig. 7 shows the transient increase in amounts of transcripts of a gene encoding a putative Ca^{2+}-binding

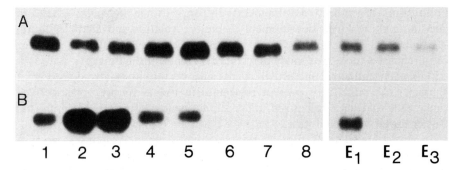

Fig. 7. Differential expression of calmodulin gene (A) and a gene (B) encoding for a putative Ca^{2+}-binding protein after auxin shock (treatment with 100 μM 2,4-D for 1 h) during transition towards embryogenesis and at various stages of somatic embryo development. Total RNA samples were prepared from microcallus suspension during the growth in hormone free medium: 1:0; 2:1 h; 3:3 h; 4:8 h; 5:1 day; 6:3 days; 7:7 days; 8:21 days. The Northern blot analysis was also carried out with RNAs isolated from somatic embryos as E_1: globular stage embryo; E_2: heart-shaped embryos; E_3: torpedo-shaped embryos. The RNA samples were fractionated in 1.4% agarose/formaldehyde gels and hybridized to ^{32}P-labelled cDNA probes. The calmodulin probe was kindly provided by K. Palme (Köln). The MsCa1 cDNA is cloned by K. Németh et al. (unpublished).

protein (MsCa1) after 2,4-D treatment. Interestingly, this gene is active in young somatic embryos. Its transcript can not be found in elongated torpedo-shaped embryos. (Nemeth et al. unpublished). In comparison, Fig. 7 also presents data about the level of the expression of calmodulin gene in the same system. Here a tobacco calmodulin probe (kindly provided by K. Palme, Köln) was used in Northern hybridization. The high level of calmodulin transcripts in somatic embryo at early developmental stages suggests a differential expression for this gene during embryogenesis. Based on expression studies with cDNA clones of various protein kinase genes, we can conclude that some of the cell cycle control genes such as cdc2 are also influenced by 2,4-D and show variable amounts of transcripts during somatic embryo development [164].

Identification of auxin-responsive cDNAs and cloning the corresponding genomic sequences should be considered as one possible experimental approach. Further information can be gained about the early molecular changes in generation of the embryogenic state with the help of transgenic plants. The gene transfer technology based on direct DNA uptake or *Agrobacterium* mediated transformation [202, 203] makes it possible to analyze the activity of various promoters linked to reporter genes such as the *E. coli* β-glucuronidase (GUS) coding sequence [204] or bacterial luciferase gene [205] in transformed cells or tissues. Based on this elegant experimental strategy, Fujii and Uchimiya [206] have studied the behavior of the *rolC* promoter in carrot tissues in the presence and absence of 2,4-D. Previously it was demonstrated that the expression of *rolC* (ORF12) gene of TL-DNA of the Ri plasmid was under tissue-specific control [207].

In transformed carrot suspension cultures, 2,4-D depressed the *rolC* promoter activity. When embryos are developing after removal of 2,4-D, the globular stage embryos show a GUS positive reaction [206]. In a different experimental system, Stefanov and coworkers (unpublished) have followed the expression of the 35S promoter of cauliflower mosaic virus (CaMV) in transformed alfalfa explants both during the initiation of callus tissues and development of somatic embryos (Fig. 8). This promoter confers a high level of expression in most of the cells in transgenic plants [208]. Deletion analyses of the 35S promoter have indicated the existence of expression modules with combinatorial control of tissue- or cell-specific function of this promoter [190]. The intact 35S promoter can show variation in level of expression in various plant organs [209]. In tobacco flowers there is considerable expression in vascular tissues and in trichomes of petals while the epidermal tissue shows weak expression [210].

The expression studies with constructs carrying 35S promoter and GUS genes during initiation and development of alfalfa somatic embryos were strongly stimulated by studies of Nagata et al. [211]. These authors provided data about variation in the expression of 35S promoter during the cell cycle with significant increase in the S phase. As is shown by Fig. 8A, the GUS activity is concentrated into a globular center zone of the developing callus mass in the presence of 2,4-D. The large elongated cells, representing a typical non-dividing cell type lack the GUS activity. We think that the detected GUS-positive tissue sector contains actively dividing cells that can be an initiation point for further steps of later embryo formation. The distribution of GUS staining in torpedo-shaped alfalfa embryo shows differential expression of 35S viral promoter (Fig. 8B). Polarity can be recognized in the activity of this regulatory element in somatic embryos on the basis of intense staining in the presumptive shoot pole region. This promoter is also active in cells along the embryogenic axis.

The presented results about the use of transformants with 35S promoter and GUS provide an example for the potentials of this approach to act as a sensitive molecular indicator to detect changes at the level of transcriptional control.

As a complementary research strategy in addition to the use of auxin-responsive or cell cycle-related genes as molecular markers for the analysis of the early inductive events, studies on carrot cell suspension culture have provided new essential information about the expression of embryo-specific genes and synthesis of stage-related proteins during somatic embryo development. It is important to recognize the fact that carrot cells or multicellular structures are already beyond the embryo initiation phase at the time of culturing in suspension culture supplemented with 2,4-D. Therefore the terminology: "proembryogenic mass" (PEM) introduced by Halperin [212] properly indicates the embryogenic nature of the carrot suspension cultures. This can explain the results of the comparative analysis between PEM grown in the presence of 2,4-D and somatic embryos developed after removal of the auxin, that have revealed only a limited number of embryo-specific polypeptides [1, 2, 213] Wilde et al. [214] have provided data about a general similarity in

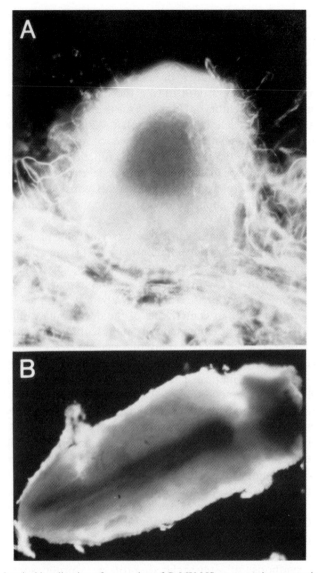

Fig. 8. Histochemical localization of expression of CaMV 35S promoter in transgenic alfalfa plant during callus induction (A) and in torpedo-shaped somatic embryo (B) (I. Stefanov et al. unpublished). *In vitro* grown plants of A2 (*Medicago varia* L.) genotype were transformed with CaMV 35S promoter – β-glucuronidase (GUS) gene fusion as described by Deák et al. [231]. Callus tissues and somatic embryos were induced from stem segments in the presence of 2,4-D. Detection of β-glucuronidase activity was carried out according to Jefferson [232].

mRNA populations from PEM and torpedo stage embryos. The temporal transcription of defined genes has made it possible to identify a set of cDNA recombinants that represent genes with differential expression in PEM or during somatic embryogenesis. The cDNA clone, Dc3 is specific for PEM or somatic embryos. The corresponding gene has been suggested to be a marker of embryogenic potential [214, 215]. Borkird et al. [4] have identified another carrot cDNA (DC8) that detects both somatic- and zygotic-specific mRNAs in Northern hybridization with increased amounts of transcripts in heart stage somatic embryos [216]. The differential screening of carrot cDNA library by Aleith and Richter [217] has resulted in the identification of clones representing genes that are induced at early embryogenic stages. These genes encode glycine rich proteins. Kiyosne et al. [218] have also identified an early marker for carrot somatic embryogenesis.

As a characteristic feature of carrot cultures, it was found that a set of extracellular proteins secreted into the culture medium could promote the acquisition of embryogenic potential [215]. As well, cDNA clones corresponding to these extracellular proteins have been identified. Based on Northern analysis one of these genes (PS48) is expressed only in embryogenic lines [219].

Considering the morphological similarities between zygotic and somatic embryos during progression of embryo development it is expected that the same genes are active in the corresponding stages. This was demonstrated also by the analysis of seed storage proteins in alfalfa and pea. All of the major and characteristic storage proteins of zygotic embryos were synthesized and processed normally in the somatic embryos; only quantitative differences were observed compared to the zygotic ones. The role of abscisic acid (ABA) in embryo maturation and seed development by regulation of gene expression has been proven by studies on a number of zygotic embryos (reviewed by Quatrano [220]). Exogenously applied ABA can significantly improve somatic embryogenesis *in vitro*. The embryo-specific gene expression in somatic embryos is supposed to be under tight developmental control, as demonstrated for zygotic embryogenesis [221]. Regulation of gene expression during plant embryogenesis has been reviewed by Goldberg et al. [222].

VI. Establishment of polarity during reorganization of cyto-architecture in dividing embryogenic cells

In auxin-treated cells, cultured *in vitro*, the transition from somatic to embryogenic cell type relies not only on the synthesis of new mRNA populations as discussed earlier, but considerable structural changes can be detected in the cellular architecture when cells are exposed to embryogenic stimuli. It seems likely that the description of transcriptional changes will not be sufficient to understand the mystery of the developmental re-direction that starts the embryogenic program in somatic cells. The organization of the microtubular cytoskeleton has undoubted importance in the processes of both zygotic and somatic embryo development [223, 224]. In a recent work Webb and Gunning [225], it was found that in fertilized *Arabidopsis thaliana* embryos the microtubules first followed random orientation in the cytoplasm of the zygote, the later distinct cortical microtubules bands were formed with transverse orientation. Furthermore the authors have detected a putative preprophase band of microtubules at the time of the first division in the zygote.

Major differences in the cellular structure between embryogenic and callus cells have become apparent from studies on both carrot and alfalfa culture systems. Cells committed to embryogenesis are highly cytoplasmic and reduced in size [224, 226]. As also shown by Fig. 1 the asymmetric pattern of the first division is an important feature that can be an indicator for the initiation of the embryogenic developmental pathway and establishment of polarity. Nuclei in alfalfa cells were situated at the cell periphery and showed irregular shape under electrical stimulation of direct embryo formation [224]. The asymmetric cell division can be the consequence of nuclear migration from the central region of the cells to the periphery.

Katsuta and Shibaoka [227] have proposed a hypothesis about the involvement of microtubules in premitotic nuclear positioning and the role of actin filaments in displacement of the nucleus in interphase cells.

As summarized by Dijak and Simmonds [224] the altered microtubule morphology, the asymmetric first division and rapid cell division with reduced cell expansion are the characteristic features of embryogenic cells. In the early stages of embryo development from alfalfa protoplasts the microtubule strands were found to be arranged in a disordered network. In carrot cultures with proembryogenic clusters the number of cortical microtubules is increased [226]. In these experiments the tubulin-protein levels either decreased or remained unchanged during early embryogenesis. Therefore the authors have concluded that the increased density of cortical microtubules in the early embryos originated from a change in the stability of the microtubules or an increase in the rate of microtubule assembly. The rearrangement of cellular structures depending on cytoskeleton can be mediated by various kinases as discussed earlier [155, 156, 183]. It might be relevant to mention that auxin has an influence on the orientation of cortical microtubules [228].

Asymmetry in the first cell division is to be considered as an initation of further polar growth that is required for embryo development. Separation of root and shoot poles can be followed by morphological changes during the subsequent stages of embryogenesis. Racusen and Sciavone [6] suggest a correlation between spatial distribution of defined gene products and changes in morphology in somatic embryos.

VII. Perspectives

Recent advances in molecular techniques have created completely new opportunities for the studies on some extraordinary characteristics of plant development. There are only few plant-specific cellular functions that can be of central interest of developmental molecular biology in general. Undoubtedly, the plastic developmental program or the flexibility of "the differentiated cell stage" in plants represents such an example of an unique biological phenomena that can be studied, with the hope of highlighting unknown mechanisms in the molecular control of cell differentiation and development in higher eukaryotes. Especially, the molecular switch in somatic cells to become embryogenic is considered as a remarkable capability that makes possible the reinitiation of a whole ontogenic cycle without sexual fertilization. In the present overview we have described a general concept that is based on the key role of the activation of a signaling system and the induction of cell division in the generation of the embryogenic stage, via the influences of various stresses, applied preferentially externally by hormones. We have pointed out several similarities in various steps of signal transduction between the fertilized egg and auxin-treated somatic cells. The embryogenic induction cannot be understood without knowing the regulatory elements in cell cycle control. The hormone-induced cell divisions open the way for reprogramming the gene expression and initiation of a coordinated cascade of synthesis of new embryogenic transcripts and subsequently, the cellular functions that are similar to those in zygotic embryos. The success in better understanding of somatic embryogenesis is highly dependent on the tissue culture systems used for molecular studies. It is expected that in addition to carrot and alfalfa, *Arabidopsis* or different mono-cotyledonous cereal species, e.g. corn will be used in research on somatic embryogenesis. In the future, the significance of developmental mutants will increase.

Analysis of mutations in genes coding for auxin receptors, signal proteins, cell cycle considerably control, and cytoskeleton functions could contribute considerably in revealing the key regulatory elements. The interpretation of information gained by cloning and structural characterization of genes involved in the control of hormone action, cell division, and cell differentiation will require functional studies as well. The analysis of transgenic plants will essentially extend our knowledge about the components responsible for coordination of gene activation that are required to control the complex changes during the reprogramming of the somatic cells.

There is also the need to find the link between altered gene expression and reorganization of cellular structure. Finally, we have to emphasize the fact that the experimental data about the molecular basis of somatic embryogenesis are essential not only to describe a special case in developmental biology, but for the improvement of tissue culture methodology as well. Somatic embryogenesis is the most direct way to regenerate plants from single cells that are the objectives of various genetic manipulation techniques such as cell fusion or DNA transformation. Therefore any progress in the research of somatic embryo-genesis can indirectly contribute to the improvement of crop species and establishment of efficient plant propagation technologies, e.g. the use of artificial seed.

298

Acknowledgements

The authors thank Béla Dusha for the excellent photographic work, and Ildikó Zsigó for preparation of the manuscript.

References

1. Sung, Z.R., and Okimoto, R., Embryonic proteins in somatic embryos of carrot, *Proc. Natl. Acad. Sci. USA*, 78, 3683, 1981.
2. Sung, Z.R., and Okimoto, R., Coordinate gene expression during somatic embryogenesis in carrots, *Proc. Natl. Acad. Sci. USA*, 80, 2661, 1983.
3. Terzi, M., Pitto, L., and Sung, Z.R., Eds., *Proc. Workshop on Somatic Embryo Genesis Carrots*, San Miniato, IPRA, 1985.
4. Borkird, C., Choi, J.H., Jin, Z.H., Franz, G., Hatzopoulos, P., Chorneau, R., Bonas, U., Pelegri, F., and Sung, Z.R., Developmental regulation of embryogenic genes in plants, *Proc. Natl. Acad. Sci. USA*, 85, 6399, 1988.
5. Komamine, A., Matsumoto, M., Tsukuhara, M., Fujiwara, A. Kawahara, R., Ito, M., Smith, J., Nomura, K., and Fujimura, T., Mechanisms of somatic embryogenesis in cell cultures-physiology, biochemistry and molecular biology, in *Progress in Plant Cellular and Molecular Biology*, Nijkamp, H.J.J., Van der Plas, L.H.W., and Van Aartrijk, J., Eds., Kluwer Academic Publishers, Dordrecht, 1990, 307.
6. Racusen, R.H., and Schiavone, F.M., Positional cues and differential gene expression in somatic embryos of higher plants, *Cell Differ. Develop.*, 30, 159, 1990.
7. Dudits, D., Bögre, L., and Györgyey, J., Molecular and cellular approaches to the analysis of plant embryo development from somatic cells in vitro, *J. Cell Science*, 99, 475, 1991.
8. Abdullah, R., Cocking, E.C., and Thompson, J.A., Efficient plant regeneration from rice protoplasts through somatic embryogenesis, *Bio/Technol.*, 4, 1087, 1986.
9. Vasil, V., Redway, F., and Vasil, I.K., Regeneration of plants from embryogenic suspension culture protoplasts of wheat (*Triticum Aestivum L.*), *Bio/Technol.*, 8, 429, 1990.
10. Mórocz, S., Donn, G., and Dudits, D., An improved system to obtain fertile regenerants via maize protoplasts isolated from a highly embryogenic suspension culture, *Theor. Appl. Genet.*, 80, 721, 1990.
11. Smith, H., Signal perception, differential expression within multigene families and the molecular basis of phenotypic plasticity, *Plant Cell Environ.*, 13, 585, 1990.
12. Brummell, D.A., and Hall, J.I., Rapid cellular responses to auxin and the regulation of growth, *Plant Cell Environ.*, 10, 523, 1987.
13. Davies, E., Action potentials as multifunctional signals in plants: a unifying hypothesis to explain apparently disparate wound responses, *Plant Cell Environ.*, 10, 623, 1987.
14. Kao, K.N., and Michayluk, M.R., Plant regeneration from mesophyll protoplasts of alfalfa, *Z. Pflanzenphysiol*, 96, 135, 1980.
15. Song, J., Sorensen, E.L., and Liang, G.H., Direct embryogenesis from single mesophyll protoplasts in alfalfa (*Medicago sativa. L*), *Plant Cell Rep.*, 9, 21, 1990.
16. Dijak, M., Smith, D.L., Wilson, T.J., and Brown, D.C.W., Stimulation of direct embryogenesis from mesophyll protoplasts of *Medicago sativa*, *Plant Cell Rep.*, 5, 468, 1986.
17. Ryan, C.A., Oligosaccharide signaling in plants, *Annu. Rev. Cell Biol.*, 3, 295, 1987.
18. Shirras, A.D., and Northcote, D.H., Molecular cloning and characterization of cDNAs complementary to mRNAs from wounded potato (*Solanum tuberosum*) tuber tissue, *Planta*, 162, 353, 1984.
19. Graham, J.S., Hall, G., Pearce, G., and Ryan, C.A., Regulation of synthesis of proteinase inhibitors I and II mRNAs in leaves of wounded tomato plants, *Planta*, 169, 399, 1986.

20. Corbin, D.R., Sauer, N., and Lamb, C.J., Differential regulation of a hydroxyproline-rich glycoprotein gene family in wounded and infected plants, *Mol. Cell. Biol.*, 7, 4337, 1987.

21. Logemann, J. Mayer, J.E., Schell, J., and Willmitzer, L., Differential expression of genes in potato tubers after wounding, *Proc. Natl. Acad. Sci. USA*, 85, 1136, 1988.

22. Hedrick, S.A., Bell, J.N., Boller, T., and Lamb, C.J., Chitinase cDNA cloning and mRNA induction by fungal elicitor, wounding, and infection, *Plant Physiol.*, 86, 182, 1988.

23. Stanford, A., Bevan, M., and Northcote, D., Differential expression within a family of novel wound-induced genes in potato, *Mol. Gen. Genet.*, 215, 200, 1989.

24. Yu, Y.B., and Yang, S.F., Auxin-induced ethylene production and its inhibition by aminoethoxylglycine and cobalt ion, *Plant Physiol.*, 64, 1074, 1979.

25. Roustan, J.P., Latche, A., and Fallot, J., Stimulation of *Daucus carota* somatic embryogenesis by inhibitors of ethylene synthesis: cobalt and nickel, *Plant Cell Rep.*, 8, 182, 1989.

26. Roustan, J.P., Latche, A., and Fallot, J., Control of carrot somatic embryogenesis by AgNO₃, an inhibitor of ethylene action: effect on arganine decarboxylase activity, *Plant Sci.*, 67, 89, 1990.

27. Purnhauser, L., Medgyesy, P., Czakó, M., Dix, P.J., and Márton, L., Stimulation of shoot regeneration in *Triticum aestivum* and *Nicotiana plumbaginifolia viv.* tissue cultures using the ethylene inhibitor AgNO₃, *Plant Cell Rep.*, 6, 1, 1987.

28. Galiba, G., and Yamada, Y., A novel method for increasing the frequency of somatic embryogenesis in wheat tissue culture by NaCl and KCl supplementation, *Plant Cell Rep.*, 7, 55, 1988.

29. Nomura, K., and Komamine, A., Embryogenesis from microinjected single cells in a carrot cell suspension culture, *Plant Sci.*, 44, 53, 1986.

30. Binh, D.Q., and Heszky, L.E., Restoration of the regeneration potential of long-term cell culture in rice (*Orysa sativa* L.) by salt pretreatment, *J. Plant Physiol.*, 136, 336, 1990.

31. Pitto, L., LoSchiavo, G., Giuliano, G., and Terzi, M., Analysis of the heat-shock protein pattern during somatic embryogenesis of carrot, *Plant Mol. Biol.*, 2, 231, 1988.

32. Zimmerman, J.L., Apuya, N., Darwish, K., and O'Carroll, C., Novel regulation of heat shock genes during carrot somatic embryo development, *Plant Cell*, 1, 1137, 1989.

33. Terzi, M., and Lo Schiavo, F., Developmental mutants in carrot, in *Progress in Plant Cellular and Molecular Biology*, Nijkamp, H.J.J., Van der Plas, L.H.W., and Van Aartrijk, J., Eds., Kluwer Academic Publishers, Dordrecht, 1990, 391.

34. Czarnecka, E., Edelman, L., Schöffl, F., and Key, J.L., Comparative analysis of physical stress responses in soybean seedlings using cloned heat shock cDNAs, *Plant Mol. Biol.*, 3, 45, 1984.

35. McClure, B.A., Hagen, G., Brown, C.S., Gee, M.A., and Guilfoyle, T.J., Transcription, organization sequence of an auxin-regulated gene cluster in soybean, *Plant Cell*, 1, 229, 1989.

36. Bond, U., and Schlesinger, M.J., Heat-shock proteins and development, *Adv. Genet.*, 24, 1, 1987.

37. Zimmerman, J.L., Petri, W., and Meselson, M., Accumulation of a specific subset of *D. melanogaster* heat shock mRNAs in normal development without heat shock, *Cell*, 32, 1161, 1983.

38. Bensaude, O., and Morange, M., Spontaneous high expression of heat shock proteins in mouse embryonal carcinoma cells and ectoderm from day 8 mouse embryo, *EMBO J.*, 2, 173, 1983.

39. Györgyey, J., Gartner, A., Németh, K., Magyar, Z., Hirt, H., Heberle-bors, E., and Dudits, D., Alfalfa heat-shock genes are differentially expressed during somatic embryogenesis, *Plant Mol. Biol.*, 16, 999, 1991.

40. Ellis, R.J., Molecular chaperones: the plant connection, *Science*, 250, 954, 1990.

41. Ingolia, T.D., and Craig, E.A., Drosophila gene related to the major heat-shock-induced gene is transcripted at normal temperatures and not induced by heat shock, *Proc. Natl. Acad. Sci. USA*, 79, 525, 1982.

42. Nuti-Ronchi, N., Bennici, A., and Martini, G., Nuclear fragmentation in dedifferentiating cells of *Nicotiana glauca* pith tissue grown *in vitro*, *Cell Differentiation*, 2, 77, 1973.

43. Parenti, R., Guille, E., Grisvard, J., Durante, M., Giorgi, L., and Buiatti, M., Transient DNA satellite in dedifferentiating pith tissue, *Nature New Biol.*, 246, 237, 1973.

44. Hase, Y., Yakura, K., and Tanifuji, J., Differential replication of satellite and main band DNA during early stages of callus formations in carrot root tissue, *Plant Cell Physiol.*, 20, 1461, 1979.
45. Durante, M., Geri, C., Grisvard, J. Guille, E., Parenti, R., and Buiatti, M., Variation in DNA complexity in *Nicotiana galuca* tissue cultures, *Protoplasma*, 114, 114, 1983.
46. Natali, L., Cavallini, A., Cremonini, R. Bassi, P, and Cionini, P.G., Amplification of nuclear DNA sequences during induced plant cell dedifferentiation, *Cell Differentiation*, 18, 157, 1986.
47. Escandon, A.S., Hopp, H.E., and Hahne, G., Differential amplification of five selected genes in callus cultures of two shrubby *Oxalis* species, *Plant Sci.*, 63, 177, 1989.
48. Nuti-Ronchi, V.N., Giorgetti, L., and Tonelli, M.G., The commitment to embryogenesis, a cytological approach, in *Progress in Plant Cellular and Molecular Biology*, Nijkamp, H.J.J., Van der Plas, L.H.W., and Van Aartrijk, J., Eds., Kluwer Academic Publishers, Dordrecht, 1990, 437.
49. Bassi, P., Cionini, P.G., Cremonini, R., and Seghizzi, P., Underrepresentation of nuclear DNA sequences in differentiating root cells of *Vicia faba*, *Protoplasma*, 123, 70, 1984.
50. Murray, L., and Christianson, M., Phylogenetic comparison of large DNA contents of differentiated cells in the roots of *Equisetum*, *Tradescantia* and *Hordeum*, *Amer. J. Bot.*, 74, 1779, 1987.
51. Altamura, M.M., Bassi, P., Cavallini, A., Cionini, G., Cremonini, R., Monacelli, B., Pasqua, G., Sassoli, O., Thanh Van, K.T., and Cionini, P.G., Nuclear DNA changes during plant development and the morphogenetic response *in vitro* of *Nicotiana tabacum* tissues, *Plant Sci.*, 53, 73, 1987.
52. Jones, P.A., and Taylor, S.M., Cellular differentiation, cytidine analogues and DNA methylation, *Cell*, 29, 85, 1980.
53. Vanyushin, B.F., and Kirnos, M.D., DNA methylation in plants, *Gene*, 74, 117, 1988.
54. Cedar, H., DNA methylation and gene activity, *Cell*, 53, 3, 1988.
55. Klaas, M, and Amasino, R.M., DNA methylation is reduced in DNaseI-sensitive regions of plant chromatin, *Plant Physiol.*, 91, 451, 1989.
56. Bashkite, E.A., Kirnos, M.D., Kiryanov, G.I., Aleksandrushkina, N.I., and Vanyushin, B.F., Replication and methylation of DNA in cells of tobacco suspension culture and the effect of auxin, *Biokhimiya*, 45, 1448, 1980.
57. Kirnos, M.D., Artyukhovskaya, N.A., Aleksandrushkina, N.I., Ashapkin, V.V., and Vanyushin, B.F., Effect of phytohormones on replicative and postreplicative methylation of nuclear DNA in the S phase of the cell cycle of cells of the first leaf of etiolated wheat seedlings, *Biokhimiya*, 51, 1875, 1986.
58. Morrish, F., and Vasil, I.K., DNA methylation and embryogenic competence in leaves and callus of Napiergrass (*Pennisetum purpureum* schum.), *Plant Physiol.*, 90, 37, 1989.
59. LoSchiavo, F., Pitto, L., Giuliano, G., Torti, G., Nuti-Ronchi, V., Marazziti, D., Vergara, R., Orselli, S., and Terzi, M., DNA methylation of embryogenic carrot cell cultures and its variations as caused by mutation, differentiation, hormones and hypomethylating drugs, *Theor. Appl. Genet.*, 77, 325, 1989.
60. Nick, H., Bowen, B., Ferl, R.J., and Gilbert, W., Detection of cytosine methylation in the maize alcohol dehydrogenase gene by genomic sequencing, *Nature*, 319, 243, 1986.
61. Epel, D., The initiation of development at fertilization, *Cell Differ. Develop.*, 29, 1, 1990.
62. Palme, K., Hesse, T., Moore, I., Campos, N., Feldwisch, J., Garbers, C., Hesse, F., and Schell, J., Hormonal modulation of plant growth: the role of auxin perception, *Mechanisms Develop.*, 33, 97, 1991.
63. Dohrmann, U., Hertel, R., and Kowalik, H., Properties of auxin binding sites in different subcellular fractions from maize coleoptiles, *Planta*, 140, 97, 1978.
64. Löbler, M., and Klambt, D., Auxin-binding protein from coleoptile membranes of corn (*Zea mays* L.) I. Purification by immunological methods and characterization, *J. Biol. Chem.*, 260, 9848, 1985.
65. Shimomura, S., Sotobayashi, T., Futai, M., and Fukui, T., Purification and properties of an auxin-binding protein from maize shoot membranes, *J. Biochem.*, 99, 1513, 1986.

66. Napier, R.M., Venis, M.A., Bolton, M.A., Richardson, L.I., and Butcher, G.W., Preparation and characterisation of monoclonal and polyclonal antibodies to maize membrane auxin-binding protein, *Planta*, 176, 519, 1988.

67. Hesse, T., Feldwisch, J., Balshüsemann, D., Bauw, G., Puype, M., Vandekerckhove, J., Löbler, M., Klambt, D., Schell, J., and Palme, K., Molecular cloning and structural analysis of a gene from *Zea mays* (L.) coding for a putative receptor for the plant hormone auxin, *EMBO J.*, 8, 2453, 1989.

68. Venis, M.A., Auxin-binding proteins in maize: purification and receptor function. in *Molecular Biology of Plant Growth Control*, Fox, E.J., Jacobs, M., Eds., Alan R. Liss, New York, 1987, 219.

69. Klämbt, D., A view about the function of auxin-binding proteins at plasma membranes, *Plant Mol. Biol.*, 14, 1045, 1990.

70. Napier, R.M., and Venis, A., Monoclonal antibodies detect an auxin-induced conformational change in the maize auxin-binding protein, *Planta*, 182, 313, 1990.

71. Barbier-Brygoo, H., Ephritikhine, G., Klambt, D., Ghislain, M., and Guern, J., Functional evidence for an auxin receptor at the plasmalemma of tobacco mesophyll protoplasts, *Proc. Natl. Acad. Sci. USA*, 86, 891, 1989.

72. Inohara, N., Shimomura, S., Fukui, T., and Futai, M., Auxin-binding protein located in the endoplasmic reticulum of maize shoots: Molecular cloning and complete primary structure, *Proc. Natl. Acad. Sci. USA*, 86, 3564, 1989.

73. Theologis, A., Rapid gene regulation by auxin, *Ann. Rev. Plant Physiol.*, 37, 407, 1986.

74. Sanders, D., Hansen, U.P., and Slayman, C.L. Role of the plasma membrane proton pump in pH regulation in non-animal cells. *Proc. Natl. Acad. Sci. USA*, 78, 5903, 1981

75. Cleland, R.E., Prins, H.B.A., Harper, J.R., and Higinbotham, N., Rapid hormone-induced hyperpolarisation of the oat coleoptile transmembrane potential, *Plant Physiol.*, 59, 395, 1977.

76. Bates, G.W., and Goldsmith, M.H.M., Rapid response of the plasma-membrane potential in oat coleoptiles to auxin and other weak acids, *Planta*, 159, 231, 1983.

77. Bögre, L., Stefanov, I., Ábrahám, M., Somogyi, I., and Dudits, D., Differences in responses to 2,4-D-dichlorophenoxy acetic acid (2,4-D) treatment between embryogenic and non-embryogenic lines of alfalfa, in *Progress in Plant Cellular and Molecular Biology*, Nijkamp, H.J.J., Van der Plas, L.H.W., and Van Aartrijk, J., Eds., Kluwer Academic Publishers, Dordrecht, 1990, 427.

78. Shen, W.H., Petit, A., Guern, J., and Tempe, J., Hairy roots are more sensitive to auxin than normal roots, *Proc. Natl. Acad. Sci. USA*, 85, 3417, 1988.

79. Berridge, M.J., Inositol trisphosphate and diacylglycerol: two interacting second messengers, *Ann. Rev. Biochem.*, 56, 159, 1987.

80. Turner, P.R., Sheetz, M.P., and Jaffe, L.A., Fertilization increases the polyphosphoinositide content of sea urchin eggs, *Nature*, 310, 414, 1984.

81. Ciapa, B., and Whitaker, M., Two phases of inositol polyphosphate and diacylglycerol production at fertilisation, *FEBS Lett.*, 195, 347, 1986.

82. Swann, K., and Whitaker, M., The part played by inositol trisphosphate and calcium in the propagation of the fertilization wave in sea urchin eggs, *J. Cell Biol.*, 103, 2333, 1986.

83. Turner, P.R., Jaffe, L.A., and Fein, A., Regulation of cortical vesicle exocytosis in sea urchin eggs by inositol 1,4,5-trisphosphate and GTP-binding protein, *J. Cell Biol.*, 102, 70, 1986.

84. Miyazaki, S., Inositol 1,4,5-trisphosphate-induced calcium release and guanine nucleotide-binding protein-mediated periodic calcium rises in golden hamster eggs, *J. Cell Biol.*, 106, 345, 1988.

85. Jaken, S., Protein kinase C and tumor promoters, *Curr. Opinion Cell Biol.*, 2, 192, 1990.

86. Poovaiah, B.W., Reddy, A.S.N., and McFadden, J.J., Calcium messenger system; role of protein phosphorylation and inositol bisphospholipids, *Plant Physiol.*, 69, 569, 1987.

87. Morse, M.J., Satter, R.L., Crain, R.C., and Coté, G.G., Signal transduction and phosphatidylinositol turnover in plants, *Physiol. Plant.*, 76, 118, 1989.

88. Guern, J. Ephritikhine, G., Imhoff, V., and Pradier, J.M., Signal transduction at the membrane level of plant cells, in *Progress in Plant Cellular and Molecular Biology*, Nijkamp, H.J.J., Van der Plas, L.H.W., and Van Aartrijk, J., Eds., Kluwer Academic Publishers, Dordrecht, 1990, 466.

89. Palme, K., Molecular analysis of plant signaling elements: the relevance of eucaryotic signal transduction models. *Int. Rev. Cytol.*, 132, 223. 1992.

90. Morré, D.J., Gripshover, B., Monroe, A., and Morré, J.T., Phosphatidylinositol turnover in isolated soybean membranes stimulated by the synthetic growth hormone 2,4-Dichlorophenoxyacetic acid, *J. Biol. Chem.*, 259, 15364, 1984.

91. Ettlinger, C., and Lehle, L., Auxin induces rapid changes in phosphatidylinositol metabolites, *Nature*, 331, 176, 1988.

92. Heim, S., and Wagner, K.G., Inositol phosphates in the growth cycle of suspension cultured plant cells, *Plant Sci.*, 63, 159, 1989.

93. Schumaker, K.S., and Sze, H., Inositol 1,4,5-trisphosphate releases Ca^{2+} from vacuolar membrane vesicles of oat roots, *J. Biol. Chem.*, 262, 3944, 1987.

94. Ranjeva, R., Carrasco, A., and Boudet, A.M., Inositol trisphosphate stimulates the release of calcium from intact vacuoles isolated from *Acer* cells, *FEBS Lett.*, 230, 137, 1988.

95. Drobak, B.K., and Ferguson, I.B., Release of Ca^{2+} from plant hypocotyl microsomes by inositol-1,4,5-trisphosphate, *Biochem. Biophys. Res. Commun.*, 130, 1241, 1985.

96. Hasunama, K., and Funadera, K., GTP-binding protein(s) in green plant, *Lenina paucicostata*, *Biochem. Biophys. Res. Commun.*, 143, 908, 1987.

97. Drobak, B.K., Allan, E.F., Comerford, J.G., Roberts, K., and Dawson, A.P., Presence of guanine nucleotide-binding proteins in a plant flypocotyl microsomal fraction, *Biochem. Biophys. Res. Commun.*, 150, 899, 1988.

98. Blum, W., Hinsch, K.D., Schultz, G., and Weiler, E.W., Identification of GTP proteins in the plasma membrane of higher plants, *Biochem. Biophys. Res. Commun.*, 156, 954, 1988.

99. Matsui, M., Sasamoto, S., Kuneida, T., Nomura, N., and Ishizaki, R., Cloning of *ara*, a putative *Arabidopsis thaliana* gene homologous to the *ras*-related family, *Gene*, 76, 331, 1989.

100. Anuntalabhochai, S., Terryn, N., Van Montagu, M., and Inze, D., Molecular characterization of an *Arabidopsis thaliana* cDNA encoding a small GTP-binding protein, Rha-1, *Plant J.*, 1, 167, 1991.

101. Palme, K., Diefenthal, T., Vingron, M., Sander, C., and Schell, J., Molecular cloning and structural analysis of genes from *Zea mays* (L.) coding for members of the ras-related ypt gene family. *Proc. Natl. Acad. Sci. USA*, 89, 787, 1992.

102. Dallmann, G., Sticher, L., Marshallsay, C., and Nagy, F., Molecular characterization of tobacco cDNAs encoding two small GTP-binding proteins, *Plant Mol. Biol.*, 19, 847, 1992.

103. Lehle, L., Phosphatidyl inositol metablosim and its role in signal transduction in growing plants, *Plant Mol. Biol.*, 15, 647, 1990.

104. Scherer, G.F.E., André, B., and Martiny-Baron, G., Hormone-activated phospholipase A2 and lysophospholipid-activated protein kinase: a new signal transduction chain and a new second messenger system in plant? *Curr. Topics Plant Biochem. Physiol.*, 9, 190, 1990.

105. André, B., and Scherer, G.F.E., Stimulation by auxin of phospholipase A in membrane vesicles from an auxin-sensitive tissue is mediated by an auxin receptor, *Planta*, 185, 209, 1991.

106. Elliot, D.C., Calcium involvement in plant hormone action, in *Molecular and Cellular Aspects of Calcium in Plant Development*, Trewavas, A.J., Ed., Plenum Press, New York, 1986, 285.

107. Poovaiah, B.W., Molecular and cellular aspects of calcium action in plants, *Hortsci.*, 23, 267, 1988.

108. Williamson, R.E., and Ashley, C.C., Free Ca^{2+} and cytoplasmic streaming in the alga *Chara*, *Nature*, 296, 647, 1982.

109. Roberts, D.M., Lukas, T.J., Harrington, H.M., and Watterson, D.M., Molecular mechanisms of calmodulin action, in *Molecular and Cellular Aspects of Calcium in Plant Development*, Trewavas, A.J., Ed., Plenum Press, New York, 1986, 11.

110. Poovaiah, B.W., and Veluthambi, K., The role of calcium and calmodulin in hormone action in plants: Importance of protein phosphorylation, in *Molecular and Cellular Aspects of Calcium in Plant Development*, Trewavas, A.J., Ed., Plenum Press, New York, 1986, 83.

111. Jaffe, L.F., The role of calcium explosions, waves and pulses in activating eggs, in *Biology of Fertilization*, Vol.3., Metz, C.B., and Monroy, A., Eds., Academic Press, New York, 1985, 127.

112. Miyazaki, S.I., Hashimoto, N., Yashimoto, Y., Kishimoto, T., Igusa, Y., and Hiramoto, Y., Temporal and spatial dynamics of the periodic increase in intracellular-free calcium at fertilization of golden hamster eggs, *Dev. Biol.*, 118, 259, 1988.

113. Speksnijder, J.A., Corson, D.W., Sardet, C., and Jaffe, L.F., Free calcium pulses following fertilization in the ascidian egg, *Dev. Biol.*, 135, 182, 1989.

114. Jansen, M.A.K., Booij, H., Schel, J.H.N., and De Vries, S.C., Calcium increases the yield of somatic embryos in carrot embryogenic suspension cultures, *Plant Cell Rep.*, 9, 221, 1990.

115. Nomura, K., Mechanisms of somatic embryogenesis in carrot suspension cultures, Ph.D. Thesis, University Tokyo, Japan, 1987.

116. Timmers, A.C.J., De Vries, S.C., and Schel, J.H.N., Distribution of membrane-bound calcium and activated calmodulin during somatic embryogenesis of carrot (*Daucus carota* L.), *Protoplasma*, 153, 24, 1989.

117. Epel, D., The role of Na^+-H^+ exchange and intracellular pH changes in fertilization, in *Na^+-H^+ Exchange*, Grinstein, S., Ed., CRC Press, Boca Raton, FL, 1988, 209.

118. Dube, F.T., Schmidt, C.H., Johnson, C.H., and Epel, D., The hierarchy of requirements for an elevated intracellular pH during early development of sea urchin embryos, *Cell*, 40, 657, 1985.

119. Sanders, D., Hansen, U.P., and Slayman, C.L., Role of the plasma membrane proton pump in pH regulation in non-animal cells, *Proc. Natl. Acad. Sci. USA*, 78, 5903, 1981.

120. Cleland, R.E., and Lomax, T., Hormonal control of H^+-excretion from oat cells, in *Regulation of Cell Membrane Activities in Plants*, Marre, E., and Ciferri, O., Eds., North-Holland Publishing Co., Amsterdam, 1977, 161.

121. Felle, H., Auxin causes oscillations of cytosolic free calcium and pH in *Zea mays* coleoptiles, *Planta*, 174, 495, 1988.

122. Felle, H., Cytoplasmic free calcium in *Riccia fluitans* L. and *Zea mays* L.: interactions of Ca^{2+} and pH? *Planta*, 176, 248, 1988.

123. Schaefer, J., Regeneration in alfalfa tissue culture, *Plant Physiol.*, 79, 584, 1985.

124. Smith, D.L., and Krikorian, A.D., Somatic proembryo production from excised, wounded zygotic carrot embryos on hormone-free medium: evaluation of the effects of pH, ethylene and activated charcoal, *Plant Cell Rep.*, 9, 34, 1990.

125. Smith, D.L., and Krikorian A.D., pH control of carrot somatic embryogenesis, in *Progress in Plant Cellular and Molecular Biology*, Nijkamp, H.J.J., Van der Plas, L.H.W., and Van Aartrijk, J., Eds., Kluwer Academic Publishers, Dordrecht, 1990, 449.

126. Murray, A.W., and Kirschner, M.W., Cyclin synthesis drives the early embryonic cell cycle, *Nature*, 339, 275, 1989.

127. Key, J.L., Hormones and nucleic acid metabolism, *Ann. Rev. Plant Physiol.*, 20, 449, 1969.

128. Hagen, G., and Guilfoyle, T.J., Rapid induction of selective transcription by auxins, *Mol. Cell. Biol.*, 5, 1197, 1985.

129. Bakó, L., Bögre, L., and Dudits, D., Protein phosphorylation in partially synchronized cell suspension culture of alfalfa, in *NATO Adv. Studies on Cellular Regulation by Protein Phosphorylation*, Heilmayer, L., Ed., Springer-Verlag, Berlin, 1991, H56, 435.

130. Yeoman, M.M., Early development in callus cultures, *Int. Rev. Cytol.*, 29, 383, 1970.

131. Yasuda, T., Yajima, Y., and Yamada, Y., Induction of DNA synthesis and callus formation from tuber tissue of Jerusalem artichoke by 2,4-dichlorophenoxyacetic acid, *Plant Cell Physiol.*, 15, 321, 1974.

132. Yeoman, M.M, and Evans, P.K., Growth and differentiation in plant tissue cultures. II. Synchronous cell division in developing callus cultures, *Ann. Bot.*, 31, 323, 1967.

133. Watanabe, A., and Imaseki, H., Induction of deoxyribonucleic acid synthesis in potato tuber tissue by cutting, *Plant Physiol.*, 51, 772, 1973.

134. Wernicke, W., Rös, M., and Jung, G., Microtubules and the first cell cycle in cultures mesophyll protoplasts of *Nicotiana*, in *Progress in Plant Cellular and Molecular Biology*, Nijkamp, H.J.J., Van der Plas, L.H.W., and Van Aartrijk, J., Eds., Kluwer Academic Publishers, Dordrecht, 1990, 538.

135. Matzk, F., A novel approach to differentiated embryos in the absence of endosperm, *Sex. Plant. Reprod.*, 4, 88, 1991.

136. Britten, E.J., Natural and induced parthenocarpy in maize and its relation to hormone production by the developing seed, *Amer. J. Bot.*, 37, 345, 1950.

137. Marshall, D.R., Molnar-Lang, M., and Ellison, F.W., Effects of 2,4-D on parthenocarpy and cross-compatibility in wheat, *Cereal Res. Commun.*, 11, 213, 1983.

138. Raghavan, V., *Experimental Embryogenesis in Angiosperms*, Cambridge, Cambridge University Press, 1986.

139. Warren, G.S., and Fowler, M.W., Cell number and cell doubling times during the development of carrot embryoids in suspension culture, *Experientia*, 34, 356, 1978.

140. Fujimura, T., and Komamine, A., The serial observation of embryogenesis in a carrot cell suspension culture, *New Phytol.*, 86, 213, 1980.

141. Nishi, A., Kato,. K., Takahashi, M., and Yoskida, R., Partial synchronization of carrot cell cultures by auxin deprivation, *Physiol. Plant.*, 39, 9, 1977.

142. Wernicke, W., and Milkovits, L., Effect of auxin on the mitotic cell cycle in cultured leaf segments at different stages of developmen in wheat, *Physiol. Plant.*, 69, 16, 1987.

143. Ahuja, P.S., Pental, D., and Cocking, E.C., Plant regeneration from leaf base callus and cell suspensions of *Triticum aestivum*, *Z. Pflanzenzüchtg.*, 81, 139, 1982.

144. Conger, B.V., Novak, F.J., Afza, R., and Erdelsky, K., Somatic embryogenesis from cultured leaf segments of *Zea mays*, *Plant Cell Rep.*, 6, 345, 1987.

145. Barcelo, P., Lazzeri, P.A., Martin, A., and Lörz, H., Competence of cereal leaf cells. I. Patterns of proliferation and regeneration capability *in vitro* of the inflorescence sheath leaves of barley, wheat and tritordeum, *Plant Sci.*, 77, 243, 1991.

146. Meyer, Y., and Chartier, Y., Hormonal control of mitotic development in tobacco protoplasts, *Plant Physiol.*, 68, 1273, 1981.

147. Hunt, T., Under arrest in the cell cycle, *Nature*, 342, 483, 1989.

148. Witters, L.A., Protein phosphorylation and dephosphorylation, *Curr. Opinion Cell Biol.*, 2, 212, 1990.

149. Draetta, G., Cell cycle control in eukaryotes: molecular mechanisms of cdc2 activation, *Trends Biochem. Sci.*, 15, 378, 1990.

150. Solomon, M.J., Glotzer, M., Lee, T.H., Philippe, M., and Kirschner, M.W., Cyclin activation of p34^{cdc2}, *Cell*, 63, 1013, 1990.

151. Johnston, L.H., and Lowndes, N.F., Cell cycle control of DNA synthesis in budding yeast. *Nucl. Acids Res.*, 20, 2403, 1992.

152. Jacobs, T., Control of the cell cycle, *Develop. Biol.*, 153, 1, 1992.

153. Broek, D., Bartlett, R., Crawford, K., and Nurse, P., Involvement of p34^{cdc2} in establishing the dependency of S phase on mitosis, *Nature*, 349, 388, 1991.

154. Moreno, S., and Nurse, P., Substrates for p34^{cdc2} *in vivo* veritas? *Cell*, 61, 549, 1990.

155. Lamb, N.J.C., Fernandez, A., Watrin, A., Labbé, J.C., and Cavadore, J.C., Microinjection of p34^{cdc2} kinase induces marked changes in cell shape, cytoskeletal organization and chromatin structure in mammalian fibroblasts, *Cell*, 60, 151, 1990.

156. Verde, F., Labbé, J.C., Dorée, M., and Karsenti, E., Regulation of microtubule dynamic by cdc2 protein kinase in cell-free extracts of Xenopus eggs, *Nature*, 343, 233, 1990.

157. Cisek, L.J., and Corden, J.L., Phosphorylation of RNA polymerase by the murine homologue of the cell-cycle control protein cdc2, *Nature*, 339, 679, 1989.

158. Minshull, J., Pines, J., Golsteyn, R., Standart, N., Mackie, S., Colman, A., Blow, J., Ruderman, J.V., Wu, M., and Hunt, T., The role of cyclin synthesis, modification and destruction in the control of cell division, *J. Cell. Sci. Suppl.*, 12, 77, 1989.

159. Reed, S.I., G$_1$-specific cyclins: in search of an S-phase-promoting factor, *Trends Genet.*, 7, 95, 1991.

160. Dudits, D., Bögre, L., Bakó, L., Dedeoglu, D., Magyar, Z., Kapros, T., Felföldi, F., and Györgyey, J. Key components of cell cycle control during auxin-induced cell division, in *Molecular and Cell Biology of the Plant Cell Cycle*, Ormrod, J.C., and Francis, D., Eds., Kluwer Academic Publishers, Dordrecht, 1993, 111.

161. Lee, M.G., and Nurse, P., Complementation used to clone a homologue of the fission yeast cell cycle control gene cdc2, *Nature*, 327, 1, 1987.

162. John, P.C.L., Sek, F.J., and Lee, M.G., A homolog of the cell cycle control protein p34^{cdc2} participates in the division cycle of *Chlamydomonas*, and a similar protein is detectable in higher plants and remote taxa, *Plant Cell*, 1, 1185, 1989.

163. Feiler, H.S., and Jacobs, T.W., Cell division in higher plants: a cdc2 gene its 34-kDa product, and histone H1 kinase activity in pea, *Proc. Natl. Acad. Sci. USA*, 87, 5397, 1990.

164. Hirt, H., Páy, A., Györgyey, J., Bakó, L., Németh, K., Bögre, L., Schweyen, R.J., Heberle-Bors, E., and Dudits, D., Complementation of a yeast cell cycle mutant by an alfalfa cDNA encoding a protein kinase homologous to p34^{cdc2}, *Proc. Natl. Acad. Sci. USA*, 88, 1636, 1991.

165. Colasanti, J., Tyers, M., and Sundaresan, V., Isolation and characterization of cDNA clones encoding a functional p34^{cdc2} homologue from Zea mays, *Proc. Natl. Acad. Sci. USA*, 88, 3377, 1991.

166. Magyar, Z., Bakó, L., Bögre, L., Dedeoglu, D., Kapros, T., and Dudits, D., Active cdc2 genes and cell cycle phase specific cdc2-related kinase complexes in hormone stimulated alfalfa cells, *The Plant J.* 4, 151, 1993.

167. Brizuela, L., Draetta, G., and Beach, D., p13^{suc1} acts in the fission yeast cell division cycle as a component of the p34^{cdc2} protein kinase, *EMBO J.*, 6, 3507, 1987.

168. Labbé, J.C., Capony, J.P., Caput, D.,Cavadore, J.C., Derancourt, J., Kaghad, M., Lelias, J.M., Picard, A., and Dorée, M., MPF from starfish oocytes at first meiotic metaphase is a heterodimer containing one molecule of cdc2 and one molecule of cyclin B, *EMBO J.*, 8, 3053, 1989.

169. Furukawa, Y., Piwnica-Worms, H., Ernst, T.J., Kanakura, Y., and Griffin, J.D., cdc2 gene expression at the G_1 to S transition in human T lymphocytes, *Science*, 250, 805, 1990.

170. Hata, S., cDNA cloning of a novel cdc2⁺/CDC28-related protein kinase from rice, *FEBS Lett.*, 279, 149, 1991.

171. Hirayama, T., Imajuku, Y., Anai, T., Matsui, M., and Oka, A., Identification of two cell-cycle-controlling cdc2 gene homologs in *Arabidopsis thaliana*, *Gene*, 105, 159, 1991.

172. Ferreira, P.C.G., Hemerly, A.S., Villaroel, R., Montagu, M.C., and Inze, D. The *Arabidopsis* functional homolog of the p34^{cdc2} protein kinase, *Plant Cell*, 3, 531, 1991.

173. Ranjeva, R., and Boudet, A.M., Phosphorylation of proteins in plants: regulatory effects and potential involvement in stimulus/response coupling, *Ann. Rev. Plant Physiol.*, 38, 73, 1987.

174. Blowers, D.P., Boss, W.F., and Trewavas, A.J., Rapid changes in plasma membrane protein phosphorylation during initiation of cell wall digestion, *Plant Physiol.*, 86, 505, 1988.

175. Bennett, J., Steinback, K.E., and Arntzen, C.J., Chloroplast phosphoproteins: regulation of excitation energy transfer by phosphorylation of thylakoid membrane polypeptides, *Proc. Natl. Acad. Sci. USA*, 77, 5253, 1980.

176. Veluthambi, K., and Poovaiah, B.W., Calcium-promoted protein phosphorylation in plants, *Science*, 223, 167, 1984.

177. Bögre, L., Oláh, Z., and Dudits, D., Ca^{2+}-dependent protein kinase from alfalfa (*Medicago varia*): partial purification and autophosphorylation, *Plant Sci.*, 58, 135, 1988.

178. Davis, J.R., and Polya, G.M., Purification and properties of a high specific activity protein kinase from a wheat germ, *Plant Physiol.*, 71, 489, 1983.

179. Murray, M.G., and Key, J.L., 2,4-dichlorophenoxyacetic acid-enhanced phosphorylation of soybean nuclear proteins, *Plant Physiol.*, 61, 190, 1978.

180. Harmon, A.C., Putnam-Evans, C., and Cormier, M.J., A calcium dependent but calmodulin-independent protein kinase from soyabean, *Plant Physiol.*, 83, 830, 1987.

181. Oláh, Z., Bögre, L., Lehel, Cs., Faragó, A., Seprödi, J., and Dudits, D., The phosphorylation site of Ca^{2+}-dependent protein kinase from alfalfa, *Plant Mol. Biol.*, 12, 453, 1989.

182. Romhányi, T., Seprödi, J., Antoni, F., Mészáros, Gy., Buday, L., and Faragó, A., The assay of the activity of protein kinase C with the synthetic oligopeptide substrate designed for histone kinase II, *Biochim. Biophys. Acta*, 888, 325, 1986.

183. Putnam-Evans, C., Harmon, A.C., and Cormier, M.J., Purification and characterization of a novel calcium-dependent protein kinase from soybean, *Biochemistry*, 29, 2488, 1990.

306

184. Goday, A., Sanchez-Martinez, D., Gomez, J., Puigdomenech, P., and Pages, M., Gene expression in developing *Zea mays* embryos: regulation by abscisic acid of a highly phosphorylated 23- to 25-kD group of proteins, *Plant Physiol.*, 88, 564, 1988.

185. Vilardell, J., Goday, A., Freire, M.A., Torrent, M., Martinez, M.C., Torné, J.M., and Pagés, M., Gene sequence, developmental expression, and protein phosphorylation of RAB-17 in maize, *Plant Mol. Biol.*, 14, 423, 1990.

186. Guilfoyle, T.J., A protein kinase from wheat germ that phosphorylates the largest subunit of RNA polymerase II, *Plant Cell*, 1, 827, 1989.

187. Datta, N., and Cashmore, A.R., Binding of a pea nuclear protein to promoters of certain photoregulated genes is modulated by phosphorylation, *Plant Cell*, 1, 1069, 1989.

188. Old, R.W., and Woodland, H.R., Histone genes: not so simple after all, *Cell*, 38, 624, 1984.

189. Schümperli, D., Multilevel regulation of replication-dependent histone genes, *Trends Genet.*, 4, 187, 1988.

190. Gigot, C., Histone genes in higher plants, in *Architecture of Eukaryotic Genes*, Kahl, G., Ed., VCH Verlagsgesellschaft, Weinheim, 1988, 229.

191. Kato, A., Fukuei, K., and Tanifuji, S., Histone synthesis during early stages of germination in *Vicia faba* embryonic axes, *Plant Cell Physiol.*, 23, 967, 1982.

192. Raghavan, V., and Olmedilla, A., Spatial patterns of histone mRNA expression during grain development and germination in rice, *Cell Differ. Develop.*, 27, 183, 1989.

193. Maxson, R., Cohn, R., and Kedes, L., Expression and organization of histone genes, *Ann. Rev. Genet.*, 17, 239, 1983.

194. Waterborg, J.H., Wicinov, I., and Harrington, R.E., Histone variants and acetylated species from the alfalfa plant *Medicago Sativa*, *Arch. Biochem. Biophys.*, 256, 167, 1987.

195. Wu, S.C., Györgyey, J., and Dudits, D., Polyadenylated H3 histone transcripts and H3 histone variants in alfalfa, *Nucl. Acids Res.*, 17, 3057, 1989.

196. Kapros, T., Bögre, L., Németh, K., Bakó, L., Györgyey, J., Wu, S.C, and Dudits, D., Differential expression of histone H3 gene variants during cell cycle and somatic embryogenesis in alfalfa, *Plant Physiol.*, 98, 621, 1992.

197. Kapros, T., Stefanov, I., Magyar, Z., Ocsovszky, I., and Dudits, D., A short histone H3 promoter from alfalfa specifies expression in S-phase cells and meristems, *In Vitro Cell. Dev. Biol.*, 29P, 27, 1993.

198. Waterborg, J.H., Harrington, R.E., and Wicinov, I., Differential histone acetylation in alfalfa (*Medicago Sativa*) due to growth in NaCl, *Plant Physiol.*, 90, 237, 1989.

199. Takahashi, Y., Kuroda, H., Tanaka, T., Machida, Y., Takebe, I., and Nagata, T., Isolation of an auxin-regulated gene cDNA expressed during the transition from G_0 to S phase in tobacco mesphyll protoplasts, *Proc. Natl. Acad. Sci. USA*, 86, 9279, 1989.

200. Ainley, W.M., Walker, J.C., Nagao, R.T., and Key, J.L., Sequence and characterization of two auxin-regulated genes from soybean, *J. Biol. Chem.*, 263, 10658, 1988.

201. Hong, J.C., Nagao, R.T., and Key, J.L., Developmentally regulated expression of soybean proline-rich cell wall protein gene, *Plant Cell*, 1, 937, 1989.

202. Klee, H.J., and Rogers, S.G., Plant gene vectors and genetic transformation: plant transformation systems based on the use of *Agrobacterium tumefaciens*, in *Cell Culture and Somatic Cell Genetics of Plants*, Vol. 6, Vasil, I.K., Ed., Academic Press, Inc., New York, 1989, 1.

203. Paszkowski, J., Saul, M.W., and Potryykus, M.W., Plant gene vectors and genetic transformation: DNA mediated direct gene transfer to plants, in *Cell Culture and Somatic Cell Genetics of Plants*, Vol. 6, Vasil, I.K., Ed., Academic Press, Inc., New York, 1989, 51.

204. Jefferson, R.A., Kavanagh, T.A., and Bevan, M.W., GUS fusions: β-glucuronidase as a sensitive and versatile gene fusion marker in higher plants, *EMBO J.*, 6, 3901, 1987.

205. Koncz, C., Olsson, O., Langridge, W.H.R., Schell, J., and Szalay, A., Expression and assembly of functional bacterial luciferase in plants, *Proc. Natl. Acad. Sci. USA*, 84, 191, 1987.

206. Fujii, N., and Uchimiya, H., Conditions favorable for the somatic embryogenesis in carrot cell culture enhance expression of the *rolC* promoter GUS fusion gene, *Plant Physiol.*, 95, 238, 1991.

207. Schmülling, T., Schell, J., and Spena, A., Promoters of the *rolA,B* and *C* genes of *Agrobacterium rhizogenes* are differentially regulated in transgenic plants, *Plant Cell*, 1, 665, 1989.
208. Odell, J.T., Nagy, F., and Chua, N.H., Identification of sequences required for activity of the cauliflower mosaic virus 35S promoter, *Nature*, 313, 810, 1985.
209. Benfey, P.N., and Chua, N.M., The cauliflower mosaic virus 35S promoter: combinatorial regulation of transcription in plants, *Science*, 250, 959, 1990.
210. Benfey, P.N., and Chua, N.M., Regulated genes in transgenic plants, *Science*, 244, 174, 1989.
211. Nagata, T., Okada, K., Kawazu, T., and Takebe, I., Cauliflower mosaic virus 35S promoter directs S phase specific expression in plant cells, *Mol. Gen. Genet.*, 207, 242, 1987.
212. Halperin, W., Alternative morphogenetic events in cell suspensions, *Amer. J. Bot.*, 53, 443, 1966.
213. Choi, J.H., and Sung, Z.R., Two-dimensional gel analysis of carrot somatic embryogenic proteins, *Plant Mol. Biol. Rep.*, 2, 19, 1984.
214. Wilde, H.D., Nelson, W.S., Booij, H., De Vries, S.C., and Thomas, T.L., Gene expression programs in embryogenic and non-embryogenic carrot cultures, *Planta*, 176, 205, 1988.
215. De Vries, S.C., Booij, H., Meyerink, P., Huisman, G., Wilde, H.D., Thomas, T.L., and Van Kammen, A., Acquisition of embryogenic potential in carrot cell-suspension cultures, *Planta*, 176, 196, 1988.
216. Franz, G., Hatzopoulos, P., Jones, T.J., Kraus, S.M., and Sung, Z.R., Molecular and genetic analysis of an embryogenic gene DC 8, from *Daucus carota* L, *Mol. Gen. Genet.*, 218, 143, 1989.
217. Aleith, F., and Richter, G., Gene expression during induction of somatic embryogenesis in carrot cell suspensions, *Planta*, 183, 17, 1990.
218. Kiyosue, T., Yamaguchi-Shinozaki, K., Shinozaki, K., Higashi, K., Satoh, S., Kamada, H., and Harada, H., Isolation and characterization of a cDNA that encodes ECP31, an embryogenic-cell protein from carrot, *Plant. Mol. Biol.*, 19, 239, 1992.
219. Booij, H., Sterk, P., Schellekens, G.A., Van Kammen, A., and De Vries, S.C., Tissue and cell-specific expression of genes encoding carrot extracellular proteins, in *Progress in Plant Cellular and Molecular Biology*, Nijkamp, H.J.J., Van der Plas, L.H.W., and Van Aartrijk, J., Eds., Kluwer Academic Publishers, Dordrecht, 1990, 398.
220. Quatrano, R.S., Regulation of gene expression by abscisic acid during angiosperm embryo development, *Oxford Surveys Plant Mol. Cell Biol.*, 3, 467, 1986.
221. Ammirato, P.V., Hormonal control of somatic embryo development from cultured cells of caraway, *Plant Physiol.*, 59, 579, 1977.
222. Goldberg, R.B., Barker, S.J., and Perez-Graun, L., Regulation of gene expression during plant embryogenesis, *Cell*, 56, 149, 1989.
223. Van Lammeren, A.A.M., Structure and function of the microtubular cytoskeleton during endosperm development in wheat: an immunofluorescence study, *Protoplasma*, 146, 18, 1988.
224. Dijak, M., and Simmonds, D.H., Microtubule organization during early direct embryogenesis from mesophyll protoplasts of *Medicago Sativa* L., *Plant Sci.*, 58, 183, 1988.
225. Webb, M.C., and Gunning, E.S., The microtubular cytoskeleton during development of the zygote, proembryo and free-nuclear endosperm in *Arabidopsis thaliana* (L.) Heynh, *Planta*, 184, 187, 1991.
226. Cyr, R.J., Bustos, M.M., Guiltinan, M.J., and Fosket, D.E., Developmental modulation of tubulin protein and mRNA levels during somatic embryogenesis in cultured carrot cells, *Planta*, 171, 365, 1987.
227. Katsuta, J., and Shibaoka, H., The roles of the cytoskeleton and the cell wall in nuclear positioning in tobacco BY-2 cells, *Plant Cell Physiol.*, 29, 403, 1988.
228. Bergfeld, R., Speth, V., and Schopfer, P., Reorientation of microfibrils and microtubules at the outer epidermal wall of maize coleoptiles during auxin-mediated growth, *Bot. Acta*, 101, 57, 1988.
229. Sussmann, M.R., and Harper, J.F., Molecular biology of the plasma membrane of higher plants, *Plant Cell*, 1, 953, 1989.
230. Csordás, A., On biological role of histone acetylation, *Biochem. J.*, 265, 23, 1990.

231. Deák, M., Kiss, Gy.B., Koncz, Cs., and Dudits, D., Transformation of *Medicago* by *Agrobacterium* mediated gene transfer, *Plant Cell Rep.*, 5, 97, 1986.
232. Jefferson, R.A., Assaying chimeric genes in plants: the GUS gene fusion system, *Plant Mol. Biol. Rep.*, 5, 387, 1987.

9. Haploid Embryogenesis

A.M.R. FERRIE, C.E. PALMER and W.A. KELLER

Contents

T.A. Thorpe (ed.), In Vitro Embryogenesis in Plants, pp. 309–344.

I. Introduction

This chapter reviews the published literature on the induction of embryogenesis and regeneration of plants from male and female gametophytic cells, i.e. haploid cells, cultured *in vitro* with emphasis placed on advances reported since 1985. Male gametophytic cells are represented by immature pollen grains or microspores and the female gametophytic cells are generally represented by the egg cells although other haploid cells are also present in the embryo sac and in some instances may be induced to undergo embryogenesis. Protocols can involve the use of anthers (containing microspores), isolated microspores, unfertilized ovaries and ovules or flower buds. Unlike microspore culture, it is not technically feasible to isolate and culture egg cells or other haploid cells in the female gametophyte and as a consequence, these are usually cultured within a mass of sporophytic tissues. There are far fewer reports of haploid embryogenesis from cultured female gametophytic cells than from male gametophytic cells. This is related to the fact that microspores are more abundant, developmentally synchronized, uniform in size and easier to manipulate physically.

The *in vitro* haploid embryogenesis developmental pathway is similar to that of *in vitro* somatic embryogenesis (see Chapters 5, 6) and *in vivo* zygotic embryogenesis (see Chapter 3) pathways. A number of studies of early stages of induction of embryogenesis have been undertaken, particularly with the isolated microspore culture system of *Brassica napus* [1–3]. The random distribution of intracellular microtubules has been reported to be characteristic of responsive microspores after inductive treatment [4]. The development of microspore-derived multicellular embryos has been characterized at the morphological as well as biochemical levels. In the case of *Brassica napus* microspore-derived embryos, general morphology as well as storage lipid content are similar to zygotic seed embryos [5].

General reviews of induction and development of haploid embryos have been written by Maheshwari et al. 1982, Dunwell 1985, Heberle-Bors 1985, Keller et al. 1987 [6–9]. The main emphasis of this chapter will be the identification and assessment of factors influencing the induction of haploid embryogenesis in higher plants. The status of plant regeneration from haploid embryos and the general applications will also be summarized.

II. Ontogeny of haploid development

Pollen development follows a precise sequence of events which have been reviewed previously [10]. A brief outline of this developmental process follows. The loss of callose in post meiotic tetrads results in the release of thin-walled, non-vacuolated microspores into the locule of the anther. Microspores increase in size and a vacuole develops. A pore is evident in the exine opposite the nucleus. DNA synthesis takes place followed by mitosis which results in a small generative cell and a large vegetative cell. The vegetative cell assumes the position adjacent to the pore where starch synthesis begins. The generative cell undergoes a second mitosis resulting in a mature pollen grain with two sperm cells. At specific stages during differentiation of microspores their developmental pathway can be altered to sporophytic development leading to the formation of haploid embryos. It is not clear whether all microspores reacquire this developmental competence after commitment to the gametophytic pathway or if a specific embryogenic fraction exist within the microspore population and require only the signals provided in culture for full expression of this capacity. The latter situation appears to apply in some species where distinct embryogenic microspores (i.e. dimorphic pollen) occur. [11, 12] Even though pollen dimorphism occurs, certain culture conditions can alter the developmental pathway of microspores. In *Nicotiana tabacum*, carbohydrate and nitrogen starvation followed by replenishment resulted in a high frequency of pollen embryos, while culture of similar microspores in a nutrient rich medium without prior starvation led to normal pollen differentiation and maturation [13, 14]. While these observations have been reported for *Nicotiana*, there is no evidence that culture conditions are responsible for pollen dimorphism and it is still an open question whether all microspores are competent to undergo embryogenesis.

Embryogenic microspores of tobacco (*Nicotiana tabacum*) have been reported to be smaller than non-embryogenic microspores [15]. The embryogenic cells are characterized by a large vacuole-like compartment, absence of starch grains, reduced size and number of plastids, reduced number of mitochondria which are condensed at later stages, clustering of organelles around the vegetative nucleus in the centre of the pollen grain, absence of polysomes and few ribosomes [15–22]. Subcellular changes occur in the embryogenic microspores. These include occurrence of a fibrillar layer on the outside of the cytoplasm, increased cytoplasmic synthesis, increase in ribosomes, redifferentiation of organelles and the appearance of starch and lipid droplets. The generative cell of the non-embryogenic microspore differs from the generative cell of the embryogenic microspore. The latter is larger in size, has more cytoplasmic membranes and increased number of mitochondria and plastids occur at all stages of development [23, 24].

Microspores from responsive species have similar plastid types. Proplastids were specific for species with embryogenic microspores while amyloplasts were specific for species where microspores did not undergo embryogenesis [25]. A number of studies in tobacco have shown that when starch accumulates in the microspore, embryogenesis does not occur [26–28]. The proplastids of *Datura* (responsive species) developed into amyloplasts then chloroplasts. Starch accumulation takes place after the first haploid mitosis. Therefore the developmental stage of pollen is important in ensuring embryogenesis. In recalcitrant species, e.g. *Antirrhinum majus*, the amyloplasts degenerated or accumulated more starch as the pollen matured [25].

Studies with a highly embryogenic genotype of *Brassica napus* indicated that in culture the uninucleate microspores were primarily embryogenic [29]. When cultured *in vitro* those microspores divided symmetrically instead of the normal asymmetric division leading to a generative and a vegetative cell which occurs *in vivo*. This equal cell division is probably an important determinant of embryogenesis [2, 3, 29].

The arrangement of microtubules and cytoskeleton seems to play a role in symmetric cell division and subsequent embryogenesis [3, 4, 30]. Inductive culture conditions such as high temperatures may trigger embryogenesis through microtubule rearrangement and nuclear migration resulting in symmetric cell division [30]. Short term exposure of microspores to anti-microtubule agents such as colchicine enhanced embryogenesis by increasing symmetric cell division [30].

The commitment to embryogenesis requires synthesis of specific proteins and macromolecules characteristic of embryogenic cells. In *Brassica napus*, an inductive high temperature treatment caused the appearance of a number of mRNAs and proteins in embryogenic microspores but not in non-embryogenic ones [31]. These appeared before cell division and may be associated with commitment to the embryogenic pathway.

In tobacco, microspore embryogenesis induced by starvation elicited the accumulation of specific mRNAs [32]. Specific phosphoproteins may also be related to the inductive process [33]. Using a similar nutrient starvation methodology, Zarsky et al. (1992) [14] demonstrated that re-entry of the vegetative cell of tobacco pollen into the G1 phase of the cell cycle was required for the induction of embryogenesis.

The developmental pattern of microspore-derived embryos resemble closely that of the zygotic embryo and globular, heart, torpedo and cotyledonary stages can be distinquished. In some cases, there may be a suspensor-like structure [34, 35]. Normal embryo maturation occurs in response to exogenous hormones and increased osmoticum and plants are readily recovered.

Biochemical studies have been conducted on microspore-derived embryos and comparisons have been made with the zygotic seed. Fatty acid biosynthesis and storage lipid accumulation were similar for both zygotic and microspore-derived embryos [5]. In microspore-derived embryos of *Brassica napus*, enzymes of lipid biosynthesis and bioassembly are comparable to those of the zygotic embryo [36]. This is also true for the major storage proteins, napin and cruciferin, which accumulate in these embryos [36, 37]. These embryos respond to increases in osmotic pressure by synthesizing abscisic acid in a manner analogous to the zygotic embryos [38]. These observations indicate that microspore-derived embryos exhibit all the biochemical competence of the zygotic counterpart. Metabolites specific to the seed accumulated in microspore-derived embryos, genes normally associated with zygotic embryo development were expressed and the microspore-derived embryos responded to abscisic acid (ABA) in a manner similar to the developing seed embryos [36].

III. *In vitro* induction of haploid embryogenesis in male gametophytic cells

A. *Factors influencing induction of embryogenesis*

1. *Donor plant growth conditions*

The physiological conditions under which donor plants are grown can significantly affect subsequent response of anthers/isolated microspores (Table 1). Anther culture studies have shown that the yield of microspore-derived embryos was affected by photoperiod and light intensity [40]. Increased microspore embryogenesis was observed in anthers from tobacco plants grown under short days [39]. Increased light intensity has resulted in increased anther culture response in *Brassica campestris* [43]. Significant increases in frequency of microspore embryogenesis in *B. napus* have been reported by Keller et al. [48] if donor plants were grown at day/night cycle of 10 °/5 °C (as opposed to 15 °/10 °C or 20 °/15 °C). Reduced growing temperatures for donor plants were also shown to be beneficial for anther and microspore cultures of *B. oleracea* [49, 50]. Species differences may, however, be significant. For example, in *B. juncea*

Table 1. Donor plant growth conditions influencing haploid embryogenesis from cultured male gametophytic cells.

Factor Studied	Species	Reference
Photoperiod	*Nicotiana tabacum*	39, 40
	Oryza sativa	41
	Triticum aestivum	42
Light intensity	*Brassica campestris*	43
	Hordeum vulgare	44
	Nicotiana tabacum	40
	Triticum aestivum	45
Field vs. greenhouse/	*Hordeum vulgare*	44
growth chamber	*Triticum aestivum*	45–47
Temperature	*Brassica napus*	48
	B. oleracea	49, 50
	Nicotiana tabacum	21, 51
	Triticum aestivum	42
Time of year	*Hordeum vulgare*	44
Nitrogen starvation	*Nicotiana tabacum*	52
Age of the plant	*Brassica napus*	53
	Nicotiana tabacum	40
	Oryza sativa	54
	Triticale	55
	Triticum aestivum	56

anther culture, higher frequencies of microspore embryogenesis were observed when anthers were taken from plants grown at 20 °/15 °C day/night cycle as opposed to lower growing temperatures (W. Keller, unpublished data). Donor plant growth temperature has also been shown to affect haploid embryogenesis in wheat [42] and tobacco [21, 51]. Other significant factors influencing haploid embryogenesis in cereals include photoperiod as well as light intensity (Table 1). Nutrient deprivation, especially nitrogen starvation of donor plants has been shown to stimulate embryogenesis in tobacco anther cultures [52].

The age of the donor plant has been shown to influence embryogenesis in some species. A higher frequency of embryogenesis was observed in young plants when compared to older plants in tobacco [40], rice [54], wheat [56] and triticale [55]. However, microspores from older plants have been shown to be more embryogenic than microspores from young plants in *B. napus* [53].

2. Pretreatments

A variety of pretreatments of spikes, buds, florets, and anthers have been employed to enhance haploid embryogenesis during subsequent *in vitro* culture (Table 2). Pretreatment of donor plants with chemicals, particularly plant growth regulators, enhanced the frequency of haploid embryogenesis in potato

Table 2. Influence of pretreatment of donor plants, buds, or anther/microspores on induction of haploid embryogenesis.

Factor	Species	Reference
Temperature		
low	*Capsicum annuum*	57
	Hordeum vulgare	58
	Nicotiana plumbaginifolia	59
	N. sylvestris	59
	Solanum tuberosum	60
	Triticum aestivum	61
elevated	*S. chacoense*	62
	S. tuberosum	63
Chemical/physical		
EMS	*N. tabacum*	64
ethanol	*B. napus*	65
irradiation	*B. napus*	65
	Malus spp.	66
N,N dimethylsuccinamic acid	*S. tuberosum*	60
Auxin	*S. tuberosum*	60
Antigibberellins	*S. tuberosum*	60

anther cultures [60]. Low temperature pretreatments have been shown to enhance embryogenesis in a number of cases with optimal temperature and duration of treatment dependent on species. For example, a 4 °C pretreatment was considered to be optimum for wheat [67, 68] and barley [69], whereas 8 °C was optimal for corn [70], rice [71] and sugarcane [72]. A 10 °C pretreatment of barley anthers on mannitol solution for four days subsequently enhanced haploid embryogenesis [73, 74]. Heat pretreatment may be beneficial in some instances with a six day treatment at 32 °C optimal for microspore embryogenesis in wheat anther cultures [75].

A pretreatment that might have widespread application involves the disruption of cytoskeleton during the initial phase of microspore culture. Zaki and Dickinson [3] reported that treatment of microspores with colchicine during the first 6–24 h of culture increased the frequencies of haploid embryogenesis. This observation provides support to the study of Simmonds et al. [4] who showed that the induction of microspore embryogenesis in B. napus is associated with reorganization of the cytoskeleton. Microspores undergoing embryogenesis had a uniform distribution of microtubules throughout the cell, whereas non-induced and non-responsive types had a polar distribution of microtubules.

Other pretreatments have included reduced atmospheric pressure [76], osmotic shock [77], gamma irradiation and ethanol stress [65]. Embryogenesis in Brassica hirta was enhanced with bud pretreatment under reduced atmospheric pressure of 500 mm of mercury for 0.5 h prior to culture [76]. Anaerobic conditions (100% N for 30 or 60 min) stimulated embryogenesis in tobacco [78].

3. Developmental stage of male gametophytic cells

The most responsive developmental stage for induction of embryogenesis varies with species (Table 3) and generally ranges from the early uninucleate to early binucleate stage. Studies with B. napus have consistently revealed that the highest frequencies of embryogenesis are achieved from microspores at the late uninucleate stage [29, 83]. The presence of binucleate microspores can result in the development of toxic conditions in culture and the consequent reduction in frequency of embryogenesis [88] and the development of abnormal, stunted embryos [89].

Inspite of these findings, it is well established that pollen embryogenesis can occur by cell division of the vegetative or generative nucleus [90, 91].

Optimal developmental stage may vary according to culture technologies employed. In tobacco, the optimal stage for isolated pollen (microspore) culture is early to mid-binucleate [13, 92] whereas in anther culture the optimal stage is less defined and is considered to be the first mitosis [22]. This may be due to inhibitory effects of the anther itself or to the anther wall acting as a barrier to nutrient flow. In an isolated report [93], it has been stated that the binucleate to trinucleate stage is more responsive for isolated microspore cultures of B. oleracea (as opposed to a more immature stage requirement for anther culture).

Table 3. Most responsive developmental stage of the male gametophyte for induction of microspore embryogenesis.

Developmental Stage	Species	Reference
Early miduninucleate	*Hordeum vulgare*	79
	Hyoscyamus niger	80
	Solanum tuberosum	60, 81
	Triticum aestivum	82
Late uninucleate – early binucleate	*Brassica napus*	29, 83
	Nicotiana otophora,	
	plumbaginifolia, sylvestris	59
	Zea mays	84
Binucleate	*Capsicum annuum*	57
	N. rustica	52, 85
	N. tabacum	13, 52, 86
	S. carolinense	87

4. Donor plant genotype

Genotype can play a major role in the embryogenic response of cultured anthers/microspores in many species. The ability to undergo *in vitro* embryogenesis is conditioned by nuclear genes [94, 95]. In barley, genotype was responsible for more than 60% of the variation observed for the frequency of embryo formation in anther cultures [94]. In *B. napus* genotypes, winter types were more embryogenic than spring types and winter by spring hybrids had a greater embryogenic capability [96]. Studies with *S. tuberosum* have indicated that the ability to undergo haploid embryogenesis *in vitro* is controlled by more than one gene and that these genes are recessive [95]. Anther culture capability or responsiveness could therefore be transferred to non-responsive lines through plant breeding procedures. In wheat, haploid production is controlled by at least three independent genes with genes on the 1D and 5BL chromosome influencing frequency of embryogenesis [97]. Other studies with anthers of wheat cultivars have shown that the most responsive possessed a 1B/1R translocation [98, 99]. In maize, plant to plant differences have been observed [100, 101] and a six-fold increase in anther culture response was observed after one cycle of selection [102]. Specific gene products related to microspore embryogenesis accumulated in the anthers of responsive maize genotypes but were not detected in non-embryogenic types [103].

The cytoplasmic background can affect embryogenesis in wheat anther cultures. Stimulatory effects have been observed in the *T. aestivum* nucleus which was associated with the *Aegilops* cytoplasm [104].

5. Composition of culture medium

Culture medium composition plays a critical role in the induction of haploid embryogenesis as well as on subsequent embryo development (Table 4). Carbon source and concentration has been shown to influence embryogenesis in a wide range of species. The most important medium component determining induction of haploid embryogenesis in *Brassica* spp is sucrose. Elevated levels of sucrose (8% or higher) are essential for inducing embryogenesis in anther cultures of *B. napus* [114], *B. campestris* [105], *B. oleracea* [116] and *B. nigra* [141], and in isolated microspore cultures of *B. napus* [115] and *B. campestris* [142]. An initial culture period in very high levels of sucrose (17–20%) followed by reduction in concentration (10–13%) has been shown to be stimulatory [117, 143, 144]. The requirement for elevated sucrose levels cannot be interpreted as a requirement for elevated osmotic pressure since other compounds such as mannitol and sorbitol have been shown to be ineffective [105]. In *Brassicas*, other sugars are apparently not superior to sucrose for induction of microspore embryogenesis with the exception of a study on *B. carinata* anther culture in which a combination of glucose and sucrose was superior to sucrose used alone [113].

Carbon source and concentration is particularly important in cereal anther culture with sucrose being the most widely used and effective carbohydrate for induction of embryogenesis. Sucrose concentrations of 8.0–9.5% was effective for induction of embryogenesis in maize [118] and barley [145]. Modification of carbohydrate type and concentration enhanced haploid embryogenesis in wheat with maltose shown to be more effective than sucrose for anther culture [109, 110, 112].

Nitrogen source and concentration have a significant effect on embryogenesis (Table 4). Glutamine has been shown to stimulate embryogenesis in both barley [123] and wheat cultures [146]. In the case of tobacco, culturing immature microspores in medium with glutamine, results in gametophytic development. However if microspores were initially cultured in medium free of glutamine then transferred to medium with sucrose and glutamine, cell division and embryogenesis occurred [13, 52].

Table 4. Media components critical to haploid embryogenesis *in vitro*.

Critical Components	Species	Reference
Source of carbon:		
sucrose superior to other carbon sources	*Brassica* spp.	105
maltose superior to sucrose	*Hordeum vulgare*	106
	Hordeum spontaneum	107
	Triticum aestivum	108, 109
glucose superior to sucrose	*T. aestivum*	110
	Triticale	111
	Hordeum vulgare	112
fructose superior to sucrose	*T. aestivum*	110
mixture of glucose, fructose superior to sucrose	*Brassica carinata*	113
	T. aestivum	110

Table 4. Continued

Critical Components	Species	Reference
Sucrose concentration:	*Brassica* spp.	105, 114–117
	Zea mays	118
	Solanum papite	119
Amino acids:		
methionine	*Datura metel*	120
glutamine	*D. metel*	121
	Nicotiana tabacum	122
	Triticale	111
	Hordeum vulgare	123
proline	*Zea mays*	124
L-cysteine	*Solanum tuberosum*	60, 125
asparagine	*Nicotiana tabacum*	122
Growth regulators:		
beneficial effect of auxin	*Solanum carolinense*	126
	Agropyron thinopyrum	127
beneficial effect of cytokinin	*B. napus*	128
beneficial effect of auxin/cytokinin	*B. napus*	114, 115
beneficial effect of ethylene inhibitors	*Solanum tuberosum*	125
	Brassica oleracea	129
beneficial effect of ethylene or ethylene promotors	*Hordeum vulgare*	130
	Datura metel	120
	Solanum carolinense	126
Mineral elements:		
calcium concentration	*S. carolinense*	87
iron concentration	*D. metel*	131
Physical factors:		
medium pH	*N. tabacum*	122
type of gelling agent	*S. tuberosum*	132
agarose concentration	*Triticum aestivum*	133
liquid superior to solid medium	*Hordeum vulgare*	106
	Zea mays	100
	Brassica napus	114
	Solanum tuberosum	119
use of membrane support rafts	*T. aestivum*	134
beneficial effect of ficoll	*Hordeum vulgare*	123, 135
anther orientation	*H. vulgare*	136
	Z. mays	137
Other beneficial additives:		
activated charcoal	*Anemone canadensis*	138
	Nicotiana tabacum	139
	Z. mays	100
	S. tuberosum	119
	Capsicum annuum	140
ascorbic acid	*S. tuberosum*	125
1-aminocyclopropane carboxylic acid	*S. carolinense*	126

Growth regulator type and composition can influence frequencies of haploid embryogenesis in many species. Cytokinins have been shown to stimulate microspore embryogenesis in *Brassica napus* [128] but most often a combination of cytokinin and auxin appears to be beneficial [114]. However, genotype and donor plant growth conditions may ultimately affect growth regulator requirements thereby leading to apparent contradictory results reported in the literature. For example, Keller et al. [48] were able to eliminate growth regulators from *B. napus* cv. Topas microspore culture medium whereas other workers have reported that elevation of growth regulator concentrations beyond levels recommended by previous researchers [147, 148].

The use of agents to block ethylene synthesis or action was beneficial in anther culture of *B. oleracea* [129, 149]. However, in the case of barley it has been shown that cultured anthers released ethylene to the medium and that specific levels of ethylene stimulated embryogenesis [130]. The addition of ethylene releasing compounds enhanced embryogenesis in genotypes that released low levels of ethylene while ethylene inhibitors enhanced embryogenesis in high ethylene producing genotypes.

Activated charcoal has been used to promote embryogenesis in anther cultures of a number of species (Table 4). These species produce phenolic substances which activated charcoal can adsorb, therefore reducing any toxic effects.

Flotation of anthers on liquid is generally more effective than plating on solid medium. Ficoll has been successfully employed to enhance flotation of anthers and anther-derived embryos [135]. Oxygen tension may also be critical as barley anthers cultured on solid medium could produce embryos if anthers were cultured with one loculus in contact with the medium [136]. Since embryoids were produced only on the upper lobes of the anthers, this suggests that oxygen availability was critical. However in rice, anther orientation was not critical to induction of haploid embryogenesis [150].

Media pH can also influence frequency of microspore embryogenesis. Generally the pH of anther/microspore culture media is in the 5.8–6.0 range. However, for *N. sylvestris*, medium with pH 6.8 was the best [151].

6. Culture environment

Physical culture conditions can significantly influence the frequency of haploid embryogenesis with culture temperature being the most significant factor. Generally, the incubation temperature is in the range of 24–27 °C. An initial culture period of elevated temperature (32–35 °C) enhanced the induction of haploid embryogenesis in *Brassica* [48, 50, 117, 152, 153] and wheat [61, 111]. It is not known why heat treatment is required. It is speculated that the high temperature inhibits protein synthesis associated with the maturing pollen grain therefore allowing sporophytic induction. In contrast, an initial culture period of 14–15 °C for four days prior to transfer to 27–28 °C enhanced embryogenesis in maize [118, 124]. Most cultures require incubation to take place in the dark. Light intensity can also influence embryogenesis with low intensities being beneficial for wheat and *Nicotiana paniculata* anther cultures [45]. CO_2 levels of 2% have stimulated embryogenesis in a number of species [154].

B. Embryogenesis protocols

There are similarities and differences between the microspore culture protocols for the species that have been evaluated. The general protocol for the Solanaceae, Gramineae, and Cruciferae are examined here. For more information on other species and protocols, refer to Huang and Keller [155].

1. Solanaceae e.g. Solanum tuberosum

Potato donor plants are grown in the greenhouse (20 °C) [132] or in the field [156]. Buds are harvested when the pollen grains are in the uninucleate stage [132, 156], this correlates to light green anthers 1.5–3.0 mm in length [157] or buds 3.5 to 5 mm in length [158]. These buds are stored at 7 °C for 72 h in the dark as a cold pretreatment [132] after which they are sterilized in 70% ethanol for two minutes then washed in sterile water. Anthers are then removed from the buds and transferred to petri plates.

Anthers have been cultured in several ways. One method involves plating the anthers on solid MS (Murashige and Skoog) [159] media supplemented with 6% sucrose, 0.5% activated charcoal and 1 mg L^{-1} benzylaminopurine (BAP). Anthers can also be placed in Erlenmeyer flasks containing half strength Linsmaier and Skoog medium [160] supplemented with 6% sucrose, 0.05% activated charcoal, 0.1 mg L^{-1} indoleacetic acid (IAA) and 2.5 mg L^{-1} BAP. Once embryos have been produced, they are transferred to solid half strength Linsmaier and Skoog media supplemented with GA_3 for further development. Another method which has given good response in terms of embryogenesis is double-layer medium [161]. The addition of gellan gum or potato starch have also been evaluated as gelling agents in anther culture protocols. Potato starch gave higher embryo yield earlier in the culture period whereas more embryos were observed on gellan gum at the end of the culture period [132]. Cultures are incubated at 25 °C with a 12 h photoperiod [60].

The media composition generally consists of 6% sucrose [162], however maltose has been used in place of sucrose which resulted in more plants per 100 anthers [158]. Growth regulators, IAA and BAP, have been used to enhance embryogenesis [157, 162, 163].

Anther cultures are incubated at 25 °C, 16 h photoperiod. [132] Macroscopic structures are observed 3–4 weeks after culture at which time they can be transferred to regeneration medium I (MS medium supplemented with 10% coconut milk, 0.3 mg L^{-1} zeatin and 3% sucrose). When embryos have developed they can be transferred to regeneration medium II (MS medium supplemented with 0.5 mg L^{-1} BAP and 3% sucrose). When roots and green shoots develop, the plantlets are transferred to soil.

2. Gramineae e.g. Hordeum vulgare

Donor plants can be grown in a growth cabinet, greenhouse or field. However microspores isolated from plants grown at low temperatures, are more embryogenic than those from plants grown at higher temperatures [164]. Generally, this is a day/night temperature of 12/10 °C with a 16 h photoperiod [74]. Spikes are harvested when pollen is in the mid to late uninucleate stage or after emergence of the flag leaf ligule. Spikes are sterilized in 70% ethanol then cultured in plates with a drop of water to maintain humidity. The petri plates are sealed with parafilm and stored at 7 °C for 14 days or 4 °C for 21–28 days in the dark [69]. Anthers are dissected from the florets and cultured at 60 to 80 anthers/mL in liquid potato medium supplemented with 9% sucrose, 1.5 mg L^{-1} 2,4-D and 0.5 mg L^{-1} kinetin [155]. For anther culture, the orientation of anthers on culture medium can influence the response of the anther. Embryos developed when a single lobe was in contact with the culture medium [136]. The cultures are incubated at 25 °C in the dark. Within 14 days, calli or embryos develop from the microspores. These embryos can be transferred to regeneration medium (MS supplemented with 2 mg L^{-1} BAP and 0.8% agar) when they reach about 1 mm in size. When green shoots develop about 7–14 days later, these shoots can be transferred to rooting medium.

The shed microspore culture technique [74] involves dissecting anthers from spikes. The cold pretreatment mentioned above can be replaced by mannitol. Anthers are cultured in media containing 0.3 M mannitol at a density of 20 anthers/mL of mannitol. These anthers are incubated at 25–28 °C for three days. After three days, the shed microspores are collected by centrifugation at 500 rpm for five minutes. The mannitol supernatant is discarded and the microspores are resuspended in 0.6 mL of ovary conditioned liquid FHG medium containing 1 mg L^{-1} IAA and 0.2 mg L^{-1} kinetin [165, 166]. Sucrose was replaced by maltose in the medium. The microspore suspension is placed in droplet form on top of 3 mL solidified FHG media. The cultures are incubated in the dark at 25 °C for 21–28 days. The microspores undergo divisions five days after initial culture. By day 7–8, the exine is ruptured to release the multicellular structures, with embryos forming by day 18. The embryos are transferred 21–28 days after culture to regeneration medium (FHG medium solidified with 0.8% Sea Plaque agarose) for further development. When plantlets are 2–4 cm in height, they are transferred to vials or flasks containing growth regulator-free MS medium (0.8% agar and 2% sucrose). Well-developed plants can be transferred to soil and grown to maturity.

Another anther culture method involves culturing anthers in 3 mL of liquid medium containing ficoll 400 and 3% sucrose and incubating the anthers in the dark for 25–30 days at 20 °C [135]. When embryos are evident, they are transferred to ficoll media with elevated levels of sucrose (4.25%) and incubated under dim light (200–300 lux) at 20 °C for 15–20 days. The mature embryos are transferred to a starch agar medium for plantlet regeneration [135].

3. Cruciferae e.g. Brassica napus

Donor plants are grown at a cool temperature, 10/5 °C and a 16 h photoperiod. This cool temperature slows the growth of the plant and allows a greater period in which to select buds. Buds (50–75) are selected when the microspores are in the mid to late uninucleate stage [48]. Buds are sterilized in sodium hypochlorite for 15 min, then washed in sterile water three times. The buds are macerated in B5 media [167] supplemented with 13% sucrose then filtered through 44 μm nylon screencloth. The crude microspore suspension is then centrifuged at 130–150 g for three minutes. The supernatant is decanted and 5 mL B5-13 is added to the pellet. This procedure is repeated twice. The number of microspores are counted and the required amount of modified Lichter medium [115] (pH 6) (NLN) supplemented with 13% sucrose and 0.83 mg L^{-1} potassium iodide, but without potato extract and hormones is added to achieve a density of 10^5 microspores/mL. These microspores are incubated at 32 °C for 72 h and then transferred to 25 °C for the remainder of three weeks at which time embryos can be counted. Embryos are allowed to turn green in the light for about one week at which time they can be transferred to regeneration medium (B5 or MS supplemented with 2% sucrose and 0.8% agar). Once a well-developed root and shoot system has been established, the plantlets can be transferred to soil.

4. Comparison of the protocols

Plants can be grown in the field or greenhouse for *Solanum*, *Hordeum*, and *Brassica*. However for optimum results, the donor plants should be grown in growth cabinets where temperature, photoperiod, light intensity and humidity can be controlled. Good fertility, water availability and good pest control are essential for healthy vigorous donor plants.

Genotypic differences have been observed in all three species [96, 106, 119]. Differences have been observed in terms of embryo production, embryo normality and regeneration ability. The most embryogenic line reported for *B. napus* is cv. Topas [48, 53, 65] and in *H. vulgare* the most responsive genotype is Igri [74, 112, 123].

For all three species, the most embryogenic developmental stage of the microspore is the uninucleate stage [29, 60, 79, 81, 83]. This has been correlated with bud size or anther size for ease of selecting buds [157, 158].

Low temperature growing conditions or pretreatment is beneficial for all three species. *S. tuberosum* requires 72 h at 7 °C [132]. For *H. vulgare*, the anthers are stored at 7 °C for 14 days or 4 °C for 21–28 days [69]. This cold period could be replaced with a 3–4 day treatment with 0.3 M mannitol [74]. For *B. napus*, the donor plants are grown at low temperatures (10/5 °C) [48] Auxins, antigibberellins, and Alar 85 (N, N, dimethylsuccinamic acid) have been used on potato plants as a pretreatment to enhance embryogenesis [60]. Ethanol and irradiation pretreatment have been used with some success in *B. napus* [65].

Carbohydrate source and concentration are critical for embryogenesis in most species. Sucrose is the most widely used carbohydrate, however improvement in embryogenic frequency in barley has been observed with

maltose, trehalose and cellobiose [112]. Sucrose at high concentrations or its breakdown products are inhibitory to embryogenesis of anthers and/or microspores [166]. Sucrose is easily hydrolyzed to glucose and fructose and it is speculated that there is a threshhold concentration above which glucose or fructose is inhibitory to microspore development [112, 166]. Maltose is degraded more slowly and may provide a metabolizable carbon source over a longer period of time. Maltose has been used in barley anther culture and was superior to sucrose [106]. Ficoll media have also been used to increase the density of the culture medium and improve embryogenesis in barley [135]. Replacement of agar by barley starch and addition of 80–120 g L^{-1} melibiose improved embryo production in barley [168, 169]. The cytokinins, benzylaminopurine and kinetin gave the best results for barley [166].

Carbohydrate concentration has been shown to influence embryogenesis in *Brassica*. For *B. napus* [114, 115], *B. campestris* [105, 142], *B. oleracea* [116], and *B. nigra* [141]. Elevated levels of sucrose (8% or more) were beneficial for inducing embryogenesis. Very high levels of sucrose (17–20%) initially, enhanced induction of embryogenesis. For continued development, a reduction in sucrose level (10–13%) was required [117, 143, 144]. Other carbohydrates have been evaluated in *Brassica* anther/microspore culture, however sucrose appears to be superior to the other carbohydrates [105].

Sucrose concentration has also been evaluated in *Solanum* species. Generally, embryo production decreased as sucrose increased [119]. The optimum concentration is 6% sucrose. Other carbohydrates have been evaluated and more embryos were produced on sucrose media. However more embryos become plants when maltose replaced sucrose in the medium [158].

Both *S. tuberosum* [132] and *H. vulgare* [74] cultures are incubated at 25 °C until embryos develop. For the *Brassica* species, an elevated temperature is required initially, e.g. is 32 °C for 72 h followed by incubation at 25 °C [48].

For *Hordeum vulgare*, regeneration of plants can be problematic with the emergence of albino plants at relatively high frequency. Several factors are thought to influence albinism in microspore-derived plants. These include genotype, developmental stage of the microspore, sucrose content, temperature and liquid vs solid medium [170]. The presence of lactic acid and other organic acids in cells under anaerobic conditions may damage the organelles resulting in albino plants [135]. A high ficoll and sucrose medium was required for obtaining a high ratio of green plants in barley [135]. Albinism is not a problem in *Solanum* and *Brassica* cultures.

IV. *In vitro* induction of haploid embryogenesis in female gametophytic cells

Gynogenetic embryos have been produced through the culture of unfertilized ovaries, ovules or flower buds of a number of species including both monocots and dicots [171, 172] The mature embryo sac with eight nuclei is usually cultured and any of the haploid cells can potentially develop into haploid embryos. In most cases, the egg cell is the origin of gynogenetic embryos e.g. in sunflower [173, 174]. Detailed histological studies of gynogenesis in *Helianthus annuus*, revealed that development was similar to that of the true zygote although culture conditions induced proliferation of some sporophytic tissues [173]. In rice, the synergids orginated the haploid embryos [175] while in *Allium tuberosum* they originate from the antipodals [176].

In gymnosperms, haploid embryogenesis and plantlet production was reported in a number of cases where the megagametophyte was cultured [177]. Embryos do not develop directly from the tissues of the megagametophyte but from a callus-like mass of cells [178]. The developmental pathway appears to be similar to that of the zygotic embryo. This topic has been the subject of a number of reviews [179, 180] and will not be covered in this review.

A. Factors influencing the induction of embryogenesis

1. Donor plant growth conditions

Studies on the effect of donor plant growth conditions on yield of embryos from cultured ovules of *Beta vulgaris* revealed that higher yields occurred when donor plants were grown in the greenhouse or growth cabinet as compared to the field [181]. The season in which donor plants are grown can greatly influence embryo yields. During the summer (May to September), *B. vulgaris* embryo yields were 3.5 to 4.5% whereas during the winter (November to February) yields were approximately 1% [181].

2. Pretreatments

A cold pretreatment of four to five days increased embryo yield in cultured *Beta vulgaris* ovules, however a pretreatment length greater than seven days was not beneficial [181]. A cold pretreatment (4 °C) for 24–48 h was also beneficial for embryo formation in *Helianthus annuus* [173].

3. Developmental stage of female gametophytic cells

Although very few documented studies are available, it would appear that ovule developmental stage influences the induction of embryogenesis. In the case of *Beta vulgaris*, ovules which were shaped like a comma were suitable for culture whereas spherical ovules degenerated [182]. In *Helianthus annuus*, ovule 3–4 days before anthesis were at the optimum stage for culture [183].

4. Donor plant genotype

A genotypic effect was observed in *B. vulgaris* [181]. Haploid plants were produced from 50% of 1300 genotypes screened and the efficiency ranged from 0 to 13% with an average of 1%. A genotypic effect (inter-and intravarietal) has been observed on frequency of embryo induction in *Beta vulgaris* [184]. Genotypic differences for embryogenesis have also been observed in *Allium cepa* [185, 186].

5. Composition of culture medium

As in the case with cultured male gametophytic cells, the medium composition is a highly significant factor influencing haploid embryogenesis in ovule/ovary cultures. Haploid embryo yields in *B. vulgaris* increased from 1.5 to 7.5% as benzylaminopurine concentration was raised from 0 to 2.0 mg L^{-1} [181]. The addition of growth regulators to media was not favourable for embryo production of *Helianthus annuus* compared to the use of media free of growth regulators; in the latter case embryos were produced and callus was inhibited [173].

Ovaries and ovules of *Allium cepa* enlarged when hormone-free media or 2,3,5-trijodobenzoic acid (TIBA)-containing media was used but degenerated when auxins, cytokinins or GA$_3$ was added to the media [185]. The frequency of ovules producing embryos was 0.28%. This response may be genotype-dependent as both auxins and cytokinins enhanced gynogenetic embryo production from cultured ovaries or flowers of *Allium cepa*. A level of 10% sucrose has been used for the initial culture medium and after four weeks, the ovules and ovaries were transferred to media with IBA, BAP and GA [187].

6. Culture environment

Limited information on the influence of physical culture environment on embryogenesis in ovule/ovary cultures is available. A cultivation temperature of 27 °C was beneficial for *Allium cepa* [185]. Low light was required for embryo development in *Carica papaya* [188].

B. Embryogenesis protocol

1. Allium cepa

A. cepa bulbs are grown in a greenhouse under natural light [186]. The flowers are collected when the microspores are in first mitosis or early binucleate stage [185]. This is the stage at which there is maximum elongation and enlargement of the flower and the ovules contain a mature embryo sac [189]. The flowers are sterilized in 3% sodium hypochlorite for 10–15 min followed by rinsing three times with sterile water. The flowers are removed and precultured on Medium I (B5 media supplemented with 10% sucrose, 2 mg L^{-1} 2,4-D and 2 mg L^{-1} BAP). After 10–14 days, the ovules are removed from the flowers and plated on to Medium II (Modified BDS media supplemented with 10% sucrose) [186].

For ovary and flower culture, the procedure is the same as above. The flowers are harvested and sterilized. For ovary culture, the ovaries are removed and plated onto Medium III (Modified BDS medium containing 10% sucrose, 1 mg L^{-1} naphthaleneacetic acid (NAA) and 2 mg L^{-1} BAP). For flower culture, the flowers are removed and cultured directly on to Medium IV (Modified BDS medium containing 10% sucrose, 2 mg L^{-1} 2,4-D and 2 mg L^{-1} BAP). After 30 days, the ovaries and flowers are subcultured on to the same media [186].

Once embryos are formed they can be transferred to MS medium with 3% sucrose for further development. When roots have formed, the plants can be transferred to soil for bulb growth.

Embryos from ovule culture required about 25 days to develop, not including the 10–14 days for preculture. For ovary and flower culture, the time required for embryos to become visible is 50–55 and 65–70 days, respectively. Ovule culture requires intensive labour therefore a more efficient method would be to culture ovaries or flowers.

V. Plant regeneration from haploid embryos

Plant regeneration has been achieved in many haploid embryo culture systems (Table 5) and has been the subject of many reviews [239–241]. Most reviews, however do not distinguish plants derived through direct embryogenesis from those derived from callus. In species such as tobacco in which simple nutrient media are used in anther culture, plants will regenerate directly on the anther culture medium. In the case of other species, such as the *Brassica* and cereal species, in which growth regulators and high level of a sugar are employed and in which embryogenesis is often induced in darkness, it has been necessary to develop an embryo culture protocol. Most embryo culture procedures involve transfer of embryos to basal nutrient media free of growth regulators with carbohydrate levels reduced to concentration levels of 2% or less. Embryos are usually cultured in long day (16 h) or continuous light conditions at temperatures in the range of 24–26 °C.

In some species, particularly in cereals (e.g. barley), albino embryos are often produced in culture and then will regenerate into albino plants. The problems of albino formation is characteristic of certain genotypes [58, 94, 242] and albino development can be reduced through modifications of culture protocol [73, 109, 135]. Cytoplasmic DNA modifications, i.e. deletions in plastid DNA have resulted in albino wheat plants [243]. Very few albinos develop from cereal ovules, this is one advantage of ovule culture over anther culture of cereals.

Plants regenerated from embryos derived from haploid gametophytic cells are generally haploid, however diploids and, occasionally, polyploid regenerants have been detected [61, 210]. Diplo- (and polyplo-)dization can occur during the early stage of induction of embryogenesis [244, 245] and also during the plant regeneration process [244]. Aneuploids have been observed in a number of species e.g. rice, maize and wheat [246–248] however these are usually found in systems where callus has been produced.

There is some evidence of genetic and molecular changes in plants derived from microspore/anther culture [239]. Morphological differences have been observed in microspore-derived lines of barley when compared to the donor plant controls [249]. Differences from the donor plant for DNA quantity and quality has been observed in microspore-derived lines of *Nicotiana tabacum* [250] and *N. sylvestris* [251]. Cytoplasmic changes may also occur [252].

Table 5. Species in which plants have been regenerated from haploid embryos derived from male gametophytic cells (since 1985).[1]

Species	Reference
Aesculus carnea	190
Anemone canadensis	191
Brassica campestris	117, 142, 192–194
Brassica carinata	113, 152
Brassica juncea	195
Brassica napus	1, 29, 48, 83, 88, 89, 128, 144, 152, 196–201
Brassica oleracea	148, 202–204
Capsicum annuum	57, 140
Citrus aurantifolia	205
Cocos nucifera	206
Datura metel	120, 121
Digitalis obscura	207
Helianthus annuus	208
Hordeum vulgare	44, 58, 74, 106, 123, 135, 136, 145, 209
Hordeum spontaneum	107
Iochroma warscewiczii	210
Lolium perenne	211
Morus	212
Nicotiana otophora	59
Nicotiana plumbaginifolia	59
Nicotiana sylvestris	59, 213
Nicotiana tabacum	214–219
Populus glandulosa	220
Raphanus sativus	193
Secale cereale	221
Sinapis alba	76, 222
Sinocalamus latiflora	223
Solanum chacoense	224, 225
Solanum papita	119
Solanum tuberosum	60, 125, 158, 226
Triticale	227
Triticum aestivum	42, 77, 98, 99, 104, 109–111, 133, 134, 227–234
Triticum turgidum	235
Zea mays	101, 236–238

[1] For a more comprehensive list of species, including those successfully regenerated prior to 1985, see the following Maheshwari et al. [6], Dunwell [7, 287], Heberle-Bors [8], Keller et al. [9], Powell [288], and Reynolds [289].

VI. Production and utilization of doubled haploids

The development of haploids or doubled haploid plants via anther, microspore, ovule or ovary culture has a number of practical advantages for plant improvement. These haploid or doubled haploid plants have been used in varietal development, genetic, mutation, genetic engineering, biochemical and physiological studies.

Doubled haploid plants can be produced through spontaneous chromosome doubling *in vitro* [253, 254] or by colchicine treatment of the microspores [255], embryos [256] or plants [201, 257]. Spontaneous chromosome doubling rate varies depending on the species and genotype [258].

A. Cultivar development

The main advantage of doubled haploids is the reduction in time to develop new cultivars. For annual self-pollinated crops it generally takes 10 or more years to develop a cultivar through a conventional plant breeding program which includes selfing and subsequent selection. The production of haploids followed by chromosome doubling to produce homozygous lines, from which superior lines are selected, can reduce the time required for varietal development by 3–4 years [259]. For cross-pollinating, heterogenous crops, the haploid system is a rapid method for producing homozygous, pure breeding lines which can be used in the development of synthetics or hybrids. This technique would be beneficial in tree improvement as it would eliminate the very long time period required to produce homozygous lines. Selection efficiency can also be improved with the production of haploids as the phenotype of the plant is not masked by dominance effects. Traits conditioned by recessive genes can be easily identified. A smaller population of doubled haploids are required when screening for desirable recombinants than would be the case for conventional diploid populations [260].

Several commercial crop cultivars have been developed through the application of haploid techniques in tobacco [261–264] and wheat [228, 229].

B. Genetic studies

Haploids can be used to detect linkage and gene interaction as well as estimate genetic variance and the number of genes for quantitative characteristics [265–267]. Haploidy can also be used for the production of genetic translocations, substitution and addition lines as shown in the case of triticale × wheat hybrids [268].

C. Mutation studies

Isolated microspores offer a haploid cell culture system which is advantageous in mutagenesis. Haploid and double haploid plants can be easily produced because of the high regenerative potential of some microspore culture systems. An additional advantage is at the plant level; recessive and dominant traits are expressed and therefore can be easily selected. Because of this, the haploid (double haploid) system is a good system for mutation and *in vitro* selection studies.

In vitro selection studies have resulted in *B. napus* plants tolerant to the herbicides, chlorsulfuron [269] and imidazolinones [270]. These plants have an altered acetohydroxyacid synthase (AHAS) enzyme. AHAS is the first enzyme in the biosynthetic pathway for leucine, isoleucine and valine which is inactivated by sulfonylureas [271, 272]. *Brassica napus* plants with elevated levels of oleic acid (18:1) and low levels of linoleic (18:2), palmitic (16:0) and stearic (18:0) in the seed oil have been selected through microspore mutagenesis [273–275]. Mutant isolation in microspore culture has been used in barley and rice [276–278].

D. Genetic engineering

The recipient material utilized for gene transfer can be microspores, protoplasts, haploid embryos or explants from whole plants. Some of the techniques used to introduce foreign genes are *Agrobacterium*-mediated, microinjection, electroporation, particle bombardment and chemical-mediated DNA uptake. The haploid embryogenic system offers an effective means of transformation as the introduced trait can be readily fixed through chromosome doubling. In addition, this system allows recovery of whole plants in one step as opposed to organogenesis where regenerated shoots must be subsequently rooted.

Agrobacterium-mediated gene transfer has been used extensively in dicotyledonous crops. Transgenic *B. napus* plants have been produced via cocultivation of microspores or microspore-derived embryos with *Agrobacterium tumefaciens* [279, 280]. *Agrobacterium rhizogenes* has been used in DNA delivery into barley anther cultures [281].

Microinjection of DNA into embryoids have also resulted in transgenic *B. napus* plants [282]. The microinjection technique is laborious, time-consuming and only a small number of cells can be conveniently handled. Therefore it can only be performed on a limited number of cells and the identification of embryogenic microspores is very important. Percoll gradient techniques [91] and morphology of microspores have been used to select the embryogenic cells [283].

Particle bombardment has been used in transformation of a number of species. For haploid culture, particle bombardment has resulted in DNA delivery into barley microspore-derived embryos [281].

Direct gene transfer procedures (PEG-mediated or electroporation) for the transformation of microspores has not been possible because the microspore is surrounded by the intine and exine. However, studies have shown that it is possible to introduce DNA into microspores by electroporation [284].

VI. Limitations and future perspectives

The induction of haploid embryos from male or female gametophytic tissues/cells is still a poorly understood process even though embryos can be readily induced in some species, notably the *Brassicas*. The response is largely genotype dependent and there are many species such as the legumes, which are recalcitrant. Even within responsive genotypes, there may be differences between individual plants, particularly for self incompatible species. There is relatively little information on the biochemistry and physiology of haploid embryogenesis even though microspores, because of their uniformity, abundance and ease of manipulation in culture, offer an ideal system for such studies. As a consequence, there are a few generalized protocols for embryo induction and the questions of why some species are responsive and not others, and what triggers the switch from gametophytic to sporophytic development remain to be answered.

What is clear, however, is that haploid embryo induction results in sustained cell division of the microspore and organization of an embryo. This undoubtedly involves growth regulators known to be associated with zygotic and somatic embryogenesis. In most cases, specific inductive conditions are required for embryogenesis and these may lead to changes in endogenous growth regulators in the microspores. Culture conditions may themselves elicit endogenous growth regulators in responsive microspores. There appears to be no specific requirements for exogenous plant growth regulators except in cereal anther culture. Recent studies have highlighted the importance of polar auxin transport in the development of zygotic embryos [285]. Since the development of microspore-derived embryos closely patterns that of the zygote, a similar role for auxin is quite likely. In some systems, specific mRNAs and proteins related to embryogenesis have been identified. Some embryoid abundant genes were shown to be temporally expressed during induction of pollen embryogenesis in wheat [286]. Although it was demonstrated that non-embryogenic microspores, or those held under non-inductive conditions did not exhibit such gene expression, it remains to be determined whether this is generally true of other systems. One question that should be addressed is whether all microspores are inherently competent for embryogenesis and only require the correct inductive stimulus, or whether this is the property of some cells. In tobacco, evidence indicates regulation of pollen embryogenesis through nutrient starvation. However, this has not been reported for other systems. The frequency of embryo production is still low compared to the number of microspores cultured. This may be improved through the development of microculture techniques where a small number of microspores can be studied under controlled conditions. Highly sychronized microspores are extremely useful in studies of the basic aspects of embryogenesis. This is more useful than the zygotic embryo which is relatively inaccessible during early development. Also mutants can be readily produced from the haploid microspores for use in studies of embryogenesis. Such embryos are well suited for studies on the physiology and biochemistry of embryo maturation and storage product biosynthesis and accumulation. They are potentially useful for germplasm multiplication and storage and perhaps the production of artificial seeds. The latter would be especially useful if microspores are treated to induce chromosome doubling before embryo development. Such treatments should not alter embryogenic capacity of the microspores.

The full potential of the female gametophyte to undergo haploid embryogenesis has not been explored, although gynogenetic haploids have been recovered from a number of species. This may be a valuable substitute for the production of homozygous lines in cases where cytoplasmic male sterility precludes the use of microspores. Another advantage is reduction in the frequency of albino plants in some species. A major limitation to the widespread use of this technique is the absence of methods to culture unfertilized ovules and ovaries under conditions where haploids can develop with minimal sporophytic tissue proliferation. Development of techniques to culture isolated embryo sacs at the early stages of development should accelerate the use of gynogenetic embryos.

References

1. Telmer, C.A., Simmonds, D.H., and Newcomb, W., Determination of developmental stage to obtain high frequencies of embryogenic microspores in *Brassica napus*, *Physiol. Plant.*, 84, 417, 1992.
2. Zaki, M.A.M., and Dickinson, H.G., Structural changes during the free divisions of embryos resulting from anther and free microspore culture in *Brassica napus*, *Protoplasma*, 156, 149, 1990.
3. Zaki, M.A.M., and Dickinson, H.G., Microspore-derived embryos in *Brassica*: the significance of division symmetry in pollen mitosis I to embryogenic development, *Sex. Plant Reprod.*, 4, 48, 1991.
4. Simmonds, D.H., Gervais, C., and Keller, W.A., Embryogenesis from microspores of embryogenic and non-embryogenic lines of *Brassica napus*, in *Proc. 8th International Rapeseed Congress*, McGregor, D.I., Ed., Saskatoon, Canada, 1991, 306.
5. Pomeroy, M.K., Kramer, J.K.G., Hunt, D.J., and Keller, W.A., Fatty acid changes during development of zygotic and microspore-derived embryos of *Brassica napus*, *Physiol. Plant.*, 81, 447, 1991.
6. Maheshwari, S.C., Rashid, A., and Tyagi, A.K., Haploids from pollen grains – retrospect and prospect, *Amer. J. Bot.*, 69, 865, 1982.
7. Dunwell, J.M., Embryogenesis from pollen *in vitro*, in *Biotechnology in Plant Science*, Zaitlin, M., Day, P., and Hollaender, A., Eds., Academic Press, New York, 1985, 49.
8. Heberle-Bors, E., *In vitro* haploid formation from pollen: a critical review, *Theor. Appl. Genet.*, 71, 361, 1985.
9. Keller, W.A., Arnison, P.G., and Cardy, B.J., Haploids from gametophytic cells – recent developments and future prospects, in *Plant Tissue and Cell Culture*, Green, C.E., Somers, D.A., Hackett, W.P., and Biesboer, D.D., Eds., Alan R. Liss Inc., New York, 1987, 223.
10. Sunderland, N., and Dunwell, J.M., Anther and pollen culture, in *Plant Tissue and Cell Culture*, Street, H.E., Ed., Blackwell Scientific Publisher, Oxford, 1977, 223.
11. Horner, M., and Street, E., Pollen dimorphism – origin and significance in pollen plant formation by anther culture, *Ann. Bot.*, 42, 763, 1978.
12. Heberle-Bors, E., and Reinert, J., Isolated pollen cultures and pollen dimorphism, *Naturwissenschaften*, 67, 311, 1980.
13. Kyo, M., and Harada, H., Control of the developmental pathway of tobacco pollen *in vitro*, *Planta*, 168, 427, 1986.
14. Zarsky, V., Garrido, D., Rihova, L., Tupy, J., Vicente, O., and Heberle-Bors, E., Derepression of the cell cycle by starvation is involved in the induction of tobacco pollen embryogenesis, *Sex. Plant Reprod.*, 5, 189, 1992.

15. Rashid, A., Siddiqui, A.W., and Reinert, J., Subcellular aspects of origin and structure of pollen embryos of *Nicotiana*, *Protoplasma*, 113, 202, 1982.
16. Sunderland, N., Anther culture: a progress report, *Sci. Prog. (London)*, 59, 527, 1971.
17. Bhojwani, S.S., Dunwell, J.M., and Sunderland, N., Nucleic acid and protein contents of embryogenic tobacco pollen, *J. Exp. Bot.*, 24, 863, 1973.
18. Dunwell, J.M., and Sunderland, N., Pollen ultrastructure in anther cultures of *Nicotiana tabacum* I. Early stages of cultures, *J. Exp. Bot.*, 25, 352, 1974.
19. Dunwell, J.M., and Sunderland, N., Pollen ultrastructure in anther cultures of *Nicotiana tabacum* II. Changes associated with embryogenesis, *J. Exp. Bot.*, 25, 363, 1974.
20. Dunwell, J.M., and Sunderland, N., Pollen ultrastructure in anther cultures of *Nicotiana tabacum* III. The first sporophytic division, *J. Exp. Bot.*, 26, 240, 1975.
21. Rashid, A., Siddiqui, A.W., and Reinert, J., Ultrastructure of embryogenic pollen of *Nicotiana tabacum* var. Badischer Burley, *Protoplasma*, 107, 375, 1981.
22. Heberle-Bors, E., Isolated pollen culture in tobacco: plant reproductive development in a nutshell, *Sex. Plant Reprod.*, 2, 1, 1989.
23. Zhou, J.Y., Pollen dimorphism and its relation to formation of pollen embryos in anther culture of wheat *Triticum aestivum*, *Acta Bot. Sin.*, 22, 117, 1980.
24. Tan, B.H., and Halloran, G.M., Pollen dimorphism and the frequency of inductive anthers in anther cultures of *Triticum monococcum*, *Biochem. Physiol. Pflanz.*, 177, 197, 1982.
25. Sangwan, R.S., and Sangwan-Norreel, B.S., Ultrastructural cytology of plastids in pollen grains of certain androgenic and non androgenic plants, *Protoplasma*, 138, 11, 1987.
26. Nitsch, J.P., Experimental androgenesis in *Nicotiana*, *Phytomorphology*, 19, 389, 1969.
27. Nitsch, J.P., and Nitsch, C., Obtention de plantes haploides a partir de pollen, *Bull. Soc. Bot. Fr.*, 117, 339, 1970.
28. Norreel, B., Etude cytologique de l'androgenese chez le *Datura innoxia* et le *Nicotiana tabacum*, *Bull. Soc. Bot. Fr.*, 117, 461, 1970.
29. Fan, Z., Armstrong, K.C., and Keller, W.A., Development of microspores *in vivo* and *in vitro* in *Brassica napus* L., *Protoplasma*, 147, 191, 1988.
30. Hause, B., Hause, G., Pechan, P., and Van Lammeren, A.A.M., Cytoskeletal changes and induction of embryogenesis in microspore and pollen cultures of *Brassica napus* L., *Cell Biol. Intl.*, 17, 153, 1993.
31. Pechan, P.M., Bartels, D., Brown, D.C.W., and Schell, J., Messenger-RNA and protein changes associated with induction of *Brassica* microspore embryogenesis, *Planta*, 184, 161, 1991.
32. Garrido, D., Eller, N., Heberle-Bors, E., Vicente, O., De novo transcription of specific mRNAs during the induction of tobacco pollen embryogenesis, *Sex. Plant Reprod.*, 6, 40, 1993.
33. Kyo, M., and Harada, H., Specific phosphoproteins in the initial period of tobacco pollen embryogenesis, *Planta*, 182, 58, 1990.
34. Sunderland, N., Roberts, M., and Evans, L.J., Multicellular pollen formation in cultured barley anthers I. Independent division of the generative and vegetative cells, *J. Exp. Bot.*, 30, 1133, 1979.
35. Rybczynski, J.J., Simonson, R.L., and Baenziger, P.S., Evidence for microspore embryogenesis in wheat anther culture, *In Vitro Cell Dev. Biol.*, 27, 168, 1991.
36. Taylor, D.C., Weber, N., Underhill, E.W., Pomeroy, M.K., Keller, W.A., Scowcroft, W.R., Wilen, R.W., Moloney, M.M., and Holbrook, L.A., Storage-protein regulation and lipid accumulation in microspore embryos of *Brassica napus* L., *Planta*, 181, 18, 1990.
37. Crouch, M.L., Non-zygotic embryos of *Brassica napus* L. contain embryo-specific storage proteins, *Planta*, 156, 520, 1982.
38. Wilen, R.W., Hormonal and environmental regulation of storage protein and oleosin gene expression in microspore-derived embryos of *Brassica napus*, Ph.D. Thesis, University of Calgary, Calgary, Canada, 1992.
39. Rashid, A., and Reinert, J. Differentiation of embryogenic pollen in cold-treated buds of *Nicotiana tabacum* var. Badischer Burley and nutritional requirements of the isolated pollen to form embryos, *Protoplasma*, 109, 285, 1981.

333

40. Dunwell, J.M., A comparative study of environmental and developmental factors which influence embryo induction and growth in cultured anthers of *Nicotiana tabacum*, *Env. Exp. Bot.*, 16, 109, 1976.
41. Lee, S.Y., Lee, Y.T., and Lee, M.S., Studies on the anther culture of *Oryza sativa* L.3. Growing environment of donor plant in anther culture – effects of photoperiod and light intensity, *The Research Reports of The Rural Development Administration*, Korea, 30, 7, 1988.
42. Jones, A.M., and Petolino, J.F., Effects of donor plant genotype and growth environment on anther culture of soft-red winter wheat (*Triticum aestivum* L.), *Plant Cell Tiss. Org. Cult.*, 8, 215, 1987.
43. Keller, W.A., Armstrong, K.C., and De la Roche, A.I., The production and utilization of microspore-derived haploids in *Brassica* crops, in *Plant Cell Cultures in Crop Improvement*, Giles, K.L. and Sen, S.K., Eds., Plenum Pub. Corp., New York, 1982, 169.
44. Luckett, D.J., and Smithard, R.A., Doubled haploid production by anther culture for Australian barley breeding, *Aust. J. Agric. Res.*, 43, 67, 1992.
45. Bjornstad, A., Opsahl-Ferstad, H.-G., and Aasmo, M., Effects of donor plant environment and light during incubation on anther cultures of some spring wheat (*Triticum aestivum* L.) cultivars, *Plant Cell Tiss. Org. Cult.*, 17, 27, 1989.
46. Ouyang, J.W., He, D.G., Feng, G.H., and Jia, S.E., The response of anther culture to culture temperature varies with growth conditions of anther-donor plants, *Plant Sci.*, 49, 145, 1987.
47. Tuvesson, I.K.D., Pedersen, S., and Andersen, S.B., Nuclear genes affecting albinism in wheat (*Triticum aestivum* L.) anther culture, *Theor. Appl. Genet.*, 78, 879, 1989.
48. Keller, W.A., Fan, Z., Pechan, P., Long, N., and Grainger, J., An efficient method for culture of isolated microspores of *Brassica napus*, in *Proc. 7th International Rapeseed Congress*, Poland, 1987, 152.
49. Muller, D., and Keller, J., Anther culture of brussel sprouts (*Brassica oleracea* var. *gemmifera*), *Arch. Zuchtungsforsch.*, 20, 219, 1990.
50. Takahata, Y., and Keller, W.A., High frequency embryogenesis and plant regeneration in isolated microspore culture of *Brassica oleracea* L., *Plant Sci.*, 74, 235, 1991.
51. Rashid, A., and Reinert, J., Factors affecting high-frequency embryo formation in *Ab initio* pollen cultures of *Nicotiana*, *Protoplasma*, 116, 155, 1983.
52. Kyo, M., and Harada, H., Studies on conditions for cell division and embryogenesis in isolated pollen culture in *Nicotiana rustica*, *Plant Physiol.*, 79, 90, 1985.
53. Takahata, Y., Brown, D.C.W., and Keller, W.A., Effect of donor plant age and inflorescence age on microspore culture of *Brassica napus* L., *Euphytica*, 58, 51, 1991.
54. Lupotto, E., Improvement in plantlet differentiation from anther culture of rice, *Genet. Agric.*, 36, 129, 1982.
55. Orlikowska, T., Induction of androgenesis *in vitro* in *Secale cereale* and Triticale, *Genet. Pol.*, 18, 51, 1977.
56. Jing, J.-K., Xi, Z.-Y., and Hu, H., Effects of high temperature and physiological condition of donor plants on induction of pollen derived plants in wheat, *Ann. Rep. Inst. Genet. Acad. Sinic.*, Beijing, 1981, 67, 1982.
57. Morrison, R.A., Koning, R.E., and Evans, D.A., Anther culture of an interspecific hybrid of *Capsicum*, *J. Plant Physiol.*, 126, 1, 1986.
58. Knudsen, S., Due, I.K., and Anderson, S.B., Components of response in barley anther culture, *Plant Breed.*, 103, 241, 1989.
59. Chen, C., Huang, C., and To, K., Anther cultures of four diploid *Nicotiana* species and chromosome numbers of regenerated plants, *Bot. Bull. Academia Sinica*, 26, 147, 1985.
60. Johansson, L., Improved methods for induction of embryogenesis in anther cultures of *Solanum tuberosum*, *Potato Res.*, 29, 179, 1986.
61. Datta, S.K., and Wenzel, G., Isolated microspore derived plant formation via embryogenesis in *Triticum aestivum* L., *Plant Sci.*, 48, 49, 1987.
62. Cappadocia, M., Cheng, D.S.K., and Ludlum-Simonette, R., Plant regeneration from *in vitro* culture of anthers of *Solanum chacoense* Bitt. and interspecific diploid hybrids *S. tuberosum* L. × *S. chacoense* Bitt., *Theor. Appl. Genet.*, 69, 131, 1984.

334

63. Batty, N.J.P., and Dunwell, J.M., The effects of various pre-treatments on the response of *Solanum tuberosum* cv. H3703 to anther culture, in *EAPR/Eucarpia Breeding and Variety Assessment Meeting*, Cambridge, 1986.

64. Medrano, H., Millo, E.P. and Guerri, J., Ethyl-methane-sulphonate effects on anther cultures of *Nicotiana tabacum*, *Euphytica*, 35, 161, 1986.

65. Pechan, P.M., and Keller, W.A., Induction of microspore embryogenesis in *Brassica napus* L. by gamma irradiation and ethanol stress, *In Vitro Cell. Develop. Biol.*, 25, 1073, 1989.

66. Zhang, Y.X., Bouvier, L. and Lespinasse, Y., Microspore embryogenesis induced by low gamma dose irradiation in apple, *Plant Breed.*, 108, 173, 1992.

67. Lazer, M.D., Schaeffer, G.W., and Baenziger, P.S., The physical environment in relation to high frequency callus and plantlet development in anther culture of wheat (*Triticum aestivum* L.) cv. Chris, *J. Plant Physiol.*, 121, 103, 1985.

68. Lazar, M.D., Schaeffer, G.W., and Baenziger, P.S., The effects of interactions of culture environment with genotype on wheat (*Triticum aestivum*) anther culture response, *Plant Cell Rep.*, 8, 525, 1990.

69. Huang, B., and Sunderland, N., Temperature-stress pretreatment in barley anther culture, *Ann. Bot.*, 49, 77, 1982.

70. Genovesi, A.D., and Collins, G.B., *In vitro* production of haploid plants of corn via anther culture, *Crop Sci.*, 22, 1137, 1982.

71. Lin, S.L., and Tsay, H.S., Effect of cold treatment on rice anther culture, *J. Agri. Assoc. China*, 127, 8, 1984.

72. Fitch, M.M., and Moore, P.H., Haploid production from anther culture of *Saccharum spontaneum* L., *Z. Pflanzenphysiol.*, 109, 197, 1983.

73. Roberts-Oehlschlager, S.L., and Dunwell, J.M., Barley anther culture: The effect of position on pollen development *in vivo* and *in vitro*, *Plant Cell Rep.*, 9, 631, 1991.

74. Ziauddin, A., Simion, E., and Kasha, K.J., Improved plant regeneration from shed microspore culture in barley (*Hordeum vulgare* L.) cv. Igri, *Plant Cell Rep.*, 9, 69, 1990.

75. Li, H., Qureshi, J.A., and Kartha, K.K., The influence of different temperature treatments on anther culture response of spring wheat (*Triticum aestivum* L.), *Plant Sci.*, 57, 55, 1988.

76. Klimaszewska, K., and Keller, W.A., The production of haploids from *Brassica hirta* Moench (*Sinapis alba* L.) anther cultures, *Z. Pflanzenphysiol.*, 109, 235, 1983.

77. Ouyang, J., Induction of pollen plants in *Triticum aestivum*, in *Haploids of Higher Plants In Vitro*, Hu, H., and Yang, H., Eds., Springer-Verlag, Berlin, 1986, 22.

78. Imamura, J., and Harada, H., Stimulation of tobacco pollen embryogenesis by anaerobic treatments, *Z. Pflanzenphysiol.*, 103, 259, 1981.

79. Wheatley, W.G., Marsolais, A.A., and Kasha, K.J., Microspore growth and anther staging in barley anther culture, *Plant Cell Rep.*, 5, 47, 1986.

80. Reynolds, T.L., Ultrastructure of anomalous pollen development in embryogenic anther cultures of *Hyoscyamus niger*, *Amer. J. Bot.*, 72, 44, 1985.

81. Powell, W., Coleman, M., and McNicol, J., The statistical analysis of potato anther culture data, *Plant Cell Tiss. Org. Cult.*, 23, 159, 1990.

82. Gang, H.D., and Ouyang, J.W., Callus and plantlet formation from cultured wheat anthers at different developmental stages, *Plant Sci. Lett.*, 33, 71, 1984.

83. Kott, L.S., Polsoni, L., and Beversdorf, W.D., Cytological aspects of isolated microspore culture of *Brassica napus*, *Can. J. Bot.*, 66, 1658, 1988.

84. Gaillard, A., Vergne, P., and Beckett, M., Optimization of maize microspore isolation and culture conditions for reliable plant regeneration, *Plant Cell Rep.*, 10, 55, 1991.

85. Harada, H., Kyo , M., and Imamura, J., Induction of embryogenesis and regulation of the developmental pathway in immature pollen of *Nicotiana* species, in *Current Topics in Developmental Biology*, Academic Press, Japan, 20, 1986, 397.

86. Aruga, K., Nakajima, T., and Yamamoto, K., Embryogenic induction in pollen grains of *Nicotiana tabacum* L., *Japan. J. Breed.*, 35, 50, 1985.

87. Reynolds, T.L., Interactions between calcium and auxin during pollen androgenesis in anther cultures of *Solanum carolinense* L., *Plant Sci.*, 72, 109, 1990.

88. Kott, L.S., Polonsi, L., Ellis, B., and Beversdorf, W.D., Autotoxicity in isolated microspore cultures of *Brassica napus*, *Can. J. Bot.*, 66, 1665, 1988.
89. Fan, Z., Holbrook, L., and Keller, W.A., Isolation and enrichment of embryogenic microspores in *Brassica napus* L. by fractionation using percoll density gradients, in *Proc. of the 7th International Rapeseed Congress*, Poland, 1988, 92.
90. Raghavan, V., From microspore to embryoid: Faces of the angiosperm pollen grain, in *Progress in Plant Cellular and Molecular Biology*, Nijkamp, H.J.J., van der Plas, L.H.W., and van Aartrijk, J., Eds., Kluwer Academic Publishers, Dordrecht, 1990, 213.
91. Pretova, A., Ruijter, N.C.A., Van Lammeren, A.A.M., and Schel, J.H.N., Structural observations during androgenesis microspore culture of 4c1 genotype of *Zea mays*, *Euphytica*, 65, 61, 1993.
92. Rashid, A., and Reinert, J., Selection of embryogenic pollen from cold-treated buds of *Nicotiana tabacum* var. Badischer Burley and their development into embryos in culture, *Protoplasma*, 105, 161, 1980.
93. Cao, M.Q., Charlot, F., and Dore, C., Embrogenese et regeneration de plantes de chou a choucroute (*Brassica oleracea* L. ssp. *capitata*) par culture *in vitro* de microspores isolees, *C.R. Acad. Sci. Paris Serie III*, 310, 203, 1990.
94. Larsen, E.T., Tuvesson, I.K.D., and Andersen, S.B., Nuclear genes affecting percentage of green plants in barley (*Hordeum vulgare* L.) anther culture, *Theor. Appl. Genet.*, 82, 417, 1991.
95. Sonnino, A., Tanaka, S., Iwanaga, M., and Schilde-Rentschler, L., Genetic control of embryo formation in anther culture of diploid potatoes, *Plant Cell Rep.*, 8, 105, 1989.
96. Cardy, B.J., Production of anther-derived doubled haploids for breeding oilseed rape (*Brassica napus* L.), Ph.D. Thesis, University of Guelph, Guelph, Ontario, Canada, 1986.
97. Agache, S., Bacheller, B., De Buyser, J., Henry, Y., and Snape, J., Genetic control of anther culture response in wheat using aneuploid, chromosome substitution and translocation lines, *Theor. Appl. Genet.*, 77, 7, 1989.
98. Foroughi-Wehr, B., and Zeller, F.J., *In vitro* microspore reaction of different German wheat cultivars, *Theor. Appl. Genet.*, 79, 77, 1990.
99. Henry, Y., and De Buyser, J., Effect of the 1B/1R translocation on anther culture ability in wheat (*Triticum aestivum* L.), *Plant Cell Rep.*, 4, 307, 1985.
100. Pace, G.M., Ried, J.N., Ho, L.C., and Fahey, J.W., Anther culture of maize and the visualization of embryogenic microspores by fluorescent microscopy, *Theor. Appl. Genet.*, 73, 863, 1987.
101. Koinuma, K., Mochizuki, N., and Inoue, Y., Embryoid and callus induction and plant regeneration by anther culture of Japanese local varieties of maize (*Zea mays* L.), *Bull. Natl. Grassl. Res. Inst.*, 43, 13, 1990.
102. Petolino, J.F., Jones, A.M., and Thompson, S.A., Selection for increased anther culture response in maize, *Theor. Appl. Genet.*, 76, 157, 1988.
103. Vergne, P., Riccardi, F., Beckert, M., and Dumas, C., Detection of androgenesis-related proteins in maize, in *Progress in Plant Cellular and Molecular Biology, Proc. of the VIIth International Congress on Plant Tissue and Cell Culture*, Nijkamp, H.J.J., van Der Plas, L.H.W., and van Aartrijk, J., Eds., Kluwer Academic Publishers, Dordrecht, 1990, 416.
104. Sagi, L., and Barnabas, B., Evidence for cytoplasmic control of *in vitro* microspore embryogenesis in the anther culture of wheat (*Triticum aestivum* L.), *Theor. Appl. Genet.*, 78, 867, 1989.
105. Hamaoka, Y., Fujita, Y., and Iwai, S., Effects of temperature on the mode of pollen development in anther culture of *Brassica campestris*, *Physiol. Plant.*, 82, 67, 1991.
106. Kuhlmann, U., and Foroughi-Wehr, B., Production of doubled haploid lines in frequencies sufficient for barley breeding programs, *Plant Cell Rep.*, 8, 78, 1989.
107. Piccirilli, M., and Arcioni, S., Haploid plants regenerated via anther culture in wild barley (*Hordeum spontaneum* C. Kock), *Plant Cell Rep.*, 10, 273, 1991.
108. Last, D.I., and Brettell, R.I.S., Embryo yield in wheat anther culture is influenced by the choice of sugar in the culture medium, *Plant Cell Rep.*, 9, 14, 1990.

336

109. Orshinsky, B.R., McGregor, L.J., Johnson, G.I.E., Hucl, P., and Kartha, K.K., Improved embryoid induction and green shoot regeneration from wheat anthers cultured in medium with maltose, *Plant Cell Rep.*, 9, 365, 1990.

110. Chu, C.C., Hill, R.D., and Brule-Babel, A.L., High frequency of pollen embryoid formation and plant regeneration in *Triticum aestivum* L. on monosaccharide containing media, *Plant Sci.*, 66, 255, 1990.

111. Eapan, S., and Rao, P.S., Factors controlling pollen embryogenesis in triticale and wheat, in *Proc. Indian Natn. Sci. Acad.*, 51, 353, 1985.

112. Hunter, C.P., Plant generation method, European Patent Application, Publication No. 0 245 898 A2, 1987.

113. Arora, R., and Bhojwani, S.S., Production of androgenic plants through pollen embryogenesis in anther cultures of *Brassica carinata* A. Braun, *Biologia Plantarum*, 30, 25, 1988.

114. Lichter, R., Anther culture of *Brassica napus* in a liquid culture medium, *Z. Pflanzenphysiol.*, 103, 229, 1981.

115. Lichter, R., Induction of haploid plants from isolated pollen of *Brassica napus*, *Z. Pflanzenphysiol.*, 105, 427, 1982.

116. Roulund, N., Andersen, S.B., and Farestveit, B., Optimal concentration of sucrose for head cabbage (*Brassica oleracea* L. convar. *capitata* (L.) Alef.) anther culture, *Euphytica*, 52, 125, 1991.

117. Baillie, A.M.R., Epp, D.J., Hutcheson, D., and Keller, W.A., *In vitro* culture of isolated microspores and regeneration of plants in *Brassica campestris*, *Plant Cell Rep.*, 11, 234, 1992.

118. Pescitelli, S.M., Johnson, C.D., and Petolino, J.F., Isolated microspore culture of maize: effects of isolation technique, reduced temperature, and sucrose level, *Plant Cell Rep.*, 8, 628, 1990.

119. Powell, W., and Uhrig, H., Anther culture of *Solanum* genotypes, *Plant Cell Tiss. Org. Cult.*, 11, 13, 1987.

120. Babbar, S.B., and Gupta, S.C., Putative role of ethylene in *Datura metel* microspore embryogenesis, *Physiol. Plant.*, 68, 141, 1986.

121. Babbar, S.B., and Gupta, S.C., Chemicals affecting androgenic response of *Datura metel*. Glutamine, glutamic acid, serine and inositol, *Beitr. Biol. Pflanzen.*, 60, 459, 1985.

122. Aruga, K., and Nakajima, T., Factors affecting the process of embryo formation from pollen grains in tobacco, *Japan. J. Breed.*, 35, 127, 1985.

123. Olsen, F.L., Induction of microspore embryogenesis in cultured anthers of *Hordeum vulgare*. The effects of ammonium nitrate, glutamine, and asparagine as nitrogen sources, *Carlsberg Res. Commun.*, 52, 393, 1987.

124. Buter, B., Schmid, J.E., and Stamp, P., Effects of L-proline and post-plating temperature treatment on maize (*Zea mays* L.) anther culture, *Plant Cell Rep.*, 10, 325, 1991.

125. Tiainen, T., The role of ethylene and reducing agents on anther culture response of tetraploid potato (*Solanum tuberosum* L.), *Plant Cell Rep.*, 10, 604, 1992.

126. Reynolds, T.L., A possible role for ethylene during IAA-induced pollen embryogenesis in anther cultures of *Solanum carolinense* L., *Amer. J. Bot.*, 74, 967, 1987.

127. Marburger, J.E., and Wang, R.R.-C., Anther culture of some perennial triticeae, *Plant Cell Rep.*, 7, 313, 1988.

128. Charne, D.G., and Beversdorf, W.D., Improving microspore culture as a rapeseed breeding tool: the use of auxins and cytokinins in an induction medium, *Can. J. Bot.*, 66, 1671, 1988.

129. Biddington, N.L., and Robinson, H.T., Ethylene production during anther culture of brussels sprouts (*Brassica oleracea* var. *gemmifera*) and its relationship with factors that affect embryo production, *Plant Cell, Tiss. Org. Cult.*, 25, 169, 1991.

130. Cho, U.-H., and Kasha, K.J., Ethylene production and embryogenesis from anther cultures of barley (*Hordeum vulgare*), *Plant Cell Rep.*, 8, 415, 1990.

131. Babbar, S.B., and Gupta, S.C., Obligatory and period specific requirement of iron for microspore embryogenesis in *Datura metel* anther cultures, *Bot. Mag. Tokyo*, 99, 225, 1986.

337

132. Calleberg, E.K., Kristjansdottir, I.S., and Johansson, L.B., Anther cultures of tetraploid *Solanum* genotypes – the influence of gelling agents and correlations between incubation temperature and pollen germination temperature, *Plant Cell Tiss. Org. Cult.*, 19, 189, 1989.

133. Ziegler, G., Dressler, K., and Hess, D., Investigations on the anther culturability of four German spring wheat cultivars and the influence of light on regeneration of green vs. albino plants, *Plant Breed.*, 105, 40, 1990.

134. Luckett, D.J., Vemnkatanagappa, S., Darvey, N.L., and Smithard, R.A., Anther culture of Australian wheat germplasm using modified C17 medium and membrane rafts, *Aust. J. Plant Physiol.*, 18, 357, 1991.

135. Kao, K.N., Saleem, M., Abrams, S., Pedras, M., Horn, D., and Mallard, C., Culture conditions for induction of green plants from barley microspores by anther culture methods, *Plant Cell Rep.*, 9, 595, 1991.

136. Hunter, C.P., The effect of anther orientation on the production of microspore-derived embryoids and plants of *Hordeum vulgare* cv. Sabarlis, *Plant Cell Rep.*, 4, 267, 1985.

137. Tsay, H.S., Miao, S.H., and Widholm, J.M., Factors affecting haploid plant regeneration from maize anther culture, *J. Plant Physiol.*, 126, 33, 1986.

138. Johansson, L., Effects of activated charcoal in anther culture, *Physiol. Plant.*, 59, 397, 1983.

139. Anagnostakis, S.L., Haploid plants from anthers of tobacco enhancement with charcoal, *Planta*, 115, 281, 1974.

140. Vagera, J., and Havranek P., *In vitro* induction of androgenesis in *Capsicum annuum* L. and its genetic aspects, *Biol. Plant. (Praha)*, 27, 10, 1985.

141. Gavel, S., Babbar, S.B., and Gupta, S.C., Plant regeneration from *in vitro* cultured anthers of black mustard (*Brassica nigra* Koch), *Plant Breed.*, 97, 64, 1986.

142. Sato, T., Nichio, T., and Hirai, M., Plant regeneration from isolated microspore cultures of Chinese cabbage (*Brassica campestris* spp. *pekinensis*), *Plant Cell Rep.*, 8, 486, 1989.

143. Chen, Z., and Chen, Z., High frequency induction of pollen-derived embryoids from anther cultures of rape (*Brassica napus*), *Kexue Tongbao*, 28, 1690, 1983.

144. Dunwell, J.M., and Thurling, N., Role of sucrose in microspore embryo production in *Brassica napus* ssp. *oleifera*, *J. Exp. Bot.*, 36, 1478, 1985.

145. Marsolais, A.A., and Kasha, K.J., Callus induction from barley microspores. The role of sucrose and auxin in a barley anther culture medium, *Can. J. Bot.*, 63, 2209, 1985.

146. Henry, Y., and De Buyser, J., Float culture of wheat anthers, *Theor. Appl. Genet.*, 60, 77, 1981.

147. Lillo, C., and Hansen, M., Anther culture of cabbage. Influence of growth temperature of donor plants and media composition on embryo yield and plant regeneration, *Norw. J. Agric. Sci.*, 1, 105, 1987.

148. Phippen, C., and Ockendon, D.J., Genotype, plant, bud size and media factors affecting anther culture of cauliflowers (*Brassica oleracea* var. *botrytis*), *Theor. Appl. Genet.*, 79, 33, 1990.

149. Biddington, N.L., Sutherland, R.A., and Robinson, H.T., Silver nitrate increases embryo production in anther culture of brussel sprouts, *Ann. Bot.*, 62, 181, 1988.

150. Mercy, S.T., and Zapata, F.J., Position of anthers at plating and its influence on anther callusing in rice, *Plant Cell Rep.*, 6, 318, 1987.

151. Rashid, A., Induction of embryos in pollen cultures of *Nicotiana sylvestris*, *Physiol. Plant.*, 56, 223, 1982.

152. Chuong, P.V., and Beversdorf, W.D., High frequency embryogenesis through isolated microspore culture in *Brassica napus* L. and *B. carinata* Braun, *Plant Sci.*, 39, 219, 1985.

153. Arnison, P.G., Donaldson, P., Ho, L.C.C., and Keller, W.A., The influence of various physical parameters on anther culture of broccoli (*Brassica oleracea* var. *italica*), *Plant Cell Tiss. Org. Cult.*, 20, 147, 1990.

154. Johansson, L., and Eriksson, T., Effects of carbon dioxide in anther cultures, *Physiol. Plant.*, 60, 26, 1984.

155. Huang, B., and Keller, W.A., Microspore culture technology, *J. Tiss. Cult. Meth.*, 12, 171, 1989.

156. Mix, G., Production of dihaploid plantlets from anthers of autotetraploid genotypes of *Solanum tuberosum* L., *Potato Res.*, 26, 63, 1983.

157. Foroughi-Wehr, B., Wilson, H.M., and Mix, G., Monohaploid plants from anthers of a dihaploid genotype of *Solanum tuberosum* L., *Euphytica*, 26, 361, 1977.

158. Batty, N., and Dunwell, J., Effect of maltose on the response of potato anthers in culture, *Plant Cell Tiss. Org. Cult.*, 18, 221, 1989.

159. Murashige, T., and Skoog, F., A revised medium for rapid growth and bioassay with tobacco tissue culture, *Physiol. Plant.*, 15, 473, 1962.

160. Linsmaier, E.M., and Skoog, F., Organic growth factor requirements of tobacco tissue cultures, *Physiol. Plant.*, 18, 100, 1965.

161. Johansson, L., Anderson, A., and Eriksson, T., Improvement of anther culture technique: Activated charcoal bound in agar medium in combination with liquid medium and elevated CO_2 concentration, *Physiol. Plant.*, 54, 24, 1982.

162. Sopory, S.K., Establishment of conditions for the induction of androgenetic embryoids in cultured anthers of dihaploid potato, in *Plant Tissue Culture, Genetic Manipulation and Somatic Hybridization*, Rao, P.S., Heble, H.R., and Chadha, M.S., Eds., Bhabha Atomic Res. Centre, Bombay, India, 1980, 85.

163. Sopory, S.K., Jacobsen, E., and Wenzel, G., Production of monohaploid embryoids and plantlets in cultured anthers of *Solanum tuberosum*, *Plant Sci. Lett.*, 12, 47, 1978.

164. Foroughi-Wehr, B., and Mix, G., *In vitro* response of *Hordeum vulgare* L. anthers cultured from plants grown under different environments, *Environ. and Exp. Bot.*, 19, 303, 1979.

165. Kohler, F. and Wenzel, G., Regeneration of isolated barley microspores in conditioned media and trials to characterize the responsible factor, *J. Plant Physiol.*, 121, 181, 1985.

166. Hunter, C.P., Plant regeneration from microspores of barley, *Hordeum vulgare*, Ph.D. Thesis, Wye College, University of London, London, 1988.

167. Gamborg, O.L., Miller, R.A., and Ojima, K., Nutrient requirements of suspension cultures of soybean root cells, *Exp. Cell Res.*, 50, 151, 1968.

168. Sorvari, S., Comparison of anther cultures of barley cultivars in barley-starch and agar gelatinized media, *Ann. Agricult. Fenn.*, 25, 127, 1986.

169. Sorvari, S., and Schieder, O., Influence of sucrose and melibiose on barley anther cultures in starch media, *Plant Breed.*, 99, 164, 1987.

170. Dunwell, J.M., Anther and ovary culture, in *Cereal Tissue and Cell Culture*, Bright, S.W.J., and Jones, M.G.K., Eds., Martinus Nijhoff/Dr. W. Junk Pub., London, 1985, 1.

171. San, L.H., and Gelebart, P., Production of gynogenetic haploids, in *Cell Culture and Somatic Cell Genetics of Plants*, Vasil, I.K., Ed., Academic Press, New York, 1986, 305.

172. Yang, H.Y., and Zhou, C., *In vitro* gynogenesis, in *Plant Tissue Culture Applications and Limitations*, Bhojwani, S.S., Ed., Elsevier, Oxford, NY, 1990, 242.

173. Yang, H.Y., Yan, H., and Zhou, C., *In vitro* production of haploids in *Helianthus*, in *Biotechnology in Agriculture and Forestry, Vol. 10: Legumes and Oilseed Crops I*, Bajaj, Y.P.S. Ed., Springer-Verlag, Berlin, 1990, 472.

174. Yan, H., Yang, H.Y., and Jensen, W.A., An electron microscope study on *in vitro* parthenogenesis in sunflower, *Sex. Plant Reprod.*, 2, 154, 1989.

175. He, C.P., and Yang, H.Y., An investigation on the stability of synergid apogamy and its condition in rice ovary culture, *J. Wuhan Bot. Res.*, 6, 203, 1988.

176. Tian, H.Q., and Yang, H.Y., Haploid embryogeny and plant regeneration in unpollinated ovary culture of *Allium tuberosum*, *Acta. Biol. Exp. Sin.*, 22, 139, 1989.

177. Von Aderkas, P., and Bonga, J.M., Formation of haploid embryoids of *Larix decudia*: early embryogenesis, *Amer. J. Bot.*, 75, 690, 1988.

178. Von Aderkas, P., Bonga, J.M., and Nagruaui, R., Promotion of embryogenesis in cultured megagametophytes of *Larix decudia*, *Can. J. Forest Res.*, 17, 1293, 1987.

179. Bonga, J.M., Von Aderkas, P., and James, D., Potential application of haploid cultures in three species, in *Genetic Manipulation of Woody Plants*, Hanover, J.W., and Keathley, D.E., Eds., Plenum Press, New York, NY, 1988, 57.

180. Rohr, R., Haploids (Gymnosperms), in *Cell and Tissue Culture in Forestry. Vol. 2.*, Bonga, J.M., and Dierjan, D.J., Eds., Martinus Nijhoff Publishers, Dordrecht, 1987, 230.

181. Lux, H., Herrmann, L., and Wetzel, C., Production of haploid sugar beet (*Beta vulgaris* L.) by culturing unpollinated ovules, *Z. Pflanzenzuchtg.*, 104, 177, 1990.

182. Van Geyt, J., Speckmann, G.J. Jr., D'Halluin, K., and Jacobs, M., *In vitro* induction of haploid plants from unpollinated ovules and ovaries of the sugarbeet (*Beta vulgaris* L.), *Theor. Appl. Genet.*, 73, 920, 1987.

183. Yang, H.Y., Zhou, C., Cai, D., Yan, H., Wu, Y., and Chen, X.M., *In vitro* culture of unfertilized ovules in *Helianthus annuus* L., in *Haploids of Higher Plants in vitro*, Hu, H., and Yang, H.Y., Eds., China Academic Pub., Springer-Verlag, Beijing, 182, 1985.

184. D'Halluin, K., and Keimer, B., Production of haploid sugarbeets (*Beta vulgaris* L.) by ovule culture, in *Genetic Manipulation in Plant Breeding*, Horn, W., Jerser, C.J., Odenbach, W., and Schieder, O., Eds., Walter de Gruyter and Co., Berlin, 1986, 307.

185. Campion, B., and Alloni, C., Induction of haploid plants in onion (*Allium cepa* L.) by *in vitro* culture of unpollinated ovules, *Plant Cell Tiss. Org. Cult.*, 20, 1, 1990.

186. Campion, B., Azzimonti, M.T., Vicini, E., Schiavi, M., and Falavigna, A., Advances in haploid plant induction in onion (*Allium cepa* L.) through *in vitro* gynogenesis, *Plant Sci.*, 86, 97, 1992.

187. Keller, J., Culture of unpollinated ovules, ovaries, and flower buds in some species of the genus *Allium* and haploid induction via gynogenesis in onion (*Allium cepa* L.), *Euphytica*, 47, 241, 1990.

188. Tsay, H.S., and Su, C.Y., Anther culture of papaya (*Carica papaya* L.), *Plant Cell Rep.*, 4, 28, 1985.

189. Guha, S., and Johri, B.M., *In vitro* development of ovary and ovule of *Allium cepa* L., *Phytomorphology*, 16, 353, 1966.

190. Radojević, L., Djordjević, N., and Tucić, B., *In vitro* induction of pollen embryos and plantlets in *Aesculus carnea* Hayne through anther culture, *Plant Cell Tiss. Org. Cult.*, 17, 21, 1989.

191. Johansson, L.B., Calleberg, E., and Gedin, A., Correlations between activated charcoal, Fe-EDTA and other organic media ingredients in cultured anthers of *Anemone canadensis*, *Physiol. Plant.*, 80, 243, 1990.

192. Burnett, L., Yarrow, S., and Huang, B., Embryogenesis and plant regeneration from isolated microspores of *Brassica rapa* L. ssp. *oleifera*, *Plant Cell Rep.*, 11, 215, 1992.

193. Lichter, R., Efficient yield of embryoids by culture of isolated microspores of different *Brassicaceae* species, *Plant Breed.*, 103, 119, 1989.

194. Sorvari, S., Production of haploids from anther culture in agriculturally valuable *Brassica campestris* L. cultivars, *Ann. Agric. Fenn.*, 24, 149, 1985.

195. Sharma, K.K., and Bhojwani, S.S., Microspore embryogenesis in anther cultures of two Indian cultivars of *Brassica juncea* (L.) Czern., *Plant Cell Tiss. Org. Cult.*, 4, 234, 1985.

196. Chuong, P.V., Pauls, K.P., and Beversdorf, W.D., High-frequency embryogenesis in male sterile plants of *Brassica napus* through microspore culture, *Can. J. Bot.*, 66, 1676, 1988.

197. Dunwell, J.M., Cornish, M., and DeCourcel, A.G.L., Influence of genotype, plant growth temperature and anther incubation temperature on microspore embryo production in *Brassica napus* ssp. *oleifera*, *J. of Exp. Bot.*, 36, 679, 1985.

198. Gland, A., Lichter, R., and Schweiger, H., Genetic and exogenous factors affecting embryogenesis in isolated microspore cultures of *Brassica napus* L., *J. Plant Physiol.*, 132, 613, 1988.

199. Huang, B., Bird, S., Kemble, R., Simmonds, D., Keller, W., and Miki, B., Effects of culture density, conditioned medium and feeder cultures on microspore embryogenesis in *Brassica napus* L. cv. Topas, *Plant Cell Rep.*, 8, 594, 1990.

200. Mandel, A., Induction of androgenetic haploids in the breeding materials of winter rape (*Brassica napus*), *Hereditas*, 106, 189, 1987.

201. Siebel, J., and Pauls, K.P., A comparison of anther and microspore culture as a breeding tool in *Brassica napus*, *Theor. Appl. Genet.*, 78, 473, 1989.

202. Arnison, P.G., and Keller, W.A., A survey of the anther culture response of *Brassica oleracea* L. cultivars grown under field conditions, *Plant Breed.*, 104, 125, 1990.

340

203. Ockendon, D.J., and Sutherland, R.A., Genetic and non-genetic factors affecting anther culture of Brussels sprouts (*Brassica oleracea* var. *gemmifera*), *Theor. Appl. Genet.*, 74, 566, 1987.

204. Roulund, N., Hansted, L., Anderson, S.B., and Farestveit, B., Effect of genotype, environment and carbohydrate on anther culture response in head cabbage (*Brassica oleracea* L. canvar. *capitata* (L.) Alef.), *Euphytica*, 49, 237, 1990.

205. Chaturvedi, H.C., and Sharma, A.K., Androgenesis in *Citrus aurantifolia* (Christm.) Swingle, *Planta*, 165, 142, 1985.

206. Monfort, S., Androgenesis of coconut: Embryos from anther culture, *Z. Pflanzenzuchtg.*, 94, 251, 1985.

207. Perez-Bermudez, P., Cornejo, M.J., and Segura, J., Pollen plant formation from anther cultures of *Digitalis obscura* L., *Plant Cell Tiss. Org. Cult.*, 5, 63, 1985.

208. Alissa, A., Serieys, H., and Jonard, R., Wild species and interspecific hybrid regeneration in *Helianthus* genus by *in vitro* androgenesis, *C.R. Acad. Sci. Paris*, 300, 25, 1985.

209. Zivy, M., Devaux, P., Blaisonneau, J., Jean, R., and Thiellement, H., Segregation distortion and linkage studies in microspore-derived double haploid lines of *Hordeum vulgare* L., *Theor. Appl. Genet.*, 83, 919, 1992.

210. Canhoto, J.M., Ludovina, M., Guimaraes, S., and Cruz, G.S., *In vitro* induction of haploid, diploid and triploid plantlets by anther culture of *Iochroma warscewiczii* Regel, *Plant Cell Tiss. Org. Cult.*, 21, 171, 1990.

211. Olesen, A., Andersen, S.B., and Due, I.K., Anther culture response in Perennial Ryegrass (*Lolium perenne* L.), *Plant Breed.*, 101, 60, 1988.

212. Shoukang, L., Dongfeng, J., and Jun, Q., *In vitro* production of haploid plants from mulberry (*Morus*) anther culture, *Scientia Sinica*, 30, 853, 1987.

213. Lespinasse, R., De Paepe, R., and Koulou, A., Induction of B chromosome formation in androgenetic lines of *Nicotiana sylvestris*, *Caryologia*, 40, 327, 1987.

214. Deaton, W.R., Collins, G.B., and Nielsen, M.T., Vigor and variation expressed by anther-derived doubled haploids of burley tobacco (*Nicotiana tabacum* L.). II. Evaluation of first- and second-cycle doubled haploids, *Euphytica*, 35, 41, 1986.

215. Kumashiro, T., and Oinuma, T., Comparison of genetic variability among anther-derived and ovule-derived doubled haploid lines of tobacco, *Japan. J. Breed.*, 35, 301, 1985.

216. Kumashiro, T., and Oinuma, T., Genetic variation of colchicine induced and spontaneous doubled haploid obtained by anther culture of an inbred tobacco, *Japan. J. Breed.*, 36, 355, 1986.

217. Ostrem, J.A., Litton, C.C., and Collins, G.B., Isolation of dark tobacco breeding lines differing in alkaloid concentration via anther derived haploids, *Z. Pflanzenzuchtg.*, 96, 224, 1986.

218. Schnell, R.J., and Wernsman, E.A., Androgenic somaclonal variation in tobacco and estimation of its value as a source of novel genetic variability, *Crop Sci.*, 26, 84, 1986.

219. Subhashini, U., and Venkateswarlu, T., Genotypes and their response to *in vitro* production of haploids in F_1 lines of *Nicotiana tabacum*, *Theor. Appl. Genet.*, 70, 225, 1985.

220. Hyun, S.K., Kim, J.H., Noh, E.W., and Park, J.I., Induction of haploid plants of *Populus* species, in *Plant Tissue Culture and its Agricultural Applications*, Withers, L.A., and Alderson, P.G., Eds., Butterworth, London, 1986, 413.

221. Flehinghaus, T., Deimling, S., and Geiger, H.H., Methodical improvements in rye anther culture, *Plant Cell Rep.*, 10, 397, 1991.

222. Jain, R.K., Brune, U., and Friedt, W., Plant regeneration from *in vitro* cultures of cotyledon explants and anthers of *Sinapis alba* and its implications on breeding of crucifers, *Euphytica*, 43, 153, 1989.

223. Tsay, H.S., Yeh, C.C., and Hsu, J.Y., Embryogenesis and plant regeneration from anther culture of bamboo (*Sinocalamus latiflora* (Munro) McClure), *Plant Cell Rep.*, 9, 349, 1990.

224. Cappadocia, M., and Ahmim, M., Comparison of two culture methods for the production of haploids by anther culture in *Solanum chacoense*, *Can. J. Bot.*, 66, 1003, 1988.

225. Rivard, S.R., Cappadocia, M., Vincent, G., Brisson, N., and Landry, B.S., Restriction fragment length polymorphism (RFLP) analyses of plants produced by *in vitro* anther culture of *Solanum chacoense* Bitt., *Theor. Appl. Genet.*, 78, 49, 1989.

226. Uhrig, H., Genetic selection and liquid medium conditions improve the yield of androgenetic plants from diploid potatoes, *Theor. Appl. Genet.*, 71, 455, 1985.

227. Hassawi, D.S., and Liang, G.H., Effect of cultivar, incubation temperature, and stage of microspore development on anther culture in wheat and triticale, *Plant Breed.*, 105, 332, 1990.

228. Daofen, H., Zhendong, Y., Yunlian, T., and Jianping, L., Jinghua No 1 – A winter wheat variety derived from pollen sporophyte, *Scientia Sinica*, 29, 733, 1986.

229. De Buyser, J., Henry, Y., Lonnet, P., Hertzog, R., and Hespel, A., "Florin": A doubled haploid wheat variety developed by the anther culture method, *Plant Breed.*, 98, 53, 1987.

230. Karsai, I., Bedo, Z., and Balla, L., The effect of repeated anther culture on *in vitro* androgenesis of wheat (*Triticum aestivum* L.), *Cereal Res. Commun.*, 19, 425, 1991.

231. Metz, S.G., Sharma, H.C., Armstrong, T.A., and Mascia, P.N., Chromosome doubling and aneuploidy in anther-derived plants from two winter wheat lines, *Genome*, 30, 177, 1988.

232. Muller, G., Vahl, U., and Wiberg, A., The use of anther culture in breeding winter wheat. II. Production of new doubled haploid lines of winter wheat with 1 AL-1 RS-translocation, *Plant Breed.*, 103, 81, 1989.

233. Rode, A., Hartmann, C., Dron, M., Picard, E., and Ouetier, F., Organelle genome stability in anther-derived doubled haploids of wheat (*Triticum aestivum* L., cv. "Moisson"), *Theor. Appl. Genet.*, 71, 320, 1985.

234. Zhou, H., Zheng, Y., and Konzak, C.F., Osmotic potential of media affecting green plant percentage in wheat anther culture, *Plant Cell Rep.*, 10, 63, 1991.

235. Ghaemi, M., Sarrafi, A., and Alibert, G., Influence of genotype and culture conditions on the production of embryos from anthers of tetraploid wheat (*Triticum turgidum*), *Euphytica*, 65, 81, 1993.

236. Hongchang, M., Liang, G.H., and Wassom, C.E., Effects of growth regulators and genotypes on callus and embryoid induction from maize anther culture, *Plant Breed.*, 106, 47, 1991.

237. Petolino, J.F., and Jones, A.M., Anther culture of elite genotypes of maize, *Crop Sci.*, 26, 1072, 1986.

238. Petolino, J.F., and Thompson, S.A., Genetic analysis of anther culture response in maize, *Theor. Appl. Genet.*, 74, 284, 1987.

239. Dunwell, J.M., Pollen, ovule and embryo culture as tools in plant breeding, in *Plant Tissue Culture and its Agricultural Applications*, Withers, L.A., and Alderson, P.G., Eds., Butterworths, London, 1986, 375.

240. Hu, H., and Zeng, J.Z., Development of new varieties via anther culture, in *Handbook of Plant Cell Cultures, Vol. 3: Crop Species*, Ammirato, P.V., Evans, D.A., Sharp, W.R., and Yamada, Y., Eds., Macmillan Pub. Co., New York, 1984, 65.

241. Morrison, R.A., and Evans, D.A., Haploid plants from tissue culture: New plant varieties in a shortened time frame, *Bio/Technol.*, 6, 684, 1988.

242. Anderson, S.B., Due, I.K., and Olesen, A., The response of anther culture in a genetically wide material of winter wheat (*Triticum aestivum* L.), *Plant Breed.*, 99, 181, 1987.

243. Day, A., and Ellis, T.H.N., Deleted forms of plastid DNA in albino plants from cereal anther culture, *Current Genetics*, 9, 671, 1985.

244. Keller, W.A., and Armstrong, K.C., Dihaploid plant production by anther culture in autotetraploid marrowstem kale (*Brassica oleracea* L. var. *Acephala* D.C.), *Can. J. Genet. Cytol.*, 23, 259, 1981.

245. Huang, B., Ultrastructural aspects of pollen embryogenesis in *Hordeum*, *Triticum*, and *Paeonia*, in *Haploids of Higher Plants In Vitro*, Hu, H., and Yang, H., Eds., Springer-Verlag, Berlin, 1986, 91.

246. Chu, Q., Zhang, Z., and Gao, Y., Cytogenetical analysis on aneuploids obtained from pollen clones of rice (*Oryza sativa* L.), *Theor. Appl. Genet.*, 71, 506, 1985.

342

247. Gu, M., Cytogenetic stability and variability of cell and cell clones originated from maize pollen and their regenerated plants, in *Haploids of Higher Plants In Vitro*, Hu, H., and Yang, H., Eds., Springer-Verlag, Berlin, 1986, 79.

248. Hu, H., Variability and gametic expression in pollen derived plants in wheat, in *Haploids of Higher Plants In Vitro*, Hu, H., and Yang, H., Eds., Springer-Verlag, Berlin, 1986, 67.

249. Powell, W., Hayter, A.M., Wood, W., Dunwell, J.M., and Huang, B., Variation in the agronomic characters of microspore-derived plants of *Hordeum vulgare* cv. Sabarlis, *Heredity*, 52, 19, 1984.

250. Dhillion, S.S., Wernsman, E.A., and Miksche, J.P., Evaluation of nuclear DNA content and heterochromatin changes in anther-derived dihaploids of tobacco (*Nicotiana tabacum*) cv. Coker 139, *Can. J. Genet. Cytol.*, 25, 169, 1983.

251. De Paepe, R., Prat, D., and Knight, J., Effects of consecutive androgenesis on morphology and fertility in *Nicotiana sylvestris*, *Can. J. Bot.*, 61, 2088, 1983.

252. Matzinger, D.F., and Burk, L.G., Cytoplasmic modification by anther culture in *Nicotiana tabacum* L., *J. Hered.*, 75, 167, 1984.

253. Charne, D.G., Pukacki, P., Kott, L.S., and Beversdorf, W.D., Embryogenesis following cryopreservation in isolated microspores of rapeseed (*Brassica napus* L.), *Plant Cell Rep.*, 7, 407, 1988.

254. Chen, J.L., and Beversdorf, W.D., Production of spontaneous diploid lines from isolated microspores following cryopreservation of spring rapeseed (*Brassica napus* L.), *Plant Breed.*, 108, 324, 1992.

255. Barnabas, B., Pfahler, P.L., and Kovacs, G., Direct effect of colchicine on microspore embryogenesis to produce dihaploid plants in wheat (*Triticum aestivum* L.), *Theor. Appl. Genet.*, 81, 675, 1991.

256. Loh, C.-S., and Ingram, D.S., The response of haploid secondary embryoids and secondary embryogenic tissues of winter oilseed rape to treatment with colchicine, *New Phytol.*, 95, 359, 1983.

257. Wong, C.K., A new approach to chromosome doubling for haploid rice plants, *Theor. Appl. Genet.*, 77, 149, 1989.

258. Beversdorf, W.D., Charne, D.G., Kott, L.S., Chuong, P.V., Polsoni, L., and Zilka, J., The utilization of microspore culture and microspore-derived doubled-haploids in a rapeseed (*Brassica napus* L.) breeding programme, in *Proc. of the 7th Intl. Rapeseed Congress*, Poland, 1987, 86.

259. Ulrich, A., Furton, W.H., and Downey, R.K., Biotechnology and rapeseed breeding: Some economic considerations, *Sci. Counc. Can. Rep.*, Ottawa, 67, 1984.

260. Rajhathy, T., Haploid flax revisited, *Z. Pflanzenzuchtg.*, 76, 1, 1976.

261. Anonymous, Success of breeding the new tobacco cultivar "Tan-Yuh no. 1", *Acta Botanica Sinica*, 16, 300, 1974.

262. Nakamura, A.T., Yamada, M., Oka, Y., Tatemichi, K., Egushi, T., Ayabe, T., and Kobayashi, K., Studies on the haploid method of breeding by anther culture in tobacco, V. Breeding of mild flue-cured variety F211 by haploid method, *Bull. Iwata Tobacco Exp. Station*, 7, 29, 1975.

263. Chaplin, J.F., Burk, L.G., Gooding, G.V., and Powell, N.T., Registration of NC744 tobacco germplasm (Reg No. GP 18), *Crop Sci.*, 20, 677, 1980.

264. Chaplin, J.F., and Burk, L.G., Registration of LMAFC 34 tobacco germplasm, *Crop Sci.*, 24, 1220, 1984.

265. Caligari, P.D.S., and Powell, W., The use of double haploids for detecting linkage and pleiotropy between quantitatively varying characters in spring barley, *J. Agric. Sci.*, 106, 75, 1986.

266. Choo, T.M., and Reinbergs, E., Doubled haploids for detecting pleiotropy and linkage of genes controlling two characters, *Genome*, 29, 584, 1987.

267. Snape, J.W., Wright, A.J., and Simpson, E., Methods for estimating gene numbers for quantitative characters using doubled haploid lines, *Theor. Appl. Genet.*, 67, 143, 1984.

268. Wang, X., and Hu, H., The chromosome constitution of plants derived from pollen of hexaploid triticale x common wheat F_1 hybrids, *Theor. Appl. Genet.*, 70, 92, 1985.

269. Swanson, E.B., Coumans, M.P., Brown, G.L., Patel, J.D., and Beversdorf, W.D., The characterization of herbicide tolerant plants in *Brassica napus* L. after *in vitro* selection of microspores and protoplasts, *Plant Cell Rep.*, 7, 83, 1988.

270. Swanson, E.B., Herrgesell, M.J., Arnoldo, M., Sippell, D.W., and Wang, R.S.C., Microspore mutagenesis and selection: canola plants with field tolerance to the imidazolinones, *Theor. Appl. Genet.*, 78, 525, 1989.

271. Ray, T.B., Site of action of chlorsulfuron, *Plant Physiol.*, 75, 827, 1984.

272. Shaner, D.L., Anderson, P.C., and Stidman, M.A., Imidazolinones – potent inhibitors of acetohydroxyacid synthase, *Plant Physiol.*, 76, 545, 1984.

273. Wong, S.-C., and Swanson, E., Genetic modifications of canola oil: high oleic acid canola, in *Fat and Cholesterol Reduced Food*, Haberstroh, C., and Morris, C.E., Eds., Gulf Pub., Houston, 1991, 154.

274. Turner, J., and Facciotti, D., High oleic *Brassica napus* from mutagenized microspores, in *Proc. of the Sixth Crucifer Genetics Workshop*, McFerson, J.R., Kresovich, S., and Dwyer, S.G., Eds., USDA-ARS, Geneva, NY, 1990, 24.

275. Huang, B., Swanson, E.B., Baszczynski, C.L., Macrae, W.D., Barbour, E., Armavil, V., Cooke, L., Arnoldo, M., Rozakis, S., Westecott, M., Keats, R.F., and Kemble, R., Application of microspore culture to canola improvement, in *Proc. 8th Intl. Rapeseed Congress*, McGregor, D.I., Ed., Saskatoon, Canada, 1991, 298.

276. Wenzel, G., Strategies in unconventional breeding for disease resistance, *Ann. Rev. Phytopathol.*, 23, 49, 1985.

277. Fadel, F., and Wenzel, G., *In vitro* selection for tolerance to fusarium in F_1 microspore populations of wheat, *Plant Breed.*, 110, 89, 1993.

278. Chen, Y., The inheritance of rice pollen plants and its application in crop improvement, in *Haploids of Higher Plants In Vitro*, Hu, H., and Yang, H., Eds., Springer-Verlag, Berlin, 1986, 118.

279. Swanson, E.B., and Erickson, L.R., Haploid transformation in *Brassica napus* using an octopine-producing strain of *Agrobacterium tumefaciens*, *Theor. Appl. Genet.*, 78, 831, 1989.

280. Oelck, M.M., Phan, C.V., Eckes, P., Donn, G., Rakow, G., and Keller, W.A., Field resistance of canola transformants (*Brassica napus* L.) to ignite (phosphinothricin), in *Proc. 8th Intl. Rapeseed Congress*, McGregor, D.I., Ed., Saskatoon, Canada, 1991, 293.

281. Creissen, G., Smith, C., Francis, R., Reynolds, H., and Mullineaux, P., *Agrobacterium*- and microprojectile-mediated viral DNA delivery into barley microspore-derived cultures, *Plant Cell Rep.*, 8, 680, 1990.

282. Neuhaus, G., Spangenberg, G., Mittelsten Scheid, O., and Schweiger, H.-G., Transgenic rapeseed plants obtained by the microinjection of DNA into microspore-derived embryoids, *Theor. Appl. Genet.*, 75, 30, 1987.

283. Bolik, M., and Koop, H.U., Identification of embryogenic microspores of barley (*Hordeum vulgare* L.) by individual selection and culture and their potential for transformation by microinjection, *Protoplasma*, 162, 61, 1991.

284. Joersbo, M., Jorgensen, R.B., and Olesen, P., Transient electropermeabilization of barley (*Hordeum vulgare* L.) microspores to propidium iodide, *Plant Cell Tiss. Org. Cult.*, 23, 125, 1990.

285. Liu, C.-M., Xu, Z.-H., and Chua, H.-H., Auxin polar transport is essential for the establishment of bilateral symmetry during early plant embryogenesis, *Plant Cell*, 5, 621, 1993.

286. Reynolds, T.L., and Kitto, S.L., Identification of embryoid abundant genes that are temporally expressed during pollen embryogenesis in wheat anther cultures, *Plant Physiol.*, 100, 1744, 1992.

287. Dunwell, J.M., Mechanisms of microspore embryogenesis, in *Reproductive Biology and Plant Breeding*, Dattee, Y.H., Dumas, C., and Gallais, A., Eds., Springer Verlag, New York, N.Y., 1992, 121.

288. Powell, W., Environmental and Genetical Aspects of Pollen embryogenesis, in *Biotechnology in Agriculture and Forestry 12; Haploids in Crop Improvement I*, Bajaj, Y.P.S., Ed., Springer-Verlag, Berlin, 1990, 45.
289. Reynolds, T.L., Ultrastructure of pollen embryogenesis, in *Biotechnology in Agriculture and Forestry 12; Haploids in Crop Improvement I*, Bajaj, Y.P.S. Ed., Springer-Verlag, Berlin, 1990, 66.

10. Somatic Embryogenesis in Herbaceous Dicots

DANIEL C.W. BROWN, KIRSTEN I. FINSTAD and EVA M. WATSON

Contents

I. Introduction

Most reports of somatic embryogenesis describe the development of adventitious bipolar structures that resemble various stages of developing zygotic embryos and which, under the appropriate conditions, can develop functional shoot and root systems. These criteria reflect those outlined by Halperin [1] and Haccius [2], but only in a few species (e.g. *Daucus carota*, *Glycine max*, *Medicago sativa*) has somatic embryo formation been widely studied, *in vitro* protocols systematically optimized and an extensive developmental and histological documentation provided. Somatic embryogenesis has been considered [2] to be a distinct developmental pathway, different from either shoot or root organogenesis, in which a single cell gives rise to a structure containing bipolar meristems and with no direct vascular connections to the maternal tissue. It has also been suggested [3] that embryogenesis is an archetypal event and that shoot organogenesis is only a modification of the process of embryo induction and development. With this view, embryogenesis could be considered to be the result of a sequence of competing organogenic events. The idea that the developmental pathways of organogenesis and embryogenesis are distinct has also been challenged by the view that their relationship is hierarchical in nature [4]. That is, the developmental processes of shoot and root organ formation are just two of several distinct genetic programs that are operative during the formation

T.A. Thorpe (ed.), In Vitro Embryogenesis in Plants, pp. 345–415.
© *1995 Kluwer Academic Publishers, Dordrecht. Printed in the Netherlands.*

of an embryo and subject to regulation by other genetic regulatory processes which may control the position of root and shoot meristems, and the size and morphogenesis proper of an embryo. It has been pointed out [5] that seemingly contradictory observations on somatic embryogenesis can be at least partially resolved if the explant cells are classed as pre-embryogenic determined cells (PEDC) or induced embryogenically determined cells (IEDC) [6, 7]. PEDCs require only permissive conditions to allow cell division and subsequent somatic embryo formation to occur. This is sometimes called direct embryo-genesis. IEDCs require more extensive manipulation for redetermination of differentiated cells, callus proliferation, induction of competence, and then induction and development of embryos. In contrast, this is called indirect embryogenesis.

It is generally accepted that somatic embryogenesis follows a developmental pattern similar to zygotic embryogenesis and that somatic embryos are similar, although not identical [3, 8], morphologically and structurally to zygotic embryos [2, 9–13]. The fact that somatic embryo formation does occur *in vivo* [12] in diverse species such as the dicot *Ranunculus sceleratus*, the orchid *Malaxis paludosa* and the succulent plants *Bryophyllum calycinum* and *Kalanchoe diagremontiana* supports the view that a somatic cell that has not been subjected to mutational change in the nucleus is able to express the full genetic potential of the organism in a similar manner to a zygote if provided with the "proper conditions". The search for the "proper conditions" has been only partially successful as a group of about 180 species of herbaceous dicotyledonous plants comprising 37 taxonomic families (see Table 1) have been reported to form somatic embryos *in vitro* out of an estimated 18,295 recognized generic names of 357 taxonomic families [14].

II. Approaches to plant regeneration

A. Research strategies

The empirical approach to the regeneration of plants has indicated that success is largely dependent on factors which are related to donor plant characteristics, culture medium composition and control of the physical environment. The most common experimental approach to achieving regeneration *in vitro* has been to systematically test one or more of these factors with the plant of interest. A history of successes starting with the first reports of somatic embryo formation in carrot has shaped our view of somatic embryo induction, strongly influenced the strategy for working with recalcitrant species and identified a large number of factors associated with the donor plant, the culture medium and the physical culture environment which influence somatic embryo induction, development and conversion into plants. The organization of the literature documented in Table 1 reflects this traditional approach as we highlighted the factors which are most widely evaluated and reported;

i.e. explant type, medium sequence used for embryo induction, the basic salt composition of the induction medium and the growth regulators, sugars and specific modifications to the medium such as vitamin or amino acid additions.

Species-, cultivar-, plant- and explant-dependant variation in the *in vitro* response has limited the successful application of the above strategy and resulted in regeneration protocols developed with a single plant line or cultivar often having a reputation for being inconsistent and unreliable. An alternate strategy to focussing primarily on empirical explant, medium and culture condition manipulation has been to select a well defined medium and a standard protocol and screen a large selection of germplasm for genotypes that are amenable to the selected protocols [e.g. 15]. A third strategy that has been used in several crop species is to use conventional breeding techniques to obtain tissue culture responsive plants [e.g. 16]. Three lines of evidence show there is a strong genetic influence on regeneration success and have led researchers to adopt this latter approach [17]. First, success in screening and selecting tissue culture responsive plants has been forthcoming in a number of species especially those that are open pollinated. Second, a strong correlation between tissue culture response and the genetic or breeding background has been noted in several species. Third, there is good evidence that *in vitro* responses are under simple genetic control. Some of the few studies that have looked at the segregation of the trait for *in vitro* embryo formation on a particular medium are with *Medicago* and suggest that the trait segregates in a pattern associated with either one or two genes in this species [18–21]. The nature of these genes has yet to be elucidated. It has been cautiously noted [3] that "the genetics associated with morphogenesis exist only in the context of the capacity to detect variation in the process" and considering that the variation in somatic embryo formation is usually evaluated under very limited *in vitro* conditions (e.g. one medium sequence) using limited parameters (e.g. embryo number), the technique of genetic analysis based on the above observations provides a view of only one dimension of the process of somatic embryogenesis and likely oversimplifies our present view of the genetic control of this developmental process.

B. The question of competence

In the early usage of the term, competence was defined by Halperin [22] and Street [23] as the ability to express inherent totipotency or morphogenic capacity; i.e. the ability to regenerate *in vitro*. What this ability consists of and how some cells or tissues have or come to have it while others do not was left for speculation. Halperin [24] proposed that epigenetic factors which comprise the differentiated state of the cells of the original explant tissues may be lost during callus formation. In this classical view, the callus cells are competent because they are undifferentiated, or "dedifferentiated". Street's view [23], on the contrary, was that cells are competent for a particular morphogenetic pathway at the time of culture initiation and that this competence may appear to diminish or change due to intercellular competition during culture. Recently, competence has been defined somewhat more specifically as the ability to receive and/or respond to a developmental signal [25, 26]. Thus, competence can be viewed as a necessary physiological, cellular or molecular state of readiness for the next step of the particular morphogenetic pathway. This definition is more compatible with Halperin's view [24] of competence as requiring a cellular change than with Street's [23] concept that competence is predetermined in the explant tissues. The concept of pre-embryogenic determined cells and induced-embryogenic determined cells reconciles these opposing views [5–7]. Recent work on morphogenetic competence and induction has been interpreted in terms of the acquisition of competence during culture. For example, Christianson and Warnick [27], Coleman and Ernst [28], Ammirato [29] and Finstad [30] all show evidence for a preculture requirement before morphogenetic induction or determination can take place. Competence, as the ability to respond to the inductive signal (usually exposure to certain growth regulators or their removal), is in these cases not present at the time of culture initiation but is acquired sometime after.

Though we can define competence as a concept and in some cases as a time period, there is virtually no data available to describe its nature, either in terms of physiology, biochemistry, cytology or molecular biology. In fact, a competent culture or tissue can only be recognized by its later behaviour following an inductive treatment, although sometimes competent sub-populations of suspension cultures can be distinguished morphologically [31, 32]. Carman [26] lists many factors influencing degree of competence such as genotype, cultural environment of both donor plant and explant, and tissue "memory" of the developmental pathways underway physically or ontogenetically proximal to the original explant. At present, however, it is not possible to distinguish effects on and variation in degree of competence from that of the subsequent differentiation, the only measure of competence available.

The question of what constitutes this competence remains untackled. It is tempting to interpret competence as simply the triggering of mitosis in culture, which may facilitate the interaction of the inductive signal with the DNA. Changes in DNA methylation and the occurrence of pseudomeiotic divisions have both been linked to the acquisition of (totipotency) competence [11]. The fact that the auxin requirements of the competence and induction stages are often distinct suggests a possible change in growth regulator receptors. However, the role of auxin in competence, if any, is not clear since competence can be acquired in its absence [30]. The acquisition of competence may not be so much the result of specific culture conditions as of the excision of the explant. Removal from neighbouring tissues and organs may release the explant from external controls regulating its developmental state. Physiological or physical isolation has long been considered a prerequisite for somatic embryogenesis [9, 33]. Although Steward [34] did not use the term competence, this idea is very reminiscent of his view that isolation of cells from their maternal tissue allows them to express their inherent totipotency.

Ranch and Pace [3] have provided a pragmatic rationale for manipulation of plant cells *in vitro*. They argue that the genetic constitution, the ontogenetic state of the donor tissue and the external culture environment should all be considered with respect to manipulation of somatic embryogenesis *in vitro*. They view the potential for embryogenesis to be determined by the genome. The ontogenetic state of the tissue determines the current action of the genome in the explant and is reflected in the explants' biochemical, physiological and cytological status. Thus, competence, as discussed above, is determined by the genome and the explant's ontogenetic state at the time of introduction *in vitro*. The external culture environment influences this biochemical and physiological state of the cells upon their introduction into culture. The first two factors are properties of cells and predispose the primary explant to a certain behaviour *in vitro* and define its competence. It is only the latter factor, the culture environment, that has been widely studied and is the tool that has been used to manipulate the *in vitro* behaviour of the tissue by influencing its biochemical and physiological state.

C. Genotype manipulation

Genotype is well-established as an important factor influencing the embryogenic response *in vitro*. Among herbaceous dicots, variability in both the occurrence and frequency of somatic embryogenesis has been observed among species (e.g. legumes [35], *Cucumis* spp. [36], *Glycine* spp. [37], *Lycopersicon* spp. [38], *Manihot* spp. [39]), and among cultivars (e.g. alfalfa [17, 40–43], carrot [44], cassava [45], cotton [46, 47], soybean [37, 48, 49]). Variation in degree of embryogenesis has also been noted among individuals of a given cultivar (e.g. alfalfa [40, 43, 50], potato [51], soybean [49, 52] and red clover [53]).

Genotype effects may be interactive with media or other culture effects. The ranking of 26 soybean cultivars according to embryogenic response changed when the auxin in the embryogenesis medium was changed [49]. Chen et al. [43] found that medium and explant source affected the response of less embryogenic alfalfa genotypes but not that of the highly embryogenic ones. Similarly, Brown [17] found that while medium choice modulated the absolute level of embryogenesis, highly regenerating alfalfa cultivars were consistently so while the relative response of lower regenerating cultivars was more subject to the medium employed. Bailey et al. [54] has reported a genotype of soybean which appears to show a superior embryogenic capacity over a range of culture conditions.

Evidence for the genetic basis of the embryogenic response comes partly from the fact that highly embryogenic cultivars can be traced to common origin(s). Brown and Atanassov [41] screened 76 alfalfa cultivars, finding that those showing the highest frequency of somatic embryogenesis had either *Medicago falcata* or "Ladak" germplasm in their background, or both. Of 33 soybean cultivars, all those showing high levels of embryogenesis were descended from one or both of two ancestral germplasm sources [48].

Inheritance studies have shown that capacity for somatic embryogenesis in some species is heritable and probably controlled by at least two loci. Variations in the degree of embryogenesis expressed in alfalfa [18, 21], cotton [47] and potato [51] is taken as evidence for the involvement of more than one gene. F_1 hybrids from parents of different embryogenic capacity may show heterosis [51] or embryogenesis at less than the midparent value but more than the non-embryogenic parent [48]. Hernandez-Fernandez and Christie [18] selfed and intercrossed three alfalfa genotypes of differing embryogenic capacities and concluded that the embryogenesis trait was controlled by two complementary loci. In F_1BC_1 families descended from a cross between an embryogenic and a non-embryogenic potato clone, two non-embryogenic parents sometimes gave rise to embryogenic offspring, suggesting either that the embryogenic alleles were recessive or that there was more than one complementary factor [51].

D. Explant choice

The choice of explant can have a great influence on the success of an embryogenesis protocol. Significant differences in embryogenic response have been observed among mature vegetative tissues [5] and among young tissues [36, 55]. However, the difference between the greater responsiveness of embryonic, seedling or immature tissues and the weaker response of mature tissues is consistently remarked upon among herbaceous dicots [e.g. 39, 46] as well as among monocots and woody species. Almost half of the protocols listed in Table 1 used embryos or seedling parts as the primary explant source. The relative responsiveness of various immature tissues seems to be species-specific; for example, cotyledons were more embryogenic than hypocotyls in cucumber [55] while hypocotyls produced more embryos than seeds or cotyledons of cotton [46]. In soybean [56, 57], only immature cotyledons of a narrow age range appear to be responsive. Occasionally, floral tissues such as ovaries, pedicels, peduncles, buds or even inflorescences are used (e.g. Table 1; [25]). Carman [26] has suggested that floral tissues may be embryogenic because of their developmental proximity to embryogenesis *in vivo*. In general, mature vegetative tissues are more convenient and more readily available than immature or floral tissues. Among herbaceous dicots, leaves and petioles are often used, with shoot tips, stems and roots being used to a lesser degree (Table 1).

The varying efficacy of explants may be a consequence of the degree of differentiation. Immature tissues, for example, may be more likely to undergo dedifferentiation and, hence, be "more embryogenic". Or, since tissues within an organ vary in their capacity to produce embryogenic callus [5], explants with varying tissue composition may be expected to vary in their embryogenic response. Chee [55] observed that cucumber hypocotyls produced a lower percentage of a particular embryogenic callus type than did cotyledons, which in turn produced more embryos. Levels of endogenous growth regulators can vary among organs, and likely affect embryogenesis [26]. Finally, unidentified epigenetic or physiological factors may influence the embryogenic response, as suggested in the interesting results of Nolan et al. [58]. This group observed significantly greater embryogenesis in explants taken from regenerated plantlets of *Medicago truncatula* than in those from the original plant.

E. Media manipulation

The history of plant tissue culture [59] and somatic embryogenesis in particular is to a large extent the story of media development. Although the first successes in *in vitro* somatic embryogenesis, the classic studies in carrot by Steward [60] and Reinert [61], used White's [62] basal medium supplemented with coconut milk, the widespread extension of somatic embryogenesis to other angiosperms did not begin until the introduction of Murashige and Skoog's [63] formulation, commonly referred to as MS medium.

Over 2000 media formulations have been documented [64, 65] but a survey of literature concerning somatic embryogenesis reveals that about half of the embryo induction medium used across all species are MS-based media. In fact, it has been suggested [66] and generally accepted that MS medium be used as a universal control medium for plant tissue cultures. MS medium was designed as a medium for optimal growth of tobacco pith tissue-derived callus and for use as a bioassay of biologically active organic substances. The medium was designed to minimize the response of the callus tissue to variations in inorganic salts, sugars, etc. commonly present in tissue extracts such that changes in the growth of the plant callus tissue reflected the presence of growth promoting organic growth substances in the medium. The surprising result was a medium of widespread applicability that tends to support the authors' view [63] that "cultures from all parts of the plant kingdom apparently have rather similar requirements, and that often their growth is limited by one or a few of a small group of common factors".

An encyclopedic documentation of plant tissue culture medium is available [64, 65] and can be consulted for details on specific formulations, species and applications. Briefly, the trends shown in Table 1 indicate that the major components for culture media for somatic embryogenesis should provide the following:

1. A carbon and energy source. Sucrose in the range of 58–88 μM is by far the most frequently used; maltose, fructose, glucose, sorbitol, mannitol, and a few other sugars are rarely used ([e.g. 10, 20]; Table 1).
2. The inorganic macro- and micronutrients essential for growth and metabolism. Potassium and iron EDTA have been shown to be required

in the medium [10]. About 75% of the reported protocols use the salts of Murashige and Skoog's [63] medium, Gamborg et al.'s B5 medium [67] or Phillips and Collins L-2 [68, 69] based medium. The more dilute White's [62] medium is now rarely used. The complex Kao medium is actually several different formulations [70–74] which are often used for protoplast culture.

3. Some level of reduced nitrogen. The quantity and quality of nitrogen has been shown to be critical in a number of species (e.g. *Daucus*, *Medicago*, *Apium*). The presence of either ammonium ion or one or more amino acids appears to be a strong requirement for embryogenesis [10, 75, 76]. The success of MS medium in somatic embryogenesis has been attributed to its high levels of reduced nitrogen [75].

4. Organic additives. Although often having an unidentified function, a number of vitamin and organic compounds have been shown empirically to stimulate growth and/or embryogenesis. MS and B5 media contain nicotinic acid, thiamine, pyridoxine and myo-inositol. Other vitamin supplements are sometimes used. Amino acids such as glutamine, glutathione, serine and adenine are often added singly or as a component of a mixture such as casein hydrolysate or yeast extract. Coconut milk or fruit juices [59, 76] are occasionally added, but with the strong trend towards fully-defined media they are often replaced with myo-inositol and/or cytokinins [22, 61, 76, 77].

5. Growth regulators, primarily auxin and cytokinin. Sequences of at least two media are very common, with exogenous growth regulators present in initial "induction" media, and reduced or removed in subsequent "differentiation" or "development" media where their continued presence appears to inhibit embryo development. The induction medium virtually always contains an auxin, the "prime controlling factor" [78] of embryogenesis. 2,4-Dichlorophenoxyacetic acid is the auxin of choice, although indole-3-acetic acid and 1-naphthalene acetic acid (primarily among the *Compositae*, *Ranunculaceae* and *Solanaceae*) are not uncommon. A cytokinin is also usually present in the induction medium, normally either kinetin, benzyladenine or 2-isopentenyl adenine. Other growth regulators such as gibberellins and abscisic acid are rare in the "induction" media but are often added to "differentiation" or "development" media. This reflects their apparent importance in embryo development and maturation.

Media are frequently optimized for a given family, species, cultivar or even genotype. The list of compositions of media [64, 65, 79] used in plant tissue culture is extensive but many recipes can be traced to a few basic formulations. Also, with the large number of modifications that have been made to media by different laboratories, the literature is filled with ambiguity and error with respect to medium composition [e.g. 64, 65, 80–82]. Systematic methods for optimizing media composition to the specific material and objective have been outlined [66, 79, 83, 84].

F. The physical environment

The environmental culture conditions have been shown to dramatically affect not only the expression of somatic embryogenesis but also the quality of the embryos produced. Much recent work has focussed on this latter effect as plant conversion of somatic embryos is a limiting factor in the commercial application of somatic embryogenesis in crop species.

In cell suspension culture, agitation rate and/or aeration was demonstrated to affect the quality of carrot embryos. High levels of ethylene in cultures have been correlated with abnormal embryo development in alfalfa. Part of the carbohydrate requirement in embryo development has been shown to be osmotic in nature [85–87] and elevated sucrose levels have been shown to reduce secondary embryogenesis in carrot [88], sunflower [89] and soybean [90] and appear to promote somatic embryogenesis in *Daucus* [91] and *Caricia* [92]. Higher sucrose levels also appear to be beneficial in inducing desiccation tolerance in carrot, alfalfa and soybean embryos. The process of desiccation has been shown to enhance embryo quality and vigour in alfalfa [92–96] and soybean embryos [56]. A short cold treatment also has been shown to improve embryo conversion rates [97–100] presumably by enhancing embryo maturation. Wild carrot somatic embryos [101] have been shown to have an electrical polarity as they develop and the exposure of alfalfa protoplasts to a weak electrical field [102] has been shown to promote embryo formation; however, there is some evidence [103] to suggest that a steep potential gradient in surrounding tissue is not a prerequisite for embryo formation but rather a consequence of the process.

III. Progress in plant families

Bhojwani et al. [104, 105] lists about 18,000 publications concerning plant tissue culture published prior to 1989 and, of that, about 900 literature citations report on the process of somatic embryogenesis. The dramatic increase in publications on plant tissue culture since 1986 to well over 1000 new titles per year is also reflected in the literature on somatic embryogenesis and, as such, all references could not be included here. The bibliographies [104, 105] on plant tissue culture along with the multiple volume works *Handbook of Plant Cell Culture* [106], *Cell Culture and Somatic Cell Genetics of Plants* [107] and *Biotechnology in Agriculture and Forestry* [108] provide an excellent source of detailed information on particular species. Examples of *in vitro* somatic embryogenesis are outlined in Table 1 and several trends are evident. Perhaps it is not surprising, but most examples cited are of species that are of agricultural or economic interest and that many plant families are represented by only a few entries or a single species. The greatest progress, as measured by literature citation, appears to have been made with the Leguminosae and Umbelliferae species and, indeed, the three most extensively characterized *in vitro* systems, *Medicago sativa*, *Glycine max* and *Daucus carota* fall within these families. The two Leguminosae species are considered in case studies and in our opinion reflect the recent progress and research trends observed in the study of *in vitro* somatic embryogenesis.

A. *Glycine* case study

Glycine max (soybean) can be regenerated with ease via shoot organogenesis or somatic embryogenesis. The present situation is in remarkable contrast to the state of the technology prior to 1985. Many pioneering studies used soybean as an experimental system. Notably, it was one of the first species to be serially propagated [67], one of the first species from which protoplasts were isolated [109], induced to divide [110] and used as a interfamily partner for successful cell hybrid production by cell fusion [111]. As well, the tissue was used in the formulation of B5 [67], Blaydes [112] and Schenk and Hildebrandt [113] media. Initially, soybean tissue cultures appeared to be recalcitrant to all known plant regeneration protocols. Low frequency shoot organogenesis from hypocotyl sections had been reported in 1973 [114] and, thereafter, high cytokinin treatments were shown in 1980 to cause multiple shoot formation from seed [115] and stem nodes [116]. A high frequency of shoot organogenesis was reported from cotyledonary nodes [117, 118] in 1986 and somatic embryo explants [119] in 1991.

Embryo-like structures had been reported in *Glycine* [120–125] cell cultures between 1977 and 1984 but these reports, although encouraging, were met with great caution as no plants were recovered from the observed structures. A report of plant recovery from embryo-like structures originating from cotyledonary tissue of *Glycine canescens* [123] was first made in 1984. However, in retrospect, the reports of partial development of plants from embryo-like structures originating in the embryonic axis of *Glycine max* [124–126] contained the key to future successes. The use of immature zygotic embryo explants on media containing relatively high 2,4-D levels appeared to lead to the production of morphologically abnormal somatic embryos. This approach was confirmed in 1985 by three groups [127–129] who reported successful plant recovery from *Gycine max* immature zygotic embryo-derived cultures. Extensive documentation of the induction, development and recovery of plants in *Glycine* is now available [48, 52, 130–152].

Explant choice and the use of high auxin concentrations in the initial culture medium appear to be the factors which most strongly affect *in vitro* response in this species. The use of immature zygotic embryos, from 2–6 mm in length [131, 143] which have been bisected or had their axis removed appear to be one of the only explants that will give a consistent and acceptable response, although the use of young somatic embryo explants has been reported to be give a superior response to zygotic embryos [147]. Levels of 5–200 μM 2,4-D, NAA, dicamba, picloram, p-CPA and 2,4,5-T have been successfully employed [52, 133, 148] with lower concentrations of 5–25 μM reported to result in higher frequencies of embryo abnormalities. 2,4-D appears to be the auxin of choice although the high levels required for embryo induction result in embryos arrested at the globular stage of development [56]. Overall, the use of MS macro and micronutrients, B5 vitamins with thiamine at 5.0 μM and nicotinic acid at 30 μM, high 2,4-D levels, low levels of sucrose in the 0.175 M range, an initiation temperature of 28 °C, lower light intensities and a medium pH range of 5.5–6.5 appear to be optimal for embryo initiation. The effects of auxin and sucrose have been reported to be interactive [151].

The origin of somatic embryos on immature zygotic embryo and somatic embryo explants has been traced largely to the epidermal and subepidermal layers of the explant [131, 134, 147] and the suggestion has been made that the site of embryo initiation may be dependent on the concentration of auxin used [147].

Genotype variation in somatic embryogenesis has been reported widely in *Glycine* [52, 54, 56, 144, 145, 147–150] with many maturity types appearing to respond *in vitro*. Attempted correlations with maturity group, seed coat colour, flower colour and disease susceptibility and *in vitro* response [52, 136] were not successful, however, there is some evidence that the early maturing, long-day photoperiod adapted genotypes [56, 150] give the most consistent response. There is also some evidence [56, 139, 150] to support a correlation between the genetic background of genotypes and their *in vitro* response and with the ability to respond *in vitro* able to be transferred to progeny in controlled crossing experiments [139, 148]. Genotype × 2,4-D [148] and genotype × sucrose [144] interaction with respect to somatic embryogenesis has also been reported to be significant. Genotype variation at the different stages of somatic embryo initiation, development and conversion has also been shown [54] but some of these effects could be overcome by protocol variation.

Somatic embryos of soybean can be recovered as whole plants with more difficulty than that observed in many other species. Success necessitates the transfer to one or more media containing higher sucrose levels or abscisic acid and partial desiccation of the embryos to promote embryo maturation [56, 141, 143, 152] with recoveries of 98% (60% being more typical) having been reported after about 50 days of maturation treatment. Observations that maturation treatments appear to be beneficial in promoting the accumulation of seed storage proteins [143, 152] and lipids [143] suggest physiological as well as morphological development is occurring. The incorporation of gibberellic acid or cytokinin in the maturation medium also appears to promote embryo maturation. It has been speculated by Ranch [57] that the historically poor recovery of plants from soybean somatic embryos may be associated with lack of physiological maturity and that criteria related to physiological status such as moisture content, respiration rate, abscisic acid titer or transcriptional activity may be a more appropriate measure of developmental progress than the morphological criteria now widely used. The factors affecting the development of isolated immature zygotic embryos of soybean have been well defined for use in wide hybridization studies [e.g. 146]; however, a systematic evaluation of all of these factors has not yet been reported for somatic embryos.

B. *Medicago* case study

Since the first report of plant regeneration in *Medicago sativa* in 1972 [153], commonly called alfalfa or lucerne, plant recovery in *Medicago* has been reported to be almost exclusively by the embryogenesis pathway [154–157]. A number of *Medicago* species have been reported to regenerate (see Table 1) but the majority of work has been with the commercially suitable *Medicago sativa-falcata* complex which includes *M. sativa*, *M. falcata*, *M. varia*, *M. caerula*, *M. glandulosa*, and *M. hemicycla* which are generally grouped as *M. sativa*. Early studies with *M. sativa* reported regeneration by organogenesis, bud formation and organ formation [16, 153, 158–160], however, closer scrutiny [161] indicated that regeneration was via embryogenesis. Presumably the shoots observed in earlier studies were actually embryos that were arrested at an early globular stage or had continued to develop after initial formation on the callus tissue and prior to isolation or transfer to a medium optimized for embryo development. Reports of organogenesis in *Medicago* are usually associated with root formation [73, 162] although there are reports of shoot formation in *M. sativa* [163, 166] and *M. arborea* [165] and one report [166] of the simultaneous formation of shoots and embryos in *M. sativa*. There is only preliminary histological evidence [163] to support the claims that regeneration can occur via organogenesis whereas good histological, biochemical and developmental evidence supports the view that regeneration is predominantly via embryo formation. Studies on leaf-derived callus, cotyledon, hypocotyl, root and petiole explants [167, 168] show that embryogenic tissue and embryos arise from the tissue surface, likely the epidermal or subepidermal layers, and appear to be single-cell in origin. In cell suspensions, it was shown by Walker et al. [160] that specific-sized cell clumps were required to induce embryo formation but it is not clear whether all the cells in the clumps participate in the formation of the embryo. The 2S, 7S and 11S seed-specific storage proteins and corresponding mRNAs have been shown to be present in developing cotyledon-staged structures (somatic embryos) *in vitro* [170] and even though the pattern of synthesis and quantitative expression of proteins vary between zygotic and somatic embryos [169, 170] their qualitative nature appears identical.

As might be expected, the biology of *Medicago* strongly affects its performance *in vitro*. Widely reported genotype variability in this genus [e.g. 14–21, 50, 154, 171–173], especially with the commercial *M. sativa-falcata* germplasm, is due to the fact that the plants of this complex are open-pollinated autotetraploids and the seed populations are highly heterogeneous. The *in vitro* responses of the self-fertile *M. truncatula* and *M. laciniata* species are far less variable as one might expect from their more homogeneous genetic makeup. The regeneration response of *M. sativa* has been correlated with the genetic background of the donor plants [15, 50] and several studies [16–21] have shown that the trait for *in vitro* somatic embryo production is heritable in *Medicago* and can be moved within a population of plants by simple breeding techniques. Segregation data from these studies suggest that the inheritance pattern most closely fits a two gene model; however, the nature of these genes is still unknown. The functional result of this work is the realization that a large component of the variation observed in regeneration response in this genus and especially in the *M. sativa-falcata* complex can be attributed to the genetic makeup of the donor plant. In theory, every plant can have a distinct genetic makeup and, therefore, *de facto* have a distinct *in vitro* response. The successful approaches [15–17] that have been adopted with this species of either selecting and/or breeding for tissue-culture-responsive plants would seem to be a good model for open-pollinated species.

A wide range of tissue explants have been successfully used in *Medicago* species. Earliest successes were reported with immature embryo-derived tissues such as the immature anther and ovary tissue [16, 153, 158, 159]. Later studies used a wide range of tissue including cotyledon [173–179], hypocotyl [173, 176–179], shoot tip explants [73], petiole [172, 176, 179, 180], leaf [171, 176, 179, 182, 187], root [174, 175, 179, 183], stem [171, 176, 179, 182], somatic embryos [177, 178] and cell suspensions [73, 184] as well as protoplasts derived from leaf [72, 74, 102, 166, 185, 186, 188–190], cotyledons [174, 175], root [162, 175] and cell suspension cultures [73, 154]. Explant type does not appear to be a critical factor in determining the somatic embryo response of this species, although it has been claimed [58] that explants taken from *M. truncatula* plants previously regenerated *in vitro* show an enhanced ability to regenerate.

The choice of culture medium has been shown to affect the regeneration response of *Medicago* species and success has been reported for at least 9 different salt formulations (see Table 1) with Murashige and Skoog [63] – based formulations being the most often used. Plant growth regulator type and concentration [158, 159] and nitrogen supply [159, 182, 192, 194] are the critical medium components that have been identified to strongly affect somatic embryogenesis. 2,4-D in the 4.5–22.5 μM range appears best for embryo induction [179]. Other related auxins such as 2,4,5-trichlorophenoxyacetic acid and (2,4-dichlorophenoxy)propanoic acid have also been shown to be very effective [30, 155]. Auxins such as 1-naphthalene acetic acid, indole-3-acetic acid and phenylacetic acid [30], which are not effective at inducing embryogenesis, have been used to propagate cell lines as callus or cell suspensions with low or zero levels of embryo induction. Transfer to a medium containing an effective embryo-inducing auxin (e.g. 2,4-dichlorophenoxyacetic acid) for a short period of 3–6 days, followed by transfer to a nitrogen-rich medium lacking growth regulators, results in large numbers of somatic embryos [30, 160, 169]. Cytokinin in combination with an auxin strongly promotes embryo induction [158–160, 179]. Kinetin in the 1–5 μM range is most often used. Although effective at inducing large numbers of embryos, exposure to high levels of 2,4-D has been reported to have a detrimental effect on embryo conversion [169, 179] and on the accumulation of storage proteins during embryo maturation [169]. Bingham et al. [154] have suggested the following guidelines with respect to growth regulator effects in this species: first, synthetic auxins and cytokinins interact to control morphogenesis; second, few structural quality differences can be seen among the cytokinins in their effects on morphogenesis; third, the structure of the auxin greatly affects the pattern of morphogensis; fourth, the response optimum for most growth regulators is broad; fifth, abnormal structures occur when non-optimized levels of growth regulators are employed.

The quantity and quality of medium-supplied nitrogen appears to be critical for embryo formation [161, 169, 193, 194] with the process of embryo development appearing to be the most sensitive to nitrogen supply [179]. Amino acid supplements of proline, alanine, glutamine, arginine, lysine, serine, asparagine and ornithgine have been shown to stimulate embryo numbers [194] especially when optimized with the NH_4^+ supply [194] in Schenck and Hildebrandt [113] based media. Optimization of nitrogen supply in a Murashige and Skoog-based medium [179], was at 60 mM, with a ratio of 1:4 NH_4^+:NO_3^- in the induction medium and a ratio of 1:2 NH_4^+:NO_3^- in the embryo development medium. When nitrogen supply was optimized, amino acid supplements of proline, glutamine or arginine were ineffective. The exception was a combination of 4.4 mM glutamine and 3.1 mM proline in the development medium that gave a 15% increase in embryo number. Interestingly, this is a similar amount glutamine and proline found in yeast extract which is frequently added to medium used for embryo development. The use of organic acids such as K^+-citrate in combination with amino acids has also been reported to increase somatic embryo formation and improve conversion frequency of embryos [195]. Sucrose supply appears to be optimal in the 50–100 mM range [182, 156] with higher levels of sucrose (116–146 mM) or maltose, malt extract, gentiobiose, maltotriose or soluble starch effective in promoting embryo quality and plant conversion [95, 156, 196]. One of the most successful treatments in improving embryo quality has been the use of abscisic acid [197] and the induction of desiccation tolerance followed by drying of the somatic embryos to dry seed levels [93–96]. Treatment of embryos in the late-cotyledon stage of development with 50 μM abscisic acid for 10 days on medium supplemented with glutamine and 5% sucrose, followed by gradual drying over a 7-day period, resulted in embryos of greater size, weight and vigor.

The use of *Medicago* as a model system to develop synthetic seed protocols [e.g. 93, 155] has encouraged the view that embryogenesis can be studied and optimized using a reductionist approach. Similar observations have been noted with respect to studies on *Daucus* [25] and *Glycine* [3]. The result is that somatic embryogenesis in *Medicago* and other species can be viewed pragmatically as a series of sequential stages (i.e. competence induction – embryo induction – embryo development – embryo maturation – embryo conversion) each of which have different *in vitro* requirements as outlined above.

Table 1. Herbaceous dicot species exhibiting somatic embryogenesis.

Family – Species (Common Name)	Explant (*Plant Recovery)[a]	Media Sequence[b]	Embryo Induction Medium				Reference
			Salts	Growth Regulators (μM)[c]	Sugars (mM)[d]	Modifications (mg/L)[e]	
Aizoaceae							
Mesembryanthemum floribundum (spinach)	leaf*	MS	MS	IAA (11.4)	Suc (88)	CM (200 mL)	198
Araliaceae							
Panax ginseng (ginseng)	root*	$MS_1 \rightarrow \frac{1}{2}MS_2$	MS_1	2,4-D (4.5)	Suc (88)	CH (1000)	199
	zyg emb	MS	MS	2,4-D (4.5)	Suc (88)		200
	som emb ptp*	$\frac{1}{2}M8P_{1(L)} \rightarrow \frac{1}{2}M8P_{2(L)} \rightarrow MS_1 \rightarrow MS_2$	MS_1	2,4-D (4.5); KIN (0.05)	Suc (88)		200
	buds*	$MS_1 \rightarrow \frac{1}{2}MS_2$	MS_1	2,4-D (4.5)	Suc (88)	MS VIT	201
	root pith	$MS_1 \rightarrow \frac{1}{2}MS_{2(F)}$ or B5	MS_1	2,4-D (4.5)	Suc (88)	CH (1000)	199
Panax quinquefolium	leaf	$LS \rightarrow LS_0$	MS	NAA (10.7); KIN (2.3)	Suc (88)	B5 VIT; several	202
	root pith*	$MS_1 \rightarrow MS_2 \rightarrow MS_3 \rightarrow MS_4$	MS_2	DIC (9.0)	Suc (88)	Thi-HCl (10); CH (100)	203
Asclepiadaceae							
Araujia sericifera	germ imm seed	M1 or M1 → M1	M1	NAA (0.5); BA (4.4)		minus NaH_2PO_4; Ad (13.5); direct embryos	204
Asclepias curassavica (milkweed)	stem	$MS_1 \rightarrow MS_2$	MS_1	2,4-D (9.0)	Suc (58)	CM (100 mL)	205
A. syriaca	stem	MS or MS → $MS_{(L)}$	MS	2,4-D (9.0); KIN (23.0)	Suc (88)	CH (1000); I (1000)	206
A. tuberosa	stem	MS or MS → $MS_{(L)}$	MS	2,4-D (2.3); BA (2.2)	Suc (58)	Ad (5); I (100); CH (400)	206

Table 1. Continued

Family – Species (Common Name)	Explant (*Plant Recovery)[a]	Media Sequence[b]	Embryo Induction Medium				Reference
			Salts	Growth Regulators (µM)[c]	Sugars (mM)[d]	Modifications (mg/L)[e]	
Hoya carnosa (wax plant)	leaf	$MS_1 \to MS_2 \to MS_3$	MS_1	2,4-D (2.3–5.0); KIN (2.5–20)	Suc (88)	Thi-HCl (0.4)	207
Pergularia minor	stem*	$MS_1 \to MS_2$	MS_1	2,4-D (9.0)	Suc (58)	CM (100 mL)	205
Tylophora indica	stem*	$W \to W_0$	W	2,4-D (4.5); 2-BTOA ($1\ mgL^{-1}$)	Suc (58)	CH (500); CM (100 mL)	208
	stem*, ptp*	$MS \to MS_0$	MS	2,4-D (9.0)	Suc (58)	CH (500); CM (100 mL); Ad (5)	209 / 210
Begoniaceae *Begonia fimbristipula* (begonia)	leaf*	$SH_{(L)} \to \frac{1}{2}SH_0$	SH	2,4-D (0.6); BA (1.1–2.2)	Suc (88)	CM (100 mL) or CH (200)	211
Berberidaceae *Dysosma pleiantha*	imm seed	$(MS_1)x \to \to MS_2;$ $(B5_1)x \to B5_2$	$MS_1;$ $B5_1$	2,4-D (4.5); 2,4-D (4.5)	Suc (88); Suc (58)	Culture in the dark	212
	mat zyg embryo leaf	$MS_1 \to (MS_2)x \to MS_3$ or $B5_1 \to (B5_2)x \to B5_3$	$MS_1;$ $MS_2;$ $B5_1;$ $B5_2$	2,4-D (4.5); 2,4-D (0.5); 2,4-D (4.5); 2,4-D (0.5)	Suc (88); Suc (58); Suc (88); Suc (58)	Culture in the dark	212; 213
	rhizome pith*	$MS_1 \to (MS_2)x \to MS_3$ or B5	$MS_1;$ MS_2	2,4-D (4.5); 2,4-D (0.5)	Suc (88)	Agar (7000)	213
	stem, pet	$(MS_1)x \to MS_2$	MS_1	2,4-D (4.5)	Suc (88)	Agar (7000)	213
Boraginaceae *Borago officinalis* (borage)	imm zyg emb	$MS_1;$ $MS_{2(L)}$	$MS_1;$ MS_2	2,4-D (0.5–4.5); 2,4-D (9.0)	Suc (88)	MS VIT	214
	imm zyg emb	MS	MS	2,4-D (4.5)	Suc (88)	CH (1000); CM (100 mL)	215

Table 1. Continued

Family – Species (Common Name)	Explant (*Plant Recovery)[a]	Media Sequence[b]	Embryo Induction Medium				Reference
			Salts	Growth Regulators (μM)[c]	Sugars (mM)[d]	Modifications (mg/L)[e]	
Campanulaceae							
Codonopsis lanceolata	cot*	MS	MS	2,4-D (4.5)	Suc (176)	MS VIT; CM (100 mL)	216
Caricaceae							
Carica papaya (papaya)	hyp*	½MS$_1$ or ½MS$_1$ → MS$_0$	MS$_1$	2,4-D (2.3–113)	Suc (176)	MS VIT; Glutm (400)	217
Chenopodiaceae							
Beta vulgaris (sugar beet)	pet*	PG$_{01}$ → PG$_{02}$ → PG$_{00}$ → PG$_{02}$ → PG$_{00}$	PG$_{02}$	NAA (5.4); BA (4.4)	Suc (58)	dark/light cycle	218
Compositae							
Brachycome lineariloba	leaf, bud	MIL$_1$ → MIL$_2$ → MIL$_{3(L)}$	MIL$_{2,3}$	NAA (2.7); KIN (0.5)			219
Cichorium endivia (endive)	mat emb cal*	MS$_{(L)}$	MS	IAA (57); KIN (0.5)	Suc (88)	MS VIT	220, 221
C. intybus (chickory)	leaf, root*	MS → MS$_0$ → ½MS$_0$	MS	BA (1.0); 2,4-D (1.0)	Suc (88)	I (100)	222
C. intybus × C. endivia (endive)	root	½MS	MS	NAA (0.1); 2iP (2.5)	Suc (30)	Glutm (248); H MICRO; KCl (753); MW VIT; Fe • EDTA (18)	223
If ptp*		MS$_{1(L)}$ → MS$_{3(L)}$ → H$_{(L)}$ → H; MS$_{1(L)}$ → MS$_2$ → M17$_{(L)}$ → H$_{(L)}$ → H; or MS$_2$ → M17$_{(L)}$ → H$_{(L)}$ → H	MS$_3$ M17	2iP (2.5); NAA (0.5) 2iP (2.5); NAA (0.1)	Suc (29–232)	½MS MACRO; H MICRO MW VIT; several	224

362

Table 1. Continued

Family – Species (Common Name)	Explant (*Plant Recovery)[a]	Media Sequence[b]	Embryo Induction Medium				Reference
			Salts	Growth Regulators (μM)[c]	Sugars (mM)[d]	Modifications (mg/L)[e]	
Gaillardia picta	leaf epider*	$B5_1 \to B5_0$	$B5_1$	NAA (1.0); BA (1.0)	Suc (58)		225
		$B5_2 \to B5_0$	$B5_2$	IBA (1.0); BA (1.0)	Suc (58)		
		$B5_3 \to B5_0$	$B5_3$	IBA (10); BA (1.0)	Suc (58)		
		$B5_4 \to B5_0$	$B5_4$	IBA (10); KIN (0.1)	Suc (58)		
Helianthus annuus (sunflower)	hyp	MS	MS	NAA (5.4); BA (4.4); GA (0.3)	Suc (88)	KNO_3 (5005); Ad (29.3); AUG (1.97)	226
	imm zyg emb*	$MS \to MS_0$	MS	2,4-D (4.5) or DIC (15)	Suc (351)	B5 VIT	227, 228
	hyp*	MS	MS	NAA (5.4); BA (4.4); GA_3 (0.3)	Suc (88)	KNO_3 (5000); I (100); Ad (40); CA (500)	229
	hyp cells	$MS_{(L)} \to B5_{(L)}$	MS	NAA (5.4); BA (4.4)	Suc (88)	CH (2000); CM (20 mL)	230
	imm emb hyp*	$MS_1 \to MS_2$	MS_1	BA (2.2–4.4)	Suc (263)	I (3900); Ala (1000); Glutm (800); Ser (160); Tryp (50); Cyst (10)	231
	hyp ptp	$MS_1 \to MS_{2(L)} \to MS_3$	MS_1	NAA (5.4); BA (4.4)	Glu (380); Man (120)	CM (20 mL)	232
	hyp ptp*	$VKM1_{(S/L)} \to$ $VKM2_{(S/L)} \to$ $VKM3_{(L)} \to MS_1 \to$ $MS_{0(L)}$	VKM2 or or	2,4-D (10) NAA (12) DIC (10)	Sor (440)	several; Glutm (1000)	233
H. annus × *H. tuberosus*	anther*	$MS_1 \to \tfrac{1}{2}MS_0$	MS_1	NAA (0.5); BA (0.9)	Suc (88)		234

Table 1. Continued

Family – Species (Common Name)	Explant (*Plant Recovery)[a]	Media Sequence[b]	Embryo Induction Medium				Reference
			Salts	Growth Regulators (μM)[c]	Sugars (mM)[d]	Modifications (mg/L)[e]	
Convolvulaceae							
Cuscuta reflexa	imm emb	W	W	IAA (5.7)	Suc (146)	CH (400)	235
Ipomoea batatas (sweet potato)	shoot apex*	MS$_1$ → MS$_0$ or MS$_1$ → MS$_2$ → MS$_0$	MS$_1$	2,4-D (10)	Suc (88)	I (90); Thi-HCl (1.7); NA (1.2); P-HCl (1.0)	236
	apical domes,	MS$_1$ → MS$_{20}$	MS$_1$	2,4-D (10); BA (1.0)	Suc (88)	I (90); NA (1.2)	237–240
	c susp	MS$_3$ → MS$_{4(L)}$ → MS$_{03}$	MS$_3$	2,4-D (1.0–5.0)	Suc (88)	Thi-HCl (1.7)	241, 242
	leaf*, sh tip*, stem, root	MS$_1$ → (MS$_2$)$_x$ → MS$_0$	MS$_1$	2,4-D (2.5–9.0)	Suc (88)	Thi-HCl (0.4)	243
	anther*	MS$_1$ → MS$_2$	MS	IAA (11.4); 2,4-D (9.0); KIN (9.2)	Suc (88)		
Cruciferae							
Brassica campestris (turnip)	pet*	SH$_1$ → SH$_2$ → SH$_3$	SH$_2$	2,4-D (0.1); KIN (1.0)	Suc (58)	minus PCPA; I (100)	244
B. juncea (mustard)	hyp ptp	KAO$_{1(L)}$ → KAO$_{2(L)}$ → K3 → MS	KAO$_1$	2,4-D (4.5); NAA (0.5); ZR (1.4)	Glu (380)		245
	lf ptp	N → MS$_1$ → MS$_2$ → MS$_3$	N	2,4-D (2.3–4.5); NAA (2.7); BA (2.2–4.4)	Suc (29) Man (550)		246
B. napus (canola, rapeseed)	lf ptp	N → MS$_1$ → MS$_2$ → MS$_3$	N	2,4-D (2.3–4.6); NAA (2.7); BA (2.2–4.4)	± Man (550)		247

Table 1. Continued

Family – Species (Common Name)	Explant (*Plant Recovery)[a]	Media Sequence[b]	Embryo Induction Medium			Modifications (mg/L)[e]	Reference
			Salts	Growth Regulators (μM)[c]	Sugars (mM)[d]		
	stem emb ptp	$N_{(L)} \to MS_{0(L)} \to PCW \to MS_2 \to MS_3$	N	2,4-D (4.5); ZR (2.8)	Suc (29); Glu (55)	CH (100)	248
	c susp	$MS_{1(L)}$ *30 days* → $MS_{1(L)} \to MS_{2(L)} \to MS_2 \to K_3$	MS_1	2,4-D (4.5)	Suc (88)	reduce inoculum	249
B. nigra (black mustard)	emb	B5	B5		Suc (58)		250
	hyp*	MS, B_5, SH, NN or B → MS_0	All	PCPA (11); NAA (2.7)	Suc (58)	Ad (0.5)	251
	hyp, lf, stem	$MS_1 \to MS_2$	MS_1	PCPA (1.6–11); ± NAA (2.7) or 2,4-D (1.4–4.5)	Suc (58)	Ad (0.5–2)	
B. oleracea (cauliflower, kale, brussel sprouts)	leaf	$MS_1 \to MS_2$	MS_1	IAA (5.7); KIN (2.3)	Suc (88)		252
Cheiranthus cheiri (wall flower)	germ seeds*	$MS_1 \to MS_{2(L)}$	MS_1	2,4-D (4.5)	Suc (88)	I (55000)	253
Eruca sativa (rocket salad)	lf ptp	$KAO_1 \to KAO_2 \to$ ½MS → MS_0	KAO_1	2,4-D (1.4); NAA (0.5); BA (0.9); GA_3 (0.3)	Glu (600)	minus Ca $(H_2PO_4)_2$	254
	imm emb*	MS → ½ MS_0	MS	NAA (2.7); BA (2.2)	Suc (88)		255
Cucurbitaceae *Citrullus lanatus* (watermelon)	imm emb*	$MS_1 \to MS_0$	MS_1	2,4-D (10); TDZ (0.5)	Suc (88)	B5 VIT; I (100)	256, 257
C. moschata	ovary*	½MS_0 → ½MS_0	MS	IBA (49 or 4.9) or 2,4-D (0.5–4.5)	Suc (88)	pH 5.2: Agar (4000) Ad (13.5); Melon	258
C. pepo (pumpkin)	hyp*	MS → MS	MS		Suc (88)	Sap (100); CH (300)	257,259, 260

Table 1. Continued

Family – Species (Common Name)	Explant (*Plant Recovery)[a]	Media Sequence[b]	Embryo Induction Medium Salts	Growth Regulators (μM)[c]	Sugars (mM)[d]	Modifications (mg/L)[e]	Reference
C. pepo (zucchini squash)	pericarp wall*	N	N	IAA (57)	Suc (147)	+Fe chelate; –Fe citrate	257, 261, 262
Cucumis melo (melon)	seed section*	MS₁ → ½MS₀	MS₁	2,4-D (0.9–45); NAA (5.4–22); BA (0.4)	Suc (88)		257, 263
	cot	MS₁ → MS₀(L) → MS₀	MS₁	2,4-D (4.5); BA (0.4)	Suc (88)		264
	hyp	N	N	2,4-D (2.3); 2,4,5-T (2.5); PCPA (2.7); IBA (2.5); DIC (2.2); plus others	Glu (222)	several	265
	cot, lf, pet, hyp	MS₁ → ½MS₂	MS₁ and or or N6	BA (0.4) 2,4-D (4.5–9.0) IAA (114–570) NAA (16–54)	Suc (88)		266
	imm cot, hyp	N6₁ → N6₀	N6	2,4-D (13.5); BA (0.4)	Suc (146)		267
C. melo utilissimus	cot	MS₁ → MS₂ → MS₀	MS₂	NAA (5.4)	Suc (88)	Ad (33.7)	268, 262
C. metuliferus × *C. anguria*	imm emb*	MS₁ → MS₂ → MS₀	MS₁	IAA (10); BA (5)	Suc (88)	I (100); CH (1000)	269
C. sativus (cucumber)	hyp*	MS₁(L) → MS₀(L) → MS₀	MS₁	2,4-D (5.0); BA (0.5–1.0)	Suc (88)		257, 270
	cot*, hyp*	MS₁ → MS₂ → MS₀	MS₁	2,4-D (9.0); KIN (2.3)	Suc (88)		271
	cot*	MS₁ → MS₀		2,4-D (4.5–9.0); KIN (2.3)	Suc (88)		272

Table 1. Continued

Family – Species (Common Name)	Explant (*Plant Recovery)[a]	Media Sequence[b]	Embryo Induction Medium			Modifications (mg/L)[e]	Reference
			Salts	Growth Regulators (μM)[c]	Sugars (mM)[d]		
	lf ptp*	$MS_1 \rightarrow MS_{1(L)} \rightarrow MS_{2(L)} \rightarrow MS_{3(L)} \rightarrow MS_4 \rightarrow MS_0$	MS_1	NAA (16.1); 2iP (14.8)	Suc (58); Man (300)	Eda (250)	273
	hyp*	$MS_{1(L)} \rightarrow MS_{0(L)}$; $\frac{1}{2}MS_{2(L)}/MSC \rightarrow \frac{1}{2}MS_0$	MS_1	2,4-D (5.0); BA (1.0)	Suc (88)	CH (100)	274
	leaf*	$(MS)_3 \rightarrow MS_{(L)} \rightarrow$ washed $\rightarrow MS_{0(L)} \rightarrow$ or MSC_0	MS	2,4,5-T (5.0); BA (4.0)	Suc (88)	Charcoal treatment	275
	leaf*	$MS_1 \rightarrow MS_1 \rightarrow MS_2$	MS_1	2,4,5-T (4.7); BA (1.1)	Suc (88)	Eda (250)	276, 277
	callus*	$MS_1 \rightarrow MS_{2(L)} \rightarrow MS_1$	MS_2	2,4,5-T (1.6); BA (1.1)	Suc (44); Glu (11)	$\frac{1}{2}$MS salts; Eda (250)	278
	cot ptp*	$DPD_{1(L)} \rightarrow DPD_{2(L)} \rightarrow MS_1 \rightarrow MS_2 \rightarrow \frac{1}{2}MS_0$	MS_1	IAA (1.1); BA (2.2)	Suc (50); Man (300)	$CaCl_2 \cdot 2H_2O$ (800); B5 VIT in MS	279
	lf*, pet*	$MS_1 \rightarrow MS_0$	MS_1	2,4-D (1.0–5.0); BA (5.0)	Suc (88)	VIT mod; dark incubation	280
	leaf	$MS_1 \rightarrow MS_{1(L)} \rightarrow MS_2 \rightarrow MS_0$	MS_1	2,4-D (5.0); BA (5.0)	Suc (88)	Try (1)	
Euphorbiaceae							
Croton bonplandianum	endospm	W	W	2,4-D (9.0); KIN (23)	Suc (58)	YE (2500)	281
Manihot esculenta (cassava)	leaf*	$MS_1 \rightarrow MSC \rightarrow \frac{1}{2}MS_0$	MS_1	2,4-D (18)	Suc (58)	$CuSO_4$ (0.3)	282
	leaf lobe*	$MS_1 \rightarrow MS_2$	MS_1	2,4-D (4.5–36)	Suc (58)		283

Table 1. Continued

Family – Species (Common Name)	Explant (*Plant Recovery)[a]	Media Sequence[b]	Embryo Induction Medium				Reference
			Salts	Growth Regulators (µM)[c]	Sugars (mM)[d]	Modifications (mg/L)[e]	
	som emb frag	$MS_{1(L)} \rightarrow MS_{2(SorL)}$; $MS_1 \rightarrow MS_2$	MS_1	2,4-D (18)	Suc (58)		284
	sh tip*, imm lf*, som emb	$MS_{1(L)} \rightarrow MS_2$	MS	2,4-D (4.5-7.2)	Suc (88)		45
	cot, lf	$MS_1 \rightarrow MS_2$	MS	2,4-D (9-54) or NAA (11); BAP (0.4)	Suc (88)		285-287
Geranaceae							
Pelargonium lortorum (geranium)	anther	$W_1 \rightarrow W_2 \rightarrow MS$	MS	NAA (2.7); KIN (11.5)	Suc (88)		288
Gesneriaceae							
Saintpaula ionantha	intact hyp	MS	MS	TDZ (1-10)	Suc (88)	B5 VIT	289
Labiatae	leaf	MS	MS	NAA (0.5-2.7); BA (0-44)	Suc (88)		290
Perilla frutescens (perilla)	leaf	$MS \rightarrow MS_0$	MS	2,4-D (4.5) or NOA (4.9)	Suc (58)		291
Lauraceae							
Persea americana (avocado)	imm zyg emb	MS	MS	PIC (0.4)	Suc (88)	I (100); Thi-HCl (0.4)	292
	imm emb*	$MS_1 \rightarrow MS_2 \rightarrow DF$	MS	PIC (0.4)	Suc (88)	Charcoal (1000); I (100); Thi-HCl (0.4)	293
Leguminosae							
Arachis hypogaea (peanut)	cot*	$B5 \rightarrow B5$	B5 or	NAA (54-108) PIC (2.1-4.2)	Suc (58)		294
	imm emb	$MS \rightarrow \frac{1}{2}MS$	MS	2,4-D (13.6)	Suc (73)		295
	leaf*	$MS_1 \rightarrow MS_2$	MS_1	2,4-D (180); KIN (0.9)	Suc (88)	B5 VIT	296

Table 1. Continued

Family – Species (Common Name)	Explant (*Plant Recovery)[a]	Media Sequence[b]	Embryo Induction Medium				Reference
			Salts	Growth Regulators (μM)[c]	Sugars (mM)[d]	Modifications (mg/L)[e]	
	cot*	MS$_1$ → MS$_2$	MS$_1$	NAA (5.4); BA (13.2)	Suc (88)		297
	imm emb*	L2$_1$ → L2$_2$	L2$_1$	PIC, NAA, 2,4-D, BA combinations	Suc (73)	CH (200); Ad (40)	298
A. paraguariensis	imm cot*, imm emb	L2$_1$ → L2$_2$	L2$_1$	PIC (0.08); ABA (0.2)	Suc (73)	Ad (40); CH (200)	298
A. pintoi	leaf*	Ms$_1$ → Ms$_2$ → Ms$_3$ → Ms$_0$	MS$_1$	NAA (10.7); BAP (4.4)	Suc (88)		299
Cicer arietinum (chickpea)	leaf*	MS$_1$ → MS$_2$	MS$_1$	2,4-D (2.3)	Suc (88)		300
Coronilla varia (crownvetch)	leaf*	UM → MS → MS$_0$	UM	2,4-D (9.0); KIN (1.2)	Suc (88)		301
			MS	2-iPA (4.9); IAA (0.57)			
	hyp*	B5$_1$ → B5$_2$ → B5$_0$ → BOi2Y or B5$_3$	B5$_2$	2,4-D (20); KIN (2.0)	Suc (58)		302, 303
	ptp*	UM → MS$_1$ → MS$_0$	MS	2iP (4.9); IAA (5.7)	Suc (88)		304
Glycine canescens (soybean)	imm cot*	MS → ½MS	MS	NAA (0.1); BA (5.0)	Suc (88)	B5 VIT; NaFe • EDTA (25)	123
	imm cot*	L2$_1$ → L2$_2$	L2$_1$	PIC (0.08); ABA (0.2) or 2,4-D (2.3); KIN (0.05)	Suc (73)	Ad (40); CH (200) Ad (2)	298
G. gracilis	imm cot*	MS$_1$ → MS$_2$ → MS$_0$ → MS$_3$ → MS$_0$	MS$_1$	NAA (54)	Suc (88)	B5 VIT; pH 7.0	57

Table 1. Continued

Family – Species (Common Name)	Explant (*Plant Recovery)[a]	Media Sequence[b]	Embryo Induction Medium				Reference
			Salts	Growth Regulators (μM)[c]	Sugars (mM)[d]	Modifications (mg/L)[e]	
G. max (soybean)	imm cot	$MS_1 \to MS_{2(L)} \to MS_0$	MS_1	2,4-D (22)	Suc (88)		138
	imm cot	$MS \to MS_0$	MS / MS_0	2,4-D (87)	Suc (88) / Suc (280)	Charcoal	141
	imm cot*	$MS_1 \to MS_2$	MS_1 / MS_0	2,4-D (45.2–180) or 2,4,5-T (22–45)	Malt (175) or Glu (175) or Suc (292)	temp 20 °C or 28 °C / light 0–2400 lux	52
		$L2_1 \to L2_2$	$L2_2$	NAA (10.6); KIN (0.05) or PIC (0.004); BA (0.6)	Suc (73) / Lac (88)	Charcoal / Ad (2)	298
	hyp	$B5 \to M_{1(L)}$	$B5$ / M_1	2,4-D (9.0) / 2,4-D (4.6); KIN (2.3); IAA (2.85)	Suc (58)		122
		$M_{2(L)}$	M_2	2,4-D (2.3); KIN (0.05–0.5)			
	hyp, epicot	$L2 \to SL2_{(L)} \to L2$	$L2$; $SL2$	2,4-D (2.2); KIN (0.46); ABA (0.1); AMO (0.1)	Suc (73)	CH (100)	120
	imm emb	$MS \to MS_0 \to Ch$ or $MS \to MS_0 \to MS_{0(L)} \to Ch$	MS	2,4-D (22.5)	Suc (58)	20 mM NH_4^+; 40 mM NO_3^-	124
	hyp	$SL_1 \to SL_{1(L)} \to SL_{0(L)} \to SL_{2(L)} \to SL_{3(L)}$	SL_2	PIC (0.25); AMO (0.025–0.25)	Suc (73)	CA (10); AdS (5.5); B5 VIT; B5 MICRO	121

Table 1. Continued

Family – Species (Common Name)	Explant (*Plant Recovery)[a]	Media Sequence[b]	Embryo Induction Medium				Reference
			Salts	Growth Regulators (μM)[c]	Sugars (mM)[d]	Modifications (mg/L)[e]	
	imm cot	L2	L2	2,4-D (5.0) or MCPA (2.5)	Suc (34)		125
	imm cot*	MS₁ → MS₂ → MS₃ → ½MS₄	MS₁	2,4-D (22.5) or NAA (53)	Suc (88)	B5 VIT	127
	imm emb	MS₁ → MS₂ → CH₀	MS₁	2,4-D (22.5)	Suc (88)		126
	imm cot*	MS₁ → MS₂ → B₅	MS₁	2,4-D (22.5–45.2)	Suc (88)	20 mM NH₄⁺; 40 mM NO₃⁻	129
	imm emb c susp*	MS₁ → MS₂(L) → MS₃ → MS₄ → MS₅(L)	MS₁	2,4-D (4.5–9.0)	Suc (58)	20 mM NH₄⁺; CM (50 mL); Ultra cold temp. treatment	128
	hyp, cot	L2₁ → L2₂(L)	L2₁	2,4-D (1.8); KIN (0.9)	Suc (73)	B5 MICRO; B5 VIT	130
			L2₂	2,4-D (1.8)	Suc (175)		
	cot	LS₁ → LS₂	LS₁	2,4-D (45); ABA (1.0)	Suc (58)		142
	imm emb*	MS₁ → MS₂ → MS₀	MS₁	NAA (43)	Suc (88)	Thi (1.7); NA (3.7)	131
	imm emb*	B5 → ½MS	B5	IBA (0.49)	Suc (58)	CM (100 mL)	135
	imm cot	MS	MS	2,4-D (43) or NAA (54)	Suc (88); Suc (44)	B5 VIT	134
	imm cot	MS₁ → MS₂	MS₂	NAA (33.6–67.1)	Suc (29–58)	B5 VIT	151
	imm cot	MS₁ → MS₁	MS₁	2,4-D (23); ± ABA (0.4)	Suc (88)	B5 VIT	133
			MS₂	NAA (25–150)			
			MS₁	2,4-D (23)			
			MS₂	NAA (50)			

Table 1. Continued

Family – Species (Common Name)	Explant (*Plant Recovery)[a]	Media Sequence[b]	Embryo Induction Medium				Reference
			Salts	Growth Regulators (μM)[c]	Sugars (mM)[d]	Modifications (mg/L)[e]	
	imm emb*	$MS_1 \rightarrow \tfrac{1}{2}MS_2 \rightarrow \tfrac{1}{2}MS_{3(L)} \rightarrow MS_4$	MS_1	NAA (54) or 2,4-D (9.0)	Suc (88)	B5 VIT	49
	imm cot	MS	MS	NAA (50)	Suc (29)	B5 VIT	139
	imm cot	$MS \rightarrow MS \rightarrow LS_0$	MS	NAA (54)	Suc (29)	B5 VIT	140
	imm cot*	$MS_1 \rightarrow MS_2$	MS_1	NAA (50)	Suc (44–175); Glu (83–166)	B5 VIT; pH 5.0	136
	imm cot	$MS \rightarrow MS$	MS	2,4-D (181)	Suc (175)	B5 VIT	90
	imm cot c susp*	$MS_1 \rightarrow MS_{2(L)} \rightarrow MS_0$	MS_1	2,4-D (181)	Suc (175)	B5 VIT; Glutm (2192); Asp (661)	137
	imm cot*	$MS_1 \rightarrow \tfrac{1}{2}MS_{(x)}$	MS_1	2,4-D (180) or NAA (46)	Malt (175); Suc (88)		143
G. soja (wild soya)	hyp, epicot	$L2 \rightarrow SL2_{(L)} \rightarrow L2$	L2	2,4-D (2.25); KIN (0.46); ABA (0.1)	Suc (73)	CH (100)	120
	hyp	$SL_1 \rightarrow SL_{1(L)} \rightarrow SL_{0(L)} \rightarrow SL_{2(L)} \rightarrow SL_{3(L)}$	SL_2	PIC (0.10–0.25); AMO (0.25–0.50)	Suc (73)	CA (10); AdS (5.5); B5 VIT; B5 MICRO	121
	imm cot*	$MS_1 \rightarrow MS_2 \rightarrow MS_3 \rightarrow \tfrac{1}{2}MS_4$	MS_1	2,4-D (23) or NAA (53)	Suc (88)	B5 VIT	127
	imm cot*	$MS_1 \rightarrow MS_2 \rightarrow B5$	MS_1 MS_2	2,4-D (22.5); 2,4-D (45.2)	Suc (88)		129
Lathyrus sativus (grass pea)	leaf	$B5_1 \rightarrow (B5_2)_5 \rightarrow B5_3$	$B5_1$	NAA (10.7); BA (2.2)	Suc (58)		305
Lens culinaris (lentil)	embryonal axes*	$B5_1 \rightarrow B5_2 \rightarrow B5_0$	$B5_2$ $B5_1$	IAA (2.7); BA (4.4) 2,4-D (4.5)	Suc (58)	NH_4NO_3 (500)	306

Table 1. Continued

Family – Species (Common Name)	Explant (*Plant Recovery)[a]	Media Sequence[b]	Embryo Induction Medium				Reference
			Salts	Growth Regulators (μM)[c]	Sugars (mM)[d]	Modifications (mg/L)[e]	
Lotus corniculatus (birdsfoot trefoil)	leaf*	$MS_1 \rightarrow \tfrac{1}{2}MS_2 \rightarrow MS_0$	MS_1	2iP (4.9); IAA (0.6)	Suc (88)		304
Lupinus albus (white lupin) L. angustifolius (blue lupin) L. mutabilis	imm emb*	$B5_1 \rightarrow B5_2$ or $MS_1 \rightarrow MS_2 \rightarrow MS_3$	$B5_1$ or MS_1	2,4-D (22.5); KIN (1.2)	Suc (58–88)		307
Medicago coerula	leaf*	$UM \rightarrow MS_0$	UM	2,4-D (9.0); KIN (1.2)	Suc (88)		186
	lf ptp*, c susp ptp*	$KAO_{1(L)} \rightarrow KAO_{2(L)} \rightarrow$ UM or $MS \rightarrow MS_0$	MS	2,4-D (4.5) BA (0.44)	Suc (88)		
	hyp*	$UM \rightarrow MS$	UM	2,4-D (9.0); KIN (1.2)	Suc (88)		173
	hyp*	$B5h \rightarrow SH \rightarrow$ BOi2Y	SH	2,4-D (50); KIN (4.6)	Suc (29)		
M. glutinosa	leaf*	$UM \rightarrow MS_0$	UM	2,4-D (9.0); KIN (1.2)	Suc (88)		186
	lf ptp*, c susp ptp*	$KAO_{1(L)} \rightarrow KAO_{2(L)} \rightarrow$ UM or $MS \rightarrow MS_0$	MS	2,4-D (4.5) ZEA (0.46)	Suc (88)		
M. lupulina (black medic)	pet	$MS_1 \rightarrow MS_2$	MS_1	2,4-D (9.0); KIN (1.2)	Suc (88)		172
M. marina	pet	$MS_1 \rightarrow MS_2$	MS_1	2,4-D (9.0); KIN (1.2)	Suc (88)		172

Table 1. Continued

Family – Species (Common Name)	Explant (*Plant Recovery)[a]	Media Sequence[b]	Embryo Induction Medium				Reference
			Salts	Growth Regulators (μM)[c]	Sugars (mM)[d]	Modifications (mg/L)[e]	
M. sativa (alfalfa)	imm an*, imm ov*, hyp*, cot*, c susp*	BII $\to$ BI2Y$_0$	B	2,4-D (2.3–36); KIN (2.3–36)	Suc (88)	YE (2000); I (100) in BI2Y	16, 153, 158
	imm ov, stem, pet	BII $\to$ BI2Y$_0$	B	2,4-D (9.0); KIN (9.0); NAA (11.0)	Suc (88)	YE (2000); I (100) in BI2Y	159
		SH$_1$ $\to$ SH$_2$ $\to$ LS$_0$, BI2Y$_0$, B5$_0$, or SH$_0$	SH$_2$	2,4-D (50); KIN (5.0)	Suc (88)		
	imm ov	SH$_1$ $\to$ SH$_0$ or	SH$_1$	2,4-D (1–100); KIN (0.1–100)	Suc (88)		160
		SH$_1$ $\to$ SH$_2$ $\to$ SH$_0$	SH$_2$	2,4-D (50); KIN (10)	Suc (88)		
	pet*	SH$_1$ $\to$ SH$_{2(L)}$ $\to$ SH$_{0(SorL)}$	SH$_2$	2,4-D (50); KIN (5)	Suc (88)		171, 179, 193, 197
	lf ptp*	KAO $\to$ KAO$_0$	KAO	ZR (0.3); 2,4-D (4.5)	Suc (73); Glu (276); Xyl (1.7)	numerous; Starch (10 gL)	72, 74
	sh tip $\to$ c susp*	KAO$_1$ $\to$ KAO$_{1(L)}$ $\to$ KAO$_{2(L)}$	KAO$_2$	2,4-D (0.5–4.5); ZR (0.3–1.4); BA (0.4–4.4); NAA (0.3–5.4)	Suc (73); Glu (276); Xyl (1.7)	numerous	73
	lf ptp*	KAO$_1$ $\to$ KAO$_{1/2(L)}$ $\to$ KAO$_{2(L)}$ $\to$ UM $\to$ MS$_1$ $\to$ MS$_0$	UM	2,4-D (9.0); KIN (1.2)	Suc (88)	numerous	162, 166
	lf $\to$ cal $\to$ ptp*	B5$_1$ $\to$ B5$_{2(L)}$ $\to$ B5$_3$ $\to$ B5$_4$ $\to$ B5$_5$	B5$_{1,4}$	2,4-D (36); KIN (37); NAA (2.7)	Suc (88)	Arg (1700)	188

Table 1. Continued

Family – Species (Common Name)	Explant (*Plant Recovery)[a]	Media Sequence[b]	Embryo Induction Medium				Reference
			Salts	Growth Regulators (μM)[c]	Sugars (mM)[d]	Modifications (mg/L)[e]	
	lf → c susp → ptp	$B5_1 \rightarrow B5_{2(L)} \rightarrow B5_{3(L)} \rightarrow B5_{4(L)} \rightarrow B5_0$	$B5_4$	BA (0.9)	Suc (146–102)	Asp (600)	188
	rt ptp*, lf ptp	$KAO_{1(L)} \rightarrow KAO_{1/2(L)} \rightarrow MS_1 \rightarrow MS_0$	KAO	2,4-D (0.9); NAA (5.4); ZEA (2.0)	several	numerous	162, 164
	lf ptp*	$KAO_{1(L)} \rightarrow KAO_2$; $SH_1 \rightarrow SH_2 \rightarrow$ B0i2Y	$SH_{1,2}$	2,4-D (36–50)	Suc (88)		185
	callus*	B → LS → SH → B	LS	IAA (5.7); KIN (27.6)	Suc (58)		85
		B → SH → SH_0 → B	SH	2,4-D (4.5); KIN (99)	Suc (58–88)		
	root*, cot*	$KAO_{1(L)} \rightarrow KAO_{1/2(L)} \rightarrow MS_{1(L)}$	KAO_1	2,4-D (0.9); ZEA (2.3); NAA (5.4)	Glu (380)	numerous	174, 175
	hyp*, lf*, cot*, pet*, stem*	$B_1 \rightarrow B_2 \rightarrow B_{3(LorS)} \rightarrow B_{30}$	B_1	2,4-D (100); KIN (5)	Suc (88)	I (100); YE (2000)	176
	leaf*, hyp	$B5_1 \rightarrow B5_{2(L)} \rightarrow B5_0$	$B5_2$	2,4-D (36); KIN (37); NAA (2.7)	Suc (58)	NH_4NO_3 (250); Arg (1700); $3 \times Fe^{++}$	188, 308
	leaf*, hyp*	UM	UM	2,4-D (9.0); KIN (1.2)	Suc (88)	CH (2000)	167
	hyp*	$MS_1 \rightarrow MS_2 \rightarrow MS_0 \rightarrow$ BOi2Y	MS_1	2,4-D (9.0); BA (1.1)	Suc (88)		177, 178

Table 1. Continued

Family – Species (Common Name)	Explant (*Plant Recovery)[a]	Media Sequence[b]	Embryo Induction Medium				Reference
			Salts	Growth Regulators (μM)[c]	Sugars (mM)[d]	Modifications (mg/L)[e]	
	leaf*, hyp*, cot*, c susp*	B5h → SH$_6$ → B0i2Y	SH$_6$	2,4-D (50); KIN (4.6)	Suc (88)	several	15, 93–96, 172, 173, 191, 309
	imm emb*	EC$_6$	EC$_6$	BA (0.22)	Suc (116)	YE (1000)	310, 311
	root, leaf, c susp → ptp*	KAO$_{1(L)}$/KAO$_{2(L)}$ → KAO$_{2(L)}$ → B5	B5	2,4-D (9.0); KIN (0.5)	Suc (58)	additions	312
	cot*, hyp*	B5$_1$ → B5$_2$ → B5$_3$ → B5$_4$	B5$_2$	2,4-D (4.5–45); KIN (0.9)	Suc (88)	CH (500); I (500); Ad (1 or 40); Glutt (10 or 50)	192
	lf ptp*	KAO$_{1(L)}$ → KAO$_{1/2(L)}$ → SH	KAO	NAA (5.4); 2,4-D (0.9); BA (2.2)	Suc (300)	numerous	102, 189, 190
	pet*, lf*, hyp*, stem*	MS → MS$_1$	MS	2,4-D (9.0); KIN (1.2)	Suc (30)	CH (2000)	179, 182
	ov*, hyp*, pet*	B → B → B0i2Y	B	2,4-D (9–90); KIN (4.6–9.2)	Suc (88)	YE (2000); I (100)	313
	lf ptp*	KAO$_{1(L)}$ → UM$_1$; KAO$_{1(L)}$ → UM$_0$	UM$_1$	2,4-D (2.7); KIN (2.8)	Suc (58); Glu (55)	CH (250); CM (20 mL)	313
	lf ptp*	SPC88$_{(L)}$ → SPC88	SPC88	2,4-D (2.3); BA (2.2)	Suc (88); Man (277)	numerous	314
	leaf*	B5$_{(L)}$ → B5$_{20(L)}$ → B5$_{30(L)}$ → MS	B5$_1$	2,4-D (18); KIN (0.9)	Suc (58); → Mal (16)	Ad (0.6); Glutt (10); PEG in B5$_2$ + B5$_3$	181
	leaf*, pet*, som emb*	B$_X$ → MS$_0$	B	IAA (11.4); NAA (10.7); KIN (9.3)	Suc (88)		180

Table 1. Continued

Family – Species (Common Name)[a]	Explant (*Plant Recovery)[a]	Media Sequence[b]	Embryo Induction Medium				Reference
			Salts	Growth Regulators (μM)[c]	Sugars (mM)[d]	Modifications (mg/L)[e]	
M. scutellata	pet	$MS_1 \rightarrow MS_2$	MS	2,4-D (9.0); KIN (1.2)	Suc (88)		172
M. truncatula (barrel medic)	leaf*	$B5_1 \rightarrow B5_2 \rightarrow B5_3 \rightarrow B5_4$	$B5_1$	NAA (10); BA (10)	Suc (58)	CH (250)	58
M. varia	root	$UM_1 \rightarrow UM_2 \rightarrow UM_0$	UM_1	2,4-D (9.0); BA (0.9)	Suc (88)		183
Onobrychis viciifolia (sainfoin)	stem, root	$LS \rightarrow LS_{(L)} \rightarrow LS \rightarrow LS_0$	LS	2,4-D (4.5); BA (4.4)	Suc (88)		315
Phaseolus coccineus (runner bean)	imm cot*	$\frac{1}{2}MS$	MS	2iP (49); NOA (0.3)	Glu (222)	minus KI; $NaH_2PO_4 \bullet H_2O$ (250); Biotin (0.05)	316
P. vulgaris (bean)	leaf	$MS_1 \rightarrow MS_2$	MS	PIC (0.3); BA (0.4)	Suc (88)	several	317, 318
P. wrightii	seed*	$MS_1 \rightarrow M2_2$	MS	BA (80)	Suc (88)	B5 VIT	319
P. arvense	stem	$MS_1 \rightarrow MS_2$	MS	PIC (0.3); BA (0.4)	Suc (88)	several	320
P. sativum (pea)	imm zyg emb*	$MS_1 \rightarrow MS_2$	MS_1	PIC (0.2–4.0); or 2,4-D (0.2–4.0)	Suc (88)	I (250); Thi (2) or B5 VIT	321, 322
	ptp cal*	$MS_1 \rightarrow MS_2$	MS_1	PIC (5.0–10); or 2,4-D (5.0–10)	Suc (58); Man (220–274)	B5 VIT; CH	323
	epicot, leaf	$MS_{1(s)} \rightarrow MS_2(L) \rightarrow MS_0$	MS_2	PIC (0.25); BA (0.44)	Suc (88)	Thi (2); I (250)	324
Psophocarpus tetragonolobus (winged bean)	stem, cot, leaf, seed	$MS_1 \rightarrow MS_2 \rightarrow \frac{1}{2}MS_0$	MS_2	NAA (.5–26); and/or 2,4-D (.5–9)	Suc (88)		325, 326
Trifolium amabile	cot*	$B5_1 \rightarrow (B5_2)_3 \rightarrow B5_3 \rightarrow W$	$B5_1$	2,4-D (13.5); KIN (2.3)	Suc (88)	CH (2)	327

Table 1. Continued

Family – Species (Common Name)	Explant (*Plant Recovery)[a]	Media Sequence[b]	Embryo Induction Medium				Reference
			Salts	Growth Regulators (μM)[c]	Sugars (mM)[d]	Modifications (mg/L)[e]	
T. ambiguum (caucasian clover)	hyp	L2 → ML$_8$ + NO$_3$ → ML$_8$	L2	PIC (0.25); BA (0.4)	Suc (73)		328
T. alpestre	hyp*, root	B5$_1$ → (B5$_2$)$_3$ → B5$_3$ → W	B5$_1$	2,4-D (13.5); KIN (2.3)	Suc (88)	CH (2)	327
T. apertum	cot*, hyp, rt	B5$_1$ → (B5$_2$)$_3$ → B5$_3$ → W	B5$_1$	2,4-D (13.5); KIN (2.3)	Suc (88)	CH (2)	327
T. batmanicum	root	B5$_1$ → (B5$_2$)$_3$ → B5$_3$ → W	B5$_1$	2,4-D (13.5); KIN (2.3)	Suc (88)	CH (2)	327
T. bocconei	hyp	B5$_1$ → (B5$_2$)$_3$ → B5$_3$ → W	B5$_1$	2,4-D (13.5); KIN (2.3)	Suc (88)	CH (2)	327
T. caucasicum	cot*, hyp	B5$_1$ → (B5$_2$)$_3$ → B5$_3$ → W	B5$_1$	2,4-D (13.5); KIN (2.3)	Suc (88)	CH (2)	327
T. cherleri	cot*, hyp	B5$_1$ → (B5$_2$)$_3$ → B5$_3$ → W	B5$_1$	2,4-D (13.5); KIN (2.3)	Suc (88)	CH (2)	327
T. echinatum	cot	B5$_1$ → (B5$_2$)$_3$ → B5$_3$ → W	B5$_1$	2,4-D (13.5); KIN (2.3)	Suc (88)	CH (2)	327
T. glanduliferum	root	B5$_1$ → (B5$_2$)$_3$ → B5$_3$ → W	B5$_1$	2,4-D (13.5); KIN (2.3)	Suc (88)	CH (2)	327
T. globosum	cot	B5$_1$ → (B5$_2$)$_3$ → B5$_3$ → W	B5$_1$	2,4-D (13.5); KIN (2.3)	Suc (88)	CH (2)	327
T. heldreichianum	cot*	B5$_1$ → (B5$_2$)$_3$ → B5$_3$ → W	B5$_1$	2,4-D (13.5); KIN (2.3)	Suc (88)	CH (2)	327
T. hirtum	hyp	B5$_1$ → (B5$_2$)$_3$ → B5$_3$ → W	B5$_1$	2,4-D (13.5); KIN (2.3)	Suc (88)	CH (2)	327
T. hybridum (2x) (alslike clover)	hyp*	B5$_1$ → (B5$_2$)$_3$ → B5$_3$ → W	B5$_1$	2,4-D (13.5); KIN (2.3)	Suc (88)	CH (2)	327

378

Table 1. Continued

Family – Species (Common Name)	Explant (*Plant Recovery)[a]	Media Sequence[b]	Embryo Induction Medium				Reference
			Salts	Growth Regulators (µM)[c]	Sugars (mM)[d]	Modifications (mg/L)[e]	
T. incarnatum (crimson clover)	cot*	$B5_1 \rightarrow (B5_2)_3 \rightarrow B5_3 \rightarrow$ W	$B5_1$	2,4-D (13.5); KIN (2.3)	Suc (88)	CH (2)	327
	hyp*	L2 → B5 or LSE → ML8	L2	PIC (0.25); BA (0.4)	Suc (73)		328
T. leucanthum	cot, hyp, rt	$B5_1 \rightarrow (B5_2)_3 \rightarrow B5_3 \rightarrow$ W	$B5_1$	2,4-D (13.5); KIN (2.3)	Suc (88)	CH (2)	327
T. michelianum	hyp	$B5_1 \rightarrow (B5_2)_3 \rightarrow B5_3 \rightarrow$ W	$B5_1$	2,4-D (13.5); KIN (2.3)	Suc (88)	CH (2)	327
T. micranthum	cot	$B5_1 \rightarrow (B5_2)_3 \rightarrow B5_3 \rightarrow$ W	$B5_1$	2,4-D (13.5); KIN (2.3)	Suc (88)	CH (2)	327
T. montanum	hyp*, root*	$B5_1 \rightarrow (B5_2)_3 \rightarrow B5_3 \rightarrow$ W	$B5_1$	2,4-D (13.5); KIN (2.3)	Suc (88)	CH (2)	327
T. obscurum	hyp	$B5_1 \rightarrow (B5_2)_3 \rightarrow B5_3 \rightarrow$ W	$B5_1$	2,4-D (13.5); KIN (2.3)	Suc (88)	CH (2)	327
T. ochroleucum	root	$B5_1 \rightarrow (B5_2)_3 \rightarrow B5_3 \rightarrow$ W	$B5_1$	2,4-D (13.5); KIN (2.3)	Suc (88)	CH (2)	327
T. pallidum	cot	$B5_1 \rightarrow (B5_2)_3 \rightarrow B5_3 \rightarrow$ W	$B5_1$	2,4-D (13.5); KIN (2.3)	Suc (88)	CH (2)	327
T. pannonicum (Hungarian clover)	cot, hyp	$B5_1 \rightarrow (B5_2)_3 \rightarrow B5_3 \rightarrow$ W	$B5_1$	2,4-D (13.5); KIN (2.3)	Suc (88)	CH (2)	327
T. pilulare	hyp	$B5_1 \rightarrow (B5_2)_3 \rightarrow B5_3 \rightarrow$ W	$B5_1$	2,4-D (13.5); KIN (2.3)	Suc (88)	CH (2)	327
T. pratense (red clover)	hyp	$B5_1 \rightarrow (B5_2)_3 \rightarrow B5_3 \rightarrow$ W	$B5_1$	2,4-D (13.5); KIN (2.3)	Suc (88)	CH (2)	327
	hyp, epicot, c susp*	L2 → $SL2_{(L)}$ → L2 or $SL2_{(L)}$ → L2	L2 and SL2	PIC (0.25); BA (0.44)	Suc (73); Suc (73)	Ad (2)	68, 69; 329, 330

Table 1. Continued

Family – Species (Common Name)	Explant (*Plant Recovery)[a]	Media Sequence[b]	Embryo Induction Medium				Reference
			Salts	Growth Regulators (μM)[c]	Sugars (mM)[d]	Modifications (mg/L)[e]	
T. pratense	c susp ptp	$KAO_{(L)} \to SL2_{(L)} \to L2 \to LSP$	KAO	2,4-D (2.3)	Suc (73); Glu (28); Xyl (1.7)	several	331
T. repens (white clover) (ladino clover)	hyp	B5	B5	2,4-D (5.7); NAA (2.7); KIN (2.3)	Suc (73)		328
	imm emb*	MS $\to$ MS	MS	BA (0.1–0.2)	Suc (117)	YE (1000)	310, 311
	imm cot, som emb	$EC_6 \to MS_0 \to \to MS_0$	EC_6	2,4-D (180)	Suc (116)	YE (1000); numerous	331–334
T. rubens	pet*	L2 $\to$ LSP	L2	PIC (0.06); BA (0.4)	Suc (73)		335
	seedling minus sh tip	$L2 \to [SL2_{(L)}]_5 \to L2 \to LSE$	L2	PIC (0.25); BA (0.50)	Suc (73)		336
	lf ptp*, c susp ptp*	$SL2_{(L)} \to SLE \to LSP$-CR2	SLE	PIC (0.25); BA (0.44)	Suc (73)		337
T. scabrum	hyp, cot, root	$B5_1 \to (B5_2)_3 \to B5_3 \to W$	$B5_1$	2,4-D (13.5); KIN (2.3)	Suc (88)	CH (2)	327
T. scutatum	root	$B5_1 \to (B5_2)_3 \to B5_3 \to W$	$B5_1$	2,4-D (13.5); KIN (2.3)	Suc (88)	CH (2)	327
T. strictum	root	$B5_1 \to (B5_2)_3 \to B5_3 \to W$	$B5_1$	2,4-D (13.5); KIN (2.3)	Suc (88)	CH (2)	327
T. spumosum	root	$B5_1 \to (B5_2)_3 \to B5_3 \to W$	$B5_1$	2,4-D (13.5); KIN (2.3)	Suc (88)	CH (2)	327
T. tomentosum	cot, hyp, root	$B5_1 \to (B5_2)_3 \to B5_3 \to W$	$B5_1$	2,4-D (13.5); KIN (2.3)	Suc (88)	CH (2)	327

Table 1. Continued

Family – Species (Common Name)	Explant (*Plant Recovery)[a]	Media Sequence[b]	Embryo Induction Medium				Reference
			Salts	Growth Regulators (μM)[c]	Sugars (mM)[d]	Modifications (mg/L)[e]	
T. vesiculosum	hyp	B5 → ML8 + NO_3 → ML8	B5	2,4-D (5.7); NAA (2.7); KIN (2.3)	Suc (73)		328
	cot*, hyp	$B5_1$ → $(B5_2)_3$ → $B5_3$ → W	$B5_1$	2,4-D (13.5); KIN (2.3)	Suc (88)	CH (2)	327
Vicia faba (field bean)	imm cot	$L2_1$ → $L2_{0(L)}$ or $L2_{0(L)}$	$L2_1$	2,4-D (2.5)	Suc (29 –73)		338
V. narbonensis (narbonne vetch)	sh tip*	MS_1 → MS_2	MS	2,4-D (4.5–45) or DIC (2.2–22) or PIC (4.2–42)	Suc (88)		339, 340
	leaf*	MS_{1-5} → MS_{01-2}	MS	2,4-D (0.9); KIN (4.6)	Suc (88)		341
Vigna aconitifolia (moth bean)	leaf c susp*	B5 → $B5_{(L)}$ → $L_{6(1-6(L)}$	L_{6-2}	2,4-D (44.5); KIN (0.4)	Suc (58); Sor (55)	B5 VIT; CH (200)	342
	lf ptp*	$GS1_{(L)}$ → $GS2_{(L)}$ → $GS3_{(L)}$ → GS3 → GS4 → GS50	GS1, 2, 3	Several	Several		343
	imm cot	MS_1 → MS_2 → MS_3 → L_6 → L_6	L_6	PIC (0.4); GA (0.03); K (2.3)	Suc (58)	B5 VIT; CH (200)	344
V. mungo (black gram)	lf ptp	$SSK8_{1(L)}$ → $SSK8_2$ → $SSK8_3$	$SSK8_2$	GA_3 (5.0); PIC (0.1); IAA (0.5); ZEA (1.0)	Man (700)		345
	imm cot	MS_1 → MS_2 → MS_3 → L_6 → L_6	L_6	PIC (0.4); GA (0.03); K (2.3)	Suc (58)	B5 VIT; CH (200)	344

Table 1. Continued

Family – Species (Common Name)	Explant (*Plant Recovery)[a]	Media Sequence[b]	Embryo Induction Medium				Reference
			Salts	Growth Regulators (μM)[c]	Sugars (mM)[d]	Modifications (mg/L)[e]	
V. radiata (mung bean)	imm cot	$MS_1 \rightarrow MS_2 \rightarrow$ $MS_3 \rightarrow L_6 \rightarrow L_6$	L_6	PIC (0.4); GA (0.03); K (2.3)	Suc (58)	B5 VIT; CH (200)	344
V. glabrescens × *V. radiata*	emb	$MS_1 \rightarrow MS_2$ $\rightarrow \frac{1}{2}MS$	MS_1	Pretreatment of pod with NAA (135); GA (2800); KIN (23)	Suc (58)	I (100); Glutm (100); NA (5); P (0.5); Thi (1)	346
Limnanthaceae							
Limnanthes alba (meadow foam)	imm zyg emb	MS	MS	2,4-D (0.1); ZEA (0.1–1.0)	Suc (205)	CM (100 mL)	347
Linaceae							
Linum usitatissimum (flax)	imm zyg emb	MON	MON	BA (0.2)	Suc (146)	Glutm (400); YE (1000)	348
Loranthaceae							
Dendrophthoe falcata (stem parasite)	embryo	W	W	IAA (14.3); KIN (46)	Suc (58)	CH (500)	349
Nuytsia floribunda (root parasite)	embryo	$W_1 \rightarrow (W_2)_{20}$	W_1 W_2	IBA (24.6); KIN (23.2); IBA (24.6); 2,4-D (9.0)	Suc (58)	CH (2000); Ad (20)	349,350
Taxillus vestitus	embryo	$W_{1-4} \rightarrow W_2$	W	IAA (14.3); KIN (46)	Suc (58)	CH (2000)	349
Malvaceae							
Abelmoschus esculentus (okra)	hyp	N	N	2,4-D (0.5); 2iP (4.9)	Glu (222)	minus suc	351
Gossypium barbadense	hyp	$MS \rightarrow MS_{0(L)}$ $\rightarrow MS_{0(L)}$	MS	KIN (2.5); 2,4-D (0.5)	Glu (166)	Thi (10); NA (1); P (1); $MgCl_2$ (750)	46

Table 1. Continued

Family – Species (Common Name)	Explant (*Plant Recovery)[a]	Media Sequence[b]	Embryo Induction Medium				Reference
			Salts	Growth Regulators (µM)[c]	Sugars (mM)[d]	Modifications (mg/L)[e]	
G. hirsutum	hyp*	MS → MS$_{0(L)}$ → MS$_{0(L)}$	MS	KIN (2.5); 2,4-D (0.5)	Glu (166)	Thi (10); NA (1); P (1); MgCl$_2$ (750)	46, 352–354
	hyp*	MS$_1$ → MS$_2$ → MS$_0$	MS$_1$	KIN (4.6); IAA (2.8–23)	Suc (88)		355
G. klotzschianum (cotton)	hyp	MS$_1$ → MS$_2$ → B5$_{2(L)}$	MS$_2$	2iP (49); NAA (5.4)	Glu (166)	Glutm (2191)	356
Hibiscus sabdariffa (roselle)	stem, hyp	N$_1$ → N$_2$	N	2,4-D (18)	Glu (222)	several	357
Onagraceae							
Fuchsia hybrida (fuchsia)	ovary	B5	B5	IAA (5.7); KIN (4.6)	Suc (58)		358
Papaveraceae							
Eschscholzia californica (California poppy)	ovary	N$_1$ → N$_2$	N		CH (500)		359–361
Macleaya cordata (plume poppy)	callus	W	W	2,4-D (4.5)		CM; CH	362
Papaver somniferum (opium poppy)	hyp	MS$_1$ → MS$_{2(L)}$ → MS$_{0(L)}$	MS$_2$	2,4-D (9.0); KIN (1.2)	Suc (88)	minus glycine	363
Polygonaceae							
Rumex acetosella	bud*	MS$_1$ → MS$_0$	MS$_1$	BA (10); IAA (1.0)	Suc (176)	Thi-HCl (1); P-HCl (1); NA (5)	364
Primulaceae							
Anagallis arvensis (pimpernel)	hyp, stem	MS$_1$ → MS$_{2(L)}$ → MS$_3$ → MS$_4$	MS$_4$	IAA (5.7); KIN (2.3)	Suc (88)	CM (100 mL)	365
Cyclamen persicum (cyclamen)	ov, ant*, pedun	MS$_1$ → ½MS$_2$ → MS$_{0(L)}$	MS$_2$	2,4-D (4.5)	Suc (88)	CM (100 mL); ½ MACRO	366
	lf*, ov*, lf stalk*	MS	MS	2,4-D (0.5–4.5); KIN (2.3–12)	Suc (176)	½ MACRO; MS VIT	367

382

Table 1. Continued

Family – Species (Common Name)	Explant (*Plant Recovery)[a]	Media Sequence[b]	Embryo Induction Medium				Reference
			Salts	Growth Regulators (μM)[c]	Sugars (mM)[d]	Modifications (mg/L)[e]	
Primula obiconia (primrose)	anther	MS	MS	IAA (2.7); 2,4-D (2.3); ZEA (9.2)	Suc (88)		368
Ranunculaceae							
Aconitum carmichaeli	imm anther*	$MS_1 \to MS_2 \to$ $MS_3 \to MS_4 \to$ $MS_5 \to MS_6$	MS_1	2,4-D (22); KIN (4.6)	Suc (88)	Culture in the dark	369
A. heterophyllum	leaf*, pet*	$MS_1 \to MS_2 \to$ $\frac{1}{4}MS_3$	MS_2	2,4-D (4.5); KIN (2.3) or NAA (26); BAP (4.4)	Suc (88); Suc (88)	CM (100 mL)	370
Coptis japonica	pedicel*	$LS_1 \to LS_2 \to LS_{(L)x}$ $\to LS_3 \to LS_4$	LS	2,4-D (1.0)	Suc (88)		371
Nigella damascens (Love-in-a-mist)	pedicel	$MS_1 \to MS_{2(L)}$	$MS_{1,2}$	2,4-D (9.0)	Suc (58)	minus CM in MS_2	372
N. damascena	endospm	MS	MS	2,4-D (5–10)	Suc (88)		373
N. sativa (black owmin)	leaf*	$W \to (MS)_8$	MS	IAA (2.8)	Suc (58)	CH (500); minus CM in MS	374
Ranunculus asiaticus	imm fl bud*	$MS_1 \to MS_0$	MS_1	2,4-D (0.5–7.2)	Suc (88)	Glutm (50); AdS (80); MW VIT	375
R. sceleratus (marsh crowfoot)	fl bud*	W	W	IAA (5.7)	Suc (58)	N MICRO; CM (100 mL)	376, 377
	lf ptp*	$W \to \frac{1}{2}MS_0$	W	BAP (4.4); NAA (16.1)	Suc (58); Man (549)	H MICRO; MW VIT; $Ca(NO_3)_2$ (200)	378

Table 1. Continued

Family - Species (Common Name)	Explant (*Plant Recovery)[a]	Media Sequence[b]	Embryo Induction Medium				Reference
			Salts	Growth Regulators (μM)[c]	Sugars (mM)[d]	Modifications (mg/L)[e]	
	lf ptp*	$MS_1 \rightarrow MS_2 \rightarrow$ $MS_3 \rightarrow MS_4 \rightarrow$ $MS_5 \rightarrow MS_6$	MS_1	BA (4.4); NAA (16.1)	Glu (166); Man (330)	Culture in the dark; CN (400); Glutm (100)	378, 379
Thalictrum urbaini	fl bud*	W	W	IAA (5.7)	Suc (88)		380, 381
	lf*, pet*	MS	MS	IAA (11.4); KIN (9.2)	Suc (88)	CM (100 mL)	382
Rosaceae							
Fragria × ananassa (pineapple strawberry)	imm emb	MS → MS	MS	2,4-D (22.5); BA (2.2)	Suc (146)	CH (500)	383
Santalaceae							
Santalum album (sandel wood)	seed	$W_1 \rightarrow W_2$	W_1	2,4-D (9.0); KIN (23)	Suc (117)	several; YE (2.5)	384
Saxifragaceae							
Chrysosplenium americanum (golden saxifrage)	lf node	¼ → ½$MS_1 \rightarrow MS_2$	MS_1	IAA (5.7) or NAA (5.4) or 2,4-D (4.5); KIN (0.04)	Suc (88)	several	385
Scrophulariaceae							
Antirrhinum majus (snap dragon)	stem*	MS → MS_0	MS	NOA (1.2–2.5)	Suc (58)	N MICRO; NN VIT; CM (1000 mL)	386
	lf ptp	MS	MS	2,4-D (4.5); BA (2.2)	Suc (60); Man (700)	several	387
Digitalis lanata	stem filament	MS_2; NMI → NMI – M → NMII – M or NMV – M	MS_2 MS	2,4-D (4.5–9.0); 2,4-D (4.5); KIN (0.1)	Suc (58) Mal (83)	several	388, 389

385

Table 1. Continued

Family – Species (Common Name)	Explant (*Plant Recovery)[a]	Media Sequence[b]	Embryo Induction Medium			Modifications (mg/L)[e]	Reference
			Salts	Growth Regulators (μM)[c]	Sugars (mM)[d]		
	fil*	$MS_1 \rightarrow MS_2$	MS_1	2,4-D (4.5–23); KIN (0.1–14)	Suc (88) →; Glu (333)	KH_2PO_4 (680)	390
D. lutea	hyp	$MS_1 \rightarrow MS_2$	MS_1	2,4-D (4.5–23); KIN (0.1–14)	Suc (88) →; Glu (333)	KH_2PO_4 (340)	390
D. obscura	hyp*	$MS_1 \rightarrow MS_0$	MS_1	IAA (4.7)	Suc (88)		391
	root, hyp	$MS_1 \rightarrow MS_2$	MS_2	KIN (4.5–9.0); IAA (0.6–3.0); or NAA (0.6–3.0)	Suc (88)		392
Solanaceae *Atropa belladonna* (night shade)	root → c susp*	$W_1 \rightarrow W_{2(L)}$	W_1	NAA (10.6); KIN (0.46)	Suc (58)	I (100); FeEDTA	393, 394
	stem → c susp → ptp*	$WB \rightarrow WB_{(L)} \rightarrow MS_{1(L)} \rightarrow MS_2 \rightarrow MS_{0(L)}$	WB and MS_1	NAA (10.7); KIN (0.46)	Suc (88); Sor (200)		395
	stem → c susp	$SSM_1 \rightarrow SSM_{2(L)} \rightarrow SSM_{0(L)}$	SSM	NAA (10.7); KIN (2.3)	Suc (58)	W VIT; $(NH_4)_2$ SO_4 (790); I (100)	396
	root → c susp	$SSM_1 \rightarrow SSM_{1(L)} \rightarrow SSM_{0(L)}$	SSM	NAA (5.4); KIN (2.3)	Suc (58)	I (100); $(NH_4)_2$ SO_4 (790)	397
Cyphomandra betacea (tree tomato)	zyg emb*, hyp*	$MS_1 \rightarrow MS_{0(L)}$	MS_1	2,4-D (9.0–2.3)	Suc (88)		398
Hyoscyamus muticus	lf ptp → c susp ptp*	$NT_{(L)} \rightarrow MSC$	MS	KIN (4.6); NAA (0.5)	Suc (88)	Charcoal (5000)	399
H. niger (henbane)	hyp*, pet*, ovary*	$BN_1 \rightarrow BN_0$	BN	2,4-D (9.0)	Suc (58)		400
Lycopersicon peruvianum (tomato)	lf ptp	$MS \rightarrow MS_0$	MS	NAA (5.4); KIN (4.6)	Suc (88); Man (500)		401

Table 1. Continued

Family – Species (Common Name)	Explant (*Plant Recovery)[a]	Media Sequence[b]	Embryo Induction Medium				Reference
			Salts	Growth Regulators (μM)[c]	Sugars (mM)[d]	Modifications (mg/L)[e]	
Nicotiana sylvestris	lf ptp*	$MS_{1(L)} \rightarrow MS_{2(L)}$ $\rightarrow MS_3$	MS_2	2,4-D (0.9); NAA (3.2); KIN (3.7)	Suc (88); Sor (176)	½MS MACRO; $CaCl_2$ (200)	402
N. tabacum (tobacco)	pet, hyp	W	W	BA (8.8); IAA (0.6)	Suc (58)	Ad (5)	403
	lf ptp	NT → WB	MS	CPA (6); KIN (0.25) or DIC (10); BA (50)	Suc (29)	NN MICRO	404
	lf ptp*	$KAO_{1(L)} \rightarrow$ $KAO_{2(L)} \rightarrow$ $KAO_{3(L)} \rightarrow MS_{1(L)}$ $\rightarrow MS_{2(L)}$	$KAO_{1,2,3}$	2,4-D (4.5); NAA (5.4); BA (2.2) ZEA (9.2)	Suc (41); Man (450–165)	several	405
Petunia inflata	stem, leaf*	MS	MS_1 MS	2,4-D (4.5); BA (0.9)	Suc (58)	NN VIT	406–408
	leaf c susp*	MS → $MS_{(L)}$ → MS_0	MS	2,4-D(0.5–4.5)	Suc (58)	NN MICRO; NN VIT	
	lf*, stem*	$MS_1 \rightarrow MS_0$	MS	2,4-D (0.5–9.0)	Suc (58)	NN MICRO; NN VIT	408
Solanum carolinense (horse-nettle)	stem*	$MS_1 \rightarrow MS_2$	MS_1; MS_2	2,4-D (45); BA (2.2)	Suc (88)	NN VIT	409
S. khasianum (bittersweet)	root*	$MS_1 \rightarrow MS_{2(L)}$	MS or	KIN (1.2); NAA (.05); BAP (1.1); NAA (.05)	Suc (88)	YE (200)	410
S. melongena (egg plant)	leaf*	MS → MS_0 or MS $\rightarrow MS_{(L)} \rightarrow MS_0$	MS	NAA (42.4–53.7)	Suc (60)	B5 VIT	411, 412

387

Table 1. Continued

Family – Species (Common Name)	Explant (*Plant Recovery)[a]	Media Sequence[b]	Embryo Induction Medium				Reference
			Salts	Growth Regulators (μM)[c]	Sugars (mM)[d]	Modifications (mg/L)[e]	
	c susp ptp*	$MS_{(L)} \to KAO_{1(L)} \to KAO_2$	MS; KAO_1	2,4-D (9.0); NAA (16.1); BAP (4.4); 2,4-D (4.5)	Glu (350)	several	413
	imm emb	$W \to W_0$; or MS $\to MS_0$	W; MS	NAA (5.4); KIN (18.6); IAA (57.0)	Suc (58)		414
	hyp*	$MS \to MS_0$	MS	NAA (42.9)	Suc (58)	VITS modified	415
	pet*, stem*	$KAO_{(L)} \to MS_1 \to MS_0$	KAO	2,4-D (0.9); ZEA (2.3); NAA (5.4)	Glu (350)		416
	leaf*	$MS_1 \to MS_{(L)} \to \frac{1}{2}MS$	MS	NAA (43.2); KIN (0.5)	Suc (88)	B5 VIT	417–419
S. tuberosum (potato)	tuber*	$MS_1 \to W_0$	MS	BAP (1.8); IAA (23); GA (1.1)	Suc (58)	NN VIT; CH (1000)	420, 421
Theaceae							
Camellia japonica	hyp*	$MS_1 \to MS_2 \to MS_3 \to MS_4$	MS_1	BA (8.8–17.2); IBA (9.8–19.6); GA_3 (2.8)	Suc (88)		422
Tiliaceae							
Corchorus capsularis (jute)	cot ptp	$KAO \to MS_1 \to MS_2$	MS_2	2,4-D (2.3); NAA (1.3)	Suc (90)	several; AdS (222)	423
Umbelliferae							
Ammi majus (bishop's weed)	hyp*	MS_1 or $MS_2 \to MS_1$	MS_1; MS_2	IAA (11.4); IAA (28.5)	Suc (88)		424

Table 1. Continued

Family – Species (Common Name)	Explant (*Plant Recovery)[a]	Media Sequence[b]	Embryo Induction Medium				Reference
			Salts	Growth Regulators (μM)[c]	Sugars (mM)[d]	Modifications (mg/L)[e]	
Anethum graveolens (dill)	ovary*	W → W	W	± IAA (5.3)	Suc (117)	(CH 100–1000) or (YE 100–1000)	425
	hyp*	MS_1 → MS_0	MS_1	2,4-D(4.5)	Suc (58)	N VIT; Fe (NITSCH)	426
	cot*	MS_3 → MS_4	MS_3; MS_4	2,4-D (10); KIN (1.0); 2,4-D (1.0)	Suc (58)	CH (1000)	
	ovary*	W	W	± IAA (5.7)	Suc (117)	YE (100–1000); CH (100–1000)	427
	infl*	W_1 → W_3; W_2 → M	$W_{1,2}$; W_2	2,4-D (2.2); KIN (2.3)	Suc (117)	CH (500); Ad (10); CM (100–200 mL)	428
Angelica acutiloba	pedicel	LS_1 → LS_2 → $LS_{(L)x}$ → LS_0	LS_2	2,4-D (1)	Suc (88)		371
A. sinensis	root*	MS_1 → MS_2	MS_1	2,4-D (3.2); NAA (27); BA (2.2)	Suc (88)	± YE (1000)	429
Apium graveolens (Chinese celery)	pet*	MS_1 → MS_1 → $MS_{1(SorL)}$ → $MS_{0(SorL)}$	MS_1	2,4-D (2.3); KIN (2.7)	Suc (88)	several	430, 431
	pet*	MS_1 → $MS_{1(L)}$ → $MS_{2(L)}$ → $MS_{1O(L)}$ → MS_2 → MS_0	MS_1; MS_2	2,4-D (2.3); KIN (2.7)	Suc (88)		432
	pet*	MS_1 → MS_2	MS_2	NAA (0.5); KIN (14)	Suc (88)		433, 434
	leaf*	MS_1 → MS_0	MS_1	2,4-D (2.3); KIN (2.7)	Suc (88)		435
	leaf	MS_1 → MS_0	MS_1	2,4-D (9.0)	Suc (88)		436
	leaf*, pet*	MS_1 → $MS_{2(L)}$ → $MS_{3(FL)}$	MS_2	2,4-D (2.3); BA (0.9)	Suc (88); Man (165)		437, 438

Table 1. Continued

Family – Species (Common Name)	Explant (*Plant Recovery)[a]	Media Sequence[b]	Embryo Induction Medium				Reference
			Salts	Growth Regulators (μM)[c]	Sugars (mM)[d]	Modifications (mg/L)[e]	
Bunium persicum	c susp*	$MS_{1(L)} \rightarrow MS_{2(L)}$ or $MS_{3(L)}$	MS_1	2,4-D (2.3); BA (0.9)	Suc (44); Man (220)		
	meric*	$MS_1 \rightarrow MS_2 \rightarrow MS_O$	MS_1; MS_2	2,4-D (9.0); KIN (18.4); 2,4-D (4.5)	Suc (88)		439
Bupleurum falcatum	root, leaf	$LS_1 \rightarrow LS_2 \rightarrow LS_{(L)} \rightarrow LS_0$	LS	ZEA (9.2)	Suc (88)		440
Carum carvi (caraway)	pet c susp	$W_{(L)} \rightarrow MS_{(L)}$	W	NAA (10.6)	Suc (58)	CH (200)	441, 442
	hyp*	$MS_1 \rightarrow MS_{0(L)} \rightarrow MS_2$	MS_1	2,4-D (1.3)	Suc (88)	NN VIT	443
Coriandrum sativum (coriander)	pet*	$MS_1 \rightarrow MS_{1(L)} \rightarrow MS_{0(L)} \rightarrow MS_0$	MS_1	2,4-D (2.3)	Suc (88)		444
	sh apex*	$MS_1 \rightarrow MS_2 \rightarrow MS_3$	MS_2	NAA (2.7); KIN (9.2)	Suc (88)	CM (150 mL)	445
Daucus carota (wild carrot)	rt, pet, pedun	several		2,4-D (0.45–9); KIN (10.2–4.6)	Suc (58–88)	± CM (1000 mL); Ad (2)	446
	hyp	$B5_{1(L)} \rightarrow B5_{0(L)}$	B5	2,4-D (2.3); BA (1.1)	Suc (58)	Pro (11500)	447, 448
	root phloem	$MS_{(L)} \rightarrow MS$	MS	2,4-D (0.5–4.5)	Suc (73)	CM (100 mL) minus KI	449
	meric	MS_O	MS	–	Suc (88)	several	450
	rt*, pet*, pedun*	several		2,4-D (0–9.0); KIN (0–4.5)	Suc (58–88)	± CM (100 mL); Ad (2)	446

Table 1. Continued

Family – Species (Common Name)	Explant (*Plant Recovery)[a]	Media Sequence[b]	Embryo Induction Medium				Reference
			Salts	Growth Regulators (μM)[c]	Sugars (mM)[d]	Modifications (mg/L)[e]	
hyp	MS	MS	IAA (57); NAA (54); IBA (4.9); 2,4-D (0.5); NOA (4.9); PCPA (5.4); MCPA (0.5); 2,4,5-T (0.4)	Suc (88)	NN VIT	451
	$MS_1 \rightarrow MS_0$	MS_1	IAA (57); NAA (54); IBA (49); 2,4-D (4.5); PCPA (5.4); MCPA (0.5); 2,4,5-T (3.9)			
hyp	$MS_1 \rightarrow MS_0$ $L_nS \rightarrow L_nS_0$ $Wm \rightarrow MS_0$	MS_1 L_nS W	2,4-D (4.5) " "	Suc (88) " "	MS MICRO NN VIT several	452
c susp	$MS_{2(L)} \rightarrow MS_{0(L)}$	MS_2	2,4-D (4.5)	Suc (88)	+KNO_3 (122); –NH_4NO_3; αAla (891); Glutm (1461); Asp (1501); AspA (1331); GlutA (1471)	
pet*	$LnS_{1(SorL)} \rightarrow LnS_2$ or $LnS_{0(SorL)}$	LnS_1; LnS_2	2,4-D (0.5–5); 2,4-D (0.05)	Suc (58–88)	reduced N_2	453, 454

Table 1. Continued

Family – Species (Common Name)	Explant (*Plant Recovery)[a]	Media Sequence[b]	Embryo Induction Medium				Reference
			Salts	Growth Regulators (μM)[c]	Sugars (mM)[d]	Modifications (mg/L)[e]	
	hyp	$B5_{1(L)} \rightarrow B5_{0(L)}$	B5	2,4-D (2.3); BA (1.1)	Suc (58)	Pro (11500)	451, 452
	rt phlm	$MS_{(L)} \rightarrow MS$	MS	2,4-D (0.5–4.5)	Suc (73)	CM (100 mL)	449
	root	$W \rightarrow MS_0$	W	NAA (10.7)	Suc (58)	CH (200)	455
	root*	$W \rightarrow W_{(L)}$	W	IAA (0.05–57)	Suc (58)	CM (150 mL)	456
	root*	$H1 \rightarrow H1_{(L)} \rightarrow H1_{(F)}$	H1	IAA (0.6–6.0)	Suc (88)	minus CM	457
	root	$MS_1 \rightarrow MS_{0(SorL)}$	MS	2,4-D (0.5–4.5); KIN (0.5)	Suc (48)	W VIT	458
	root*	MS	MS	2,4-D (0.02)	Suc (58)	W VIT; $-NH_4NO_3$; $+KNO_3$ (4155)	459
		W	W	2,4-D (0.02)	Suc (58)	$\pm NH_4NO_3$ (1650); $\pm KNO_3$ (4155)	
	root	$MS_1 \rightarrow MS_0 \rightarrow$	MS	2,4-D (0.02)	Suc (58)	W VIT	460
		$W \rightarrow MS_0 \rightarrow$	W	″	″		
	rt/pet $\rightarrow$	$B5_1 \rightarrow B5_{1(L)}$	$B5_1$	2,4-D (0.5)	Suc (58)		461
	c susp	$B5_2 \rightarrow B5_0$	$B5_2$	2,4-D (0.5)	Glu (50); Man (230)		
	pet $\rightarrow$ c susp	$LnS_1 \rightarrow LnS_{1(L)} \rightarrow LnS_{0(L)}$ or $LnS_{2(L)}$	LnS_1; LnS_2	2,4-D (0.5); ZEA (0.1)	Suc (58)	KNO_3 (5555); NH_4Cl (267); Thi-HCl (3); NA (5)	462, 463
	c susp	$LnS_0 \rightarrow LnS_1$	LnS_1	2iP (0.5–2.5)	Suc (88)		464
	c susp ptp	$B5_{(L)} \rightarrow Kao_{(L)} \rightarrow B5_{(L)}$	B5; KAO	2,4-D (4.5); NAA (1.0); ZEA (0.5)	Suc (58); Glu (380); D-rib (1.0)	CM (50 mL); B5 MICRO + VIT	465
	root	$H \rightarrow H_0$	H	IAA (1.06)	Glu (166)		466

Table 1. Continued

Family – Species (Common Name)	Explant (*Plant Recovery)[a]	Media Sequence[b]	Embryo Induction Medium				Reference
			Salts	Growth Regulators (μM)[c]	Sugars (mM)[d]	Modifications (mg/L)[e]	
	hyp c susp	$LnS_{1(L)} \to LnS_{2(1)}$	LnS_1	2,4-D (0.5); ZEA (1.0)	Man (200)		31
	hyp c susp	$MS_{1(L)} \to MS_{2(L)}$	MS_1	2,4-D (4.5)	Suc (88)	minus NH_4NO_3; alanine (890)	467
	hyp	$MS_{(L)} \to MS \to MS_{0(L)}$	MS	2,4-D (10)	Suc (88)		468
	root c susp	$MS_{1(L)} \to MS_{2(L)}$	MS_1	2,4-D (0.45); ZEA (0.1)	Suc (73)		469, 470
	root*	$W \to W$	W	IAA (5.7 or 57)	Suc (58)	YE (1)	471
	root	$MS_{(L)} \to MS_{(L)}$	MS	NAA (16.1)	Suc (73)	CM (50) minus KI; Ad (3)	472
	wounded zyg emb	DS – 5	DS – 5	–	Suc (88)	NH_4Cl (80); Thi (1)	473, 474
	c susp	$LnS_{1(L)} \to LnS_{20(L)}$	LnS_1	2,4-D (2.3)	several	WCM; pH maintained $\pm$ K^+	475–478, 732, 734
	hyp	$DS_1 \to DS_0 \to DS_{10}$	DS_1	2,4-D (4.5)	Suc (58)	DS – 5a; $DS_0 \to$ pH 4.5	479
Foeniculum vulgare (fennel)	root	$MS_1 \to MS_0$	MS	2,4-D (4.5)	Suc (88)		480
	pet $\to$ c susp $\to$ ptp*	$LS_1 \to LS_{1(L)} \to LS_2 \to LS_0$ or $LS_1 \to LS_{1(L)} \to LS_{20}$	LS_1; LS_2	2,4-D (1.0); KIN (1.0); 2,4-D (1.0); KIN (1.0)	Suc (88); Glu (350–650)		481
	stem	$W \to N_1 \to N_2$	N_2	IAA (1.0); KIN (1.0)	Suc (58)	CM (150 mL); YE (100)	482
	pet*	$LS_1 \to LS_{1(L)} \to LS_0 \to \frac{1}{4} LS_0$	LS_1	2,4-D (1.0); KIN (1.0)	Suc (88)		483

Table 1. Continued

Family – Species (Common Name)	Explant (*Plant Recovery)[a]	Media Sequence[b]	Embryo Induction Medium				Reference
			Salts	Growth Regulators (μM)[c]	Sugars (mM)[d]	Modifications (mg/L)[e]	
Petroselinum hortense (parsley)	pet callus*	$MS_1 \to MS_2$	$MS_{1,2}$	IAA (57); KIN (0.2)	Suc (88)	AdS (15–20) in MS_2	210, 484
(parsley)	pet*	$MS_1 \to MS_2$ or MS_0	MS_1; MS_2	2,4-D (4.5) 2iP (5–10)	Suc (88)		485
Pimpinella anisum (anise)	hyp c susp	$B5_1 \to B5_2$	$B5_1$; $B5_2$	2,4-D (4.4); Ado or 2iP (0.05–0.5)	Suc (58)		486–488
Selinum candolii	pet*	$MS_1 \to MS_{1(L)}$; $MS_{2+oil} \to MS_2$	MS_1	2,4-D (4.5); KIN (1.2)	Suc (88)	Oil overlay	489
Sium suave (water parsnip)	embryo	$W \to MS_0$	W	NAA (10.7)	Suc (58)	CM (100 mL); CH (200)	455
Trachyspermum ammi (borage)	hyp*	$MS_1 \to MS_2 \to$ $MS_3 \to MS_0$	MS_2	2,4-D (2.3); KIN (2.3)	Suc (88)		490
	hyp*	$MS_1 \to MS_x$	MS	2,4-D (9.0); KIN (2.3)	Suc (88)		491
Valerianaceae *Nardostachys jatamansi*	pet	$MS_1 \to MS_2 \to$ $MS_3 \to MS_4 \to$ $MS_5 \to MS_6$	MS_1; MS_6	NAA (16); KIN (1.2); NAA (1.3); KIN (9.3)	Suc (88)	I (100)	566

[a] Code for explant source: ant = anther; cal = callus; cot = cotyledon; c susp = cell suspension; emb = embryo; embal = embryonal; endospm = endosperm; epicot = epicotyl; epider = epidermus; fil = filament; fl = flower; frag = fragmented; germ = germinated; hyp = hypocotyl; infl = inflorescences; imm = immature; lf = leaf; mat = mature; meric = mericarp; ov = ovary; ped = pedicels; pedun = peduncles; pet = petiole; phlm = phloem; ptp = protoplast; rt = root; sh = shoot; som = somatic; zyg = zygotic.
[b] Code for media sequence: B = Blaydes; BI1, BOi2Y, BI2Y = modified Blaydes; B5 = Gamborg; BN = Bourgain and Nitsch; Ch = Cheng; DF = Dixon and Fuller; DPD = Durand, Potrykus and Dom; DS = Smith and Krikorian; EC_6 = Maheswaran and Williams; GS = Shekhawat and Galston; H = Heller; H_1 = Hildebrandt; K3 = Nagy and Maliga; KAO = modified KAO's; L_6 = Kumar, Gamborg and Nabors; LnS = Lin and Staba; LS = Linsmaier and Skoog; LSE = Philip and Collins; L2 = Philip and Collins; LSP = Phillips, Collins and Taylor; M1 = Pecaut, Dumas de Vaulx and Lot; MIL = Miller; MON = Monnier; MS = Murashige and Skoog; MSC = MS medium with charcoal;

MW = Morel and Wetmore; M8P = Glimelius, Djupsjöbacka and Feldegg; N = Nitsch; N6 = Chu; NN = Nitsch and Nitsch; NT = Nagata and Takebe; PCW = Chinese potato extract; SH = Shenk and Hildebrandt; SL2, SLE = Phillip and Collins; SPC88 = Song, Sorensen and Liang; SSM = standard synthetic medium; SSK8 = Sinha, Das and Sen; UM = Uchimiya and Murashige; VKM = Binding and Nehls; W = White; Wm = White's modified; WB = Wood and Braun.

(F) = medium moistened filter paper; (L) = liquid medium; (S/L) = cells embedded in solid medium immersed in liquid med; (X) = several subcultures.

[c] Code for growth regulators: ABA = Absisic Acid; AMO 1618 = 2-Isoproply-4-dimethylamino-5-methyl-phenyl-1-piperidine carboxymethyl chloride; BAP/BA = 6-Benzylaminopurine/ Benzyladenine; 2-BTOA = 2-Benzothiazoleacetic acid; 2,4-D = 2,4-Dichlorophenoxyacetic acid; GA = Gibberellic Acid; IAA = Indole-3-acetic Acid; IBA = Indole-3-butyric acid; 2-iP = 6γγ-Dimethylallylamino-purine; KIN = Kinetin; MCPA = 2-Methyl-4-chlorophenoxyacetic acid; NAA = β-Naphthaleneacetic acid; NOA = 2-Naphthoxyacetic acid; PCPA = p-Chlorophenozyacetic acid; PIC = Picloram; 2,4,5-T = 2,4,5-Trichlorophenoxyacetic acid; TDZ = Thidiazuron; ZEA = Zeatin; ZR = Zeatin riboside.

[d] Code for sugars: D-rib = D-ribose; Glu = glucose; Lac = lactose; Malt = maltose; Man = mannitol; Sor = sorbitol; Suc = sucrose; Xyl = xylose.

[e] Code for modifications: Ad = adenine; AdS = adenine sulfate; Ala = alanine; Arg = arginine; Asp = asparagine; AspA = Aspartic acid; CA = casamino acids; CH = casean hydrosolate; CM = coconut milk; CuSO4 = Cupric sulphate; Cyst = cysteine; Eda = edamine; Glut A = glutamic acid; Glutm = glutamine; Glutt = glutathione; gly = glycine; I = inositol; NA = nicotinic acid; P = pyridoxine; P-HCl = pyrodoxine-HCl; Pro = proline; Ser = Serine; Thi = Thiamine; Thi-HCl = thiamine-HCl; Try = tryptone; Tryp = tryptophane; VIT = vitamines; YE = yeast extract.

References

1. Halperin, W., Morphogenesis in cell cultures, *Ann. Rev. Plant Physiol.*, 20, 395, 1969.
2. Haccius, B., Question of unicellar origin of non-zygotic embryos in callus cultures. *Phytomorphology*, 28, 74, 1978.
3. Ranch, J., and Pace, G.M., Science in the art of plant regeneration from cultured cells: An essay and a proposal for a conceptual framework, *Iowa State J. Res.*, 62, 537, 1988.
4. Kitano, H., Tamura, Y., Satoh, H., and Nagato, Y., Hierarchical regulation of organ differentiation during embryogenesis in rice, *Plant J.*, 3, 607, 1993.
5. Williams, E.G., and Maheswaran, G., Somatic embryogenesis: Factors influencing coordinated behaviour of cells as an embryogenic group, *Ann. Bot.*, 57, 443, 1986.
6. Sharp, W.R., Evans, D.A., and Sondahl, M.R., Application of somatic embryogenesis to crop improvement, in *Plant Tissue Culture 1982*, Fujiwara, A., Ed., Japanese Assoc. Plant Tissue Culture, Tokyo, 1982, 759.
7. Sharp, W.R., Cadas, L.S., and Maraffa, S.B., The physiology of *in vitro* asexual embryogenesis, *Hort. Rev.*, 2, 286, 1980.
8. Xu, N., and Bewley, J.D., Contrasting pattern of somatic and zygotic embryo development in alfalfa (*Medicago sativa* L.) as revealed by scanning electron microscopy, *Plant Cell Rep.*, 11, 279, 1992.
9. Brown, D.C.W., and Thorpe, T.A., Vasil, I.K., Eds. Plant regeneration by organogenesis, in *Cell Culture and Somatic Cell Genetics of Plants*, Vol. 1, Academic Press, New York, 1986, 1.
10. Ammirato, P.V., Embryogenesis, in *Handbook of Plant Cell Culture*, Vol. 1. Evans, D.A., Sharp, W.R., Ammirato, P.V., Yamada, Y., Eds., Macmillan, New York, 1983, 82.
11. Terzi, M., and Loschiavo, F., Somatic embryogenesis, in *Plant Tissue Culture: Applications and Limitations*, Bhojwani, S.S., Ed., Elsevier, Amsterdam, 1990, 54.
12. Steeves, T.A., and Sussex, I.M., *Patterns in Plant Development*, Second Edition, Cambridge University Press, Cambridge, 1989.
13. Ammirato, P.V., Recent progress in somatic embryogenesis, *IAPTC Newslett.*, 57, 2, 1989.
14. Brummitt, R.K., *Vascular Plant Families and Genera*, Royal Botanic Gardens, Kew, 1992.
15. Brown, D.C.W., and Atanassov, A., Role of genetic background in somatic embryogenesis in *Medicago, Plant Cell Tiss. Org. Cult.*, 4, 111, 1985.
16. Bingham, E.T., Hurley, L.V., Kaatz, D.M., and Saunders, J.W., Breeding alfalfa which regenerates from callus tissue in culture, *Crop Sci.*, 15, 719, 1975.
17. Brown, D.C.W., Germplasm determination of *in vitro* somatic embryogenesis in alfalfa, *HortSci.*, 23, 526, 1988.
18. Hernandez-Fernandez, M.M., and Christie, B.R., Inheritance of somatic embryogenesis in alfalfa *Medicago sativa* L., *Genome*, 32, 318, 1989.
19. Wan, Y., Sorenson, E.L., and Liang, G.H., Genetic control of *in vitro* regeneration in alfalfa (*Medicago sativa* L.), *Euphytica*, 39, 3, 1988.
20. Ray, I.M., and Bingham, E.T., Breeding diploid alfalfa for regeneration from tissue culture, *Crop Sci.*, 29, 1545, 1989.
21. Seitz Kris, M.H., and Bingham, E.T., Interactions of highly regenerative genotypes of alfalfa (*Medicago sativa*) and tissue culture protocols, *In Vitro Cell. Develop. Biol.*, 24, 1047, 1988.
22. Halperin, W., Population density effects on embryogenesis in carrot cell cultures, *Exp. Cell Res.*, 48, 170, 1967.
23. Street, H.E., Embryogenesis and chemically induced organogenesis. in *Plant Cell and Tissue Culture Principles and Applications*, Sharp, W.R., Larsen, P.O., Paddock, E.F., and Raghavan, V., Eds., Ohio State University Press, Columbus, 1979, 123.
24. Halperin, W., The use of cultured tissue in studying developmental problems, *Can. J. Bot.*, 51, 1801, 1973.
25. Choi, J.H., and Sung, Z.R., Induction, commitment, and progression of plant embryogenesis, *Plant Biotechnol.*, 11, 141, 1989.

396

26. Carman, J.G., Embryogenic cells in plant tissue cultures: occurrence and behavior, *In Vitro Cell. Develop. Biol.*, 26, 746, 1990.

27. Christianson, M.L., and Warnick, D.A., Competence and determination in the process of *in vitro* shoot organogenesis, *Dev. Biol.*, 95, 288, 1983.

28. Coleman, G.D., and Ernst, S.G., Shoot induction competence and callus determination in *Populus deltoides*, *Plant Sci.*, 71, 83, 1990.

29. Ammirato, P.V., Patterns of development in culture, in *Tissue Culture in Forestry and Agriculture*, Henke, R.R., Hughes, K.W., Constantin, M.J., and Hollaender, A., Eds., Plenum Press, New York, 1985, 9.

30. Finstad, K., Biochemical and developmental markers of induction of somatic embryogenesis in alfalfa tissue culture, Ph.D. Thesis, Carleton University, Ottawa, 1992.

31. Nomura, K., and Komamine, A., Identification and isolation of single cells that produce somatic embryos at a high frequency in a carrot suspension culture, *Plant Physiol.*, 79, 988, 1985.

32. Finer, J.J., Plant regeneration from somatic embryogenic suspension cultures of cotton (*Gossypium hirsutum* L.), *Plant Cell Rep.*, 7, 399, 1988.

33. Meins, J. Jr., Determination and morphogenetic competence in plant tissue culture, in *Botanical Monographs (Oxford), Vol. 23, Plant Cell Culture Technology*, Yeoman, M.M. Ed., Blackwell Scientific Publications, Oxford, 1986, 7.

34. Steward, F.C., Mapes, M.O., Kent, A.E., and Holsten, R.D., Growth and development of cultured plant cells, *Science*, 143, 20, 1964.

35. Oelck, M.M., and Schieder, O., Genotypic differences in some legume species affecting the redifferentiation ability from callus to plants, *Z. Pflanzenzüchtg.*, 91, 312, 1983.

36. Punja, Z.K., Abbas, N., Sarmento, G.G., and Tang, F.A., Regeneration of *Cucumis sativus* var. *sativus* and *C. sativus* var. *hardwickii*, *C. melo*, and *C. metuliferus* from explants through somatic embryogenesis and organogenesis, *Plant Cell Tiss. Org. Cult.*, 21, 93, 1990.

37. Komatsuda, T., and Ko, S.-W., Screening of soybean [*Glycine max* (L.) Merrill] genotypes for somatic embryo production from immature embryos, *Japan. J. Breed.*, 40, 249, 1990.

38. Koornneef, M., Manhart, C., Jongsma, M., Toma, I., Weide, R., Zabel, P., and Hille, J., Breeding of a tomato genotype readily accessible to genetic manipulation, *Plant Sci.*, 45, 201, 1986.

39. Szabados, L., Hoyos, R., and Roca, W., *In vitro* somatic embryogenesis and plant regeneration of cassava, *Plant Cell Rep.*, 6, 248, 1987.

40. Matheson, S.L., Nowak, J., and MacLean, N.L., Selection of regenerative genotypes from highly productive cultivars of alfalfa, *Euphytica*, 45, 105, 1990.

41. Brown, D.C.W., and Atanassov A., Role of genetic background in somatic embryogenesis in *Medicago*, *Plant Cell Tiss. Org. Cult.*, 4, 11, 1985.

42. Takamizo, T., Suginobum K.-I., and Ohsugi, R., Somatic embryogenesis in a recalcitrant cultivar of alfalfa *Medicago sativa* L. in an improved medium, *Bull Natl. Grassl. Res. Inst.*, 44, 15, 1991.

43. Chen, T.H.H, Marowitch, J., and Thompson, B.G., Genotypic effects on somatic embryogenesis and plant regeneration from callus cultures of alfalfa, *Plant Cell Tiss. Org. Cult.*, 8, 73, 1987.

44. Feirer, R.P., and Simon, P.W., Biochemical differences between carrot inbreds differing in plant regeneration potential, *Plant Cell Rep.* 10, 152, 1991.

45. Szabados, L., Hoyos, R., and Roca, W., *In vitro* embryogenesis and plant regeneration of cassava, *Plant Cell Rep.*, 6, 248, 1987.

46. Trolinder, N.L., and Chen, X., Genotype specificity of the somatic embryogenesis response in cotton, *Plant Cell Rep.*, 8, 133, 1989.

47. Gawel, N.J., and Robacker, C.D., Genetic control of somatic embryogenesis in cotton petiole callus cultures, *Euphytica*, 49, 249, 1990.

48. Parrott, W.A., Williams, E.G., Hildebrand, D.F., and Collins, G.B., Effect of genotype on somatic embryogenesis from immature cotyledons of soybean, *Plant Cell Tiss. Org. Cult.*, 11, 111, 1989.

49. Komatsuda, T., and Ohyama, K., Genotypes of high competence for somatic embryogenesis and plant regeneration in soybean *Glycine max*, *Theor. Appl. Genet.*, 75, 695, 1989.
50. Mitten, D.H., Sato, S.J., and Skokut, T.A., *In vitro* regenerative potential of alfalfa germplasm sources, *Crop Sci.*, 24, 943, 1984.
51. Sonnino, A., Tanaka, S., Iwanaga, M., and Schilde-Rentschler, L., Genetic control of embryo formation in anther culture of diploid potatoes, *Plant Cell Rep.*, 8, 105, 1989.
52. Ranch, J.P., Ogelsby, L., and Zielinski, A.C., Plant regeneration from tissue cultures of soybean by somatic embryogenesis, in *Cell Culture and Somatic Cell Genetics of Plants*, Vol. 3, Vasil, I.K., Ed., Academic Press, New York, 1986, 97.
53. Keyes, G.T., Collins, G.B., and Taylor. N.L., Genetic variation in tissue cultures of red clover, *Theor. Appl. Genet.*, 58, 265, 1980.
54. Bailey, M.A., Boerma, H.R., and Parrott, W.A., Genotype effects on proliferative embryogenesis and plant regeneration of soybean, *In Vitro Cell. Develop. Biol.*, 29, 102, 1993.
55. Chee, P.P., High frequency of somatic embryogenesis and recovery of fertile cucumber plants, *HortSci.*, 25, 792, 1990.
56. Ranch, J.P., The potential for synthetic soybean seed, in *Synseeds: Applications of Synthetic Seeds to Crop Improvement*, Redenbaugh, K., Ed., CRC Press, Boca Raton, FL, 1993, 329.
57. Komatsuda, T., Lee, W., and Oka, S., Maturation and germination of somatic embryos as affected by sucrose and plant growth regulators in soybeans *Glycine gracilis* Skvortz and *Glycine max* (L.) Merr., *Plant Cell Tiss. Org. Cult.*, 28, 103, 1992.
58. Nolan, K.E., Rose, R.J., and Gorst, J.R., Regeneration of *Medicago truncatula* from tissue culture: Increased somatic embryogenesis using explants from regenerated plants, *Plant Cell Rep.*, 8, 278, 1989.
59. Gautheret, R.J., History of plant tissue and cell culture, A personal account, in *Cell Culture and Somatic Cell Genetics of Plants*, Vol. 2, Vasil, I.K., Ed., Academic Press, Orlando, 1985, 1.
60. Stewart, F.C., Mapes, M.O., and Mears, K., Growth and organised development of cultured cells. II. Organization in cultures grown freely from suspended cells, *Am. J. Bot.*, 445, 705, 1958.
61. Reinert, J., Über die Kontrolle der Morphogenese und die Induktion von Adventiveembryonnen an Gewebekulturen aus Karotten, *Planta*, 58, 318, 1959.
62. White, P.R., *A Handbook of Plant Tissue Culture*, The Ronald Press Co., New York, 1943.
63. Murashige, T., and Skoog, F., A revised medium for rapid growth and bioassay with tobacco tissue cultures, *Physiol. Plant.*, 15, 475, 1962.
64. George, E.F., Puttock, D.J.M., and George, H.J., *Plant Culture Media, Vol. 1, Formulations and Uses*, Exegetics Ltd., Eversley, 1987.
65. George, E.F., Puttock, D.J.M., and George, H.J., *Plant Culture Media, Vol. 2, Commentary and Analysis*, Exegetics Ltd., Eversley, 1988.
66. Gamborg, O.L., Murashige, T., Thorpe, T.A., and Vasil, I.K., Plant tissue culture media, *In Vitro*, 12, 473, 1976.
67. Gamborg, O., Miller, R., and Ojima, K., Nutrient requirements of suspension cultures of soybean root cells, *Exp. Cell Res.*, 50, 151, 1968.
68. Phillips, G.C., and Collins, G.B., Somatic embryogenesis from cell suspension cultures of red clover, *Crop Sci.*, 20, 323, 1980.
69. Phillips, G.C., and Collins, G.B., *In vitro* tissue culture of selected legumes and plant regeneration from callus cultures of red clover, *Crop Sci.*, 19, 59, 1979.
70. Kao, K.N., Constabel, F., Michayluk, M.R., and Gamborg, O.L., Plant protoplast fusion and growth of intergeneric hybrid cells, *Planta*, 120, 215, 1974.
71. Kao, K.N., Chromosomal behaviour in somatic hybrids of soybean-*Nicotiana glauca*, *Molec. Gen. Genet.*, 150, 225, 1977.
72. Kao, K.N., and Michayluk, M.R., Plant regeneration from mesophyll protoplasts of alfalfa, *Z. Pflanzenphysiol.*, 96, 135, 1980.
73. Kao, K.N., and Michayluk, M.R., Embryoid formation in alfalfa cell suspension cultures from different plants, *In Vitro*, 17, 645, 1991.
74. Kao, K.N., and Michayluk, M.R., Nutritional requirements for growth of *Vicia hajastana* cells and protoplasts at a very low population density in liquid media, *Planta*, 126, 105, 1975.

75. Kohlenbach, H.W., Basic aspects of differentiation and plant regeneration from cell and tissue cultures, in *Plant Tissue Culture and Its Bio-technological application*, Barz, W., Reinhard, E., and Zenk, M.H., Eds., Springer-Verlag, Berlin, 1977, 355.

76. Stewart, F.C., Ammirato, P.V., and Mapes, M.O., Growth and development of totipotent cells: Some problems, procedures, and perspectives, *Ann. Bot.*, 34: 761, 1970.

77. Murashige, T., Nutrition of plant cells and organs *in vitro*, *In Vitro*, 9, 81, 1973.

78. Kohlenbach, H.W., Comparative somatic embryogenesis, in *Frontiers of Plant Tissue Culture*, Thorpe, T.A., Ed., University of Calgary, Calgary, 1978, 59.

79. George, E.F., and Sherrington, P.D., *Plant Propagation by Tissue Culture*, Exegetics Ltd., Eversley, 1984.

80. Singh, M., and Krikorian, A.D., White's standard nutrient solution, *Ann. Bot.*, 47, 133, 1981.

81. Singh, M., and Krikorian, A.D., Chelated iron in culture media, *Ann. Bot.*, 46, 807, 1980.

82. Owen, H.R., and Miller, R., An examination and correction of plant tissue culture basal medium formulations, *Plant Cell Tiss. Org. Cult.*, 28, 147, 1992.

83. Bhojwani, S.S., and Razdan, M.K., *Plant Tissue Culture: Theory and Practice*, Elsevier, Amsterdam, 1983.

84. Albrecht, C., Optimization of tissue culture media, *Proc. Int. Plant Prop. Soc.*, 35, 196, 1985.

85. Stavarek, S.J., Croughan, T.P., and Rains, D.W., Regeneration of plants from long-term cultures of alfalfa cells, *Plant Sci. Lett.*, 19, 253, 1980.

86. Nadel, B.L., Altman, A., and Meira, Z., Regulation of somatic embryogenesis in celery cell suspensions, *Plant Cell Tiss. Org. Cult.*, 18, 181, 1989.

87. Finkelstein, R.R., and Crouch, M.L., Rapeseed embryo development in culture on high osmoticum is similar to that in seeds, *Plant Physiol.*, 81, 907, 1986.

88. Kamada, H., Koyosus, T., and Harada, H., New methods for somatic embryo induction and their use for synthetic seed production, *In Vitro Cell. Develop. Biol.*, 24, 71A, 1988.

89. Finer, J.J., Direct somatic embryogenesis and plant regeneration from immature embryos of hybrid sunflower (*Helianthus annuus* L.) on a high sucrose-containing medium, *Plant Cell Rep.*, 6, 372, 1987.

90. Finer, J.J., Apical proliferation of embryogenic tissue of soybean (*Glycine max* (L.) Merrill), *Plant Cell Rep.*, 7, 238, 1988.

91. Wetherell, D.F., Enhanced adventive embryogenesis resulting from plasmolysis of cultured wild carrot cells, *Plant Cell Tiss. Org. Cult.*, 3, 221, 1984.

92. Litz, R.E., and Conover, R.A., High frequency somatic embryogenesis from *Carica* suspension cultures, *Ann. Bot.*, 51, 683, 1983.

93. Senaratna, T., McKersie, B.D., and Bowley, S.R., Desiccation tolerance of alfalfa (*Medicago sativa* L.) somatic embryos, influence of abscisic acid, stress pretreatments and drying rates, *Plant Sci.*, 65, 253, 1989.

94. McKersie, B.D., Senaratna, T., Bowley, S.R., Brown, D.C.W., Krochko, J.E., and Bewley, J.D., Application of artificial seed technology in the production of hybrid alfalfa (*Medicago sativa* L.), *In Vitro Cell. Develop. Biol.*, 25, 1183, 1989.

95. Anandarajah, K., and McKersie, B.D., Enhanced vigor of dry somatic embryos of *Medicago sativa* L. with increased sucrose, *Plant Sci.*, 71, 261, 1990.

96. Anandarajah, K., and McKersie, B.D., Manipulating the desiccation tolerance and vigor of dry somatic embryos of *Medicago sativa* L. with sucrose, heat shock and abscisic acid, *Plant Cell Rep.*, 9, 451, 1990.

97. Kott, L.S., and Beversdorf, W.F., Enhanced plant regeneration from microspore-derived embryos of *Brassica napus* by chilling, partial desiccation and age selection, *Plant Cell Tiss. Org. Cult.*, 23, 187, 1990.

98. Huang, B., Bird, S., Kemble, B., and Keller, W., Plant regeneration from microspore-derived embryos of *Brassica napus*: Effect of embryo age, culture temperature, osmotic pressure, and abscisic acid, *In Vitro Cell. Develop. Biol.*, 27, 28, 1991.

99. Redenbaugh, K., Fugii, J.A., and Slade, D., Encapsulated plant embryos, in *Biotechnology in Agriculture*, Mizrahi, A., Ed., Alan R. Liss, Inc., New York, 1988, 225.

100. Kitto, S.L., and Janick, J., Hardening treatment increased survival of synthetically-coated asexual embryos of carrot, *J. Amer. Soc. Hort. Sci.*, 110, 283, 1985.

101. Rathore, K.S., Hodges, T.K., and Robinson, K.R., Ionic basis of currents in somatic embryos of *Daucus carota*, *Planta*, 175, 280, 1988.

102. Dijak, M., Smith, D.L., Wilson, T.J., and Brown, D.C.W., Stimulation of direct embryogenesis from mesophyll protoplasts of *Medicago sativa*, *Plant Cell Rep.*, 5, 468, 1986.

103. Goldsworthy, A., and Lago, A., Electrical control of differentiation in callus natural electric potentials, *Plant Cell Tiss. Org. Cult.*, 30, 221, 1992.

104. Bhojwani, S.S., Dhawan, V., and Arora, R., Somatic embryogenesis, in *Plant Tissue Culture, A Classified Bibliography*, Chapter 7, Elsevier, Amsterdam, 1986, 171.

105. Bhojwani, S.S., Dhawan, V., and Arora, R., Somatic embryogenesis, in *Plant Tissue Culture, A Classified Bibliography 1985–1989*, Chapter 7, Elsevier, Amsterdam, 1990, 57.

106. Evans, D., and Sharp, W.R., Eds., *Handbook of Plant Cell Culture, Vol. 1, Techniques for Propagation and Breeding*, McGraw-Hill Publishing Co., New York, 1983.

107. Vasil, I.K., Ed., *Cell Culture and Somatic Cell Genetics of Plants*, Academic Press, Orlando, Vol. 1, 1984.

108. Bajaj, Y.P.S., Ed., *Biotechnology in Agriculture and Forestry*, Springer-Verlag, Berlin, Vol. 1, 1986.

109. Schenk, R.U., and Hildebrandt, A.C., Production of protoplasts from plant cells in liquid culture using purified commercial cellulases, *Crop Sci.*, 9, 629, 1969.

110. Kao, K.N., Keller, W.A., and Miller, R.A., Cell division in newly formed cells from protoplasts of soybean, *Exptl. Cell. Res.*, 62, 338, 1970.

111. Kao, K.N., Chromosomal behaviour in somatic hybrids of soybean-*Nicotiana glauca*, *Molec. Gen. Genet.*, 150, 225, 1977.

112. Blaydes, D.F., Interaction of kinetin and various inhibitors in the growth of soybean tissue, *Physiol. Plant.*, 19, 748, 1966.

113. Schenk, R.U., and Hildebrandt, A.C., Medium and techniques for induction and growth of monocotyledonous and dicotyledonous plant cell cultures, *Can. J. Bot.*, 50, 199, 1972.

114. Kimball, S.L., and Bingham, E.T., Adventitious bud development of soybean hypocotyl sections in culture, *Crop Sci.*, 13, 758, 1973.

115. Cheng, T.Y., Saka, H., and Voqui-Dinh, T.H., Plant regeneration from soybean cotyledonary nodes in culture, *Plant Sci. Lett.*, 19, 91, 1980.

116. Saka, H., Voqui-Dinh, T.H., and Cheng, T.Y., Stimulation of multiple shoot formation on soybean stem nodes in culture, *Plant Sci. Lett.*, 19, 193, 1980.

117. Wright, M.S., Koehler, S.M., Hinchee, M.A., and Carnes, M.G., Plant regeneration by organogenesis in *Glycine max* L. Merr.; plants from tissue-cultured epicotyls, *Plant Cell Rep.*, 5, 150, 1986.

118. Wright, M.S., Carnes, M.G., Hinchee, M.A., Davis, G.C., Keohler, S.M., Williams, M.H., Colburn, S.M., and Pierson, P.E., Plant regeneration from tissue cultures of soybean by organogenesis, in *Cell Culture and Somatic Cell Genetics of Plants*, Vol. 3, Vasil, I.K., Ed., Academic Press, Orlando, 1986, 97.

119. Wright, M.S., Launis, K.L., Novitzky, R., Deusing, J.H., and Harms, C.T., A simple method for the recovery of multiple fertile plants from individual somatic embryos of soybean (*Glycine max* (L.) Merrill), *In Vitro Cell. Develop. Biol.*, 27, 153, 1991.

120. Phillips, G.C., and Collins, G.B., Induction and development of somatic embryos from cell suspension cultures of soybean, *Plant Cell Tiss. Org. Cult.*, 1, 123, 1981.

121. Gamborg, O.L., Davis, B.P., and Stahlhut, R.W., Somatic embryogenesis in cell cultures of *Glycine* species, *Plant Cell Rep.*, 2, 209, 1983

122. Beversdorf, W.D., and Bingham, E.D., Degrees of differentiation obtained in tissue cultures of *Glycine* species, *Crop Sci.*, 17, 303, 1977.

123. Grant, J.E., Plant regeneration from cotyledonary tissue of *Glycine canescens*, a perennial wild relative of soybean, *Plant Cell Tiss. Org. Cult.*, 3, 169, 1984.

124. Christianson, M.L., Warnick, D.A., and Carlson, P.S., A morphogenetically competent soybean suspension culture, *Science*, 222, 632, 1983.

125. Lippmann, B., and Lippmann, G., Induction of somatic embryos in cotyledonary tissue of soybean, *Glycine max* L. Merr., *Plant Cell Rep.*, 3, 215, 1984.
126. Christianson, M.L., An embryogenic culture of soybean: towards a general theory of somatic embryogenesis, in *Tissue Culture in Forestry and Agriculture*, Henke, R.R., Hughes, K.W., Constantin, M.J., Hollaender, A., and Wilson, C.M., Eds., Plenum Press, New York, 1985, 83.
127. Lazzeri, P.A., Hildebrand, D.F., and Collins, G.B., A procedure for plant regeneration from immature cotyledon tissue of soybean, *Plant Mol. Biol. Rep.*, 3, 160, 1985.
128. Li, B.J., Langridge, W.H.R., and Szalay, A.A., Somatic embryogenesis and plantlet regeneration in the soybean *Glycine max*, *Plant Cell Rep.*, 4, 344, 1985.
129. Ranch, J.P., Oglesby, L., and Zielinski, A.C., Plant regeneration from embryo-derived tissue cultures of soybeans, *In Vitro Cell. Develop. Biol.*, 21, 653, 1985.
130. Kerns, H.R., Barwale, U.B., Meyer, M.M. Jr., and Widholm, J.M., Correlation of cotyledonary node shoot proliferation and somatic embryoid development in suspension cultures of soybean (*Glycine max* L. Merr.), *Plant Cell Rep.*, 5, 140, 1986.
131. Barwale, U.B., Kerns, H.R., and Widholm, J.M., Plant regeneration from callus cultures of several soybean genotypes via embryogenesis and organogenesis, *Planta*, 167, 473, 1986.
132. Ghazi, T.D., Cheema, H.V., and Nabors, M.W., Somatic embryogenesis and plant regeneration from embryogenic callus of soybean, *Glycine max* L., *Plant Cell Rep.*, 5, 452, 1986.
133. Lazzeri, P.A., Hildebrand, D.F., and Collins, G.B., Soybean somatic embryogenesis: Effects of hormones and culture manipulations, *Plant Cell Tiss. Org. Cult.*, 10, 197, 1987.
134. Hartweck, L.M., Lazzeri, P.A., Cui, D., Collins, G.B., and Williams, E.G., Auxin-orientation effects of somatic embryogenesis from immature soybean cotyledons, *In Vitro Cell. Develop. Biol.*, 24, 821, 1988.
135. Hammatt, N., and Davey, M.R., Somatic embryogeneis and plant regeneration from cultured zygotic embryos of soybean (*Glycine max* L. Merr.), *J. Plant Physiol.*, 128, 219, 1987.
136. Lazzeri, P.A., Hildebrand, D.F., and Collins, G.B., Soybean somatic embryogenesis: Effects of nutritional, physical and chemical factors, *Plant Cell Tiss. Org. Cult.*, 10, 209, 1987.
137. Finer, J.J., and Nagasawa, A., Development of an embryogenic suspension culture of soybean (*Glycine max* Merrill), *Plant Cell Tiss. Org. Cult.*, 15, 125, 1988.
138. Hepher, A., Boulter, M.E., Harris, N., and Nelson, R.S., Development of a superficial meristem during somatic embryogenesis for immature cotyledons of soybean (*Glycine max* L.), *Ann. Bot.*, 62, 513, 1988.
139. Parrott, W.A., Willliams, E.G., Hildebrand, D.F., and Collins, G.B., Effect of genotype on somatic embryogenesis from immature cotyledons of soybean, *Plant Cell Tiss. Org. Cult.*, 16, 15, 1989.
140. Parrott, W.A., Dryden, G., Vogt, S., Hildebrand, D.F., Collins, G.B., and Williams, E.G., Optimization of somatic embryogenesis and embryo germination in soybean, *In Vitro Cell. Develop. Biol.*, 24, 817, 1988.
141. Buchheim, J.A., Colburn, S.M., and Ranch, J.P., Maturation of soybean somatic embryos and the transition of plantlet growth, *Plant Physiol.*, 89, 768, 1989.
142. Ghazi, T.D., Cheema, H.V., and Nabors, M.W., Somatic embryogenesis and plant regeneration from embryogenic callus of soybean, *Glycine max* L., *Plant Cell Rep.*, 5, 452, 1986.
143. Slawinska, J., and Obendorf, R.L., Soybean somatic embryo maturation: composition, respiration and water relations, *Seed Sci. Res.*, 1, 251, 1991.
144. Komatsuda, T., Ability of soybean (*Glycine max* L. Merr.) genotypes to produce somatic embryos on a medium containing a low concentration of sucrose, *Japan. J. Breed.*, 40, 371, 1990.
145. Dhir, S.K., Dhir, S., and Widholm, J.M., Regeneration of fertile plants from protoplasts of soybean (*Glycine max* L. Merr.): genotypic differences in culture response, *Plant Cell Rep.*, 11, 285, 1992.
146. Lippmann, B., and Lippmann, G., Soybean embryo culture: factors influencing plant recovery from isolated embryos, *Plant Cell Tiss. Org. Cult.*, 32, 83, 1993.

147. Liu, W. Moore, P.J., and Collins, G.B., Somatic embryogenesis in soybean via somatic embryo cycling, *In Vitro Cell. Develop. Biol.*, 28, 153, 1992.

148. Shoemaker, R.C., Amberger, L.A., Palmer, R.G., Oglesby, L., and Ranch, J.P., Effect of 2,4-dichlorophenoxyacetic acid concentration on somatic embryogenesis and heritable variation in soybean (*Glycine max* (L.) Merr.), *In Vitro Cell. Develop. Biol.*, 27, 84, 1991.

149. Lambert, M., Embryogenèse somatique chez le Soja (*Glycine max* L. Merrill), Ph.D. Thesis, Université de Paris-Sud, Centre D'Orsay, 1991.

150. Tian, L., Brown, D.C.W., Voldeng, H., and Webb, J., *In vitro* response and pedigree analysis for somatic embryogenesis of long-day photoperiod adapted soybean, *Plant Cell Tiss. Org. Cult.*, 1993 (in press).

151. Lazzeri, P.A., Hildebrand, D.F., Sunega, J., Williams, E.G., and Collins, G.B., Soybean somatic embryogenesis: interactions between sucrose and auxin, *Plant Cell Rep.*, 7, 517, 1988.

152. Komatsuda, T., Research on somatic embryogenesis and plant regeneration in soybean, *Bull. Natl. Inst. Agriobiol. Resour.*, 7, 1, 1992.

153. Saunders, J.W., and Bingham, E.T., Production of alfalfa plants from callus tissue. *Crop Sci.*, 12, 804, 1972.

154. Bingham, E.T., McCoy, T.J., and Walker, K.A., Alfalfa tissue culture, in *Alfalfa and Alfalfa Improvement*, Hanson, A.A., Ed., Amer. Soc. of Agronomy, Madison, Wisconsin, 1988, 903.

155. Redenbaugh, K., and Walker, K., Role of artificial seeds in alfalfa breeding, in *Plant Tissue Culture: Applications and Limitations*, Bhojwani, S.S., Ed., Elsevier, Amsterdam, 1990, 102.

156. Arcioni, S., Damiani, F., Pezzotti, M., and Lupotto, E., Alfalfa, Lucerne (*Medicago* spp.), in *Biotechnology in Agriculture and Forestry, Vol. 10, Legumes and Oilseed Crops I*, Bajaj, Y.P.S., Ed., Springer-Verlag, Berlin, 1990, 197.

157. Gilmour, D.M., Golds, T.J., and Davey, M.R., *Medicago* protoplasts: fusion, culture and plant regeneration, in *Biotechnology in Agriculture and Forestry, Vol. 8, Plant Protoplasts and Genetic Engineering I*, Bajaj, Y.P.S., Ed., Springer-Verlag, Berlin, 1989, 370.

158. Saunders, J.W., and Bingham, E.T., Growth regulator effects on bud initiation in callus cultures of *Medicago sativa*, *Amer. J. Bot.*, 62, 850, 1975.

159. Walker, K.A., Yu, P.C., Sato, S.J., and Jaworski, E.G., The hormonal control of organ formation in callus of *Medicago sativa* L. cultured *in vitro*, *Amer. J. Bot.*, 65, 654, 1978.

160. Walker, K.A., Wendeln, M.L., and Jaworski, E.G., Organogenesis in callus tissue of *Medicago sativa*. The temporal separation of induction processes from differentiation processes, *Plant Sci. Lett.*, 16, 23, 1979.

161. Walker, K.A., and Sato, S.J., Morphogenesis in callus tissue of *Medicago sativa*: the role of ammonium ion in somatic embryogenesis, *Plant Cell Tiss. Org. Cult.*, 1, 109, 1981.

162. Xu, Z.-H., Davey, M.R., and Cocking, E.C., Organogenesis from root protoplasts of the forage legumes *Medicago sativa* and *Trigonella foenum-graecum*, *Z. Pflanzenphysiol.*, 4, 107, 231, 1982.

163. Kim, D.M., Choi, K.S., and Chung, W.I., Induction of multishoots and plant regeneration from protoplasts of alfalfa *Medicago sativa* L., *Korean J. Bot.*, 32, 313, 1989.

164. Téoulé, E., and Dattée, Y., Recherche d'une méthode fiable de culture de protoplastes, d'hybridation somatique et de régénération chez *Medicago*, *Agronomie*, 7, 575, 1987.

165. Mariotti, D., Arcioni, S., and Pezzotti, M., Regeneration of *Medicago arborea* L. plants from tissue and protoplast cultures of different organ origin, *Plant Sci. Lett.*, 37, 149, 1984.

166. Dos Santos, A.V.P., Outka, D.E., Cocking, E.C., and Davey, M.R., Organogenesis and somatic embryogenesis in tissues derived from leaf protoplasts and leaf explants of *Medicago sativa*, *Z. Pflanzenphysiol.*, 99, 261, 1980.

167. Dos Santos, A.V.P., Cutter, E.G., and Davey, M.R., Origin and development of somatic embryos in *Medicago sativa* L. (Alfalfa), *Protoplasma*, 117, 107, 1983.

168. Wenzel, C.L., and Brown, D.C.W., Histological events leading to somatic embryo formation in cultured petioles of alfalfa, *In Vitro Cell. Develop. Biol.*, 27, 190, 1991.

169. Stuart, D.A., Nelson, J., McCall, C.M., Strickland, S.G., and Walker, K.A., Physiology of the development of somatic embryos in cell cultures of alfalfa and celery, in *Biotechnology in Plant Science*, Zaitlin, M., Day, P., and Hollaender, A., Eds., Academic Press, Inc., New York, 1985, 35.

170. Krochko, J.E., Pramanik, S.K., and Bewley, J.D., Contrasting storage protein synthesis and messenger RNA accumulation during development of zygotic and somatic embryos of alfalfa (*Medicago sativa* L.), *Plant Physiol.*, 99, 46, 1992.

171. Bianchi, S., Flament, P., and Dattée, Y., Embryogenèse somatique et organogenèse *in vitro* chez la luzerne: évaluation des potentialités de divers génotypes, *Agronomie*, 8, 121, 1988.

172. Walton, P.D., and Brown, D.C.W., Screening of *Medicago* wild species for callus formation and the genetics of somatic embryogenesis, *J. Genet.*, 67, 95, 1988.

173. Meijer, E.G.M., and Brown, D.C.W., Screening of diploid *Medicago sativa* germplasm for somatic embryogenesis, *Plant Cell Rep.*, 4, 285, 1985.

174. Lu, D.Y., Pental, D., and Cocking, E.C., Plant regeneration from seedling cotyledon protoplasts, *Z. Pflanzenphysiol.*, 107, 59, 1982.

175. Lu, D.Y., Davey, M.R., Pental, D., and Cocking, E.C., Forage legume protoplasts: somatic embryogenesis from protoplasts of seedling cotyledons and roots of *Medicago sativa*, in *Plant Tissue Culture 1982*, Fujiwara, A., Ed., Japanese Assoc. Plant Tissue Culture, Tokyo, 1982, 597.

176. Novak, F.J., and Konecna, D., Somatic embryogenesis in callus and cell suspension cultures of alfalfa (*Medicago sativa* L.), *Z. Pflanzenphysiol.*, 105, 279, 1982.

177. Lupotto, E., Propagation of an embryogenic culture of *Medicago sativa* L. *Z. Pflanzenphysiol.*, 111, 95, 1983.

178. Lupotto, E., The use of single somatic embryo culture in propagating and regenerating lucerne (*Medicago sativa* L.), *Ann. Bot.*, 57, 19, 1986.

179. Meijer, E.G.M., and Brown, D.C.W., A novel system for rapid high frequency somatic embryogenesis in *Medicago sativa*, *Physiol. Plantarum*, 69, 591, 1987.

180. Parrott, W.A., and Bailey, M.A., Characterization of recurrent somatic embryogenesis of alfalfa on auxin-free medium, *Plant Cell Tiss. Org. Cult.*, 32, 60, 1993.

181. Denchev, P., Velcheva, M., and Atanassov, A., A new approach to direct somatic embryogenesis in *Medicago*, *Plant Cell Rep.*, 10, 338, 1991.

182. Meijer, E.G.M., and Brown, D.C.W., Role of exogenous reduced nitrogen and sucrose in rapid high frequency somatic embryogenesis in *Medicago sativa*, *Plant Cell Tiss. Org. Cult.*, 10, 11, 1987.

183. Feher, F., Tarczy, M.H., Bocsa, I., and Dudits, D., Somaclonal chromosome variation in tetraploid alfalfa, *Plant Sci.*, 60, 91, 1989.

184. McCoy, T.J., and Bingham, E.T., Regeneration of diploid alfalfa plants from cells grown in suspension culture, *Plant Sci. Lett.*, 10, 56, 1977.

185. Johnson, L.B., Stuteville, D.L., Higgins, R.K., and Skinner, D.Z., Regeneration of alfalfa plants from protoplasts of selected Regen S clones, *Plant Sci. Lett.*, 20, 297, 1981.

186. Arcioni, A., Davey, M.R., Dos Santos, A.V.P., and Cocking, E.C., Somatic embryogenesis in tissues from mesophyll and cell suspension protoplasts of *Medicago coerulea* and *M. glutinosa*, *Z. Pflanzenphysiol.*, 106, 105, 1982.

187. Denchev, P.D., Kuklin, A.I., Atanassov, A.I. and Scragg, A.H., Kinetic studies of embryo development and nutrient utilization in an alfalfa direct somatic embryogenesis system, *Plant Cell Tiss. Org. Cult.*, 33, 67, 1993.

188. Mezentsev, A.V., Control of somatic embryogenesis and mass regeneration of plants from alfalfa cells and protoplasts, *Sel' skokhozyaistvennaya biologiya*, 16, 253, 1981.

189. Dijak, M., and Brown, D.C.W., Patterns of direct and indirect embryogenesis from mesophyll protoplasts of *Medicago sativa*, *Plant Cell Tiss. Org. Cult.*, 9, 121, 1987.

190. Dijak, M., and Brown, D.C.W., Donor tissue and culture condition effects on mesophyll protoplasts of *Medicago sativa*, *Plant Cell Tiss. Org. Cult.*, 9, 217, 1987.

191. Atanassov, A., and Brown, D.C.W., Plant regeneration from suspension culture and mesophyll protoplasts of *Medicago sativa L*, *Plant Cell Tiss. Org. Cult.*, 3, 149, 1984.

192. Atanassov, A., Vlakhova, V., Denchev, P., and Dragoeva, A., *In vitro* embryogenesis and cell selection for alfalfa genetics and breeding, in *Nuclear Techniques and In Vitro* Culture for Plant Improvement, International Atomic Energy Agency, Vienna, 1986, 35.

193. Stuart, D.A., and Strickland, S.G., Somatic embryogenesis from cell cultures of *Medicago sativa* L. I. The role of amino acid additions to the regeneration medium, *Plant Sci. Lett.*, 34, 165, 1984.

194. Stuart, D.A., and Strickland, S.G., Somatic embryogenesis from cell cultures of *Medicago sativa* L. II. The interaction of amino acids with ammonium, *Plant Sci. Lett.*, 34, 175, 1984.

195. Nichol, J.W., Slade, D., Viss, P., and Stuart, D.A., Effect of organic acid pretreatment on the regeneration and development(conversion) of whole plants from callus cultures of alfalfa, *Medicago sativa* L., *Plant Sci.*, 79, 181, 1991.

196. Strickland, S., Nichols, J., McCall, C., and Stuart, D., Effect of carbohydrate source on alfalfa somatic embryogenesis, *Plant Sci.*, 48, 113, 1987.

197. Fujii, J.A., Slade, D., Olsen, R., Ruzin, S.E., and Redenbaugh, K., Alfalfa somatic embryo maturation and conversion of plants, *Plant Sci.*, 72, 93, 1990.

198. Mehra, A., and Mehra, P.N., Differentiation in callus cultures of *Mesembryanthemum floribundum*, *Phytomorphology*, 22, 171, 1972.

199. Chang, W.C., and Hsing, Y.I., Plant regeneration through somatic embryogenesis in root-derived callus of ginseng (*Panax ginseng* C.A. Meyer), *Appl. Genet.*, 57, 133, 1980.

200. Arya, S., Liu, J.R., and Eriksson, T., Plant regeneration from protoplasts of *Panax ginseng* (C.A. Meyer) through somatic embryogenesis, *Plant Cell Rep.*, 10, 277, 1991.

201. Shoyama, Y., Kamura, K., and Nishioka, I., Somatic embryogenesis and clonal multiplication of *Panax ginseng*, *Planta Medica*, 54, 155, 1988.

202. Cellarova, E., Rychlova, M., and Vranova, E., Histological characterization of *in vitro* regenerated structures of *Panax ginseng*, *Plant Cell Tiss. Org. Cult.*, 30, 165, 1992.

203. Wang, A.S., Callus induction and plant regeneration of American ginseng, *HortSci.*, 25, 571, 1990.

204. Torné, J.M., Claparols, I., and Santos, M.A., Somatic embryogenesis in *Araujia sericifera*, *Plant Cell Tiss. Org. Cult.*, 29, 269, 1992.

205. Prabhudesai, V.R. and Narayanaswamy, S., Organogenesis in tissue cultures of certain Asclepiads, *Z. Pflanzenphysiol.*, 71, 181, 1974.

206. Dunbar, K.B., Wilson, K.J., Petersen, B.H., and Biesboer, D.D., Identification of laticifers in embryoids derived from callus and suspension cultures of *Asclepias* species (Asclepiadaceae), *Amer. J. Bot.*, 73, 847, 1986.

207. Maraffa, S.B., Sharp, W.R., Tayama, H.K., and Fretz, T.A., Apparent asexual embryogenesis in cultured leaf sections of *Hoya*, *Z. Pflanzenphysiol.*, 102, 45, 1981.

208. Rao, P.S., and Narayanaswamy, S., Morphogenetic investigations in callus cultures of *Tylophora indica*, *Physiol. Plant.*, 27, 271, 1972.

209. Mhatre, M., Bapat, V.A., and Rao, P.S., Plant regeneration in protoplast cultures of *Tylophora indica*, *J. Plant Physiol.*, 115, 231, 1984.

210. Rao, P.S., Narayanaswamy, S., and Benjamin, B.D., Differentiation *ex ovula* of embryos and plantlets in stem tissue cultures of *Tylophora indica*, *Physiol. Plant.*, 23, 140, 1970.

211. Zhang, L.Y., Li, G.G., and Guo, J.Y., Study on the somatic embryogenesis from leaf of *Begonia fimbristipula* hance *in vitro*, *Acta Bot. Sin.*, 30, 134, 1988.

212. Chuang, M.J., and Chang, W.C., Somatic embryogenesis and plant regeneration in callus culture derived from immature seeds and mature zygotic embryos of *Dysosma pleiantha* (Hance) Woodson, *Plant Cell Rep.*, 6, 484, 1987.

213. Chuang, M.J., and Chang, W.C., Embryoid formation and plant regeneration in callus cultures derived from vegetative tissues of *Dysosma pleiantha* (Hance) Woodson, *J. Plant Physiol.*, 128, 279, 1987.

214. Quinn, J., Simon, J.E., and Janick, J., Recovery of y-linolenic acid from somatic embryos of borage, *J. Amer. Soc. Hort. Sci.*, 114, 511, 1989.

215. Janick, J., Simon, J.E., and Whipkey, A., *In vitro* propagation of borage, *HortSci.*, 22, 493, 1987.

216. Min, S.R., Yang, S.G., Liu, J.R., Choi, P.S., and Soh, W.Y., High frequency somatic embryogenesis and plant regeneration in tissue cultures of *Codonopsis lanceolata*, *Plant Cell Rep.*, 10, 621, 1992.

404

217. Fitch, M.M.M., High frequency somatic embryogenesis and plant regeneration from papaya hypocotyl callus, *Plant Cell Tiss. Org. Cult.*, 32, 205, 1993.

218. Tétu, T., Sangwan, R.S., and Sangwan-Norreel, B.S., Hormonal control of organogenesis and somatic embryogenesis in *Beta vulgaris* callus, *J. Ex. Bot.*, 38, 506, 1987.

219. Gould, A.R., Diverse pathways of morphogenesis in tissue cultures of the Composite *Brachycome lineariloba* (2n = 4), *Protoplasma*, 97, 125, 1978.

220. Vasil, I.K., Hildebrandt, A.C., and Riker, A.J., Endive plantlets from freely suspended cells and cell groups grown *in vitro*, *Science*, 146, 76, 1964.

221. Vasil, I.K., and Hildebrandt, A.C., Variations of morphogenetic behavior in plant tissue cultures I. *Cichorium endivia*, *Amer. J. Bot.*, 53, 860, 1966.

222. Schoofs, J., and De Langhe, E., Chicory (*Cichorium intybus* L.), in *Biotechnology in Agriculture and Forestry, Vol. 6, Crops II*, Bajaj, Y.P.S., Ed., Springer-Verlag, Berlin, 1988, 294.

223. Dubois, T., Guedira, M., Dubois, J., and Vasseur, J., Direct somatic embryogenesis in roots of *Cichorium*: Is Callose an Early Marker? *Ann. Bot.*, 65, 539, 1990.

224. Sidikou-Seyni, R., Rambaud, C., Dubois, J., and Vasseur, J., Somatic embryogenesis and plant regeneration from protoplasts of *Cichorium intybus* L. × *Cichorium endivia* L., *Plant Cell Tiss. Org. Cult.*, 29, 83, 1992.

225. Pillai, K.G., Rao, I.U., Rao, I.V.R., and Ram, H.Y.M., Induction of division and differentiation of somatic embryos in the leaf epidermis of *Gaillardia picta*, *Plant Cell Rep.*, 10, 599, 1992.

226. Paterson-Robinson, K.E., and Adams, D.O., The role of ethylene in the regeneration of *Helianthus annuus* (sunflower) plants from callus, *Physiol. Plant.*, 71, 151, 1987.

227. Finer, J.J., Direct somatic embryogenesis and plant regeneration from immature embryos of hybrid sunflower (*Helianthus annuus* L.) on a high sucrose-containing medium, *Plant Cell Rep.*, 6, 372, 1987.

229. Paterson, K.E., and Everett, N.P., Regeneration of *Helianthus annuus* inbred plants from callus, *Plant Sci.*, 42, 125, 1985.

228. Robinson, K.E.P., and Everett, N.P., Sunflower (*Helianthus annuus* L.): Establishment of cultures, transformation, and the regeneration of plants, in *Biotechnology in Agriculture and Forestry, Vol. 10 Legumes and Oilseed Crops I*, Bajaj, Y.P.S., Ed., Springer-Verlag, Berlin, 1990, 434.

230. Pelissier, B., Bouchefra, O., Pepin, R., and Freyssinet, G., Production of isolated somatic embryos from sunflower thin cell layers, *Plant Cell Rep.*, 9, 47, 1990.

231. Freyssinet, M., and Freyssinet, G., Fertile plant regeneration from sunflower (*Helianthus annuus* L.) immature embryos, *Plant Sci.*, 56, 177, 1988.

232. Moyne, A.L., Thor, V., Pelissier, B., Bergounioux, C., Freyssinet, G., and Gadal, P., Callus and embryoid formation from protoplasts of *Helianthus annuus*, *Plant Cell Rep.*, 7, 437, 1988.

233. Krasnyanski, S., and Menczel, L., Somatic embryogenesis and plant regeneration from hypocotyl protoplasts of sunflower (*Helianthus annuus* L.), *Plant Cell Rep.*, 12, 260, 1993.

234. Pugliesi, C., Megale, P., Cecconi, F., and Baroncelli, S., Organogenesis and embryogenesis in *Helianthus tuberosus* and in the interspecific hybrid *Helianthus annuus* × *Helianthus tuberosus*, *Plant Cell Tiss. Org. Cult.*, 33, 187, 1993.

235. Maheshawri, P., and Baldev, B., Artificial production of buds from the embryos of *Cuscuta reflexa*, *Nature*, 191, 197, 1961.

236. Chée, R.P., and Cantliffe, D.J., Somatic embryony patterns and plant regeneration in *Ipomoea batatas* Poir., *In Vitro Cell. Develop. Biol.*, 24, 955, 1988.

237. Chée, R.P., and Cantliffe, D.J., Selective enhancement of *Ipomoea batatas* Poir. embryogenic and non-embryogenic callus growth and production of embryos in liquid culture, *Plant Cell Tiss. Org. Cult.*, 15, 149, 1988.

238. Chée, R.P., Schultheis, J.R., and Cantliffe, D.J., Plant recovery from sweet potato somatic embryos, *HortSci.*, 25, 795, 1990.

239. Chée, R.P., and Cantliffe, D.J., Inhibition of somatic embryogenesis in response to 2,3,5-triiodobenzoic acid and 2,4-dichlorophenoxyacetic acid in *Ipomoea batatas* (L.) Lam. cultured *in vitro*, *J. Plant Physiol.*, 135, 398, 1989.

240. Chée, R.P., and Cantliffe, D.J., Composition of embryogenic suspension cultures of *Ipomoea batatas* Poir. and production of individualized embryos, *Plant Cell Tiss. Org. Cult.*, 17, 39, 1989.

241. Lui, J.R., and Cantliffe, D.J., Somatic embryogenesis and plant regeneration in tissue cultures of sweet potato (*Ipomea batatas* Poir.), *Plant Cell Rep.*, 3, 112, 1984.

242. Jarret, R.L., Salazar, S., and Fernandez, R., Somatic embryogenesis in sweet potato, *HortSci.*, 19, 397, 1984.

243. Tsay, H.S., and Tseng, M.T., Embryoid formation and plantlet regeneration from anther callus of sweet potato, *Bot. Bull. Academia Sin.*, 20, 117, 1979.

244. Bhattacharya, N.M., and Sen, S.K., Production of plantlets through somatic embryogenesis in *Brassica campestris*, *Z. Pflanzenphysiol.*, 99, 357, 1980.

245. Kirti, P.B., and Chopra, V.L., Rapid plant regeneration through organogenesis and somatic embryogenesis from cultured protoplasts of *Brassica juncea*, *Plant Cell Tiss. Org. Cult.*, 20, 65, 1990.

246. Eapen, S., Abraham, V., Gerdemann, M., and Schieder, O., Direct somatic embryogenesis, plant regeneration and evaluation of plants obtained from mesophyll protoplasts of *Brassica juncea*, *Ann. Bot.*, 63, 369, 1989.

247. Li, L.-C., and Kohlenbach, H.W., Somatic embryogenesis in quite a direct way in cultures of mesophyll protoplasts of *Brassica napus* L., *Plant Cell Rep.*, 1, 209, 1982.

248. Kohlenbach, H.W., Wenzel, G., and Hoffmann, F., Regeneration of *Brassica napus* plantlets in cultures from isolated protoplasts of haploid stem embryos as compared with leaf protoplasts, *Z. Pflanzenphysiol.*, 105, 131, 1982.

249. Kranz, E., Somatic embryogenesis in stationary phase suspension cultures derived from hypocotyl protoplasts of *Brasssica napus* L., *Plant Cell Tiss. Org. Cult.*, 12, 141, 1988.

250. Keller, W.A., and Armstrong, K.C., Embryogenesis and plant regeneration in *Brassica napus* anther cultures, *Can. J. Bot.*, 55, 1383, 1977.

251. Narasimhulu, S.B., Kirti, P.B., Prakash, S., and Chopra, V.L., Somatic embryogenesis in *Brassica nigra* (Koch), *J. Ex. Bot.*, 43, 1203, 1992.

252. Pareek, L.K., and Chandra, N., Somatic embryogenesis in leaf callus from cauliflower (*Brassica oleracea var.* Botrytis), *Plant Sci. Lett.*, 11, 311, 1978.

253. Khanna, P., and Staba, E.J., *In vitro* physiology and morphogenesis of *Cheiranthus cheiri* var. Clott of gold and *C. cheiri* var., *Goliath, Botanical Gazette*, 131, 1, 1970.

254. Sikdar, S.R., Chatterjee, G., Das, S., and Sen, S.K., Regeneration of plants from mesophyll protoplasts of the wild crucifer *Eruca sativa* Lam., *Plant Cell Rep.*, 6, 486, 1987.

255. Ahloowalia, B.S., Somatic embryogenesis and plant regeneration in *Eruca sativa*, *Crop Sci.*, 27, 813, 1987.

256. Compton, M.E., and Gray, D.J., Somatic embryogenesis and plant regeneration from immature cotyledons of watermelon, *Plant Cell Rep.*, 12, 61, 1993.

257. Debeaujon, I., and Branchard, M., Somatic embryogenesis in *Cucurbitaceae*, *Plant Cell Tiss. Org. Cult.*, 34, 91, 1993.

258. Kwack, S.N., and Fujieda, K., Somatic embryogenesis in cultured unfertilized ovules of *Cucurbita moschata*, *J. Japan. Soc. Hort. Sci.*, 57, 34, 1988.

259. Jelaska, S., Embryogenesis and organogenesis in pumpkin explants, *Physiol. Plant.*, 31, 257, 1974.

260. Jelaska, S., Embryoid formation by fragments of cotyledons and hypocotyls in *Cucurbita pepo*, *Planta (Berl.)*, 103, 278, 1972.

261. Schroeder, C.A., Adventive embryogenesis in fruit pericarp tissue *in vitro*, *Bot. Gaz.*, 129, 374, 1968.

262. Jelaska, S., Cucurbits, in *Biotechnology in Agriculture and Forestry, Vol. 2, Crops I*, Bajaj, Y.P.S., Ed., Springer-Verlag, Berlin/Heidelberg, 1986, 371.

263. Homma, Y., Sugiyama, K., and Oosawa, K., Improvement in production and regeneration of somatic embryos from mature seed of melon (*Cucumis melo* L.) on solid media, *Japan. J. Breed.*, 41, 543, 1991.

264. Oridate, T., and Oosawa, K., Somatic embryogenesis and plant regeneration from suspension callus culture in melon (*Cucumis melo* L.), *Japan J. Breed.*, 36, 424, 1986.

265. Blackmon, W.J., Reynolds, B.D., and Postek, C.E., Production of somatic embryos from callused cantaloupe hypocotyl plants, *HortSci.*, 16, 87, 1981.

266. Tabei, Y., Kanno, T., and Nishio, T., Regulation of organogenesis and somatic embryogenesis by auxin in melon, *Cucumis melo* L., *Plant Cell Rep.*, 10, 225, 1991.

267. Oridate, T., Atsumi, H., Ito, S., and Araki, H., Genetic difference in somatic embryogenesis from seeds in melon (*Cucumis melo* L.), *Plant Cell Tiss. Org. Cult.*, 29, 27, 1992.

268. Halder, T., and Gadgil, V.N., Morphogenesis in some plant species of the family *Cucurbitaceae*, in *Proc. COSTED Symp. on Tissue Culture of Economically Important Plants*, Rao, A.N., Ed., Singapore, 1981, 98.

269. Fassuliotis G., and Nelson B.V., Interspecific hybrids of *Cucumis metuliferus* × *C. anguria* obtained through embryo culture and somatic embryogenesis, *Euphytica*, 37, 53, 1968.

270. Rajasekaran, K., Mulins, M.G., and Nair, Y., Flower formation *in vitro* by hypocotyl explants of cucumber (*Cucumis sativus* L.), *Ann. Bot.*, 52, 417, 1983.

271. Chee, P.P., High frequency of somatic embryogenesis and recovery of fertile cucumber plants, *HortSci.*, 25, 792, 1990.

272. Cade, R.M., Wehner, T.C., and Blazich, F.A., Somatic embryos derived from cotyledons of cucumber, *J. Amer. Soc. Hort. Sci.*, 115, 691, 1990.

273. Orczyk, W., and Malepszy, S., *In vitro* culture of *Cucumis sativus* L. V. Stabilizing effect of glycine on leaf protoplasts, *Plant Cell Rep.*, 4, 269, 1985.

274. Ziv, M., and Gadasi, G., Enhanced embryogenesis and plant regeneration from cucumber (*Cucumis sativus* L.) callus by activated charcoal in solid/liquid double layer cultures, *Plant Sci.*, 47, 115, 1986.

275. Chee, P.P., and Tricoli, D.M., Somatic embryogenesis and plant regeneration from cell suspension cultures of *Cucumis sativus* L., *Plant Cell Rep.*, 7, 274, 1988.

276. Malepszy, S., Cucumber (*Cucumis sativus* L.), *Biotechnology in Agriculture and Forestry, Vol. 6, Crops II*, Bajaj, Y.P.S., Ed., Springer-Verlag, Berlin, 277, 1988.

277. Malepszy, S., and Nadolska-Orczyk, A., *In vitro* culture of *Cucumis sativus* I. Regeneration of plantlets from callus formed by leaf explants, *Z. Pflanzenphysiol.*, 111, 273, 1983.

278. Malepszy, S., and Solarek, E., *In vitro* culture of *Cucumis sativus* L. IV. conditions for cell suspension, *Genetica Polonica*, 27, 249, 1986.

279. Jia, H.R., Fu, Y.Y., and Lin, Y., Embryogenesis and plant regeneration from cotyledon protoplast culture of cucumber (*Cucumis sativus* L.), *J. Plant Physiol.*, 124, 393, 1986.

280. Bergervoet, J.H.W., Van der Mark, F., and Custers, J.B.M., Organogenesis versus embryogenesis from long-term suspension cultures of cucumber (*Cucumis sativus* L.), *Plant Cell Rep.*, 8, 116, 1989.

281. Bhojwani, S.S., Morphogenetic behaviour of mature endosperm of *Croton bonplandianum* Baill. in culture, *Phytomorphology*, 16, 349, 1966.

282. Matthews, H., Schopke, C., Carcamo, R., Chavarriaga, P., Fauquet, C., and Beachy, R.N., Improvement of somatic embryogenesis and plant recovery in cassava, *Plant Cell Rep.*, 12, 328, 1993.

283. Raemakers, C.J.J.M., Bessembinder, J.J.E., Staritsky, G., Jacobsen, E., and Visser, R.G.F., Induction, germination and shoot development of somatic embryos in cassava, *Plant Cell Tiss. Org. Cult.*, 33, 151, 1993.

284. Raemakers, C.J.J.M., Schavemaker, C.M., Jacobsen, E., and Visser, R.G.F., Improvements of cyclic somatic embryogenesis of cassava (*Manihot esculenta* Crantz), *Plant Cell Rep.*, 12, 226, 1993.

285. Stamp, J.A., Somatic embryogenesis in cassava: The anatomy and morphology of the regeneration process, *Ann. Bot.*, 59, 451, 1987.

286. Stamp, J.A., and Henshaw, G.G., Secondary somatic embryogenesis and plant regeneration in cassava, *Plant Cell Tiss. Org. Cult.*, 10, 227, 1987.
287. Stamp, J.A., and Henshaw, G.G., Somatic embryogenesis from clonal leaf tissues of cassava, *Ann. Bot.*, 59, 445, 1987.
288. Abo, E.N., Mostafa, M., and Hildebrandt, A.C., Geranium plant differentiation from anther callus, *Amer. J. Bot*, 58, 475, 1971.
289. Qureshi, Javed A., and Saxena, P.K., Adventitious shoot induction and somatic embryogenesis with intact seedlings of several hybrid seed geranium (*Pelargonium* × *hortorum* Bailey) varieties, *Plant Cell Rep.*, 11, 443, 1992.
290. Zhang, P., Ni, D., Wang, Q., and Wang, K., Hormonal control of somatic embryogenesis and callus growth of *Saintpaulia ionantha in vitro*, *Acta Bot. Sin.*, 28, 446, 1986.
291. Tanimoto, S., and Harada, H., Hormonal control of morphogenesis in leaf explants of *Perilla frutescens* Britton var. crispa Decaisne f. viridi-crispa Makino, *Ann. Bot.*, 45, 321, 1980.
292. Pliego-Alfaro, F., and Murashige, T., Somatic embryogenesis in avocado (*Persea americana* Mill.) *in vitro*, *Plant Cell Tiss. Org. Cult.*, 12, 61, 1988.
293. Mooney, P.A., and Van Staden, J., Induction of embryogenesis in callus from immature embryos of *Persea americana*, *Can. J. Bot.*, 65, 622, 1987.
294. Ozias-Akins, P., Plant regeneration from immature embryos of peanut, *Plant Cell Rep.*, 8, 217, 1989.
295. Hazra, S., Sathaye, S.S., and Mascarenhas, A.F., Direct somatic embryogenesis in peanut (*Arachis hypogea*), *Biotechnology*, 7, 949, 1989.
296. Baker, C.M., and Wetzstein, H.Y., Somatic embryogenesis and plant regeneration from leaflets of peanut, *Arachis hypogaea*, *Plant Cell Rep.*, 11, 71, 1992.
297. Banerjee, S., Bandyopadhyay, S., and Ghosh, P.D., Cotyledonary node culture and multiple shoot formation in peanut: Evidences for somatic embryogenesis, *Curr. Sci.*, 57, 252, 1988.
298. Sellars, R.M., Southward, G.M., and Phillips, G.C., Adventitious somatic embryogenesis from cultured immature zygotic embryos of peanut and soybean, *Crop Sci.*, 30, 408, 1990.
299. Burtnik, O.J., and Mroginski, L.A., Regeneracion de plantas de *Arachis pintoi* (*Leguminosae*) por cultivo *in vitro* de tejidos foliares, *Oléagineux*, 40, 609, 1985.
300. Rao, B.G., and Chopra, V.L., Regeneration in chickpea (*Cicer arietinum* L.) through somatic embryogenesis, *J. Plant Physiol.*, 134, 637, 1975.
301. Mariotti, D., and Sergio, A., Callus culture of *Coronilla varia* L. (crownvetch), plant regeneration through somatic embryogenesis, *Plant Cell Tiss. Org. Cult.*, 2, 103, 1983.
302. Moyer, B.G., and Gustine, D.L., Regeneration of *Coronilla varia* L. (crownvetch) plants from callus culture, *Plant Cell Tiss. Org. Cult.*, 3, 143, 1984.
303. Gustine, D.L., and Moyer, B.G., Crownvetch (*Coronilla varia* L.), in *Biotechnology in Agriculture and Forestry, Vol. 10, Legumes and Oilseed Crops I*, Bajaj, Y.P.S., Ed., Springer-Verlag, Berlin, 1990, 341.
304. Arcioni S., Mariotti, D., Damiani, F., and Pezzotti, M., Birdsfoot trefoil (*Lotus corniculatus* L.), crownvetch (*Coronilla varia* L.) and sainfoin (*Onobrychis viciifolia* Scop.), in *Biotechnology in Agriculture and Forestry, Vol. 6, Crops II*, Bajaj, Y.P.S., Ed., Springer-Verlag, Berlin, 1988, 548.
305. Gharyal, P.K., and Maheshwari, S.C., Genetic and physiological influences on differentiation in tissue cultures of a legume, *Lathyrus sativus*, *Theor. Appl. Genet.*, 66, 123, 1983.
306. Saxena, P.K., and King, J., Morphogenesis in lentil: plant regeneration from callus cultures of *Lens culinaris* Medik. via somatic embryogenesis, *Plant Sci.*, 52, 223, 1987.
307. Nadolska-Orczyk, A., Somatic embryogenesis of agriculturally important lupin species (*Lupinus angustifolius, L. albus, L. mutabilis*), *Plant Cell Tiss. Org. Cult.*, 28, 19, 1992.
308. Mezentsev, A.V., Experimental morphogenesis and regeneration of alfalfa plants from somatic cells, *Transactions of the USSR Academy of Sciences, Biological Series*, 2, 190, 1982.
309. Atanassov, A.I., and Brown, D.C.W., Plant regeneration from suspension culture and mesophyll protoplasts of alfalfa, *Experienta Suppl.*, 45, 40, 1983.
310. Maheswaran, G., and Williams, E.G., Direct somatic embryoid formation on immature

408

embryos of *Trifolium repens, T. pratense* and *Medicago sativa*, and rapid clonal propagation of *T. repens, Ann. Bot.*, 54, 201, 1984.

311. Maheswaran, G., and Williams, E.G., Origin and development of somatic embryoids formed directly on immature embryos of *Trifolium repens in vitro, Ann. Bot.*, 56, 619, 1985.

312. Pezzotti, M., Arcioni, A., and Mariotti, D., Plant regeneration from mesophyll, root and cell suspension protoplasts of *Medicago sativa* cv. Adriana, *Genet. Agr.*, 38, 195, 1984.

313. Nam, L.S., and Heszky, L.E., Screening for plant regeneration in callus and protoplast cultures of alfalfa (*Medicago sativa* L.) germplasms, *Acta Botanica Hungarica*, 33, 387, 1987.

314. Song, J., Sorensen, E.L., and Liang, G.H., Direct embryogenesis from single mesophyll protoplasts in alfalfa (*Medicago sativa* L.), *Plant Cell Rep.*, 9, 21, 1990.

315. Gu, Z., Callus culture of sainfoin (*Onobrychis viciifolia*) and plant regeneration through somatic embryogenesis, *Ann. Bot.*, 60, 309, 1987.

316. Angelini, R.R., and Allavena, A., Plant regeneration from immature cotyledon explant cultures of bean (*P. coccineus* L.), *Plant Cell Tiss. Org. Cult.*, 19, 167, 1989.

317. Breuer, R., Kysely, W., and Jacobsen, H.J., Recent results on somatic embryogenesis in pea and bean, in *Genetic Manipulation in Plant Breeding*, Horn, W., Jensen, C.J., Odenbach, W., and Schieder, O., Eds., Walter de Gruyter, Berlin, 1986, 279.

318. Jacobsen, H.J., and Kysely, W., Induction of *in vitro*-regeneration via somatic embryogenesis in pea (*Pisum sativum*) and bean (*Phaseolus vulgaris*), in *Genetic Manipulation in Plant Breeding*, Horn, W., Jensen, C.J., Odenbach, W., and Schieder, O., Eds., Walter de Gruyter, Berlin, 1986, 445.

319. Malik, K.A., and Saxena, P.K., Somatic embryogenesis and shoot regeneration from intact seedlings of *Phaseolus acutifolius* A., *P. aureus* (L.) Wilczek, *P. coccineus* L., and *P. wrightii* L., *Plant Cell Rep.*, 11, 163, 1992.

320. Griga, M., and Novak, F.J., Pea (*Pisum sativum* L.), in *Biotechnology in Agriculture and Forestry, Vol. 10, Legumes and Oilseed Crops I*, Bajaj, Y.P.S., Ed., Springer-Verlag, Berlin, 1990, 65.

321. Kysely, W., and Jacobsen, H.J., Somatic embryogenesis from pea embryos and shoot apices, *Plant Cell Tiss. Org. Cult.*, 20, 7, 1990.

322. Kysely, W., Myers, J.R., Lazzeri, P.A., Collins, G.B., and Jacobsen, H.J., Plant regeneration via somatic embryogenesis in pea (*Pisum sativum* L.), *Plant Cell Rep.*, 6, 305, 1987.

323. Lehminger-Mertens, R., and Jacobsen, H.J., Plant regeneration from pea protoplasts via somatic embryogenesis, *Plant Cell Rep.*, 8, 379, 1989.

324. Jacobsen, H.-J., and Kysely, W., Induction of somatic embryos in pea, *Pisum sativum* L., *Plant Cell Tiss. Org. Cult.*, 3, 319, 1984.

325. Venketeswaran, S., Winged Pea (*Psophocarpus tetragonolobus* (L.) D.C.), in *Biotechnology in Agriculture and Forestry, Vol. 10, Legumes and Oilseed Crops I*, Bajaj, Y.P.S., Ed., Springer-Verlag, Berlin, 1990, 170.

326. Tran Thanh Van, K., Lie-Schricke, H., Marcotte, J.L., and Trinh, T.H., Winged Bean (*Psophocarpus tetragonolobus* (L.) DC.), in *Biotechnology in Agriculture and Forestry, Vol. 2, Crops I*, Bajaj, Y.P.S., Ed., Springer-Verlag, Berlin, 1986, 556.

327. Yamada, T., and Higuchi, S., *In vitro* culture of genus *Trifolium* germplasm and plant regeneration, *J. Jap. Soc. Grassland Sci.*, 36, 47, 1990.

328. Pederson, G.A., *In vitro* culture and somatic embryogenesis of four *Trifolium* species, *Plant Sci.*, 45, 101, 1986.

329. Collins, G.B., and Phillips, G.C., *In vitro* tissue culture and plant regeneration in *Trifolium pratense* L., in *Variability in Plants Regenerated from Tissue Culture*, Earle, E.D., and Demarly, Y., Eds., Praeger Publishers, New York, 1982, 22.

330. Williams, E.G., Collins, G.B., and Myers, J.R., Clovers (*Trifolium* spp.), in *Biotechnology in Agriculture and Forestry, Vol. 10, Legumes and Oilseed Crops I*, Bajaj, Y.P.S., Ed., Springer-Verlag, Berlin, 1990, 242.

331. Myers, J.R., Grosser, J.W., Taylor, N.L., and Collins, G.B., Genotype-dependent whole plant regeneration from protoplasts of red clover (*Trifolium pratense* L.), *Plant Cell Tiss. Org. Cult.*, 19, 113, 1989.

332. Parrott, W.A., Auxin-stimulated somatic embryogenesis from immature cotyledons of white clover, *Plant Cell Rep.*, 10, 17, 1991.

333. Weissinger, A.K. II, and Parrott, W.A., Repetitive somatic embryogenesis and plant recovery in white clover, *Plant Cell Rep.*, 12, 125, 1993.

334. Maheswaran, G., and Williams, E.G., Direct secondary somatic embryogenesis from immature sexual embryos of *Trifolium repens* cultures *in vitro*, *Ann. Bot.*, 57, 109, 1986.

335. Cui, D., Myers, J.R., Collins, G.B., and Lazzeri, P.A., *In vitro* regeneration in *Trifolium*. 1. Direct somatic embryogenesis in *T. rubens* (L.), *Plant Cell Tiss. Org. Cult.*, 15, 33, 1988.

336. Parrott, W.A., and Collins, G.B., Callus and shoot-tip culture of eight *Trifolium* species *in vitro* with regeneration via somatic embryogenesis of *T. rubens*, *Plant Sci. Lett.*, 28, 189, 1983.

337. Grosser, J.W., and Collins, G.B., Isolation and culture of *Trifolium rubens* protoplasts with whole plant regeneration, *Plant Sci. Lett.*, 37, 165, 1984.

338. Griga, M., Kubalakova, M., and Tejklova, E., Somatic embryogenesis in *Vicia faba* L., *Plant Cell Tiss. Org. Cult.*, 9, 167, 1987.

339. Pickardt, T., Schieder, O., Saalbach, I., Tegeder, M., Machemehl, F., Saalbach, G., Kohn, H., and Muntz, K., Regeneration and transformation in the legume *Vicia narbonensis*, in *Plant Tissue Culture and Gene Manipulation for Breeding and Formation of Phytochemicals*, Oono, K., Hirabayashi, T., Kikuchi, S., Handa, H., and Kajiwara, K. Eds., National Institute of Agrobiological Resources, Tsukuba, Japan, 1992, 89.

340. Pickardt, T., Huancaruna Perales, E., and Schieder, O., Plant regeneration via somatic embryogenesis in *Vicia narbonensis*, *Protoplasma*, 149, 5, 1989.

341. Albrecht, C., and Kohlenbach, H.W., Induction of somatic embryogenesis in leaf-derived callus of *Vicia narbonensis* L., *Plant Cell Rep.*, 8, 267, 1989.

342. Kumar, A.S., Gamborg, O.L., and Nabors, M.W., Plant regeneration from cell suspension cultures of *Vigna aconitifolia*, *Plant Cell Rep.*, 7, 138, 1988.

343. Shekhawat, N.S., and Galston, A.W., Isolation, culture and regeneration of moth bean *Vigna aconitifolia* leaf protoplasts, *Plant Sci. Lett.*, 32, 43, 1983.

344. Eapen, S., and George, L., Ontogeny of somatic embryos of *Vigna mungo* and *Vigna radiata*, *Ann. Bot.*, 66, 219, 1990.

345. Sinha, R.R., Das, K., and Sen, S.K., Embryoids from mesophyll protoplasts of *Vinga mungo* L. Hepper, a seed legume crop plant, in *Plant Cell Culture in Crop Improvement*, Sen, S.K. and Giles, K.L., Eds., Plenum Press, New York, 1983, 209.

346. Chen, H.K., Mok, M.C., and Mok, D.W.S., Somatic embryogenesis and shoot organogenesis from interspecific hybrid embryos of *Vigna glabrescens* and *V. radiata*, *Plant Cell Rep.*, 9, 77, 1990.

347. Southworth, D., and Kwiatkowski, S., Somatic embryogenesis from immature embryos in meadowfoam (*Limnanthes alba*), *Plant Cell Tiss. Org. Cult.*, 24, 193, 1991.

348. Pretova, A., and Williams, E.G., Direct somatic embryogenesis from immature zygotic embryos of flax (*Linum usitatissimum* L.), *J. Plant Physiol.*, 126, 155, 1986.

349. Nag, K.K., and Johri, B.M., Experimental morphogenesis of the embryo of *Dendrophtoe*, *Taxillus*, and *Nuytia*, *Bot. Gaz.*, 137, 378, 1976.

350. Nag, K.K., and Johri, B.M., Organogenesis and chromosomal constitution in embryo callus of *Nuytsia floribunda*, *Phytomorphology*, 19, 405, 1969.

351. Reynolds, B.D., Blackmon, W.J., and Postek, C.E., Production of somatic embryos from callused okra hypocotyl explants, *HortSci.*, 16, 87, 1981.

352. Trolinder, N.L., and Goodin, J.R., Somatic embryogenesis in cotton (*Gossypium*) I. Effects of source of explant and hormone regime, *Plant Cell Tiss. Org. Cult.*, 12, 31, 1988.

353. Trolinder, N.L., and Goodin, J.R., Somatic embryogenesis in cotton (*Gossypium*). II. Requirements for embryo development and plant regeneration, *Plant Cell Tiss. Org. Cult.*, 12, 43, 1988.

354. Trolinder, N.L., and Goodin, J.R., Somatic embryogenesis and plant regeneration in cotton (*Gossypium hirsutum* L.), *Plant Cell Rep.*, 6, 231, 1987.

355. Zhang, D., and Wang, Z., Tissue culture and embryogenesis of *Gossypium hirsutum* L., *Acta Bot. Sin.*, 31, 161, 1989.

410

356. Price, H.J., and Smith, R.H., Somatic embryogenesis in suspension cultures of *Gossypium klotzschianum* Anderss, *Planta*, 45, 305, 1979.

357. Reynolds, B.D., Blackmon, W.J., and Postek, C.E., Production of somatic embryos from roselle callus, *HortSci.*, 15, 432, 1980.

358. Coumans, M., Kevers, C., Coumans-Gilles, M.F., and Gaspar T., Fuchsia, in *Handbook of Plant Cell Culture, Vol. 5, Ornamental Species*, Ammirato, P.V., Evans, D.R., Sharp, W.R., and Bajaj, Y.P.S., Eds., Macmillan, New York, 1990, 429.

359. Kavathekar, A.K., and Ganapathy, P.S., Embryoid differentiation in *Eschscholzia californica*, *Current Sci.*, 42, 671, 1973.

360. Kavathekar, P.S., Ganapathy, P.S., and Johri, B.M., Chilling induces development of embryoids into plantlets in *Eschscholzia*, *Z. Pflanzenphysiol.*, 81, 358, 1977.

361. Kavathekar, A.K., Ganapathy, P.S., and Johri, B.M., *In vitro* responses of embryoids of *Eschscholzia californica*, *Biol. Plant.*, 20, 98, 1978.

362. Kohlenbach, H.W., Über organisierte Bilbungen aus *Macleaya cordata* Kallus, *Planta*, 64, 37, 1965.

363. Nessler, C.L., Somatic embryogenesis in the opium poppy, *Papaver somniferum*, *Physiol. Plant.*, 55, 453, 1982.

364. Culafic, L., Budimir, S., Vujicic, R., and Neskovic, M., Induction of somatic embryogenesis and embryo development in *Rumex acetosella* L., *Plant Cell Tiss. Org. Cult.*, 11, 133, 1987.

365. Bajaj, Y.P.S., and Mader, M., Growth and morphogenesis in tissue cultures of *Anagallis arvensis*, *Physiol. Plant.*, 32, 43, 1974.

366. Kiviharju, E. Tuominen, U., and Tormala, T., The effect of explant material on somatic embryogenesis of *Cyclamen persicum* Mill, *Plant Cell Tiss. Org. Cult.*, 28, 187, 1992.

367. Wicart, G., Mouras, A., and Lutz, A., Histological study of organogenesis and embryogenesis in *Cyclamen persicum* Mill. tissue cultures: Evidence for a single organogenetic pattern, *Protoplasma*, 119, 159, 1984.

368. Bajaj, Y.P.S., Regeneration of plants from ultra-low frozen anthers of *Primula obconica*, *Sci. Hort.*, 14, 93, 1981.

369. Hatano, K., Shoyama, Y., and Nishioka, I., Somatic embryogenesis and plant regeneration from the anther of *Aconitum carmichaeli* Debx., *Plant Cell Rep.*, 6, 446, 1987.

370. Giri, A., Ahuja, P.S., and Kumar, P.V.A., Somatic embryogenesis and plant regeneration from callus cultures of *Aconitum heterophyllum* Wall, *Plant Cell Tiss. Org. Cult.*, 32, 213, 1993.

371. Nakagawa, K., Miura, Y., Fukui, H., and Tabata, M., Clonal propagation of medicinal plants through the induction of somatic embryogenesis from the cultured cells, in *Plant Tissue Culture 1982*, Fujiwara, A., Ed., Japanese Assoc. Plant Tissue Culture, Tokyo, 1982, 701.

372. Raman, K., and Greyson, R.I., *In vitro* induction of embryoids in tissue cultures of *Nigella damascena*, *Can. J. Bot.*, 52, 1988, 1974.

373. Sethi, M., and Rangaswamy, N.S., Endosperm embryoids in cultures of *Nigella damascena*, *Curr. Sci.*, 45, 109, 1976.

374. Banerjee, S., and Gupta, S., Embryogenesis and differentiation in *Nigella sativa* leaf callus *in vitro*, *Physiol. Plant.*, 38, 115, 1976.

375. Beruto, M., and Debergh, P., Somatic embryogenesis in *Ranunculus asiaticus* L. hybr. thalamus cultivated *in vitro*, *Plant Cell Tiss. Org. Cult.*, 29, 161, 1992.

376. Konar, R.N., Thomas, E., and Street, H.E., Origin and structure of embryoids arising from epidermal cells of the stem of *Ranunculus sceleratus* L., *J. Cell Sci.*, 11, 77, 1972.

377. Konar, R.N., and Nataraja, K., Morphogenesis of isolated floral buds of *Ranunculus sceleratus* L., *Acta Bot. Neerl.*, 18, 680, 1969.

378. Dorion, N., Chupeau, Y., and Bourgin, J.P., Isolation, culture and regeneration into plants of *Ranunculus sceleratus* L. leaf protoplasts, *Plant Sci. Lett.*, 5, 325, 1975.

379. Dorion, N., Godin, B., and Bigot, C., Embryogenèse somatique à partir de cultures issues de protoplastes foliaires de *Ranunculus sceleratus*, *Can. J. Bot.*, 62, 2345, 1984.

380. Konar, R.N., and Nataraja, K., *In vitro* control of floral morphogenesis in *Ranunculus sceleratus* L., *Phytomorphology*, 14, 558, 1964.

381. Konar, R.N., and Nataraja, K., Experimental studies in *Ranunculus sceleratus* L. plantlets from freely suspended cells and cell groups, *Phytomorphology*, 15, 206, 1965.

382. Yang, Y.W., and Chang, W.C., Embryoid formation and subsequent plantlet regeneration from callus culture of *Thalictrum urbaini* Hayata (Ranunculaceae), *Z. Pflanzenphysiol.*, 97, 19, 1982.

383. Wang, D., Wergin, W.P., and Zimmerman, R.H., Somatic embryogenesis and plant regeneration from immature embryos of strawberry. *HortSci.*, 19, 71, 1984.

384. Rao, P.S., *In vitro* induction of embryonal proliferation in *Santalum album* L., *Phytomorphology*, 15, 175, 1965.

385. Bala, R., and Ibrahim, R., Induction of callus culture on *Chrysosplenium americanum* (Saxifragaceae), *Plant Physiol.*, 69, 33, 1982.

386. Sangwan, R.S., and Harada, H., Chemical regulation of callus growth, organogenesis, plant regeneration, and somatic embryogenesis in *Antirrhinum majus* tissue and cell cultures, *J. Exp. Bot.*, 26, 868, 1975.

387. Piorier-Hamon, S., Rao, P.S., and Harada, H., Culture of mesophyll protoplasts stem segments of *Antirrhinum majus* (Snapdragon): growth and organization of embryoids, *J. Exp. Bot.*, 25, 752, 1974.

388. Kuberski, C., Scheibner, H., Steup, C., Dietrich, B., and Luckner, M., Embryogenesis and cardenolide formation in tissue cultures of *Digitalis lanata*, *Phytochemistry*, 23, 1407, 1984.

389. Scheibner, H. Diettrich, B., Schulz, U., and Luckner, M., Somatic embryos of *Digitalis lanata* synchronization of development and cardenolide biosynthesis, *Biochem. Physiol. Pflanzen.*, 184, 311, 1989.

390. Tewes, A., Wappler, A., Peschke, E.M., Garve, R., and Nover, L., Morphogenesis and embryogenesis in long-term cultures of *Digitalis*, *Z. Pflanzenphysiol.*, 106, 311, 1982.

391. Arrillaga, L., Brisa, M.C., and Segura, J., Somatic embryogenesis and plant regeneration from hypocotyl cultures of *Digitalis obscura* L., *J. Plant Physiol.*, 124, 425, 1986.

392. Brisa, M.C., and Segura, J., Morphogenic potential of mechanically isolated single cells from *Digitalis obscura* L. callus, *Plant Cell Tiss. Org. Cult.*, 19, 129, 1989.

393. Konar, R.N., Thomas, E., and Street, H.E., The diversity of morphogenesis in suspension cultures of *Atropa belladonna* L., *Ann. Bot.*, 36, 249, 1972.

394. Thomas, E., and Street, H.E., Organogenesis in cell suspension cultures of *Atropa belladonna* L. and *Atropa belladonna* cultivar *lutea* Doll, *Ann. Bot.*, 34, 657, 1970.

395. Gosch, G., Bajaj, Y.P.S., and Reinert, J., Isolation, culture, and induction of embryogenesis in protoplasts from cell-suspensions of *Atropa belladonna*, *Protoplasma*, 86, 405, 1975.

396. Rashid, A., and Street, H.E., Growth, embryogenic potential and stability of a haploid cell culture of *Atropa belladonna* L., *Plant Sci. Lett.*, 2, 89, 1974.

397. Thomas, E., and Street, H.E., Factors influencing morphogenesis in excised roots and suspension cultures of *Atropa belladonna*, *Ann. Bot.*, 36, 239, 1972.

398. Guimaraes, M.L.S., Cruz, G.S., and Montezuma-de-Carvalho, J.M., Somatic embryogenesis and plant regeneration in *Cyphomandra betacea* (Cav.) Sendt, *Plant Cell Tiss. Org. Cult.*, 15, 161, 1988.

399. Lorz, H., Wernicke, W., and Potrykus, I., Culture and plant regeneration of *Hyoscyamus* protoplasts, *J. Medicinal Plant Res.*, 36, 21, 1979.

400. Cheng, J., and Raghavan, V., Somatic embryogenesis and plant regeneration in *Hyoscyamus niger*, *Amer. J. Bot.*, 72, 580, 1987.

401. Zapata, F.J., and Sink, K.C., Somatic embryogenesis from *Lycopersicon peruvianum* leaf mesophyll protoplasts. *Theor. Appl. Genet.*, 59, 265, 1981.

402. Facciotti, D., and Pilet, P.-E., Plants and embryoids from haploid *Nicotiana sylvestris* protoplasts, *Plant Sci. Lett.*, 15, 1, 1979.

403. Prabhudesai, V.R., and Narayanaswamy, S., Differentiation of cytokinin-induced shoot buds and embryoids on excised petioles of *Nicotiana tabacum*, *Phytomorphology*, 23, 133, 1973.

404. Lorz, H., Potrykus, I., and Thomas, E., Somatic embryogenesis from tobacco protoplasts, *Naturwissenschaften*, 64, 439, 1977.

412

405. Rao, K.S., and Gunasekari, K., Control of morphogenesis in tobacco protoplast cultures: organogenesis vs. embryogenesis, *Indian J. Biochem. Biophys.*, 28, 467, 1991.
406. Handro, W., Rao, P.S., and Harada, H., A histological study of the development of buds, roots, and embryos in organ cultures of *Petunia inflata* R. Fries, *Ann. Bot.*, 37, 817, 1973.
407. Rao, P.S., Handro, W., and Harada, H., Bud formation and embryo differentiation in *in vitro* cultures of Petunia, *Z. Pflanzenphysiol.*, 69, 87, 1973.
408. Rao, P.S., Handro, W., and Harada, H., Hormonal control of differentiation of shoots, roots and embryos in leaf and stem cultures of *Petunia inflata* and *Petunia hybrida*, *Physiol. Plant.*, 28, 458, 1973.
409. Reynolds, T.L., Somatic embryogenesis and organogenesis from callus cultures of *Solanum carolinense*, *Amer. J. Bot.*, 73, 914, 1986.
410. Chaturvedi, H.C., and Sinha, M., Mass clonal progagation of *Solanum khasianum* through tissue culture, *Indian J. Exp. Biol.*, 17, 153, 1979.
411. Gleddie, S., Keller, W., and Setterfield, G., Somatic embryogenesis and plant regeneration from leaf explants and cell suspension of *Solanum melongena* (eggplant), *Can. J. Bot.*, 61, 656, 1983.
412. Hinata, K., Eggplant (*Solanum melongena* L.), in *Biotechnology in Agriculture and Forestry, Vol. 2, Crops I*, Bajaj, Y.P.S., Ed., Springer-Verlag, Berlin, 1986, 363.
413. Gleddie, S., Keller, W.A., and Setterfield, G., Somatic embryogenesis and plant regeneration from cell suspension derived protoplasts of *Solanum melongena* (eggplant), *Can. J. Bot.*, 64, 355, 1984.
414. Yamada, T., Nakagawa, H., and Sinoto, Y., Studies on the differentiation in cultured cells I. Embryogenesis in three strains of *Solanum* callus, *Bot. Mag. Tokyo*, 80, 68, 1967.
415. Matsuoka, H., and Hinata, K., NAA-induced organogenesis and embryogenesis in hypocotyl callus of *Solanum melongena* L., *J. Exp. Bot.*, 30, 363, 1979.
416. Sihachakr, D., and Ducreux, G., Cultural behavior of protoplasts from different organs of eggplant (*Solanum melongena* L.), and plant regeneration, *Plant Cell Tiss. Org. Cult.*, 11, 179, 1987.
417. Rao, P.V.L., and Singh, B., Plantlet regeneration from encapsulated somatic embryos of hybrid *Solanum melongena* L., *Plant Cell Rep.*, 10, 7, 1991.
418. Gleddie, S.C., *In vitro* morphogenesis from tissues, cells and protoplasts of *Solanum melongena*, M.Sc. Thesis, Carleton University, 1981.
419. Gleddie, S., Keller, W.A., and Setterfield, G., Eggplant, in *Handbook of Plant Cell Culture, Vol. 4, Techniques and Applications*, Evans, D.A., Sharp, W.R., and Ammirato, P.V., Eds., Macmillan Publishing Co., New York, 1985, 500.
420. Skirvin, R.M., Lam, S.L., and Janick, J., Plantlet formation from potato callus *in vitro*, *HortSci.*, 10, 413, 1975.
421. Lam, S.L., Shoot formation in potato tuber discs in tissue culture, *Amer. Potato J.*, 52, 103, 1975.
422. Kato, M., Polyploids of camellia through culture of somatic embryos, *HortSci.*, 24, 1023, 1989.
423. Saha, T., and Sen S.K., Somatic embryogenesis in protoplast derived calli of cultivated jute, *Corchorus capsularis* L., *Plant Cell Rep.*, 10, 633, 1992.
424. Grewal, S., Sachdeva, U., and Atal, C.K., Regeneration of plants by embryogenesis from hypocotyl cultures of *Ammi majus* L., *Indian J. Exp. Biol.*, 14, 216, 1976.
425. Johri, B.M., and Sehgal, C.B., *In vitro* production of neomorphs in *Anethum graveolens* L., *Naturwiss.*, 50, 47, 1963.
426. Ratnamba, S.P., and Chopra, R.N., *In vitro* induction of embryoids from hypocotyls and cotyledons of *Anethum graveolens*, seedlings, *Z. Pflanzenphysiol.*, 73, 452, 1974.
427. Sehgal, C.B., *In vitro* development of neomorphs in *Anethum graveolens* L., *Phytomorphology*, 18, 509, 1968.
428. Sehgal, C.B., Differentiation of shoot buds and embryoids from inflorescence of *Anthum graveolens* in cultures, *Phytomorphology*, 28, 291, 1978.

429. Zhang, S., and Zheng, G., Induction of embryogenic callus and histocytological study on embryoid development of *Angelica sinensis* (Oliv.) Diels, *Acta Bot. Sin.*, 28, 241, 1986.

430. Williams, L., and Collin, H.A., Embryogenesis and plantlet formation in tissue cultures of celery, *Ann. Bot.*, 409, 325, 1976.

431. Al-Abta, S., and Collin, H.A., Control of embryoid development in tissue cultures of celery, *Ann. Bot.*, 42, 773, 1978.

432. Zee, S.-Y., and Wu, S.C., Embryogenesis in the petiole explants of chinese celery, *Z. Pflanzenphysiol.*, 93, 325, 1979.

433. Chen, C.H., Vegetative progagation of the celery plant by tissue culture, *Proc. S.D. Acad. Sci.*, 55, 44, 1976.

434. Browers, M.A., and Orton, T.J., Celery (*Apium graveolens* L.), in *Biotechnology in Agriculture and Forestry, Vol 2, Crops I*, Bajaj, Y.P.S., Ed., Springer-Verlag, Berlin, 1986, 405.

435. Zee, S.-Y., and Wu, S.C., Somatic embryogenesis in the leaf explants of Chinese celery, *Aust. J. Bot.*, 28, 429, 1980.

436. Kim, Y.H., and Janick, J., Origin of somatic embryos in celery tissue culture, *HortSci.*, 24, 671, 1989.

437. Nadel, B.L., Altman, A., and Ziv, M., Regulation of somatic embryogenesis in celery cell suspensions 1. Promoting effects of mannitol on somatic embryo development, *Plant Cell Tiss. Org. Cult.*, 18, 181, 1989.

438. Nadel, B.L., Altman, A., and Ziv, M., Regulation of somatic embryogenesis in celery cell suspensions 2. Early detection of embryogenic potential and the induction of synchronized cell cultures, *Plant Cell Tiss. Org. Cult.*, 20, 119, 1990.

439. Wakhlu, A.K., Nagari, S., and Barna, K.S., Somatic embryogenesis and plant regeneration from callus cultures of *Bunium persicum* Boiss, *Plant Cell Rep.*, 9, 137, 1990.

440. Wang, P.J., and Huang, C.I., Production of saikosaponins by callus and redifferentiated organs of *Bupleurum falcatum* L., in *Plant Tissue Culture 1982*, Fujiwara, A., Ed., Maruzen Co., Tokyo, 1982, 71.

441. Ammirato, P.V., Hormonal control of somatic embryo development from cultured cells of caraway, *Plant Physiol.*, 59, 579, 1977.

442. Ammirato, P.V., The effects of abscisic acid in the development of somatic embryos from cells of caraway (*Carum carvi* L.), *Bot. Gaz.*, 135, 328, 1974.

443. Furmanowa, M., Oledzka, H., and Sowinska, D., Regeneration of plants by embryogenesis with callus cultures of *Carum carvi* L., *J. Plant Physiol.*, 115, 209, 1984.

444. Zee, S.-Y., Studies on adventive embryo formation in the petiole explants of Coriander (*Coriandrum sativum*), *Protoplasma*, 107, 21, 1981.

445. Kumar, K.B., Pareek, N., Pillai, S.K., and Pillai, A., Callus development, shoot formation and somatic embryogenesis in *Coriandrum sativum* L. *in vitro*, *Beitr. Biol. Pflanzen*, 57, 369, 1982.

446. Halperin, W., and Wetherell, D.F., Adventive embryony in tissue cultures of the wild carrot, *Daucus carota*, *Amer. J. Bot.*, 51, 274, 1964.

447. Ronchi, V.N., Caligo, M.A., Nozzolini, M., and Luccarini, G., Stimulation of carrot somatic embryogenesis by proline and serine, *Plant Cell Rep.*, 3, 210, 1984.

448. Caligo, M.A., Ronchi, V.N., and Nozzolini, M., Proline and serine affect polarity and development of carrot somatic embryos, *Cell Differ.*, 17, 193, 1985.

449. Jones, L.H., Factors influencing embryogenesis in carrot cultures (*Daucus carota* L.), *Ann. Bot.*, 38, 1077, 1974.

450. Smith, D.L., and Krikorian, A.D., Production of somatic embryos from carrot tissues in hormone-free medium, *Plant Sci.*, 58, 103, 1988.

451. Kamada, H., and Harada, H., Studies on the organogenesis in carrot tissue cultures I. Effects of growth regulators on somatic embryogenesis and root formation, *Z. Pflanzenphysiol.*, 91, 255, 1979.

452. Kamada, H., and Harada, H., Studies on the organogenesis in carrot tissue cultures II. Effects of amino acids and inorganic nitrogenous compounds on somatic embryogenesis, *Z. Pflanzenphysiol.*, 91, 453, 1979.

414

453. Halperin, W., Alternative morphogenetic events in cell suspensions, *Amer. J. Bot.*, 53, 443, 1966.
454. Halperin, W., Morphogenetic studies with partially synchronized cultures of carrot embryos, *Science*, 146, 408, 1964.
455. Ammirato, P.V., and Steward, C.F., Some effects of environment on the development of embryos from cultured free cells, *Bot. Gaz.*, 132, 149, 1971.
456. Wetherell, D.F., and Halperin, W., Embryos derived from callus tissue cultures of the wild carrot, *Nature*, 200, 1336, 1963.
457. Sussex, I.M., and Frei, K.A., Embryoid development in long-term tissue cultures of carrot, *Phytomorphology*, 18, 339, 1968.
458. Smith, S.M., and Street, H.E., The decline of embryogenic potential as callus and suspension cultures of carrot (*Daucus carota* L.) are serially subcultured, *Ann. Bot.*, 38, 223, 1974.
459. Reinert, J., Some aspects of embryogenesis in somatic cells of *Daucus carota*, *Phytomorphology*, 17 ,510, 1967.
460. Reinert, J., and Backs, D., Control of totipotency in plant cells growing *in vitro*, *Nature*, 220, 1340, 1968.
461. Grambow, H.J., Kao, K.N., Miller, R.A., and Gamborg, O.L., Cell division and plant development from protoplasts of carrot cell suspension cultures, *Planta*, 103, 348, 1972.
462. Fujimura, T.T., and Komamine, A., Effects of various growth regulators on the embryogenesis in a carrot cell suspension culture, *Plant Sci. Lett.*, 5, 359, 1975.
463. Fujimura, T., and Komamine, A., Synchronization of somatic embryogenesis in a carrot cell suspension culture, *Plant Physiol.*, 64, 162, 1979.
464. Sung, Z.R., Smith, R., and Horowitz, J., Quantitative studies of embryogenesis in normal and 5-methyltryptophan-resistant cell lines of wild carrot, *Planta*, 147, 236, 1979.
465. Dudits, D., Kao, K.N., Constabel, F., and Gamborg, O.L., Embryogenesis and formation of tetraploid and hexaploid plants from carrot protoplasts, *Can. J. Bot.*, 54, 1063, 1976.
466. Homès, J.L.A., and Guillaume, M., Phénomènes d'organogenèse dans les cultures *in vitro* de tissus de carotte (*Caucus carota* L.), *Bull. Soc. Roy. Bot. Belgique Tome*, 100, 239, 1967.
467. Kamada, H., and Harada, H., Studies on nitrogen metabolism during somatic embryogenesis in carrot. I. Utilization of α-alanine as a nitrogen source, *Plant Sci. Lett.*, 33, 7, 1984.
468. Schiavone, F.M., and Cooke, T.J., A geometric analysis of somatic embryo formation in carrot cell cultures, *Can. J. Bot.*, 63, 1573, 1985.
469. Warren, G.S., and Fowler, M.W., Cell number and cell doubling times during the development of carrot embryoids in suspension culture, *Experientia*, 34, 356, 1978.
470. Warren, G.S., and Fowler, M.W., Physiological interactions during the initial stages of embryogenesis in cultures of *Daucus carota* L., *New Phytol.*, 87, 481, 1981.
471. Kato, H., and Takeuchi, M., Morphogenesis *in vitro* starting from single cells of carrot root, *Plant Cell Physiol.*, 4, 243, 1963.
472. Kessell, R.H.J., Goodwin, C., and Philp, J., The relationship between dissolved oxygen concentration, ATP and embryogenesis in carrot (*Daucus carota*) tissue cultures, *Plant Sci. Lett.*, 10, 265, 1977.
473. Smith, D.L., and Krikorian, A.D., Release of somatic embryogenic potential from excised zygotic embryos of carrot and maintenance of proembryonic cultures in hormone-free medium, *Amer. J. Bot.*, 76, 1832, 1989.
474. Smith, D.L., and Krikorian, A.D., Somatic proembryo production from excised, wounded zygotic carrot embryos on hormone-free medium: evaluation of the effects of pH, ethylene and activated charcoal, *Plant Cell Rep.*, 9, 34, 1990.
475. Dougall, D.K., and Verma, D.C., Growth and embryo formation in wild-carrot suspension cultures with ammonium ion as a sole nitrogen source, *In Vitro*, 14, 180, 1978.
476. Verma, D.C., and Dougall, D.K., Influence of carbohydrates on quantitative aspects of growth and embryo formation in wild carrot suspension cultures, *Plant Physiol.*, 59, 81, 1977.
477. Brown, S., Wetherell, D.F., and Dougall, D.K., The potassium requirement for growth and embryogenesis in wild carrot suspension cultures, *Physiol. Plant.*, 37, 73, 1976.

478. Wetherell, D.F., and Dougall, D.K., Sources of nitrogen supporting growth and embryogenesis in cultured wild carrot tissue, *Physiol. Plant.*, 37, 97, 1976.
479. Smith, D.L., and Krikorian, A.D., Low external pH replaces 2,4-D in maintaining and multiplying 2,4-D-initiated embryogenic cells of carrot, *Physiol. Plant.*, 80, 329, 1990.
480. Ammirato, P.V., Carrot, in *Handbook of Plant Cell Culture, Vol. 4, Techniques and Applications*, Evans, D.A., Sharp, W.R., and Ammirato, P.V., Eds., Macmillan Publishing Co., New York, 1985, 457.
481. Miura, Y., and Tabata, M., Direct somatic embryogenesis from protoplasts of *Foeniculum vulgare*, *Plant Cell Rep.*, 5, 310, 1986.
482. Maheshwari, S.C., and Gupta, G.R.P., Production of adventitious embryoids *in vitro* from stem callus of *Foeniculum vulgare*, *Planta*, 67, 384, 1965.
483. Vasil, I.K., and Hildebrandt, A.C., Variations of morphogenetic behavior in plant tissue cultures II. *Petroselinum hortense*, *Amer. J. Bot.*, 53, 869, 1966.
483. Miura, Y., Fukui, H., and Tabata, M., Clonal propagation of chemically uniform fennel plants through somatic embryoids, *Planta Medica*, 53, 92, 1986.
485. Watin, C., and Bigot, C., Régénération de plantes fertiles à partir de cals embryogènes de persil commun (*Petroselinum hortense* Hoffm.), *C.R. Acad. Sci. Paris*, 309, 653, 1989.
486. Ernst, D., Schafer, W., and Oesterhelt, D., Isolation and quantitation of isopentenyladenosine in an anise cell culture by single-ion monitoring, radioimmunoassay and bioassay, *Planta*, 159, 216, 1983.
487. Ernst, D., and Oesterhelt, D., Effect of exogenous cytokinins on growth and somatic embryogenesis in anise cells (*Pimpinella anisum* L.), *Planta*, 161, 246, 1984.
488. Huber, J., Constabel, F., and Gamborg, O.L., A cell counting procedure applied to embryogenesis in cell suspension cultures of anise (*Pimpinella anisum* L.), *Plant Sci. Lett.*, 12, 209, 1978.
489. Mathur, J., Enhanced somatic embryogenesis in *Selinum candolii* DC. under a mineral oil overlay, *Plant Cell Tiss. Org. Cult.*, 27, 23, 1991.
490. Jasrai, Y.T., Barot, S.M., and Mehta, A.R., Plant regeneration through somatic embryogenesis in hypocotyl explants of *Trachyspermum ammi* (L.) Sprague, *Plant Cell Tiss. Org. Cult.*, 29, 57, 1992.
491. Mathur, J., Somatic embryogenesis from callus cultures of *Nardostachys jatamansi*, *Plant Cell Tiss. Org. Cult.*, 33, 163, 1993.

11. Somatic Embryogenesis in Herbaceous Monocots

SANKARAN KRISHNARAJ and INDRA K. VASIL

Contents

I. Introduction

The seed is the primary organ for the perpetuation of germplasm and propagation in all flowering plants. It contains the embryo which develops from the zygote following the fertilization of the egg cell by one of the male gametes. Gene recombination during the formation, as well as the fusion, of the gametes plays a critical role in the evolution of the angiosperms. During its development the embryo is nourished by the endosperm, which itself is formed as a result of the fusion of the second male gamete with two nuclei (in most of the angiosperms) of the central cell of the embryo sac. During the course of evolution many species have developed special means of propagation to overcome physical, environmental and genetic factors that prevent flowering and seed formation. These include the formation of perennial structures such as tubers, rhizomes, bulbs, stolons, etc (see Sharma and Thorpe, this volume). An advanced form of perennation is found in many apomictic species which produce embryos from unfertilized gametic or somatic cells. Like the zygotic embryos, the apomictic embryos develop within the seed and are protected and dispersed like natural seed. Apomictic embryos of all genotypes – irrespective of heterozygosity – developing from unreduced and unfertilized somatic or gametic cells (agamospermy, adventive embryony, apospory, diplospory) breed

417

T.A. Thorpe (ed.), In Vitro Embryogenesis in Plants, pp. 417–470.

418

true and give rise to clonal populations. Apomixis is known to be under genetic control, but environmental conditions such as temperature and photoperiod may also promote a switch from sexual to apomictic development [1, 2].

Although apomictic adventive embryony occurs in nature in many species of angiosperms, its experimental induction in cultured cells and tissues, commonly described as somatic embryogenesis and characterized by the formation of somatic embryos, is a rather recent phenomenon. The first clear demonstration of the formation of somatic embryos in tissue cultures was made by Steward et al. [3] and Reinert [4] in carrot. Since then this phenomenon has been described in scores of species, and has been studied in exhaustive detail particularly in carrot. Although much progress has been made in the elucidation of the factors that control somatic embryogenesis *in vitro*, the molecular basis for the dramatic switch in the developmental potential of somatic cells to the embryogenic pathway is not fully understood.

The bulk of the information presented in this volume is on somatic embryogenesis in dicotyledonous species. The reason for this is that although the phenomenon has been described in many monocotyledonous species for a long time (Table 1), it has not been studied in much detail except in the Gramineae [5, 6]. This chapter describes the principal features of somatic embryogenesis in herbaceous monocots, particularly in the Gramineae.

II. Somatic embryogenesis in herbaceous monocots

Somatic embryogenesis has been induced in many herbaceous species belonging to at least the following eleven families of monocots:

Alliaceae	Araceae
Asparagaceae	Dioscoreaceae
Gramineae	Hemerocallidaceae
Iridaceae	Liliaceae
Musaceae	Orchidaceae
Zingiberaceae	

Representative examples from each of the families, in which clear evidence of the formation of somatic embryos has been provided, are listed in Table 1. Although there is an extensive body of literature on tissue culture propagation of orchids, much of it does not include clear references to or descriptions of somatic embryogenesis. Nevertheless, for all practical purposes the formation of protocorms, which closely mimic the development of zygotic embryos, in tissue cultures of orchids should be considered to be a manifestation of somatic embryogenesis.

Table 1. Somatic embryogenesis in herbaceous monocot species.

Species (Common name)	Explant (*Success in Regeneration)	Media Sequence[a] (Salts)	SE Induction Medium		SE Development Medium		Reference
			Growth Regulators[b] (μM)	Others[c] (mM/g)	Growth Regulators[b] (μM)	Others[c] (mM/g)	
Alliaceae							
Allium carinatum	Seedling roots*	K+ → K+ → K	2,4-D (5) + KN (10) + IAA (10) → KN (50) + IAA (10)	-	-	-	7
Allium cepa (Onion)	Shoot meristem* and basal plate-scale*	K+ → K+	Piclo (3.1) + BA (6.7–8.9)	-	Piclo (0.1–0.2) + BA (1.6–4.4)	-	8
Allium fistulosum (Japanese bunching onion)	Mat. seed*	K+ → K+ → K+	2,4-D (4.5) ± KN (4.6)	-	KN (7) + ADS (217) → IBA (0.5)	-	9
Allium fistulosum × A. cepa (Beltsville bunching onion)	Mat. seed*	K+ → K+ → K+	2,4-D (4.5) ± KN (4.6)	-	KN (7) + ADS (217) → IBA (0.5)	-	9
	Basal plate* and young inflores.*	K+ → K+ → 1/2K	Piclo (3.1) + BA (8.9)	L-Pro (21.7)	Piclo (0.1) + 2-iP (1.6)	L-Pro (21.7)	10
Allium porrum (Leek)	Mat. embryo*	MS+ → MS+	2,4-D (5)	Phytagel (3 g) + Suc (87.6)	KN (5.1) + ABA (1)	-	11
	Mat. embryo*	MS+ → MS+	2,4-D (5)	Phytagel (3 g) + Suc (87.6)	KN (5.1) + ABA (1)	-	11
Allium sativum (Garlic)	Receptacle tissue*	K+ → MS+	NAA (1) + BA (10)	CNS (100)	NAA (0–5) + BA (0–15)	-	12
	Stem segments*	AZ+ → AZ+	CPA (10) + KN (0.5) + 2,4-D (2)	-	IAA (10) + KN (20)	-	13

Table 1. Continued

Species (Common name)	Explant (*Success in Regeneration)	Media Sequence[a] (Salts)	SE Induction Medium		SE Development Medium		Reference
			Growth Regulators[b] (μM)	Others[c] (mM/g)	Growth Regulators[b] (μM)	Others[c] (mM/g)	
Araceae							
Anthurium andraeanum (Flamingo plant/Painters Palette/Arum Lily)	Leaf blade*	1/2macMS+ → 1/2macMS+	2,4-D (4.5–18.1) + KN (1.5–4.6)	Suc (58.4) + Glu (55.5) + Gelrite (1.8 g) + my-inos (0.6)	BA (0.9)	Suc (58.4)	14
Asparagaceae							
Asparagus cooperi	Spear disc*	MS+ → MS+ → MS+	NAA (5.4) + KN (4.6)	KNO_3 (28.7) → L-Glut (2.1) + CH (1 g) + NH_4NO_3 (15.0)	Z (4.6) + GA_3 (2.9)	–	15
Asparagus officinalis (Asparagus)	Cell culture*	MS+ → MS/MS+	2,4-D (5)	–	± 2,4-D (5)	–	16
	Cladophylls*	MS+ → MS → MS+	NAA (50) + BA (1)	Suc (300)	NAA (50) + BA (1)	Suc (300)	17
	Lateral buds*	MS+ → MS+ → MS MS+ → MS	NAA (0.5) + KN (0.1) → NAA (8.1) + KN (0.5)	–	NAA (0.3) + KN (0.5)	Glu (111)	18
	Lateral stem bud*	MS+ → MS+	2,4-D (6.8) + 2-iP (1.5)	Glu/Fru (111–333)/Suc (58–175)	NAA (0.4) + 2-iP (1.0)	Glu/Fru (277.5–333)/Suc (146–175)	19
	Leaf cell protoplasts	MS+ → MS → MS+	Z (8) + NAA (3)/ BA (5) + NAA (5)	Adenine (30)	IAA (1) + Z (1/2/3)	–	20
	Seedling crown* and lateral buds*	LS+ → LS → 1/2LS	2,4-D (5)	Suc (58.4)	–	–	21
	Shoot and cladode*	LS+/MS+ → LS+/MS+	NAA (5.4) + KN (4.7)	–	IAA (0.5–5.7) + BA (0.44–17.7)	–	22 23

Table 1. Continued

Species (Common name)	Explant (*Success in Regeneration)	Media Sequence[a] (Salts)	SE Induction Medium		SE Development Medium		Reference
			Growth Regulators[b] (μM)	Others[c] (mM/g)	Growth Regulators[b] (μM)	Others[c] (mM/g)	
	Spear segments* and lateral buds* and in vitro crowns*	MS+ → MS+	2,4-D (4.5–45.3)/NAA (5.4–53.7) + KN (0–4.7)	–	NAA (5.4–53.7)	–	24
	Stem cutting*	MS+ → 1/2MS → 1/2MS+	2,4-D (4.5)	–	IBA (4.9) + GA₃ (2.9)	Suc (29.2)	25
	Stem sections*	MS+ → MS+ → MS+ → MS+	NAA (0.54) + 2-iP (1.4) → NAA (54–107)	–	NAA (0.54) + 2-iP (0.98)	Suc (180)/Glu/Fru (330)	26
Dioscoreaceae							
Dioscorea deltoidea (Yam)	Tuber	MS+	2,4-D (9) + KN (41.8)	–	n.r.	n.r.	27
Dioscorea floribunda (Medicinal Yam)	Embryo*	MS+ → MS+/MS	2,4-D (4.5)	–	± ABA (0.1)	± Glut (3.4)	28
Dioscorea rotundata (Dioscorea)	Mat. embryo*	MS+ → MS+	NAA (6)	CH (1 g)	NAA (0.6)	CH (0.5 g)	29
Gramineae (Poaceae)							
Agropyron repens × Bromus inermis (Intergeneric hybrid AGROMUS)	Inflores. primordia*	MS+ → MS; MS+ → MS	GA₃ (5.8) → 2,4-D (2.3)	Suc (58.4) + Gly (26.6) + my-inos (0.6)	–	Suc (58.4) + Gly (26.6) + my-inos (0.6)	30
Agrostis palustris (Creeping bent grass)	Mature seed*	MS+ → 1/2MS	Dica (30) + BA (2.25)	CH (0.5 g)	–	CH (0.5 g)	31

Table 1. Continued

Species (Common name)	Explant (*Success in Regeneration)	Media Sequence[a] (Salts)	SE Induction Medium		SE Development Medium		Reference
			Growth Regulators[b] (µM)	Others[c] (mM/g)	Growth Regulators[b] (µM)	Others[c] (mM/g)	
Avena sativa (Oat)	Imm. embryo*	MS →MS	-	Suc (175.3)	-	Suc (58.4)	32
	Imm. embryo*	MS+ → → MS+	2,4-D (9.1)	-	2,4-D (2.3)	-	33
	Imm. embryo*	MS+ → MS → MS	2,4-D (9.1)	Aspn (1.1) + Suc (58.4)	-	Suc (175.3) → Suc (58.4)	34
	Mat. and imm. embryo* and mesocotyls*	LS+ → LS+ → LS+	2,4,5-T (3.9)	-	2,4-D/2,4,5-T (0–4) → 2,4,5-T (0–20)	-	35
Bambusa beecheyana (Bamboo)	Young inflores.*	MS+ → MS/MS+	2,4-D (13.6) + KN (9.3)	CH (1 g) + Suc (175.3)	-	-	36
Bambusa oldhamii (Green bamboo)	Young inflores.*	MS+ → MS	2,4-D (13.6) + KN (9.3)	Suc (175.3) + CH (1 g)	-	-	37
Bothriochloa caucasica (Caucasian bluestem)	Mat. seed*	MS+ → MS	2,4-D (5)	L-Pro (12) + B5 vits + Suc (58.4)	-	B5 vits	38
Bothriochloa ischaemum (Blue stem grass)	Young inflores.*	1/2MS+ → 1/2MS+ → 1/2MS	2,4-D (13.6) + IAA (5.7)	-	2,4-D (2.3) + Z (4.6)	-	39
Bromus inermus (Brome grass)	Mesocotyl tissue*	B5+ → B5	2,4-D (4.5)	-	-	-	40
Cenchrus ciliaris (Buffel grass/african foxtail)	Young inflores.*	MS+ → MS → 1/2MS+	2,4-D (27.1/63.3)	-	± NAA (0.5)	-	41
Cenchrus setigerus	Young inflores.*	MS+ → MS → 1/2MS+	2,4-D (27.1/63.3)	-	+NAA (0.5)	-	41
Coix lacryma-jobi (Job's tears)	Young inflores.*	N6+ → MS+	2,4-D (4.5–9)	Suc (87.6–146.1)	KN (2.3) + NAA (0.05)	-	42

Table 1. Continued

Species (Common name)	Explant (*Success in Regeneration)	Media Sequence[a] (Salts)	SE Induction Medium		SE Development Medium		Reference
			Growth Regulators[b] (μM)	Others[c] (mM/g)	Growth Regulators[b] (μM)	Others[c] (mM/g)	
Cymbopogon winterianus (Java citronella)	Shoot buds*	MS+/SH+ → MS → MS (l)	2,4-D (2.3–18.1)/NAA (2.7–21.5)/IAA (2.9–22.8) + KN (2.3–9.3)/BAP (2.2–8.9)	–	–	Suc (2.9)	43
	Young inflores.*	MS+ → MS+/MS → MS+	2,4-D (4.5)	–	IAA (5.7)/NAA (5.4)/KN (4.7)/BAP (4.4)	–	44
Cynodon dactylon (Bermuda grass)	Imm. inflores.*	1/2MS+ → 1/2MS+	2,4-D (13.6)	CH (0.2 g)	2,4-D (2.3) + Z (5.5)	–	45
	Young inflores.*	1/2MS+ → 1/2MS+ → 1/2MS	2,4-D (13.6) + IAA (5.7)	–	2,4-D (2.3) + Z (4.6)	–	39
	Young inflores.*	N6+ → N6	2,4-D (4.5)	Suc (175.3)	–	–	46
	Young inflores.*	N6+ → N6+ → N6	2,4-D (4.5) → 2,4-D (2.3)	–	2,4-D (1.1) → 2,4-D (0)	–	47
Dactylis glomerata (Orchard grass)	Anthers* and pistils (unpollinated)	SH+ → SH	Dica (30)	Suc (87.6)	–	–	48
	Anthers*	SH+ → SH	Dica (30)	Suc (0.3)	–	–	49
	Leaf base	SH+	Dica (30)	CH (3 g)	n.d.	n.d.	50
	Leaf base*	SH+ (l) → SH (s)	Dica (30)	CH (3 g)	–	–	51
	Leaf bases*	SH+ → SH	Dica (30)	Suc (87.6) + CH (3 g)+ L-Pro (12.5) + L-Ser (12.5)	–	–	52

423

Table 1. Continued

Species (Common name)	Explant (*Success in Regeneration)	Media Sequence[a] (Salts)	SE Induction Medium Growth Regulators[b] (μM)	Others[c] (mM/g)	SE Development Medium Growth Regulators[b] (μM)	Others[c] (mM/g)	Reference
	Leaf segment	SH+	Dica (30)	Suc (87.6) + my-inos (5.6)	n.d.	n.d.	53
	Leaf segment*	SH+ (l) →SH (l)	Dica (30/40/60)	CH (3 g/4 g)	-	CH (0–4 g)	54
	Leaf segment*	SH+ (l) → SH+ (s) → SH (l/s)	Dica (30)	± CH (1.5 g)	-	-	55
	Leaf segment*	SH+ → SH	Dica (30) + 2,4-D (10)	-	-	-	56
	Leaf segment*	SH+ → SH	Dica (30)	-	-	-	57
	Mat. embryo	LS+ → LS+	2,4-D (20) → 2,4-D (1)	-	n.d.	n.d.	58
	Young leaves*	SH+ → SH	Dica (30)	-	n.d.	n.d.	59
	Young leaves*	SH+ → SH+ → 1/2SH	Dica (30)	-	Dica (1)	-	60
Echinochloa crusgalli (Barnyard grass)	Mesocotyl plate tissue	MS+ → MS+	2,4-D (22.6/45.2) + BAP (8.9)	-	2,4-D (< 4.5)	-	61
	Young inflores.*	MS+ → MS+/MS	2,4-D (2.3–22.6) + BA (2.2)	CH (0.5 g)	± (2,4-D (2.3) + BA (2.2)	-	62
	Seedling mesocotyl plate tissue*	MS+ → MS+ → 1/2MS (l)	2,4-D (22.6/45.2) + BAP (8.9)	Suc (73)	2,4-D (≤ 22.6)	-	63
Echinochloa frumentacea	Young inflores.* and imm. embryo*	MS+ → N6+	2,4-D (10/20/50)	-	2,4-D (10/20)	-	64
	Young inflores.*	MS+ → MS+	2,4-D (22.4) + KN (2.3)	my-inos (0.6) + Gly (33.3)	KN (2.3)	-	65

Table 1. Continued

Species (Common name)	Explant (*Success in Regeneration)	Media Sequence[a] (Salts)	SE Induction Medium Growth Regulators[b] (µM)	SE Induction Medium Others[c] (mM/g)	SE Development Medium Growth Regulators[b] (µM)	SE Development Medium Others[c] (mM/g)	Reference
Echinochloa glabrescens	Young inflores.* and young leaf	MS+ → MS+ → LS	2,4-D (9) + BAP (0.9)	–	BAP (17.8) + IAA (2.9) + NAA (2.7)	CH (0.5 g)	66
Echinochloa muricata	Seedling mesocotyl plate tissue*	MS+ → MS+ → 1/2MS (1)	2,4-D (22.6/45.2) + BAP (8.9)	Suc (73)	2,4-D (≤ 22.6)	–	63
	Mesocotyl plate tissue	MS+ → MS+	2,4-D (22.6/45.2) + BAP (8.9)	–	2,4-D (< 4.5)	–	61
Echinochloa oryzicola	Leaf sheath*	MS+ → MS+ → MS/MS+	2,4-D (22.6) → 2,4-D (18.1)	–	KN (0–46.5)/BA (0.4–44.4) → NAA (0.5–53.7)	–	67
Eleusine coracana (Finger millet)	Mat. seed* and shoot apex*	MS+ → MS	Piclo (8.3) + KN (0.5) → Piclo (0.4) + KN (4.7)/Z (4.6)	–	–	–	68
	Mat. seed*	MS+ → MS+ → MS+	2,4-D (4.5/13.6) → 2,4-D (0.9)	–	GA$_3$ (2.9)/IBA (4.9)	–	69
	Mat. seed*	MS+ → MS	Piclo (16.6) + KN (2.3) → Piclo (8.3) + KN (0.5)	–	–	Cefo. (0.05–1.0 g)/Carben. (0.5–1.0 g)/Strept.(0.05 g)	70
	Young inflores.*	MS+ → MS	Piclo (8.3) + KN (0.5)	Suc (87.6/175.3/262.9)	–	Suc (87.6)	71
Eleusine indica (Yard/Goose grass)	Young inflores.*	N6+ → N6	2,4-D (9.1)	CH (0.5 g)	–	–	72
Elymus canadensis (Canada wildrye)	Imm. embryo* and imm. inflores.*	MS+ → MS+/MS	2,4-D (9.1)	–	± (2,4-D (2.3) + GA$_3$ (0.9))	–	73

425

Table 1. Continued

Species (Common name)	Explant (*Success in Regeneration)	Media Sequence[a] (Salts)	SE Induction Medium Growth Regulators[b] (μM)	SE Induction Medium Others[c] (mM/g)	SE Development Medium Growth Regulators[b] (μM)	SE Development Medium Others[c] (mM/g)	Reference
Festuca arundinacea (Tall fescue)	Leaf base*	MS+ → MS	2,4-D (30)	-	-	-	74
	Mat. and imm. embryo*	MS+/MS+ → 1/2MS+	2,4-D (9 → 31.7) + BAP (0.9)/2,4-D (45.2 → 22.6)	CH (1 g)	BAP (0.4)	Suc (58.4)	75
Festuca rubra (Red fescue)	Mat. caryopses*	MS+ → 1/2MS	2,4-D (20)	-	-	-	76
	Mat. embryo callus*	MS+ → 1/2MS	2,4-D (22.5)	Suc (120) + B5 vits	-	B5 vits + Suc (135/180)	77
	Mat. embryo*	1/2MS+ → 1/2MS+	2,4-D (20)	-	BA (0.4)	-	78
	Mat. seed*	1/2MS+ → 1/2MS	2,4-D (18.1)	B5 vits + CH (3 g)	-	B5 vits	79
Hordeum bulbosum × Triticum aestivum hybrid	Imm. haploid embryo*	B5+ (s) → AA+ (1) → B5/B5+	2,4-D (9) → 2,4-D (9)	-	± 2,4-D (9)	-	80
Hordeum chilense × Triticum turgidum hybrid	Imm. embryo* and young inflores.*	MS+ → MS	2,4-D (2.3/4.5/9)	-		-	81
Hordeum vulgare (Barley)	Apical meristem*	MS+ → MS	NAA (10) + 2,4-D (15) + 2-iP (1.5/4.5)	± Trypt (0.24)	-	-	82
	Imm. embryo* and young leaf tissue*	MS+ → MS	2,4-D (9)	Suc (87.6)	-	-	83
	Imm. embryo* and young inflores.*	MS+ → MS+/MS	2,4-D (10)	-	± ABA (0.4)	-	84

Table 1. Continued

Species (Common name)	Explant (*Success in Regeneration)[a]	Media Sequence[a] (Salts)	SE Induction Medium		SE Development Medium		Reference
			Growth Regulators[b] (µM)	Others[c] (mM/g)	Growth Regulators[b] (µM)	Others[c] (mM/g)	
	Imm. embryo*	B5+ → B5+ → B5+	2,4-D (11.3/22.6)	–	ABA (0.04/0.19) → GA$_3$ (2.9)/KN (0.46)	–	85
	Imm. embryo*	MS+/CC+ → CC+ → 1/2MS	2,4-D (9)	CH (1 g)	Z (0.2) + IAA (5.7)	–	86
	Mat. embryo*	MS+ → MS+ → MS/MS+	2,4-D (4.5) → 2,4-D (0–2.3)	–	2,4-D (0.1)	–	87
	Mat. embryo*	MS+ → MS+	2,4-D (9/22.6)	Suc (131.5) + my-inos (2.8)	TIBA (3)	Suc (58.4)	88
	Mat. seed*	SH+ → SH+	2,4-5-T (5)	–	2,4,5-T (1)	–	89
	Seedling tissues*	MS+ → MS+	2,4-D (7.8) + BA (2.2)	Suc (131.5) + L-Pro (10)	TIBA (3)	Suc (58.4)	90
	Seedling*	MS+ → MS+ → MS+	2,4-D (7.8) + BA (2.2)	Suc (131.5)	ABA (0.4)/TIBA (3) → IBA (4.9)	Suc (131.5) → Suc (87.6)	91
	Seedling*	MS+ → MS+	2,4-D/2,4,5-T (19.5/31.3) ± BA (2.2/4.4)	–	TIBA (3)	–	92
Hordeum vulgare × Triticum aestivum hybrid	Young inflores.*	N6+ → N6+ → N6+ → N6	2,4-D (9) → 2,4-D (0–4.5)	–	± 2,4-D (0.5)/NAA (0.5) + KN (4.6)	–	93
Lasiurus scindicus (Shevan grass)	Mat. embryo*	MS+ → MS+ → MS → 1/2MS	2,4-D (27.2) → 2,4-D (9.1)	Asc acid (0.25) + Suc (131.5)	–	–	94
Lolium multiflorum (Italian rye grass)	Imm. embryo*	MS+ → MS	2,4-D (9) + BA (0.9)	–	–	–	95

Table 1. Continued

Species (Common name)	Explant (*Success in Regeneration)	Media Sequence[a] (Salts)	SE Induction Medium		SE Development Medium		Reference
			Growth Regulators[b] (μM)	Others[c] (mM/g)	Growth Regulators[b] (μM)	Others[c] (mM/g)	
	Imm. inflores.*	MS+ → MS+	2,4-D (33.9)	–	2,4-D (33.9)	–	96
	Mat. and imm. embryo*	MS+/MS+ → 1/2MS+	2,4-D (9 → 31.7) + BAP (0.9)/2,4-D (45.2 → 22.6)	CH (1 g)	BAP (0.4)	Suc (58.4)	75
Lolium multiflorum × L. perenne hybrid	Imm. embryo*	MS+ → 1/2MS+ → 1/2MS+ → 1/2MS	2,4-D (6.8) + KN (10.2) + IAA (37.1)	–	2,4-D (3.2) + Z (4.6) + IAA (21.1) → 2,4-D (2.3)	Suc (43.8)	97
Lolium perenne (Perennial rye grass)	Imm. embryo* and imm. inflores.*	MS+ → MS/MS+ → MS	2,4-D (22.6)	Suc (87.6)	± 2,4-D (9.1)	–	96
	Mat. seed*	1/2MS+ (l) → 1/2MS+	2,4-D (27.2)	CH (3 g) + B5 vits	BA (2.2)	Fluri (0.5 mg) + B5 vits	98
Lolium temulentum (Darnel)	Mat. and imm. embryo* and shoot apices*	MS+ → MS+ → 1/2MS+ → MS	2,4-D (9) → 2,4-D (9)	–	2,4-D (2.3) + IAA (11.4) + KN (4.6)	–	99
	Shoot tip suspension*	MS+ → MS+ → M155 → MS+	2,4-D (27.1) → (13.6)	Suc (87.6)	2,4-D (0.5) + BAP (0.4) → KN (0.9)	Glu (379.7) + Gelrite (3 g) → Suc (87.6)	100
Oryza longistaminata (African wild rice)	Mat. embryo*	MS+ → MS+	2,4-D (9)	–	NAA (0.27) + BAP (2.2)	N6 vits	101
Oryza perennis (Wild rice)	Mat. seed* and young inflores.*	MS+ → MS+	2,4-D (9) + BAP (8.9)	–	IAA (2.9) + NAA (2.7) + BAP (17.8)	CH (0.5 g)	102

Table 1. Continued

Species (Common name)	Explant (*Success in Regeneration)	Media Sequence[a] (Salts)	SE Induction Medium Growth Regulators[b] (µM)	SE Induction Medium Others[c] (mM/g)	SE Development Medium Growth Regulators[b] (µM)	SE Development Medium Others[c] (mM/g)	Reference
Oryza sativa (Rice)	Imm. embryo*	MS+ → MS	2,4-D (4.5)	L-Pro (25)	–	–	103
	Imm. embryo*	N6+ (s) → N6+ (s)	2,4-D (4.5)	Suc (87.6) + CH (0.3 g) + L-Pro (10)	NAA (5.4) + KN (23.2)	Suc (87.6)	104
		N6+ (s) → 1/2N6+ (l)	2,4-D (4.5)	Suc (87.6) + CH (0.3 g) + L-Pro (10)	NAA (0.05) + 4-PU (0.1 mg)	Suc (131.5)	
	Imm. leaf*	MS+ → MS	2,4-D (4.7)	–	–	–	105
	Leaf base callus	KPR → N6 → MS	n.r.	n.r.	–	Suc (233.7)	106
	Mat. and imm. embryo* protoplasts*	MS+/CC+ → MS+/ CC+ → CC → 1/2MS	2,4-D (9)	CH (1 g)	Z (0.2) + IAA (5.7)/IAA (2.9) + NAA (2.7) + BAP (17.8)	–	107
	Mat. and imm. embryo*	R2+ → N6+ → N6+ → N6	2,4-D (9) → 2,4-D (9)	Suc (400) → Suc (175.3)	KN (9.3/23.2)/Z (9.1/22.8)	Suc (175.3)	108
	Mat. caryopses*	MS+ → MS+ → MS+ → MS	2,4-D (9) + BA (1.1) → 2,4-D (4.5) + BA (1.1)	–	BA (1.1)	–	109
	Mat. embryo derived callus protoplasts*	N6+ → MS+ → MS	2,4-D (6.8)	L-Pro (8.7) + Suc (400 → 175)	KN (9.3) + NAA (5.4)	Suc (87.6) → Suc (58.4)	110
	Mat. seed*	LS+ → LS+	2,4-D (4.5/11.3/22.6)	L-Trypt (0.5)	BA (0.4-2.2)/TIBA (0.2-1.0)	–	111
	Mat. seed*	LS+ → MS+ → MS	2,4-D (11.3)	Suc (87.6)	NAA (0.3) + BAP (2.2)	–	112

Table 1. Continued

Species (Common name)	Explant (*Success in Regeneration)	Media Sequence[a] (Salts)	SE Induction Medium		SE Development Medium		Reference
			Growth Regulators[b] (µM)	Others[c] (mM/g)	Growth Regulators[b] (µM)	Others[c] (mM/g)	
	Mat. seed*	MS+ →, MS+ →, MS+	2,4-D (4.5) + KN (4.6) → 2,4-D (2.3) + ABA (0.5)	Suc (175.3) → Suc (87.6)	NAA (10.7) + KN (9.3)	Suc (87.6)	113
	Mat. seed*	MS+ →, MS+ → MS	2,4-D (9.1)	–	2,4-D (4.5)	–	114
	Mat. seed*	MS+ →, MS+	2,4-D (10)	Suc (58.4)	IAA (2.8) + KN (23)	CW (10%) + L-Pro (3–12)/L-Trypt (0.24)	115
	Mat. seed*	MS → B5+ → N6+	2,4-D (9)	CH (0.3g) + Suc (58.4)	BA (8.9) + NAA (2.7)	CH (1 g) + Suc (87.6)	116
	Root segment*	MS+ →, MS+ →, MS+	2,4-D (9) → 2,4-D (4.5–9)	CH (2 g) → CH (1 g) + Suc (58.4)	KN (2.3)	CH (2 g)	117
	Root tissue*	MS+ →, MS+ → MS	2,4-D (1–10) + NAA (1–10)	my-inos (0.6) + CH (0.2 g)	2,4-D (2.5) + KN (1)	–	118
	Root tissue*	MS+ →, MS+ →, MS+	2,4-D (13.6) → 2,4-D (9)	CH (2 g) + Suc (87.6)	2,4-D (0.1) + KN (46.5)	–	119
	Root tissue*	MS+ →, MS+	2,4-D (13.6)	CH (2 g) + my-inos (1.1)	2,4-D (0–0.9) + KN (0–46.5)	CH (2 g) + my-inos (1.1)	120
	Seedling*	MS+ → MS → MS	2,4-D (10/20) + BA (0–0.25)/KN (0–0.25)	CH (0.2 g)	–	–	121

Table 1. Continued

Species (Common name)	Explant (*Success in Regeneration)	Media Sequence[a] (Salts)	SE Induction Medium Growth Regulators[b] (μM)	SE Induction Medium Others[c] (mM/g)	SE Development Medium Growth Regulators[b] (μM)	SE Development Medium Others[c] (mM/g)	Reference
Oryza sativa × Oryza latifolia hybrid	Imm. inflores.*	HE+ → HE+/MS+ → HE+/MS+	2,4-D (9) ± NAA (10.7) ± KN (13.9)	± YE (1.36 g) ± CH (0.3 g) + Suc (87.6–175.3)	NAA (10.7) + KN (9.3)	Suc (87.6/175.3)	122
Oryza sativa (Rice)	Young inflores.*	MS + (l) → MS+	2,4-D (4.5–31.7)	CM (5%)	BA (4.4)	–	123
	Young inflores.* and mat. seed*	MS+ → MS+	KN (9.3)/NAA (10.7) + 2,4-D (9.1)	Suc (87.6)	ABA (0.5)	Suc (87.6)	124
	Young inflores.*	LS+ → LS+ → 1/2LS → 1/4LS	2,4-D (4.5/9/11.3)	Suc (87.6)	IAA (2.3–2.9) + KN (9.3)/BAP (4.4)	± CM (2.5%)	125
	Young inflores.*	N6+ → N6	2,4-D (9)	CH (0.5 g)	–	–	126
	Young inflores.*, stem node*, anthers*	MS+/N6+ → MS+	2,4-D (9.1) ± KN (4.6)	Suc (58.4/87.6) ± Man (164.7) + CM (10%)	NAA (5.4/10.7) + KN (9.3/2.3) ± BA (8.9)	Suc (87.6)	127
	Young leaf base*	LS+ → MS+ → KPR+ → MS/N6 → MS	2,4-D (11.3) → NAA (0.3) + BAP (2.2) → 2,4-D (2.3)	Glu (555.1)	–	Suc (233.7)	128
Otatea acuminata aztecorum (Mexican weeping bamboo)	Mat. embryo*	MS+ → B5	2,4-D (13.6) + BAP (2.2)	Suc (58.4)	–	–	129

Table 1. Continued

Species (Common name)	Explant (*Success in Regeneration)	Media Sequence[a] (Salts)	SE Induction Medium		SE Development Medium		Reference
			Growth Regulators[b] (μM)	Others[c] (mM/g)	Growth Regulators[b] (μM)	Others[c] (mM/g)	
Panicum bisulcatum	Mesocotyl*	MS+ → MS+ → MS/MS+	2,4-D (11.3–45.2) → 2,4-D (0.5)	Suc (87.6) → Suc (29.2)	± Z (9.1)	Suc (58.4)	130
Panicum maximum (Guinea grass)	Imm. embryo	MS+	2,4-D (22.6)	CM (5%)	n.d.	n.d.	131
	Imm. embryo* and young inflores.*	MS+ (l) → KM+ (l) → MS+ (s) → MS	2,4-D (9) → 2,4-D (1.1–22.6)	CM (5%)	2,4-D (1.1–4.5)	–	132
	Imm. embryo* and young inflores.*	MS+ (l) → MS+ (s) → MS → 1/2MS	2,4-D (4.5)	CW (2.5%)	2,4-D (0.9)/NAA (1.1)	–	133
	Imm. embryo* and young inflores.*	MS+ → MS+ → 1/2MS	2,4-D (11.3–45.2)	–	± 2,4-D (0.9)	CM (15%)	134
	Imm. embryo*	MS+ → MS+ → MS+ → 1/2MS	2,4-D (45.2) → 2,4-D (4.5)	CW (10%) → CW (5%)	GA$_3$ (2.9) + KN (4.6)	CW (5%)	135
	Leaf tissue*	MS+ → MS+ → 1/2MS	2,4-D (11.3–45.2)	CW (5–15%)	GA (2.9)	–	136
Panicum miliaceum (Proso millet)	Mat. and imm. embryo*	LS+ → KM+ → LS	2,4-D (11.3) → 2,4-D (4.5) + BAP (4.4)	–	–	–	137

Table 1. Continued

Species (Common name)	Explant (*Success in Regeneration)	Media Sequence[a] (Salts)	SE Induction Medium Growth Regulators[b] (µM)	Others[c] (mM/g)	SE Development Medium Growth Regulators[b] (µM)	Others[c] (mM/g)	Reference
	Mat. seed*	KM+ (l) → LS+ (s) → LS	2,4-D (4.5) + BAP (4.4) → 2,4-D (11.3)	Suc (116.9)	–	–	138
	Young inflores.*	MS+ → MS+ → 1/2MS	2,4-D (11.3)	CW (5%)	2,4-D (0.9)	Suc (58.4)	139
Panicum miliare (Little millet)	Young inflores.*	MS+ → MS+ → 1/2MS	2,4-D (11.3)	CW (5%)	2,4-D (0.9)	Suc (58.4)	139
Paspalum notatum (Bahia grass)	Caryopses*	MS+ → MS	2,4-D (4.5)	Suc (58.4)	–	–	140
	Mat. and imm. embryo*	MS+ → MS+ → MS	2,4-D (4.5)	–	NAA (0.5/5.4) + BAP (4.4/13.3)	Suc (233.7)	141
	Mat. seed*	MS+ → MS	2,4-D (9.1)	Suc (58.4)	–	–	142
Paspalum scrobiculatum (Kodo millet)	Young inflores. and young leaf derived protoplasts	KM+ → M2 → MS+ → MS+ → MS+ → 1/2MS	2,4-D (4.5) → 2,4-D (9.1) + NAA (5.4) + BAP (4.4) → 2,4-D (2.3)	CM (2%) → CM (5%)	BAP (8.9–22.2) + NAA (1.3–2.7)	–	143
	Young inflores.* and young leaf base*	MS+ (l) → MS+ (s) → MS+	2,4-D (9.1)	CM (5%)	NAA (1.3–2.7) + BAP (8.9)	–	144
Pennisetum americanum (Pearl millet)	Imm. embryo	MS+	2,4-D (1.1–11.3)	± CW (5%)	n.d.	n.d.	145
	Imm. embryo*	LS+ → LS	2,4-D (11.3)	CW (5%)	–	–	146

Table 1. Continued

Species (Common name)	Explant (*Success in Regeneration)	Media Sequence[a] (Salts)	SE Induction Medium		SE Development Medium		Reference
			Growth Regulators[b] (μM)	Others[c] (mM/g)	Growth Regulators[b] (μM)	Others[c] (mM/g)	
	Imm. embryo*	MS+ (s/l) → MS+ → MS	2,4-D (11.3)	CM (5%)	ABA (0.04–0.08)/Z (4.6)/2-iP (4.9)	CM (5%)	147
	Imm. embryo*	MS+ → MS+ → MS+	2,4-D (11.3)	-	IAA (5.7) → IAA (1.2) + KN (4.6)	Adenine (0.3)	148
	Young inflores.*	MS+ → MS+ → MS	2,4-D (11.3)	± CM (5%)	ABA (0.04)	Suc (175.3) + Amm. malate (0.4 g)	148
	Mat. embryo*	MS+ → MS+ → MS+ → 1/2MS+	2,4-D (22.6/45.2/67.9)	CM (5, 10, 15%)	2,4-D (0.9) → 2,4-D (0.9) + KN (0.9) → NAA (1.1) + 2,4-D (0.9)	-	149
Pennisetum americanum × *P. purpureum* hybrid	Young inflores.*	MS+ → MS+ → MS	2,4-D (11.3)	± CM (5%)	ABA (0.04)	Suc (175.3) + Amm. malate (0.4 g)	148
Pennisetum americanum (Pearl millet)	Young inflores.	MS+ → MS+ → MS+	2,4-D (2.3–22.6)	± CM (5%)	2,4-D (0.9)	-	150
	Young inflores.*	MS+ (l) → MS+ (s) → MS+ (s)	2,4-D (11.3) → 2,4-D (1.1/2.3/4.5)	CM (5%)	ABA (0.04/0.4)	-	151
	Young inflores.*	MS+ → MS+ → 1/2MS+	2,4-D (1.1) + KN (1.2)	-	2,4-D (0.6) → NAA (1.1)	-	152
	Imm. inflores.*	MS+ → MS+ → MS+ → 1/2MS+	2,4-D (22.6) + KN (4.6) → 2,4-D (2.3) + BA (2.2) + NAA (5.4)	CW (5%)	BA (4.4) + IAA (0.57) → NAA (0.5)	-	153

Table 1. Continued

Species (Common name)	Explant (*Success in Regeneration)	Media Sequence[a] (Salts)	SE Induction Medium Growth Regulators[b] (μM)	SE Induction Medium Others[c] (mM/g)	SE Development Medium Growth Regulators[b] (μM)	SE Development Medium Others[c] (mM/g)	Reference
Pennisetum purpureum (Napier/Elephant grass)	Inflores. derived protoplasts	KM+ → MS+ → MS+ → MS+	2,4-D (2.3) + Z (1.1) ± NAA (5.4)	Glu (400)	2,4-D (0.5) + BAP (0.4) + NAA (1.1/2.7/5.4) → NAA (1.1)	–	154
	Leaf tissue*	MS+ → MS+ → 1/2MS	2,4-D (2.3) + BAP (2.2) + NAA (5.4)	CM (5%)	GA (0.3)	–	155
	Leaf tissue*	MS+ → MS+ → MS	2,4-D (2.3–45.2)	CM (5%)	2,4-D (2.3) + BAP (2.2) + NAA (5.4)	CM (5%)	156
	Anther*	MS+ → MS+ → MS	2,4-D (2.3) + BAP (2.2) + NAA (5.4)	–	2,4-D (2.3) + BAP (2.2) + NAA (5.4)	CM (5%)	156
	Young inflores.*	N6+ → MS+ → MS/MS+	2,4-D (4.5) + BAP (2.2) + NAA (5.4) → 2,4-D (11.3) + GA (2.9)	CM (5%)	± GA (2.9)	–	157
Pennisetum typhoides (Bulrush millet)	Young inflores.*	N6+ → N6	2,4-D (20/50)	Suc (58.4)	–	–	158
Poa pratensis (Kentucky blue grass)	Mat. seed* and young inflores.*	MS+ → MS+ → MS	2,4-D (9.1) → 2,4-D (0.9)	–	–	–	159
Saccharum berberi (Wild sugarcane)	Leaf section* and young stem*	MS+ → MS+ → MS	2,4-D (2.3–13.6)	CM (5%) + CH (0.5 g) + Argn (0.3)	2,4-D (4.5) → 2,4-D (9.1) + KN (2.3)	CM (5%) + CH (0.5 g) + Act. char (1 g) + Argn (0.3) → CH (2 g)	160
Saccharum officinarum (Sugarcane)	Leaf segment*	MS+ → MS+ → MS	2,4-D (31.7) → 2,4-D (4.5)	–	–	–	161

Table 1. Continued

Species (Common name)	Explant (*Success in Regeneration)	Media Sequence^a (Salts)	SE Induction Medium Growth Regulators^b (μM)	SE Induction Medium Others^c (mM/g)	SE Development Medium Growth Regulators^b (μM)	SE Development Medium Others^c (mM/g)	Reference
	Leaf tissue derived protoplasts	MS+ → MS+	2,4-D (9.1)	Suc (87.6–146.1)	BA (8.9) + IAA (2.9) + IBA (2.5) + NAA (2.7–5.4)	–	162
	Leaf section* and young stem*	MS+ → MS+ → MS	2,4-D (2.3–13.6)	CM (5%) + CH (0.5 g) + Argn (0.3)	2,4-D (4.5) → 2,4-D (9.1) + KN (2.3)	CM (5%) + CH (0.5 g) + Act. char (1 g) + Argn (0.3) → CH (2 g)	160
Saccharum officinarum × Pennisetum americanum (Intergeneric hybrid)	Cell suspension protoplasts	MS+ → MS+	2,4-D (0.5–1.1) → 2,4-D (0.02–0.23) ± BAP (0.4–2.2)	Act. char (10 g)	n.d.	n.d.	163
Saccharum officinarum (Sugarcane)	Young leaf* and apical meristem*	MS+ → MS+ (s/l) → 1/2MS	2,4-D (13.6) → 2,4-D (2.3)	CW (10%)	–	CW (10%)	164
	Young leaf*	MS+ → MS+ → 1/2MS	2,4-D (2.3–13.6)/ 2,4-D (1.1–2.3)	CM (5%) + Suc (87.6–233.7)	Z (0.01–0.09)/ GA(2.9)	Suc (175.3)	165
	Young leaf*	MS+ → MS+ → 1/2MS	2,4-D (9–13.6) → 2,4-D (1.1–9)	CH (0.5 g) + CW (5%)	–	Suc (175.3)	166
	Young leaf*	MS+ → MS+ → MS+	2,4-D (1.1–4.5)	Act. char (1%)	BA (2.2) ± Z (1.1–2.3)/2-iP (1.2–2.5) → BA (2.2)	Fluri (0.5 mg)	167
	Young leaf*	MS+/N6+ → MS+/N6+ → MS	2,4-D (12.5→4.5)	+ Suc (5%)	–	CW (10%)	168

Table 1. Continued

Species (Common name)	Explant (*Success in Regeneration)	Media Sequence[a] (Salts)	SE Induction Medium Growth Regulators[b] (μM)	SE Induction Medium Others[c] (mM/g)	SE Development Medium Growth Regulators[b] (μM)	SE Development Medium Others[c] (mM/g)	Reference
Saccharum robustum (Wild sugarcane)	Leaf section* and young stem*	MS+ → MS+ → MS	2,4-D (2.3–13.6)	CM (5%) + CH (0.5 g) + Argn (0.3)	2,4-D (4.5) → 2,4-D (9.1) + KN (2.3)	CM (5%) + CH (0.5 g) + Act. char (1 g) + Argn (0.3) → CH (2 g)	160
Saccharum sinense (Natural hybrid sugarcane)	Leaf section* and young stem*	MS+ → MS+ → MS	2,4-D (2.3–13.6)	CM (5%) + CH (0.5 g) + Argn (0.3)	2,4-D (4.5) → 2,4-D (9.1) + KN (2.3)	CM (5%) + CH (0.5 g) + Act. char (1 g) + Argn (0.3) → CH (2 g)	160
Saccharum spontaneum (Wild perennial sugarcane)	Leaf section* and young stem*	MS+ → MS+ → MS	2,4-D (2.3–13.6)	CM (5%) + CH (0.5 g) + Argn (0.3)	2,4-D (4.5) → 2,4-D (9.1) + KN (2.3)	CM (5%) + CH (0.5 g) + Act. char (1 g) + Argn (0.3) → CH (2 g)	160
Saccharum spp. (Sugarcane)	Spindle leaf*	MS+ → MS+ → 1/2MS	2,4-D (13.6)	Argn (0.34) + Suc (58.4)	KN (4.6) + NAA (5.4)	CM (10%) + Suc (58.4)	169
Schizachyrium scoparium (Little bluestem)	Young inflores.*	LS+ → LS → LS+	2,4-D (22.6)	–	KN (4.7–23.2)	–	170
Secale africanum (Wild rye)	Imm. embryo*	MS+ → MS+ → MS+	2,4-D (2.8–22.6) → 2,4-D (2.3–9)	CH (0.75 g)	(Z (4.6) + KN (4.6))/GA$_3$ (0.3)	–	171
Secale ancestrale (Wild rye)	Imm. embryo*	MS+ → MS+ → MS+	2,4-D (2.8–22.6) → 2,4-D (2.3–9)	CH (0.75 g)	(Z (4.6) + KN (4.6))/GA$_3$ (0.3)	–	171
Secale cereale (Rye)	Imm. embryo*	CC+ → MS+/CC+	Dica (30)	CW (5%) + Suc (87.6)	KN (4.5)	–	172

Table 1. Continued

Species (Common name)	Explant (*Success in Regeneration)	Media Sequence[a] (Salts)	SE Induction Medium		SE Development Medium		Reference
			Growth Regulators[b] (µM)	Others[c] (mM/g)	Growth Regulators[b] (µM)	Others[c] (mM/g)	
	Imm. embryo*	MS+ → MS+ → MS+ MS/MS+	2,4-D (11.3) → 2,4-D (3.4–9.1)	–	± (BAP (4.4)/KN (4.6))/GA$_3$ (2.9)/2-iP (4.9)	–	173
	Imm. embryo*	MS+ → MS+	2,4-D (4.5)	–	2,4-D (4.5) + Z (4.6) CH (1 g) + Glut (5) + KN (0.5–4.6) + BAP (2.2–8.9)	–	174
	Leaf tissue*	MS+ → MS+/MS	2,4-D (2.3–4.5)	–	± 2,4-D (2.3–4.5)	–	175
	Young inflores.*	MS+ → MS+	2,4-D (2.3/4.5/6.8/9)	–	–	–	176
Secale kuprianovii (Wild rye)	Imm. embryo*	MS+ → MS+ → MS+	2,4-D (2.8–22.6) → 2,4-D (2.3–9)	CH (0.75 g)	(Z (4.6) + KN (4.6))/GA$_3$ (0.3)	–	171
Secale segetale (Wild rye)	Imm. embryo*	MS+ → MS+ → MS+	2,4-D (2.8–22.6) → 2,4-D (2.3–9)	CH (0.75 g)	(Z (4.6) + KN (4.6))/GA$_3$ (0.3)	–	171
Secale vavilovii (Wild rye)	Imm. inflores.*	MS+ → MS	2,4-D (2.3/9.1)	–	–	–	177
	Imm. embryo*	MS+ → MS	2,4-D (4.5/9.1)	–	–	–	177
	Young leaf*	MS+ → MS	2,4-D (4.5/9.1/22.6)	–	–	–	177
	Imm. embryo*	MS+ → MS+ → MS+	2,4-D (2.8–22.6) → 2,4-D (2.3–9)	CH (0.75 g)	(Z (4.6) + KN (4.6))/GA$_3$ (0.3)	–	171
Setaria italica (Foxtail millet)	Young inflores.*	MS+ → MS+ → MS+	2,4-D (9) + BAP (0.9–2.2)/KN (0.9–2.3)	–	BAP (8.9) + NAA (2.7)	–	178
Sinocalamus latiflora (Bamboo)	Mat. embryo*	MS+ → MS+	2,4-D (27.1) + KN (13.9)	PVP (0.25 g) + Suc (146.1)	2,4-D (13.6) + KN (9.3)	–	179

Table 1. Continued

Species (Common name)	Explant (*Success in Regeneration)	Media Sequence[a] (Salts)	SE Induction Medium Growth Regulators[b] (μM)	SE Induction Medium Others[c] (mM/g)	SE Development Medium Growth Regulators[b] (μM)	SE Development Medium Others[c] (mM/g)	Reference
Sorghum almum (Columbus grass)	Young inflores.*	MS+ → MS+ → 1/2MS+	2,4-D (9) + Z (0.5)	–	KN (2.3)/BA (4.4) → NAA (0.5)	–	180
Sorghum arundinaceum (Sudan grass)	Young inflores.*	MS+ → MS+ → 1/2MS+	2,4-D (0.05–22.6)	CW (5%)	2,4-D (4.5) → NAA (2.7)	CW (5%) → Suc (58.4)	181
Sorghum bicolor (Sorghum)	Imm. embryo* and mat. embryo*	MS+ → MS	2,4-D (11.6–23.4)	± CW (10%)	–	–	182
	Imm. inflores.*	MS+ → MS	2,4-D (0.9–4.5) + BA (0.3–4.7) + KN (0.5–4.7)	–	–	–	183
	Imm. inflores.*	MS+ → MS+ → MS+ → 1/2MS+	2,4-D (9.1) + Z (0.5) → (2,4-D (0.5)/TIBA (0.2)) + KN (23.2/46.5)/Z (4.6/9.1)/BA (4.4/8.9)	–	KN (0.5/2.3)/BA (4.4) → NAA (0.5)	–	184
	Imm. leaf*	MS+ → MS+/MS	2,4-D (9)	Suc (116.9)	± KN (0.5)/BAP (0.4)	–	185
	Leaf segment*	MS+ → MS+/MS	2,4-D (9) + KN (0.5)	Suc (116.9)	± KN (0.5)	–	186
	Leaf tissue*	MS+ → N6+ → 1/2MS	2,4-D (9.1) → NAA (0.5) + Z (2.3)	Suc (58.4)	–	–	187
	Mat. and imm. embryo*	LS+ → LS+ → LS+	2,4-D (9) + KN (2.3)	Suc (116.9)	IAA (5.7) + BA (2.2) → IBA (14.8)	–	188

Table 1. Continued

Species (Common name)	Explant (*Success in Regeneration)	Media Sequence^a (Salts)	SE Induction Medium		SE Development Medium		Reference
			Growth Regulators^b (μM)	Others^c (mM/g)	Growth Regulators^b (μM)	Others^c (mM/g)	
	Seedling*	MS+ → MS	2,4-D (11.3) + KN (1.2–46.5)/BA (1.1–44.4)	–	–	–	189
	Shoot tip*	MS+ → MS	2,4-D (11.3) + BA (2.2, 11.1, 22.2)	–	–	–	189
Sorghum halepense (Johnson grass)	Imm. inflores.*	MS+ → MS+ → 1/2MS+	2,4-D (9.1) + Z (0.5)	–	KN (0.5)/BA (0.4) → NAA (0.5)	–	190
Tripsacum dactyloides (Gama grass)	Mat. embryo*	MS+ → MS+ → MS	Dica (10/20) → Dica (10)	Suc (87.6/175.3) → Suc (87.6)	–	Suc (58.4)	191
Triticosecale (Triticale)	Imm. embryo*	MS+ → MS+	2,4-D (4.5)	–	2,4-D (4.5) + Z (4.6) + KN (0.5–4.6) + BAP (2.2–8.9)	CH (1 g) + Glut (5)	174
	Imm. embryo*	MS+/Ka+ → MS+ → 1/2MS	2,4-D (36)	L-Pro (25) + L-Glut, L-Ala, L-Argn (7.5) + CM (1%/20%)	KN (4.6)	–	192
Triticum aestivum (Wheat)	Imm. embryo derived protoplasts*	MS+ → MS+ (I) → MS+/KM+ → MS → MS+/ 1/2MS+	2,4-D (9.1) → 2,4-D (2.3/4.5)	my-inos (0.56) → MES (0.6 g) + Glu (4.11)	IAA (5.7) + Z (4.6) → ± IAA (2.9/5.7) ± KN (0.1)	± Act. char (1%)	193
	Imm. embryo	MS+ → MS+	2,4-D (2.3/4.5)	L-Aspn (1.1) + MgCl$_2$ (3.7)	2,4-D (0.5)	L-Aspn (1.1) + MgCl$_2$ (3.7)	194
	Imm. embryo	MS+	2,4-D (4.5) ± ABA (0.4)	CM (10%)	n.d.	n.d.	195

441

Table 1. Continued

Species (Common name)	Explant (*Success in Regeneration)	Media Sequence[a] (Salts)	SE Induction Medium Growth Regulators[b] (μM)	Others[c] (mM/g)	SE Development Medium Growth Regulators[b] (μM)	Others[c] (mM/g)	Reference
	Imm. embryo	MS+/2MS+ /1/2MS+ → 1/2macMS+ /1/2macMS	2,4-D (10)	Suc (87.6)	± 2,4-D (0.01-0.05)	–	196
	Imm. embryo* and young inflores.*	MS+ → MS	2,4-D (4.5)	CM (10%)	–	–	197
	Imm. embryo* and young inflores.*	MS+ → MS+ → 1/2MS	2,4-D (9)	Suc (58.4)	2,4-D (0.9) + 2-iP (0.5)/Z (4.6) + IAA (5.7)	–	198
	Imm. embryo*	1/2macMS+ → 1/2mac-MS+	2,4-D (1–40)	–	± NAA (2) ± KN (3)	–	199
	Imm. embryo*	1/2macMS+ → 1/2mac-MS+	2,4-D (10)	–	2,4-D (0.01)	–	200
	Imm. embryo*	1/2macMS+ → 1/2mac-MS+ → 1/2macMS+	2,4-D (10) → 2,4-D (3)	–	2,4-D (0.01)	–	201
Transgenic wheat	Imm. embryo*	2MS+ → MS+ → MS+	2,4-D (9.1) → (4.5)	Suc (87.6) + CH (0.1 g) + Glut (3.4) → my-inos (0.6) + Suc (58.4)	Z (4.6) + IAA (5.7)	–	202
Triticum aestivum (Wheat)	Imm. embryo*	2MS+ → MS+	2,4-D (4.5–9)	Suc (58.4) + CH (0.1–0.2 g) + Glut (3.4)	ABA (0.1)	–	203

Table 1. Continued

Species (Common name)	Explant (*Success in Regeneration)	Media Sequence[a] (Salts)	SE Induction Medium Growth Regulators[b] (μM)	Others[c] (mM/g)	SE Development Medium Growth Regulators[b] (μM)	Others[c] (mM/g)	Reference
	Imm. embryo*	2MS+ → MS+	2,4-D (9.1)	Suc (87.6) + CH (0.1 g) + Glut (3.4)	IAA (5.7) + Z (4.6)	-	204
	Inflorescence*	MS+ → MS+	2,4-D (4.5)	Suc (87.6)	2,4-D (0.9) + 2-iP (4.9)	-	204
	Anthers*	P2 → MS+	-	-	NAA (2.7) + KN (2.3)	-	204
	Imm. embryo*	MS+ → MS	2,4-D (4.5)	Suc (87.6)	-	Suc (29.2)	205
	Imm. embryo*	MS+ → MS+ → 1/2MS	2,4-D (9.1)	-	KN (4.7)	Suc (58.4)	206
	Imm. embryo*	MS+ → MS+ → MS	2,4-D (4.5/13.5)	Suc (175.3) + Aspn (1130)	2,4-D (0.45)/KN (23)	-	207
	Imm. embryo*	MS+ → MS → MS	2,4-D (9.1)/2,4-D (22.6) → NAA (26.9)	± Adenine (0.4)	-	-	208
	Imm. embryo*	MS+ → MS+	2,4-D (0.9/1.8) + BAP (0.9)	Suc (58.4) + CH (0.1 g)	2,4-D (0.9)	Suc (58.4)	209
	Imm. embryo*	MS+ → MS+	2,4-D (9) ± ABA/ABA anal (1.9)	± DMSO (1-3 g)	2,4-D (2.3)	-	210
	Imm. embryo*	MS+ → MS+/MS	2,4-D (4.5)	NaCl (40)/KCl (40)	+ 2,4-D (2.3/4.5)	-	211
	Imm. embryo*	N6+ → N6+	Dica (1/10 mg)	-	Dica (1/10 mg)	-	212
	Imm. inflores.	N6+ → N6+	2,4-D (9)	Suc (146.1)	NAA (1.1) + KN (2.3)/2,4-D (0.5-2.3)	Suc (146.1)	213
	Leaf base*	MS+ → MS	2,4-D (1-50)	-	-	-	214
	Mat. embryo*	LS+ → LS	2,4-D (90.5)	-	-	-	215

Table 1. Continued

Species (Common name)	Explant (*Success in Regeneration)	Media Sequence[a] (Salts)	SE Induction Medium		SE Development Medium		Reference
			Growth Regulators[b] (μM)	Others[c] (mM/g)	Growth Regulators[b] (μM)	Others[c] (mM/g)	
	Mat. embryo*	MS+ → MS	2,4-D (10)	-	-	± AGN (0.06)	216
	Shoot meristem*	MS+ → MS/MS+	2,4-D (9)	-	± BAP (0.4–4.4)	-	217
Triticum aestivum × Leymus angustus hybrids (TLF1 hybrids)	Inflorescence* and leaf segment*	MS+ → MS+	2,4-D (2.3/4.5/9.1)	CM (5%)	2,4-D (2.3)	CM (5%)	218
	Leaf and inflores. cell suspension*	MS+ → MS	2,4-D (2.3/4.5)	CH (5 g) + CW (5%) + Suc (87.6)	± 2,4-D (1.1/0.5/0.05)	-	219
Triticum aestivum (Wheat)	Young inflores. derived protoplasts*	MS+ → N6 → MS+ → MS+	2,4-D (9.1)	Gly (26.6) + Glut (1.0) + CH (0.2 g)	2,4-D (0.9–9.1) → Z (4.6) + IAA (5.7) → IAA (2.9)	-	220
	Young inflores.*	MS+ → MS/MS+	2,4-D (0.45)	CH (0.02 g) + CM (10%)	2,4-D (0.25/5) ± KN (0.1)	-	221
	Imm. embryo	MS+	2,4-D (3.62)/Dica (9.05) + KN (2.56/4.65)	L-Aspn (1) + Suc (0.06)	-	-	222
	Imm. embryo*	2MS+ → MS+	2,4-D (4.52)	CH (0.1 g) + Suc (58.43)	2,4-D (4.52) + KN (0.46)	L-Aspn (1) + L-Arg (0.23) + L-Trypt (0.2) + Suc (58.43)	223
Triticum durum (Durum wheat)	Imm. embryo*	MS+ → MS+ → MS	2,4-D (9) → 2,4-D (4.5)	my-inos (1.1) + L-Aspn (1.5) + Suc (73)	-	-	224
	Imm. embryo*	MS+ → MS+ → 1/2MS	2,4-D (9.1)	-	KN (4.7)	Suc (58.4)	206

Table 1. Continued

Species (Common name)	Explant (*Success in Regeneration)	Media Sequence[a] (Salts)	SE Induction Medium		SE Development Medium		Reference
			Growth Regulators[b] (μM)	Others[c] (mM/g)	Growth Regulators[b] (μM)	Others[c] (mM/g)	
Triticum durum × Elytrigia intermedium hybrid (Trititigia)	Imm. inflores.*	KM+ → MS+ → N6+/B5+ → MS	2,4-D (4.5) + KN (0.9) → 2,4-D (9.1) → 2,4-D (4.5/2.3) + KN (0.9/9.3) ± BA (2.2) ± IAA (5.7)	Glu (650.2) + Suc (14.6) + CH (0.1 g) → Suc (87.6) + CH (0.3 g) + Act. Char (1 g) + Glut (1.4)	-	CH (0.3 g) + Glut (1.4)	225
Triticum aestivum (Wheat)	Imm. embryo*	MS → MS+ → MS+ → 1/2MS/1/2M-S+	2,4-D (9.1)	CH (0.1 g) + Glut (3.4)	IAA (5.7) + Z (4.6)	Suc (58.4) → Suc (29.2)	226
Triticum turgidum (Wheat)	Imm. embryo* and young inflores.*	MS+ → MS	2,4-D (2.3/4.5/9)	-	-	-	81
Triticum turgidum × Hordeum chilense hybrid (Tritordeum)	Imm. embryo* and young inflores.*	MS+ → MS	2,4-D (2.3/4.5/9)	-	-	-	81
Urochloa panicoides (Fodder grass)	Imm. inflores.*	MS+ → MS+ → 1/4MS	2,4-D (18.1)	KM vits + L-Glut (0.5) + Asc acid (0.13) + Suc (146.1)	KN (4.6)	-	227
Zea diploperennis	Imm. embryo* and imm. leaf*	MS+ → MS+ → MS+	2,4-D (4.5-27.1)	Suc (58.4/175.3) + CM (5%)	2,4-D (0/1.1) ± KN (1.2) → NAA (1.3)	Suc (58.4)	228
Zea mays (Maize)	Imm. embryo	KM+	2,4-D (0.5-0.9) + ABA (0.8)	-	-	-	229
	Imm. embryo	MS+ → N6+	2,4-D (2.3-4.5) → 2,4-D (6.8)	Suc (58.4-175.3) → CH (0.2 g) + Suc (58.4)	n.r.	n.r.	230
	Imm. embryo	MS+	2,4-D (9)	Suc (175.3)	n.d.	n.d.	231

Table 1. Continued

Species (Common name)	Explant (*Success in Regeneration)	Media Sequence[a] (Salts)	SE Induction Medium Growth Regulators[b] (μM)	Others[c] (mM/g)	SE Development Medium Growth Regulators[b] (μM)	Others[c] (mM/g)	Reference
	Imm. embryo	MS+/N6+	2,4-D (4.5)	Suc (58.4/175.3) ± CA (0.1 g) ± L-Pro (6)	n.d.	n.d.	232
	Imm. embryo	MS+/N6+	2-CPA (100)/3-CPA (10)/4-CPA (1)	–	n.r.	n.r.	233
	Imm. embryo*	CC+ → YP+	2,4-D (9)	Suc (58.4) + CW (10%) + Man (36.4)	2,4-D (4.5)	CH (0.5g) + L-Pro (3.5) + Suc (350.6)	234
	Imm. embryo*	MS+ → 1/2MS/MS/ MS+ → 1/2MS	2,4-D (1.1–9)	Suc (175.3/350.6)	± GA (2.9)	Suc (58.4/87.6)	235
	Imm. embryo*	MS+ → MS	2,4-D (2.3)	–	–	–	236
	Imm. embryo*	MS+ → MS+ → 1/2MS	2,4-D (1.1–2.3)	Suc (175.3/262.9/ 350.6)	GA (2.9)	–	237
	Imm. embryo*	MS+ → MS+ → MS	2,4-D (4.5)	CM (5%) + Suc (175.3) → CH (0.5 g) + Suc (58.4)	–	–	238
	Imm. embryo*	MS+ → MS+ → MS	2,4-D (22.6–45.2) + KN (0–2.3)	Suc (146.1) + CH (0.5 g) + Cyste (0.1)	KN (13.9)/Z (13.7)/ GA_3 (2.9) + IAA (5.7)	Suc (58.4)	239
	Imm. embryo*	MS+ → MS+ → MS+ → MS/N6	2,4-D (9) → 2,4-D (9)	Man (200) → Man (300–800) + Suc (58.4) + CW (2%) + Glut (1.7)	2,4-D (9)	Whites vits + Suc (58.4)	240

Table 1. Continued

Species (Common name)	Explant (*Success in Regeneration)	Media Sequence[a] (Salts)	SE Induction Medium		SE Development Medium		Reference
			Growth Regulators[b] (µM)	Others[c] (mM/g)	Growth Regulators[b] (µM)	Others[c] (mM/g)	
	Imm. embryo*	MS+ → MS+	2,4-D (2.3–4.5)	Suc (175.3–350.6)	2,4-D (0.5) + BAP (0.4)/KN (0.5)	Suc (58.4)	241
	Imm. embryo*	MS+ → MS+	2,4-D (3.4)	L-Aspn (1)	n.r.	n.r.	242
	Imm. embryo*	MS+/YP+ → MS	2,4-D (4.5)	L-Pro (3.5) + Suc (350.6)	–	Suc (58.4)	243
	Imm. embryo*	N6+ → MS	2,4-D (4.5)	L-Pro (6) + CA (0.1 g) + Suc (58.4)	–	–	244
	Imm. embryo*	N6+ → MS+ → MS+/N6+ → MS+ → 1/2MS+	2,4-D (9.1/13.6) + NAA (10.7)	Pro (4.3) + CH (0.5 g) + Suc (87.6)	2,4-D (0.9) + KN (18.6/46.5)	CH (0.5 g) + Suc (175.3) → Suc (58.4) + Act. Char (5 g)	245
	Imm. embryo*	N6+ → MS → 1/2MS	2,4-D (2.5/10)	Suc (87.6/175.3)	–	Suc (58.4)	246
	Imm. embryo*	N6+ → N6 → MS	2,4-D (2.3–4.5)	L-Aspn (1.1) + CA (0.1 g) + Suc (58.4)	–	Suc (175.3) → Suc (58.4)	247
	Imm. embryo*	N6+ → N6 → MS+	2,4-D (2.5)	–	IAA (2) + NAA (3)	Suc (175.3) → Suc (58.4)	248
	Imm. embryo*	N6+ → N6 → MS+	2,4-D (2.5)	Suc (350.6)	IAA (2) + NAA (3)	Suc (350.6)	249
	Imm. embryo*	N6+ → N6	2,4-D (2.3)	Suc (58.4) + CH (0.2 g) + L-Pro (20)	–	Suc (175.3)	250
	Imm. embryo*	N6+ → N6	2,4-D (2.5)	Suc (350.6)	–	Suc (175.3)	251
	Imm. embryo*	N6+ → N6	2,4-D (4.5)	L-Pro (20) + Suc (58.4)	–	Suc (175.3)	252
	Imm. embryo*	N6+ → N6	2,4-D (11.3)	Suc (350.6)	–	Suc (175.3)	253

Table 1. Continued

Species (Common name)	Explant (*Success in Regeneration)	Media Sequence[a] (Salts)	SE Induction Medium Growth Regulators[b] (μM)	Others[c] (mM/g)	SE Development Medium Growth Regulators[b] (μM)	Others[c] (mM/g)	Reference
	Imm. embryo*	N6+ → N6/N6+	2,4-D (5.7)	L-Pro (575) + CH (0.1 g) + Suc (58.4)	2,4-D (0.5–0.9)	Suc (116.9–175.3) → Suc (58.4)	254
	Imm. embryo*	N6+/MS+ → N6/MS	2,4-D (2.3) → 2,4-D (4.5/9.1)	CH (0.2 g) + L-Pro (20) + Suc (87.6)	-	Suc (175.3)	255
	Imm. glumes*	MS+ → MS	2,4-D (9.1)	NH_4NO_3 (36)	-	-	256
	Imm. inflores.*	MS+	2,4-D (4.5)	Suc (58.4) + L-Aspn (1.0)	-	-	257
	Imm. embryo*	N6+	2,4-D (5–20)/Chlora/DCBA (10)/ABA (0.5–1)	Suc (262.9–350.6)	n.r.	n.r.	258
	Leaf segment*	SH+ → MS+	Dica (30)	-	NAA (1) + IAA (1) + 2-iP (2)	Suc (175.3)	259
	Mat. embryo*	N6+ → N6+ (l/s) → MS → MS+	2,4-D (9)	L-Pro (6) + CH (0.1 g) + Suc (58.4) + Man (164.7)	-	Suc (175.3) → Suc (58.4)	260
	Root tissue	MS+	2,4-D (4.5)	Suc (58.4)	n.r.	n.r.	261
	Suspension protoplasts* (cryopreserved)	N6+ → N6+	2,4-D (9.1)	Suc (87.6)	BA (4.4) + NAA (1.1)	Act. char (5 g) + Suc (87.6)	262
	Young inflores.*	MS+ → MS+ → MS+	2,4-D (9.1) + ABA (0.003)	CH (0.1 g) + L-Pro (24) + AgN (0.2)	BAP (15.5) → NAA (26.9)	-	263
	Young leaf*	MS+ → MS+ → MS+	2,4-D (9) → 2,4-D (36.2)	-	2,4-D (0.01)/ABA (0.004)	-	264
	Zygotic embryo*	MS+/N6+ → MS	2,4-D (2.3)	Suc (58.4/87.6)	-	Suc (175.3)	265

Table 1. Continued

Species (Common name)	Explant (*Success in Regeneration)	Media Sequence^a (Salts)	SE Induction Medium Growth Regulators^b (µM)	Others^c (mM/g)	SE Development Medium Growth Regulators^b (µM)	Others^c (mM/g)	Reference
Zizania latifolia (Wild rice)	Young inflores.*	MS+ → B5+ → MS/B5	2,4-D (4.5) → KN (4.6) + NAA (0.5) + BA (4.4)	Aspn (0.8) + CLP (0.025 g)/Spermine (20 mg)	–	–	266
Zoysia japonica (Japanese lawn grass)	Mat. seed*	1/2MS+ → N6+ → 1/2MS	Piclo (12.4)	CH (0.5 g) + Suc (58.4)	–	–	267
Hemerocallidaceae							
Hemerocallis spp. (Daylily)	Shoot tip*	MS + (1) → mDS-5a	2,4-D (9)	Suc (87.6) + CW (10%) + my-inos (0.56)	–	Suc (67) + MES (25)	268
	Ovary*	MS+/W+/S-H+ → MS/W/SH	2,4-D (9)	CW (10%)	–	–	269
Iridaceae							
Freesia refracta (Freesia)	Young inflores.* and young leaves*	N6+/MS+ → N6+ → N6	IAA (11.4) + BAP (13.3/2.2)/NAA (2.7)	–	IAA (11.4) + BAP (13.3)	–	270
Gladiolus sp. (Gladiolus)	Inflores. stalk*	MS+ → MS	NAA (53.6)	–	–	–	271
	Cormel slices*	MS+ → MS	2,4-D (2.2/4.5)	–	–	–	271
Iris pseudacorus	Root segment*	MS+ →	2,4-D (4.5/22.5) + NAA (5.4) + KN (0.5)	–	BA (9/22)/KN (5) + TIBA (2)/TIBA (4) + BA (9)	–	272
I. setosa		MS+ →					
I. versicolor (Iris)		MS+					
Iris spp. (Iris)	Shoot apex*	MS+ →MS+	2,4-D (4.5) + BA (0.4)	–	IAA (0.6)	–	22, 23

Table 1. Continued

Species (Common name)	Explant (*Success in Regeneration)	Media Sequence[a] (Salts)	SE Induction Medium Growth Regulators[b] (μM)	SE Induction Medium Others[c] (mM/g)	SE Development Medium Growth Regulators[b] (μM)	SE Development Medium Others[c] (mM/g)	Reference
Liliaceae							
Bellevalia romana (Hyacinthus romanus)	Cotyledon* and mesocotyl*	MS+ (s) → MS+ (l) →MS	2,4-D (4.5/13.6/22.6) → 2,4-D (0.9)	–	–	–	273
Brimeura amethystina	root*, leaf*, bulb scale* and mesocotyls*	MS+ → MS	NAA (10.8) + BAP (2.2)	–	–	Act. Char (1 g)	274
Gasteria verrucosa	Leaf tissue*	MS+ → 1/4MS+/1/4 MS → MS	Piclo (2.1–12.4) + KN (1.2/4.6)	Suc (58.4)	Piclo (0.4) + KN (1.2)	–	275
Haworthia fasciata	Leaf tissue*	MS+ → 1/4MS+/1/4 MS → MS	Piclo (2.1–12.4) + KN (1.2/4.6)	Suc (58.4)	Piclo (0.4) + KN (1.2)	–	275
Scilla indica (Squill)	Imm. anthers*	MS+ → MS+ → MS → 1/2MS	2,4-D (9.1) → NAA (10.7)	CM (15%)	–	–	276
Urginea indica (Indian squill)	Bulb scale*	MS+ → MS+ → MS+ → MS (l)	2,4-D (9)	± CM (15%)	BAP (0.2–0.4) → NAA (0.05) + KN (0.2)	–	277
	Bulb scale*	MS+ → MS+ → MS+ → MS (l)	2,4-D (9)	CM (15%)	BAP (0.2–0.4) → NAA (0.05) + KN (0.2)	CM (10%/15%)	278

Table 1. Continued

Species (Common name)	Explant (*Success in Regeneration)	Media Sequence^a (Salts)	SE Induction Medium Growth Regulators^b (μM)	Others^c (mM/g)	SE Development Medium Growth Regulators^b (μM)	Others^c (mM/g)	Reference
	Bulbs and inflores. segments*	MS+ → MS+	2,4-D (4.5–9.1) ± KN (4.6)	–	± NAA (0.05–0.3) ± KN (0.2)	–	279
Musaceae							
Musa acuminata M. balbisiana & hybrids (Banana/Plantains)	Imm. embryo*	MS+ → MS+	Piclo (7.5)/ Dica (9)	Suc (126)	NAA (5.3)	–	280
Musa ornata (Ornamental banana)	Imm. embryo*	MS+ → SH → MS → 1/2SH	2,4-D (2.25/4.5/9)	CW (5%)	–	± CW (5%)	281
Musa sp. (Banana/Plantains)	Corm segments*	SH+ → MS (l)	Dica (30–120)	–	–	–	282
	Shoot tips	MS+ → MS+ (l)	BAP (22.2) → 2,4-D (4.5)/2,4,5-T (3.9) + BAP (0.4–4.4)	–	n.s.	n.s.	283
	Shoot tips*	MS+ → MS+ → MS+	BAP (22.2) → 2,4,5-T (3.9)	Suc (116.9) → CW (5%) + Suc (58.4)	BAP (22.2)	Suc (58.4)	284
Orchidaceae Bletilla striata	Imm. seed*	HX+/VW+ → VW+ → VW	2,4-D (13.6) + KN (4.6)	–	–	–	285
Cattleya nobilior × C. loddigesii (Hybrid orchid)	Root tip	VW+	2,4-D (0.5) + NAA (2.7) ± BA (0.2)	–	n.d.	n.d.	286

Table 1. Continued

Species (Common name)	Explant (*Success in Regeneration)	Media Sequence[a] (Salts)	SE Induction Medium		SE Development Medium		Reference
			Growth Regulators[b] (μM)	Others[c] (mM/g)	Growth Regulators[b] (μM)	Others[c] (mM/g)	
Zingiberaceae							
Zingiber officinale (Ginger)	Rhizome*	MS+	n.r.	n.r.	BA (4.4)	–	287
	Young leaf segment*	MS+ →	Dica (2.7)	–	BA (8.9)	–	288
		MS+					

[a] Media sequence code (include respective media and their modifications): MS = Murashige and Skoog [289]; N6 = Chu et al. [290]; SH = Schenk and Hildebrandt [291]; B5 = Gamborg et al. [292]; LS = Linsmaier and Skoog [293]; P2 = Chuang et al. [294]; K = Dunstan and Short [295]; AZ = Abo El-Nil and Zettler [296]; HX = Hyponex/Kano [298]; KM = Kao and Michayluk [299]; Ka = Kao [300]; YP = Yu-Pei/Ku et al. [301]; HE = Ling et al. [302]; AA = Muller and Grafe [303]; R2 = Ohira et al. [304]; CC = Potrykus et al. [305]; mDS-5a = Smith and Krikorian [306]; VW = Vacin and Went [307]; W = White's [308]; M2 = Nayak and Sen [144]; KPR = Thompson et al. [112]. Medium symbols: + = medium containing plant growth regulators; 1/2 = half-strength of regular salt concentrations; 1/4 = one-fourth strength of regular salt concentrations; 2 = double strength of regular salt concentrations; 1/2mac = half-strength of macro nutrients; (l) = liquid medium; (s) = semi-solid / agar based medium; media without (+) symbols are respective basal media composition.

[b] Growth regulator code: 2,4-D = 2,4-dichlorophenoxyacetic acid; KN = kinetin; Piclo = 4-amino-3,5,6-trichloropicolinic acid (picloram); NAA = napthaleneacetic acid; 2-iP = 6-γ-dimethylallylamino purine (2-isopentenyladenine); BA = benzyladenine; BAP = 6-benzylaminopurine; ABA = abscisic acid; ABA anal = abscisic acid analoges; IAA = indole-3-acetic acid; IBA = indole-3-butyric acid; GA/GA₃ = gibberellic acid; Dica = 3,6 dichloro-2-methoxybenzoic acid (dicamba); CPA = para-chlorophenoxyacetic acid; TIBA = 2,3,5-triiodobenzoic acid; Z = Zeatin; 4-PU = 4-pyridyl urea; Chlora = chloramben; DCBA = 2,5-dichlorobenzoic acids; 2-, 3-, 4-, CPA = 2-, 3-, 4-, chlorophenoxyacetic acids, respectively; 2,4,5-T = 2,4,5-trichlorophenoxyacetic acid; ADS = adenine sulfate.

[c] Other supplements code: Glu = glucose; Fru = fructose; Suc = sucrose; Man = mannitol; CW = coconut water (%); CM = coconut milk (%); YE = yeast extract; Act. char = activated charcoal; CNS = cane sugar; CH = casein hydrolysate; Gly = glycine; my-inos = myo-inositol; CA = casamino acids; KNO₃ = potassium nitrate; NH₄NO₃ = ammonium nitrate; Amm. malate = ammonium malate; MgCl₂ = magnesium chloride; AGN = silver nitrate; CLP = calcium pantothenate; PVP = polyvinyl pyrrolidone; Fluri = fluridone; Cefo. = cefotaxime; Carben. = carbenicillin; Strept. = streptomycin sulfate; Asc. acid = ascorbic acid; DMSO = dimethyl sulfoxide; MES = (2-[N-Morpholino]ethanesulfonic acid; L-Pro = L-Proline; Glut/L-Glut = L-Glutamine; Trypt/L-Trypt = L-Tryptophan; L-Ser = L-Serine; Argn/L-Argn = L-Arginine; L-Ala = L-Alanine; Aspn/L-Aspn = L-Asparagine; Cyste = cysteine; B5 vits = Gamborg et al. [292] vitamins; KM vits = Kao and Michayluk [299] vitamins; Whites vits = White's [308] vitamins.

n.r. = not recorded; n.s. = not specified; n.d. = not determined; ± = with or without; → = transfered onto media or media containing.

III. Factors affecting somatic embryogenesis

A. Explant and genotype

The choice of explant is perhaps the most important factor in tissue culture work and determines the success or failure of most experiments. This is particularly so in monocots where cells differentiate early and rapidly, followed by the loss of mitotic and morphogenetic ability. As a consequence, only meristematic and undifferentiated cells that are developmentally uncommited undergo sustained divisions to form callus and retain their morphogenetic competence. The reasons for the early loss of totipotency are not known, but may be linked to the activity of genes that regulate the synthesis and metabolism of endogenous plant growth regulators [309, 310]. In this context, it is noteworthy that embryogenic cells are often formed in the vicinity of differentiating vascular tissues, which are thought to contain high levels of endogenous plant growth regulators [311].

It is thus natural that almost all the work on somatic embryogenesis in herbaceous monocots is based on the culture of tissue or organ explants that contain meristematic and undifferentiated cells. These include immature embryos or seeds, leaf bases (Gramineae) or tips (Orchidaceae) meristems, bulb scales (Liliaceae), lateral buds, etc. Both the stage of development and the physiological condition of the explant are critical in eliciting the desired response. Developmental gradients present in the plants/explants are reflected in gradients of response obtained *in vitro* [309, 312]. Vigorously growing and healthy plants provide the best tissues for culture. The explants may contain both differentiated as well as undifferentiated cells, but only the latter form embryogenic callus.

Genotype is considered by many to be an important factor in tissue culture response. This commonly accepted dogma is often based on inadequate information and has tended to divert attention from the critical role of the developmental and physiological status of the explants. Only in some instances, such as in maize [242] and wheat [313], has the high regeneration ability *in vitro* been shown to be under genetic control, and sexually transferable to less responsive genotypes through hybridization. It is significant, however, that manipulation of the physiological and developmental phases of donor tissues, as well as nutrient media, can help to overcome genotype specificity in a wide variety of species [311, 314]. It is thus not uncommon to see reports of regeneration in genotypes which had been previously described as recalcitrant [233, 315].

B. Nutrient medium

The Murashige and Skoog's [289] nutrient solution (MS medium) is the most commonly used medium for gramineous species. In some instances the B5 [292] and the N6 [290] media have also been used. In most instances the addition of the synthetic auxin 2,4-dichlorophenoxyacetic acid (2,4-D) was found to be sufficient to induce the formation of embryogenic tissues. In addition, picloram and dicamba have been found to be useful for some species. Generally no other additives are needed, but supplementing the basal medium with casein hydrolysate, coconut water, low levels of cytokinins, etc., can improve the efficiency of embryogenic callus formation and plant regeneration. Somatic embryos of grass species have a tendency to germinate precociously *in vitro*. This problem can be generally alleviated by the addition of abscisic acid to the medium which supports embryo maturation.

Various modifications of the Vacin and Went [307], Knudson [316] and Murashige and Skoog [289] media are used for tissue cultures of orchids [317]. In many instances the medium is supplemented with complex natural substances such as coconut water or extracts of banana, potato, etc. Agar and liquid media can be used with equal advantage.

In general, good results can be obtained in most herbaceous monocots with any of the three chemically defined media: MS, B5 and N6. Embryogenic cultures can be induced by the addition of strong auxins such as 2,4-D, and sometimes by high levels of cytokinins. The use of complex organic additives, though common, is not recommended unless absolutely necessary. Further details on nutrient media and additives used for particular species can be found in George et al. [318, 319].

C. The nature of embryogenic cultures

The most common type of embryogenic cultures described in herbaceous monocots are callus cultures grown on agar media. Characteristically, the embryogenic calluses are compact, organized, generally white in color and grow slowly [5]. They are comprised of small, highly cytoplasmic, non-vacuolated, starch and lipid containing meristematic cells. Embryogenic cultures are heterogeneous in nature and contain some non-embryogenic cells. The latter arise either from the division of non-induced cells present in the original explant or from the differentiation of embryogenic cells caused by declining levels of exogenous plant growth regulators in the medium owing to their rapid metabolism and/or degradation. Rapid subculture, but particularly the removal of non-embryogenic cells/sectors during subculture, is useful for the maintenance of embryogenic cultures.

Embryogenic suspension cultures are very difficult to establish and maintain, and have been obtained only in a few species of the Gramineae [5] and some Amaryllidaceae [320]. Typically the embryogenic suspension cultures are finely dispersed, rapidly growing, and comprise groups of small highly cytoplasmic, non-vacuolated embryogenic cells [321]. They are especially valued as a unique source of totipotent protoplasts, and can be conveniently and efficiently preserved for long periods of time by cryopreservation [322–324].

IV. Histology of somatic embryogenesis

Somatic embryos arise from single cells, often indirectly through the formation of proembryogenic masses [114, 145]. Direct formation of somatic embryos, without any intervening callus phase, has also been described in several species of the Gramineae. The embryos are characterized by a closed vascular system. The formation of multiple or fused cotyledons, and the suppression of shoot or root meristems are common problems associated with somatic embryogenesis *in vitro*. Improved culture conditions, especially addition of abscisic acid and increased osmotica, are useful in obtaining well defined somatic embryos.

V. Analysis of regenerants

In comparison to the high mortality of shoots with adventitious roots, plantlets derived from somatic embryos can be easily transferred to soil and grown to maturity because of the presence of a well established root-shoot axis and primary roots. Perhaps one of the most useful features of plants regenerated from somatic embryos is that they are generally non-chimeral in nature and are free of any major genetic or cytological changes. This has been attributed to their single cell origin and stringent selection in favor of non-aberrant cells during the development of somatic embryos [152, 274, 325–328].

VI. Applications

There are many practical uses of plant regeneration through somatic embryogenesis. The most useful aspect of immediate benefit is its efficiency and potential for mass propagation, and the uniformity of plants derived from somatic embryos. This should be of particular interest and benefit to the micropropagation industry which relies on large scale clonal propagation. In addition, embryogenic cultures are well suited to growth in bioreactors and lend themselves easily to automation and the production of synthetic seed [329].

Embryogenic cultures have proven to be of considerable value in the genetic manipulation of herbaceous monocots, particularly the Gramineae, as they provide the only source of totipotent protoplasts for this group of important food crops. Protoplasts isolated from embryogenic suspension cultures have been used to obtain somatic hybrids and male sterile cybrids [330–332], as well as transgenic plants [333–337]. Transgenic plants have also been recovered from the direct delivery of DNA into embryogenic cell suspension [338, 339], or callus [202] or directly into the scutellar tissues of immature embryos [340].

Acknowledgements

The authors would like to thank Mrs. Rangathilakam KrishnaRaj for her help in compiling and cross-checking the references, and Mr. Lisheng Kong for translating some references.

References

1. Knox, R.B., Apomixis: seasonal and population differences in a grass, *Science*, 157, 325, 1967.
2. Bashaw, E.C., and Hanna, W.W., Apomictic reproduction, in *Reproductive Versatility in the Grasses*, Chapman, G.P., Ed., Cambridge University Press, 1990, 100.
3. Steward, F.C., Mapes, M.O., and Mears, K., Growth and organised development of cultured cells. II. Organisation in cultures grown from freely suspended cells, *Am. J. Bot.*, 45, 705, 1958.
4. Reinert, J., Morphogenese und ihre kontrolle an Gewebekulturen aus Karroten, *Naturwiss.*, 45, 344, 1958.
5. Vasil, I.K., and Vasil, V., Regeneration in cereal and other grass species, in *Cell Culture and Somatic Cell Genetics of Plants, Vol. 3, Plant Regeneration and Genetic Variability*, Vasil, I.K., Ed., Academic Press, Orlando, 1986, 121.
6. Morrish, F.M., Vasil, V., and Vasil, I.K., Developmental morphogenesis and genetic manipulation in tissue and cell cultures of the Gramineae, *Adv. Genet.*, 24, 431, 1987.
7. Havel, L., and Novak, F.J., Regulation of somatic embryogenesis and organogenesis in *Allium carinatum* L., *J. Plant Physiol.*, 132, 373, 1988.
8. Phillips, G.C., and Luteyn, K.J., Effects of Picloram and other auxins on onion tissue cultures, *J. Amer. Soc. Hort. Sci.*, 108, 948, 1983.
9. Shahin, E.A., and Kaneko, K., Somatic embryogenesis and plant regeneration from callus cultures of nonbulbing onions, *HortSci.*, 21, 294, 1986.
10. Lu, C.-C., Currah, L., and Peffley, E.B., Somatic embryogenesis and plant regeneration in diploid *Allium fistulosum* × *A. cepa* F1 hybrid onions, *Plant Cell Rep.*, 7, 696, 1989.
11. Van der Valk, P., Scholten, O.E., Verstappen, F., Jansen, R.C., and Dons, J.J.M., High frequency somatic embryogenesis and plant regeneration from zygotic embryo-derived callus cultures of three *Allium* species, *Plant Cell Tiss. Org. Cult.*, 30, 181, 1992.
12. Xue, H.E., Araki, H., Shi, L., and Yakuwa, T., Somatic embryogenesis and plant regeneration in basal plate and receptacle derived-callus cultures of garlic (*Allium sativum* L.), *J. Japan Soc. Hort. Sci.*, 60, 627, 1991.
13. Abo El-Nil, N.M., Organogenesis and embryogenesis in callus cultures of garlic (*Allium sativum* L.), *Plant Sci. Lett.*, 9, 259, 1977.
14. Kuehnle, A.R., Chen, F.-C., and Sugii, N., Somatic embryogenesis and plant regeneration in *Anthurium andraeanum* hybrids, *Plant Cell Rep.*, 11, 438, 1992.
15. Ghosh, B., and Sen, S., Plant regeneration through somatic embryogenesis from spear callus culture of *Asparagus cooperi* Baker, *Plant Cell Rep.*, 9, 667, 1991.
16. Chin, C.-K., and Khunachak, A., Effect of ancymidol on asparagus somatic embryo development, *HortSci.*, 19, 79, 1984.
17. Uragami, A., Sakai, A., Nagai, M., and Takahashi, T., Survival of cultured cells and somatic embryos of *Asparagus officinalis* cryopreserved by vitrification, *Plant Cell Rep.*, 8, 418, 1989.
18. Levi, A., and Sink, K.C., Histology and morphology of asparagus somatic embryos, *HortSci.*, 26, 1322, 1991.
19. Levi, A., and Sink, K.C., Differential effects of sucrose, glucose and fructose during somatic embryogenesis in Asparagus, *J. Plant Physiol.*, 137, 184, 1990.
20. Bui Dang Ha, D., Norreel, B., and Masset, A., Regeneration of *Asparagus officinalis* L. through callus cultures derived from protoplasts, *J. Exptl. Bot.*, 26, 263, 1975.
21. Saito, T., Nishizawa, S., and Nishimura, S., Improved culture conditions for somatic

456

embryogenesis from *Asparagus officinalis* L. using an aseptic ventilative filter, *Plant Cell Rep.*, 10, 230, 1991.

22. Reuther, G., Adventitious organ formation and somatic embryogenesis in callus cultures of *Asparagus* and *Iris* and its possible application, *Acta Hort.*, 78, 217, 1977.

23. Reuther, G., Embryoide differenzierungsmuster im Kalus der Gattungen *Iris* und *Asparagus*, *Ber. Dtsch. Bot. Ges.*, 90, 417, 1977.

24. Levi, A., and Sink, K.C., Somatic embryogenesis in asparagus: the role of explants and growth regulators, *Plant Cell Rep.*, 10, 71, 1991.

25. Kunitake, H., and Mii, M., Somatic embryogenesis and plant regeneration from protoplasts of Asparagus (*Asparagus officinalis* L.), *Plant Cell Rep.*, 8, 706, 1990.

26. Levi, A., and Sink, K.C., Asparagus somatic embryos: Production in suspension culture and conversion to plantlets on solidified medium as influenced by carbohydrate regime, *Plant Cell Tiss. Org. Cult.*, 31, 115, 1992.

27. Singh, J.P., Respiratory metabolism of somatic embryogenesis in callus cultures of *Dioscorea deltoidea* Wall, *Curr. Sci. (Banglore)*, 51, 618, 1982.

28. Ammirato, P.V., Somatic embryogenesis and plantlet development in suspension cultures of the medicinal yam, *Dioscorea floribunda*, *Am. J. Bot.*, 65, 89, 1978.

29. Osifo, E.O., Somatic embryogenesis in *Dioscorea*, *J. Plant Physiol.*, 133, 378, 1988.

30. Gyulai, G., Janovszky, J., Kiss, E., Csillag, A., and Heszky, L.E., Callus initiation and plantlet regeneration from inflorescence primordia of the intergeneric hybrid *Agropyron repens* (L.) Beauv. × *Bromus inermis* Leyss. cv. *nanus* on a modified nutritive medium, *Plant Cell Rep.*, 11, 266, 1992.

31. Zhong, H., Srinivasan, C., and Sticklen, M.B., Plant regeneration via somatic embryogenesis in creeping bentgrass (*Agrostis palustris* Huds.), *Plant Cell Rep.*, 10, 453, 1991.

32. Bregitzer, P., Somers, D.A., and Rines, H.W., Development and characterization of friable, embryogenic oat callus, *Crop Sci.*, 29, 798, 1989.

33. Nassuth, A., Fife, C., Fedak, G., and Altosaar, I., Somatic embryogenesis in oat cultivars, *J. Cell. Biochem.*, suppl. 11B., 27, 1987.

34. Bregitzer, P., Bushnell, W.R., Rines, H.W., and Somers, D.A., Callus formation and plant regeneration from somatic embryos of Oat (*Avena sativa* L.), *Plant Cell Rep.*, 10, 243, 1991.

35. Heyser, J.W., and Nabors, M.W., Long term plant regeneration, somatic embryogenesis and green spot formation in secondary oat (*Avena sativa*) callus, *Z. Pflanzenphysiol.*, 107, 153, 1982.

36. Yeh, M.-L., and Chang, W.-C., Somatic embryogenesis and subsequent plant regeneration from inflorescence callus of *Bambusa beecheyana* Munro var. beecheyana, *Plant Cell Rep.*, 5, 409, 1986.

37. Yeh, M.-L. and, Chang, W.-C., Plant regeneration through somatic embryogenesis in callus culture of green bamboo (*Bambusa oldhamii* Munro), *Theor. Appl. Genet.*, 73, 161, 1986.

38. Franklin, C.I., Trieu, T.N., and Gonzales, R.A., Plant regeneration through somatic embryogenesis in the forage grass Caucasian bluestem (*Bothriochloa caucasica*), *Plant Cell Rep.*, 9, 443, 1990.

39. Artunduaga, I.R., Taliaferro, C.M., and Johnson, B.L., Effects of auxin concentration on induction and growth of embryogenic callus from young inflorescence explants of Old World bluestem (*Bothriochloa* spp.) and bermuda (*Cynodon* spp.) grasses, *Plant Cell Tissue Organ Cult.*, 12, 13, 1988.

40. Gamborg, O.L., Constabel, F., and Miller, R.A., Embryogenesis and production of albino plants from cell cultures of *Bromus inermis*, *Planta*, 95, 355, 1970.

41. Kackar, A., and Shekhawat, N.S., Plant regeneration from cultured immature inflorescences of *Cenchrus setigerus* and *Cenchrus ciliaris*, *Ind. J. Exptl. Biol.*, 29, 62, 1991.

42. Sun, C.S., and Chu, C.C., Somatic embryogenesis and plant regeneration from immature inflorescence segments of *Coix lacryma-jobi*, *Plant Cell Tiss. Org. Cult.*, 5, 175, 1986.

43. Barthakur, M., and Bordoloi, D.N., *In vitro* regeneration of Java citronella (*Cymbopogon winterianus* Jowitt), *Herba Hungarica*, 28, 21, 1989.

44. Sreenath, H.L., and Jagadishchandra, K.S., Somatic embryogenesis and plant regeneration from inflorescence culture of Java citronella (*Cymbopogon winterianus*), *Ann. Bot. (London)*, 64, 211, 1989.

45. Artunduaga, I.R., Taliaferro, C.M., and Johnson, B.B., Induction and growth of callus from immature inflorescences of "Zebra" bermuda grass as affected by casein hydrolysate and 2,4-D concentration, *In Vitro*, 25, 753, 1989.

46. Ahn, B.J., Huang, F.H., and King, J.W., Plant regeneration through somatic embryogenesis in common bermudagrass tissue culture, *Crop Sci.*, 25, 1107, 1985.

47. Ahn, B.J., Huang, F.H., and King, J.W., Regeneration of bermuda cultivars and evidence of somatic embryogenesis, *Crop Sci.*, 27, 594, 1987.

48. Songstad, D.D., and Conger, B.V., Direct embryogenesis from cultured anthers and pistils of *Dactylis glomerata*, *Amer. J. Bot.*, 73, 989, 1986.

49. Songstad, D.D., and Conger, B.V., Factors influencing somatic embryo induction from orchardgrass anther cultures, *Crop Sci.*, 28, 1006, 1988.

50. Conger, B.V., Hovanesian, J.C., Trigiano, R.N., and Gray, D.J., Somatic embryo ontogeny in suspension cultures of orchardgrass, *Crop Sci.*, 29, 448, 1989.

51. Trigiano, R.N., Conger, B.V., and Songstad, D.D., Effects of mefluidide and dicamba on *in vitro* growth and embryogenesis of *Dactylis glomerata* (Orchard grass), *J. Plant Growth Regul.*, 6, 133, 1987.

52. Trigiano, R.N., and Conger, B.V., Regulation of growth and somatic embryogenesis by proline and serine in suspension cultures of *Dactylis glomerata*, *J. Plant Physiol.*, 130, 49, 1987.

53. Trigiano, R.N., Gray, D.J., Conger, B.V., and McDaniel, J.K., Origin of direct somatic embryos from cultured leaf segments of *Dactylis glomerata*, *Bot. Gaz.*, 150, 72, 1989.

54. Gray, D.J., and Conger, B.V., Influence of dicamba and casein hydrolysate on somatic embryo number and culture quality in cell suspensions of *Dactylis glomerata* (Gramineae), *Plant Cell Tiss. Org. Cult.*, 4, 123, 1985.

55. Gray, D.J., Conger, B.V., and Hanning, G.E., Somatic embryogenesis in suspension and suspension-derived callus cultures of *Dactylis glomerata*, *Protoplasma*, 122, 196, 1984.

56. Hanning, G.E., and Conger, B.V., Factors influencing somatic embryogenesis from cultured leaf segments of *Dactylis glomerata*, *J. Plant Physiol.*, 123, 23, 1986.

57. Gavin, A.L., Conger, B.V., and Trigiano, R.N., Sexual transmission of somatic embryogenesis in *Dactylis glomerata*, *Plant Breed.*, 103, 251, 1989.

58. McDaniel, J.K., Conger, B.V., and Graham, E.T., A histological study of tissue proliferation, embryogenesis and organogenesis from tissue cultures of *Dactylis glomerata* L., *Protoplasma*, 110, 121, 1982.

59. Conger, B.V., Hanning, G.E., Gray, D.J., and McDaniel, J.K., Direct embryogenesis from mesophyll cells of orchard grass, *Science*, 221, 850, 1983.

60. Hanning, G.E., and Conger, B.V., Embryoid and plantlet formation from leaf segments of *Dactylis glomerata* L., *Theor. Appl. Genet.*, 63, 155, 1982.

61. VanderZee, D., Cobb, B.G., Loescher, W.H., and Kennedy, R.A., Plantlet regeneration via somatic embryogenesis in the grasses *Echinochloa crus-galli* var. *oryzicola* and *E. muricata* from subcultured callus, *Plant Physiol.*, 72, 47, 1983.

62. Wang, D.-Y., and Yan, K., Somatic embryogenesis in *Echinochloa crusgalli*, *Plant Cell Rep.*, 3, 88, 1984.

63. Cobb, B.G., Vanderzee, D., Loescher, W.H., and Kennedy, R.A., Evidence for plantlet regeneration via somatic embryogenesis in the grasses *Echinochloa muricata* and *E. crus-galli* var. Oryzicola, *Plant Sci.*, 40, 121, 1985.

64. Talwar, M., and Rashid, A., Somatic embryo formation from unemerged inflorescences and immature embryos of a graminaceous crop *Echinochloa*, *Ann. Bot. (London)*, 64, 195, 1989.

65. Sankhla, A., Davis, T.D., Sankhla, D., Sankhla, N., Upadhyaya, A., and Joshi, S., Influence of growth regulators on somatic embryogenesis, plantlet regeneration, and post-transplant survival of *Echinochloa frumentacea*, *Plant Cell Rep.*, 11, 368, 1992.

66. Wang, M., and Zapata, F.J., Somatic embryogenesis and plant regeneration in tissue culture of *Echinochloa glabrescens* Munro ex Hook. F., *J. Plant Physiol.*, 130, 79, 1987.

67. Takahashi, A., Sakuragi, Y., Kamada, H., and Ishizuka, K., Plant regeneration through somatic embryogenesis in barnyard grass, *Echinochloa oryzicola* Vasing, *Plant Sci. Lett.*, 36, 161, 1984.

458

68. Eapen, S., and George, L., High frequency plant regeneration through somatic embryogenesis in finger millet (*Eleusine coracana* Gaertn.), *Plant Sci.*, 61, 127, 1989.

69. Sivadas, P., Kothari, S.L., and Chandra, N., High frequency embryoid and plantlet formation from tissue cultures of the finger millet- *Eleusine coracana* (L.) Gaertn., *Plant Cell Rep.*, 9, 93, 1990.

70. Eapen, S., and George, L., Influence of phytohormones, carbohydrates, amino acids, growth supplements and antibiotics on somatic embryogenesis and plant differentiation in finger millet, *Plant Cell Tiss. Org. Cult.*, 22, 87, 1990.

71. George, L., and Eapen, S., High frequency plant-regeneration through direct shoot development and somatic embryogenesis from immature inflorescence cultures of finger millet (*Eleusine coracana* Gaertn.), *Euphytica*, 48, 269, 1990.

72. Cheng, J.-C., and Zhon, J.-Y., Somatic embryogenesis and plant regeneration from inflorescence segments of *Eleusine indica* (L.) Grertn., *Int. Bot. Congr. Abstr.*, 17, 121, 1987.

73. Park, C.H., and Walton, P.D., Embryogenesis and plant regeneration from tissue culture of Canada wildrye, *Elymus canadensis* L., *Plant Cell Rep.*, 8, 289, 1989.

74. Kearney, J.F., Parrott, W.A., and Hill, N.S., Infection of somatic embryos of tall fescue with *Acremonium coenophialum*, *Crop Sci.*, 31, 979, 1991.

75. Rajoelina, S.R., Alibert, G., and Planchon, C., Continuous plant regeneration from established embryogenic cell suspension cultures of italian ryegrass and tall fescue, *Plant Breed.*, 104, 265, 1990.

76. Torello, W.A., Rufner, R., and Symington, A.G., The ontogeny of somatic embryos from long-term callus cultures of red fescue, *HortSci.*, 20, 938, 1985.

77. Zaghmout, O.M.F., and Torello, W.A., Restoration of regeneration potential of long-term cultures of red fescue (*Festuca rubra* L.) by elevated sucrose levels, *Plant Cell Rep.*, 11, 142, 1992.

78. Torello, W.A., Symington, A.G., and Rufner, R., Callus initiation, plant regeneration, and evidence of somatic embryogenesis in red fescue, *Crop Sci.*, 24, 1037, 1984.

79. Zaghmout, O.M.F., and Torello, W.A., Somatic embryogenesis and plant regeneration from suspension cultures of red fescue, *Crop Sci.*, 29, 815, 1989.

80. Inagaki, M., De Buyser, J., and Henry, Y., Occurrence of somatic embryoids in suspension callus cultures initiated from immature haploid wheat embryos, *Japan J. Breed.*, 38, 103, 1988.

81. Barcelo, P., Vazquez, A., and Martin, A., Somatic embryogenesis and plant regeneration from *Tritordeum*, *Plant Breed.*, 103, 235, 1989.

82. Weigel, R.C. Jr., and Hughes, K.W., Long term regeneration by somatic embryogenesis in barley (*Hordeum vulgare* L.) tissue cultures derived from apical meristem explants, *Plant Cell Tiss. Org. Cult.*, 5, 151, 1985.

83. Ruiz, M.L., Rueda, J., Pelaez, M.I., Espino, F.J., Candela, M., Sendino, A.M., and Vazquez, A.M., Somatic embryogenesis, plant regeneration and somaclonal variation in barley, *Plant Cell Tiss. Org. Cult.*, 28, 97, 1992.

84. Thomas, M.R., and Scott, K.J., Plant regeneration by somatic embryogenesis from callus initiated from immature embryos and immature inflorescences of *Hordeum vulgare*, *J. Plant Physiol.*, 121, 159, 1985.

85. Kott, L.S., and Kasha, K.J., Initiation and morphological development of somatic embryoids from barley cell cultures, *Can. J. Bot.*, 62, 1245, 1984.

86. Luhrs, R., and Lorz, H., Plant regeneration *in vitro* from embryogenic cultures of spring- and winter- type barley (*Hordeum vulgare* L.) varieties, *Theor. Appl. Genet.*, 75, 16, 1987.

87. Gozukirmizi, N., Ari, S., Oraler, G., Okatan, Y., and Unsal, N., Callus induction, plant regeneration and chromosomal variations in barley, *Acta Bot. Neerl.*, 39, 379, 1990.

88. Rengel, Z., Embryogenic callus induction and plant regeneration from cultured *Hordeum vulgare* mature embryos, *Plant Physiol. Biochem.*, 25, 43, 1987.

89. Coppens, L., and Dewitte, D., Esterase and peroxidase zymograms from barley (*Hordeum vulgare* L.) callus as a biochemical marker system of embryogenesis and organogenesis, *Plant Sci.*, 67, 97, 1990.

90. Rengel, Z., and Jelaska, S., The effect of L-Proline on somatic embryogenesis in long-term callus culture of *Hordeum vulgare*, *Acta Bot. Croat.*, 45, 71, 1986.

91. Rengel, Z., Effect of abscisic acid on plant development from *Hordeum vulgare* embryogenic callus, *Biochem. Physiol. Pflanzen.*, 181, 605, 1986.

92. Rengel, Z., and Jelaska, S., Somatic embryogenesis and plant regeneration from seedling tissues of *Hordeum vulgare* L., *J. Plant Physiol.*, 124, 385, 1986.

93. Chu, C.C., Sun, C.S., Chen, X., Zhang, W.X., and Du, Z.H., Somatic embryogenesis and plant regeneration in callus from inflorescences of *Hordeum vulgare* x *Triticum aestivum* hybrids, *Theor. Appl. Genet.*, 68, 375, 1984.

94. Kackar, A., and Shekhawat, N.S., Regeneration of *Lasiurus scindicus* from tissue culture, *Ann. Bot. (London)*, 64, 455, 1989.

95. Dale, P.J., Embryoids from cultured immature embryos of *Lolium multiflorum*, *Z. Pflanzenphysiol.*, 100, 73, 1980.

96. Creemers-Molenaar, T., Loeffen, J.P.M., and Zaal, M.A.C., Somatic embryogenesis and plant regeneration from different type of explants of *Lolium perenne* and *Lolium multiflorum*, *Int. Bot. Congr. Abstr.*, 17, 121, 1987.

97. Ahloowalia, B.S., Spectrum of variation in somaclones of triploid ryegrass, *Crop Sci.*, 23, 1141, 1983.

98. Faiz Zaghmout, O.M., and Torello, W.A., Somatic embryogenesis and plant regeneration from embryogenic suspension cultures of perennial ryegrass, *In Vitro Cell. Dev. Biol.*, 26, 419, 1990.

99. Arumuganathan, K., Dale, P.J., and Cooper, J.P., Vernalisation in *Lolium temulentum* L.: Responses of *in vitro* cultures of mature and immature embryos, shoot apices and callus, *Ann. Bot. (London)*, 67, 173, 1991.

100. Dalton, S.J., and Thomas, I.D., A statistical comparison of various factors on embryogenic proliferation, morphogenesis and regeneration in *Lolium temulentum* cell suspension colonies, *Plant Cell Tiss. Org. Cult.*, 30, 15, 1992.

101. Boissot, N., Valdez, M., and Guiderdoni, E., Plant regeneration from leaf and seed-derived calli and suspension cultures of the African perennial wild rice, *Oryza longistaminata*, *Plant Cell Rep.*, 9, 447, 1990.

102. Wang, M.-S., Zapata, F.J., and De Castro, D.C., Plant regeneration through somatic embryogenesis from mature seed and young inflorescences of wild rice (*Oryza perennis* Moench), *Plant Cell Rep.*, 6, 294, 1987.

103. Maggioni, L., Lusardi, M.C., and Lupotto, E., Effects of cultural conditions on callus induction and plant regeneration from mature and immature embryos of rice varieties (*Oryza sativa* L.), *J. Genet. Breed.*, 43, 99, 1989.

104. Ozawa, K., and Komamine, A., Establishment of a system of high-frequency embryogenesis from long-term cell suspension cultures of rice (*Oryza sativa* L.), *Theor. Appl. Genet.*, 77, 205, 1989.

105. Wernicke, W., Brettell, R., Wakizuka, T., and Potrykus, I., Adventitious embryoid and root formation from rice leaves, *Z. Pflanzenphysiol.*, 103, 361, 1981.

106. Zhang, H.M., Yang, H., Rech, E.L., Golds, T.J., Davis, A.S., Mulligan, B.J., Cocking, E.C., and Davey, M.R., Transgenic rice plants produced by electroporation-mediated plasmid uptake into protoplasts, *Plant Cell Rep.*, 7, 379, 1988.

107. Hartke, S., and Lorz, H., Somatic embryogenesis and plant regeneration from various indica rice (*Oryza sativa* L.) genotypes, *J. Genet. Breed.*, 43, 205, 1989.

108. Kyozuka, J., Hayashi, Y., and Shimamoto, K., High frequency plant regeneration from rice protoplasts by novel nurse culture methods, *Mol. Gen. Genet.*, 206, 408, 1987.

109. Jones, T.J., Somatic embryogenesis and plant regeneration from four varieties of rice (*Oryza sativa* L.), *Amer. J. Bot.*, 72, 804, 1985.

110. Datta, K., Potrykus, I., and Datta, S.K., Efficient fertile plant regeneration from protoplasts of the Indica rice breeding line IR72 (*Oryza sativa* L.), *Plant Cell Rep.*, 11, 229, 1992.

111. Siriwardana, S., and Nabors, M.W., Tryptophan enhancement of somatic embryogenesis in rice, *Plant Physiol.*, 73, 142, 1983.

460

112. Thompson, J.A., Abdullah, R., and Cocking, E.C., Protoplast culture of rice (*Oryza sativa* L.) using media solidified with agarose, *Plant Sci.*, 47, 123, 1986.

113. Ling, D.-H., Brar, D.S., and Zapata, F.J., Cytology and histology in somatic embryogenesis of indica rice, *Acta Bot. Sinica*, 30, 485, 1988.

114. Jones, T.J., and Rost, T.L., The developmental anatomy and ultrastructure of somatic embryos from rice (*Oryza sativa* L.) scutellum epithelial cells, *Bot. Gaz.*, 150, 41, 1989.

115. Chowdhry, C.N., Tyagi, A.K., Maheshwari, N., and Maheshwari, S.C., Effect of L-proline and L-tryptophan on somatic embryogenesis and plantlet regeneration of rice (*Oryza sativa* L. cv. Pusa 169), *Plant Cell Tiss. Org. Cult.*, 32, 357, 1993.

116. Binh, D.Q., Heszky, L.E., Guylai, G., and Csillag, A., Plant regeneration of NaCl-pretreated cells from long-term suspension culture of rice (*Oryza sativa* L.) in high saline conditions, *Plant Cell Tiss. Org. Cult.*, 29, 75, 1992.

117. Abe, T., and Futsuhara, Y., Plant regeneration from suspension cultures of rice (*Oryza sativa* L.), *Japan J. Breed.*, 36, 1, 1986.

118. Sticklen, M.B., Direct somatic embryogenesis and fertile plants from rice root cultures, *J. Plant Physiol.*, 138, 577, 1991.

119. Abe, T., and Futsuhara, Y., Varietal difference of plant regeneration from root callus tissues in rice, *Japan J. Breed.*, 34, 147, 1984.

120. Abe, T., and Futsuhara, Y., Efficient plant regeneration by somatic embryogenesis from root callus tissues of rice (*Oryza sativa* L.), *J. Plant Physiol.*, 121, 111, 1985.

121. Sticklen, M.B., Rumpho, M.E., and Kennedy, R.A., Somatic embryogenesis in rice seedlings, *Plant Physiol.*, 83, 75, 1987.

122. Ling, D.H., Chen, W.Y., Chen, M.F., and Ma, Z.R., Somatic embryogenesis and plant regeneration in an interspecific hybrid of *Oryza*, *Plant Cell Rep.*, 2, 169, 1983.

123. Gengguang, L., Lanying, Z., Ruzhu, C., and Kailian, L., Somatic embryogenesis and plant regeneration from suspension cultures of young inflorescences of indica rice, *Acta Botanica Yunnanica*, 12, 317, 1990.

124. Ling, D.-H., and Yosida, S., The study of some factors affecting somatic embryogenesis in IR lines of rice, *Acta Bot. Sinica*, 29, 1, 1987.

125. Chen, T.-H., Lam, L., and Chen, S.-C., Somatic embryogenesis and plant regeneration from cultured young inflorescences of *Oryza sativa* L. (rice), *Plant Cell Tiss. Org. Cult.*, 4, 51, 1985.

126. Chen, C.J-., and Zhou, J.-Y., Somatic embryogenesis and plant regeneration from inflorescence of paddy rice, *Amer. J. Bot.*, 75, 55, 1988.

127. Wu, Y.-Z., and Ling, D.-H., *In vitro* culture of different explants from Hubei photoperiod sensitive genic male-sterile rice, *Acta Bot. Sinica*, 33, 489, 1991.

128. Abdullah, R., Cocking, E.C., and Thompson, J.A., Efficient plant regeneration from rice protoplasts through somatic embryogenesis, *Bio/Technol.*, 4, 1087, 1986.

129. Woods, S.H., Phillips, G.C., Woods, J.E., and Collins, G.B., Somatic embryogenesis and plant regeneration from zygotic embryo explants in mexican weeping bamboo, *Otatea acuminata aztecorum*, *Plant Cell Rep.*, 11, 257, 1992.

130. Fladung, M., and Hesselbach, J., Callus induction and plant regeneration in *Panicum bisulcatum* and *Panicum milioides*, *Plant Cell Rep.*, 3, 169, 1986.

131. Lu, C.-Y., and Vasil, I.K., Histology of somatic embryogenesis in *Panicum maximum* (Guinea grass), *Amer. J. Bot.*, 72, 1908, 1985.

132. Lu, C.-Y., Vasil, V., and Vasil, I.K., Isolation and culture of protoplasts of *Panicum maximum* Jacq. (Guinea grass): Somatic embryogenesis and plantlet formation, *Z. Pflanzenphysiol.*, 104, 311, 1981.

133. Lu, C.-Y., and Vasil, I.K., Somatic embryogenesis and plant regeneration from freely-suspended cells and cell groups of *Panicum maximum* Jacq., *Ann. Bot. (London)*, 48, 543, 1981.

134. Lu, C.-Y., and Vasil, I.K., Somatic embryogenesis and plant regeneration in tissue cultures of *Panicum maximum* Jacq, *Amer. J. Bot.*, 69, 77, 1982.

135. Akashi, R., and Adachi, T., High frequency somatic embryo formation in cultures of immature embryos of guinea grass, *Panicum maximum* Jacq., *Japan J. Breed.*, 41, 85, 1991.

136. Lu, C.-Y., and Vasil, I.K., Somatic embryogenesis and plant regeneration from leaf tissues of *Panicum maximum* Jacq., *Theor. Appl. Genet.*, 59, 275, 1981.

137. Heyser, J.W., Isolation and culture of totipotent protoplasts of Proso millet (*Panicum miliaceum*), *Plant Physiol.*, 72, 143, 1983.

138. Heyser, J.W., Callus and shoot regeneration from protoplasts of Proso millet (*Panicum miliaceum* L.), *Z. Pflanzenphysiol.*, 113, 293, 1984.

139. Rangan, T.S., and Vasil, I.K., Somatic embryogenesis and plant regeneration in tissue cultures of *Panicum miliaceum* L. and *Panicum miliare* Lamk., *Z. Pflanzenphysiol.*, 109, 49, 1983.

140. Marousky, F.J., and West, S.H., Somatic embryos from bahiagrass caryopses cultured *in vitro*, *HortSci.*, 22, 94, 1987.

141. Bovo, O.A., and Mroginski, L.A., Somatic embryogenesis and plant regeneration from cultured mature and immature embryos of *Paspalum notatum* (Gramineae), *Plant Sci.*, 65, 217, 1989.

142. Marousky, F.J., and West, S.H., Somatic embryogenesis and plant regeneration from cultured mature caryopses of bahiagrass (*Paspalum notatum* Flugge), *Plant Cell Tiss. Org. Cult.*, 20, 125, 1990.

143. Nayak, P., and Sen, S.K., Plant regeneration through somatic embryogenesis from suspension culture-derived protoplasts of *Paspalum scrobiculatum* L., *Plant Cell Rep.*, 10, 362, 1991.

144. Nayak, P., and Sen, S.K., Plant regeneration through somatic embryogenesis from suspension cultures of a minor millet, *Paspalum scrobiculatum*, *Plant Cell Rep.*, 8, 296, 1989.

145. Vasil, V., and Vasil, I.K., The ontogeny of somatic embryos of *Pennisetum americanum* (L.) K. Schum. I. in cultured immature embryos, *Bot. Gaz.*, 143, 454, 1982.

146. Vasil, V., and Vasil, I.K., Isolation and culture of cereal protoplasts. II. Embryogenesis and plantlet formation from protoplasts of *Pennisetum americanum*, *Theor. Appl. Genet.*, 56, 97, 1980.

147. Vasil, V., and Vasil, I.K., Somatic embryogenesis and plant regeneration from suspension cultures of Pearl millet (*Pennisetum americanum*), *Ann. Bot. (London)*, 47, 669, 1981.

148. Vasil, V., and Vasil, I.K., Somatic embryogenesis and plant regeneration from tissue cultures of *Pennisetum americanum*, and *P. americanum* × *P. purpureum* hybrid, *Amer. J. Bot.*, 68, 864, 1981.

149. Botti, C., and Vasil, I.K., Plant regeneration by somatic embryogenesis from parts of cultured mature embryos of *Pennisetum americanum* (L.) K. Schum, *Z. Pflanzenphysiol.*, 111, 319, 1983.

150. Botti, C., and Vasil, I.K., Ontogeny of somatic embryos of *Pennisetum americanum*. II. In cultured immature inflorescences, *Can. J. Bot.*, 62, 1629, 1984.

151. Vasil, V., and Vasil, I.K., Characterization of an embryogenic cell suspension culture derived from cultured inflorescences of *Pennisetum americanum* (Pearl millet, Gramineae), *Amer. J. Bot.*, 69, 1441, 1982.

152. Swedlund, B., and Vasil, I.K., Cytogenetic characterization of embryogenic callus and regenerated plants of *Pennisetum americanum* (L.) K. Schum, *Theor. Appl. Genet.*, 69, 575, 1985.

153. Pius, J., George, L., Eapen, S., and Rao, P.S., Enhanced plant regeneration in pearl millet (*Pennisetum americanum*) by ethylene inhibitors and cefotaxime, *Plant Cell Tiss. Org. Cult.*, 32, 91, 1993.

154. Vasil, V., Wang, D.-Y., and Vasil, I.K., Plant regeneration from protoplasts of Napier Grass (*Pennisetum purpureum* Schum.), *Z. Pflanzenphysiol.*, 111, 233, 1983.

155. Chandler, S.F., and Vasil, I.K., Optimization of plant regeneration from long term embryogenic callus cultures of *Pennisetum purpureum* Schum. (Napier grass), *J. Plant Physiol.*, 117, 147, 1984.

156. Haydu, Z., and Vasil, I.K., Somatic embryogenesis and plant regeneration from leaf tissues and anthers of *Pennisetum purpureum* Schum., *Theor. Appl. Genet.*, 59, 269, 1981.

157. Wang, D., and Vasil, I.K., Somatic embryogenesis and plant regeneration from inflorescence segments of *Pennisetum purpureum* Schum. (Napier or Elephant grass), *Plant Sci. Lett.*, 25, 147, 1982.

158. Talwar, M., and Rashid, A., Factors affecting formation of somatic embryos and embryogenic

462

callus from unemerged inflorescences of a graminaceous crop *Pennisetum*, *Ann. Bot. (London)*, 66, 17, 1990.

159. Van der Valk, P., Zaal, M.A.C.M., and Creemers-Molenaar, J., Somatic embryogenesis and plant regeneration in inflorescence and seed derived callus cultures of *Poa pratensis* L. (Kentucky bluegrass), *Plant Cell Rep.*, 7, 644, 1989.

160. Chen, W.H., Davey, M.R., Power, J.B., and Cocking, E.C., Control and maintenance of plant regeneration in sugarcane callus cultures, *J. Exptl. Bot.*, 39, 251, 1988.

161. Guiderdoni, E., and Demarly, Y., Histology of somatic embryogenesis in cultured leaf segments of sugarcane plantlets, *Plant Cell Tiss. Org. Cult.*, 14, 71, 1988.

162. Yan, Q.-S., Zhang, X.-Q., and Gu, M.-G., Somatic embryogenesis from sugarcane protoplasts, *Acta Bot. Sinica*, 29, 242, 1987.

163. Tabaeizadeh, Z., Ferl, R.J., and Vasil, I.K., Somatic hybridization in the Graminae: *Saccharum officinarum* (sugarcane) and *Pennisetum americanum* (L.) K. Schum. (pearl millet), *Proc. Natl. Acad. Sci. USA*, 83, 5616, 1986.

164. Ahloowalia, B.S., and Maretzki, A., Plant regeneration via somatic embryogenesis in sugarcane, *Plant Cell Rep.*, 2, 21, 1983.

165. Ho, W.-J., and Vasil, I.K., Somatic embryogenesis in sugarcane (*Saccharum officinarum* L.) I. The morphology and physiology of callus formation and the ontogeny of somatic embryos, *Protoplasma*, 118, 169, 1983.

166. Ho, W.-J., and Vasil, I.K., Somatic embryogenesis in sugarcane (*Saccharum officinarum* L.): Growth and plant regeneration from embryogenic cell suspension cultures, *Ann. Bot. (London)*, 51, 719, 1983.

167. Srinivasan, C., and Vasil, I.K., Plant regeneration from protoplasts of sugarcane (*Saccharum officinarum* L.), *J. Plant Physiol.*, 126, 41, 1986.

168. Fitch, M.M.M., and Moore, P.H., Long-term culture of embryogenic sugarcane callus, *Plant Cell Tiss. Org. Cult.*, 32, 335, 1993.

169. Lee, T.S.G., Micropropagation of sugarcane (*Saccharum* spp.), *Plant Cell Tiss. Org. Cult.*, 10, 47, 1987.

170. Songstad, D.D., Chen, C.H., and Boe, A.A., Plant regeneration in callus cultures derived from young inflorescences of Little Bluestem, *Crop Sci.*, 26, 827, 1986.

171. Rybczynski, J.J., and Zdunczyk, W., Somatic embryogenesis and plantlet regeneration in the genus *Secale*. 1. Somatic embryogenesis and organogenesis from cultured immature embryos of five wild species of rye, *Theor. Appl. Genet.*, 73, 267, 1986.

172. Zimny, J., and Lorz, H., High frequency of somatic embryogenesis and plant regeneration of rye (*Secale cereale* L.), *Plant Breed.*, 102, 89, 1989.

173. Lu, C.-Y., Chandler, S.F., and Vasil, I.K., Somatic embryogenesis and plant regeneration from cultured immature embryos of rye (*Secale cereale* L.), *J. Plant Physiol.*, 115, 237, 1984.

174. Bebeli, P., Karp, A., and Kaltsikes, P.J., Plant regeneration and somaclonal variation from cultured immature embryos of sister lines of rye and triticale differing in their content of heterochromatin I. Morphogenetic response, *Theor. Appl. Genet.*, 75, 929, 1988.

175. Linacero, R., and Vazquez, A.M., Somatic embryogenesis and plant regeneration from leaf tissues of rye (*Secale cereale* L.), *Plant Sci.*, 44, 219, 1986.

176. Linacero, R., and Vazquez, A.M., Somatic embryogenesis from immature inflorescences of rye, *Plant Sci. (Limmerick)*, 72, 253, 1990.

177. Vazquez, A.M., Espino, F.J., Rueda, J., Candela, M., and Sendino, A.M., A comparative study of somatic embryogenesis in *Secale vavilovii*, *Plant Cell Rep.*, 10, 265, 1991.

178. Zhi-hong, X., Da-yuan, W., Li-jun, Y., and Zhi-ming, W., Somatic embryogenesis and plant regeneration in cultured immature inflorescences of *Setaria italica*, *Plant Cell Rep.*, 3, 149, 1984.

179. Yeh, M.-L., and Chang, W.-C., Plant regeneration via somatic embryogenesis in mature embryo-derived callus culture of *Sinocalamus latiflora* (Munro) McClure, *Plant Sci.*, 51, 93, 1987.

180. George, L., and Eapen, S., Plant regeneration by somatic embryogenesis from immature inflorescence culture of *Sorghum almum*, *Ann. Bot. (London)*, 61, 589, 1988.

181. Boyes, C.J., and Vasil, I.K., Plant regeneration by somatic embryogenesis from cultured young inflorescences of *Sorghum arundinaceum* (Desv.) Stapf. var. Sudanense (Sudan grass), *Plant Sci. Lett.*, 35, 153, 1984.

182. Thomas, E., King, P.J., and Potrykus, I., Shoot and embryo-like structures from cultured tissues of *Sorghum bicolor*, *Naturwiss.*, 64, 587, 1977.

183. Brettell, R.I.S., Wernicke, W., and Thomas, E., Embryogenesis from cultured immature inflorescence of *Sorghum bicolor*, *Protoplasma*, 104, 141, 1980.

184. George, L., Eapen, S., and Rao, P.S., High frequency somatic embryogenesis and plant regeneration from immature inflorescence cultures of two Indian cultivars of sorghum (*Sorghum bicolor* L. Moench), *Proc. Indian Acad. Sci. (Plant Sci.)*, 99, 405, 1989.

185. Wernicke, W., Potrykus, I., and Thomas, E., Morphogenesis from cultured leaf tissue of *Sorghum bicolor* – The morphogenetic pathways, *Protoplasma*, 111, 53, 1982.

186. Wernicke, W., and Brettell, R., Somatic embryogenesis from *Sorghum bicolor* leaves, *Nature*, 287, 138, 1980.

187. Jabri, A., Chaussat, R., Jullien, M., and Le Deunff, Y., Callogenesis and somatic embryogenesis ability of leaf portions of three varieties of sorghum (*Sorghum bicolor* L. Moench) with and without tannins, *Agronomie*, 9, 101, 1989.

188. MacKinnon, C., Gunderson, G., and Nabors, M.W., Plant regeneration by somatic embryogenesis from callus cultures of sweet sorghum, *Plant Cell Rep.*, 5, 349, 1986.

189. Bhaskaran, S., and Smith, R.H., Control of morphogenesis in sorghum by 2,4-Dichlorophenoxyacetic acid and cytokinins, *Ann. Bot. (London)*, 64, 217, 1989.

190. George, L., Eapen, S., and Rao, P.S., Plant regeneration by somatic embryogenesis from inflorescence cultures of *Sorghum halepense* L. Pers. [Johnson Grass], *Ind. J. Exp. Biol.*, 29, 16, 1991.

191. Furini, A., and Jewell, D.C., Somatic embryogenesis and plant regeneration of *Tripsacum dactyloides* L., *Euphytica*, 55, 111, 1991.

192. Stolarz, A., and Lorz, H., Somatic embryogenesis, *in vitro* multiplication and plant regeneration from immature embryo explants of hexaploid *Triticale* (x *Triticosecale* Wittmack), *Z. Pflanzenzuchtg.*, 96, 353, 1986.

193. Vasil, V., Redway, F., and Vasil, I.K., Regeneration of plants from embryogenic suspension culture protoplasts of wheat (*Triticum aestivum* L.), *Bio/Technol.*, 429, 1990..

194. Ou, G., Wang, W.C., and Nguyen, H.T., Inheritance of somatic embryogenesis and organ regeneration from immature embryo cultures of winter wheat, *Theor. Appl. Genet.*, 78, 137, 1989.

195. Morris, P.C., Maddock, S.E., Jones, M.G.K., and Bowles, D.J., Lectin levels in tissues of cultured immature wheat embryos, *Plant Cell Rep.*, 5, 460, 1986.

196. He, D.G., Yang, Y.M., and Scott, K.J., A comparison of scutellum callus and epiblast callus induction in wheat: The effect of genotype, embryo age and medium, *Plant Sci.*, 57, 225, 1988.

197. Maddock, S.E., Lancaster, V.A., Risiott, R., and Franklin, J., Plant regeneration from cultured immature embryos and inflorescences of 25 cultivars of wheat (*Triticum aestivum*), *J. Exptl. Bot.*, 34, 915, 1983.

198. Ozias-Akins, P., and Vasil, I.K., Plant regeneration from cultured immature embryos and inflorescences of *Triticum aestivum* L. (Wheat): Evidence for somatic embryogenesis, *Protoplasma*, 110, 95, 1982.

199. He, D.G., Tanner, G., and Scott, K.J., Somatic embryogenesis and morphogenesis in callus derived from the epiblast of immature embryos of wheat (*Triticum aestivum*), *Plant Sci.*, 45, 119, 1986.

200. He, D.G., Yang, Y.M., and Scott, K.J., The effect of macroelements in the induction of embryogenic callus from immature embryos of wheat (*Triticum aestivum* L.), *Plant Sci.*, 64, 251, 1989.

201. He, D.G., Yang, Y.M., Bertram, J., and Scott, K.J., The histological development of the regenerative tissue derived from cultured immature embryos of wheat (*Triticum aestivum* L.), *Plant Sci. (Limmerick)*, 68, 103, 1990.

202. Vasil, V., Castillo, A.M., Fromm, M.E., and Vasil, I.K., Herbicide resistant fertile transgenic

464

wheat plants obtained by microprojectile bombardment of regenerable embryogenic callus, *Bio/Technol.*, 10, 667, 1992.

203. Ozias-Akins, P., and Vasil, I.K., Improved efficiency and normalization of somatic embryogenesis in *Triticum aestivum* (Wheat), *Protoplasma*, 117, 40, 1983.

204. Redway, F.A., Vasil, V., Lu, D., and Vasil, I.K., Identification of callus types for long-term maintenance and regeneration from commercial cultivars of wheat (*Triticum aestivum* L.), *Theor. Appl. Genet.*, 79, 609, 1990.

205. Nambisan, P., and Chopra, V.L., Somatic embryogenesis in *Triticum aestivum* L. Morphological observations on germination, *Indian J. Exp. Biol.*, 30, 12, 1992.

206. Bapat, S.A., Joshi, C.P., and Mascarenhas, A.F., Occurrence and frequency of precocious germination of somatic embryos is a genotype-dependent phenomenon in wheat, *Plant Cell Rep.*, 7, 538, 1988.

207. Magnusson, I., and Bornman, C.H., Anatomical observations on somatic embryogenesis from scutellar tissues of immature zygotic embryos of *Triticum aestivum*, *Physiol. Plant.*, 63, 137, 1985.

208. Bennici, A., Caffaro, L., Dameri, R.M., Gastaldo, P., and Profumo, P., Callus formation and plantlet regeneration from immature *Triticum durum* Desf. embryos, *Euphytica*, 39, 255, 1988.

209. Ozias-Akins, P., and Vasil, I.K., Proliferation of and plant regeneration from the epiblast of *Triticum aestivum* (Wheat: Gramineae) embryos, *Amer. J. Bot.*, 70, 1092, 1983.

210. Qureshi, J.A., Kartha, K.K., Abrams, S.R., and Steinhauer, L., Modulation of somatic embryogenesis in early and late-stage embryos of wheat (*Triticum aestivum* L.) under the influence of (±)-abscisic acid and its analogs, *Plant Cell Tiss. Org. Cult.*, 18, 55, 1989.

211. Galiba, G., and Yamada, Y., A novel method for increasing the frequency of somatic embryogenesis in wheat tissue culture by NaCl and KCl supplementation, *Plant Cell Rep.*, 7, 55, 1988.

212. Hunsinger, H., and Schauz, K., The influence of dicamba on somatic embryogenesis and frequency of plant regeneration from cultured immature embryos of wheat (*Triticum aestivum* L.), *Plant Breed.*, 98, 119, 1987.

213. Chu, C.C., Wang, Y.X., and Sang, J.L., Long term plant regeneration from callus cultures of haploid wheat, *Plant Cell Tiss. Org. Cult.*, 11, 221, 1987.

214. Rajyalakshmi, K., Grover, A., Maheshwari, N., Tyagi, A.K., and Maheshwari, S.C., High frequency regeneration of plantlets from the leaf-bases via somatic embryogenesis and comparison of polypeptide profiles from morphogenic and non-morphogenic calli in wheat (*Triticum aestivum*), *Physiol. Plant.*, 82, 617, 1991.

215. Heyser, J.W., Nabors, M.W., MacKinnon, C., Dykes, T.A., Demott, K.J., Kautzman, D.C., and Mujeeb-Kazi, A., Long-term, high-frequency plant regeneration and the induction of somatic embryogenesis in callus cultures of wheat (*Triticum aestivum* L.), *Z. Pflanzenzuchtg.*, 94, 218, 1985.

216. Kim, S.C., and Kim, S.-G., Plant regeneration from single cell culture of wheat (*Triticum aestivum* L.), *Korean J. Bot.*, 32, 227, 1989.

217. Wernicke, W., and Milkovits, L., The regeneration potential of wheat shoot meristems in the presence and absence of 2,4-Dichlorophenoxyacetic acid, *Protoplasma*, 131, 131, 1986.

218. Tabaeizadeh, Z., Plourde, A., and Comeau, A., Somatic embryogenesis and plant regeneration in *Triticum aestivum* × *Leymus angustus* F1 hybrids and the parental lines, *Plant Cell Rep.*, 9, 204, 1990.

219. Tabaeizadeh, Z., and Campeau, N., Embryogenic cell suspensions of *Triticum aestivum* × *Leymus angustus* F1 hybrids: characterization and plant regeneration, *Plant Cell Rep.*, 11, 81, 1992.

220. Li, Z.-Y., Xia, G.-M., and Chen H.-M., Somatic embryogenesis and plant regeneration from protoplasts isolated from embryogenic cell suspensions of wheat (*Triticum aestivum* L.), *Plant Cell Tiss. Org. Cult.*, 28, 79, 1992.

221. Rajyalakshmi, K., Dhir, S.K., Maheshwari, N., and Maheshwari, S.C., Callusing and regeneration of plantlets via somatic embryogenesis from inflorescence cultures of *Triticum aestivum* L.: Role of genotype and long-term retention of morphogenic potential, *Plant Breed.*, 101, 80, 1988.

222. Carman, J.G., Jefferson, N.E., and Campbell, W.F., Induction of embryogenic *Triticum aestivum* L. calli. II. Quantification of organic addenda and other culture variable effects, *Plant Cell Tiss. Org. Cult.*, 10, 115, 1987.
223. Carman, J.G., Jefferson, N.E., and Campbell, W.F., Induction of embryogenic *Triticum aestivum* L. calli. I. Quantification of genotype and culture medium effects, *Plant Cell Tiss. Org. Cult.*, 10, 101, 1987.
224. Borrelli, G.M., Lupotto, E., Locatelli, F., and Wittmer, G., Long-term optimized embryogenic cultures in durum wheat (*Triticum durum* Desf.), *Plant Cell Rep.*, 10, 296, 1991.
225. Wang, T.-B., Chien, Y.C., Li, J.-L., Qu, G.-P., and Tsai, C.K., Plant regeneration from protoplast of *Trititrigia* (*Triticum* sect. *Trititrigia* Mackey), *Acta Bot. Sinica*, 32, 329, 1990.
226. Redway, F.A., Vasil, V., and Vasil, I.K., Characterization and regeneration of wheat (*Triticum aestivum* L.) embryogenic cell suspension cultures, *Plant Cell Rep.*, 8, 714, 1990.
227. Kackar, A., and Shekhawat, N.S., Plant regeneration from cultured immature inflorescence of *Urochloa panicoides* (Beauv.), *Indian J. Exp. Biol.*, 29, 331, 1991.
228. Swedlund, B., and Locy, R.D., Somatic embryogenesis and plant regeneration in two-year old cultures of *Zea diploperennis*, *Plant Cell Rep.*, 7, 144, 1988.
229. Vasil, V., and Vasil, I.K., Formation of callus and somatic embryos from protoplasts of a commercial hybrid of maize (*Zea mays* L.), *Theor. Appl. Genet.*, 73, 793, 1987.
230. Vasil, V., Lu, C.-Y., and Vasil, I.K., Histology of somatic embryogenesis in cultured immature embryos of maize (*Zea mays* L.), *Protoplasma*, 127, 1, 1985.
231. Fransz, P.F., and Schel, J.H.N., An ultrastructural study on early callus development from immature embryos of the maize strains A188 and A632, *Acta Bot. Neerl.*, 36, 247, 1987.
232. Fransz, P.F., and Schel, J.H.N., Cytodifferentiation during the development of friable embryogenic callus of maize, *Can. J. Bot.*, 69, 26, 1991.
233. Close, K.R., and Gallagher-Ludeman, L.A., Structure-activity relationships of auxin-like plant growth regulators and genetic influences on the culture induction response in maize (*Zea mays* L.), *Plant Sci.*, 61, 245, 1989.
234. Fahey, J.W., Reed, J.N., Readdy, T.L., and Pace, G.M., Somatic embryogenesis from three commercially important inbreds of *Zea mays*, *Plant Cell Rep.*, 5, 35, 1986.
235. Lu, C.-Y., Vasil, V., and Vasil, I.K., Improved efficiency of somatic embryogenesis and plant regeneration in tissue cultures of maize (*Zea mays* L.), *Theor. Appl. Genet.*, 66, 285, 1983.
236. Willman, M.R., Schroll, S.M., and Hodges, T.K., Inheritance of somatic embryogenesis and plantlet regeneration from primary (type1) callus in maize, *In Vitro Cell. Dev. Biol.*, 25, 95, 1989.
237. Lu, C., Vasil, I.K., and Ozias-Akins, P., Somatic embryogenesis in *Zea mays*, *Theor. Appl. Genet.*, 62, 109, 1982.
238. Vasil, V., Vasil, I.K., and Lu, C.-Y., Somatic embryogenesis in long-term callus cultures of *Zea mays* L. (Gramineae), *Amer. J. Bot.*, 71, 158, 1984.
239. Radojević, L., Tissue culture of Maize *Zea mays* [Cudu] I. Somatic embryogenesis in the callus tissue, *J. Plant Physiol.*, 119, 435, 1985.
240. Kamo, K.K., Cahng, K.L., Lynn, M.E., and Hodges, T.K., Embryogenic callus formation from maize protoplasts, *Planta*, 172, 245, 1987.
241. Vasil, V., Lu, C.-Y., and Vasil, I.K., Proliferation and plant regeneration from the nodal region of *Zea mays* L. (Maize, Gramineae) embryos, *Amer. J. Bot.*, 70, 951, 1983.
242. Tomes, D.T., and Smith, O.S., The effect of parental genotype on initiation of embryogenic callus from elite maize (*Zea mays* L.) germplasm, *Theor. Appl. Genet.*, 70, 505, 1985.
243. Rapela, M.A., Organogenesis and somatic embryogenesis in tissue cultures of Argentine maize (*Zea mays* L.), *J. Plant Physiol.*, 121, 119, 1985.
244. Fransz, P.F., and Schel, J.H.N., An ultrastructural study on the early development of *Zea mays* somatic embryos, *Can. J. Bot.*, 69, 858, 1991.
245. Shi, J.-C., Liu, J.-H., and Guo, Z.-C., Plant regeneration from the protoplast culture of multitiller and multispike forage maize and seeding of the regenerated plant, *Acta Bot. Sinica*, 33, 409, 1991.

466

246. Prioli, L.M., and da Silva, W.J., Somatic embryogenesis and plant regeneration capacity in tropical maize inbreds, *Rev. Brasil. Genet.*, 12, 553, 1989.

247. Armstrong, C.L., and Green, C.E., Establishment and maintenance of friable, embryogenic maize callus and the involvement of L-Proline, *Planta*, 164, 207, 1985.

248. Novak, F.J., Daskalov, S., Brunner, H., Nesticky, M., Afza, R., Dolezelova, M., Lucretti, S., Herichova, A., and Hermelin, T., Somatic embryogenesis in maize and comparison of genetic variability induced by gamma radiation and tissue culture techniques, *Plant Breed.*, 101, 66, 1988.

249. Nesticky, M., Herichova, A., Novak, F.J., and Dolezelova, M., Somaclonal variation in plants derived from somatic embryos of maize (*Zea mays* L.), *Scientia Agriculturae Bohemoslovaca*, 19, 253, 1987.

250. Kamo, K.K., Becwar, M.R., and Hodges, T.K., Regeneration of *Zea mays* L. from embryogenic callus, *Bot. Gaz.*, 146, 327, 1985.

251. Lazanyi, J., Variation in the R3 generation of CHI-31 maize inbred line after *in vitro* regeneration of immature embryos, *Cereal Res. Commun.*, 16, 251, 1988.

252. McCain, J.W., and Hodges, T.K., Anatomy of somatic embryos from maize embryo cultures, *Bot. Gaz.*, 147, 453, 1986.

253. Lazanyi, J., Novak, F.J., Brunner, H., and Hermelin, T., Somaclonal variation in the R3-generation of a maize inbred line, *Acta Agronomica Hungarica*, 39, 101, 1990.

254. Vasil, V., and Vasil, I.K., Plant regeneration from friable embryogenic callus and cell suspension cultures of *Zea mays* L., *J. Plant Physiol.*, 124, 399, 1986.

255. McCain, J.W., Kamo, K.K., and Hodges, T.K., Characterization of somatic embryo development and plant regeneration from friable maize callus cultures, *Bot. Gaz.*, 149, 16, 1988.

256. Rao, K.V., Suprasanna, P., and Reddy, G.M., Somatic embryogenesis from immature glume calli of *Zea mays* L., *Indian J. Exp. Biol.*, 28, 531, 1990.

257. Xie, Y., Chen, L., and Dai, J., Study of somatic embryogenesis and plant regeneration in maize, *Acta Genetica Sinica*, 13, 113, 1986.

258. Close, K.R., and Ludeman, L.A., The effect of auxin-like plant growth regulators and osmotic regulation on induction of somatic embryogenesis from elite maize inbreds, *Plant Sci.*, 52, 81, 1987.

259. Conger, B.V., Novak, F.J., Afza, R., and Erdelsky, K., Somatic embryogenesis from cultured leaf segments of *Zea mays*, *Plant Cell Rep.*, 6, 345, 1987.

260. Emons, A.M.C., and Kieft, H., Histological comparison of single somatic embryos of maize from suspension culture with somatic embryos attached to callus cells, *Plant Cell Rep.*, 10, 485, 1991.

261. Woodward, B.R., Furze, M.J., and Cresswell, C.F., Callus initiation and somatic embryogenesis in root cultures of the maize inbred line 21A-6, *Afr. J. Bot.*, 56, 695, 1990.

262. Lu, T.-G., Sun, C.S., Jian, L.-C., and Sun, L.-H., Plant regeneration from cryopreserved protoplasts of maize (*Zea mays* L.), *Chinese J. Bot.*, 1, 165, 1989.

263. Pareddy, D.R., and Petolino, J.F., Somatic embryogenesis and plant regeneration from immature inflorescences of several elite inbreds of maize, *Plant Sci.*, 67, 211, 1990.

264. Ray, D.S., and Ghosh, P.D., Somatic embryogenesis and plant regeneration from cultured leaf explants of *Zea mays*, *Ann. Bot. (London)*, 66, 497, 1990.

265. Hodges, T.K., Kamo, K.K., Imbrie, C.W., and Becwar, M.R., Genotype specificity of somatic embryogenesis and regeneration in maize, *Bio/Technol.*, 4, 219, 1986.

266. Jong, T.-M., and Chang, W.-C., Somatic embryogenesis and plant regeneration from inflorescence callus of *Zizania latifolia* Turcz., *J. Plant Physiol.*, 130, 67, 1987.

267. Asano, Y., Somatic embryogenesis and protoplast culture in Japanese lawngrass (*Zoysia japonica*), *Plant Cell Rep.*, 8, 141, 1989.

268. Smith, D.L., and Krikorian, A.D., Growth and maintenance of an embryogenic cell culture of Daylily (*Hemerocallis*) on hormone-free medium, *Ann. Bot. (London)*, 67, 443, 1991.

269. Krikorian, A.D., and Kann, R.P., Plantlet production from morphogenetically competent cell suspensions of daylily, *Ann. Bot. (London)*, 47, 679, 1981.

270. Wang, L., Huang, B., He, M., and Hao, S., Somatic embryogenesis and its hormonal regulation in tissue cultures of *Freesia refracta*, *Ann. Bot. (London)*, 65, 271, 1990.

271. Kamo, K., Chen, J., and Lawson, R., The establishment of cell suspension cultures of *Gladiolus* that regenerate plants, *In Vitro Cell. Dev. Biol.*, 26, 425, 1990.

272. Laublin, G., Saini, H.S., and Cappadocia, M., *In vitro* plant regeneration via somatic embryogenesis from root culture of some rhizomatous irises, *Plant Cell Tiss. Org. Cult.*, 27, 15, 1991.

273. Lupi, C., Bennici, A., and Gennai, D., *In vitro* culture of *Bellevalia romana* (L.) Rchb. I. Plant regeneration through adventitious shoots and somatic embryos, *Protoplasma*, 125, 185, 1985.

274. Cavallini, A., and Natali, L., Cytological analyses of *in vitro* somatic embryogenesis in *Brimeura amethystina* Salisb. (Liliaceae), *Plant Sci.*, 62, 255, 1989.

275. Beyl, C.A., and Sharma, G.C., Picloram induced somatic embryogenesis in *Gasteria* and *Haworthia*, *Plant Cell Tiss. Org. Cult.*, 2, 123, 1983.

276. Chakravarty, B., and Sen, S., Regeneration through somatic embryogenesis from anther explants of *Scilla indica* (Roxb.) Baker, *Plant Cell Tiss. Org. Cult.*, 19, 71, 1989.

277. Jha, S., Cytological analysis of embryogenic callus and regenerated plants of *Urginea indica* Kunth., Indian Squill, *Caryologia*, 42, 165, 1989.

278. Jha, S., and Sen, S., Development of Indian squill (*Urginea indica* Kunth.) through somatic embryogenesis from long term culture, *J. Plant Physiol.*, 124, 431, 1986.

279. Jha, S., Sahu, N.P., and Mahato, S.B., Callus induction, organogenesis and somatic embryogenesis in three chromosomal races of *Urginea indica* and production of bufadienolides, *Plant Cell Tiss. Org. Cult.*, 25, 85, 1991.

280. Escalant, J.V., and Teisson, C., Somatic embryogenesis and plants from immature zygotic embryos of the species *Musa acuminata* and *Musa balbisiana*, *Plant Cell Rep.*, 7, 665, 1989.

281. Cronauer-Mitra, S.C., and Krikorian, A.D., Plant regeneration via somatic embryogenesis in the seeded diploid banana *Musa ornata* Roxb., *Plant Cell Rep.*, 7, 23, 1988.

282. Novak, F.J., Afza, R., Perea-Dallos, M., Van Duren, M., Conger, B.V., and Tang, X., Formation of somatic embryos in cultured tissues of diploid and triploid dessert bananas (*Musa* AA and AAA) and triploid cooking bananas (*Musa* ABB), *Int. Bot. Congr. Abstr.*, 17, 190, 1987.

283. Cronauer, S.S., and Krikorian, A.D., Response levels of *Musa* to various aseptic culture techniques, *Plant Physiol.*, 75, 14, 1984.

284. Cronauer, S., and Krikorian, A.D., Somatic embryos from cultured tissues of Triploid Plantains (*Musa*'ABB'), *Plant Cell Rep.*, 2, 289, 1983.

285. Lee, J.S., Kim, Y.J., Cheong, H.S., and Hwang, B., Studies on the induction of transformation and multiplication in orchid plants I. Formation of somatic embryos and regeneration from immature seeds of *Bletilla striata*, *Korean J. Bot.*, 33, 271, 1990.

286. Kerbauy, G.B., *In vitro* conversion of *Cattleya* root tip cells into protocorm like bodies, *J. Plant Physiol.*, 138, 248, 1991.

287. Hosoki, T., and Sagawa, Y., Clonal propagation of ginger (*Zingiber officinale* Roscoe) through tissue culture, *HortSci.*, 12, 451, 1977.

288. Kackar, A., Bhat, S.R., Chandel, K.P.S., and Malik, S.K., Plant regeneration via somatic embryogenesis in ginger, *Plant Cell Tiss. Org. Cult.*, 32, 289, 1993.

289. Murashige, T., and Skoog, F., A revised medium for rapid growth and bioassays with tobacco tissue cultures, *Physiol. Plant.*, 15, 473, 1962.

290. Chu, C.C., Wang, C.C., Sun, C.S., Hsu, C., Yin, K.C., Chu, C.Y., and Bi, F.Y., Establishment of an efficient medium for anther culture of rice through comparative experiments on the nitrogen sources, *Sci. Sin.*, 18, 659, 1975.

291. Schenk, R.U., and Hildebrandt, A.C., Medium and techniques for induction and growth of monocotyledonous and dicotyledonous plant cell cultures, *Can. J. Bot.*, 50, 199, 1972.

292. Gamborg, O.L., Miller, R.A., and Ojima, K., Nutrient requirements of suspension cultures of soyabean root cells, *Exp. Cell Res.*, 50, 151, 1968.

293. Linsmaier, E.M., and Skoog, F., Organic growth factor requirements of tobacco tissue culture, *Physiol. Plant.*, 18, 100, 1965.

294. Chuang, C.C., Ouyang, J.W., Chia, H., Chou, S.M., and Ching, C.K., A set of potato media for wheat anther culture, in *Proc. Symp. Plant Tiss. Cult.*, Science Press, Peking, 1978, 52.

295. Dunstan, D.I., and Short, K.C., Improved growth of tissue cultures of the onion, *Allium cepa*, *Physiol. Plant.*, 41, 70, 1977.

296. Abo El-Nil, M.M., and Zettler, F.W., Callus initiation and organ differentiation from shoot tip cultures of *Colacasia esculenta*, *Plant Sci. Lett.*, 6, 401, 1976.

297. Dalton, S.J., Plant regeneration from cell suspension protoplasts of *Festuca arundinaceae* Shreb. (tall fescue) and *Lolium perenne* L. (perennial ryegrass), *J. Plant Physiol.*, 132, 170, 1988.

298. Kano, K., Studies on the media for orchid seed germination, *Mem. Fac. Agri. Kagawa Univ.*, 20, 1, 1962.

299. Kao, K.N., and Michayluk, M.R., Nutritional requirements for growth of *Vicia hajastana* cells and protoplasts at a very low population density in liquid media, *Planta*, 126, 105, 1975.

300. Kao, K.N., Chromosomal behavior in somatic hybrids of soyabean-*Nicotiana glauca*, *Molec. Gen. Genet.*, 150, 225, 1977.

301. Ku, M.K., Cheng, W.C., Kuo, L.C., Kuan, Y.L., An, H.P., and Huang, C.H., Induction factors and morphocytological characteristics of pollen-derived plants in maize (*Zea mays*), in *Proc. Symp. Plant Tiss. Cult.*, Science Press, Peking, 1978, 35.

302. Ling, D.H., Xian, W.N., and Zeng, B.L., in *Proc. Symp. on Anther Culture*, Science Press, Peking, 1977, 265.

303. Muller, A.J., and Grafe, R., Isolation and characterization of cell lines of *Nicotiana tabacum* lacking nitrate reductase, *Mol. Gen. Genet.*, 161, 67, 1978.

304. Ohira, K., Ojima, K., and Fujiwara, A., Studies on the nutrition of rice cell culture 1. A simple, defined medium for rapid growth in suspension culture, *Plant Cell Physiol.*, 14, 1113, 1973.

305. Potrykus, I., Harms, C.T., and Lorz, H., Callus formation from cell culture protoplasts of corn (*Zea mays* L.), *Theor. Appl. Genet.*, 54, 209, 1979.

306. Smith, D.L., and Krikorian, A.D., Release of somatic embryogenic potential from excised zygotic embryos of carrot and maintenance of proembryonic cultures in hormone-free medium, *Amer. J. Bot.*, 76, 1834, 1989.

307. Vacin, E.F., and Went, F.W., Some pH changes in nutrient solutions, *Bot. Gaz.*, 110, 605, 1949.

308. White, P.R., Ed., *The Cultivation of Animal and Plant Cells*, 2nd Ed., Ronald Press, New York, 1963.

309. Rajasekaran, K., Hein, M.B., Davis, G.C., Carnes, M.G., and Vasil, I.K., Endogenous plant growth regulators in leaves and tissue cultures of Napier grass (*Pennisetum purpureum* Schum.), *J. Plant Physiol.*, 130, 13, 1987.

310. Rajasekaran, K., Hein, M.B., and Vasil, I.K., Endogenous abscisic acid and indole-3-acetic acid and somatic embryogenesis in cultured leaf explants of *Pennisetum purpureum* Schum.: effects *in vivo* and *in vitro* of glyphosate, fluridone and paclobutrazol, *Plant Physiol.*, 84, 47, 1987.

311. Vasil, I.K., Developing cell and tissue culture systems for the improvement of cereal and grass crops, *J. Plant Physiol.*, 128, 193, 1987.

312. Franco, M.A., and Vasil, I.K., Development of somatic embryos and plants from stem internode thin layers and discs of *Pennisetum purpureum* Schum. (Napier grass) (submitted).

313. Higgins, P., and Mathias, R.J., The effect of the 4B chromosomes of hexaploid wheat on the growth and regeneration of callus cultures, *Theor. Appl. Genet.*, 74, 439, 1987.

314. Vasil, I.K., Progress in the regeneration and genetic manipulation of cereal crops, *Bio/Technol.*, 6, 397, 1988.

315. Duncan, D.R., Williams, M.E., Zehr, B.E., and Widholm, J.M., The production of callus capable of plant regeneration from immature embryos of numerous *Zea mays* (L.) genotypes, *Planta*, 165, 322, 1985.

316. Knudson, L., Nutrient solutions for orchids, *Bot. Gaz.*, 112, 528, 1951.

317. Sagawa, Y., and Kunisaki, J.T., Clonal propagation: Orchids, in *Cell Culture and Somatic Cell Genetics of Plants, Vol. 1, Laboratory Procedures and their Applications*, Vasil, I.K., Ed., Academic Press, Orlando, 1984, 61.

318. George, E.F., Puttock, D.J.M., and George, H.J., *Plant Culture Media, Vol. 1, Formulations and Uses*, Exegetics Limited, Edington, 1987.

319. George, E.F., Puttock, D.J.M., and George, H.J., *Plant Culture Media, Vol. 2, Commentary and Analysis*, Exegetics Limited, Edington, 1988.

320. Krikorian, A.D., Kann, R.P., and Fitter-Corbin, M.S., Daylily, in *Handbook of Plant Cell Culture, Vol. 5, Ornamental Species*, Ammirato, P.V., Evans, D.R., Sharp, W.R., and Bajaj, Y.P.S., Eds., McGraw-Hill, New York, 1990, 375.

321. Karlsson, S.B., and Vasil, I.K., Morphology and ultrastructure of embryogenic cell suspension cultures of *Panicum maximum* Jacq. (Guinea grass) and *Pennisetum purpureum* Schum. (Napier grass), *Amer. J. Bot.*, 73, 894, 1986.

322. Chen, T.H.H., Kartha, K.K., and Gusta, L.V., Cryopreservation of wheat suspension culture and regenerable callus, *Plant Cell Tiss. Org. Cult.*, 4, 101, 1985.

323. Gnanapragasam, S., and Vasil, I.K., Plant regeneration from a cryopreserved embryogenic cell suspension of a commercial sugarcane hybrid (*Saccharum* sp.), *Plant Cell Rep.*, 9, 419, 1990.

324. Gnanapragasam, S., and Vasil, I.K., Cryopreservation of immature embryos, embryogenic callus and cell suspension cultures of gramineous species, *Plant Sci.*, 83, 205, 1992.

325. Cavallini, A., Lupi, M.C., Cremonini, R., and Bennici, A., *In vitro* culture of *Bellevalia romana* (L.) Rchb. III. Cytological study of somatic embryos, *Protoplasma*, 139, 66, 1987.

326. Morrish, F.M., Hanna, W.W., and Vasil, I.K., The expression and perpetuation of inherent somatic variation in regenerants from embryogenic cultures of *Pennisetum glaucum* (L.) R.Br. (Pearl millet), *Theor. Appl. Genet.*, 80, 409, 1990.

327. Gmitter, F.G. Jr., Ling, X., Cai, C., and Grosser, J.W., Colchicine induced polyploidy in *Citrus* embryogenic cultures, somatic embryos and regenerated plants, *Plant Sci.*, 74, 135, 1991.

328. Shenoy, V.B., and Vasil, I.K., Biochemical and molecular analysis of plants derived from embryogenic tissue cultures of napier grass (*Pennisetum purpureum* K. Schum.), *Theor. Appl. Genet.*, 83, 947, 1992.

329. Datta, S.K., and Potrykus, I., Artificial seeds in barley: encapsulation of microspore-derived embryos, *Theor. Appl. Genet.*, 77, 820, 1989.

330. Hayashi, Y., Kyozuka, J., and Shimamoto, K., Hybrids of rice (*Oryza sativa* L.) and wild *Oryza* species obtained by cell fusion, *Molec. Gen. Genet.*, 214, 6, 1988.

331. Kyozuka, J., Taneda, K., and Shimamoto, K., Production of cytoplasmic male sterile rice (*Oryza sativa* L.) by cell fusion, *Bio/Tecnol.*, 7, 1171, 1989.

332. Takamizo, T., Spagenberg, G., Suginobu, K., and Ptrykus, I., Intergeneric somatic hybridization in Gramineae: somatic hybrid plants between tall fescue (*Festuca arundinacea* Schreb. and Italian ryegrass (*Lolium multiflorum* Lam.), *Molec. Gen. Genet.*, 231, 1, 1991.

333. Horn, M.E., Shillito, R.D., Conger, B.V., and Harms, C.T., Transgenic plants of Orchardgrass (*Dactylis glomerata* L.) from protoplasts, *Plant Cell Rep.*, 7, 469, 1988.

334. Shimamoto, K., Terada, R., Izawa, T., and Fujimoto, H., Transgenic rice plants regenerated from transformed protoplasts, *Nature*, 338, 274, 1989.

335. Donn, G., Nilges, M., and Morocs, S., Stable transformation of maize with a chimeric modified phosphinothricin-acetyltransferase gene from *Streptomyces viridochromogenes*, in *Progress in Plant Cellular and Molecular Biology*, Nijkamp, H.J.J., Van der Plas, L.H.W., and Aartrij, J., Eds., Kluwer Academic Publishers, Dordrecht, 1990, 53.

336. Bower, R., and Birch, R.G., Transgenic sugarcane plants via microprojectile bombardment, *Plant J.*, 2, 409, 1992.

337. Wang, Z., Takamizo, T., Iglesias, V.A., Osusky, M., Nagel, J., Potrykus, I., and Spangenberg, G., Transgenic plants of tall fescue (*Festuca arundinacea* Schreb.) obtained by direct gene transfer to protoplasts, *Bio/Technol.*, 10, 691, 1992.

338. Fromm, M.E., Morrish, F., Armstrong, C., Williams, R., Thomas, J., and Klein, T.M., Inheritance and expression of chimeric genes in the progeny of transgenic maize plants, *Bio/Technol.*, 8, 833, 1990.

339. Gordon-Kamm, W.J., Spencer, T.M., Mangano, M.L., Adams, T.R., Daines, R.J., Start,

470

W.G., O'Brien, J.V., Chambers, S.A., Adams, W.R., Willets, N.G., Rice, T.B., Mackey, C.J., Krueger, R.W., Kausch, A.P., and Lemaux, P.G., Transformation of maize cells and regeneration of fertile transgenic plants, *Plant Cell*, 2, 603, 1990.

340. Christou, P., Ford, T.L., and Kofron, M., Production of transgenic rice (*Oryza sativa* L.) plants from agronomically important Indica and Japonica varieties via electric discharge particle accelaration of exogenous DNA into immature zygotic embryos, *Bio/Technol.*, 9, 957, 1991.

12. Somatic Embryogenesis in Woody Plants

DAVID I. DUNSTAN, THOMAS E. TAUTORUS
and TREVOR A. THORPE

Contents

I. Introduction

A. Woody plants

Woody plants may be defined as trees and shrubs "whose stems live for a number of years and increase in diameter each year by addition of woody tissue" [1]. This contrasts with the description for herbaceous plants "with a relatively short-lived aboveground stem that is comparatively thin, soft, and nonwoody" [1]. Included in this definition of woody plants are those vines which are woody, i.e. "plants whose stems are not self-supporting, but either trail on the ground

471

T.A. Thorpe (ed.), In Vitro Embryogenesis in Plants, pp. 471–538.
© 1995 *Kluwer Academic Publishers, Dordrecht. Printed in the Netherlands.*

or climb by attaching to other plants or supports" [1]. In this context species such as *Manihot esculenta* and *Carica papaya*, included as woody by others [2, 3] are regarded as herbaceous here.

Woody plants therefore have a number of common features, for example because they are long-lived they are exposed to and must survive the numerous environmental changes (air movement, precipitation, drought, temperature, nutrient availability) that can occur over many years. Included in this are the prolonged exposures to solar radiation and to the prevailing gaseous atmosphere. This long-lived attribute presumably has an evolutionarily competitive advantage in some respects, i.e. the plant is already there occupying a niche. The woodiness and the size of many woody plants also must be evolutionarily advantageous in their ability to survive conditions inclement to non-woody plants and in their ability to shade other plants. Woody plants are invariably seed plants, i.e. gymnosperms and angiosperms (monocotyledons and dicotyledons). However, the range of morphologies of woody plants is broad and it is hard to see further similarities among such diverse examples as can be seen from Tables 1 and 2. Not surprisingly therefore somatic embryogenesis and the conditions under which it has been achieved are very variable.

Woody plant somatic embryogenesis is briefly explored in Section II, principally by reference to the tables. Following that are three specific examples of woody species; *Citrus sinensis*, *Elaeis guineensis*, and *Picea glauca*; which have been selected because of their socio-economic importance and because of the considerable differences that they present in both the manner in which somatic embryogenesis has been achieved and in the research emphasis with each. In addition to these species, it is worth noting that among the other species listed in the tables research with the following is prominent: *Gossypium hirsutum*, *Hevea brasiliensis*, *Phoenix dactylifera*, *Picea abies*, *Santalum album*, *Theobroma cacao*, and *Vitis vinifera*.

II. Woody plant somatic embryogenesis

A. Historical notes

Somatic embryogenesis with woody plants dates back to the 1960s, the first reports occurring with *Citrus spp.* [4–6], *Biota orientalis* [7], *Santalum album* [8], and *Zamia integrifolia* [9, 10]. However, it was not until the 1980s that woody plant somatic embryogenesis was more widely reported (over 180 citations for angiosperms, and over 35 citations for gymnosperms; 1985 was the first year in which conifer somatic embryogenesis was reported). To date somatic embryogenesis has been reported for approximately 150 woody species and related hybrids, presenting a considerable diversity (Tables 1 and 2).

Table 1. Induction of somatic embryogenesis and production of plantlets in woody angiosperms.

Species	Explant	Induction Media[a]	Induction Phytohormones μM[b]	Plantlet Production[b]	References
Acanthopanax senticosus	mature zygotic embryo	MS	2.3 2,4-D	Yes	11
Achras sapota	various	White's I	21.5 NAA, 1.9 KIN	Yes	12
Actinidia chinensis	anther	LS or B5	24.6 2-iP, 1.7–5.7 IAA	Yes	13
Actinidia deliciosa	anther	LS or B5	24.6 2-iP, 1.7 IAA	No	13
	young leaf	MS	9 ZEA	No	433
Aesculus hippocastanum	young leaf	MS	9 2,4-D, 10.8 NAA, 9.3 KIN	Yes	14, 15
	immature zygotic embryo	MS	13.6 2,4-D, 4.7 KIN	Yes	16
	filament	WPM	1) 2.5, 5 2,4-D[c], 2.5, 5 BA; 2) 0.75 BA	No	17
	cotyledon	MS	0.5 2,4-D	Yes	18
	immature zygotic embryo filament	MS	4.4–8.8 2,4-D, 5.4 NAA	Yes	451
Albizzia lebbeck	hypocotyl	B5	none	Yes	19
Albizzia richardiana	hypocotyl	B5	1) 1 BA;[c] 2) none	Yes	20
Bactris gasipaes	shoot tip	MS	1) 0.3 PIC,[c] 22.2 BA; 2) 0.1 PIC, 22.2 BA; 3) none	Yes	21
Betula pendula	seed seedlings	N7	9 2,4-D, 2.3 KIN	No	22
	leaf	N7	9 2,4-D, 2.3 KIN	Yes	22
	leaf	N7	1) 9 2,4-D[c], 2.3 KIN; 2) 0.5 2,4-D, 0.1 KIN		
Brahea dulcis	mature zygotic embryo	MS	452.5 2,4-D, 14.8 2-iP	No	23
Camellia chrysantha	cotyledon	MS	BA,[d] NAA[d]	No	24
Camellia japonica	immature, mature zygotic embryo	MS	none	Yes	25
	root	MS	none	Yes	26
	cotyledonary stage somatic embryo	MS	4.4 BA, 0.5 IBA	No	27
Camellia reticulata	immature zygotic embryo	MS	2.2 BA, 0.5 IBA	Yes	27
Camellia sasanqua	cotyledon	MS	4.4 BA, 1.1–2.7 NAA	No	430
				Yes	28

Table 1. Continued

Species	Explant	Induction Media[a]	Induction Phytohormones µM[b]	Plantlet Production[b]	References
Camellia sinensis	cotyledon	MS	44.4 BA, 2.5 IBA	Yes	434
Carya illinoensis	immature zygotic embryo	WPM	1) 9 2,4-D,[c] 1.1 BA; 2) none	No	29
Castanea dentata	immature zygotic embryo	WPM	1) 9 2,4-D,[c] 1.1 BA; 2) none	Yes	30
	ovule	WPM	1) 1.1 BA,[c] 32.3 NAA; 2)	No	436
Castanea sativa × Castanea crenata (hybrid)	cotyledon	MS	none	No	31
	immature zygotic embryo	MS	9 2,4-D	No	32
Cercis canadensis	immature zygotic embryo	SH	2.3 or 4.5 2,4-D	No	33
	immature zygotic embryo	WPM	13.6 2,4-D	Yes	34
Chamaedorea costaricana	ovule	MS	5 2,4-D	No	35
Citrus sp.	nucellus	White's I	452.5 2,4-D, 4.9 2-iP	Yes	4
Citrus aurantifolia	ovary	MS	9.3 KIN	Yes	36
	ovule	MS	2.3 KIN, 1.4 IAA	Yes	36
	ovule	MT	0.5 KIN	Yes	37
Citrus aurantifolia, Citrus sinensis (donor plants)	protoplasts (embryogenic suspension,[e] leaf protoplasts)	MT	none	Yes (somatic hybrid)	38
Citrus aurantium	ovule	MT	none	No	39–41
	ovule	MT	none	Yes	42
	protoplasts (embryogenic tissue[f])	MT	none	Yes	42
Citrus aurantium, Citrus limon (donor plants)	mature zygotic embryo	MT	1) 9 2,4-D,[c] 22.2 BA; 2) none	Yes	43
	protoplasts (γ-irradiated embryogenic tissue,[f] iodoacetate tissue[f])	MT	none	Yes (cybrid)	44
Citrus clementina	nucellus	MS	none	Yes	45
Citrus grandis	nucellus	MS	2.7 NAA	Yes	6

Table 1. Continued

Species	Explant	Induction Media[a]	Induction Phytohormones μM[b]	Plantlet Production[b]	References
	stem	MS	1) 1.2 KIN,[c] 13.4 NAA, 1.1 2,4-D; 2) 1.2 KIN, 2.7 NAA, 1.1 2,4-D; 3) 1.1 ZEA, 0.5 NAA	No	46
Citrus jambhiri	ovule	MT	none	No	41
Citrus limon	nucellus	MS	2.7 NAA	Yes	6
	ovule	MT	none	No	40, 41
	protoplasts (embryogenic tissue[f])	MT	none	Yes	42
Citrus madurensis	ovule	MT	none	Yes	37, 47
	anther	MS	11.4 IAA or 1.1 NAA, 0.1 KIN	Yes	48
	protoplasts (embryogenic tissue[g])	MT	none	Yes	48
Citrus microcarpa	nucellus	White's I	none	Yes	5
	ovule	White's I	5.7 IAA	Yes	49
Citrus mitis	ovule	MT	none	Yes	50
	protoplasts (embryogenic suspension[f])	MT	none	Yes	50
Citrus nobilis	ovule	MT	none	No	40
Citrus paradisi	ovule, nucellus	MT	none	Yes	51
	stem, leaf segments	MT	1) 2.7 NAA;[c] 2) 2.3 2,4-D; 3) 2.2 BA, 0.8 NAA	No	52
	ovule	MT	none	Yes	42
	protoplasts (embryogenic tissue[f])	MT	none	Yes	42
	ovule	MT	none	No	41
	ovule	MT	0.1 DAM	Yes	47
	ovule	MT	0.05 2,4-D	Yes	37

Table 1. Continued

Species	Explant	Induction Media[a]	Induction Phytohormones μM[b]	Plantlet Production[b]	References
	stem, root segments	1) MS;[c] 2)½MS	1) 53.8 NAA,[c] 0.9 KIN; 2) 22.2 BA	Yes	53
Citrus paradisi × *Citrus reticulata* (hybrid)	nucellus	MT	none	No	54
Citrus reticulata	ovule	MT	none	Yes	37
	nucellus	MT	none	Yes	55
	nucellus	MS	none	No	56
	ovule	MT	none	Yes	42, 47
	protoplasts (embryogenic tissue[f])	MT	none	Yes	42
	ovule	MT	none	No	41
	immature zygotic embryo	MS	50 KIN	No	57
	protoplasts (embryogenic tissue[h])	MS	1 IAA, 1 ZEA, 10 GA$_3$	Yes	57
Citrus reticulata, Citropis gilletiana (donor plants)	protoplasts (embryogenic suspension,[e] leaf protoplasts)	MT	none	Yes (somatic hybrid)	58
Citrus reticulata × *Citrus paradisi* (hybrid)	ovule	MT	none	Yes	59
Citrus reticulata × *Citrus sinensis* (hybrid)	nucellus	MS	2.7 NAA	Yes	6
Citrus sinensis	ovule	MS or White's	none	Yes	60
	ovule	MT	none	Yes	37, 42, 47, 51, 59, 61, 64
	nucellus	MT	none	Yes	51
	ovary	MS	2.3 KIN, 1.4 IAA	Yes	36
	ovule	MT	none	No	39–41, 65, 66, 69, 70
	protoplasts (embryogenic tissue[g])	MT	none	No	71
	protoplasts (embryogenic tissue[f])	MT	none	Yes	42, 69, 72

Table 1. Continued

Species	Explant	Induction Media[a]	Induction Phytohormones μM[b]	Plantlet Production[b]	References
	immature zygotic embryo	MS	50 KIN	No	57
	protoplasts (embryogenic tissue[h])	MS	1 IAA, 1 ZEA, 10 GA$_3$	Yes	57
Citrus sinensis, Citrus paradisi (donor plants)	ovule	MS	none	Yes	73
	ovule	MT	44.4 BA	Yes	74, 75
	protoplasts (embryogenic suspension,[f] leaf protoplasts)	MT	none	Yes (somatic hybrid)	76
Citrus sinensis, Citrus unshiu (donor plants)	protoplasts (embryogenic suspension,[f] leaf protoplasts)	MT	none	Yes (somatic hybrid)	77
Citrus sinensis, "Murcott" tangor (donor plants)	protoplasts (embryogenic suspension,[f] leaf protoplasts)	MT	none	Yes (somatic hybrid)	78
Citrus sinensis, Poncirus trifoliata (donor plants)	protoplasts (embryogenic suspension,[f] leaf protoplasts)	MT	none	Yes (somatic hybrid)	79
	protoplasts (embryogenic suspension,[i] leaf protoplasts)	MT	none	Yes (somatic hybrid)	80
	mature zygotic embryo	MT	1) 9 2,4-D[c] 22.2 BA; 2) none	Yes	43
Citrus sinensis, Severinia disticha (donor plants)	protoplasts (embryogenic suspension,[e] epicotyl tissue protoplasts)	MT	none	Yes (somatic hybrid)	81
Citrus sudachi, Citrus aurantifolia (donor plants)	protoplasts (embryogenic suspension,[f] leaf protoplasts)	MT	none	Yes (somatic hybrid)	82
Citrus unshiu	ovule	MT	none	No	83
	protoplasts (embryogenic tissue[f])	MT	none	Yes	83
Citrus yuko	immature zygotic embryo	MS	50 KIN	No	57
	protoplasts (embryogenic tissue[h])	MS	1 IAA, 1 ZEA, 10 GA$_3$	Yes	57
Cladastris lutea	immature zygotic embryo	SH	1) 9 2,4-D;[c] 2) 25 NAA	Yes	84

Table 1. Continued

Species	Explant	Induction Media[a]	Induction Phytohormones μM[b]	Plantlet Production[b]	References
Cocos nucifera	stem segments rachillae	MS	1) 5 BA,[c] 5 2-iP, 100 2,4-D; 2) 5 BA, 5 2-iP, 0.01 2,4-D[j]	No	85
	rachillae	MS	1) 5 BA,[c] 5 2-iP, 100 2,4-D; 2) 5 BA, 5 2-iP, 0.01 2,4-D[j]	Yes	86
	immature zygotic embryo	KP	12–20 2,4-D	No	452
	young leaf	KP	12–20 2,4-D	No	453
	young leaf	Y3	140–270 2,4-D[j]	Yes	454
Coffea arabica	mature leaf	1) MS;[c] 2) ½MS; 3) ½MS	1) 18.4 KIN,[c] 4.5 2,4-D; 2) as above; 3) 2.3 KIN, 0.3 NAA	Yes	87, 88
	mature leaf	1) MS;[c] 2) ½MS	1) 20 KIN,[c] 5 2,4-D; 2) 2.5 KIN, 0.25 NAA	No	89
	young leaf	MS	5 BA	Yes	90
	leaf	½MS	35.6 BA, 4.5 2,4-D	Yes	91
	young leaf	¼MS	4.4 BA	No	92
	young leaf	MS	1) 18.6 KIN,[c] 4.5 2,4-D; 2) 2.3 KIN, 0.3 NAA	No	447
Coffea canephora	internode stem sections	LS	0.5 KIN, 0.5 2,4-D	Yes	93
	internode stem sections	MS	24.6 IBA, 4.4 BA	No	94
	ovule	½MS	0.2 2,4-D, 1 NAA	Yes	95
	young leaf	½MS	4.9 2-iP, 24.6 IBA	No	96
	protoplasts (embryogenic suspension[k])	B5	1) 2.3 2,4-D,[c] 2.7 NAA, 2.3 KIN; 2) none	Yes	97
Cornus florida	leaf discs	¼MS	5 2-iP	Yes	98
Corylus avellana	leaf	MS	1.4 2,4-D, 7 KIN	Yes	456
	immature zygotic embryo	MS or SH	13.6 2,4-D	Yes	99
	immature zygotic embryo	MS	4.5 2,4-D, 4.7 KIN	Yes	100
	immature zygotic embryo	MS	4.5 2,4-D, 4.7 KIN	No	101

Table 1. Continued

Species	Explant	Induction Media[a]	Induction Phytohormones µM[b]	Plantlet Production[b]	References
	cotyledonary nodes	K (h)	1) 0.5 BA,[c] 5 IBA; 2) 5 BA, 0.5 IBA	Yes	102, 103
Cyclamen persicum	leaf blade, leaf stalk, ovary	MS	0.5–5.7 NAA, IAA, or KIN, 2.3–11.6 KIN	Yes	104
Diospyros kaki	leaf segments	MS	10 NAA, 1 BA	No	105
Elaeis guineensis	mature zygotic embryo	Heller's	1) 1 2,4-D,[c] 0.5–1 KIN; 2) 1 IAA	No	106
	leaf	MS	1) 9 2,4-D;[c] 2) 2.3–9 2,4-D, 2.7–10.8 NAA	No	107
	young leaf	1) MS;[c] 2) ½MS	1) 226.2–316.7 2,4-D;[c] 2) 45.2 2,4-D, 0.05 2-iP; 3) 0.05–2.5 2-iP	Yes	108
	mature zygotic embryo, young leaf	½MS	268.8 NAA	Yes	109
Ephedra gerardiana	young leaf	MS	auxin[d]	No	110
	callus[i]	MS	1) 0.5 KIN,[c] 53.8 NAA; 2) 23.3 KIN, 5.4 NAA	No	111
Eucalyptus citriodora	mature zygotic embryo	B5	16.1 NAA	Yes	112, 113
Eucalyptus grandis	young leaf	MS	1.1–4.5 2,4-D	Yes	435
Eucalyptus teichow	seedling	B5	1) 3.7–4.7 KIN,[c] 9 2,4-D; 2) 0.9–4.4 BA, 2.7 NAA	Yes	114
Eugenia jambos	ovule	½MS	4.5, 9 2,4-D	No	115
Eugenia malaccensis	ovule	½MS	9 2,4-D	No	115
Euphoria longan	young leaf	B5	4.5 2,4-D, 9.2 KIN	Yes	116
Euterpe edulis	immature, mature zygotic embryo	LS	452.5 2,4-D	Yes	117
Fagus sylvatica	anther	MS	5 2,4-D, 2.5 BA	No	118
Ficus religiosa	young leaf	MS	1) 2.3 2,4-D;[c] 2) 0.4 BA, 0.3 NAA	Yes	119

Table 1. Continued

Species	Explant	Induction Media[a]	Induction Phytohormones μM[b]	Plantlet Production[b]	References
Fraxinus americana	seed	DKW	1) 5 NAA;[c] 5 BA; 2) none	Yes	120
Fuchsia sp.	ovary	B5	1) 4.5 2,4-D or 5.7 IAA;[c] 2) KIN[d]	Yes	121
	ovary	B5	1) 4.5 2,4-D;[c] 2) 5.7 IAA, 4.7 KIN	No	122
Gossypium barbadense	hypocotyl	MS	0.5 2,4-D, 2.3 KIN	No	123
Gossypium barbadense ×	hypocotyl	MS	0.5 2,4-D, 0.5, 2.3 KIN	No	123
Gossypium hirsutum (hybrid)					
Gossypium hirsutum	cotyledon	LS	1) 10.8 NAA,[c] 4.7 KIN; 2) 5.4 NAA, 2.3 KIN; 3) none	Yes	124
	leaf disc	LS	2.7 NAA, 4.7 KIN, 0.3 GA$_3$	No	125
	petiole	MS	21.5 NAA, 4.7 KIN	Yes	125, 126
	hypocotyl	MS	11.4 IAA, 4.7 KIN	Yes	127
	hypocotyl	MS	11.4 IAA, 4.7 KIN	No	128
	hypocotyl	MS	0.5 2,4-D, 0.5 KIN	Yes	123, 129, 130
	hypocotyl	MS	1) 0.5 2,4-D,[c] 0.5 KIN; 2) none	Yes	131
	cell suspensions[l]	MS	2.1 PIC or 0.5 2,4-D	Yes	132, 133
	stem	MS	9.9 NAA, 0.5 KIN	No	134
	leaf disc	MS	0.5 2,4-D, 4.9 2-iP	No	134
	various seedling sections	MS	0.5 2,4-D, 2.3 KIN	No	134
	petiole	MS	21.5 NAA, 4.7 KIN	No	135
	cotyledon	LS	1) 0.5 2,4-D,[c] 0.5 KIN; 2) 21.5 NAA, 4.7 KIN	No	444
Gossypium klotzschianum	hypocotyl	MS	11.4 IAA, 2.3 KIN	Yes	432
Hedera helix	cell suspensions[k,l]	B5	+/- 0.5 2,4-D	No	136, 137
	stem segments (mature)	1) MS; 2) White's II	1) 21.5 NAA;[c] 9.3 KIN; 2) 5.4 NAA, 2.3 KIN	Yes	138
Hevea brasiliensis	anther	MS	4.5 2,4-D, 4.7 KIN	Yes	455

Table 1. Continued

Species	Explant	Induction Media[a]	Induction Phytohormones μM[b]	Plantlet Production[b]	References
	seed inner integument	1) MH1; 2) MH2; 3) MH3	1) 1.4 2,4-D,[c] 5.7 IAA, 4.4–22.2 BA; 2) as above; 3) 2.5 NOA, 2.2 BA	No	139
	seed inner integument	1) MH1; 2) MH2	1) 9 3,4-D,[c] 9 BA; 2) 0.45 NOA, 0.45 BA	No	140
	seed inner integument	1) MH1; 2) MH3	1) 9 3,4-D,[c] 4.4–22.2 BA; 2) 2.5 NOA, 2.2 BA	No	141
	seed inner integument	1) MH1; 2) MH3	1) 9 2,4-D,[c] 9 BA; 2) 2 2,4-D, 2 BA	No	142
	seed inner integument	1) MH1; 2) MH1; 3) MH2	1) 4.5 3,4-D,[c] 4.5 BA; 2) 0.45 3,4-D, 0.45 BA; 3) 0.45 NOA, 0.45 BA	Yes	143
	seed inner integument	MS	1) 4.5 3,4-D,[c] 4.5 BA; 2) 0.45 3,4-D, 0.45 BA	No	144
	seed inner integument	MS	1) 3 3,4-D,[c] 3 BA; 2) 1.4 3,4-D, 1.4 BA, 0.05 ABA	Yes	448
Hibiscus acetosella	young leaf	Medium H	1) 0.6 CHL,[c] 0.3 NOA, 49.3 2-iP; 2) 4.5 2,4-D, 4.7 KIN; 3) 0.6 CHL, 18.1 2,4-D	Yes	145
	seedling root, cotyledon	Medium H	1) 4.5 2,4-D,[c] 4.7 KIN; 2) 0.6 CHL, 18.1 2,4-D	Yes	145
Howeia forsteriana	ovule	MS	1) 452.5 2,4-D,[c] 4.9 2-iP; 2) none	No	35
Ilex aquifolium	immature zygotic	LS	none	No	146
	immature zygotic embryo	LS	none	Yes	68, 147
Juglans hindsii	immature cotyledon	DKW	1) 4.4 BA,[c] 9.3 KIN, 0.05 IBA; 2) none	Yes	148
Juglans major	immature cotyledon	DKW	1) 4.4 BA,[c] 9.3 KIN, 0.05 IBA; 2) none	Yes	457

Table 1. Continued

Species	Explant	Induction Media[a]	Induction Phytohormones μM[b]	Plantlet Production[b]	References
Juglans nigra	immature cotyledon	DKW	1) 4.4 BA,[c] 9.3 KIN, 0.05 IBA; 2) none	Yes	457
Juglans nigra × Juglans regia (hybrid)	immature cotyledon	DKW	1) 4.4 BA,[c] 9.3 KIN, 0.05 IBA; 2) none	Yes	457
	immature cotyledon	DKW	1) 4.4 BA,[c] 9.3 KIN, 0.05 IBA; 2) none	Yes	457, 458
Juglans regia	immature cotyledon	DKW	1) 4.4 BA,[c] 9.3 KIN, 0.05 IBA; 2) none	Yes	148–150
	endosperm	DKW	1) 4.4 BA,[c] 9.3 KIN, 0.05 IBA; 2) 0.4 BA, 4.9 IBA, 0.9 KIN; 3) same as 1)	Yes	151
Leucosceptrum canum	young leaf	MS	2.3 2,4-D	Yes	152
Liquidambar styraciflua	hypocotyl	Blaydes	1) 5.4 NAA,[c] 8.9 BA; 2) none	Yes	153
Liriondendron tulipifera	immature zygotic embryo	Blaydes	1) 9 2,4-D,[c] 22.2 BA; 2) 9 2,4-D, 1.1 BA	Yes	154
	protoplasts (embryogenic suspension[h])	Blaydes	1) 4.5 2,4-D,[c] 1.1 BA; 2) none	Yes	155
Livistona decipiens	immature zygotic embryo	Blaydes	9 2,4-D, 1.1 BA	No	156
	mature zygotic embryo	MS	452.5 2,4-D, 14.8 2-iP	No	24
Malus domestica	nucellus	MS	none	No	157
	young leaf	MS	44.4 BA, 16.1 NAA	Yes	158
Malus pumila	nucellus, endosperm	LS	none	No	159 (see also 228)
	immature zygotic embryo	LS	4.4 BA, 0.5 NAA	No	159
	various	1) White's I; 2) Nitsch I	1) 21.5 NAA,[c] 9.3 KIN, 2.9 GA3; 2) 10.8 NAA, 8.9 BA	No	160
Mangifera indica	ovule	½MS	4.4, 8.9 BA	No	161
	ovule	½MS	4.5 2,4-D	No	162–164

Table 1. Continued

Species	Explant	Induction Media[a]	Induction Phytohormones μM[b]	Plantlet Production[b]	References
	nucellus	½MS	1) 9 2,4-D;[c] 2) none	No	165
	ovule	½MS	4.5 2,4-D	Yes	166
Metroxylon sp.	shoot tip	MS	1) 452.5 2,4-D,[c] 14.8 2-iP; 2) none	No	172
Microcitrus sp. (M. australis × M. australasica) (hybrid)	ovule	MT	1) 15 IAA-1-alanine,[c] 8.9 BA; 2) none	No	173
	protoplasts (embryogenic tissue[e])	MT	none	Yes	173
Microcitrus sp., Citrus aurantium (donor plants)	protoplasts (γ-irradiated embryogenic tissue,[e] iodo-acetate embryogenic tissue[f])	MT		Yes (cybrid)	174
Microcitrus sp., Citrus jambhiri (donor plants)	protoplasts (γ-irradiated embryogenic tissue,[e] iodo-acetate embryogenic tissue[f])	MT	none	Yes (cybrid)	174
Myrciaria cauliflora	ovule	1) ½MS;[c] 2) ½MS	1) 9 2,4-D;[c] 2) none	No	175
Olea europaea	immature zygotic embryo	½MS	0.5 BA	Yes	176
Paulownia tomentosa	ovule with placenta	MS	4.5 2,4-D, 4.7 KIN	No	101, 177
	mature zygotic embryo	MS	5.7 IAA, 4.7 KIN	Yes	177
Persea americana	immature zygotic embryo	MS	0.4 PIC	Yes	178
	immature zygotic embryo	MS	0.4 PIC	No	179
Phoenix canariensis	shoot tip	MS	1) 452.5 2,4-D,[c] 14.8 2-iP; 2) none	No	172
Phoenix dactylifera	ovule	MS	1) 452.5 2,4-D,[c] 4.9 2-iP; 2) none	No	35
	various	MS	1) 452.5 2,4-D,[c] 14.8 2-iP; 2) none	Yes	180
	axillary bud (from offshoots)	MS	1) 135.7 2,4-D,[c] 14.8 2-iP; 2) none	Yes	181

Table 1. Continued

Species	Explant	Induction Media[a]	Induction Phytohormones μM[b]	Plantlet Production[b]	References
	axillary bud (from offshoots)	MS	1) 452.5 2,4-D,[c] 14.8 2-iP; 2) 0.5 NAA	Yes	182
	various	MS	1) 452.5 2,4-D,[c] 14.8 2-iP; 2) 0.5 NAA	Yes	183
	axillary bud, shoot tips (from offshoots)	MS	1) 452.5 2,4-D,[c] 22.2 BA; 2) none	Yes	184, 185
	mature zygotic embryo	MS	452.5 2,4-D, 14.8 2-iP	No	24
	shoot tips (from offshoots)	MS	1) 452.5 2,4-D,[c] 14.8 2-iP; 2) none	No	172
	cotyledonary sheaths	SH	4.5 2,4-D, 0.4 BA	No	186, 187
	ovule	MS	1) 445 2,4-D,[c] 9.5 KIN; 2) 200 PIC, 9.5 KIN; 3) none	Yes	188
Phoenix pusilla	mature zygotic embryo	MS	452.5 2,4-D, 14.8 2-iP	No	24
Phoenix roebelenii	shoot tip	MS	1) 452.5 2,4-D,[c] 14.8 2-iP; 2) none	No	172
Phoenix sylvestris	mature zygotic embryo	MS	452.5 2,4-D, 14.8 2-iP	No	24
	shoot tip	MS	1) 452.5 2,4-D,[c] 22.2 BA; 2) none	Yes	189
Poncirus trifoliata ("Poorman" × *Poncirus trifoliata*), *Citrus aurantium* (donor plants)	mature zygotic embryo	MT	1) 9 2,4-D,[c] 22.2 BA; 2) none	Yes	43
	protoplasts (γ-irradiated embryogenic tissue,[f] iodo-acetate embryogenic tissue[f])	MT	none	Yes (cybrid)	44
("Poorman" × *Poncirus trifoliata*), *Citrus limon* (donor plants)	protoplasts (γ-irradiated embryogenic tissue,[f] iodo-acetate embryogenic tissue[f])	MT	none	Yes (cybrid)	44
Populus alba × *Populus grandidentata* (hybrid)	leaf disc	MS	22.6 2,4-D, 0.2 BA	No	190
Populus ciliata	young leaf	MS	2.3 2,4-D	Yes	191

Table 1. Continued

Species	Explant	Induction Media[a]	Induction Phytohormones μM[b]	Plantlet Production[b]	References
Populus nigra × *Populus maximowiczii* (hybrid)	punctured leaf	MS	2.3 2,4-D, 0.4 BA	Yes	192
Prestoea sp.	mature zygotic embryo	MS	452.5 2,4-D, 14.8 2-iP	No	24
Prunus persica	cotyledon	MS	22.6 2,4-D, 9.3 KIN, 8.9 BA	Yes	193
	immature zygotic embryo	MS	1) 22 2,4-D,[c] 9.5 BA, 9.5 KIN; 2) 4 2,4-D, 2 BA, 2 KIN	No	446
Punica granatum	immature zygotic embryo	½MS	1) none;[c] 2) 5 2,4-D, 2 BA, 1 KIN	Yes	194
Pyrus communis	various	1) White's I; 2) Nitsch I	1) 21.5 NAA,[c] 9.3 KIN, 2.9 GA3; 2) 10.8 NAA, 8.9 BA	Yes	195
Quercus alba	immature zygotic embryo	MS	4.5 2,4-D, 4.4 BA	No	196
Quercus bicolor	catkin	MS	1) 4.5 2,4-D;[c] 2) 4.4 BA; 3) none	Yes	443
Quercus ilex	leaf	MS	17.8 BA, 2.7 NAA	No	197
Quercus petraea	anther	MS	1) 2.5,5 2,4-D,[c] 0–10 BA; 2) 2.5 BA	No	118
Quercus robur	immature zygotic embryo	MS or WPM	4.4 BA, 2.9 GA3	Yes	198
Quercus rubra	immature zygotic embryo	MS	none	Yes	196
Quercus serrata	immature zygotic embryo	MS (liquid)	0.1 2,4-D, 0.1 BA	No	440
Quercus suber	internode fragments	MS	10 IBA, 8.9 BA	No	199
	internode fragments	MS	10 IBA, 8.9 BA	No	200
	immature, mature zygotic embryo	SM	2.3 2,4-D	Yes	438
Ribes nigrum	ovule	Miller's	10.8 NAA	Yes	201
Ribes rubrum	ovule	Miller's	10.8 NAA	Yes	202
Robinia pseudoacacia	immature zygotic embryo	MS	1) 18.1 2,4-D,[c] 1.1 BA; 2) none	Yes	203
Rosa sp.	leaf	½MS	0.5 KIN, 0.5 NOA	Yes	449

Table 1. Continued

Species	Explant	Induction Media[a]	Induction Phytohormones μM[b]	Plantlet Production[b]	References
Rosa hybrida	leaf stem internode	½MS	1) 2.2 BA,[c] 5.4 NAA, 2.2 2,4-D; 2) 2.2 BA, 0.05 NAA, 0.3 GA$_3$	No	437
Rosa persica × *Rosa xanthina* (hybrid)	filament	B5	9 2,4-D, 6.8 ZEA	Yes	450
	protoplasts (embryogenic suspenion[m])	1) KM;[c] 2) SH	1) 10.8 NAA,[c] 4.4 BA; 2) 13.6 2,4-D	Yes	204
Sabal minor	mature zygotic embryo	MS	452.5 2,4-D, 14.8 2-iP	No	24
Santalum album	endosperm containing zygotic embryo	White's I	5.7 IAA, 9 2,4-D, 23.3 KIN	No	8
	hypocotyl	BM	4.4 BA	Yes	205
	shoot segment	MS	1) 4.5 2,4-D,[c] 0.9–2.3 KIN; 2) 1 mg/l GA[d]	Yes	206
	endosperm	MS	1) 4.5 2,4-D;[c] 2) 1–2 mg/l GA[d]	Yes	207
	hypocotyl	BM	1) 4.4 BA;[c] 2) 2.9, 5.7 IAA	Yes	208
	internodal stem segments	MS	1) 4.5 2,4-D;[c] 2) 5.7 IAA, 4.4 BA	Yes	209
	protoplasts[n]	1) V47;[c] 2) ½MS	1) 4.5 2,4-D,[c] 5.4 NAA, 4.4 BA; 2) 5.7 IAA	Yes	210
	protoplasts (embryogenic suspension[m])	1) V47;[c] 2) MS	1) 2.3 2,4-D,[c] 2.2 BA; 2) 5.7 IAA, 4.4 BA	Yes	211
Sapindus trifoliatus	internodal stem segments	MS	4.5 2,4-D	Yes	212
	young leaf	MS	1) 9 2,4-D,[c] 2.3 KIN; 2) 2.3 2,4-D, 2.3 KIN	Yes	213
	young leaf	MS	1) 9 2,4-D,[c] 2.3 KIN; 2) 2.3 2,4-D, 2.3 KIN	No	214
Sassafras randaiense	mature zygotic embryo	½MS	50 2,4-D	Yes	215
Simmondsia chinensis	immature zygotic embryo	MS	4.5 2,4-D, 0.4 BA	No	216
Thea sinensis	cotyledon	MS	2.3 2,4-D, 0.2 KIN	Yes	217

Table 1. Continued

Species	Explant	Induction Media[a]	Induction Phytohormones μM[b]	Plantlet Production[b]	References
Theobroma cacao	immature zygotic embryo	MS	8.1 NAA	No	218, 219
	immature zygotic embryo	MS	80 IAA	No	220, 221 (see also 222–226)
	immature zygotic embryo	MS	none	No	229–231
	cotyledon, embryo axes (immature zygotic embryo)	½MS	10 NAA	Yes	232
	immature zygotic embryo	MS	10.8 NAA or 4.4 BA, 0.6 IAA, 0.03 GA$_3$	Yes	233
	immature zygotic embryo	MS	16.1 NAA	Yes	234
	immature zygotic embryo	MS	1 NAA	Yes	235
	cotyledon	MS	13.3 BA, 5.4 NAA	Yes	431
Tilia cordata	immature zygotic embryo	MS	2.3 2,4-D	Yes	198
Veitchia merrilli	mature zygotic embryo	MS	1) 5–25 2,4-D;[c] 2) none	Yes	62
Vitis sp.	flower cluster, petiole, internodal stem segments	MS	1) 4.5 2,4-D,[c] 0.4 BA; 2) 10.8 NAA, 0.4 BA	Yes	236
Vitis longii	anther	Medium H	4.5 2,4-D, 1.1 BA	Yes	237
	anther, ovary	MS	5 2,4-D, 1 BA	Yes	238
Vitis riparia × *Vitis rupestris* (hybrid)	anther	Medium H	1) 1 2,4-D,[c] 0.25 BA; 2) 1 IAA, 0.5 BA	Yes	239
Vitis rupestris	anther	Medium H	1) 1 2,4-D[c] 0.25 BA; 2) 1 IAA, 0.5 BA	Yes	239
	anther, stamen attached to ovary	½MS	1) 4.5 2,4-D,[c] 1.1 BA; 2) 0.6 IAA, 1.1 BA	Yes	240
Vitis rupestris × *Vitis berlandieri* (hybrid)	anther	MS	none	Yes	241
	anther	Medium H	1) 1 2,4-D,[c] 0.25 BA; 2) 1 IAA, 0.5 BA	Yes	239
Vitis vinifera	ovule	Nitsch I	1) 5 BA,[c] 2) 5 BA, 5 NOA; 3) 2.5 BA, 5 NOA	Yes	242

Table 1. Continued

Species	Explant	Induction Media[a]	Induction Phytohormones μM[b]	Plantlet Production[b]	References
	ovule	Nitsch II	1) 5 2,4-D,[c] 1 BA; 2) 10 NOA, 1 BA	No	243
	anther	Medium H	1) 1 2,4-D,[c] 0.25 BA; 2) 1 IAA, 0.5 BA	Yes	239
	ovule	Medium H	1) 1 BA,[c] 5 2,4-D; 2) none	No	244
	anther	½MS	1) 4.5 2,4-D,[c] 1 BA; 2) 0.5 NAA, 1 BA	Yes	245
	leaf disc	Medium H	1) 5 2,4-D,[c] 5,10 KT-30; 2) 1 2,4-D	Yes	246
	anther	Medium H (liquid)	1) 4.5 2,4-D,[c] 1.1 BA; 2) none	No	445
Vitis vinifera × Vitis rupestris (hybrid)	anther	Medium H	1) 5 2,4-D,[c] 1 BA; 2) none	Yes	247
	ovule	Nitsch II	1) 5 2,4-D,[c] 1 BA; 2) 10 NOA, 1 BA	No	243
	anther	Medium H (liquid)	1) 1 2,4-D,[c] 0.25 BA; 2) 1 IAA, 0.5 BA	Yes	239

[a] Some of these basal media (macroelements) have been modified, refer to particular references for details; B5 = Gamborg et al. [248]; Blaydes = Witham et al. [249]; BM = Rao and Bapat [250]; DKW = Driver and Kuniyuki [251]; Heller's = Heller [252]; K (h) = Cheng [253]; KM = Kao and Michayluk [254]; KP = Karunaratne and Periyapperuma [452]; LS = Linsmaier and Skoog [255]; Medium H = Nitsch and Nitsch [256]; MH = Carron and Enjalric [139]; Miller's = Miller [257]; MS = Murashige and Skoog [258]; MT = Murashige and Tucker [259]; Nitsch I = Nitsch [260]; Nitsch II = Nitsch et al. [261]; SH = Schenk and Hildebrandt [262]; SM = Sommer et al. [439]; White's I = White [263]; White's II = White [264]; N7 = Simola [265]; V47 = Binding [266]; WPM = Woody Plant Medium [267]; Y3 = Eeuwens [268].

Abbreviations: BA = N[6]-benzylaminopurine; CHL = chlormequat; 2,4-D = 2,4-dichlorophenoxyacetic acid; 3,4-D = 3,4-dichlorophenoxyacetic acid; DAM = daminozide; GA = gibberellic acid; IAA = indole-3-acetic acid; IBA = 3-indolebutyric acid; 2-iP = 6-γ,γ-dimethylallylaminopurine; KIN = kinetin; KT-30 = N-(2-chloro-4-pyridyl)-N′-phenylurea (or 4PU-30), Kyowa Hakko Kogyo Co., Japan; NAA = naphthaleneacetic acid; NOA = β-naphthoxyacetic acid; PIC = picloram; ZEA = zeatin.

[b] In some cases to obtain further development, changes to phytohormones have been made; refer to particular references for details.

[c] Multi-step induction process.

[d] Type or concentration not reported.

[e] Derived from ovule.

[f] Derived from nucellus.

[g] Derived from anther.

[h] Derived from immature zygotic embryo.

[i] Source not reported.

[j] Serially subcultured to reduced 2,4-D concentration.

[k] Derived from leaf tissue.

[l] Derived from various seedling sections.

[m] Derived from roots or shoots.

[n] Derived from non-differentiated stem tissue.

Table 2. Induction of somatic embryogenesis and production of plantlets in gymnosperms

Species	Explant	Induction Media[a]	Induction Phytohormones μM[b]	Plantlet Production[b]	References
Abies alba	female gametophyte[c]	SH	4.4 BA	No	269
Abies nordmanniana	female gametophyte,[c] immature zygotic embryo	½MS	10 BA	No	270
Biota orientalis	immature zygotic embryo	BY	none	No	7
Ceratozamia hildae	immature zygotic embryo	B5	9.3 KIN	No	459
Ceratozamia mexicana	pinnae	B5	4.7 KIN, 4.5 2,4-D	Yes	417
	immature zygotic embryo	B5	4.5 2,4-D	No	459
Larix decidua	immature zygotic embryo	MS	9 2,4-D, 2.2 BA	No	271
	female gametophyte[c]	MS	45.2 2,4-D, 23.3 KIN, 22.2 BA	No	272
L. decidua × L. leptolepis	immature zygotic embryo	DCR	9 2,4-D, 2.2 BA	Yes[d]	273
	immature zygotic embryo	MS	9 2,4-D, 2.2 BA	No	271
Larix leptolepis	immature zygotic embryo	MS	9 2,4-D, 2.2 BA	No	271
L. leptolepis × L. decidua	immature zygotic embryo	MS	9 2,4-D, 2.2 BA	Yes[d]	273
	immature zygotic embryo	MS	9 2,4-D, 2.2 BA	No	271
Larix × eurolepis	protoplasts (embryogenic tissue, suspension)[e]	MS	9 2,4-D, 2.2 BA	Yes	274
Larix occidentalis	immature zygotic embryo	½LM	9 2,4-D, 2.2 BA	Yes	469
Picea abies	immature zygotic embryo	MS	1) 4.5–18 2,4-D,[f] 4.4 BA; 2) 0.9 2,4-D, 2.2 BA	Yes	275
	mature zygotic embryo	MS	1) 9 2,4-D,[f] 4.4 BA; 2) 0.9 2,4-D, 2.2 BA	Yes	275
	immature zygotic embryo	LP	10 2,4-D, 5 BA	No	276, 277 (see also 278, 279)
	immature zygotic embryo	LP	10 2,4-D, 5 BA	Yes	280, 281
	mature zygotic embryo	½MS	50 2,4-D, 20 BA, 20 KIN	Yes	282, 283
	cotyledon	½MS	1) 1 BA;[f] 2) 50 2,4-D, 25 BA, 25 KIN	No	284
	mature zygotic embryo	LP	10 2,4-D, 5 BA	No	285

Table 2. Continued

Species	Explant	Induction Media[a]	Induction Phytohormones μM[b]	Plantlet Production[b]	References
	cotyledon	MS	1) 4.5 BA,[f] 0.05 NAA; 2) 10.7, 21.5 NAA, 0.45 BA	No	286, 287
	mature zygotic embryo	½LP	10 NAA, 5 BA	Yes	288, 289
	mature zygotic embryo	LP	10 2,4-D, 5 BA	Yes	290
	immature zygotic embryo	½LP	10 NAA, 5 BA	Yes	289
	immature zygotic embryo	LP	9 2,4-D, 4.4 BA	Yes	291
	mature zygotic embryo	½LP	9 2,4-D, 4.4 BA	Yes	292, 293
	stage 3 (cotyledonary) somatic embryo	½LP	9 2,4-D, 4.4 BA	Yes	292
	mature zygotic embryo	½BLG	10.7 NAA, 4.5 BA	Yes	294
	immature and mature zygotic embryo	½LP	10 NAA, 5 BA	No	295
	female gametophyte[c]	N6	1) 9.1 2,4-D,[f] 2.3 KIN; 2) 10 2,4-D, 2.5 KIN	No	296
	seedling	½LP	9 2,4-D, 4.4 BA	Yes	293
	mature zygotic embryo	LP	10 2,4-D, 5 BA	No	425
	immature zygotic embryo	LP	5 2,4-D, 5 BA	Yes	470
	mature zygotic embryo	½LP	10 NAA, 5 BA	Yes	470
	protoplasts (embryogenic suspension)[e]	LP	9 2,4-D, 4.4 BA	No	297
Picea glauca	immature zygotic embryo	LP	10 2,4-D, 5 BA	Yes	298 (see also 352)
	immature zygotic embryo	LP	5 PIC, 5 BA	Yes	299
	protoplasts (embryogenic suspension)[e]	LP	10 2,4-D, 5 BA	No	300
	protoplasts (embryogenic suspension)[e]	LP	10 2,4-D, 5 BA	Yes	301
	cotyledon	LP	9 2,4-D, 4.5 BA	No	302

Table 2. Continued

Species	Explant	Induction Media[a]	Induction Phytohormones μM[b]	Plantlet Production[b]	References
	cotyledon	MS	1) 4.5 BA,[f] 0.05 NAA; 2) 21.5 NAA, 0.45 BA	Yes	303
Picea glauca engelmannii complex[g]					
	mature zygotic embryo	½LM	10 2,4-D, 5 BA	Yes	304
	immature zygotic embryo	LP	10 2,4-D, 5 BA	No	305
	immature zygotic embryo	LP	10 2,4-D, 5 BA	Yes	306, 307
	immature zygotic embryo	LP	5 2,4-D, 2 BA	Yes	308, 309
	stage 3 (cotyledonary) somatic embryo	LP	5 2,4-D, 2 BA	No	310
Picea glehnii	mature zygotic embryo	LEP	10 NAA, 5 BA	Yes	427
Picea jezoensis	mature zygotic embryo	LEP	10 NAA, 5 BA	Yes	427
Picea mariana	immature zygotic embryo	LP	10 2,4-D, 5 BA	Yes	298
	cotyledon	LP	9 2,4-D, 4.5 BA	Yes	302
	cotyledon	MS	1) 4.5 BA,[f] 0.05 NAA; 2) 9.1 2,4-D, 4.5 BA	Yes	303
	immature zygotic embryo	LP	9 2,4-D, 4.5 BA	No	311
	mature zygotic embryo	½LM	9 2,4-D, 4.5 BA	No	311 (see also 352)
	protoplasts (embryogenic suspension)[h]	½LM	9 2,4-D, 4.5 BA	No	311
	immature zygotic embryo	½LM	10 2,4-D, 5 BA	No	312
	mature zygotic embryo	½LM	2 2,4-D, 0.5 BA	Yes	424
Picea pungens	mature zygotic embryo	½LM	2 NAA, 10 BA	Yes	415
Picea rubens	mature zygotic embryo	½LM	10 2,4-D, 5 BA	No	312 (see also 426)
Picea sitchensis	mature zygotic embryo	LP	10 NAA, 10 BA	Yes	414
	immature zygotic embryo	MS	50 2,4-D, 25 BA, 25 KIN	Yes	313
	mature zygotic embryo	½LP	10 2,4-D, 5 BA	No	314
	mature zygotic embryo	MS	50 2,4-D, 25 BA, 25 KIN	Yes	315
	immature zygotic embryo	LP	10 2,4-D, 5 BA	Yes	316

Table 2. Continued

Species	Explant	Induction Media[a]	Induction Phytohormones μM[b]	Plantlet Production[b]	References
Picea wilsonii	immature zygotic embryo	LP	10 2,4-D, 4.9 BA	Yes	317
Pinus caribaea	female gametophyte[c]	½LP	10 2,4-D, 5 BA	Yes	318
	protoplasts (embryogenic suspension)[e]	C	10 NAA, 7 BA	No	318
Pinus elliottii	immature zygotic embryo	WPM	20 2,4-D, 5 BA	No	319
Pinus lambertiana	immature and mature zygotic embryo	DCR	13.5 2,4-D	Yes[i]	320
Pinus serotina	female gametophyte[c]	MS	n.r.[j]	No	321
Pinus strobus	female gametophyte[c], immature zygotic embryo	DCR	9 2,4-D, 4.4 BA	No	322
Pinus taeda	female gametophyte[c]	½MS	50 2,4-D, 20 BA, 20 KIN	Yes	323 (see also 67)
	protoplasts (embryogenic suspension)[e]	½MS	5 2,4-D, 2 BA, 2 KIN	No	324
	female gametophyte[c]	MSG	none	No	325
	immature zygotic embryo	DCR	13.6 2,4-D, 2.2 BA	No	325
	female gametophyte[c]	DCR	13.6 2,4-D, 2.2 BA	No	63
Pseudotsuga menziesii	immature and mature zygotic embryo	½MS	50 2,4-D, 20 BA, 20 KIN	Yes[i]	326
	protoplasts (embryogenic suspension)[e]	DCR	5 2,4-D, 2 BA, 2 KIN	No	327
Sequoia sempervirens	mature zygotic embryo, hypocotyl cotyledon	MS	2.5–10 2,4-D, 2 BA, 2 KIN	Yes[i]	328
Zamia fischeri	female gametophyte[c]	B5	13.9 KIN, 0.5 2,4-D	No	460
Zamia furfuracea	immature zygotic embryo, female gametophyte[c]	B5	4.7 KIN	No	460
Zamia integrifolia	mature zygotic embryo	White's I	none	No	9

Table 2. Continued

Species	Explant	Induction Media[a]	Induction Phytohormones μM[b]	Plantlet Production[b]	References
	immature zygotic embryo	1) White's I;[f] 2) LS; 3) White's I	1) 0.05–0.5 2,4-D;[f] 0.2–2.3 KIN; 2) 4.5 2,4-D, 4.7 KIN; 3) none	No	10
Zamia pumila (= Z. integrifolia)	mature zygotic embryo	MS	1) 5.4 NAA,[f] 4.4 BA; 2) none	No	329
	immature zygotic embryo, female gametophyte[c]	B5	9.3 KIN	No	460

[a] Some of these basal media have been modified, refer to particular references for details; B5 = Gamborg et al. [248]; BLG = modified Brown and Lawrence [330] as per Ammerson et al. [331]; BY = Butenko and Yakovleva [332]; C = Lainé et al. [333]; DCR = Douglas fir cotyledon revised medium [334]; LM = Litvay medium [335]; LEP = Lepoivre medium [428]; LP = von Arnold and Eriksson medium [336]; LS = Linsmaier and Skoog [255]; MS = Murashige and Skoog [258]; MSG = modified MS as per Becwar et al. [321]; N6 = Chu et al. [337]; SH = Schenk and Hildebrandt [262]; White's I = White [263]; WPM = Lloyd and McCown medium [267].

[b] In some cases to obtain further development, changes to phytohormones have been made; refer to particular references for details.

[c] Containing zygotic embryo.

[d] Authors did not indicate which reciprocal cross of L. leptolepis × L. decidua was used for plantlet production.

[e] Derived from immature zygotic embryo.

[f] Multi-step induction process.

[g] Picea glauca-engelmannii complex (interior spruce) represents a mixture of Picea glauca (Moench) Voss and Picea engelmannii Parry, from the interior of British Columbia where their ranges overlap [338].

[h] Derived from mature zygotic embryo.

[i] Authors did not indicate which original explant was used for plantlet production.

[j] n.r. = not reported.

B. Somatic embryogenesis

Somatic embryogenesis is a distinctive process, describing the production of bipolar structures nominally capable of germination without separate shoot and root induction phases. Sometimes the distinction between structures derived by organogenesis and somatic embryogenesis can be blurred, especially when it is necessary to give shoot and/or root development treatments to the latter. For this reason it is necessary to present histological information at least once for each species, to verify that somatic embryos have been produced from an explant. Histological evidence has been presented in support of somatic embryogenesis with for example *Acanthopanax* [11], *Aesculus* [14], *Castanea* [31], *Cammellia* [430], *Cercis* [33, 34], *Citrus* [339], *Cocos* [85, 454], *Coffea* [87, 92, 94, 96, 340], *Corylus* [101], *Cyclamen* [104], *Elaeis* [110], *Fuschia* [121], *Gossypium* [127], *Hevea* [141, 448], *Larix* [271, 435], *Paulownia* [101], *Persea* [178], *Phoenix* [181], *Picea* [276, 293, 341–343], *Pinus* [320, 322], *Sequoia* [328], *Thea* [217], and *Vitis* [241]. Some authors have critically compared somatic and zygotic embryogenesis [474], an essential element in regulation of terminology. However, for many genera, or species there is no convincing histological evidence presented.

C. Explant

The most frequently used explant type (see Tables 1 and 2) has been developmentally juvenile, such as the developing zygotic embryo (e.g. *Aesculus, Camellia, Corylus, Liriodendron, Malus, Olea, Paulownia, Persea, Picea, Pinus, Theobroma*) and young seedling parts e.g. hypocotyl and cotyledon (e.g. *Aesculus, Gossypium, Picea*). However there have been a number of reports of somatic embryogenesis with older explant material, such as leaf parts (e.g. *Betula, Citrus, Cocos, Coffea, Gossypium, Populus, Sapindus*) and shoot segments (e.g. *Cocos, Coffea, Gossypium, Phoenix, Santalum*). For some genera particular explants have proven to be highly responsive, such as the nucellus with *Citrus* species, the inner integument with *Hevea brasiliensis*, the anther wall with *Vitis* species, and the fertilized female gametophyte with *Pinus* species. In some instances the development of somatic embryogenesis may be due to the ability of a species to undergo polyembryony e.g. with nucellus culture in *Citrus sinensis*. With *Picea glauca* polyembryogenesis is not normally expressed during zygotic embryogenesis though polyembryony exists in many coniferous species.

The usefulness of juvenile explants for true-to-type clonal propagation is more limited for species which are outbreeders and that have lengthy juvenile phases which delay phenotypic selections, e.g. conifers. In this connection, examples exist in which strategies have been developed for storage of somatic embryos in culture [344, 345], for partial drying of somatic embryos, [30, 308] and for cryopreservation [346, 347], which may be the basis for methods of longer term storage during clonal evaluation of somatic embryo-derived plantlets. Species which have juvenile vegetative growth on mature individuals present special opportunities for true-to-type clonal propagation, as for *Phoenix dactylifera* [181] for example. With *Aesculus hippocastanum* a system of somatic embryogenesis has been developed which uses the filament as the initial explant [17, 451]. Such a technique makes it possible to clonally propagate from mature individuals, a desirable goal which remains elusive for the majority of woody plant species.

D. Medium and other culture conditions

The most extensively used medium formulation has been that of Murashige and Skoog (MS [258]). Its use, and the use of related media such as Linsmaier and Skoog (LS [255]), and Murashige and Tucker (MT [259]) account for over 180 examples of somatic embryogenesis with woody species. Medium usage for some angiosperm genera or species appears to be very characteristic, e.g. N7 [265] medium is used with *Betula pendula* [22, 23], MH medium [139] with *Hevea brasiliensis* [139–142], Blaydes [249] with *Liriodendron tulipifera* [154, 156], Nitsch and Nitsch medium [256] (and variants) with *Vitis* species. [239, 242–244, 246]. MT medium is most widely used with *Citrus* and related genera. Medium use with conifer species reflects the same variation, with the addition of two other widely used medium formulations, LP [336] with *Picea* species, and DCR [334] with *Pinus* species. Medium use can often be a personal preference, but because experimentation with new medium formulations, or with modifications to existing medium formulations is time-consuming, researchers invariably follow the patterns of earlier successes with a particular species or explant. Modifications to carbohydrate content [426] and nitrogenous compounds [477] have been investigated. It might be appropriate to completely re-evaluate medium use with each new species or from time-to-time during routine experimentation, for an example of such investigations see the works of Teasdale and colleagues [475, 476].

Many authors use agar-solidified media for the induction and maintenance of somatic embryos, a culture milieu which in itself can produce inconsistencies resulting from variance in agar type and purity [477, 478]. Alternatively, some authors have investigated the use of liquid suspension culture [227, 376, 423, 446, 461, 467, 468]. This technique offers a convenient experimental system for analyses of nutrient utilization [227, 423, 468], and an avenue to large scale culture in bioreactors [423, 467]. However, this culture method is not without problems also, e.g., in the establishment of liquid cultures [370], or with the effect of long-term maintenance on subsequent development of embryos [465, 466].

Control of the ambient culture atmosphere has produced interesting results when investigated. For example, pCO_2 and pO_2 had significant effects on somatic embryogenic tissue induction with *Picea abies*, the stimulatory effects of high pCO_2 or of low pO_2 [425], could be related to their inhibition of ethylene synthesis. This is similar to the observation made with embryogenic suspension cultures of *Picea glauca* for which reduced growth was coincident with high ambient ethylene levels [399]. The effects of temperature and light availability [323, 477], involving duration of light period, light intensity and spectra, have rarely been intensively investigated. A complete range of such cultural conditions can be found in papers detailing woody plant somatic embryogenesis.

E. Phytohormones

Induction of somatic embryogenesis with many woody species has resulted predominantly from the use of 2,4-dichlorophenoxyacetic acid (2,4-D), usually at concentrations under 10 μM, but occasionally as high as 452 μM (e.g. with *Cocos nucifera* [348], and *Phoenix dactylifera* [35]). Other auxins have been used, often as substitutes to avoid the potential toxicity of 2,4-D. Cytokinins have also proved useful in inducing somatic embryogenesis with some species, e.g. *Abies alba* [269], *Actinidia deliciosa* [13], *Coffea arabica* [87], *Juglans regia* [148]. Distinct differences can occur therefore in phytohormone requirement for induction within seemingly related plant groups. For example within the Coniferae induction of somatic embryogenesis with *Picea* species often uses 9–10 μM 2,4-D and 2–5 μM benzyl adenine (BA), whereas with *Abies* no 2,4-D is used. In general however a combination of an auxin with a cytokinin has been used, usually with predominance of one type over the other. With some *Citrus* species somatic embryogenesis can be obtained from nucellar explants in the absence of exogenous phytohormones. With these species the addition of malt extract was important. Useful information about the effects of phytohormones on the *in vitro* propagation of woody species is contained in the review by Schwarz [349].

F. Progress toward plantlet production

Somatic embryo development with many angiosperm species appears to occur spontaneously, i.e. somatic embryos mature in a relatively asynchronous manner under routine maintenance conditions. However, with some species development is triggered by alterations to the culture conditions, e.g. with the use of gibberellic acid (GA$_3$) with some *Citrus* species. Abscisic acid (ABA) has rarely been used with woody angiosperms, for example to prevent precocious germination of *Mangifera indica* somatic embryos [166]. The latter contrasts with the more extensive use of ABA in examples of conifer somatic embryo cultures primarily for promotion of maturation [302, 350–352, 413]. The ABA which has been used is a racemic compound, ($\pm$)-ABA, which is a synthetic mixture having a 1:1 ratio of the natural compound (+)-ABA and its unnatural (−)-ABA enantiomer. The (−)-ABA enantiomer may or may not interfere with processes involving (+)-ABA, or it may itself be effective. For this reason authors should specify the composition of the ABA they have used. ABA is most often used as the sole exogenous phytohormone for conifer somatic embryo maturation, although IBA has been included by some [307], and BA has been found to have a synergistic effect when added with ABA [470]. It is possible that osmotic treatments such as use of PEG 4000 [421, 422] may stimulate endogenous (+)-ABA. Germination from matured conifer somatic embryos occurs after removal of ABA. Studies such as performed with *Vitis* somatic embryos, on the measurement of endogenous ABA during maturation and germination [353], are needed to elucidate the impact of exogenous treatments with the racemic compound, e.g. during conifer somatic embryo maturation. One key group of plants for which plantlet development appears to be problematic are the agronomically important tropical species, e.g. *Cocos nucifera*, *Hevea brasiliensis*, and *Theobroma cacao*.

The propagation of many woody species is seed based, emphasising the need for development of reliable systems for somatic embryo production based on (storable) artificial seed. Artificial (synthetic) seed of *Santalum album* [212] have been created by encapsulation in sodium alginate [354, 355]. A scheme for propagation involving encapsulation of somatic embryos is shown for spruce in Figure 1 (see also [486]). Encapsulation may offer a means to provide a surrogate nutritive support in lieu of the endosperm or megagametophyte found in angiosperm or gymnosperm seeds respectively. This may be critical because, for example, somatic embryos of "interior spruce" [338] (*P. glauca engelmannii* complex [310], and see Section V) were found to hydrolyze storage protein and triacylglyceride reserves during germination sooner than intact germinating seed [356]. An understanding of the seed nutrient reserves and their mobilization during germination is indispensible in this regard, such as the studies made for *P. glauca* [471, 472]. Artificial seed may require desiccation for reliable storage. Desiccation was recently reported for *P. glauca* [433, 447], involving the use of gradual reduction in relative humidity of the ambient environment by use of a series of saturated solutions. There is also evidence that partial drying, under atmospheres of high relative humidity, is achievable and is beneficial [30, 308, 316].

It is not always possible for laboratory researchers to transplant tissue-cultured plantlets into greenhouse facilities in a horticulturally acceptable way. For this reason reports of plantlet production from laboratory experiments may not describe the most effective means for plantlet development, acclimatization, transplantation or continued growth. For limited scale plantlet production such methods may suffice, however they may be misleading in terms of practical use with larger scale propagation. Information on large scale somatic embryo plantlet production is rare, and may indicate problems with the processes being developed in the research laboratory.

G. Cryopreservation

Somatic embryos provide a unique material suitable for cryopreservation, a technique which has relevance both as a method of preserving genetic resource, and as a safety precaution in the routine maintenance of valuable tissue cultures [346, 357], see also Fig. 1). Surprisingly research with cryopreservation of

498

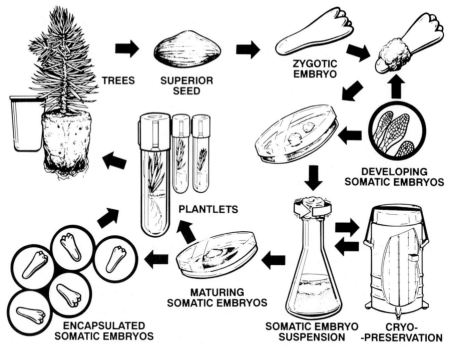

TREES   SUPERIOR   ZYGOTIC
           SEED   EMBRYO

PLANTLETS

DEVELOPING
SOMATIC EMBRYOS

MATURING
SOMATIC EMBRYOS

ENCAPSULATED   SOMATIC EMBRYO   CRYO-
SOMATIC EMBRYOS   SUSPENSION   -PRESERVATION

Fig. 1. A schematic representation of plantlet production through spruce somatic embryogenesis. The production of plantlets following encapsulation of somatic embryos has recently been reported in the literature [486]. For further details refer to Section V.

somatic embryos of woody plant species has been reported very rarely. Notable examples of successful cryopreservation include *Citrus* [75], *Elaeis* [347], *Larix* [418], *Phoenix* [358], *Picea* [346, 359, 418, 487], *Pinus* [416], *Theobroma* [234], and *Vitis* [441, 442].

H. Protoplast culture

Protoplast culture methods have been developed for many woody plant species. Such methods are useful in providing material e.g. for cell biology investigations, genetic transformation, and for single cell regeneration to plantlets with or without protoplast fusion. Species which represent the best evidence of regenerable protoplast cultures include the following: *Citrus sinensis* [57] (and see Section III), *Coffea canephora* [97], *Larix × eurolepis* [274], *Liriodendron tulipifera* [155], *Picea glauca* [301] (and see Section V), *Rosa persica × xanthina* [204], and *Santalum album* [210, 211]. In many cases the source material used for protoplast culture with these species was an embryogenic culture, similar to observations made with the Gramineae, [360] although leaf-material was used as a source of protoplasts with embryogenic competence for *Actinidia deliciosa* [433]. Protoplast fusion with woody plants is represented prominently in research with various species of *Citrus* (e.g. Saito et al. [82]) and related genera, as shown in Table 1 and Section III.

I. Genetic transformation

There are comparatively few reports which detail methods for transient gene expression in somatic embryo cultures of woody species, or protoplasts derived from them. Among the examples is research with *Picea* spp. [324, 334, 361, 408, 480, 481, 488], (and as detailed in Section V) and *Larix* × *eurolepis* [479, 481]. The successful development of methods for the recovery of stably transformed plants is currently restricted to *Gossypium hirsutum* [131, 133], *Juglans regia* [149, 150, 362] and very recently *Picea glauca* [482, 489]. These examples used either *Agrobacterium* [131, 149, 150, 362] or particle bombardment [133, 482, 489] to achieve stable transformation.

J. Other features

Notable among the reports of somatic embryogenesis in woody plants are the following aspects: the production of variant culture lines with or without mutagenic treatment, e.g. with *Phoenix dactylifera* [188], *Citrus* spp [363, 462]; electrophoretic analyses of protein content for the optimization of explant selection with *Liriodendron tulipifera* [156], *Picea* spp. [305], or for the characterization of storage proteins during somatic embryo maturation with *Gossypium hirsutum* [128], *Larix* × *eurolepis* [429], *Picea* spp. [307, 343, 420, 473]; the heritability of somatic embryogenesis potential with *Gossypium hirsutum* [126]; the occurrence of multiple paternal genotypes in embryogenic tissue from individual seeds of *Pinus taeda* [63]; cell biology investigations with *Corylus avellana* [364], *Picea glauca* [341, 365], and *Picea mariana* [366, 367, 419]; the relationship between occurrence of extracellular proteins and somatic embryogenesis in *Citrus aurantium* [463, 464], and *Vitis* sp. [483], and between the occurrence of free polyamines and of putrescine and spermidine and somatic embryo development in *Vitis vinifera* [445].

Several published texts contain interesting reviews relevant to the topic of this chapter, these and related works should be consulted for further information [167–171].

III. *Citrus sinensis* osb.

A. Introduction

The sweet orange is one member of a large family of woody fruit species, grouped together under the heading of *Citrus*. The commercial significance of this plant group has led to the establishment of plantations throughout the tropical and subtropical regions of the world. Its principal value lies in its contribution of vitamin C to the human diet. It is consumed as fresh fruit, as juice, dried and canned product, and as preserves. There are several *Citrus* species which have been successfully cultured *in vitro*, but the most detail available is for *C. sinensis* (Table 1).

B. Development of methods

Citrus sinensis has a unique advantage for somatic embryogenesis, in its inherent ability to produce more than one embryo per seed. Unlike polyembryony in conifer species for example, the adventive embryos of polyembryonic *Citrus* species are produced from nucellar tissue adjacent to the developing zygotic embryo. The culture of nucellus tissue of polyembryonic and monoembryonic *Citrus* species was a major focus of the initial attempts to initiate somatic embryogenesis [4, 6, 49]. With *C. sinensis* Button and Bornman [60] and Kochba et al. [51] obtained adventive embryogenesis from explanted nucellar tissue, cultured on phytohormone-free modified Murashige and Skoog (MS) medium [258] or Murashige and Tucker (MT) medium [259] respectively. Explanted nucellus tissue had no nucellar or zygotic embryos prior to culture [51]. Malt extract (400 [60] or 500 [51] mg L $^{-1}$) was reported to be stimulatory for the production of adventive nucellar embryos.

Subsequently methods were developed for the production of tissue cultures from explanted unfertilized ovules of cv. Shamouti, using medium containing 500 mg L^{-1} malt extract [36, 61]. Tissue culture growth was promoted by transfer to MT medium containing 4.7 μM kinetin (KIN) and 5.7 μM indole acetic acid (IAA) with or without malt extract. Subsequent somatic embryo production was stimulated by transfer to MT medium containing 1000 mg L^{-1} malt extract, and plantlet development was reported to occur when somatic embryos were transferred to MT medium with 2.9 μM GA$_3$ [61]. In later work these authors do not refer to the use of malt extract in culture maintenance [368, 369].

C. Somatic embryogenesis

With *C. sinensis* the embryogenic tissue was found to be highly organized, being composed of numerous somatic embryos in various stages of development [339]. Button et al. [339] described structures, a few cells in size, which they called "proembryoids". These could give rise to further proembryoids in a repetitive process, commencing with single cells usually on their periphery. The term "callus" therefore may be inappropriate to describe this tissue proliferation because it does not agree with the accepted definition of a callus [370].

D. Further methods

Conditions which favoured tissue growth (i.e. regular subculture to fresh phytohormone-free MT medium every 5 weeks) resulted in a loss of embryogenic potential. Extending the culture period to 14 weeks had the effect of aging the tissue; when aged tissue was then transferred to fresh phytohormone-free MT medium a 100-fold increase in embryogenesis was recorded [368]. Tissue aging was also found to be stimulatory by Button and Rijkenberg [371]. A similar, but transient effect was observed with sucrose starvation of tissue [368]. This observation caused Kochba et al. [40,372] to investigate the effects of alternative carbohydrate sources. Galactose and lactose at concentrations above 32 mM markedly stimulated somatic embryogenesis in habituated tissue, and the development of embryos was improved most by lactose (from 80 mM and above). The stimulatory affect of galactose and related sugars might result from presumptive antagonism with auxin, or might be related to growth retardation, i.e. similar to the observation with sucrose deprivation [372]. A number of other compounds or treatments have also been found to be stimulatory to somatic embryogenesis in tissue cultures of *C. sinensis*, including ABA [373], glycerol [41], and irradiation [65, 70], though the precise reasons for such stimulation remain unclear.

As noted earlier some tissue lines developed by Kochba and colleagues [51, 61] were found to be "habituated", that is they lacked a requirement for exogenous auxin and cytokinin. Indeed, the addition of exogenous auxin and cytokinin was inhibitory to embryogenesis. Interestingly, inhibitors of auxin synthesis (hydroxy nitrobenzyl-bromide, and 7-aza-indole) and presumptive antagonists to endogenous cytokinin activity (8-aza-guanine) had the effect of stimulating embryogenesis [369]. It was also found that embryogenic tissue lines had a reduced endogenous auxin content compared with non-embryogenic lines, suggestive that auxin concentration was a critical factor in the embryogenesis process [374].

Tissue culture procedures with or without the use of mutagenic agents can provide alternative ways of producing novel plants for breeding programs. In this connection, a method was developed to obtain single cells from the embryogenic tissue which were capable of regeneration to whole plants. This process involved the use of enzymatic and osmotic treatment to facilitate the separation of single cells [375]. Also in this context, the use of fluorescein diacetate as an indicator of cell viability in suspension cultures [376] is a useful aid in determining the relative lethality of mutation treatments, and of the post-treatment selection process. Salt-tolerance has been one focus in mutation breeding with cv. Shamouti [363], and salt-tolerant cultures have been recovered after selection by extended exposure to 0.1 M NaCl [39, 66]. Such cultures required different conditions than non-selected material for plant regeneration, in particular the presence of NaCl [363]. More recently Gmitter and colleagues [59, 64] have used colchicine (0.01%) to produce non-chimaeric tetraploid plants derived via somatic embryogenesis from unfertilized ovules [59]. If fertile, such plants may be hybridized with diploids; triploids in the progeny could possess seedless phenotypes [59].

E. Plant regeneration

Proembryoids could enlarge, becoming macroscopic globular somatic embryos, referred to as pseudobulbils [49]. Pseudobulbils 2 mm in diameter could subsequently become ovoid and then heart-shaped. Mature bicotyledonary somatic embryos were occasionally observed from which plantlets could develop. Somatic embryo development could be improved by addition of 2.9 μM GA$_3$ and 0.15 mM adenine sulphate [339]. It appeared that the GA$_3$ stimulated root meristem initiation in smaller pseudobulbils, and adenine sulphate promoted subsequent root development [339]. Development of heart-shaped and cotyledonary somatic embryos could also be improved by use of the GA$_3$/adenine sulphate combination [377]. Fasciated and other aberrant somatic embryo morphologies were frequently observed with cv. Shamouti [377] and cv. Hamlin [37]. Aberrant embryo morphology and imbalanced root-shoot development during germination have been cited as major obstacles to successful plant production [37]. When germination in the presence of 2.9 μM GA$_3$ resulted in plantlets of cv. Hamlin with balanced shoot and root development, 60% or more survived acclimatization to produce plants of normal appearance. This occurred whether from short or long-term tissue cultures [37]. The report of Navarro et al. [45] presents a more cautionary note, at least for monoembryonic *Citrus* cultivars. These authors observed that nucellus culture resulted in the production of 29% aberrant phenotypes.

F. Cryopreservation

Several recent reports detail methods for the cryopreservation of *C. sinensis* tissue. The first report of plantlet regeneration after 5 min storage of somatic embryos at -196 °C [73] employed a pretreatment with 10% DMSO followed by slow cooling at 0.5 °C min^{-1} down to -42 °C, and slow thawing at room temperature. However, survival of somatic embryos was poor, and plantlets which were regenerated were derived by a process of secondary embryogenesis. A similar process, using 5% DMSO pretreatment, a cooling rate of 0.5 °C down to -40 °C, and rapid thaw at $+40$ °C, led to culture survival after 1 h at -196 °C that was 70% of control non-frozen material. Frozen cultures were regrown and cotyledonary somatic embryos and plantlets were regenerated [74]. These authors subsequently reported the use of a rapid freezing technique. This employed a pretreatment with 2 M glycerol, a rapid freeze by exposing suspension cultures held in transparent straws to -30 °C for 30 min prior to immersion in liquid nitrogen, and a rapid thaw at $+40$ °C. Survival after 30 min at -196 °C was about 90% that of non-frozen material. Frozen cultures were regrown, and cotyledonary somatic embryos and plantlets were regenerated [75].

G. Protoplast culture

Regeneration of somatic embryos and plants, from protoplasts of a non-regenerative culture from an ovule of cv. Shamouti, was first reported by Vardi et al. [71] These authors used an enzyme treatment with separate incubations in Pectinase (1% w/v) and Cellulase (1% w/v). They found that protoplast-derived tissue cultures were stimulated to produce embryos following X-ray treatment. This protoplast method was subsequently improved, using an enzyme mix containing 0.2% (w/v) Cellulase, 0.1% (w/v) Driselase, and 0.3% (w/v) Macerozyme, osmotic stabilization during protoplast regeneration (0.6 M sucrose), and, after 6–10 weeks, transfer of regenerated tissue cultures with immature somatic embryos to basal MT medium. Treatment with 28.9 μM GA_3 was necessary to promote somatic embryo axis elongation [42]. Essentially similar results were achieved with cv. Trovita [69]. These authors subsequently developed procedures for the almost direct production of somatic embryos from individual protoplast-derived cells, using a combination of low plating density (e.g. 4×10^4 protoplasts mL^{-1}) and low mannitol concentration (e.g. 0.4 M) achieved by a gradual dilution of 0.05 M at each dilution commencing with 0.7 M mannitol used for protoplast isolation [72].

The production of novel plants through tissue culture was one of the early objectives of *Citrus* tissue culturists [61, 65]. Protoplast culture contributes to this goal in two ways, as a source of protoplasts with potential to regenerate from single cells, and as a source of protoplasts for use in somatic hybridization. The latter aspect has been extensively examined with *Citrus* because of the opportunities it presents to circumvent problems experienced during conventional genetic improvement programmes, e.g. due to polyembryony Ohgawara et al. [79] and Grosser et al. [80] reported the production of somatic hybrid plants between *C. sinensis* and *Poncirus trifoliata* by recovery of protoplast fusion products through somatic embryogenesis. Both research groups took advantage of differences in regenerability of protoplast cultures derived from *C. sinensis* somatic embryogenic tissue (regenerative) and *P. trifoliata* leaf tissue (non-regenerative), and of the probability that asexual hybrids would have the distinct trifoliate leaf which was characteristic of sexual hybrids of the same cross. For fusion each group used polyethylene glycol (PEG, M.W.6000, 40%), though post-fusion treatments differed. Fusion products were regenerated via somatic embryogenesis to plants, either by somatic embryo maturation or after induction of organogenesis from somatic embryos. Subsequent fusion research has been predominantly from these groups and has used substantially similar experimental strategies. Ohgawara and Kobayashi have recently reviewed protoplast fusion in *Citrus* [378]. Parasexual hybridizations are shown in Table 1. Analytical methods for the identification of somatic hybrids have included the use of isozyme banding patterns [379] (and references of Grosser and colleagues [80]), and endonuclease treatment of nuclear ribosomal DNA [383] (and references of Ohgawara and colleagues [79]). As expected somatic hybrids will have amphidiploid chromosome numbers. Kobayashi et al. [78] found that somatic hybrids from fusions between protoplasts of *C. sinensis* and Murcott tangor had chloroplastic DNA from either one parent or the other, though they all had the mitochondrial DNA of *C. sinensis*. Somatic hybrids can be expected to be unpredictable in their contributions to breeding programs, indeed fertility of somatic hybrids is commonly lost [378]. An alternative approach to somatic hybridization used with *Citrus* is donor-recipient protoplast fusion (asymmetric hybridization), which has the advantage of producing diploid fusion products with the nuclear genome of one donor (cybrids). Donor-recipient fusions have been reported by Vardi and colleagues [44, 174], who used gamma irradiation to arrest the nuclear divisions of donor protoplasts, and iodoacetate to transiently arrest metabolism, and thence cell divisions, in non-fused recipient protoplasts. Various fusions were accomplished resulting in cybrid plants from fusions between e.g. Poorman × *Poncirus trifoliata* with *Citrus limon* [44], and *Microcitrus* sp. with *C. jambhiri* [174]. Although cybrid production involving *C. sinensis* has not been reported, the system has potential for exploring organellar inheritance in fusion products, and for the production of novel cybrid trees for use in breeding programs [174].

Citrus sinensis somatic embryogenesis has been the foundation for considerable advances in our knowledge of the maleability of woody plant tissue cultures, particularly owing to the work of Kochba, Spiegel-Roy and colleagues on metabolism, and the work of various research groups on protoplast fusions and cybrid production. Other reviews on *Citrus* should be consulted for further details [381, 382].

IV. *Elaeis guineensis* Jacq.

A. Introduction

The oil palm is one of a group of monocotyledonous plantation trees with which somatic embryogenesis has been obtained, and is perhaps the best developed example but least well described. Oil palm is endemic to West Africa but as a result of its commercial importance it is grown in plantations in many tropical regions of the world [384]. As its name implies the value of oil palm lies principally in its ability to supply palm oil and palm kernel oil. The related species *E. oleifera*, which differs in its commercial desirability from *E. guineensis*, is endemic to South America, and is useful in hybridization programmes with *E. guineensis*. Oil palm has a solitary vegetative shoot, which restricts the possibilities for vegetative propagation.

B. Development of methods

Somatic embryogenesis in oil palm owes its origins to the work of Staritsky [385], and Rabechault and colleagues [106], who attempted to obtain regenerative tissue cultures from shoot apices and zygotic embryos, respectively. A method for oil palm somatic embryogenesis is attributed to Jones [386], though that reference does not provide a method nor is the term somatic embryo used to describe the structures which were presented. Similarly, no detail of the process nor of the embryo-like features of the regenerated structures was provided in the later work of Corley et al. [387] or Ahee et al. [388]. The latter authors [388] describe the development of a fast growing, friable and subculturable callus (FGC, or CCR [cultures à croissance rapide]) originated from explants taken from young leaves. The related work of Pannetier et al. [389] briefly describes the production of embryo-like structures from callus derived from such leaf explants. Hanower and Hanower [107] subsequently noted that callus obtained from culture of leaf explants on MS medium with 9 μM 2,4-D, gave rise after varying periods to FGC when subcultured onto MS medium with 2–9 μM 2,4-D and 3–11 μM NAA. Embryogenesis was observed when FGC were transferred to medium with 5.4–10.8 μM NAA and 4.4–13.3 μM BA [107].

A process for the induction of embryo-like structures from leaf-derived callus was also partially described by Thomas and Rao [108]. Their procedure involved the use of meristematic zones from bases of immature leaves excised from 6-month-old plants. About 50% of explants showed subculturable callus production after 8 weeks on 226–317 μM 2,4-D. Callus was subcultured with a sequential reduction in 2,4-D concentration (not specified). The use of 45 μM 2,4-D, 0.05 μM 2iP with half-strength MS salts, and an increased concentration of sodium dihydrogen orthophospate, maintained the callus as a source for the continuous production of embryo-like structures. The latter could be observed after 1 week culture on 0.05–2.5 μM 2-iP. Shoots which developed from the embryo-like structures were cultured on filter-paper bridges in MS medium with 5.4 μM NAA and 3 μM GA$_3$, to stimulate root development.

Dormant intact zygotic embryos have also been used as explants for the induction of somatic embryogenesis [109]. When cultured on modified MS medium with 23–45 μM 2,4-D or 27–54 μM NAA, a compact nodular callus developed, from which elongate embryos were produced by further development of the constituent nodules. Shoot and root development on embryos could be stimulated by brief passage on medium with 108 μM NAA and 3 μM GA$_3$ in the presence of activated charcoal [109].

C. Somatic embryogenesis

A histological analysis of somatic embryogenesis from leaf-derived callus [110], showed that the callus resulted after the development of a cambium-like zone derived from perivascular tissue of leaf veins. The callus lacked an epidermis and was nodular, and subsequently gave rise to other nodules by continued callogenesis from the cambium-like zone. With further growth a discontinuity in the latter zone led to the eventual development of protruberances, distinguishable from the callus nodules in their possession of epidermis, procambium, and storage lipid. These differentiated into conical somatic embryos, with shoot and root apices. Ferdinando et al. [390] undertook an intensive analysis of the biochemical and morphological events occurring during zygotic embryogenesis in oil palm. Such information is useful for comparative purposes in describing somatic embryogenesis. These authors detail a process for the immunological recognition of zygotic embryo-specific proteins. Immunochemical recognition of embryo-specific proteins would be invaluable in the verification of somatic embryogenesis as a process within oil palm callus culture systems, provided that such a system could be applied stringently i.e. with no cross recognition of proteins produced in other regeneration pathways. See also the study of Turnham and Northcote [391] described in Section-D.

D. Further methods

After long-term culture (about 5 years) it was found that FGC, derived according to Hanower and Hanower [107], varied in their embryogenic potential, from highly embryogenic, through partially embryogenic, to non-embryogenic, largely independent of callus genotype. These authors found that non-embryogenic FGC could prevent embryogenesis in control embryogenic FGC when co-cultivated together. Similarly, partially-embryogenic FGC could retard the development of control embryogenic FGC, leading the authors to hypothesize that inhibitory substances were excreted from the non- or partially-embryogenic lines. Embryogenesis in partially-embryogenic lines could be enhanced by use of the anti-auxin 2-o-chlorophenoxyisobutyric acid (1 μM), presumably by interference with the activity of endogenous auxin. The greatest stimulation of somatic embryogenesis resulted from the use of 1 mM phloridzine [107], a flavanoid compound from apple.

Turnham and Northcote [391] investigated the occurrence of biochemical indicators useful in the prediction of embryogenic potential. These authors showed that it was possible to correlate the activity of acetyl-CoA carboxylase with the degree of embryogenesis in oil palm callus culture. This was based on the catalysis of the ATP-dependent carboxylation of acetyl-CoA to malonyl CoA, the first committed step in triacylglycerol production. Triacylgylcerol was a specific storage product of embryogenic cultures. It was proposed that an assay for acetyl-CoA carboxylase activity could be used efficiently in the screening of media for their effect on embryogenic potential. Further, Hughes et al. [392] describe the use of ^{31}P NMR to analyse cell metabolism during the growth cycle in oil palm cell suspension culture. Such work has relevance in assisting the optimisation of culture maintenance conditions. Although it is unclear how their suspension cultures were produced, and whether they had embryogenic potential, these authors provide a method for suspension culture maintenance, based on the use of modified MS medium and 9 μM 2,4-D [392]. This suspension was used in the development of a protoplast culture and cell regeneration procedure, which could find use in the development of methods for transformation, especially relevant if used in conjunction with cultures regenerable to somatic embryos and plants.

Suspension cultures have also been derived from the nodular callus produced from immature leaf fragments of adult trees [107, 461]. Callus was placed in culture medium containing approximately 163 μM adenine sulphate, 4.4 μM benzyl adenine, and either 362 or 452 μM 2,4-D. After two passages each of 4–6 weeks, suspensions of embryogenic nodule masses were obtained, these could be maintained in this proliferating state by continued culture on the same medium. Further development and plant regeneration were occasionally obtained following removal of 2,4-D in agar-solidified culture medium [461].

E. Plant regeneration

The development of plantlets from oil palm somatic embryos is not well described in the literature. There appears to be a critical requirement for the sequential use of precise auxin concentrations to maintain a system for continuous somatic embryo production and germination [393]. As noted previously, the use of GA_3 to stimulate root development has been reported by some authors [108, 109]. Suspension culture-derived embryos could develop shoot poles when embryos were cultured on 2,4-D-free medium, though root pole development was more sporadic. Rooted plants were obtained from such embryos [461].

The requirements for transplantation and acclimatization are not well described, though it appears that a high humidity treatment extending over 3 weeks is needed [394].

There has been concern that plantlets regenerated from the callus-based embryogenesis system would show a high frequency of genetic aberration. There are however only brief reports related to studies on genotypic stability of *in vitro* cultures or regenerants. These reports indicate very little genetic instability at either level of organization [394]. There is more substantial information regarding field performance of regenerated oil palms. Data indicate that the variability recorded for traits such as petiole characteristics and fruit composition, are much less in the clonal trees than in seedlings [394]. Recent reports of a low frequency abnormality in flower development of some regenerated oil palm clones [380] may be attributable to physiological imbalances introduced during *in vitro* culture [395], though this has not been characterized further.

F. Cryopreservation

Engelmann and colleagues [344, 347, 396] have investigated ways to preserve somatic embryos of oil palm, for use in the reduced maintenance of stock cultures, and as a culture repository. A method for reduced maintenance involves the use of air-tight plastic boxes with a controlled atmosphere of 1% oxygen and 99% nitrogen. Cultures with somatic embryos could be maintained with minimal growth on modified MS medium without phytohormones for 4 months without subculture [344]. Longer term storage is also possible through the use of cryopreservation in liquid nitrogen. A simple technique has been described, involving culture for 2 months on 0.3 M sucrose to obtain suitable somatic embryos, followed by 7 days on 0.75 M sucrose and rapid (200 °C min^{-1}) freezing by direct immersion in liquid nitrogen. Survival of somatic embryos after rapid thaw (1 min at 40 °C) and gradual reduction in sucrose concentration was 50–80%. In the context of such cold treatments it is of interest to note that oil palm is particularly sensitive to chilling injury, and is injured at temperatures below 18 °C. To characterize chilling injury Corbineau et al. [397] investigated the relative abundance of ethylene in cultures of somatic embryos which had been exposed to various degrees of non-freezing chilling temperatures. These authors found that the amount of ethylene released from cultures was inversely related to the amount of injury, i.e. the greater the injury the lower the amount of ethylene released when cultures were returned to optimal conditions.

Progress with somatic embryogenesis research with *Elaeis guineensis* is unclear because of the comparative lack of information in many of the reports, especially, and perhaps understandably, those from industrial research groups. Ascribing the original scientific report of somatic embryogenesis in oil palm remains a judgement call. It is highly desirable that further research be carried out to develop efficient methods for the production of somatic embryos and plantlets of this valuable species, and for a more complete characterization of the process. Until this is done the effectiveness of the somatic embryogenesis process as a means to propagate oil palm will remain uncertain. For further information other reviews on *Elaeis guineensis* should be consulted [384, 393, 395].

V. *Picea glauca* (Moench) Voss

A. Introduction

Picea glauca, white spruce, is a coniferous forest species which grows in the boreal forest region of northern North America. In the Pacific northwest of North America *P. glauca* hybridizes with *P. engelmannii* where their ranges overlap, the resulting hybrids have been called "interior spruce" [338] and have also been referred to as *P. glauca engelmannii* complex [310]. For simplicity interior spruce will be referred to as *P. glauca engelmannii* in this text. White spruce is a long-lived, commercially important species. It is used in the production of lumber for building construction, and of plywood sheets and wall panelling. The whiteness of its wood and its fibre qualities make it a desirable species for use in pulp and paper production.

B. Development of methods

Somatic embryogenesis with white spruce has been induced from immature i.e. developing [298, 299, 306, 342], and mature i.e. quiescent or resting [304] zygotic embryos, and from shoot portions of young seedlings up to 30-days-old [302]. In addition somatic embryogenesis with *P. glauca engelmannii* has been achieved by re-induction from matured cotyledonary somatic embryos [310] (Table 2). In most cases explants have been placed directly on von Arnold and Eriksson's LP medium [336] containing 9–10 μM 2,4-D and 4–5 μM BA. Some texts refer to this medium as AE [299] and VE [306]. Alternatively, Tremblay [304] has used a modified Litvay medium (LM) [335]. The von Arnold and Eriksson and Litvay media are rich in nitrogen-containing compounds each having different emphases on organic and inorganic nitrogen constituents, and white spruce cultures can be grown satisfactorily on either salt formulation with agar-solidified media. Embryogenic tissue can also be maintained and rapidly multiplied in liquid shake flasks [227, 370, 398, 410], and see following Section D.

C. Somatic embryogenesis

Typically, within 2 to 4 weeks of culture an embryogenic tissue is produced, particularly from the hypocotyl region of whole zygotic embryo explants [299, 304, 306], This tissue has been described as mucilaginous and white, and possesses translucent cells [298, 299, 302, 342]. The composition of this mucilaginous matrix has not been reported.

The appearance of white spruce somatic embryos, and the developmental sequence in somatic embryology has been described by various authors [341, 342, 398], Most authors have referred to the embryogenic tissue as a callus. However, as discussed with *Citrus sinensis*, this term may be inappropriate because the tissue contains highly organized somatic embryos [298]. In addition other cell types can occur in the tissue, including isodiametric cells isolated or in clusters, and free suspensor-like cells [398]. Hakman et al. [341] noted three mechanisms by which somatic embryos could originate. Somatic embryos appear to arise from single cells by initial asymmetric divisions, the products of which give rise to the embryonal mass and suspensor system. Meristematic cells within the suspensor system also appear capable of developing into somatic embryos. Such meristematic cells may be remnants of the embryonal mass which failed to develop, or may be the products of asymmetric divisions of suspensor cells within the suspensor system. Somatic embryos appear also to undergo cleavage, resulting in multiple embryos which are frequently attached to one another.

In some cases the developmental sequence of somatic embryos has been described using terms borrowed from spruce zygotic embryology. For example Nagmani et al. [342] have described some of the structures in the embryogenic tissue as "somatic proembryos". In white spruce zygotic embryology "proembryo" refers to a stage of development ending with the production of a 16-celled proembryo, arranged in four tiers each of four cells, within the archegonium [338]. The use of the term "somatic proembryo" could therefore lead to confusion, if not applied in a very restricted sense to the earliest events in the derivation of white spruce somatic embryos. The term may have been misapplied in describing a structure with an embryonal mass of 16–32 cells [342] and in describing even later developmental states [350, 399]. Later stages in zygotic embryology are described as having an embryonal mass atop a suspensor system which has been contributed to by the suspensor tier (primary suspensor) and by the embryonal mass (secondary suspensor) [338]. It is normally difficult to identify such an hierarchy in tissue cultures, and the suspensor system can be very complex with tiered suspensor cells and less regular cells similar to embryonal tubes. "Embryonal mass" and "suspensor system" are terms consistent with conifer terminology and with the usual state of most embryos in embryogenic tissue cultures. Simple nomenclature systems (e.g. stage 1, 2, etc.) such as that described for *P. abies* [289] have also been used with *P. glauca* [350, 351] and may be the most useful way of comparison amongst different research groups.

D. Further methods

Various factors have been found to influence the frequency of induction of embryogenic tissue from explants *in vitro*. These have included the time of collection during zygotic embryo maturation, early cotyledonary embryos being preferred [298, 299, 306]. For *P. glauca engelmannii* this was correlated to the period prior to deposition of storage protein [305]. The length of time in storage for resting seeds was influential, induction decreased with increasing years of storage [304]. Assuming that such seed contain storage proteins, the observation made on protein deposition with immature zygotic embryo explants [305] might be interpreted in terms of changes in endogenous phytohormonal status during zygotic embryo maturation. Interacting with these influences are effects of genotype (i.e. the explant donor) [304, 306].

Embryogenic tissue can be maintained routinely on agar-solidified medium. The use of liquid shake flasks has also been reported [227, 370, 398, 410]. Such liquid cultures offer possibilities for rapid multiplication of genotypes and for analyses of nutrient utilization for the determination of requirements for optimal growth [227]. In this connection, Lulsdorf et al. [227] evaluated the usefulness of various biotic and abiotic parameters as measures of growth with somatic embryo cultures of *P. glauca engelmannii*. These authors indicated that osmolarity was the most suitable non-destructive parameter closely correlated to culture growth. Ammonium, nitrate and potassium ions (in LP medium) were found not to be limiting, but carbohydrate supply (30 mM sucrose) was limiting.

Bioreactor-culture of somatic embryo cultures of *P. glauca engelmannii* has also been reported [423, 490]. Bioreactors have the capacity to produce a more homogeneous culture than shake flasks owing to the increased working volume and greater control over physical parameters. Embryo yield was reduced on a per ml basis compared to shake flasks, the yield of immature embryos was approximately 13.5 million in the 6 l of bioreactor medium over the 6–8 day culture period. Medium conductivity was found to be the most appropriate parameter with close correlation to culture growth [423].

E. Plant regeneration

Maturation of white spruce somatic embryos can be promoted by culture on medium with racemic ABA [302, 350–352, 413, 422]. In a recent study the maximum production of cotyledonary embryos was achieved on day 42, after using 36–60 μM racemic ABA during the first 35 days [413]. This is similar to findings for *P. glauca engelmannii* [307, 309]. However, a broad range of racemic ABA concentrations has been reported in the literature for use in maturation of white spruce somatic embryos [302, 350–352, 413, 422]. The variety of provenances which have been used, and differences in culture conditions among laboratories, could account for some of the variations in ABA usage. In general it has been difficult to obtain germinable somatic embryos when these have been matured on low concentrations of racemic ABA, e.g. 7.6 μM [351], or 12 μM [350, 422].

Somatic embryo maturation can be problematic, e.g. in the occurrence of aberrant embryo forms [413], and with the variability in response evident with long-term suspension-grown somatic embryos of *P. glauca* [465]. Evidence suggests that there is a strong genotype influence regarding the requirements for adequate maturation conditions, and that these requirements are affected by prior maintenance conditions [466]. These observations may be attributable to physiological and biochemical differences, and altered gene expression.

During racemic ABA-stimulated maturation there is deposition of storage product, e.g. protein, triacylglycerides and carbohydrate [343]. "Storage" protein deposition appeared to be influenced by the racemic ABA concentration used in the maturation phase and was related to the subsequent ability of *P. glauca engelmannii* somatic embryos to germinate; "storage" protein deposition and germination were enhanced with high racemic ABA concentrations (e.g. 40 μM) in the maturation medium [307, 308, 491]. Similarly, triacylglycerol (TAG) biosynthesis was stimulated in *P. glauca* somatic embryo maturation by use of high racemic ABA concentration (16–24 μM) [422]. TAG biosynthesis was also promoted by additional treatment with polyethylene glycol (7.5% PEG, 4000) The combination of treatments had the effect of preventing precocious germination and enhancing desiccation tolerance [422].

The potential of biologically active analogues of (+)-ABA to act as substitutes for racemic ABA in white spruce somatic embryo maturation has been compared [350, 413]. Comparison of the molecular configuration of ABA-like compounds indicated that analogues possessing an ABA-like side chain with two double bonds, were more effective in the white spruce somatic embryo maturation process than compounds with a triple bond in the side chain. However, only abscisyl alcohol was as effective as racemic ABA in the promotion of somatic embryo maturation.

It is not known if (+)-ABA, (−)-ABA, or their respective metabolic products are responsible for the observed enhancement of somatic embryo maturation in white spruce. Using somatic embryo suspension cultures it was observed that there was a quantitative conversion of (+)-ABA to phaseic acid (PA) in the

E   E   E   S   S   S

Fig. 2. Picea glauca-engelmannii somatic embryo-derived plants (E) and seedling-derived plants (S) after one growing season in a forest seedling nursery in British Columbia, Canada. The growth curves, final heights, time of bud set, and root and shoot morphologies were similar for the two groups. Reproduced with permission from Webster et al. [309].

conifer culture medium. The (−)-ABA enantiomer was not metabolized and was not transformed to 7'-hydroxyabscisic acid or to ABA conjugates. The enzyme system which oxidizes (+)-ABA to PA does not appear to hydroxylate the (−)- form of the phytohormone in this culture [400]. Recently it has been suggested that ABA metabolism seems to occur essentially independently of a culture's ability to undergo somatic embryo maturation [492].

Somatic embryo germination and transplantation of plantlets to non-sterile soil conditions have presented problems [350–352], possibly because of suboptimal concentration of racemic ABA in the maturation medium, or to other suboptimal conditions during maturation and germination. Germination and plantlet vigour was improved with *P. glauca engelmannii* somatic embryos by use of a partial drying treatment, in an atmosphere at relatively high humidity [308, 309]. Attree et al. [421, 422] showed that PEG (7.5%, 4000), used as a non-permeating osmoticum in conjunction with racemic ABA, enhanced desiccation tolerance during use of a series of saturated solutions with decreasing relative humidity environments (see also [493]). The acquisition of desiccation tolerance represents a significant developmental stage in between embryo development and germination. Somatic embryos which were not matured under appropriate PEG/ABA treatment underwent precocious germination. To date the best example of successful transfer to soil has been obtained with plantlets derived from *P. glauca engelmannii* somatic embryos [309], and see Fig. 2. Such material appears comparable to seedlings in physiological response and morphological development [495-497].

Early evidence, with *P. glauca engelmannii*, suggests that somatic embryogenesis is a conservative process; using isozyme analyses no somaclonal variants were detected among 1500 clonally related cotyledonary somatic embryos [310].

F. Cryopreservation

Somatic embryos of white spruce can be cryopreserved in liquid nitrogen, using osmotic pre-culture, followed by a treatment with a cryoprotectant combination of 0.4 M sorbitol with 5% DMSO, and controlled freezing down to -35 °C prior to storage in liquid nitrogen [346]. Such material has been thawed after several years storage and used in subsequent experimentation [352, 413] apparently unaltered. This basic method [346] has recently been used to cryopreserve over 300 embryogenic lines of *P. glauca engelmannii* [487]. An alternative storage method involves the placement of embryogenic tissue on agar-solidified culture medium in serum-capped culture flasks, maintained at 24 °C [345]. The latter method is probably most useful as a procedure for intermediate term storage, leaving the cryopresevation method as an ultimate repository. A scheme for the propagation of spruce by somatic embryogenesis, in which cryopreserved samples are periodically thawed for culture renewal, is shown in Figure 1.

G. Protoplast culture

Protoplasts have been obtained from white spruce somatic embryo cultures [401], regenerated back into somatic embryos [297, 300] and to plantlets [301]. Protoplasts can be isolated by incubation for 3–5 h in an enzyme mixture consisting of 1% (w/v) Cellulase, 0.25% (w/v) each of Rhozyme, Pectinase and Driselase, with 5 mM $CaCl_2$, and 0.44 M mannitol. Following initial culture in modified LP medium [300] protoplast-derived cultures had to be carefully returned to the original osmotic strength of the maintenance medium using a series of gradual dilutions over a 35 day period [297, 301]. Cultures regenerated from protoplasts were subsequently maintained in liquid suspension culture, aliquots containing immature somatic embryos were cultured on agar-solidified medium with racemic ABA, leading to the production of mature, germinable, somatic embryos [301].

Freshly isolated protoplasts occurred in uninucleate and multinucleate forms, in each case having extensive networks of randomly arranged cortical microtubules [365]. For either of the two protoplast forms regeneration to somatic embryos was preceded by the establishment of parallel arrays of cortical microtubules during initial cell wall formation [365]. Uninucleate protoplasts showed cell divisions within 24 h of protoplast culture, repeated divisions led to the formation of cell clusters, from which suspensor cells elongated. This was followed by the development of somatic embryos after 12 days. The nuclei in multinucleate protoplasts underwent synchronous division, with the development of multiple phragmoplasts, becoming multicellular by 5 days. Embryo-like structures were observed from these protoplasts by day 7 [365].

The uptake of membrane-impermeable macromolecules into protoplasts of white spruce has been studied using chemical permeabilization with polyethylene glycol (PEG, M.W. 3350) and electropermeabilization [402]. Electropermeabilization was more effective than PEG in stimulating the uptake of the membrane impermeable fluorescent dye (calcein, M.W. 622). Using impermeable fluorescent molecules of different molecular weights it was found that the larger molecules were not internalized as effectively as the calcein dye, suggestive of a limitation due to molecular weight or configuration.

H. Genetic transformation

Electroporation [403] and PEG (M.W. 8000) [404] have been used to study gene expression of *cat* (encoding chloramphenicol acetyl transferase, CAT) and *uidA* (encoding β-glucuronidase, β-GUS) in assays with transiently transformed protoplasts of white spruce. In each case the white spruce protoplasts were capable of regulating *cat* expression under the control of the cauliflower mosaic virus 35S constitutive promoter. Background levels in fluorescent assays for β-GUS-like activity were observed with either permeabilization system, but were most problematic with electroporation [403]. With a tandem duplicated form of the 35S promoter, CAT and β-GUS activities were greatly enhanced in electroporated protoplasts [405]. The electroporation procedure has also been used to investigate the regulation by white spruce protoplasts of heterologous gene constructs under the control of inducible promoters. Good et al. [406] demonstrated the abilities of white spruce to accurately regulate *Adh* (encoding alcohol dehydogenase, ADH) constructs, containing *Adh* promoters from pea, *Arabidopsis*, or maize when each was linked to the *cat* gene. Induction of an *Adh* promoter by anaerobiosis was detected by transient expression of the *cat* gene only in protoplasts electroporated with the the maize construct [406]. Ellis et al. [407] investigated transient gene expression regulated by heterologous inducible promoters in embryogenic cultures exposed to electrically discharged particles. These authors also showed that inducible promoters from angiosperms were regulated in the conifer. White spruce protoplasts are also capable of regulating an ABA-inducible promoter [494], which was isolated from wheat and is contained in the plasmid pBM113Kp [409]. Recently there have been two reports confirming the integration of heterologous genes in white spruce [482, 489]. Both methods used particle bombardment (electrical discharge [482], and helium pressure [489]) to introduce reporter genes. Two different methods were used to obtain transformed tissues, one, by selection for *nptII* expression based on resistance to 5 mg l^{-1} kanamycin [482] and, two, by successive propagation of tissue sectors, periodically assessed for positive reaction to histochemical GUS assay, without the use of kanamycin [489].

In general greater progress has been made in somatic embryogenesis with spruce species than with other conifer species, including pine, although the latter are more widely distributed and prominent world-wide [370]. This is partly attributable to the preponderance of research using *Picea abies*, and partly to the inherent responsiveness of *Picea* species *in vitro*. Progress with the various aspects of *Picea glauca* somatic embryogenesis presented in the preceeding description has been rapid, and it has been a highly responsive species with which to explore conifer biotechnology. Other reviews on conifer biotechnology should also be consulted [3, 370, 410–412].

VI. Conclusions

As a grouping, woody plants are associated with one another by virtue of their woodiness, and additionally in the perception that they are a difficult group with which to work. It is rewarding therefore to see the extent to which somatic embryogenesis has been developed with this group. In some instances research with somatic embryogenesis has proceeded at a tremendous pace.

The driving forces for such developments are as varied as the procedures which have been developed. As an example, consider the three species which have been chosen to illustrate somatic embryogenesis with woody plants. With *Citrus*, research has focussed on the production of variants for use in breeding programmes. This has involved the development of processes such as protoplast fusion. Much information has been accumulated about *Citrus* somatic embryogenesis during this endeavour, and progress has been steady. Comparatively, research with *Elaeis* has concentrated on somatic embryogenesis as a procedure for clonal propagation, progress appears to have been relatively slow. Research with *Picea* initially concerned the characterization of the somatic embryogenesis induction process, and the regeneration of somatic embryos to plantlets. However, cell biology, protoplast culture and transient transformation feature prominently. Progress has been very rapid.

As illustrated in the text, somatic embryogenesis can be the basis for other research endeavours, such as plant breeding through protoplast hybridization and through genetic transformation. It is to be hoped that this review will lead to further stimulation of such endeavours, as well as to the cross-fertilization of ideas, based on the variety of procedures used among the woody species with which somatic embryogenesis has been induced.

It is disturbing that the processes underlying the induction of somatic embryogenesis, or the maturation and germination of somatic embryos, remain relatively unclear. Progress tends to be the result of empirical research. This does not assist in the resolution of problems that from time-to-time arise, such as with the variation in maturation potential experienced with *Picea* spp. [465, 466, 484] somatic embryo cultures. It is highly desirable that research into the physiological/biochemical and/or molecular events associated with these phases is stimulated.

For many researchers the goal of a somatic embryogenesis method is its use as a propagation tool. In this connection it is appropriate that more research effort be directed toward understanding seed maturation and germination. This knowledge would be of great assistance for those interested in somatic embryo maturation and desiccation processes and in the development of synthetic seed.

There is relatively little information about clonal fidelity and about field performance of somatic embryo-derived plants. These aspects are essential to verify that the somatic embryo process for each particular species is not causing aberrations, such as noted with *Elaeis* [380, 395]. Unfortunately, field performance data for woody plants is accumulated often over long periods, making such work unattractive. Never-the-less the importance of such information mandates the establishment of field trials accompanied, for example, by ecophysiological assessments of survivability and growth potential relative to seedlings [485].

We hope in writing this chapter to have dispelled some of the mysticism associated with woody plants, principally relating to their recalcitrance to *in vitro* experimentation. There remains considerable potential for scientific progress. It is our belief that the investment of research effort into this group of plants will reward the investigator and the broader scientific community.

References

1. Ray, P.M., Steeves, T.A., and Fultz, S.A., *Botany*, 1st Ed., Saunders College Publishing, New York, 1983.
2. Tulecke, W., Somatic embryogenesis in woody perennials, in *Cell and Tissue Culture in Forestry*, Vol. 2, Bonga, J.M., and Durzan, D.J. Eds., Martinus Nijhoff Publishers, Dordrecht, 1987, 61.
3. Wann, S.R., Somatic embryogenesis in woody species, *Hort. Rev.*, 10, 153, 1988.
4. Ranga Swamy, N.S., Culture of nucellar tissue of *Citrus in vitro*, *Experientia*, 14, 111, 1958.
5. Maheshwari, P., and Ranga Swamy, N.S., Polyembryony and *in vitro* culture of embryos of *Citrus* and *Mangifera*, *Ind. J. Hort.*, 15, 275, 1958.
6. Rangan, T.S., Murashige, T., and Bitters, W.P., *In vitro* initiation of nucellar embryos in monoembryonic *Citrus*, *HortSci.*, 3, 126, 1968.
7. Konar, R.N., and Oberoi, Y.P., *In vitro* development of embryoids on the cotyledons of *Biota orientalis*, *Phytomorphology*, 15, 137, 1965.
8. Rao, P.S., *In vitro* induction of embryonal proliferation in *Santalum album* L, *Phytomorphology*, 15, 175, 1965.
9. Norstog, K., Induction of apogamy in megagametophytes of *Zamia integrifolia*, *Amer. J. Bot.*, 52, 993, 1965.
10. Norstog, K., and Rhamstine, E., Isolation and culture of haploid and diploid cycad tissues, *Phytomorphology*, 17, 374, 1967.
11. Gui, Y., Guo, K., Ke, S., and Skirvin, R.M., Somatic embryogenesis and plant regeneration in *Acanthopanax senticosus*, *Plant Cell Rep.*, 9, 514, 1991.
12. Sachdeva, S., and Mehra, P.N., *In vitro* studies on *Achras sapota* (Chikoo) I. callus and early embryogenesis, *Phytomorphology*, 36, 315, 1986.
13. Fraser, L.G., and Harvey, C.F., Somatic embryogenesis from anther-derived callus in two *Actinidia* species, *Sci. Hort.*, 29, 335, 1986.

14. Dameri, R.M., Caffaro, L., Gastaldo, P., and Profumo, P., Callus formation and embryogenesis with leaf explants of *Aesculus hippocastanum* L, *J. Plant Physiol.*, 126, 93, 1986.

15. Profumo, P., Gastaldo, P., Dameri, R.M., and Caffaro, L., Histological study of calli and embryoids from leaf explants of *Aesculus hippocastanum* L, *J. Plant Physiol.*, 126, 97, 1986.

16. Radojevic, L., Plant regeneration of *Aesculus hippocastanum* L. (Horse chestnut) through somatic embryogenesis, *J. Plant Physiol.*, 132, 322, 1988.

17. Jörgensen, J., Somatic embryogenesis in *Aesculus hippocastanum* L. by culture of filament callus, *J. Plant Physiol.*, 135, 240, 1989.

18. Profumo, P., Gastaldo, P., Bevilacqua, L., and Carli, S., Plant regeneration from cotyledonary explants of *Aesculus hippocastanum* L, *Plant Sci.*, 76, 139, 1991.

19. Gharyal, P.K., and Maheshwari, S.C., *In vitro* differentiation of somatic embryoids in a leguminous tree – *Albizzia lebbeck* L, *Naturwiss.*, 68, 379, 1981.

20. Tomar, U.K., and Gupta, S.C., Somatic embryogenesis and organogenesis in callus cultures of a tree legume – *Albizzia richardiana* King, *Plant Cell Rep.*, 7, 70, 1988.

21. Valverde, R., Arias, O., and Thorpe, T. A., Picloram-induced somatic embryogenesis in pejibaye palm (*Bactris gasipaes* H.B.K.), *Plant Cell Tiss. Org. Cult.*, 10, 149, 1987.

22. Kurten, U., Nuutila, A.M., Kauppinen, V., and Rousi, M., Somatic embryogenesis in cell cultures of birch (*Betula pendula* Roth.), *Plant Cell Tiss. Org. Cult.*, 23, 101, 1990.

23. Nuutila, A.M., Kurtén, U., and Kauppinen, V., Optimization of sucrose and inorganic nitrogen concentrations for somatic embryogenesis of birch (*Betula pendula* Roth.) callus cultures: a statistical approach, *Plant Cell Tiss. Org. Cult.*, 24, 73, 1991.

24. Zaid, A., and Tisserat, B., Survey of the morphogenetic potential of excised palm embryos *In vitro*, *Crop Res. (Hort. Res.)*, 24, 1, 1984.

25. Chengji, Z., and Hanxing, L., Somatic embryogenesis and plantlets formation in cotyledon cutlure of *Camellia chrysantha*, *Acta Bot. Yunnanica*, 7, 446, 1985.

26. Vieitez, A.M., and Barciela, J., Somatic embryogenesis and plant regeneration from embryonic tissues of *Camellia japonica* L, *Plant Cell Tiss. Org. Cult.*, 21, 267, 1990.

27. Vieitez, A.M., San-José, C., Vieitez, J., and Ballester, A., Somatic embryogenesis from roots of *Camellia japonica* plantlets cultured *in vitro*, *J. Amer. Soc. Hort. Sci.*, 116, 753, 1991.

28. Chengji, Z., Jinyu, D., and Jiankui, Z., Somatic embryogenesis and plantlet regeneration of *Camellia sasanqua*, *Acta Bot. Yunnanica*, 10, 241, 1988.

29. Merkle, S.A., Wetzstein, H.Y., and Sommer, H.E., Somatic embryogenesis in tissue cultures of pecan, *HortSci.*, 22, 128, 1987.

30. Wetzstein, H.Y., Ault, J.R., and Merkle, S.A., Further characterization of somatic embryogenesis and plantlet regeneration in pecan (*Carya illinoensis*), *Plant Sci.*, 64, 193, 1989.

31. Gonzalez, M.L., Vieitez, A.M., and Vieitez, E., Somatic embryogeneisis from chestnut cotyledon tissue cultured *in vitro*, *Sci. Hort.*, 27, 97, 1985.

32. Vieitez, F.J., San-Jose, M.C., Ballester, A., and Vieitez, A.M., Somatic embryogenesis in cultured immature zygotic embryos chestnut, *J. Plant Physiol.*, 136, 253, 1990.

33. Trigiano, R.N., Beaty, R.M., and Graham, E.T., Somatic embryogenesis from immature embryos of redbud (*Cercis canadensis*), *Plant Cell Rep.*, 7, 148, 1988.

34. Geneve, R.L., and Kester, S.T., The initiation of somatic embryos and adventitious roots from developing zygotic embryo explants of *Cercis canadensis* L. cultured *in vitro*, *Plant Cell Tiss. Org. Cult.*, 22, 71, 1990.

35. Reynolds, J.F., and Murashige, T., Asexual embryogenesis in callus cultures of palms, *In Vitro Cell. Dev. Biol.*, 15, 383, 1979.

36. Mitra, G.C., and Chaturvedi, H.C., Embryoids and complete plants from unpollinated ovaries and from ovules of *in vivo*-grown emasculated flower buds of *Citrus* spp, *Bull. Torrey Bot. Club*, 99, 184, 1972.

37. Gmitter, F. Jr., and Moore, G.A., Plant regeneration from undeveloped ovules and embryogenic calli of *Citrus*: Embryo production, germination, and plant survival, *Plant Cell Tiss. Org. Cult.*, 6, 139, 1986.

38. Grosser, J.W., Moore, G.A., and Gmitter, J.F.G., Interspecific somatic hybrid plants from the

518

fusion of "Key" lime (*Citrus aurantifolia*) with "Valencia" sweet orange (*Citrus sinensis*) protoplasts, *Sci. Hort.*, 39, 23, 1989.

39. Kochba, J., Ben-Hayyim, G., Spiegel-Roy, P., Saad, S., and Neumann, H., Selection of stable salt-tolerant callus cell lines and embryos in *Citrus sinensis* and *C. aurantium*, *Z. Pflanzenphysiol.*, 106, 111, 1982.

40. Kochba, J., Spiegel-Roy, P., Neumann, H., and Saad, S., Effect of carbohydrates on somatic embryogenesis in subcultured nucellar callus of *Citrus* cultivars, *Z. Pflanzenphysiol.*, 105, 359, 1982.

41. Ben-Hayyim, G., and Neumann, H., Stimulatory effect of glycerol on growth and somatic embryogenesis in *Citrus* callus cultures, *Z. Pflanzenphysiol.*, 110, 331, 1983.

42. Vardi, A., Spiegel-Roy, P., and Galun, E., Plant regeneration from *Citrus* protoplasts: variability in methodological requirements among cultivars and species, *Theor. Appl. Genet.*, 62, 171, 1982.

43. Beloualy, N., Plant regeneration from callus culture of three *Citrus* rootstocks, *Plant Cell Tiss. Org. Cult.*, 24, 29, 1991.

44. Vardi, V., Breiman, A., and Galun, E., *Citrus* cybrids: production by donor-recipient protoplast-fusion and verification by mitochondrial-DNA restriction profiles, *Theor. Appl. Genet.*, 75, 51, 1987.

45. Navarro, L., Ortiz, J.M., and Juarez, J., Aberrant citrus plants obtained by somatic embryogenesis of nucelli cultured *in vitro*, *HortSci.*, 20, 214, 1985.

46. Chaturvedi, H.C., and Mitra, G.C., A shift in morphogenetic pattern in *Citrus* callus tissue during prolonged culture, *Ann. Bot.*, 39, 683, 1975.

47. Moore, G.A., Factors affecting *in vitro* embryogenesis from undeveloped ovules of mature *Citrus* fruit, *J. Amer. Soc. Hort. Sci.*, 110, 66, 1985.

48. Ling, J.-T., Nito, N., and Iwamasa, M., Plant regeneration from protoplasts of calamondin (*Citrus madurensis* Lour.), *Sci. Hort.*, 40, 325, 1989.

49. Ranga Swamy, N.S., Experimental studies on female reproductive structures of *Citrus microcarpa* Bunge, *Phytomorphology*, 11, 109, 1961.

50. Sim, G.-E., Loh, C.-S., and Goh, C.-J., Direct somatic embryogenesis from protoplasts of *Citrus mitis* Blanco, *Plant Cell Rep.*, 7, 418, 1988.

51. Kochba, J., Spiegel-Roy, P., and Safran, H., Adventive plants from ovules and nucelli in *Citrus*, *Planta*, 106, 237, 1972.

52. Raj Bhansali, R., and Arya, H.C., Differentiation in explant of *Citrus paradisi* Macf. (grapefruit) grown in culture, *Indian J. Exp. Biol.*, 16, 409, 1977.

53. Dhillon, B.S., Raman, H., and Brar, D.S., Somatic embryogenesis in *Citrus paradisi* and characterization of regenerated plants, *Acta Hort.*, 239, 113, 1989.

54. Gavish, H., Lewinsohn, E., Vardi, A., and Fluhr, R., Production of flavanone-neohesperidosides in *Citrus* embryos, *Plant Cell Rep.*, 8, 391, 1989.

55. Juarez, J., Navarro, L., and Guardiola, J.L., Obtention de plants nucellaires de divers cultivars de clémentinier au moyen de la culture de nucelle *in vitro*, *Fruits*, 31, 751, 1976.

56. Tisserat, B., and Murashige, T., Probable identity of substances in *Citrus* that repress asexual embryogenesis, *In Vitro Cell. Dev. Biol.*, 13, 785, 1977.

57. Hidaka, T., and Kajiura, I., Plantlet differentiation from callus protoplasts induced from *Citrus* embryo, *Sci. Hort.*, 34, 85, 1988.

58. Grosser, J.W., Gmitter, F.G. Jr., Tusa, N., and Chandler, J.L., Somatic hybrid plants from sexually incompatible woody species: *Citrus reticulata* and *Citropsis gilletiana*, *Plant Cell Rep.*, 8, 656, 1990.

59. Gmitter, F.G. Jr., and Ling, X., Embryogenesis *in vitro* and nonchimeric tetraploid plant recovery from undeveloped *Citrus* ovules treated with colchicine, *J. Amer. Soc. Hort. Sci.*, 116, 317, 1991.

60. Button, J., and Bornman, C.H., Development of nucellar plants from unpollinated and unfertilised ovules of the Washington navel orange *in vitro*, *J. S. Afr. Bot.*, 37, 127, 1971.

61. Kochba, J., and Spiegel-Roy, P., Effect of culture media on embryoid formation from ovular callus of "Shamouti" orange (*Citrus sinensis*), *Z. Pflanzenzüchtg.*, 69, 156, 1973.

62. Srinivasan, C., Litz, R.E., Barker, J., and Norstog, K., Somatic embryogenesis and plantlet formation from christmas palm callus, *HortSci.*, 20, 278, 1985.

63. Becwar, M.R., Blush, T.D., Brown, D.W., and Chesick, E.E., Multiple paternal genotypes in embryogenic tissue derived from individual immature loblolly pine seeds, *Plant Cell Tiss. Org. Cult.*, 26, 37, 1991.

64. Gmitter, F.G. Jr., Ling, X., Cai, C., and Grosser, J.W., Colchicine-induced polyploidy in *Citrus* embryogenic cultures, somatic embryos, and regenerated plantlets, *Plant Sci.*, 74, 135, 1991.

65. Spiegel-Roy, P., and Kochba, J., Stimulation of differentiation in orange (*Citrus sinensis*) ovular callus in relation to irradiation of the media, *Radiation. Bot.*, 13, 97, 1973.

66. Ben-Hayyim, G., and Kochba, J., Growth characteristics and stability of tolerance of citrus callus cells subjected to NaCl stress, *Plant Sci. Lett.*, 27, 87, 1982.

67. Gupta, P.K., and Pullman, G.S., Method for reproducing coniferous plants by somatic embryogenesis, U.S. Patent, 4,957,866, 1990.

68. Hu, C.Y., Holly (*Ilex spp.*), in *Trees II*, Vol. 5, Bajaj, Y.P.S., Ed., Springer-Verlag, Berlin, 1989, Chapter I.22.

69. Kobayashi, S., Uchimiya, H., and Ikeda, I., Plant regeneration from "Trovita" orange protoplasts, *Japan. J. Breed*, 33, 119, 1983.

70. Nito, N., Ling, J.-T., Iwamasa, M., and Katayama, Y., Effects of α-irradiation on growth and embryogenesis of *Citrus* callus, *J. Japan. Soc. Hort. Sci.*, 58, 283, 1989.

71. Vardi, A., Spiegel-Roy, P., and Galun, E., *Citrus* cell culture: isolation of protoplasts, plating densities, effect of mutagens and regeneration of embryos, *Plant Sci. Lett.*, 4, 231, 1975.

72. Kobayashi, S., Ikeda, I., and Uchimiya, H., Conditions for high frequency embryogenesis from orange (*Citrus sinensis* Osb.) protoplasts, *Plant Cell Tiss. Org. Cult.*, 4, 249, 1985.

73. Marin, M.I., and Duran-Vila, N., Survival of somatic embryos and recovery of plants of sweet orange (*Citrus sinensis* (L.) Osb.) after immersion in liquid nitrogen, *Plant Cell Tiss. Org. Cult.*, 14, 51, 1988.

74. Kobayashi, S., and Oiyama, I., Cryopreservation in liquid nitrogen of cultured navel orange (*Citrus sinensis* Osb.) nucellar cells and subsequent plant regeneration, *Plant Cell Tiss. Org. Cult.*, 23, 15, 1990.

75. Sakai, A., Kobayashi, S., and Oiyama, I., Cryopreservation of nucellar cells of navel orange (*Citrus sinensis* Osb.) by a simple freezing method, *Plant Sci.*, 74, 243, 1991.

76. Ohgawara, T., Kobayashi, S., Ishii, S., Yoshinaga, K., and Oiyama, I., Somatic hybridization in *Citrus*: navel orange (*C. sinensis* Osb.) and grapefruit (*C. paradisi* Macf.), *Theor. Appl. Genet.*, 78, 609, 1989.

77. Kobayashi, S., Ohgawara, T., Ohgawara, E., Oiyama, I., and Ishii, S., A somatic hybrid plant obtained by protoplast fusion between navel orange (*Citrus sinensis*) and satsuma mandarin (*C. unshiu*), *Plant Cell Tiss. Org. Cult.*, 14 63, 1988.

78. Kobayashi, S., Ohgawara, T., Fujiwara, K., and Oiyama, I., Analysis of cytoplasmic genomes in somatic hybrids between navel orange (*Citrus sinensis* Osb.) and "Murcott" tangor, *Theor. Appl. Genet.*, 82, 6, 1991.

79. Ohgawara, T., Kobayashi, S., Ohgawara, E., Uchimiya, H., and Ishii, S., Somatic hybrid plants obtained by protoplast fusion between *Citrus sinensis* and *Poncirus trifoliata*, *Theor. Appl. Genet.*, 71, 1, 1985.

80. Grosser, J.W., Gmitter, F.G. Jr., and Chandler, J.L., Intergeneric somatic hybrid plants of *Citrus sinensis* cv. Hamlin and *Poncirus trifoliata* cv. Flying Dragon, *Plant Cell Rep.*, 7, 5, 1988.

81. Grosser, J.W., Gmitter, F.G. Jr., and Chandler, J.L., Intergeneric somatic hybrid plants from sexually incompatible woody species: *Citrus sinensis* and *Severinia disticha*, *Theor. Appl. Genet.*, 75, 397, 1988.

82. Saito, W., Ohgawara, T., Shimizu, J., and Ishii, S., Acid citrus somatic hybrids between sudachi (*Citrus sudachi* Hort. ex Shirai) and lime (*C. aurantifolia*) produced by electrofusion, *Plant Sci.*, 77, 125, 1991.

83. Kunitake, H., Kagami, H., and Mii, M., Somatic embryogenesis and plant regeneration from protoplasts of "Satsuma" mandarin (*Citrus unshiu* Marc.), *Sci. Hort.*, 47, 27, 1991.

520

84. Weaver, L.A., and Trigiano, R.N., Regeneration of *Cladrastis lutea* (Fabaceae) via somatic embryogenesis, *Plant Cell Rep.*, 10, 183, 1991.
85. Branton, R.L., and Blake, J., Development of organized structures in callus derived from explants of *Cocos nucifera* L, *Ann. Bot.*, 52, 673, 1983.
86. Branton, R.L., and Blake, J., Clonal propagation of coconut palm, in *Proc. Int. Conf. Cocoa and Coconuts*, 771, 1984.
87. Söndahl, M.R., and Sharp, W.R., High frequency induction of somatic embryos in cultured leaf explants of *Coffea arabica* L, *Z. Pflanzenphysiol.*, 81, 395, 1977.
88. Söndahl, M.R., Spahlinger, D.A., and Sharp, W.R., A histological study of high frequency and low frequency induction of somatic embryos in cultured leaf explants of *Coffea arabica* L., *Z. Pflanzenphysiol.*, 94, 101, 1979.
89. Söndahl, M.R., Salisbury, J.L., and Sharp, W.R., SEM characterization of embryogenic tissue and globular embryos during high frequency somatic embryogenesis in coffee callus cells, *Z. Pflanzenphysiol.*, 94, 185, 1979.
90. Yasuda, T., Fujii, Y., and Yamaguchi, T., Embryogenic callus induction from *Coffea arabica* leaf explants by benzyladenine, *Plant Cell Physiol.*, 26, 595, 1985.
91. De Garcia, E., and Menendez, A., Embriogénesis somática a partir de explants foliares de cafeto *Catimor*, *Café Cacao Thé*, 31, 15, 1987.
92. Michaux-Ferrière, N., Bieysse, D., Alvard, D., and Dublin, P., Étude histologique de l'"embryogenèse somatique chez *Coffea arabica*, induite par culture sur milieux uniques de fragments foliaires de génotypes différents, *Café Cacao Thé*, 33, 207, 1989.
93. Staritsky, G., Embryoid formation in callus tissues of coffee, *Acta Bot. Neerl.*, 19, 509, 1970.
94. Nassuth, A., Wormer, T.M., Bouman, F., and Staritsky, G., The histogenesis of callus in *Coffea canephora* stem explants and the discovery of early embryoid initiation, *Acta Bot. Neerl.*, 29, 49, 1980.
95. Lanaud, C., Production de plantules de *C. canephora* par embryogenèse somatique réalisée à partir de culture *in vitro* d'ovules, *Café Cacao Thé*, 25, 231, 1981.
96. Pierson, E.S., Van Lammeren, A.M., Schel, J.H.N., and Staritsky, G., *In vitro* development of embryoids from punched leaf discs of *Coffea canephora*, *Protoplasma*, 115, 208, 1983.
97. Schöpke, C., Müller, L.E., and Kohlenbach, H.-W., Somatic embryogenesis and regeneration of plantlets in protoplast cultures from somatic embryos of coffee (*Coffea canephora* P. ex Fr.), *Plant Cell Tiss. Org. Cult.*, 8, 243, 1987.
98. Hatanaka, T., Arakawa, O., Yasuda, T., Uchida, N., and Yamaguchi, T., Effect of plant growth regulators on somatic embryogenesis in leaf cultures of *Coffea canephora*, *Plant Cell Rep.*, 10, 179, 1991.
99. Trigiano, R.N., Beaty, R.M., and Dietrich, J.T., Somatic embryogenesis and plant regeneration in *Cornus florida*, *Plant Cell Rep.*, 8, 270, 1989.
100. Radojević, L., Vujicić, R., and Nesković, M., Embryogenesis in tissue culture of *Corylus avellana* L, *Z. Pflanzenphysiol.*, 77, 33, 1975.
101. Radojević, L., Kovoor, J., and Zylberberg, L., Étude anatomique et histochimique des cals embryogènes du *Corylus avellana* L. et du *Paulownia tomentosa* STEUD, *Rev. Cytol. Biol. végét.-Bot.*, 2, 155, 1979.
102. Pérez, C., Fernàndez, B., and Rodriguez, R., *In vitro* plantlets regeneration through asexual embryogenesis in cotyledonary segments of *Corylus avellana* L, *Plant Cell Rep.*, 2, 226, 1983.
103. Pérez, C., Rodriguez, R., and Tamés, R.S., Regulation of asexual embryogenesis in filbert cotyledonary nodes. Morphological variability, *Plant Sci.*, 45, 59, 1986.
104. Wicart, G., Mouras, A., and Lutz, A., Histological study of organogenesis and embryogenesis in *Cyclamen persicum* Mill. tissue cultures: evidence for a single organogenetic pattern, *Protoplasma*, 119, 159, 1984.
105. Fukui, H., Nishimoto, K., Murase, I., and Nakamura, M., Somatic embryogenesis from the leaf tissues of continuously subcultured shoots of Japanese Persimmon (*Diospyros kaka* Thunb.), *Japan. J. Breed*, 38, 465, 1988.
106. Rabéchault, H., Ahée, J., and Guénin, G., Colonies cellulaires et formes embryoides obtenues

in vitro à partir de cultures d'embryons de palmiere à huile (*Elaeis guineensis* Jacq. var. dura Becc.), *C.R. Acad. Sci. Paris Sér. D*, 270, 3067, 1970.

107. Hanower, J., and Hanower, P., Inhibition et stimulation, en culture *in vitro* de l'embryogenèse des souches issues d'explants foliaires de palmier à huile, *C.R. Acad. Sci. Paris, Sér. III*, 298, 45, 1984.

108. Thomas, V., and Rao, P.S., *In vitro* propagation of oil palm (*Elaeis guineensis* Jacq. var tenera) through somatic embryogenesis in leaf-derived callus, *Curr. Sci.*, 54, 184, 1985.

109. Nwankwo, B.A., and Krikorian, A.D., Morphogenetic potential of embryo- and seedling-derived callus of *Elaeis guineensis* Jacq. var. *pisifera* Becc, *Ann. Bot.*, 51, 65, 1986.

110. Schwendiman, J., Pannetier, C., and Michaux-Ferriere, N., Histology of somatic embryogenesis from leaf explants of the oil palm *Elaeis guineensis*, *Ann. Bot.*, 62, 43, 1988.

111. Ramawat, K.G., and Arya, H.C., Growth and morphogenesis in callus cultures of *Ephedra gerardiana*, *Phytomorphology*, 26, 395, 1976.

112. Muralidharan, E.M., and Mascarenhas, A.F., *In vitro* plantlet formation by organogenesis in *E. camaldulensis* and by somatic embryogenesis in *Eucalyptus citriodora*, *Plant Cell Rep.*, 6, 256, 1987.

113. Muralidharan, E.M., Gupta, P.K., and Mascarenhas, A.F., Plantlet production through high frequency somatic embryogenesis in long term cultures of *Eucalyptus citriodora*, *Plant Cell Rep.*, 8, 41, 1989.

114. Quan, O., Hai-zhong, P., and Qi-quan, L., Studies on the development of embryoid from *Eucalyptus* callus, *Sci. Silvae Sin.*, 17, 1, 1981.

115. Litz, R.E., *In vitro* responses of adventitious embryos of two polyembronic *Eugenia* species, *HortSci.*, 19, 720, 1984.

116. Litz, R.E., Somatic embryogenesis from cultured leaf explants of the tropical tree *Euphoria longan* Stend, *J. Plant Physiol.*, 132, 190, 1988.

117. Guerra, M.P., and Handro, W., Somatic embryogenesis and plant regeneration in embryo cultures of *Euterpe edulis* mart. (palmae), *Plant Cell Rep.*, 7, 550, 1988.

118. Jörgensen, J., Embryogenesis in *Quercus petraea* and *Fagus sylvatica*, *J. Plant Physiol.*, 132, 638, 1988.

119. Lal, M., Narayan, P., and Jaiswal, V.S., Induction of somatic embryogenesis and associated changes in peroxidase activity in leaf callus cultures of *Ficus religiosa* L, *Proc. Indian Natl. Sci. Acad.*, 54, 171, 1988.

120. Preece, J.E., Zhao, J., and Kung, F.H., Callus production and somatic embryogenesis from white ash, *HortSci.*, 24, 377, 1989.

121. Bouharmont, J., and Dabin, P., Origine et évolution des embryons somatiques chez les fuchsias, *Cellule*, 74, 307, 1989.

122. Dabin, P., and Faranna, G., Callogenèse, étalement et embryogenèse chez une lignée embryogène de fuchsia, *Bull. Soc. Roy. Belg.*, 122, 170, 1989.

123. Trolinder, N.L., and Xhixian, C., Genotype specificity of the somatic embryogenesis response in cotton, *Plant Cell Rep.*, 8, 133, 1989.

124. Davidonis, G.H., and Hamilton, R.H., Plant regeneration from callus tissue of *Gossypium hirsutum* L, *Plant Sci. Lett.*, 32, 89, 1983.

125. Gawel, N.J., Rao, A.P., and Robacker, C.D., Somatic embryogenesis from leaf and petiole callus cultures of *Gossypium hirsutum* L, *Plant Cell Rep.*, 5, 457, 1986.

126. Gawel, N.J., and Robacker, C.D., Genetic control of somatic embryogenesis in cotton petiole callus cultures, *Euphytica*, 49, 249, 1990.

127. Shoemaker, R.C., Couche, L.J., and Galbraith, D.W., Characterization of somatic embryogenesis and plant regeneration in cotton (*Gossypium hirsutum* L.), *Plant Cell Rep.*, 3, 178, 1986.

128. Shoemaker, R.C., Christofferson, S.E., and Galbraith, D.W., Storage protein accumulation patterns in somatic embryos of cotton (*Gossypium hirsutum* L.), *Plant Cell Rep.*, 6, 12, 1987.

129. Trolinder, N.L., and Goodin, J.R., Somatic embryogenesis and plant regeneration in cotton (*Gossypium hirsutum* L.), *Plant Cell Rep.*, 6, 231, 1987.

522

130. Trolinder, N.L., and Goodin, J.R., Somatic embryogenesis in cotton (*Gossypium*). II. Requirements for embryo development and plant regeneration, *Plant Cell Tiss. Org. Cult.*, 12, 43, 1988.

131. Umbeck, P., Johnson, G., Barton, K., and Swain, W., Genetically transformed cotton (*Gossypium hirsutum* L.) plants, *Bio/Technol.*, 5, 263, 1987.

132. Finer, J.J., Plant regeneration from somatic embryogenic suspension cultures of cotton (*Gossypium hirsutum* L.), *Plant Cell Rep.*, 7, 399, 1988.

133. Finer, J.J., and McMullen, M.D., Transformation of cotton (*Gossypium hirsutum* L.) via particle bombardment, *Plant Cell Rep.*, 8, 586, 1990.

134. Trolinder, N.L., and Goodin, J.R., Somatic embryogenesis in cotton (*Gossypium*) I. Effects of source explant and hormone regime, *Plant Cell Tiss. Org. Cult.*, 12, 31, 1988.

135. Gawel, N.J., and Robacker, C.D., Somatic embryogenesis in two *Gossypium hirsutum* genotypes on semi-solid versus liquid proliferation media, *Plant Cell Tiss. Org. Cult.*, 23, 201, 1990.

136. Price, H.J., and Smith, R.H., Somatic embryogenesis in suspension cultures of *Gossypium klotzschianum* Anderss, *Planta*, 145, 305, 1979.

137. Finer, J.J., and Smith, R.H., Initiation of callus and somatic embryos from explants of mature cotton (*Gossypium klotzschianum* Anderss.), *Plant Cell Rep.*, 3, 41, 1984.

138. Banks, M.S., Plant regeneration from callus from two growth phases of English ivy, *Hedera helix* L, *Z. Pflanzenphysiol.*, 92, 349, 1979.

139. Carron, M.-P., and Enjalric, F., Somatic embryogenesis from inner integument of the seed of *Hevea brasiliensis* (Kunth., Müll. Arg.), *C.R. Acad. Sci. Paris Sér. III*, 300, 653, 1985.

140. El Hadrami, I., Michaux-Ferrière, N., Carron, M.-P., and d'Auzac, J., Polyamines, a possible limiting factor in somatic embryogenesis of *Hevea brasiliensis*, *C.R. Acad. Sci. Paris Sér. III*, 308, 205, 1989.

141. Michaux-Ferrière, N., and Carron, M.-P., Histology of early somatic embryogenesis in *Hevea brasiliensis*: The importance of the timing of subculturing, *Plant Cell Tiss. Org. Cult.*, 19, 243, 1989.

142. Auboiron, E., Carron, M.-P., and Michaux-Ferrière, N., Influence of atmospheric gases, particularly ethylene, on somatic embryogenesis of *Hevea brasiliensis*, *Plant Cell Tiss. Org. Cult.*, 21, 31, 1990.

143. El Hadrami, I., Carron, M.P., and d'Auzac, J., Influence of exogenous hormones on somatic embryogenesis in *Hevea brasiliensis*, *Ann. Bot.*, 67, 511, 1991.

144. Etienne, H., Berger, A., and Carron, M.P., Water status of callus from *Hevea brasiliensis* during induction of somatic embryogenesis, *Physiol. Plant.*, 82, 213, 1991.

145. Reynolds, B.D., and Blackmon, W.J., Embryogenesis and plantlet regeneration from callus of *Hibiscus acetosella*, *J. Amer. Soc. Hort. Sci.*, 108, 307, 1983.

146. Hu, C.Y., and Sussex, I.M., *In vitro* development of embryoids on cotyledons of *Ilex aquifolium*, *Phytomorphology*, 21, 103, 1971.

147. Hu, C.Y., Ochs, J.D., and Mancini, F.M., Further observations on *Ilex* embryoid production, *Z. Pflanzenphysiol.*, 89, 41, 1978.

148. Tulecke, W., and McGranahan, G., Somatic embryogenesis and plant regeneration from cotyledons of walnut, *Juglans regia* L, *Plant Sci.*, 40, 57, 1985.

149. McGranahan, G.H., Leslie, C.A., Uratsu, S.L., Martin, L.A., and Dandekar, A.M., *Agrobacterium*-mediated transformation of walnut somatic embryos and regeneration of transgenic plants, *Bio/Technol.*, 6, 800, 1988.

150. Polito, V.S., McGranahan, G., Pinney, K., and Leslie, C., Origin of somatic embryos from repetitively embryogenic cultures of walnut (*Juglans regia* L.): implications for *Agrobacterium*-mediated transformation, *Plant Cell Rep.*, 8, 219, 1989.

151. Tulecke, W., McGranahan, G., and Ahmadi, H., Regeneration by somatic embryogenesis of triploid plants from endosperm of walnut, *Juglans regia* L. cv Manregian, *Plant Cell Rep.*, 7, 301, 1988.

152. Pal, A., Banerjee, A., and Dhar, K., *In vitro* organogenesis and somatic embryogenesis from leaf explants of *Leucosceptrum canum* sm, *Plant Cell Rep.*, 4, 281, 1985.

153. Sommer, H.E., and Brown, C.L., Embryogenesis in tissue cultures of sweetgum, *Forest Sci.*, 26, 257, 1980.

154. Merkle, S.A., and Sommer, H.E., Somatic embryogenesis in tissue cultures of *Liriodendron tulipifera*, *Can. J. For. Res.*, 16, 420, 1986.

155. Merkle, S.A., and Sommer, H.E., Regeneration of *Liriodendron tulipifera* (family Magnoliaceae) from protoplast culture, *Amer. J. Bot.*, 74, 1317, 1987.

156. Sotak, R.J., Sommer, H.E., and Merkle, S.A., Relation of the developmental stage of zygotic embryos of yellow-poplar to their somatic embryogenic potential, *Plant Cell Rep.*, 10, 175, 1991.

157. Eichholtz, D.A., Robitaille, H.A., and Hasegawa, P.M., Adventive embryony in apple, *HortSci.*, 14, 699, 1979.

158. Liu, J.R., Sink, K.C., and Dennis, J.F.G., Adventive embryogenesis from leaf explants of apple seedlings, *HortSci.*, 18, 871, 1983.

159. James, D.J., Passey, A.J., and Deeming, D.C., Adventitious embryogenesis and the *in vitro* culture of apple seed parts, *J. Plant Physiol.*, 115, 217, 1984.

160. Mehra, P.N., and Sachdeva, S., Embryogenesis in apple *in vitro*, *Phytomorphology*, 34, 26, 1984.

161. Litz, R.E., Knight, R.L., and Gazit, S., Soamtic embryos from cultured ovules of polyembryonic *Mangifera indica* L, *Plant Cell Rep.*, 1, 264, 1982.

162. Litz, R.E., *In vitro* somatic embryogenesis from nucellar callus of monoembryonic mango, *HortSci.*, 19, 715, 1984.

163. Litz, R.E., and Schaffer, B., Polyamines in adventitious and somatic embryogenesis in mango (*Mangifera indica* L.), *J. Plant Physiol.*, 128, 251, 1987.

164. DeWald, S.G., Litz, R.E., and Moore, G.A., Optimizing somatic embryo production in mango, *J. Amer. Soc. Hort. Sci.*, 114, 712, 1989.

165. Litz, R.E., Knight, R.J. Jr., and Gazit, S., *In vitro* somatic embryogenesis from *Mangifera indica* L. callus, *Sci. Hort.*, 22, 233, 1984.

166. DeWald, S.G., Litz, R.E., and Moore, G.A., Maturation and germination of mango somatic embryos, *J. Amer. Soc. Hort. Sci.*, 114, 837, 1989.

167. Conger, B.V., Ed., *Cloning Agricultural Plants Via In Vitro Techniques*, CRC Press, Boca Raton, 1981.

168. Evans, D.A., Sharp, W.R., Ammirato, P.V., and Yamada, Y., Eds., *Handbook of Plant Cell Culture, Volume 1, Techniques for Propagation and Breeding*, Macmillan Publishing Co., New York, 1983.

169. Vasil, I.K., Ed., *Cell Culture and Somatic Cell Genetics of Plants, Volume 1, Laboratory Procedures and Their Applications*, Academic Press, New York, 1984.

170. Bajaj, Y.P.S., Ed., *Biotechnology in Agriculture and Forestry 1, Trees Volume I*, Springer-Verlag, Berlin, 1986.

171. Bonga, J.M., and Durzan, D.J., Eds., *Cell and Tissue Culture in Forestry, Volume 1, General Principles and Biotechnology*, Martinus Nijhoff Publishers, Dordrecht, 1987.

172. Gabr, M.F., and Tisserat, B., Propagating palms *in vitro* with special emphasis on the date palm (*Phoenix dactylifera* L.), *Sci. Hort.*, 25, 255, 1985.

173. Vardi, A., Hutchison, D.J., and Galun, E., A protoplast-to-tree system in *Microcitrus* based on protoplasts derived from a sustained embryogenic callus, *Plant Cell Rep.*, 5, 412, 1986.

174. Vardi, A., Arzee-Gonen, P., Frydman-Shani, A., Bleichman, S., and Galun, E., Protoplast-fusion-mediated transfer of organelles from *Microcitrus* into *Citrus* and regeneration of novel alloplasmic trees, *Theor. Appl. Genet.*, 78, 741, 1989.

175. Litz, R.E., *In vitro* somatic embryogenesis from callus of jaboticaba, *Myrciaria cauliflora*, *HortSci.*, 19, 62, 1984.

176. Rugini, E., Somatic embryogenesis and plant regeneration in olive (*Olea europaea* L.), *Plant Cell Tiss. Org. Cult.*, 14, 207, 1988.

177. Radojević, L., Somatic embryos and plantlets from callus cultures of *Paulownia tomentosa* Steud, *Z. Pflanzenphysiol.*, 91, 57, 1979.

524

178. Mooney, P.A., and Van Staden, J., Induction of embryogenesis in callus from immature embryos of *Persea americana, Can. J. Bot.*, 65, 622, 1987.
179. Pliego-Alfaro, F., and Murashige, T., Somatic embryogenesis in avocado (*Persea americana* Mill.) *in vitro, Plant Cell Tiss. Org. Cult.*, 12, 61, 1988.
180. Tisserat, B., Propagation of date palm (*Phoenix dactylifera* L.) *in vitro, J. Exp. Bot.*, 30, 1275, 1979.
181. Tisserat, B., and DeMason, D.A., A histological study of development of adventive embryos in organ cultures of *Phoenix dactylifera* L, *Ann. Bot.*, 46, 465, 1980.
182. Tisserat, B., Factors involved in the production of plantlets from date palm callus cultures, *Euphytica*, 31, 201, 1982.
183. Zaid, A., and Tisserat, B., Morphogenetic responses obtained from a variety of somatic explant tissues of date palm, *Bot. Mag. Tokyo*, 96, 67, 1983.
184. Sharma, D.R., Dawra, S., and Chowdhury, J.B., Somatic embryogenesis and plant regeneration in date palm (*Phoenix dactylifera* Linn.) cv. "Khadravi" through tissue culture, *Indian J. Exp. Biol.*, 22, 596, 1984.
185. Sharman, D.R., Deepak, S., and Chowdhury, J.B., Regeneration of plantlets from somatic tissues of the date palm *Phoenix dactylifera* Linn, *Indian J. Exp. Biol.*, 24, 763, 1986.
186. Calero, N., Actions de radiations rouges et bleues sur l'embryogenèse somatique du Plamer dattier (*Phoenix dactylifera* L.) en culture *in vitro* et sur sa teneur en leucoanthocyanes, *C.R. Soc. Biol. Paris*, 183, 307, 1989.
187. Calero, N., Blanc, A., and Benbadis, A., Effets conjugués de la BAP et de la lumière rouge sur l'embryogenèse somatique dans de gaines cotylédonaires de Palmier dattier (*Phoenix dactylifera* L.) en culture *in vitro, Bull. Soc. bot. Fr. Actual. Bot.*, 137, 13, 1990.
188. Omar, M.S., and Novak, F.J., *In vitro* plant regeneration and ethylmethanesulphonate (EMS) uptake in somatic embryos of date palm (*Phoenix dactylifera* L.), *Plant Cell Tiss. Org. Cult.*, 20, 185, 1990.
189. Sharma, D.R., Yadav, N.R., and Chowdhury, J.B., Somatic embryogenesis and plant regeneration from shoot tip calli of wild date palm *Phoenix sylvestris* Roxb, *Indian J. Exp. Biol.*, 26, 854, 1988.
190. Michler, C.H., and Bauer, E.O., High frequency somatic embryogenesis from leaf tissue of *Populus* spp, *Plant Sci.*, 77, 111, 1991.
191. Cheema, G.S., Somatic embryogenesis and plant regeneration from cell suspension and tissue cultures of mature himalayan poplar (*Populus ciliata*), *Plant Cell Rep.*, 8, 124, 1989.
192. Park, Y.G., and Son, S.H., *In vitro* organogenesis and somatic embryogenesis from punctured leaf discs of *Populus nigra* × *P. maximowiczii*, *Plant Cell Tiss. Org. Cult.*, 15, 95, 1988.
193. Raj Bhansali, R., Driver, J.A., and Durzan, D.J., Rapid multiplication of adventitious somatic embryos in peach and nectarine by secondary embryogenesis, *Plant Cell Rep.*, 9, 280, 1990.
194. Raj Bhansali, R., Somatic embryogenesis and regeneration of plantlets in pomegranate, *Ann. Bot.*, 66, 249, 1990.
195. Mehra, P.N., and Jaidka, K., Experimental induction of embryogeneis in pear, *Phytomorphology*, 35, 1, 1985.
196. Gingas, V.M., and Lineberger, R.D., Asexual embryogenesis and plant regeneration in *Quercus, Plant Cell Tiss. Org. Cult.*, 17, 191, 1989.
197. Féraud-Keller, C., and Espagnac, H., Conditions d'apparition d'une embryogénèse somatique sur des cals issus de la culture de tissus foliaires de chêne vert (*Quercus ilex*), *Can. J. Bot.*, 67, 1066, 1989.
198. Chalupa, V., Plant regeneration by somatic embryogenesis from cultured immature embryos of oak (*Quercus robur* L.) and linden (*Tilia cordata* Mill.), *Plant Cell Rep.*, 9, 398, 1990.
199. El Maâtaoui, M., Espagnac, H., and Michaux-Ferrière, N., Histology of callogenesis and somatic embryogenesis induced in stem fragments of cork oak (*Quercus suber*) cultured *in vitro, Ann. Bot.*, 66, 183, 1990.
200. El Maâtaoui, M., and Espagnac, H., Neoformation of somatic embryo-like structures from cork-oak (*Quercus suber* L.) tissue cultures, *C.R. Acad. Sci. Paris Sér. III*, 304, 83, 1987.

201. Zatykó, J.M., Kiss, F., and Szalay, F., Induction of adventive embryony in cultured ovules of black currant (*Ribes nigrum* L.), *Hort. Res.*, 21, 99, 1981.

202. Zatykó, J.M., Simon, I., and Szabó, C.S., Induction of polyembryony in cultivated ovules of red currant, *Plant Sci. Lett.*, 4, 281, 1975.

203. Merkle, S.A., and Wiecko, A.T., Regeneration of *Robinia pseudoacacia* via somatic embryogenesis, *Can. J. For. Res.*, 19, 285, 1989.

204. Matthews, D., Mottley, J., Horan, I., and Roberts, A.V., A protoplast to plant system in roses, *Plant Cell Tiss. Org. Cult.*, 24, 173, 1991.

205. Bapat, V.A., and Rao, P.S., Somatic embryogenesis and plantlet formation in tissue cultures of sandalwood (*Santalum album* L.), *Ann. Bot.*, 44, 629, 1979.

206. Lakshmi Sita, G., Raghava Ram, N.V., and Vaidyanathan, C.S., Differentiation of embryoids and plantlets from shoot callus of sandalwood, *Plant Sci. Lett.*, 15, 265, 1979.

207. Sita, G.L., Ram, N.V.R., and Vaidyanathan, C.S., Triploid plants from endosperm cultures of sandalwood by experimental embryogenesis, *Plant Sci. Lett.*, 20, 63, 1980.

208. Bapat, V.A., and Rao, P.S., Regulatory factors for *in vitro* multiplication of sandalwood tree (*Santalum album* Linn.). I. Shoot bud regeneration and somatic embryogenesis in hypocotyl cutures, *Proc. Indian Acad. Sci.*, 93, 19, 1984.

209. Rao, P.S., Bapat, V.A., and Mhatre, M., Regulatory factors for *in vitro* multiplication of sandalwood tree (*Santalum album* Linn.). II. Plant regeneration in nodal and internodal stem explants and occurrence of somaclonal variations in tissue culture raised plants, *Proc. Indian Natl. Sci. Acad.*, 50, 196, 1984.

210. Bapat, V.A., Gill, R., and Rao, P.S., Regeneration of somatic embryos and plantlets from stem callus protoplasts of sandalwood tree (*Santalum album* L.), *Curr. Sci.*, 54, 978, 1985.

211. Rao, P.S., and Ozias-Akins, P., Plant regeneration through somatic embryogenesis in protoplast cultures of sandalwood (*Santalum album* L), *Protoplasma*, 124, 80, 1985.

212. Bapat, V.A., and Rao, P.S., Sandalwood plantlets from "synthetic seeds", *Plant Cell Rep.*, 7, 434, 1988.

213. Desai, H.V., Bhatt, P.N., and Mehta, A.R., Plant regeneration of *Sapindus trifoliatus* L. (soapnut) through somatic embryogenesis, *Plant Cell Rep.*, 3, 190, 1986.

214. Unnikrishnan, S.K., Prakash, L., Josekutty, P.C., Bhatt, P.N., and Mehta, A.R., Effect of NaCl salinity on somatic embryo development in *Sapindus trifoliatus* L, *J. Exp. Bot.*, 42, 401, 1991.

215. Chen, M.-H., and Wang, P.-J., Somatic embryogenesis and plant regeneration on *Sassafras randaiense* (Hay.) rehd, *Bot. Bull. Academia Sinica*, 26, 1, 1985.

216. Wang, Y.-C., and Janick, J., Somatic embryogenesis in jojoba, *J. Amer. Soc. Hort. Sci.*, 111, 281, 1986.

217. Bano, Z., Rajarathnam, S., and Mohanty, B.D., Somatic embryogenesis in cotyledon cultures of tea (*Thea sinensis*, L.), *J. Hort. Sci.*, 66, 465, 1991.

218. Pence, V.C., Hasegawa, P.M., and Janick, J., Asexual embryogenesis in *Theobroma cacao* L, *J. Amer. Soc. Hort. Sci.*, 104, 145, 1979.

219. Pence, V.C., Hasegawa, P.M., and Janick, J., *In vitro* cotyledonary development and anthocyanin synthesis in zygotic and asexual embryos of *Theobroma cacao*, *J. Amer. Soc. Hort. Sci.*, 106, 381, 1981.

220. Pence, V.C., Hasegawa, P.M., and Janick, J., Initiation and development of asexual embryos of *Theobroma cacao* L. in vitro, *Z. Pflanzenphysiol.*, 98, 1, 1980.

221. Pence, V.C., Hasegawa, P.M., and Janick, J., Sucrose-mediated regulation of fatty acid composition in asexual embryos of *Theobroma cacao*, *Physiol. Plant.*, 53, 378, 1981.

222. Kononowicz, H., and Janick, J., Response of embryogenic callus of *Theobroma cacao* L. to gibberellic acid and inhibitors of gibberellic acid synthesis, *Z. Pflanzenphysiol.*, 113, 359, 1984.

223. Kononowicz, A.K., and Janick, J., The influence of carbon source on the growth and development of asexual embryos of *Theobroma cacao*, *Physiol. Plant.*, 61, 155, 1984.

224. Elhag, H.M., Whipkey, A., and Janick, J., Induction of somatic embryogenesis from callus in *Theobroma cacao* in response to carbon source and concentration, *Rev. Theobroma*, 17, 153, 1987.

526

225. Kononowicz, H., and Janick, J., Somatic embryogenesis via callus of *Theobroma cacao* L. II. Total protein content in nucleus, nucleolus and cytoplasm, *Acta Physiol. Plant.*, 10, 107, 1988.

226. Kononowicz, H., and Janick, J., Somatic embryogenesis via callus of *Theobroma cacao* L. I. Cell cycle, DNA content, RNA synthesis, and DNA template activity, *Acta Physiol. Plant.*, 10, 93, 1988.

227. Lulsdorf, M.M., Tautorus, T.E., Kikcio, S.I., and Dunstan, D.I., Growth parameters of embryogenic suspension cultures of interior spruce (*Picea glauca-engelmannii* complex) and black spruce (*Picea mariana* Mill.), *Plant Sci.*, 82, 27, 1992.

228. James, D.J., Passey, A.J., MacKenzie, K.A.D., Jones, O.P., and Menhinick, E.C., Regeneration of temperate fruit trees *in vitro* via organogenesis and embryogenesis, in *Genetic Manipulation in Plant Breeding*, Horn, W., Ed., Walter de Gruyter & Co., New York, 1986, 433.

229. Wright, D.C., Janick, J., and Hasegawa, P.M., Temperature effects on *in vitro* lipid accumulation in asexual embryos of *Theobroma cacao* L, *Lipids*, 18, 863, 1983.

230. Wright, D.C., Kononowicz, A.K., and Janick, J., Factors affecting *in vitro* fatty acid content and composition in asexual embryos of *Theobroma cacao*, *J. Amer. Soc. Hort. Sci.*, 109, 77, 1984.

231. Wang, Yi-C., and Janick, J., Inducing precocious germination in asexual embryos of cacao, *HortSci.*, 19, 839, 1984.

232. Adu-Ampomah, Y., Novak, F.J., Afza, R., Van Duren, M., and Perea-Dallos, M., Initiation and growth of somatic embryos of cocoa (*Theobroma cacao* L.), *Café Cacao Thé*, 32, 187, 1988.

233. Duhem, K., Le Mercier, N., and Boxus, P., Données nouvelles sur l'induction et le développement d'embryons somatiques chez *Theobroma cacao* L, *Café Cacao Thé*, 33, 9, 1989.

234. Pence, V.C., Cryopreservation of immature embryos of *Theobroma cacao*, *Plant Cell Rep.*, 10, 144, 1991.

235. Wen, M.C., and Kinsella, J.E., Somatic embryogenesis and plantlet regeneration of *Theobroma cacao*, *Food Biotechnol.*, 5, 119, 1991.

236. Krul, W.R., and Worley, J.F., Formation of adventitious embryos in callus cultures of "Seyval", a french hybrid grape, *J. Amer. Soc. Hort. Sci.*, 102, 360, 1977.

237. Netzer, M.H., Guellec, V., and Branchard, M., Behaviour of grapevine rootstocks during somatic embryogenesis, under conditions of calcareous chlorosis, *Agron.*, 11, 125, 1991.

238. Gray, D.J., and Mortensen, J.A., Initiation and maintenance of long term somatic embryogenesis from anthers and ovaries of *Vitis longii* "Microsperma", *Plant Cell Tiss. Org. Cult.*, 9, 73, 1987.

239. Bouquet, A., Piganeau, B., and Lamaison, A.-M., Genotypic effect on *in vitro* production of callus, embryoids, and plantlets from cultured anthers in *Vitis*, *C.R. Acad. Sci. Paris Sér. III*, 295, 569, 1982.

240. Monnier, M., Faure, O., and Sigogneau, A., Somatic embryogenesis in *Vitis*, *Bull. Soc. bot. Fr. Actual. Bot.*, 137, 35, 1990.

241. Faure, O., Embryons somatiques de *Vitis rupestris* et embryons zygotiques de *Vitis* sp.: morphologie, histologie, histochimie et développement, *Can. J. Bot.*, 68, 2305, 1990.

242. Mullins, M.G., and Srinivasan, C., Somatic embryos and plantlets from an ancient clone of the grapevine (cv. Cabernet-Sauvignon) by apomixis *in vitro*, *J. Exp. Bot.*, 27, 1022, 1976.

243. Srinivasan, C., and Mullins, M.G., High-frequency somatic embryo production from unfertilized ovules of grapes, *Sci. Hort.*, 13, 245, 1980.

244. Rajasekaran, K., and Mullins, M.G., Somatic embryo formation by cultured ovules of cabernet sauvignon grape: effects of fertilization and of the male gametocide toluidine blue, *Vitis*, 24, 151, 1985.

245. Mauro, M.C., Nef, C., and Fallot, J., Stimulation of somatic embryogenesis and plant regeneration from anther culture of *Vitis vinifera* cv. Cabernet Sauvignon, *Plant Cell Rep.*, 5, 377, 1986.

246. Matsuta, N., and Hirabayashi, T., Embryogenic cell lines from somatic embryos of grape (*Vitis vinifera* L.), *Plant Cell Rep.*, 7, 684, 1989.

247. Rajasekaran, K., and Mullins, M.G., Embryos and plantlets from cultured anthers of hybrid grapevines, *J. Exp. Bot.*, 30, 399, 1979.

248. Gamborg, O.L., Miller, R.A., and Ojima, K., Nutrient requirements of suspension cultures of soybean root cells, *Exp. Cell Res.*, 50, 151, 1968.

249. Witham, F.H., Blaydes, D.F., and Devlin, R.M., *Experiments in Plant Physiology*, Van Nostrand-Reinhold Co., New York, 1971.

250. Rao, P.S., and Bapat, V.A., Vegetative propagation of sandalwood plants through tissue culture, *Can. J. Bot.*, 56, 1153, 1978.

251. Driver, J.A., and Kuniyuki, A.H., *In vitro* propagation of paradox walnut rootstock, *HortSci.*, 19, 507, 1984.

252. Heller, R., Recherches sur la nutrition minerale des tissues vegetaux *in vitro*, *Ann. Sci. Nat. Bot., Biol. Veg.*, 14, 1, 1953.

253. Cheng, T.-Y., Adventitious bud formation in culture of Douglas fir (*Pseudotsuga menziesii* (Mirb.) Franco), *Plant Sci. Lett.*, 5, 97, 1975.

254. Kao, K.N., and Michayluk, M.R., Nutritional requirements for growth of *Vicia hajastana* cells and protoplasts at a very low population density in liquid media, *Planta*, 126, 105, 1975.

255. Linsmaier, E.M., and Skoog, F., Organic growth factor requirements of tobacco tissue cultures, *Physiol. Plant.*, 18, 100, 1975.

256. Nitsch, J.P., and Nitsch, C., Haploid plants from pollen grains, *Science*, 163, 85, 1969.

257. Miller, C.O., Cytokinins in *Zea mays*, *Ann. N.Y. Acad. Sci.*, 144, 251, 1967.

258. Murashige, T., and Skoog, F., A revised medium for rapid growth and bio assays with tobacco tissue cultures, *Physiol. Plant.*, 15, 473, 1962.

259. Murashige, T., and Tucker, D.P.H., Growth factor requirements of *Citrus* tissue culture, in *Proc. of the First International* Citrus *Symposium*, Vol. 3, Chapman, H.D. Ed., University of California, Riverside, 1969, 1155.

260. Nitsch, J.P., Growth and development *in vitro* of excised ovaries, *Amer. J. Bot.*, 38, 566, 1951.

261. Nitsch, J.P., Nitsch, C., and Hamon, S., Réalisation expérimentale de l'androgénèse chez divers *Nicotiana*, *C.R. Soc. Biol. Paris*, 162, 369, 1968.

262. Schenk, R.U., and Hildebrandt, A.C., Medium and techniques for induction and growth of monocotyledonous and dicotyledonous plant cell cultures, *Can.J.Bot.*, 50, 199, 1972.

263. White, P.R., *A Handbook of Plant Tissue Culture*, The Jacquese Catell Press, Lancaster, PA, 1943.

264. White, P.R., *The Cultivation of Plant and Animal Cells*, 2nd Ed., Roland Press, New York, 1963.

265. Simola, L.K., Propagation of plantlets from leaf callus of *Betula pendula* F. *purpurea*, *Sci. Hort.*, 26, 77, 1985.

266. Binding, H., Regeneration von haploiden und diploiden Pflanzen aus protoplasten von *Petunia hybrida* L, *Z. Pflanzenphysiol.*, 74, 327, 1974.

267. Lloyd, G., and McCown, B., Commercially-feasible micropropagation of mountain laurel, *Kalmia latifolia*, by use of shoot-tip culture, *Proc. Int. Plant Propagators' Soc.*, 30, 421, 1981.

268. Eeuwens, C.J., Mineral requirements for growth and callus initiation of tissue explants excised from mature coconut palms (*Cocos nucifera*) and cultured *in vitro*, *Physiol. Plant.*, 36, 23, 1976.

269. Schuller, A., Reuther, G., and Geier, T., Somatic embryogenesis from seed explants of *Abies alba*, *Plant Cell Tiss. Org. Cult.*, 17, 53, 1989.

270. Nørgaard, J.V., and Krogstrup, P., Cytokinin induced somatic embryogenesis from immature embryos of *Abies nordmanniana* Lk, *Plant Cell Rep.*, 9, 509, 1991.

271. Von Aderkas, P., Klimaszewska, K., and Bonga, J.M., Diploid and haploid embryogenesis in *Larix leptolepis*, *L. decidua*, and their reciprocal hybrids, *Can. J. For. Res.*, 20, 9, 1990.

272. Cornu, D., and Geofrrion, C., Aspects de l'embryogenèse somatique chez le mélèze, *Bull. Soc. bot. Fr. Actual. Bot.*, 137, 25, 1990.

273. Klimaszewska, K., Plantlet development from immature zygotic embryos of hybrid larch through somatic embryogenesis, *Plant Sci.*, 63, 95, 1989.

274. Klimaszewska, K., Recovery of somatic embryos and plantlets from protoplast cultures of *Larix* × *eurolepis*, *Plant Cell Rep.*, 8, 440, 1989.

528

275. Chalupa, V., Somatic embryogenesis and plantlet regeneration from cultured immature and mature embryos of *Picea abies* (L.) Karst, *Comm. Inst. Forest. Cech.*, 14, 57, 1985.
276. Hakman, I., Fowke, L.C., and Eriksson, T., The development of somatic embryos in tissue cultures initiated from immature embryos of *Picea abies* (Norway spruce), *Plant Sci.*, 38, 35, 1985.
277. Feirer, R.P., Conkey, J.H., and Verhagen, S.A., Triglycerides in embryogenic conifer calli: a comparison with zygotic embryos, *Plant Cell Rep.*, 8, 207, 1989.
278. Wann, S.R., Johnson, M.A., Noland, T.L., and Carlson, J.A., Biochemical differences between embryogenic and non-embryogenic callus of *Picea abies* (L.) Karst, *Plant Cell Rep.*, 6, 39, 1987.
279. Becwar, M.R., Noland, T.L., and Wann, S.R., A method for quantification of the level of somatic embryogenesis among Norway spruce callus lines, *Plant Cell Rep.*, 6, 35, 1987.
280. Hakman, I., and Von Arnold, S., Plantlet regeneration through somatic embryogenesis in *Picea abies* (Norway spruce), *J. Plant Physiol.*, 121, 149, 1985.
281. Becwar, M.R., Noland, T.L., and Wann, S.R., Somatic embryo development and plant regeneration from embryogenic Norway spruce callus, *Tech. Assoc. Pulp. Paper. Ind. J.*, 70, 155, 1987.
282. Gupta, P.K., and Durzan, D.J., Plantlet regeneration via somatic embryogenesis from subcultured callus of mature embryos of *Picea abies* (Norway spruce), *In Vitro Cell. Dev. Biol.*, 22, 685, 1986.
283. Boulay, M.P., Gupta, P.K., Krogstrup, P., and Durzan, D.J., Development of somatic embryos from cell suspensions cultures of Norway spruce (*Picea abies* Karst.), *Plant Cell Rep.*, 7, 134, 1988.
284. Krogstrup, P., Embryolike structure from cotyledons and ripe embryos of Norway spruce (*Picea abies*), *Can. J. For. Res.*, 16, 664, 1986.
285. Von Arnold, S., and Hakman, I., Effect of sucrose on initiation of embryogenic callus cultures from mature zygotic embryos of *Picea abies* (L.) Karst. (Norway spruce), *J. Plant Physiol.*, 122, 261, 1986.
286. Lelu, M.-A., Boulay, M., and Arnaud, Y., Obtention de cals embryogénes á partir de cotylédons de Picea abies (L.) Karst. prélevés sur de jeunes plantes âgées de 3 â 7 jours aprés germination, *C.R. Acad. Sci. Paris Sér. III*, 305, 105, 1987.
287. Lelu, M.-A., Boulay, M.P., and Bornman, C.H., Somatic embryogenesis in cotyledons of *Picea abies* is enhanced by an adventitious bud-inducing treatment, *New For.*, 4, 125, 1990.
288. Von Arnold, S., Improved efficiency of somatic embryogenesis in mature embryos of *Picea abies* (L.) Karst, *J. Plant Physiol.*, 128, 233, 1987.
289. Von Arnold, S., and Hakman, I., Regulation of somatic embryo development in *Picea abies* by abscisic acid (ABA), *J. Plant Physiol.*, 132, 164, 1988.
290. Jain, M.S., Newton, R.J., and Soltes, E.J., Enhancement of somatic embryogenesis in Norway spruce (*Picea abies*), *Theor. Appl. Genet.*, 76, 501, 1988.
291. Becwar, M.R., Noland, T.L., and Wyckoff, J.L., Maturation, germination, and conversion of Norway spruce (*Picea abies* L.) somatic embryos to plants, *In Vitro Cell. Dev. Biol.*, 25, 575, 1989.
292. Mo, L.H., Von Arnold, S., and Lagercrantz, U., Morphogenic and genetic stability in longterm embryogenic cultures and somatic embryos of Norway spruce (*Picea abies*) L. Karst), *Plant Cell Rep.*, 8, 375, 1989.
293. Mo, L.H., and Von Arnold, S., Origin and development of embryogenic cultures from seedlings of Norway spruce (*Picea abies*), *J. Plant Physiol.*, 138, 223, 1991.
294. Verhagen, S.A., and Wann, S.R., Norway spruce somatic embryogenesis: high frequency initiation from light cultured mature embryos, *Plant Cell Tiss. Org. Cult.*, 16, 103, 1989.
295. Hakman, I., Stabel, P., Engström, P., and Eriksson, T., Storage protein accumulation during zygotic and somatic embryo development in *Picea abies* (Norway spruce), *Physiol. Plant.*, 80, 441, 1990.
296. Simola, L.K., and Santanen, A., Improvement of nutrient medium for growth and

embryogenesis of megagametophyte and embryo callus lines of *Picea abies*, *Physiol. Plant.*, 80, 27, 1990.

297. Attree, S.M., Bekkaoui, F., Dunstan, D.I., and Fowke, L.C., Regeneration of somatic embryos from protoplasts isolated from an embryogenic suspension culture of white spruce (*Picea glauca*), *Plant Cell Rep.*, 6, 480, 1987.

298. Hakman, I., and Fowke, L.C., Somatic embryogenesis in *Picea glauca* (white spruce) and *Picea mariana* (black spruce), *Can. J. Bot.*, 65, 656, 1987.

299. Lu, C.-Y., and Thorpe, T.A., Somatic embryogenesis and plantlet regeneration in cultured immature embryos of *Picea glauca*, *J. Plant Physiol.*, 128, 297, 1987.

300. Attree, S.M., Dunstan, D.I., and Fowke, L.C., Initiation of embryogenic callus and suspension cultures, and improved embryo regeneration from protoplasts, of white spruce (*Picea glauca*), *Can. J. Bot.*, 67, 1790, 1988.

301. Attree, S.M., Dunstan, D.I., and Fowke, L.C., Plantlet regeneration from embryogenic protoplasts of white spruce (*Picea glauca*), *Bio/Technol.*, 7, 1060, 1989.

302. Attree, S.M., Budimir, S., and Fowke, L.C., Somatic embryogenesis and plantlet regeneration from cultured shoots and cotyledons of seedlings from stored seeds of black and white spruce (*Picea mariana* and *Picea glauca*), *Can. J. Bot.*, 68, 30, 1990.

303. Lelu, M.-A., and Bornman, C.H., Induction of somatic embryogenesis in excised cotyledons of *Picea glauca* and *Picea mariana*, *Plant Physiol. Biochem.*, 28, 785, 1990.

304. Tremblay, F.M., Somatic embryogenesis and plantlet regeneration form embryos isolated from stored seeds of *Picea glauca*, *Can. J. Bot.*, 68, 236, 1990.

305. Roberts, D.R., Flinn, B.S., Webb, D.T., Webster, F.B., and Sutton, B.C.S., Characterization of immature embryos of interior spruce by SDS-PAGE and microscopy in relation to their competence for somatic embryogenesis, *Plant Cell Rep.*, 8, 285, 1989.

306. Webb, D.T., Webster, F., Flinn, B.S., Roberts, D.R., and Ellis, D.D., Factors influencing the induction of embryogenic and caulogenic callus from embryos of *Picea glauca* and *P. engelmannii*, *Can. J. For. Res.*, 19, 1303, 1989.

307. Roberts, D.R., Flinn, B.S., Webb, D.T., Webster, F.B., and Sutton, B.C.S., Abscisic acid and indole-3-butyric acid regulation of maturation and accumulation of storage proteins in somatic embryos of interior spruce, *Physiol. Plant.*, 78, 355, 1990.

308. Roberts, D.R., Sutton, B.C.S., and Flinn, B.S., Synchronous and high frequency germination of interior spruce somatic embryos following partial drying at high relative humidity, *Can. J. Bot.*, 68, 1086, 1990.

309. Webster, F.B., Roberts, D.R., McInnis, S.M., and Sutton, B.C.S., Propagation of interior spruce by somatic embryogenesis, *Can. J. For. Res.*, 20, 1759, 1990.

310. Eastman, P.A.K., Webster, F.B., Pitel, J.A., and Roberts, D.R., Evaluation of somaclonal variation during somatic embryogenesis of interior spruce (*Picea glauca engelmannii* complex) using culture morphology and isozyme analysis, *Plant Cell Rep.*, 10, 425, 1991.

311. Tautorus, T.E., Attree, S.M., Fowke, L.C., and Dunstan, D.I., Somatic embryogenesis from immature and mature zygotic embryos, and embryo regeneration from protoplasts in black spruce (*Picea mariana* Mill.), *Plant Sci.*, 67, 115, 1990.

312. Tremblay, L., and Tremblay, F.M., Effects of gelling agents, ammonium nitrate, and light on the development of *Picea mariana* (Mill) B.S.P. (black spruce) and *Picea rubens* Sarg. (red spruce) somatic embryos, *Plant Sci.*, 77, 233, 1991.

313. Krogstrup, P., Eriksen, E.N., Møller, J.D., and Rouland, H., Somatic embryogenesis in Sitka spruce (*Picea sitchensis* (Bong.) Carr), *Plant Cell Rep.*, 7, 594, 1988.

314. Von Arnold, S., and Woodward, S., Organogenesis and embryogenesis in mature zygotic embryos of *Picea sitchensis*, *Tree Physiol.*, 4, 291, 1988.

315. Krogstrup, P., Effect of culture densities on cell proliferation and regeneration from embryogenic cell suspensions of *Picea sitchensis*, *Plant Sci.*, 72, 115, 1990.

316. Roberts, D.R., Lazaroff, W.R., and Webster, F.B., Interaction between maturation and high relative humidity treatments and their effects on germination of sitka spruce somatic embryos, *J. Plant Physiol.*, 138, 1, 1991.

530

317. Ying-hong, L., and Zhong-shen, G., Somatic embryogenesis and plantlet formation of *Picea wilsonii* Mast. in different conditions, *Acta Bot. Sin.*, 32, 568, 1990.
318. Lainé, E., and David, A., Somatic embryogenesis in immature embryos and protoplasts of *Pinus caribaea*, *Plant Sci.*, 69, 215, 1990.
319. Jain, S.M., Dong, N., and Newton, R.J., Somatic embryogenesis in slash pine (*Pinus elliotii*) from immature embryos cultured *in vitro*, *Plant Sci.*, 65, 233, 1989.
320. Gupta, P.K., and Durzan, D.J., Somatic polyembryogenesis from callus of mature sugar pine embryos, *Bio/Technol.*, 4, 643, 1986.
321. Becwar, M.R., Wann, S.R., Johnson, M.A., Verhagen, S.A., and Feirer, R.P., Development and characterization of *in vitro* embryogenic systems in conifers, in *Somatic Cell Genetics of Woody Plants*, Ahuja, M.R., Ed., Kluwer Academic Publishers, Dordrecht, 1988, 1.
322. Finer, J.J., Kriebel, H.B., and Becwar, M.R., Initiation of embryogenic callus and suspension cultures of eastern white pine (*Pinus strobus* L.), *Plant Cell Rep.*, 8, 203, 1989.
323. Gupta, P.K., and Durzan, D.J., Biotechnology of somatic polyembryogenesis and plantlet regeneration in loblolly pine, *Bio/Technol.*, 5, 147, 1987.
324. Gupta, P.K., and Durzan, D.J., Somatic embryos from protoplasts of loblolly pine proembryonal cells, *Bio/Technol.*, 5, 710, 1987.
325. Becwar, M.R., Nagmani, R., and Wann, S.R., Initiation of embryogenic cultures and somatic embryo development in loblolly pine (*Pinus taeda*), *Can. J. For. Res.*, 20, 810, 1990.
326. Durzan, D.J., and Gupta, P.K., Somatic embryogenesis and polyembryogenesis in Douglas-fir cell suspension cultures, *Plant Sci.*, 52, 229, 1987.
327. Gupta, P.K., Dandekar, A.M., and Durzan, D.J., Somatic proembryo formation and transient expression of a luciferase gene in Douglas fir and loblolly pine protoplasts, *Plant Sci.*, 58, 85, 1988.
328. Bourgkard, F., and Favre, J.M., Somatic embryos from callus of *Sequoia sempervirens*, *Plant Cell Rep.*, 7, 445, 1988.
329. Webb, D.T., Rivera, M.E., Starszak, E., and Matos, J., Callus initiation and organized development from *Zamia pumila* embryo explants, *Ann. Bot.*, 51, 711, 1983.
330. Brown, C.L., and Lawrence, R.H., Culture of pine callus on a defined medium, *Forest Sci.*, 14, 62, 1968.
331. Amerson, H.V., Frampton, L.J. Jr., McKeand, S.E., Mott, R.L., and Weir, R.J., Loblolly pine tissue culture: laboratory, greenhouse, and field studies, in *Tissue Culture in Forestry and Agriculture*, Henke, R.R., Hughes, K.W., Constantin, M.J., and Hollaender, A., Eds., Plenum Press, New York, 1985, 271.
332. Butenko, R.G., and Yakovleva, S.M., Controlled organogenesis and regeneration of a whole plant in a culture of non-differentiated plant tissue, *Izv. Akad. Nauk SSSR. Biol. Ser.*, 2, 230, 1962.
333. Lainé, E., David, H., and David, A., Callus formation from cotyledon protoplasts of *Pinus oocarpa* and *Pinus patula*, *Physiol. Plant.*, 72, 374, 1988.
334. Gupta, P.K., and Durzan, D.J., Shoot multiplication from mature trees of Douglas-fir (*Pseudotsuga menziesii*) and sugar pine (*Pinus lambertiana*), *Plant Cell Rep.*, 4, 177, 1985.
335. Litvay, J.D., Johnson, M.A., Verma, D., Einspahr, D., and Weyrauch, K., *Conifer Suspension Culture Medium Development Using Analytical Data From Developing Seeds*, Institute Paper Chem. Tech. Paper Series 115, Appleton, Wisconsin, 1981.
336. Von Arnold, S., and Eriksson, T., *In vitro* studies of adventitious shoot formation in *Pinus contorta*, *Can. J. Bot.*, 59, 870, 1981.
337. Chu, C.C., Wang, C.C., Sun, C.S., Hsü, C., Yin, K.C., Chu, C.Y., and Bi, F.Y., Establishment of an efficient medium for anther culture of rice through comparative experiments on the nitrogen sources, *Sci. Sinica*, 18, 658, 1975.
338. Owens, J.N., and Molder, M., *The Reproductive Cycle of Interior Spruce*, Information Services Branch, B.C. Ministry of Forests, Victoria, B.C., 1984.
339. Button, J., Kochba, J., and Bornman, C.H., Fine structure of and embryoid development from embryogenic ovular callus of "Shamouti" orange (*Citrus sinensis* Osb.), *J. Exp. Bot.*, 25, 446, 1974.

340. Michaux-Ferrière, N., Dublin, P., and Schwendiman, J., Histological study of somatic embryogenesis from foliar explants of *Coffea arabica* L, *Café Cacao Thé*, 31, 112, 1987.

341. Hakman, I., Rennie, P., and Fowke, L., A light and electron microscope study of *Picea glauca* (white spruce) somatic embryos, *Protoplasma*, 140, 100, 1987.

342. Nagmani, R., Becwar, M.R., and Wann, S.R., Single-cell origin and development of somatic embryos in *Picea abies* (L.) karst. (Norway spruce) and *P. glauca* (Moench) Voss (white spruce), *Plant Cell Rep.*, 6, 157, 1987.

343. Joy, R.W. IV, Yeung, E.C., Kong, L., and Thorpe, T.A., Development of white spruce somatic embryos: I. Storage product deposition, *In Vitro Cell. Dev. Biol.*, 27P, 32, 1991.

344. Engelmann, F., Utilisation d'atmosphères à teneur en oxygène réduite pour la conservation de cultures d'embryons somatiques de palmier à huile (*Elaeis guineensis* Jacq.), *C.R. Acad. Sci. Paris Sér III*, 310, 679, 1990.

345. Joy, R.W. IV, Kumar, P.P., and Thorpe, T.A., Long term storage of somatic embryogenic white spruce tissue at ambient temperature, *Plant Cell Tiss. Org. Cult.*, 25, 53, 1991.

346. Kartha, K.K., Fowke, L.C., Leung, N.L., Caswell, K.L., and Hakman, I., Induction of somatic embryos and plantlets from cryopreserved cell cultures of white spruce (*Picea glauca*), *J. Plant Physiol.*, 231, 529, 1988.

347. Engelmann, F., Intérêt de la cryoconservation des organes végétaux: cas des embryons somatiques de palmier à huile (*Elaeis guineensis* Jacq.), *Int. J. Refrig.*, 13, 26, 1990.

348. Gupta, P.K., Kendurkar, S.V., Kulkarni, V.M., Shirgurkar, M.V., and Mascarenhas, A.F., Somatic embryogenesis and plants from zygotic embryos of coconut (*Cocos nucifera* L.) *in vitro*, *Plant Cell Rep.*, 3, 222, 1984.

349. Schwarz, O.J., Plant growth regulator effects in the *in vitro* propagation of three hardwood tree genera: *Castanea*, *Juglans*, and *Quercus*, in *Hormonal Control of Tree Growth*, Kossuth, S.V., and Ross, S. Eds., Martinus Nijhoff, Dordrecht, 1987, 113.

350. Dunstan, D.I., Bekkaoui, F., Pilon, M., Fowke, L.C., and Abrams, S.R., Effects of abscisic acid and analogues on the maturation of white spruce (*Picea glauca*) somatic embryos, *Plant Sci.*, 58, 77, 1988.

351. Hakman, I., and Von Arnold, S., Somatic embryogenesis and plant regeneration from suspension cultures of *Picea glauca* (white spruce), *Physiol. Plant.*, 72, 579, 1988.

352. Attree, S.M., Tautorus, T.E., Dunstan, D.I., and Fowke, L.C., Somatic embryo maturation, germination, and soil establishment of plants of black and white spruce (*Picea mariana* and *Picea glauca*), *Can. J. Bot.*, 68, 2583, 1990.

353. Rajasekaran, K., Vine, J., and Mullins, M.G., Dormancy in somatic embryos and seeds of *Vitis*: changes in endogenous abscisic acid during embryogeny and germination, *Planta*, 154, 139, 1982.

354. Redenbaugh, K., Application of artificial seed to tropical crops, *HortSci.*, 25, 251, 1990.

355. Gray, D.J., and Purohit, A., Somatic embryogenesis and development of synthetic seed technology, *Crit. Rev. Plant Sci.*, 10, 33, 1991.

356. Cyr, D.R., Webster, F.B., and Roberts, D.R., Biochemical events during germination and early growth of somatic embryos and seed of interior spruce (*Picea glauca engelmannii* complex), *Seed Sci. Res.*, 1, 91, 1991.

357. Chen, T.H.H., and Kartha, K.K., Cryopreservation of woody species, in *Cell and Tissue Culture in Forestry*, Vol. 2, Bonga, J.M., and Durzan, D.J., Eds., Martinus Nijhoff, Dordrecht, 1987, 305.

358. Tisserat, B., Ulrich, J.M., and Finkle, B.J., Cryogenic preservation and regeneration of date palm tissue, *HortSci.*, 16, 47, 1981.

359. Bercetche, J., Galerne, M., and Dereuddre, J., Efficient regeneration of plantlets from embryogenic callus of *Picea abies* (L.) Karst after freezing in liquid nitrogen, *C. R. Acad.Sci. Paris Sér. III*, 310, 357, 1990.

360. Vasil, I.K., Progress in the regeneration and genetic manipulation of cereal crops, *Bio/Technol.*, 6, 397, 1988.

361. Tautorus, T.E., Bekkaoui, F., Pilon, M., Datla, R.S.S., Crosby, W.L., Fowke, L.C., and Dunstan, D.I., Factors affecting transient gene expression in electroporated black spruce

(*Picea mariana*) and jack pine (*Pinus banksiana*) protoplasts, *Theor. Appl. Genet.*, 78, 531, 1989.

362. McGranahan, G.H., Leslie, C.A., Uratsu, S.L., and Dandekar, A.M., Improved efficiency of the walnut somatic embryo gene transfer system, *Plant Cell Rep.*, 8, 512, 1990.

363. Ben-Hayyim, G., and Goffer, Y., Plantlet regeneration from an NaCl-selected salt-tolerant callus culture of Shamouti orange (*Citrus sinensis* L. Osbeck), *Plant Cell Rep.*, 7, 680, 1989.

364. Vujicić, R., Radojević, L., and Nesković, M., Orderly arrangement of ribosomes in the embryogenic callus tissues of *Corylus avellana* L, *J. Cell Biol.*, 69, 686, 1969.

365. Fowke, L.C., Attree, S.M., Wang, H., and Dunstan, D.I., Microtubule organization and cell division in embryogenic protoplast cultures of white spruce (*Picea glauca*), *Protoplasma*, 158, 86, 1990.

366. Wang, H., Cutler, A.J., and Fowke, L.C., Microtubule organization in culture soybean and black spruce cells: interphase-mitosis transition and spindle morphology, *Protoplasma*, 162, 46, 1991.

367. Tautorus, T.E., Tissue and cell culture studies of black spruce (*Picea mariana* Miller B.S.P.) and jack pine (*Pinus banksiana* Lambert), Ph.D. Thesis, University of Saskatchewan, 1990.

368. Kochba, J., and Button, J., The stimulation of embryogenesis and embryoid development in habituated ovular callus from the "Shamouti" orange (*Citrus sinensis*) as affected by tissue age and sucrose concentration, *Z. Pflanzenphysiol.*, 73, 415, 1974.

369. Kochba, J., and Spiegel-Roy, P., The effects of auxins, cytokinins and inhibitors on embryogenesis in habituated ovular callus of the "Shamouti" orange (*Citrus sinensis*), *Z. Pflanzenphysiol.*, 81, 283, 1977.

370. Tautorus, T.E., Fowke, L.C., and Dunstan, D.I., Somatic embryogenesis in conifers, *Can. J. Bot.*, 69, 1873, 1991.

371. Button, J., and Rijkenberg, F.H.J., The effect of subculture interval on organogenesis in callus cultures of *Citrus sinensis*, *Acta Hort.*, 78, 225, 1977.

372. Kochba, J., Spiegel-Roy, P., Saad, S., and Neumann, H., Stimulation of embryogenesis in *Citrus* tissue culture by galactose, *Naturwiss.*, 65, 261, 1978.

373. Kochba, J., Spiegel-Roy, P., Neumann, H., and Saad, S., Stimulation of embryogenesis in citrus ovular callus by ABA, ethephon, CCC and Alar and its suppression by GA_3, *Z. Pflanzenphysiol.*, 89, 427, 1978.

374. Epstein, E., Kochba, J., and Neumann, H., Metabolism of indoleacetic acid by embryogenic and non-embryogenic callus lines of "Shamouti" orange (*Citrus sinensis* Osb.), *Z. Pflanzenphysiol.*, 85, 263, 1977.

375. Button, J., and Botha, C.E.J., Enzymic maceration of *Citrus* callus and the regeneration of plants from single cells, *J. Exp. Bot.*, 26, 723, 1975.

376. Nadel, B.L., Use of fluorescein diacetate in *Citrus* tissue cultures for the determination of cell viability and the selection of mutants, *Sci. Hort.*, 39, 15, 1989.

377. Kochba, J., Button, J., Spiegel-Roy, P., Bornman, C.H., and Kochba, M., Stimulation of rooting of *Citrus* embryoids by gibberellic acid and adenine sulphate, *Ann. Bot.*, 38, 795, 1974.

378. Ohgawara, T., and Kobayashi, S., Application of protoplast fusion to *Citrus* breeding, *Food Biotechnol.*, 5, 169, 1991.

379. Ben-Hayyim, G., Shani, A., and Vardi, A., Evaluation of isozyme systems in *Citrus* to facilitate identification of fusion products, *Theor. Appl. Genet.*, 64, 1, 1982.

380. Corley, R.H.V., Lee, C.H., Law, I.H., and Wong, C.Y., Abnormal flower development in oil palm clones, *Planter, Kuala Lumpur*, 62, 233, 1986.

381. Barlass, M., and Skene, K.G.M., *Citrus* (*Citrus* species), in *Biotechnology in Agriculture and Forestry*, Vol. 1, Bajaj, Y.P.S., Ed., Springer-Verlag, New York, 1986, 207.

382. Spiegel-Roy, P., and Vardi, A., *Citrus*, in *Handbook of Plant Cell Culture*, Vol. 3, Ammirato, P.V., Evans, D.A., Sharp, W.R., and Yamada, Y., Eds., Macmillan Publishing Co., New York, 1984, 355.

383. Uchimiya, H., Ohgawara, T., Kato, H., Akiyma, T., and Harada, H., Detection of two different nuclear genomes in parasexual hybrids by ribosomal RNA gene analysis, *Theor. Appl. Genet.*, 64, 117, 1983.

384. Paranjothy, K., Oil palm, in *Handbook of Plant Cell Culture*, Vol. 3, Ammirato, P.V., Evans, D.A., Sharp, W.R., and Yamada, Y., Eds., Macmillan Publishing Company, New York, 1984, 591.

385. Staritsky, G., Tissue culture of the oil palm (*Elaeis guineensis* Jacq.) as a tool for its vegetative propagation, *Euphytica*, 19, 288, 1970.

386. Jones, L.H., Propagation of clonal oil palms by tissue culture, *Oil Palm News*, 17, 1, 1974.

387. Corley, R.H.V., Barrett, J.N., and Jones, L.H., Vegetative propagation of oil palm via tissue culture, *Oil Palm News*, 22, 2, 1977.

388. Ahée, J., Arthuis, P., Cas, G., Duval, Y., Guénin, G., Hanower, J., Hanower, P., Lievoux, D., Lioret, C., Malaurie, B., Pannetier, C., Raillot, D., Varechon, C., and Zuckerman, L., Vegetative propagation of the oil palm *in vitro* by somatic embryogenesis, *Oléagineux*, 36, 115, 1981.

389. Pannetier, C., Arthuis, P., and Lievoux, D., Neoformation of young *Elaeis guineensis* plants from primary calluses obtained on leaf fragments cultured *in vitro*, *Oléagineux*, 36, 121, 1981.

390. Ferdinando, D., Hulme, J., and Hughes, W.A., Oil palm embryogenesis: A biochemical and morphological study, in *Experimental Manipulation of Ovule Tissues*, Chapman, G.P., Mantell, S.H., and Daniels, R.W., Eds., Longman, London, 1985, 135.

391. Turnham, E., and Northcote, D.H., The use of acetyl-CoA carboxylase activity and changes in wall composition as measures of embryogenesis in tissue cultures of oil palm (*Elaeis guineensis*), *Biochem. J.*, 208, 323, 1982.

392. Hughes, W.A., Bociek, S.M., Barrett, J.N., and Ratcliffe, R.G., An investigation of the growth characteristics of oil palm (*Elaeis guineensis*) suspension cultures using ^{31}P NMR, *Biosci. Rep.*, 3, 1141, 1983.

393. Brackpool, A.L., Branton, R.L., and Blake, J., Regeneration in palms, in *Plant Regeneration and Genetic Variability*, Vol. 3, Vasil, I.K., Ed., Harcourt Brace Jovanovich, New York, 1986, 207.

394. Choo, W.K., Yew, W.C., and Corley, R.H.V., Tissue culture of palms – a review, in *Proc. COSTED Symp. Tissue Culture of Economically Important Plants*, 1981, 138.

395. Jones, L.H., and Hughes, W.A., Oil palm (*Elaeis guineensis* Jacq.), in *Trees II*, Vol. 5, Bajaj, Y.P.S., Ed., Springer-Verlag, Berlin, 1989, Chapter I.10.

396. Engelmann, F., Duval, Y., and Dereuddre, J., Survie et prolifération d'embryons somatiques de Palmier à huile (*Elaeis guineensis* Jacq.) après congélation dans l'azote liquide, *C.R. Acad. Sci. Paris Sér. III*, 301, 111, 1985.

397. Corbineau, F., Engelmann, F., and Côme, D., Ethylene production as an indicator of chilling injury in oil palm (*Elaeis guineensis* Jacq.) somatic embryos, *Plant Sci.*, 71, 29, 1990.

398. Hakman, I., and Fowke, L.C., An embryogenic cell suspension culture of *Picea glauca* (white spruce), *Plant Cell Rep.*, 6, 20, 1987.

399. Kumar, P.P., Joy, R.W. IV, and Thorpe, T.A., Ethylene and carbon dioxide accumulation, and growth of cell suspension cultures of *Picea glauca* (white spruce). *J. Plant Physiol.* 135, 592, 1989.

400. Dunstan, D.I., Bock, C.A., Abrams, G.D., and Abrams, S.R., Metabolism of (+)- and (−)-abscisic acid by suspension cultures of white spruce (*Picea glauca*) somatic embryos, *Phytochemistry*, 31, 1451, 1991.

401. Bekkaoui, F., Saxena, P.K., Attree, S.M., Fowke, L.C., and Dunstan, D.I., The isolation and culture of protoplasts from an embryogenic cell suspension culture of *Picea glauca* (Moench) voss, *Plant Cell Rep.*, 6, 476, 1987.

402. Bekkaoui, F., and Dunstan, D.I., Permeabilization of *Picea glauca* protoplasts to macromolecules, *Can. J. For. Res.*, 19, 1316, 1989.

403. Bekkaoui, F., Pilon, M., Lainé, E., Raju, D.S.S., Crosby, W.L., and Dunstan, D.I., Transient gene expression in electroporated *Picea glauca* protoplasts, *Plant Cell Rep.*, 7, 481, 1988.

404. Wilson, S.M., Thorpe, T.A., and Moloney, M.M., PEG-mediated expression of GUS and CAT genes in protoplasts from embryogenic suspension cultures of *Picea glauca*, *Plant Cell Rep.*, 7, 704, 1989.

405. Bekkaoui, R., Datla, R.S.S., Pilon, M., Tautorus, T.E., Crosby, W.L., and Dunstan, D.I., The

534

effects of promoter on transient expression in conifer cell lines, *Theor. Appl. Genet.*, 79, 353, 1990.

406. Good, A.G., Bekkaoui, F., Pilate, G., Dunstan, D.I., and Crosby, W.L., Anaerobic induction in conifers: Expression of endogenous and chimeric anaerobically-induced genes, *Physiol. Plant.*, 78, 441, 1990.

407. Ellis, D.D., McCabe, D., Russell, D., Martinell, B., and McCown, B.H., Expression of inducible angiosperm promoters in a gymnosperm, *Picea glauca* (white spruce), *Plant Mol. Biol.*, 17, 19, 1991.

408. Duchesne, L.C., and Charest, P.J., Transient expression of the β-glucuronidase gene in embryogenic callus of *Picea mariana* following microprojection, *Plant Cell Rep.*, 10, 191, 1991.

409. Marcotte, W.R. Jr., Bayley, C.C., and Quatrano, R.S., Regulation of a wheat promoter by abscisic acid in rice protoplasts, *Nature*, 335, 454, 1988.

410. Dunstan, D.I., Prospects and progress in conifer biotechnology, *Can. J. For. Res.*, 18, 1497, 1988.

411. Gupta, P.K., Timmis, R., Pullman, G., Yancey, M., Kreitinger, M., Carlson, W., and Carpenter, C., Development of an embryogenic system for automated propagation of forest trees, in *Cell Culture and Somatic Cell Genetics of Plants*, Vol. 8, Vasil, I.K., Ed., Academic Press, San Diego, 1991, 75.

412. Attree, S.M., Dunstan, D.I., and Fowke, L.C., White spruce [*Picea glauca* (Moench) Voss] and black spruce [*Picea mariana* (Mill) B.S.P.], in *Biotechnology in Agriculture and Forestry*, Vol. 16, Bajaj, Y.P.S., Ed., Springer-Verlag, Berlin, 1991, 423.

413. Dunstan, D.I., Bethune, T.D., and Abrams, S.R., Racemic abscisic acid and abscisyl alcohol promote maturation of white spruce (*Picea glauca*) somatic embryos, *Plant Sci.*, 76, 219, 1991.

414. Harry, I.S., and Thorpe, T.A., Somatic embryogenesis and plant regeneration from mature zygotic embryos of red spruce, *Bot Gaz.*, 152, 446, 1991.

415. Afele, J.C., Senaratna, T., McKersie, B.D., and Saxena, P.K., Somatic embryogenesis and plant regeneration from zygotic embryo culture in blue spruce (*Picea pungens* Engelman.), *Plant Cell Rep.*, 11, 299, 1992.

416. Lainé, E., Bade, P., and David, A., Recovery of plants from cryopreserved embryogenic cell suspensions of *Pinus caribaea*, *Plant Cell Rep.*, 11, 295, 1992.

417. Chavez, V.M., Litz, R.E., Moon, P.A., and Norstog, K., Somatic embryogenesis from leaf callus of mature plants of the gymnosperm *Ceratozamia mexicana* var. Robusta (Miq.) Dyer (Cycadales), *In Vitro Cell Dev. Biol.*, 28P, 59, 1992.

418. Klimaszewska, K., Ward, C., and Cheliak, W.M., Cryopreservation from embryogenic cultures of larch (*Larix x eurolepis*) and black spruce (*Picea mariana*), *J. Exp. Bot.*, 43, 73, 1992.

419. Tautorus, T.E., Wang, H., Fowke, L.C., and Dunstan, D.I., Microtubule pattern and the occurrence of pre-prophase bands in embryogenic cultures of black spruce (*Picea mariana* Mill.) and non-embryogenic cultures of jack pine (*Pinus banksiana* Lamb.), *Plant Cell Rep.*, 11, 419, 1992.

420. Flinn, B.S., Roberts, D.R., and Taylor, I.E.P., Evaluation of somatic embryos of interior spruce. Characterization and developmental regulation of storage proteins, *Physiol. Plant.*, 82, 624, 1991.

421. Attree, S.M., Moore, D., Sawhney, V.K., and Fowke, L.C., Enhanced maturation and desiccation tolerance of white spruce (*Picea glauca* [Moench] Voss) somatic embryos: effects of a non-plasmolysing water stress and abscisic acid, *Ann. Bot.*, 68, 519, 1991.

422. Attree, S.M., Pomeroy, M.K., and Fowke, L.C., Manipulation of conditions for the culture of somatic embryos of white spruce for improved triacylglycerol biosynthesis and desiccation tolerance, *Planta*, 187, 395, 1992.

423. Tautorus, T.E., Lulsdorf, M.M., Kikcio, S.I., and Dunstan, D.I., Bioreactor culture of *Picea mariana* Mill. (black spruce) and the species complex *Picea glauca-engelmannii* (interior spruce) somatic embryos. Growth parameters, *Appl. Microbiol. Biotechnol.*, 38, 46, 1992.

424. Cheliak, W.M., and Klimaszewska, K., Genetic variation in somatic embryogenic response in open-pollinated families of black spruce, *Theor. Appl. Genet.*, 82, 185, 1991.

425. Kvaalen, H., and Von Arnold, S., Effects of various partial pressures of oxygen and carbon dioxide on different stages of somatic embryogenesis in *Picea abies, Plant Cell Tiss. Org. Cult.*, 27, 49, 1991.

426. Tremblay, L., and Tremblay, F.M., Carbohydrate requirements for the development of black spruce (*Picea mariana* [Mill.] B.S.P.) and red spruce (*P. rubens* Sarg.) somatic embryos, *Plant Cell Tiss. Org. Cult.*, 27, 95, 1991.

427. Ishii, K., Somatic embryo formation and plantlet regeneration through embryogenic callus from mature zygotic embryos of *Picea jezoensis*, and *Picea glehnii, J. Jpn. For. Soc.*, 73, 24, 1991.

428. Aitken-Christie, J., and Thorpe, T.A., Clonal propagation: Gymnosperms, in *Cell Culture and Somatic Cell Genetics of Plants*, Vol. 1, Vasil, I.K., Ed., Academic Press, Orlando, 1984, 82.

429. Pitel, J.A., Yoo, B.Y., Klimaszewska, K., and Charest, P.J., Changes in enzyme activity and protein patterns during the maturation phase of somatic embryogenesis in hybrid larch (*Larix* × *eurolepis*), *Can. J. For. Res.*, 22, 553, 1992.

430. Plata, E., Ballester, A., and Vieitez, A.M., An anatomical study of secondary embryogenesis in *Camellia reticulata, In Vitro Cell Dev. Biol.*, 27P, 183, 1991.

431. Aguilar, M.E., Villalobos, V.M., and Vasquez, N., Production of cocoa plants (*Theobroma cacao* L.) via micrografting of somatic embryos, *In Vitro Cell Dev. Biol.*, 28P, 15, 1992.

432. Voo, K.S., Rugh, C.L., and Kamalay, J.C., Indirect somatic embryogenesis and plant recovery from cotton (*Gossypium hirsutum* L.), *In Vitro Cell Dev. Biol.*, 27P, 117, 1991.

433. Oliveira, M.M., and Pais, M.S.S., Somatic embryogenesis in leaves and leaf-derived protoplasts of *Actinidia deliciosa* var. *deliciosa* cv. Hayward (kiwifruit), *Plant Cell Rep.*, 11, 314, 1992.

434. Jha, T.B., Jha, S., and Sen, S.K., Somatic embryogenesis from immature cotyledons of an elite Darjeeling tea clone, *Plant Sci.*, 84, 209, 1992.

435. Watt, M.P., Blakeway, F., Cresswell, C.F., and Herman, B., Somatic embryogenesis in *Eucalyptus grandis, Suid-Afrikaanse Bosboutydskrif*, 157, 59, 1991.

436. Merkle, S.A., Wiecko, A.T., and Watson-Pauley, B.A., Somatic embryogenesis in American chestnut, *Can. J. For. Res.*, 21, 1698, 1991.

437. Rout, G.R., Debata, B.K., and Das, P., Somatic embryogenesis in callus cultures of *Rosa hybrida* L. cv. Landora, *Plant Cell Tiss. Org. Cult.*, 27, 65, 1991.

438. Bueno, M.A., Astorga, R., and Manzanera, J.A., Plant regeneration through somatic embryogenesis in *Quercus suber, Physiol. Plant.*, 85, 30, 1992.

439. Sommer, H.E., Brown, C.L., and Kormanik, P.P., Differentiation of plantlets in longleaf pine (*Pinus palustris* Mill.) tissue cultured *in vitro, Bot. Gaz.*, 136, 196, 1975.

440. Sasamoto, H., and Hosoi, Y., Callus proliferation from the protoplasts of embryogenic cells of *Quercus serrata, Plant Cell Tiss. Org. Cult.*, 29, 241, 1992.

441. Dussert, S., Mauro, M.C., Deloire, A., Hamon, S., and Engelmann F., Cryopreservation of grape embryogenic cell suspensions: 1. Influence of pretreatment, freezing and thawing conditions, *Cryo-lett.*, 12, 287, 1991.

442. Dussert, S., Mauro, M.C., and Engelmann F., Cryopreservation of grape embryogenic cell suspensions: 2. Influence of post-thaw culture conditions and application to different strains, *Cryo-lett.*, 13, 15, 1992.

443. Gingas, V.M., Asexual embryogenesis and plant regeneration from male catkins of *Quercus, HortSci.*, 26, 1217, 1991.

444. Liu, C.-m., and Yao, D.-y., Studies on somatic embryogenesis and histological observation in *Gossypium hirsutum* L., cv. Coker 312, *Acta Bot. Sinica*, 33, 378, 1991.

445. Faure, O., Mengoli, M., Nougarede, A., and Bagni, N., Polyamine pattern and biosynthesis in zygotic and somatic embryo stages of *Vitis vinifera, J. Plant. Physiol.*, 138, 545, 1991.

446. Bhansali, R.R., Driver, J., and Durzan, D.J., Somatic embryogenesis in cell suspension cultures of *Prunus persica* (L.), *J. Hort. Sci.*, 66, 601, 1991.

447. Neuenschwander, B., and Baumann, T.W., A novel type of somatic embryogenesis in *Coffea arabica, Plant Cell Rep.*, 10, 608, 1992.

536

448. Michaux-Ferrière, Grout, H., and Carron, M.P., Origin and ontogenesis of somatic embryos in *Hevea brasiliensis* (Euphorbiaceae), *Am. J. Bot.*, 79, 174, 1992.
449. De Wit, J.C., Esendam, H.F., Honkanen, J.J., and Tuominen, U., Somatic embryogenesis and regeneration of flowering plants in rose, *Plant Cell Rep.*, 9, 456, 1990.
450. Noriega, C., and Söndahl, M.R., Somatic embryogenesis in hybrid tea roses, *Biotech.*, 9, 991, 1991.
451. Kiss, J., Heszky, L.E., Kiss, E., and Gyulai, G., High efficiency adventive embryogenesis on somatic embryos of anther, filament and immature proembryo origin in horse-chestnut (*Aesculus hippocastanum* L.) tissue culture, *Plant Cell Tiss. Org. Cult.*, 30, 59, 1992.
452. Karunaratne, S., and Periyapperuma, K., Culture of immature embryos of coconut, *Cocos nucifera* L: callus proliferation and somatic embryogenesis, *Plant Sci.*, 62, 247, 1989.
453. Karunaratne, S., Gamage, C., and Kovoor, A., Leaf maturity, a critical factor in embryogenesis, *J. Plant Physiol.*, 139, 27, 1991.
454. Buffard-Morel, J., Verdeil, J.-L., and Pannetier, C., Embryogenèse somatique du cocotier (*Cocos nucifera* L.) à partir d'explants foliaires: Étude histologique, *Can J. Bot.*, 70, 735, 1992.
455. Wang, Z., Wu, H., Zeng, X., Chen, C., and Li, Q., Embryogeny and origin of anther plantlet of *Hevea brasiliensis*, *Chin. J. Trop. Crops*, 5, 9, 1984.
456. Zamarripa, A., Ducos, J.P., Bollon, H., Dufour, M., and Petiard, V., Production of somatic embryos of coffee in liquid medium: effects of inoculation density and renewal in the medium. *Café Thé* (Paris), 35, 233, 1991.
457. Cornu, D., Somatic embryogenesis in tissue cultures of walnut (*Juglans nigra, J. major* and hybrids *J. nigra* × *J. regia*), in *Somatic Cell Genetics of Woody Plants*, Ahuja, M.R. Ed., Kluwer Academic Publishers, Dordrecht, 1987, 45.
458. Deng, M.-D., and Cornu, D., Maturation and germination of walnut somatic embryos, *Plant Cell Tiss. Org. Cult.*, 28, 195, 1992.
459. Chavez, V.M., Litz, R.E., and Norstog, K., *In vitro* morphogenesis of *Ceratozamia hildae* and *C. mexicana* from megagametophytes and zygotic embryos, *Plant Cell Tiss. Org. Cult.*, 30, 93, 1992.
460. Chavez, V.M., Litz, R.E., and Norstog, K., Somatic embryogenesis and organogenesis in *Zamia fischeri, Z. furfuracea* and *Z. pumila*, *Plant Cell Tiss. Org. Cult.*, 30, 99, 1992.
461. De Touchet, B., Duval, Y., and Pannetier, C., Plant regeneration from embryogenic suspension cultures of oil palm (*Elaeis guineensis* Jacq.), *Plant Cell Rep.*, 10, 529, 1991.
462. Beloualy, N., and Bouharmont, J., NaCl tolerant plants of *Poncirus trifoliata* regenerated from tolerant cell lines, *Theor. Appl. Genet.*, 83, 509, 1992.
463. Gavish, H., Vardi, A., and Fluhr, R., Extracellular proteins and early development in *Citrus* nucellar cell cultures, *Physiol. Plant.*, 82, 606, 1991.
464. Gavish, H., Vardi, A., and Fluhr, R., Suppression of somatic embryogenesis in *Citrus* cell cultures by extracellular proteins, *Planta*, 186, 511, 1992.
465. Dunstan, D.I., Bethune, T.D., and Bock, C.A., Somatic embryo maturation from long-term suspension cultures of white spruce (*Picea glauca*), *In Vitro Cell. Dev. Biol.*, 29P, 109, 1993.
466. Tautorus, T.E., Kikcio, S.I., Lulsdorf, M.M., and Dunstan, D.I., Effect of genotype and culture conditions on maturation of the species complex *Picea glauca-engelmannii* (interior spruce) somatic embryos, *3rd Canadian Workshop on Plant Tissue Culture and Genetic Engineering*, Guelph, Ontario, June, 1992.
467. Bapat, V.A., Fulzele, D.P., Heble, M.R., and Rao, P.S., Production of sandalwood somatic embryos in bioreactors, *Curr. Sci.*, 59, 746, 1990.
468. Nuutila, A.M., and Kauppinen, V., Nutrient uptake and growth of an embryogenic and a non-embryogenic cell line of birch (*Betula pendula* Roth.) in suspension culture, *Plant Cell Tiss. Org. Cult.*, 30, 7, 1992.
469. Thompson, R.G., and Von Aderkas, P., Somatic embryogenesis and plant regeneration from immature embryos of western larch, *Plant Cell Rep.*, 11, 379, 1992.
470. Bozhkov, P.V., Lebedenko, L.A., and Shiryaeva, G.A., A pronounced synergistic effect of abscisic acid and 6-benzyladenine on Norway spruce (*Picea abies* L. Karst) somatic embryo maturation, *Plant Cell Rep.*, 11, 386, 1992.

471. Gifford, D.J., and Tolley, M.C., The seed proteins of white spruce and their mobilization following germination, *Physiol. Plant.*, 77, 254, 1989.

472. Misra, S., and Green, M.J., Developmental gene expression in conifer embryogenesis and germination. I. Seed proteins and protein body composition of mature embryo and the megagametophyte of white spruce (*Picea glauca* [Moench] Voss.), *Plant Sci.*, 68, 163, 1990.

473. Roberts, D.R., Abscisic acid and mannitol promote early development, maturation and storage protein accumulation in somatic embryos of interior spruce, *Physiol. Plant.*, 83, 247, 1991.

474. Von Aderkas, P., Bonga, J., Klimaszewska, K., and Owens, J., Comparison of larch embryogeny *in vivo* and *in vitro*, in *Woody Plant Biotechnology*, Ahuja, M.R., Ed., Plenum Press, New York, 1991, 139.

475. Teasdale, R.D., and Richards, D.K., Boron deficiency in cultured pine cells: quantitative studies of the interaction with Ca and Mg, *Plant Physiol.*, 93, 1071, 1990.

476. Teasdale, R.D., Dawson, P.A., and Woolhouse, H.W., Mineral nutrient requirements of a loblolly pine (*Pinus taeda*) cell suspension culture, *Plant Physiol.*, 4, 82, 942, 1986.

477. Tremblay, L., and Tremblay, F.M., Effects of gelling agents, ammonium nitrate, and light on the development of *Picea mariana* (Mill) B.S.P. (black spruce) and *Picea rubens* Sarg. (red spruce) somatic embryos, *Plant Sci.*, 77, 233, 1991.

478. Debergh, P.C., Effects of agar brand and concentration on the tissue culture medium, *Physiol. Plant.*, 59, 270, 1983.

479. Charest, P.J., Devantier, Y., Ward, C., Jones, C., Schaffer, U., and Klimaszewska, K., Transient expression of foreign chimeric genes in the gymnosperm hybrid larch following electroporation, *Can. J. Bot.*, 69, 1731, 1991.

480. Newton, R.J., Yibrah, H.S., Dong, N., Clapham, D.H., and Von Arnold, S., Expression of an abscisic acid responsive promoter in *Picea abies* (L.) Karst. following bombardment from an electric discharge particle accelerator, *Plant Cell Rep.*, 11, 188, 1992.

481. Duchesne, L.C., and Charest, P.J., Effect of promoter sequence on transient expression of the β-glucuronidase gene in embryogenic calli of *Larix x eurolepis* and *Picea mariana* following microprojection, *Can. J. Bot.*, 70, 175, 1992.

482. Ellis, D.D., McCabe, D.E., McInnis, S., Ramachandran, R., Russell, D.R., WAllace, K.M., Martinell, B.J., Roberts, D.R., Raffa, K.F., and McCown, B.H., Stable transformation of *Picea glauca* by particle acceleration. *Bio/Technol.*, 11, 84, 1993.

483. Coutos-Thevenot, P., Goebel-Tourand, I., Mauro, M-C., Jouanneau, J-P, Boulay, M., Deloire, A., and Guern, J., Somatic embryogenesis from grapevine cells. I-Improvement of embryo development by changes in culture conditions, *Plant Cell Tiss. Org. Cult.*, 29, 125, 1992.

484. Jalonen, P., and Von Arnold, S., Characterization of embryogenic cell lines of *Picea abies* in relation to their competence for maturation, *Plant Cell Rep.*, 10, 384, 1991.

485. Grossnickle, S.C., Roberts, D.R., Major, J.E., Folk, R.S., Webster, F.B., and Sutton, B.C.S., Integration of somatic embryogenesis into operational forestry. Comparison of interior spruce emblings and seedlings during production of 1+0 stock, *U.S.D.A. For. Serv. Gen. Tech. Rep.* No. RM-211, 1991, 106.

486. Lulsdorf, M.M., Tautorus, T.E., Kikcio, S.I., and Dunstan, D.I., Germination of encapsulated embryos of interior spruce (*Picea glauca engelmannii* complex) and black spruce (*Picea mariana* Mill.), *Plant Cell Rep.*, 12, 385, 1993.

487. Cyr, D.R., Lazaroff, W.R., Grimes, S.M.A., Quan, G., Bethune, T.D., Dunstan, D.I., and Roberts, D.R., Cryopreservation of interior spruce embryogenic cultures, *Plant Cell Rep.*, 13, 574, 1994.

488. Bommineni, V.R., Datla, R.S.S., and Tsang, E.W.T., Expression of GUS in somatic embryo cultures of black spruce after microprojectile bombardment, *J. Exp. Bot.*, 45, 491, 1994.

489. Bommineni, V.R., Chibbar, R.N., Datla, R.S.S., and Tsang, E.W.T., Transformation of white spruce (*Picea glauca*) somatic embryos by microprojectile bombardment, *Plant Cell Rep.*, 13, 17, 1993.

538

490. Tautorus, T.E., Lulsdorf, M.M., Kikcio, S.I., and Dunstan, D.I., Nutrient utilization during bioreactor culture, and evaluation of maintenance regime on maturation of somatic embryo cultures of *Picea mariana* and the species complex *Picea glauca-engelmannii, In Vitro Cell. Dev. Biol.*, 30P, 58, 1994.

491. Flinn, B.S., Roberts, D.R., Newton, C.H., Cyr, D.R., Webster, F.B., and Taylor, I.E.P., Storage protein gene expression in zygotic and somatic embryos of interior spruce, *Physiol. Plant.*, 89, 719, 1993.

492. Dunstan, D.I., Berry, S., and Bock, C.A., ABA consumption in Norway spruce (*Picea abies*) and white spruce (*Picea glauca*) somatic embryo cultures, *In Vitro Cell. Dev. Biol.*, 30, 156, 1994.

493. Misra, S., Attree, S.M., Leal, I., and Fowke, L.C., Effect of abscisic acid, osmoticum, and desiccation on synthesis of storage proteins during the development of white spruce somatic embryos, *Ann. Bot.*, 71, 11, 1993.

494. Dong, J-Z., Pilate, G., Abrams, S.A., and Dunstan, D.I., Induction of a wheat Em promoter by optically pure ABA and ABA analogs, in white spruce (*Picea glauca*) protoplasts, *Physiol. Plant.*, 90, 513, 1994.

495. Grossnickle, S.C., Major, J.E., and Folk, R.S., Interior spruce seedlings compared with emblings produced from somatic embryogenesis. I. Nursery development, fall acclimation, and over-winter storage, *Can. J. For. Res.*, 24, 1376, 1994.

496. Grossnickle, S.C., and Major, J.E., Interior spruce seedlings compared with emblings produced from somatic embryogenesis. II. Stock quality assessment prior to field planting, *Can. J. For. Res.*, 24, 1385, 1994.

497. Grossnickle, S.C., and Major, J.E., Interior spruce seedlings compared with emblings produced from somatic embryogenesis. III. Physiological response and morphological development on a reforestation site, *Can. J. For. Res.*, 24, 1397, 1994.

List of Contributors

BAKÓ, L., *Institute of Plant Physiology, Biological Research Center, Hungarian Academy of Sciences, 6701 Szeged, P.O. Box 521, Hungary*
(Chapter 8, with D. Dudits, J. Györgyey and L. Bögre

BÖGRE, L., *Institute of Plant Physiology, Biological Research Center, Hungarian Academy of Sciences, 6701 Szeged, P.O. Box 521, Hungary*
(Chapter 8, with D. Dudits, J. Györgyey and L. Bakó)

BROWN, D.C.W., *Plant Biotechnology Program, Plant Research Centre, Agriculture Canada, Ottawa, Ontario, Canada K1A 0C6*
(Chapter 10, with K.I. Finstad and E.M. Watson)

DUDITS, D., *Institute of Plant Physiology, Biological Research Center, Hungarian Academy of Sciences, 6701 Szeged, P.O. Box 521, Hungary*
(Chapter 8, with J. Györgyey, L. Bögre and L. Bakó)

DUNSTAN, D.I., *National Research Council Canada, Plant Biotechnology Institute, 110 Gymnasium Place, Saskatoon, Saskatchewan, Canada S7N 0W9*
(Chapter 12, with T.E. Tautorus and T.A. Thorpe)

FERRIE, A.M.R., *Plant Biotechnology Institute, National Research Council, 110 Gymnasium Place, Saskatoon, SK, Canada S7N 0W9*
(Chapter 9, with C.E. Palmer and W.A. Keller)

FINSTAD, K.I., *Plant Biotechnology Program, Plant Research Centre, Agriculture Canada, Ottawa, Ontario, Canada K1A 0C6*
(Chapter 10, with D.C.W. Brown and E.M. Watson)

FLINN, B.S., *Department of Botany, The University of Georgia, Athens, GA 30602, U.S.A.*
(Chapter 5, with S.A. Merkle and B.S. Flinn)

GYÖRGYEY, J., *Institute of Plant Physiology, Biological Research Center, Hungarian Academy of Sciences, 6701 Szeged, P.O. Box 521, Hungary*
(Chapter 8, with D. Dudits, L. Bögre and L. Bakó)

HALPERIN, W., *Botany Department, University of Washington, Seattle, Washington, 98195-0001, U.S.A.*
(Chapter 1)

540

KELLER, W.A., *Plant Biotechnology Institute, National Research Council, 110 Gymnasium Place, Saskatoon, SK, Canada S7N 0W9*
(Chapter 9, with A.M.R. Ferrie and C.E. Palmer)

KOMAMINE, A., *Biological Institute, Faculty of Science, Tohoku University, Sendai, Japan*
(Chapter 7, with K. Nomura)

KRISHNARAJ, S., *Department of Biological Sciences, University of Calgary, Calgary, Alberta, Canada T2N 1N4*
(Chapter 11, with I.K. Vasil)

MERKLE, S.A., *D.B. Warnell School of Forest Resources, The University of Georgia, Athens, GA 30602, U.S.A.*
(Chapter 5, with W.A. Parrott and B.S. Flinn)

MONNIER, M., *Université Pierre et Marie Curie, Laboratoire d'Histophysiologie Végétale, 12, Rue Cuvier, 75005 Paris, France*
(Chapter 4)

NOMURA, K., *Institute of Agriculture and Forestry, Tsukuba University, Tennodai, Tsukuba, Japan*
(Chapter 7, with A. Komamine)

PALMER, C.E., *Department of Plant Science, University of Manitoba, Winnipeg, MB, Canada R3T 2N2*
(Chapter 9, with A.M.R. Ferrie and W.A. Keller)

PARROTT, W.A., *Department of Crop and Soil Sciences, The University of Georgia, Athens, GA 30602, U.S.A.*
(Chapter 5, with S.A. Merkle and B.S. Flinn)

RAGHAVAN, V., *Department of Plant Biology, The Ohio State University, Columbus, Ohio 43210, U.S.A.*
(Chapter 3, with K.K. Sharma)

SHARMA, K.K., *Plant Physiology Research Group, Department of Biological Sciences, University of Calgary, Calgary, Alberta, Canada T2N 1N4, Present Address:* Legumes Cell Biology Unit, ICRISAT, Patancheru 502324, Andhra Pradesh, India
(Chapter 2, with T.A. Thorpe)
(Chapter 3, with V. Raghavan)

TAUTORUS, T.E., *National Research Council Canada, Plant Biotechnology Institute, 110 Gymnasium Place, Saskatoon, Saskatchewan, Canada S7N 0W9*
(Chapter 12, with D.I. Dunstan and T.A. Thorpe)

THORPE, T.A., *Plant Physiology Research Group, Department of Biological Sciences, University of Calgary, Calgary, Alberta, Canada T2N 1N4*
(Chapter 2, with K.S. Sharma)
(Chapter 12, with D.I. Dunstan and T.E. Tautorus)

VASIL, I.K., *Laboratory of Plant Cell and Molecular Biology University of Florida, Gainesville, FL 326711-0690, U.S.A.*
(Chapter 11, with S. KrishnaRaj)

541

WATSON, E.M., *Plant Biotechnology Program, Plant Research Centre, Agriculture Canada, Ottawa, Ontario, Canada K1A 0C6*
(Chapter 10, with D.C.W. Brown and K.I. Finstad)

YEUNG, E.C., *Department of Biological Sciences, The University of Calgary, Calgary, Alberta, Canada T2N 1N4*
(Chapter 6)

Subject Index